Planets, Stars and Stellar Systems

Astronomical Techniques, Software, and Data

Terry D. Oswalt (Editor-in-Chief)

Howard E. Bond (Volume Editor)

Planets, Stars and Stellar Systems

Volume 2: Astronomical Techniques, Software, and Data

With 121 Figures and 32 Tables

Editor-in-Chief
Terry D. Oswalt
Department of Physics & Space Sciences
Florida Institute of Technology
University Boulevard
Melbourne, FL, USA

Volume Editor
Howard E. Bond
Labrador Lane
Cockeysville, MD, USA

ISBN 978-94-007-5617-5 ISBN 978-94-007-5618-2 (eBook)
ISBN 978-94-007-5619-9 (print and electronic bundle)
DOI 10.1007/978-94-007-5618-2

This title is part of a set with
Set ISBN 978-90-481-8817-8
Set ISBN 978-90-481-8818-5 (eBook)
Set ISBN 978-90-481-8852-9 (print and electronic bundle)

Springer Dordrecht Heidelberg New York London

Library of Congress Control Number: 2012953926

© Springer Science+Business Media Dordrecht 2013
This work is subject to copyright. All rights are reserved by the Publisher, whether the whole or part of the material is concerned, specifically the rights of translation, reprinting, reuse of illustrations, recitation, broadcasting, reproduction on microfilms or in any other physical way, and transmission or information storage and retrieval, electronic adaptation, computer software, or by similar or dissimilar methodology now known or hereafter developed. Exempted from this legal reservation are brief excerpts in connection with reviews or scholarly analysis or material supplied specifically for the purpose of being entered and executed on a computer system, for exclusive use by the purchaser of the work. Duplication of this publication or parts thereof is permitted only under the provisions of the Copyright Law of the Publisher's location, in its current version, and permission for use must always be obtained from Springer. Permissions for use may be obtained through RightsLink at the Copyright Clearance Center. Violations are liable to prosecution under the respective Copyright Law.
The use of general descriptive names, registered names, trademarks, service marks, etc. in this publication does not imply, even in the absence of a specific statement, that such names are exempt from the relevant protective laws and regulations and therefore free for general use.
While the advice and information in this book are believed to be true and accurate at the date of publication, neither the authors nor the editors nor the publisher can accept any legal responsibility for any errors or omissions that may be made. The publisher makes no warranty, express or implied, with respect to the material contained herein.

Printed on acid-free paper

Springer is part of Springer Science+Business Media (www.springer.com)

Series Preface

It is my great pleasure to introduce "Planets, Stars, and Stellar Systems" (PSSS). As a "Springer Reference", PSSS is intended for graduate students to professionals in astronomy, astrophysics and planetary science, but it will also be useful to scientists in other fields whose research interests overlap with astronomy. Our aim is to capture the spirit of 21st century astronomy – an empirical physical science whose almost explosive progress is enabled by new instrumentation, observational discoveries, guided by theory and simulation.

Each volume, edited by internationally recognized expert(s), introduces the reader to a well-defined area within astronomy and can be used as a text or recommended reading for an advanced undergraduate or postgraduate course. Volume 1, edited by Ian McLean, is an essential primer on the tools of an astronomer, i.e., the telescopes, instrumentation and detectors used to query the entire electromagnetic spectrum. Volume 2, edited by Howard Bond, is a compendium of the techniques and analysis methods that enable the interpretation of data collected with these tools. Volume 3, co-edited by Linda French and Paul Kalas, provides a crash course in the rapidly converging fields of stellar, solar system and extrasolar planetary science. Volume 4, edited by Martin Barstow, is one of the most complete references on stellar structure and evolution available today. Volume 5, edited by Gerard Gilmore, bridges the gap between our understanding of stellar systems and populations seen in great detail within the Galaxy and those seen in distant galaxies. Volume 6, edited by Bill Keel, nicely captures our current understanding of the origin and evolution of local galaxies to the large scale structure of the universe.

The chapters have been written by practicing professionals within the appropriate subdisciplines. Available in both traditional paper and electronic form, they include extensive bibliographic and hyperlink references to the current literature that will help readers to acquire a solid historical and technical foundation in that area. Each can also serve as a valuable reference for a course or refresher for practicing professional astronomers. Those familiar with the "Stars and Stellar Systems" series from several decades ago will recognize some of the inspiration for the approach we have taken.

Very many people have contributed to this project. I would like to thank Harry Blom and Sonja Guerts (Sonja Japenga at the time) of Springer, who originally encouraged me to pursue this project several years ago. Special thanks to our outstanding Springer editors Ramon Khanna (Astronomy) and Lydia Mueller (Major Reference Works) and their hard-working editorial team Jennifer Carlson, Elizabeth Ferrell, Jutta Jaeger-Hamers, Julia Koerting, and Tamara Schineller. Their continuous enthusiasm, friendly prodding and unwavering support made this series possible. Needless to say (but I'm saying it anyway), it was not an easy task shepherding a project this big through to completion!

Most of all, it has been a privilege to work with each of the volume Editors listed above and over 100 contributing authors on this project. I've learned a lot of astronomy from them, and I hope you will, too!

January 2013

Terry D. Oswalt
General Editor

Preface to Volume 2

Volume 2 of *Planets, Stars, and Stellar Systems* is entitled "*Astronomical Techniques, Software, and Data*." When I began my astronomical career in the 1960s, astronomical techniques, at least for optical observers, consisted mostly of exposing and developing photographic plates, or obtaining single-channel photometry with a photomultiplier tube. Software, if used at all, was run using punched cards that you took over to the computer center and came back 24 hours later to pick up the output (which often consisted of pointing out the typographical errors in your FORTRAN code). Computations were done with a slide rule or a mechanical Friden calculator (the most advanced model could actually calculate a square root, although it took about 15 seconds to do so!). Data consisted of photographic plates or strip-chart recordings of your photometry or hand-written columns of numbers or plots prepared with a Leroy lettering set. If you needed data from a collaborator, the plates had to be shipped to you in a sturdy wooden box or you had to travel to your colleague's institution or the tables of numbers had to be mailed to you.

The advances in astronomical methods in recent decades have come in steady steps, but as I look back from our contemporary viewpoint to 40 or 50 years ago, they are all but inconceivable. I hold in my hand a computing device orders of magnitude more powerful than the room-sized computer of 1965 (and it can even make telephone calls!). What's that bright thing next to the moon? Just start up the app, aim the phone at the sky, and it will tell you. In the old days you made a finding chart by laboriously pulling out a Palomar Sky Survey print and photographing it with a Polaroid camera; now, in a few seconds, and from any mountaintop observatory in the world, I can display a Digitized Sky Survey image of any point in the sky, and read off the coordinates just by moving the cursor. Do you want the spectral-energy distribution of your source from the far-UV, through the optical, to the near- and mid-IR? You can find all of that in a few moments now.

The venerable Chicago *Stars and Stellar Systems* had two volumes dedicated to "*Astronomical Techniques*" and "*Basic Astronomical Data*." The new volume captures the basic spirit of those SSS volumes, in terms of introducing the reader to some of the basic observing and data-analysis techniques, but of course many of the actual topics are vastly different from, or didn't exist at all, five decades ago.

Volume 2 starts with two articles (Stetson, and Massey & Hanson) describing modern techniques of astronomical photometry and spectroscopy, primarily at optical wavelengths. The next two articles (Tokunaga, Vacca, & Young, and Snik & Leller) move to the realms of techniques of infrared astronomy and polarimetry of astrophysical sources. As I have already mentioned, the availability of multi-wavelength sky surveys has transformed modern observational astronomy, and the amazing breadth of survey data available now (or in the near future) is comprehensively reviewed by Djorgovski, Mahabal, Drake, Graham, & Donalek.

Moving to still longer wavelengths, Wilson reviews the techniques of radio astronomy, and then Monnier & Allen reveal the methods of interferometry at both radio and optical frequencies. To understand your data, you usually have to calibrate them to absolute physical units, and these techniques are explained by Deustua, Kent, & Smith.

The new science of astroinformatics is reviewed by Borne. Statistical methods of particular utility in astronomy are discussed by Feigelson & Babu, and the volume closes with a review of modern numerical techniques in astronomy by Wood.

This volume would not have been possible without the contributions of the authors, the guiding influence of the other volume editors and the editor-in-chief, and the staff at Springer. I thank all of them, and I hope that the new PSSS volumes will have as much influence on contemporary astronomy as the old and still cherished Chicago SSS volumes did.

Howard E. Bond
Cockeysville, MD
USA

Editor-in-Chief

Dr. Terry D. Oswalt
Department Physics & Space Sciences
Florida Institute of Technology
150 W. University Boulevard
Melbourne, Florida 32901
USA
E-mail: toswalt@fit.edu

Dr. Oswalt has been a member of the Florida Tech faculty since 1982 and was the first professional astronomer in the Department of Physics and Space Sciences. He serves on a number of professional society and advisory committees each year. From 1998 to 2000, Dr. Oswalt served as Program Director for Stellar Astronomy and Astrophysics at the National Science Foundation. After returning to Florida Tech in 2000, he served as Associate Dean for Research for the College of Science (2000–2005) and interim Vice Provost for Research (2005–2006). He is now Head of the Department of Physics & Space Sciences. Dr. Oswalt has written over 200 scientific articles and has edited three astronomy books, in addition to serving as Editor-in-Chief for the six-volume Planets, Stars, and Stellar Systems series.

Dr. Oswalt is the founding chairman of the Southeast Association for Research in Astronomy (SARA), a consortium of ten southeastern universities that operates automated 1-meter class telescopes at Kitt Peak National Observatory in Arizona and Cerro Tololo Interamerican Observatory in Chile (see the website www.saraobservatory.org for details). These facilities, which are remotely accessible on the Internet, are used for a variety of research projects by faculty and students. They also support the SARA Research Experiences for Undergraduates (REU) program, which brings students from all over the U.S. each summer to participate one-on-one with SARA faculty mentors in astronomical research projects. In addition, Dr. Oswalt secured funding for the 0.8-meter Ortega telescope on the Florida Tech campus. It is the largest research telescope in the State of Florida.

Dr. Oswalt's primary research focuses on spectroscopic and photometric investigations of very wide binaries that contain known or suspected white dwarf stars. These pairs of stars, whose separations are so large that orbital motion is undetectable, provide a unique opportunity to explore the low luminosity ends of both the white dwarf cooling track and the main sequence; to test competing models of white dwarf spectral evolution; to determine the space motions, masses, and luminosities for the largest single sample of white dwarfs known; and to set a lower limit to the age and dark matter content of the Galactic disk.

Volume Editor

Howard E. Bond
9615 Labrador Lane
Cockeysville, MD 21030
USA
E-mail: bond@stsci.edu

Dr. Bond received his Ph.D. in astronomy from the University of Michigan in 1969.

From 1970 to 1984, he was a faculty member in the Department of Physics and Astronomy at Louisiana State University. In 1984, he moved to the Space Telescope Science Institute in Baltimore, Maryland, which manages the scientific programs of the Hubble Space Telescope. At STScI, Dr. Bond helped develop the peer review procedures for Hubble observers. He managed the Hubble Postdoctoral Fellowship Program from 1994 to 2002. He was a cofounder of the Hubble Heritage Project, which makes the most spectacular HST images available to the public, and has received both the Klumpke-Roberts Award of the Astronomical Society of the Pacific and the Education Prize of the American Astronomical Society. Dr. Bond was involved in the development of the Wide-Field Camera 3 for the Hubble telescope, serving on the WFC3 Scientific Oversight Committee from 1998 to the present.

Dr. Bond's research interests are in observational stellar astronomy. He has published over 500 scientific papers, concentrating particularly on planetary nebulae and their central stars, white dwarfs, stellar chemical compositions, binary stars, and transient astrophysical phenomena. He is an active user of the Hubble Space Telescope, having received observing time on the telescope in all 20 peer-review cycles.

Bond served as Councilor of the American Astronomical Society from 1987 to 1989, and was Managing Editor of Publications of the Astronomical Society of the Pacific from 1991 to 1997. In 2008, he became Astronomer Emeritus at the Space Telescope Science Institute, and is currently continuing his research programs as an independent contractor and consultant.

Table of Contents

Volume 2

List of Contributors

Ronald J. Allen
Science Mission Office
Space Telescope Science Institute
Baltimore, MD
USA

G. Jogesh Babu
Department of Statistics
The Pennsylvania State University
University Park, PA
USA
and
Center for Astrostatistics
The Pennsylvania State University
University Park, PA
USA

Kirk Borne
George Mason University
Fairfax, VA
USA

Susana Deustua
Instruments Division
Space Telescope Science Institute
Baltimore, MD
USA

S. George Djorgovski
California Institute of Technology
Pasadena, CA
USA

Ciro Donalek
California Institute of Technology
Pasadena, CA
USA

Andrew Drake
California Institute of Technology
Pasadena, CA
USA

Eric D. Feigelson
Department of Astronomy & Astrophysics
The Pennsylvania State University
University Park, PA
USA
and
Center for Astrostatistics
The Pennsylvania State University
University Park, PA
USA

Matthew Graham
California Institute of Technology
Pasadena, CA
USA

Margaret M. Hanson
Department of Physics
University of Cincinnati
Cincinnati, OH
USA

Christoph U. Keller
Sterrewacht Leiden
Universiteit Leiden
Leiden
The Netherlands

Stephen Kent
Fermi National Accelerator Laboratory
Batavia, IL
USA

Ashish Mahabal
California Institute of Technology
Pasadena, CA
USA

Philip Massey
Lowell Observatory
Flagstaff, AZ
USA

John D. Monnier
Astronomy Department
Experimental Astrophysics
University of Michigan
Ann Arbor, MI
USA

J. Allyn Smith
Austin Peay State University
Clarksville, TN
USA

Frans Snik
Sterrewacht Leiden
Universiteit Leiden
Leiden
The Netherlands

Peter B. Stetson
Dominion Astrophysical Observatory
Herzberg Institute of Astrophysics
National Research Council Canada
Victoria, BC
Canada

Alan T. Tokunaga
Institute for Astronomy
University of Hawaii
Honolulu, HI
USA

William D. Vacca
SOFIA
NASA Ames Research Center
Moffett Field, CA
USA

T. L. Wilson
Naval Research Laboratory
Washington, DC
USA

Matt Wood
Department of Physics and Space Sciences
Florida Institute of Technology
Melbourne, FL
USA

Erick T. Young
SOFIA
NASA Ames Research Center
Moffett Field, CA
USA

1 Astronomical Photometry

Peter B. Stetson
Dominion Astrophysical Observatory, Herzberg Institute of Astrophysics, National Research Council Canada, Victoria, BC, Canada

T.D. Oswalt, H.E. Bond (eds.), *Planets, Stars and Stellar Systems. Volume 2: Astronomical Techniques, Software, and Data*, DOI 10.1007/978-94-007-5618-2_1, © Springer Science+Business Media Dordrecht 2013

1 Introduction

Astronomers use the term "photometry" to refer to the precise measurement of the apparent brightness of astronomical objects in particular specified ranges of electromagnetic wavelength in and near the optically visible band. Historically, this task has been most commonly carried out with the human eye, photographic plates, photomultiplier tubes, and – most recently as of this writing – charge-coupled devices. At wavelengths significantly shorter or longer than the optical region, different detector technologies must be used, and some other term than "photometry" is often used to name the process.

The basic unit of astronomical photometry is the magnitude. The modern magnitude scale is logarithmic, with a flux ratio of 100 defined to be precisely five magnitudes; a photometric difference of one magnitude, therefore, corresponds to a flux ratio of $\sqrt[5]{100} \approx 2.512$. Magnitudes increase numerically as the objects become apparently fainter. The conversions between flux F and magnitude m are $m = constant - 2.5 \log F$ and $F \propto 10^{0.4(constant-m)}$. For quick mental calculations, it is useful to remember that a factor of two in flux is about three-quarters of a magnitude, and differences of one, two, three, and four magnitudes correspond to flux ratios of about two and a half, six and a quarter, sixteen, and forty. Also, $2.5 \log 1.01 \approx 0.0108 = 1.08\%$, and most photometrists tend to use the terms "about one one-hundredth of a magnitude" and "about one percent" interchangeably.

This magnitude system – a legacy from the ancient days of subjective, naked-eye brightness estimates – often causes bewilderment or scorn among some of our colleagues in the physics community, who feel that more fundamental units, such as $\mathrm{erg\,cm^{-2}\,s^{-1}\,nm^{-1}}$, would be of more practical use and greater scientific rigor. For example, at wavelengths much longer than we are concerned with here, the "jansky" (defined as $10^{-28}\,\mathrm{W\,m^{-2}\,Hz^{-1}}$) is the most commonly used basic unit of astronomical flux. For our purposes, however, the traditional magnitude system still provides some valuable advantanges, most notably in helping to compartmentalize the observational error budget.

This is because the most widely used methods of astronomical photometry are fundamentally *differential* in nature: the measuring instrument, properly used, observes the brightness *ratio* between two astronomical objects. As we have seen, this ratio can be expressed as a magnitude difference, and under ideal experimental conditions it can in principle be measured with a precision ultimately limited only by the Poisson statistics of the detected photons. Throughout history, the precision of possible *flux-ratio* measurements has exceeded the accuracy of the available magnitude-to-flux conversions (the "*constant*" in the above equations, which is actually a function of wavelength) by several orders of magnitude. This means that photometric measurements expressed as magnitude differences (= flux ratios) relative to some specified reference network of standard stars can retain their quantitative meaning and scientific validity for many decades or centuries, while subsequent generations of more fundamental experiments progressively refine the estimated magnitude-to-absolute-flux conversion. Most astronomical problems relying on photometry for their solution can benefit from the one percent, or millimagnitude, or micromagnitude precision possible from a well-designed photometric experiment without being adversely affected by the typically few percent uncertainty in the estimated magnitude-to-flux conversion factor at any given wavelength (e.g., Massey et al. 1988; Hamuy et al. 1992). When photometric measurements are published on a well-defined magnitude scale, future investigators will be able to apply their own contemporary notion of the magnitude-to-flux conversion to historical results, without having to remember and correct for the conversion that was considered to be valid at the time when each different set of observations was made.

Astronomical photometry is a measuring process, and the observations themselves contain both random and systematic measurement errors that are usually at least as large as the uncertainties we ultimately want in our derived results. Accordingly, it is hardly ever sufficient to imagine the most obvious way of performing a particular measurement. At each step of the way it is essential to consider all possible sources of random and systematic error, and much of the experimental design is aimed at finding clever ways to reduce the random uncertainties, to induce systematic errors to cancel themselves out, and to obtain reliable quantitative estimates of those unavoidable uncertainties that remain. Numerous examples of these basic principles will appear below.

Bessell (2005) has provided an up-to-date, comprehensive survey of modern photometric systems in both historical and contemporary astrophysical context. I will not attempt to duplicate his information (or his reference list) here. Instead, the purpose of this chapter is to provide an overview of the basic principles and techniques of astronomical photometry. The discussion will relate to the optical range of wavelengths, and as far into the near infrared and the near ultraviolet as the same techniques may be said to be appropriate. That is to say, the discussion will stop at that point in the infrared where the thermal emission of the equipment, and the relatively poor stability of both the terrestrial atmosphere and the currently available generation of detectors become (for now, at least) the dominant impediment to photometric precision. The relevance of the present discussion will also cease at that point in the near ultraviolet where the opacity of the atmosphere prevents useful observations from our planet's surface. I shall also concern myself primarily with photometry of stars. Photometry of extended objects like galaxies adds an entirely different class of problem, such as defining the spatial extent of individual entities of various morphologies in some meaningful and homogeneous way, and disentangling the contributions of overlapping objects. These problems have no clear and obvious solution, and should be treated separately.

I am much better at remembering the past than predicting the future. The subject matter of this chapter will therefore be dominated by collective lessons learned by astronomers during, roughly speaking, the six decades from about 1950 to 2010. This interval is neatly bisected by the date 1980, which is approximately when the charge-coupled device, or CCD, began to have a serious impact on astronomical observations (e.g., Leach et al. 1980). In the future, innovations like queue-scheduled observing and dedicated telescopes for large-scale surveys will increasingly and profoundly change the way in which astronomical photometry is done. It can only be anticipated that the consumers of astronomical photometry will become ever more decoupled from those who produce it. I hope that the present chapter will be interesting and educational for the former, and a useful introduction to the field for those hoping to become the latter.

2 General Properties of Photometric Detectors

Before the discussion proceeds further, we should understand that the blanket term *stellar photometry* embraces a number of recognizable subtopics. First, there is what I will call "absolute" or "all-sky" photometry, where the purpose is to measure the magnitude differences (or flux ratios) among objects widely separated on the sky. This may be contrasted with "relative" photometry, which seeks to measure the magnitude differences among objects that can be observed simultaneously, and "time-domain" photometry, which wants to track changes of brightness in the same object observed at different times. Second, we should recognize that there are some practical differences between measuring magnitudes (e.g., for estimating a distance, or defining a variable-star light curve) and colors (= magnitude differences or flux ratios between different wavelength bands for a given object in order to estimate, e.g., temperature, reddening, or other crude properties of its spectral-energy distribution). Third, different types of measuring equipment produce data requiring different types of treatment to provide useful photometric results. Among these, the photographic plate, the photomultiplier tube, and the CCD may be taken as prototypes representing three very different classes of photometric detector.

2.1 Photographic Plates

Historically, the first of these detectors to see astronomical use was the photographic plate. This is an area detector, capable of recording an image of a significant area of sky and acquiring positional and photometric information for hundreds or thousands of targets simultaneously. The quantum efficiency of plates is low, typically of order 1% (e.g., Latham 1976; Kaye 1977), but for some applications (for instance, color-magnitude diagrams of star clusters) the multiplex advantage of recording many targets at once more than makes up for this shortcoming. The plate itself requires careful chemical processing to develop the latent astronomical image and render it permanent, but once this has been done the plate can be analyzed in detail later, and it serves as its own durable data record.

The principal shortcoming of the photographic plate is that it is a *non-linear* photometric detector: there is no simple or universally applicable relationship between a star's apparent brightness on the sky and the properties of its recorded image on the plate. The photographic image of a faint star is a faint, gray, fuzzy patch whose size is set by the astronomical seeing, telescope aberrations, and the scattering of photons within the photographic emulsion. The images of slightly brighter stars are not much larger than those of fainter stars, but become progressively blacker for stars that are brighter and brighter. For stars that are brighter yet, the photographic image saturates, meaning the central blackness remains roughly constant, while the apparent diameter of the star image on the plate grows with increasing stellar brightness. At extremely high flux levels, the photographic emulsion can "solarize": it actually de-exposes itself and the centers of bright star images turn from black back into gray.

The index of photometric brightness that one extracts from a photographic plate, therefore, is a conflation of the blackness and diameter of the star image. This relationship is not necessarily constant even for plates from the same emulsion batch that have been exposed and developed on the same night. Therefore, to obtain the best possible photometry from a photographic plate, it is important to have independently determined magnitudes for a sequence of stars on that plate, spanning the full brightness range of the target stars. These can be used to calibrate an empirical relationship between some photographic magnitude index, however defined, and stellar magnitude on a true photometric scale.

The light-sensitive component of the photographic emulsion consists of grains of silver-halide crystal, of diverse sizes, suspended in a jelly-like substrate. In general, the more sensitive the emulsion the coarser the grain size, but grain dimensions of order 0.1 μm to a few μm are typical. Statistical variations in the distribution of these grains within the emulsion substrate typically limit the precision achievable with a single photometric measurement to not better than 2–5% (e.g., Stetson 1979).

2.2 Photomultipliers

The photomultiplier tube is a technology of World War II vintage. The light of a star, collected by the telescope, is projected through a filter and onto a photocathode where some fraction (typically ~10%; e.g., Giannì et al. 1975) of the photons will cause the ejection of photoelectrons. These are accelerated through a static electric field until they hit a dynode, where the augmented energy of each individual photoelectron causes the ejection of a number of low-energy secondary electrons. These are in turn accelerated through the electric field until they hit another dynode, producing a still larger cloud of low-energy electrons. This process continues through several more stages, resulting in a final amplification factor that can be as great as 100 million, and producing a final burst of electrons large enough to be detected by macroscopic laboratory equipment. In the earlier years of photoelectric photometry, the amount of DC current emerging from the photomultiplier, as displayed in analog form on a dial or on a paper-chart recorder, was the measure of the target star's brightness. Later, pulse-counting circuitry was developed which can detect the output current spike produced by each initial photoelectron. The number of such spikes counted in a fixed amount of time (of order, for instance, 10 s) is a reliable, *digital* measure of the flux from the target star.

Inside the photoelectric photometer, an aperture – usually a simple hole in a piece of metal – is positioned in the focal plane of the telescope. The astronomer centers the image of the target of interest in this aperture, and the light passing through the hole is optically relayed through the filter and a "Fabry," or "field" lens to the photomultiplier (e.g., Johnson 1962). The purpose of the field lens is to produce an image of the entrance pupil (the telescope's primary mirror or lens) on the photocathode of the photomultiplier. Presuming that the primary mirror or lens and the photometer and its individual parts are rigidly attached to the telescope itself, the image of the pupil is fixed in location on the photocathode, and does not move either as the telescope is pointed to different places on the sky or as the star image wanders in the entrance aperture due to imperfect tracking. Furthermore, the image of the pupil occupies a finite area that can be matched in size to the photocathode. Therefore, any positional variations in the sensitivity of the photocathode itself are both averaged out and invariant during the observations. As a result, the photomultiplier can be an almost perfect linear photometric detector, subject only to Poisson statistical variance in the production of photoelectrons, as long as the photon arrival rate is low enough that multiple pulses in the output are not often blended together and counted as one. A simple and reliable correction formula can be applied when the number of coincident pulses is a small fraction of the total (perhaps a correction of a few percent at a count rate of 1 MHz, e.g., Giannì et al. 1975), thus somewhat increasing the useful dynamic range of the instrument. There are no other important instrumental noise sources.

The major practical shortcoming of a photomultiplier is that it is only one "pixel" and therefore can measure only one object at a time. The "pixel" – that is to say, the fixed entrance aperture – must be large not only to contain the wings of the stellar profile; it must also allow for the star image to wander due to seeing and imperfect telescope tracking while the observation is being made. Typically aperture diameters of order ten or tens of arcseconds are used. This large an aperture also lets in a significant quantity of diffuse sky light – terrestrial skyglow, scattered moonlight, zodiacal light, and unrecognized astronomical objects – whose contribution can be estimated from separate offset observations of "blank" sky regions. This both increases the necessary observing time and limits the faintness of objects that can be practically observed: at a good, dark site on a moonless night, the sky brightness is typically apparent magnitude

20 or 21 per square arcsecond at blue and green wavelengths, and still brighter in the red and near infrared. With a measuring aperture of order ten arcseconds in diameter (~100 square arcseconds in area), then, the apparent sky brightness would contribute comparably with a target star of apparent magnitude 15 or 16, while a target star of apparent magnitude 20 or 21 would represent a mere 1% perturbation on top of the sky contamination.

2.3 CCDs

The charge-coupled device, or CCD, combines many of the best properties of the photographic plate and the photomultiplier, and then some. First, it is an area detector, capable of simultaneously recording many science targets in its field of view. Second, when properly adjusted, it is a fairly *linear* detector: the output signal is very nearly proportional to the flux received. Third, it has high quantum efficiency: modern chips record nearly 100% of the incident photons at visible wavelengths (e.g., Leach et al. 1980; Suntzeff and Walker 1996; Burke et al. 2005).

CCDs are physically small compared to photographic plates. Plates with dimensions up to 0.35 m and even 0.5 m square have been used for astronomical imaging while, in contrast, individual CCDs have dimensions of at most a few centimeters. Some modern mosaic cameras containing arrays of CCDs with overall dimensions as large as ~25 cm are in routine operation (e.g., Baade et al. 1999), and still larger arrays are being developed. The financial cost of such cameras, as well as the computing facilities required to handle the large volumes of digital data, can be considerable.

Unlike photomultipliers, early generations of CCDs were plagued by high "readout" noise, a constant uncertainty in the conversion of detected analog signal to digital output data caused by intrinsic amplifier noise. This could be much larger than the Poisson statistics of photon arrivals and a significant limiting factor in low-flux-level applications. Early CCDs also tended to have fabrication defects that could corrupt the data from small patches, or columns, or rows of pixels, sometimes amounting to as much as a few percent of the total detector area. Modern design and fabrication techniques have largely eliminated readout noise and detector blemishes as major causes for concern. Finally, CCDs are highly effective at detecting charged particles as well as photons. In a photon-counting photomultiplier, the arrival of a charged particle (which might be a cosmic ray or a radioactive decay product from the environment near the detector) produces a single pulse in the output: it has the same effect as just one more photon in the diffuse sky light. In a CCD, an energetic particle impact can liberate thousands of electrons and invalidate the signal from one, or a few, or many pixels. Long-term exposure to energetic charged particles eventually leads to deterioration of the detector.

In comparison to photomultiplier data, CCD images require appreciable computer processing to remove instrumental signatures. Unlike the case with the photoelectric photometer with its field lens and its single photocathode, different scientific targets impinge on different parts of the detector. In principle, each of the thousands or millions of pixels that constitute a CCD can have its own individual zero point and scale for flux measurements.

Since the digital image produced by a CCD camera contains some irreducible noise level originating in the readout and amplification electronics, the device is tuned to provide some non-zero output signal in the presence of zero input signal, so that the noise fluctuations cannot ever produce an apparently negative output when an amplified low-level input signal is presented to the analog-to-digital converter. This produces some positive, roughly constant, zero-point offset in the relationship between input flux and output digital signal. The mean value of this offset can be determined by programing the electronics to read out more pixels than are physically present in the device; these phantom pixels contain only the bias offset in the flux scale (plus noise). It can also happen that the reaction of the device to being read out can impose some constant pattern in the zero-signal bias across the face of the device. For a well-behaved detector, the effective flux-measurement zero point of each pixel can be relatively easily accounted for by subtraction of a suitable "bias" frame. This is constructed by averaging together many (to reduce the net effects of readout noise) individual digital images obtained with zero exposure time.

Normalization of the digital image to a common flux *scale* for each pixel requires division by a suitable "flat-field" image. This usually turns out to be the most delicate of the basic calibration steps, and is often a dominant source of inaccuracy in the final scientific photometry. Separate flat-field images are required for the different filters employed by the observer, because the throughput of each filter and the spectral sensitivity of the detector itself may both vary as functions of position in the focal plane. A flat-field image is produced by averaging a large number of exposures to a uniform source of illumination. Usually this is either a white target placed in front of the top end of the telescope tube (or the inside surface of the dome itself) and illuminated by electric lamps ("dome flats"), or the twilight sky observed after sunset or before sunrise, when it is not dark enough to observe astronomical targets ("sky flats").

Dome flats have the advantage that they can be obtained in the daytime when astronomical observations are impossible, so a large number of individual exposures in each of the different filters can be taken and averaged to reduce the Poisson and readout noise to insignificant levels. In contrast, for sky flats, only a limited amount of time is available between when the sky is dark enough not to saturate the detector in the shortest practical exposure time, and when it is no longer bright enough to overwhelm any background astronomical objects. This makes it difficult to obtain high-quality sky flats for a large number of filters on any one occasion.

On the other hand, a dome flat is relatively vulnerable to scattered light, because the inside of the dome is illuminated. Particularly with an open telescope tube (such as a Serrurier truss design), light entering the side of the tube can be scattered off of the inside of the tube components, the telescope's top end, and the inside of the primary mirror hole, and reach the CCD without having passed through the normal optical chain. This can produce a pattern of additional diffuse illumination across the detector which has nothing to do with the throughput of the actual optical system. This compromises the use of the flat field as a measure of the spatial sensitivity variations of the camera. (Grundahl and Sørensen (1996) present a simple and clever way to diagnose scattered light problems with CCD cameras.) In the case of sky flats, with the dome shutter and wind screen maximally restricting the light illuminating the inside of the dome, the diffuse illumination reaching the camera can be dominated by light following the canonical optical path, and scattered-light problems can be minimized.

Sky flats have the additional advantage that the overall spectral-energy distribution of scattered sunlight is likely to more closely resemble that of other astronomical objects than the light emitted by incandescent light bulbs. Even when photographic color-balance filters are employed to make the incandescent light bluer in the optical regime, these can still pass light in the near-infrared range (~700–1,000 nm), which can be a serious problem with a short-wavelength filter that has a significant red leak of its own. Red leaks are a common problem with near-ultraviolet U filters (e.g., Argue 1963; Shao and Young 1965). Designed to transmit primarily light at ~350 nm, they are often insufficiently opaque in the near-IR – which was not a problem with photocathodes that were themselves insensitive at the longer wavelengths, but is a serious concern with CCDs that are often quite sensitive at those same wavelengths. Available incandescent lamps produce many orders of magnitude more photons at ~800 nm than at ~350 nm, so even with a red leak that seems small in percentage terms, short-wavelength dome flats can be dominated by photons of completely inappropriate wavelength (see, e.g., Stetson 1989a, ➋ Sect. 1, for an example). This is less of a problem with sky flats, where short-wavelength photons are abundant compared with long-wavelength ones.

Optimum flat-fielding is clearly contingent on the specifics of the dome, the telescope, the camera, the filters, the detector, and the scientific goals, so it is impractical to provide a universal set of best practices. On a personal level, I have had generally good results by combining dome flats and sky flats in the following way. The mean sky flat in a given filter is divided by the mean dome flat in the same filter. The resulting ratio image is then smoothed using a kernel of a few pixels and, finally, the mean dome flat is multiplied by the smoothed ratio image:

$$\text{Flat} = \text{Dome} \times \left\langle \frac{\text{Sky}}{\text{Dome}} \right\rangle_{\text{smoothed}} .$$

The dome flat, which is the average of a very large number of individual well-exposed, diffusely illuminated images, contains the best information on the relative quantum efficiencies of individual pixels. The sky flat is typically based on a smaller number of images, and since the sky brightness is varying rapidly with time during twilight, it is difficult to obtain the optimum exposure level in each individual image; therefore, individual pixels in the sky flats can be subject to Poisson statistics. However, the sky flat contains the purest information on the global sensitivity variations across the face of the filter/detector combination. Multiplying the dome flat by the smoothed ratio of Sky:Dome retains the excellent pixel-to-pixel Poisson statistics of the dome flats, but restores the large-scale illumination pattern of the sky flats.

If the scientific goal of the observation is to measure the surface brightness of an extended astronomical object, it may also be desirable to subtract a "dark" frame to account for the fact that over time the CCD can accumulate thermal signal from within the detector itself. This dark signal is not necessarily uniform over the detector, especially when – as sometimes happens – some components within the CCD itself act as light-emitting diodes (e.g., Suntzeff and Walker 1996). Typically, a dark frame is constructed by averaging a number of exposures obtained under ordinary observing conditions, but with the camera completely sealed off from any incident astronomical or environmental light by a dark slide. The dark frames may be obtained with the same integration time as the science frames, or – if the detector system is well behaved so the dark signal accumulates linearly over time – they may be proportionately scaled to different integration times. For most purposes of stellar photometry, the dark current can be considered merely another component of the incident diffuse sky light, and dark frames can be omitted.

Finally, it can happen that the actual time that the camera shutter is open can differ from the time intended by the observer. A finite amount of time, typically of order 100 ms can pass between when the control computer instructs the mechanical shutter to open and when it is in fact completely open. Similarly, some amount of time passes between the instruction to close and when the shutter is actually closed. These two time delays need not be numerically the same. Furthermore, if the shutter itself is placed closer to the focal plane than to a pupil, vignetting by the shutter aperture as it is opening and closing can cause different parts of the imaged field to receive different total exposure times. If the shutter timing error is of order 100 ms, then in a 10-s exposure this can cause a systematic, possibly position dependent, 1% error in the fluxes inferred for the science targets. Obviously, the systematic error becomes worse for shorter exposures, and may be negligible for much longer exposures. However, shutter-timing corrections, whether constant timing offsets or two-dimensional correction maps, can readily be estimated from nighttime observations of bright stars (e.g., Massey and Jacoby 1992), or an appropriate series of dome-flat exposures of different integration times (e.g., Stetson 1989a, ➲ Sect. 2).

The rectification of the raw images generally takes place in the following order:

1. For the bias, dark, flat, and science images, the average bias level is determined from the overscan region (phantom pixels) and subtracted from the overall images. Either this is subtracted as a constant value from the entire image, or it can be determined and subtracted on a row-by-row basis if the bias level fluctuates perceptibly during readout. Any pixels lying outside the light-sensitive parts of the data array are then cropped from the digital images.
2. The mean bias frame is constructed, and subtracted from the dark, flat, and science images.
3. If desired, the mean dark frame is constructed, and subtracted from the flat and science images after scaling to the appropriate exposure times.
4. If necessary, the flat-field images are individually corrected for positional shutter-timing errors. They are then averaged together, and the resulting mean image is normalized to an average value of unity.
5. Each science image is then divided by the appropriate flat-field image. If the shutter-timing error shows positional dependence, it can be corrected now; if the timing error is constant across the image, it can simply be applied to the requested integration time to produce the true integration time in a later stage of analysis (see below).

The relatively complex readout and amplification electronics of a CCD camera system and its cabling can also be prone to maladjustment and sensitive to ambient electric noise. Furthermore, it can be difficult to achieve a sufficiently uniform distribution of incident light with appropriate spectral characteristics for the production of completely satisfactory flat-field corrections. Massey and Jacoby (1992) have provided some common sense and practical guidelines for recognizing, diagnosing, and in some cases rectifying, some typical problems.

All told, however, the shortcomings of CCDs are minor compared to their advantanges, and these detectors have now almost entirely supplanted photographic plates and photomultipliers in contemporary photometric studies. In fact, CCDs are so nearly ideal as detectors that even if they are themselves soon replaced by some improved technology, most of the following discussion that relates directly to CCD-based techniques will likely be applicable to the next generation of detectors, which can at best be only marginally more capable and convenient.

3 The General Photometric Problem

In what follows, I shall adopt all-sky "absolute" photometry with CCDs as a standard photometric reference mission. I shall also try, as appropriate, to point out ways in which different types of photometric program may require different approaches or allow shortcuts to be taken.

The type of measurement that I am calling absolute photometry requires what astronomers call "photometric" observing conditions. This means that the properties of the terrestrial atmosphere (and of the instrumentation) must be effectively constant over the length of time and range of direction required for the measurement of the science targets as well as a number of reference standard objects of known photometric properties. For "effectively constant," I mean that the percentage variation in the throughput of the atmosphere and in the sensitivity of the equipment on relevant time scales must not be vastly larger than the percentage uncertainty desired in the final photometric results.

As stated in the introduction above, astronomical photometry is a measuring process, subject to both random and systematic measurement errors that can be large compared to the accuracy we desire in our final results. Therefore, a statistical approach to the subject is essential, and it is normal to vastly overdetermine the solution of the problem. That is to say, we try to obtain many independent measurements of the science targets *and* of standard stars of well-defined photometric properties, and then estimate the final desired quantities through a statistical procedure of optimization such as the method of least squares. As a starting point for the present discussion, a minimalist set of formulae representing the astronomical photometry problem typically looks something like the following[1]:

$$v \doteq V + \alpha + \beta(B{-}V) + \gamma X, \tag{1.1}$$

$$b \doteq B + \delta + \zeta(B{-}V) + \eta X, \ldots \tag{1.2}$$

In these equations, I use the symbol "≐" to indicate that here we are not dealing with equations in the strict definition of the term, but rather with a network of many relationships of this form, for which we intend to find an optimum albeit inexact solution, for example, in a maximum-likelhood sense. Here I have used lower-case Roman letters to represent the *instrumental* magnitudes in the *visual* (v) and *blue* (b) bands – that is, the raw observations themselves, which are presumed to be subject to observational uncertainty – defined as follows:

$$(\textit{instrumental magnitude}) \equiv (\textit{arbitrary constant}) - 2.5\log\left[\frac{(\textit{integrated signal})}{(\textit{integration time})}\right].$$

In this equation, "signal" may refer to electron pulse counts produced by a photomultiplier or to digital data numbers ("DN") produced by a CCD, but in either case they are presumed to come from the science target only, after removal of the instrumental signature and correction for any contribution of sky photons. Furthermore, it is essential to note that "integration time" here refers to true integration time, corrected for any shutter-timing errors. Upper-case Roman letters in (➋ 1.1) and (➋ 1.2) represent the corresponding photometric magnitudes on the standard system – those that you would find in the literature, and those that you would hope to publish. Finally, upper- and lower-case Greek letters are parameters that describe the instantaneous state of the instrumentation and atmosphere at the time the observations were made: (a) zero points α and δ describe the overall throughput of the system in each bandpass; (b) coefficients β and ζ represent *first-order Taylor expansions* of the mismatch between the observer's photometric bandpasses and those that were used to establish the standard photometric system; and (c) coefficients γ and η describe the first-order extinction of the incident flux that occurs in the terrestrial atmosphere. The upper-case Greek letter chi, X, represents the path length of the observation through Earth's atmosphere, which is commonly referred to as the "airmass." Each of these points requires some further explication, and I will take them in reverse order.

[1]Purely for purposes of illustration, I have chosen a notation that implies we care considering the visual and blue (V and B) bands of the venerable (Johnson and Morgan 1953) photometric system. In fact, the discussion to be presented will have far greater generality.

3.1 Atmospheric Extinction

We know from radiative transfer that the intensity of a beam of light passing through a uniform layer of some absorbing/scattering medium diminishes as $I_\circ\, e^{-\tau}$, where τ is the optical thickness of the semi-opaque layer, and $I_\circ$ is the beam's initial intensity before entering the medium. The atmosphere over the observatory can be thought of as an arbitrarily large number of arbitrarily thin horizontal layers, each having some vertical thickness Δh_i and average opacity κ_i. Each of these layers reduces the intensity of a vertically incident beam by a factor $e^{-\tau_i} = e^{-\kappa_i \Delta h_i}$. The final intensity of the beam at the telescope, I_f, is then given by

$$I_f = I_\circ \prod_i (e^{-\tau_i}) = I_\circ e^{-\sum_i \tau_i} = I_\circ e^{-\tau_T},$$

where $\tau_T \equiv \sum_i \tau_i$ is the total optical thickness of the atmosphere above the observatory. But a magnitude is a logarithm of an intensity, so if m is the star's apparent magnitude at the observatory,

$$\begin{aligned} m &= constant - 2.5 \log I_f \\ &= constant - 2.5 \log I_\circ - 2.5 \log e^{-\tau_T} \\ &= m_\circ + 2.5 \log e^{\tau_T} \\ &= m_\circ + (2.5 \log e)\, \tau_T, \end{aligned}$$

where $m_\circ$ is its apparent magnitude outside the atmosphere, and $+(2.5 \log e)\, \tau_T$ is the net extinction along the line of sight, now expressed as a magnitude difference rather than a flux ratio. On a magnitude scale, the extinction produced by the atmosphere is a linear function of the total opacity along the line of sight.

In a homogeneous, isotropic, plane-parallel stratified atmosphere, the path length of an *inclined* line of sight through any given layer of the atmosphere is simply $\Delta h_i \sec z = \frac{\Delta h_i}{\cos(z)}$, where z represents the zenith distance of the observation: the angle between the zenith of the observatory and the direction to the science target at the instant the observation was made. From spherical trigonometry,

$$\sec z = (\sin L \sin D + \cos L \cos D \cos h)^{-1}, \tag{1.3}$$

(e.g., Smart 1977) where L is the observer's terrestrial latitude, D is the target's declination, and h is the target's hour angle at the moment of observation. The integrated opacity along this line of sight, then, is $(2.5 \log e)\, \tau_T \sec z$, and we may define $(2.5 \log e)\, \tau_T \equiv \gamma$ or ζ, while $\sec z \equiv X$. Thus, by definition, an observation is made through "one atmosphere" when looking straight up ($X = \sec(0) = 1$), and to very good approximation the extinction increases proportional to $\sec z$ at other angles relative to the vertical: $X \approx 1.15$ at a zenith distance of 30°, ≈ 1.4 at 45°, and ≈ 2 at 60°, and it becomes infinite when the target is on the horizon. It should be noted that when tracking a target at any given declination the airmass is at a minimum and changes most slowly with time at the meridian ($|h| \approx 0$) – especially when the target passes near the zenith ($h \sim 0$ and $L \sim D$) – and changes rapidly for angles far from the vertical.

In actuality, of course, the layers of even a homogeneous and isotropic atmosphere are curved along with the underlying surface of the Earth, and due to refraction the true and the apparent zenith distance will differ by a small amount. Several improvements to (➋ 1.3) have been proposed to better describe the true atmospheric path length of an observation. Among these are

$$X = \sec z - 0.0018167(\sec z - 1) - 0.002875(\sec z - 1)^2 - 0.0008083(\sec z - 1)^3$$

(e.g., Hardie 1962) and

$$X = (1 - 0.0012 \tan^2 z) \sec z$$

(Young 1974). However, the differences between these elaborations and the simple $X = \sec z$ are small compared to the typical inherent variability – both temporal and spatial – of the terrestrial atmosphere. Furthermore, the true correction from a plane-parallel approximation to a spherically symmetric atmosphere depends upon the vertical distribution of the opacity sources. Therefore, the more complex formulae are really unnecessary in all but very extraordinary experiments beyond the scope of the present discussion.

For an observatory at an intermediate elevation (say, 2,500 m above sea level) on a clear night, typical extinction values are ~0.5 mag airmass^{-1} in the near ultraviolet (~350 nm), 0.25 in the blue (450 nm), 0.14 in the visual (550 nm), 0.10 in the red (625 nm), and 0.07 in the near infrared (825 nm). These values are just order-of-magnitude approximations (compare them to the average values given by Landolt (1992) for Cerro Tololo: 0.591, 0.276, 0.152, 0.108, and 0.061; or those given by Sung and Bessell (2000) for Siding Spring: 0.538, 0.309, 0.160, 0.108, and 0.086). They should definitely not be regarded as correct for every observatory, or for every night at any observatory; the actual values could easily be different by a factor of order two (and possibly variable over time and/or direction) under some weather conditions. Landolt (1992) states that he experienced V-band extinction coefficients ranging from +0.099 to +0.250 mag airmass^{-1} during his observing runs at Cerro Tololo. I have measured V-band extinction values as high as 0.6–0.7 mag airmass^{-1} on La Palma, when strong, dust-laden winds were coming off the Sahara desert. The eruption of the Pinatubo volcano on June 15, 1991, produced spectacular sunsets as well as atmospheric extinction effects that could be detected around the world for several years afterward (e.g., Forbes et al. 1995).

3.2 Bandpass Mismatch

The terms proportional to $B-V$ color in (➋ 1.1) and (➋ 1.2) allow for the fact that with available filters and detectors it is virtually impossible to perfectly reproduce the spectral response of the equipment that was originally used to define the standard photometric system. For example, the solid curves in ➋ *Fig. 1-1* represent the throughput as a function of wavelength for the filters in Cerro Tololo's *UBVRI* filter set No. 3 that Landolt (1992) used to define his version of the U, B, and V photometric bandpasses of Johnson (1955; also Johnson and Morgan 1953, and Johnson and Harris 1954) and the R and I bandpasses of Cousins (1976; also Kron et al. 1953). For clarity, these have all been normalized to a peak throughput of unity. The dashed curves represent the typical transparency of the terrestrial atmosphere as a function of wavelength (labeled "atm"; these data have been taken from Allen 1973, p. 126); the reflectivity of *one* typical aluminum surface ("Al," Allen 1973, p. 108); and the quantum efficiency of Cerro Tololo's RCA 3103A photomultiplier tube ("PMT," as tabulated in Landolt 1992; here the curve has been scaled up by a factor of 5 for clarity). ➋ *Figure 1-2* shows the effect of multiplying the filter curves by the atmospheric transmission, the aluminum reflectivity squared (to account for the primary and secondary mirrors of a typical Cassegrain photoelectric photometer), and the detector quantum

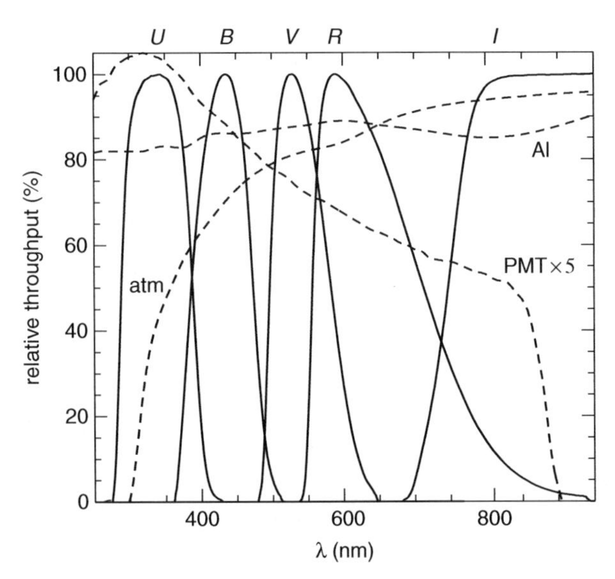

Fig. 1-1

The *solid curves* show the transmission as a function of wavelength for the five *UBVRI* filters in CTIO Filter Set #3 that were used by Landolt (1992) to establish his network of equatorial standard stars. For clarity, these have all been normalized to unity. The *dashed curves* show the typical transparency of the terrestrial atmosphere ("atm"), the reflectivity of one average aluminum surface ("Al"), and the spectral sensitivity of CTIO's RCA 3103A photomultiplier, also used by Landolt. The photomultiplier sensitivity has been scaled up by a factor of five for clarity

efficiency. Note that the short-wavelength side of the U bandpass is defined by the terrestrial atmosphere, *not* by the filter, and the long-wavelength side of the I bandpass is defined by the detector, *not* the filter. The effects of the atmosphere, mirrors, and detector quantum efficiency on the other bandpasses are comparatively minor.

In *Figure 1-3*, the dashed curve represents one filter/CCD combination designed at Cerro Tololo to approximate as nearly as possible the standard V bandpass, Landolt's version of which is repeated here as the solid curve. The two curves have been normalized to the same area. It is evident that the match between the two bandpass curves, while good, is not perfect. In particular, the Cerro Tololo bandpass is shifted toward the blue end of the spectrum by a few percent, relative to the standard bandpass. The two dotted curves in *Figure 1-3* represent spectral-energy distributions for the blue star 60 Cygni (spectral class B1v, $B{-}V = -0.27$) and the red star HD 151288 (spectral class K7v, $B{-}V = +1.35$) from the Gunn and Stryker (1983) atlas. From this, it is apparent that if the stars appear to be exactly equally bright when measured in the standard V bandpass, then the blue star will be measured somewhat brighter than the red one in the Cerro Tololo V bandpass. Conversely, in another V filter that was shifted to a slightly longer wavelength than the standard bandpass, the red star would be measured somewhat brighter than the blue one.

In (1.1) above, then, the term $\beta(B{-}V)$ plays the role of the first-order term in a Taylor expansion describing the differences between the two bandpasses:

$$v(observed) \sim V(standard) + (\lambda_o - \lambda_s)\frac{d(-2.5 \log F_\lambda)}{d\lambda},$$

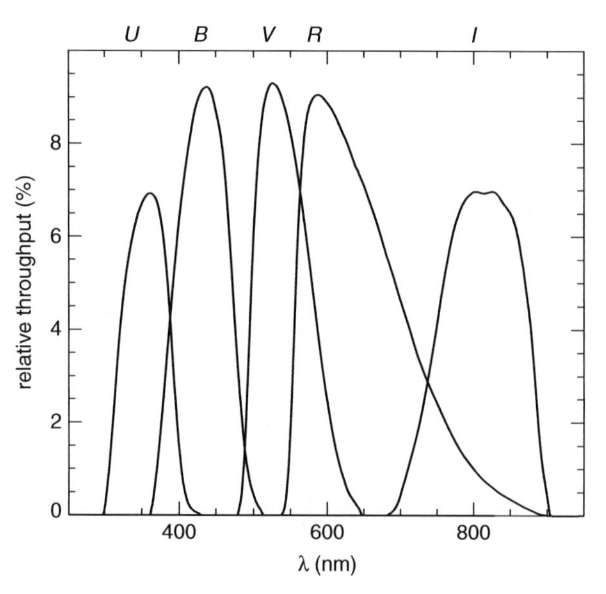

Fig. 1-2
The throughput as a function of wavelength for the five filters shown in *Fig. 1-1* after including the transmission of the atmosphere, two reflections from aluminum surfaces, and the quantum efficiency of the photomultiplier

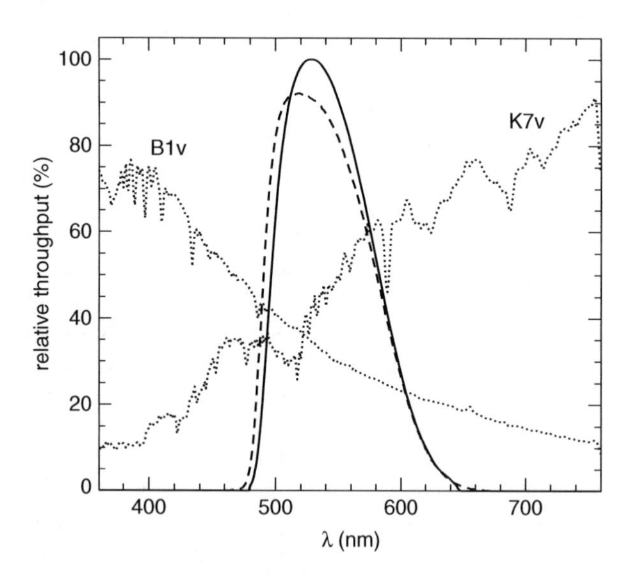

Fig. 1-3
The *solid curve* shows the same *V* bandpass as in *Fig. 1-2*, while the *dashed curve* shows the bandpass produced by the *V* filter from CTIO filter set #2 combined with the RCA #1 CCD. The two curves have been normalized to the same total area. The two *dotted curves* show the spectral-energy distributions of the blue star 60 Cygni (spectral class B1V, $B–V = -0.27$) and the red star HD 151288 (spectral class K7V, $B–V = +1.35$) from the Gunn and Stryker (1983) atlas

where λ_o and λ_s represent the effective wavelengths of the observed and standard photometric bandpasses, respectively. In the empirical correction, β is $\lambda_o - \lambda_s$ and $B-V$ is $-2.5\, d\,(\log F)/d\,\lambda$ in suitable units. We use the $B-V$ color of the star for this purpose simply because it is a conveniently available, albeit crude, empirical measure of the spectral-energy gradient in the stars we are observing. The $V-R$ color, the $V-I$ color, or even the Morgan-Keenan spectral class could be just as easily used, provided they are available. However, it is clearly convenient and appropriate to use a color based upon filters having wavelengths near the spectral region where one is actually observing, because then the observed spectral-energy gradient will be most suitable for the Taylor expansion. Not quite as obviously, it is important that the color be based upon the photometric bandpasses that one is actually employing for the observations, because it will then be possible to define that color for science target stars that do *not* already have the necessary photometric information, as well as for the standard stars whose colors can be taken from the literature (see Sect. 3.4 below).

The numerical value of the correction for bandpass mismatch is derived from observations of stars of known photometric properties rather than, for instance, filter transmissions and detector quantum-efficiency curves measured in the laboratory, because the laboratory cannot adequately reproduce actual observing conditions. A color correction based upon direct observations of standard stars includes the spectrum-distorting effects of variable factors like the dust aloft in the atmosphere on the night in question, the current state of tarnish on the telescope's reflective surfaces, and changes in the filter throughput caused by age and the night's relative humidity, as well as the more static physical properties of the filter and detector. If the filter/detector combinations are well designed, the necessary bandpass-mismatch corrections amount to only a few percent, and if those corrections themselves can be estimated to within a relative precision of a few percent then the residual systematic errors will be at the millimagnitude level, which is respectable for all-sky absolute photometry and quite good enough for comparison with theoretical predictions of star colors.

It is vital to note that this empirical correction works adequately for most purposes because the spectral-energy distributions of normal stars depend strongly on a single parameter – temperature – and much more weakly on secondary characteristics such as surface gravity, chemical abundance, and rotation velocity. Different stars with the same $B-V$ color (for instance) will usually have very similar spectral-energy distributions in general – at least within the wavelength range spanned by the B and V filters – and a correction derived from a set of normal standard stars will be valid for similarly normal science target stars. If a given set of filters is poorly matched to the standard set, however, serious systematic errors can occur if color corrections based upon ordinary standard stars are applied blindly to objects with strikingly dissimilar spectral-energy distributions, such as supernovae, quasars, or composite stellar populations. In such circumstances, it is important to investigate the consequences of such mismatches through numerical experiments with actual filter/detector sensitivity curves from the laboratory and representative spectral-energy distributions. Even then, it may not be possible to properly characterize the true photometric properties of extragalactic targets whose Doppler shifts are unknown.

In addition, severe bandpass mismatch can produce systematic errors even for normal stars if they have strong spectral features. For instance, the spectra of M stars and carbon stars contain molecular bands with equivalent widths that are an appreciable fraction of the effective width of typical photometric bandpasses. In such cases, bandpass mismatch may produce systematic calibration errors as a function of surface gravity, chemical abundance, or small differences in temperature. The same concern is particularly severe for near-ultraviolet bandpasses that include the Balmer convergence and Balmer jump (roughly 370–390 nm). In stars of spectral classes late B and A, where the hydrogen absorption spectrum is at its strongest, the relation between an instrumental u magnitude and a standard U magnitude can be a seriously complicated function of temperature and gravity. Considerations like these represent an absolute limit to the precision with which observations obtained with one equipment setup may be compared to data obtained with another, or to theoretical models.

3.3 Zero Points

The constant terms α and δ in (➋ 1.1) and (➋ 1.2) embody the cumulative effect of all the multiplicative scaling factors that govern how much light ultimately arrives at the detector: the aperture of the telescope, the reflectivity of mirror coatings, and losses due to dust and unwanted reflections at the surfaces of lenses, windows, and other optical elements, as well as the throughput of the filter and the quantum efficiency of the detector at the wavelengths of interest. When the calibration formulae are expressed in this way, however, these zero points do *not* include the overall transparency of the atmosphere. This is entirely encapsulated within the extinction terms: if X is unity at the zenith and increases as sec z at other zenith distances, then α in (➋ 1.1) and δ in (➋ 1.2) are the differences between the observed and the true magnitude when the calibration formulae are extrapolated to $X = 0$, that is, they represent the calibration formulae that would have been obtained had the observer been outside the atmosphere. To the extent that the transparency of the atmosphere is the only factor that is likely to change significantly over the course of a few nights (if the increasing deterioration of the reflective mirror coatings, for instance, is imperceptible over that span of time), then these zero points should be effectively constant for a given telescope and instrument from one night to the next.

3.4 Methodology

➋ Equations 1.1 and ➋ 1.2 are used in two distinct ways. In the first instance, in what one might call "training" mode, the formulae are applied to observations of standard stars of known photometric properties. Here, the numerical values of V, B, and X are regarded as perfectly known quantities; the values of v and b are observed quantities with their associated standard errors; and the numerical values of the fitting parameters $\alpha, \beta, \ldots, \zeta$ are initially unknown, but can be estimated by a statistical procedure such as least squares. The second use of the equations, in what one might call "application" mode, treats $\alpha, \beta, \ldots, \zeta$ and X as parameters whose values are now known with negligible uncertainty, v and b are again observed quantities subject to measurement errors, and V and B are now the unknowns to be estimated by statistical analysis.

As mentioned above, to obtain final photometric results whose uncertainties are small compared to the typical random measuring error associated with a single observation, it is desirable to grossly overdetermine this network of equations in both the training and application modes. In training mode, that is to say, observations of a large number of standard stars spanning a suitably broad range of color $B–V$ and of airmass X are inserted into (➋ 1.1) and (➋ 1.2) to estimate a single set of best values for the parameters $\alpha, \beta, \ldots, \zeta$ by a statistical optimization. In the application mode, many different observations of a given target star are similarly subjected to statistical optimization to arrive at final values for V and B that are improved by the canonical factor $\sim \sqrt{N}$ relative to the precision of a single observation.

There are some features to note and ponder in this formulation. For example, (1) in both the training and application modes, the quantities that are subject to the greatest observational uncertainty, namely, the instrumental magnitudes v and b, are isolated on the left side of the equals sign as is appropriate for a least-squares adjustment. The right side of the equation is supposed to consist only of fitting parameters and numerical quantities presumed to have negligible uncertainties. (2) The airmass of the observation – the atmospheric path length X – is treated as one of these perfectly known quantities in both the training and application stages; defined purely geometrically, for an instantaneous observation its value can be calculated directly without appreciable uncertainty, but the appropriate numerical value to use for an observation of extended duration is not necessarily obvious (although Stetson 1989a, Sect. 3, suggests an approximate numerical solution to the problem). (3) In the application mode, the initially unknown quantity $B–V$ appears in both equations; the numerical value that is to be assigned to $B–V$ is a direct consequence of the values of V and B, which are themselves initially unknown, and are subsequently estimated with some degree of uncertainty. In a standard least-squares approach to statistical optimization, this requires that the solution of equations be iterated: given an initial guess at the $B–V$ color of the target star (which could be zero, but would even better be set to the typical color of stars generally resembling the target star), provisional values for V and B are derived. These are differenced to yield a new estimate of the $B–V$ color of the target star and the solution is performed again. This process is repeated until the answer stops changing. As long as $|\beta|$ and $|\zeta|$ are both $\ll 1$ (as would be the case for even a moderately appropriate filter/detector combination), this process converges rapidly to a solution. However, it remains the case that the numerical feedback inherent in requiring a value for $B–V$ in order to compute values for V and B can potentially introduce some bias in the final results. It is therefore useful to reduce multiple observations of the target star simultaneously. In this way, it is possible to estimate optimum values of V, B, and $B–V$ whose global uncertainties are small compared to the uncertainty of a single measurement of v or b. This, combined with numerical values near zero for β and ζ minimizes the potential for bias in the photometric results; and the fact that $B–V$ is estimated rather than known with arbitrary precision becomes a matter of little concern.

The photometric problem has not always been solved in this way. In particular, for ease of computation some researchers adopt an alternative approach typified by equations of the form

$$V = v + \alpha' + \beta'(b–v) - \gamma' X, \tag{1.4}$$

$$B = b + \delta' + \zeta'(b–v) - \eta' X, \tag{1.5}$$

where $b–v$ is now an "instrumental color" defined as a difference between instrumental magnitudes from a near-simultaneous pair of v, b observations. Equation 1.5 is sometimes replaced by

$$B–V = \delta' + \zeta'(b–v) - \eta' X, \tag{1.6}$$

where now $\zeta' \approx 1$. This approach has the advantage that, since the value of the instrumental $b–v$ color is available for each suitably matched pair of observations – whether of standard or of target star – no iteration is required either for the training or for the application of the calibration. This approach does have some problems, however.

1. The instrumental color $b–v$ is not necessarily a well defined quantity. If v and b are measured at different times, it is possible that they are measured at significantly different airmasses. An instrumental $b–v$ color for one star measured in v and b at one pair of airmasses will

not be on strictly the same photometric system as the color of another star measured at a different pair of airmasses.

2. The instrumental colors $b-v$ for a set of stars contain random measuring errors; in the case of CCD observations, in particular, these will be at least as large as the uncertainties in the v and b magnitudes from which they are constructed. When error-containing quantities appear on the right-hand side of the equations in a standard least-squares adjustment, a systematic bias can result. This fact is known to anyone who has tried fitting a linear regression of y as a function of x and then x as a function of y for some noisy data set, and discovered that the two slopes so derived are not negative inverses of each other. More sophisticated approaches to optimization that ameliorate or eliminate this problem exist,[2] but they are neither conceptually trivial nor widely used by photometrists.
3. This approach restricts the available observing strategies. In particular, the magnitudes v and b should always be observed in air-mass-matched pairs. Otherwise the instrumental magnitudes must be individually corrected for the difference in extinction and any other relevant variables before the instrumental color can be inferred. In practice, this means either manual intervention by the person reducing the data, or particularly intricate software. It can be difficult to apply this method to archival data, or to data obtained by a colleague who did not strictly adhere to the desired observing protocol.

It should be stressed that these objections apply most directly to data obtained with CCDs, and not so much to data from a photoelectric photometer. A well-designed photoelectric observing program usually ensures that each observation cycle is symmetric in time (e.g., Landolt 1992, Sect. 2). For instance, assuming that integration times are short (say, of order 10 s, e.g., Giannì et al. 1975) and that filter changes are practically instantaneous, the astronomer could observe the star (along with the diffuse sky light that is also in the aperture) and an offset blank-sky position in the following order: $vbbv$(star plus sky)–$vbbv$(blank sky)–$vbbv$(star plus sky). This observing sequence can be repeated as many times as neceesary to achieve the desired Poisson statistics. Under these circumstances, an instrumental color

$$b-v = 2.5\log\left[\frac{(v \text{ count rate star+sky } \textit{minus } v \text{ count rate sky alone})}{(b \text{ count rate star+sky } \textit{minus } b \text{ count rate sky alone})}\right]$$

is a well-defined quantity. Even if the atmospheric transmission or the sky brightness fluctuates significantly during the observing, these variations will cancel to first order in each individual symmetric observing sequence, and will average out as $\sqrt{N}$ if the sequence is repeated N times. In such a case, it is quite possible that the instrumental $b-v$ color can be known with considerably higher precision than either of the instrumental v or b magnitudes inidividually.

However, several factors conspire to render this approach impractical when observing with CCDs. It takes time to read out and digitize the data from the many pixels of a CCD for each individual exposure; new photons cannot be collected while this readout is taking place. Furthermore, each digitized image requires storage space and subsequent processing effort. Repeated short exposures are subject to higher net readout noise, relative to the Poisson photon statistics, than single, longer exposures. Finally, astronomers typically try to observe the faintest objects they can; otherwise they would use a smaller, less oversubscribed telescope. These factors militate against observing sequences that consist of many short integrations.

[2]One example may be found in Stetson (1989b), lecture 3A.

In contrast, when the reduction scheme typified by (➋ 1.1) and (➋ 1.2) is used, it is not important whether the observations in the different filters are closely matched in time and airmass. It is not necessary that they be obtained on the same night, or even with the same telescope or camera. As long as appropriate standard stars of known photometric properties are observed on each night, it is possible to use those observations in training mode to derive the parameters $\alpha, \ldots, \gamma$ and/or $\delta, \ldots, \eta$ that characterize the instantaneous properties of the atmosphere and instrumentation in the relevent photometric bandpass(es). In application mode, for any given target object *all* the observed instrumental magnitudes v or b, each one accompanied by the values of $\alpha, \ldots, \gamma$ or $\delta, \ldots, \eta$ appropriate for the telescope, camera, and night on which it was obtained, can be included in a single statistical optimization to obtain the unique values of V, B, and $B-V$ that best explain the entire corpus of data for that target.[3]

3.5 Higher-Order Effects

As mentioned above, (➋ 1.1) and (➋ 1.2) are minimal: they illustrate the most important problems inherent in the calibration of astronomical photometry, and represent first-order solutions to those problems. It is often necessary to expand the above equations to deal with more subtle effects and achieve the desired accuracy in the final results.

1. When the bandpass of a filter is wide, say $\gtrsim 100$ nm in effective width for wavelengths in the optical regime (i.e., a spectral resolution $R \lesssim 5$), it is not necessarily sufficient to assume that the atmospheric opacity is perfectly gray across wavelength range of the filter. For a blue star observed in a wide filter, most of the transmitted flux will lie toward the blue end of the bandpass; for a red star observed in the same filter, most of the flux will lie toward the red end. Since the atmospheric opacity generally increases toward shorter wavelengths, the measured flux of the blue star will experience greater extinction than the flux of the red star in the same filter under otherwise identical circumstances. Numerically, this can be modeled by replacing ζ in (➋ 1.2) by $[\zeta_1 + \zeta_2(B-V)]$, for instance, and similarly for other bandpasses. In typical broad-band filters (~100 nm) this effect is small but not completely absent: An order-of-magnitude estimate is $\zeta_2 \sim -0.02$ mag per airmass per one magnitude change in (say) $B-V$ color (cf. Landolt 1992, Table 1). This correction is sufficiently small that it is often better to estimate it from laboratory spectral-sensitivity curves and measured stellar spectral-energy distributions than from direct observations of standard stars. The effect is obviously most pronounced at the shortest wavelengths, where the atmospheric opacity is high in general and changes most rapidly with wavelength. It can be usually be neglected at wavelengths longer than 500 nm, and for bandpasses of a few tens of nanometers or less in breadth.
2. If the bandpass mismatch is appreciable, and if photometric results of the highest absolute accuracy are desired, a first-order Taylor expansion in a color may not be adequate to match the observed instrumental magnitudes to the standard ones. If the filter being used is broader (or narrower) than the standard filter, for instance, the equipment will measure *both* blue stars and red stars as appearing brighter (or fainter) than yellow stars of the same standard magnitude. Or, when a near-ultraviolet bandpass is being calibrated it may be found

[3]This, of course, presupposes that the photometric properties of the target object do not vary with time. The diligent astronomer will certainly test the validity of this assumption by examination of the fitting residuals.

that one correction term proportional to $B-V$ color combined with another term proportional to $U-B$ color provides a better fit than either term alone: because of the Balmer jump, A-, F-, and G-type stars can all have the same $U-B$ colors despite the fact they occupy quite different ranges of effective temperature and of $B-V$ color. The term in $B-V$ allows the calibration to provide different corrections to the U-band magnitudes for the different spectral classes.[4]

3. There is no guarantee that on any given occasion the properties of the atmosphere will be isotropic, homogeneous, or constant with time. The mean properties of the atmospheric dust may change as the wind increases or decreases, or as the vehicle traffic in the neighborhood of the observatory drops off after sunset. Cirrus too thin to see with the naked eye may form or dissipate. For an observatory with a coastal plain and broad ocean on one side and high desert mountains on the other, it is possible that the atmospheric extinction may change as a function of direction. Smooth, gradual variations of this sort can be modeled by the addition of suitable terms to the calibrating equations. For instance, γX could be replaced by $(\gamma_A + \gamma_B t)X$, where t is the time of the observation expressed, say, in ($\pm$) hours from midnight. Directional terms could be added; functional forms involving the quantity $(X-1)$, such as

$$(X-1)\cos(azimuth); \quad (X-1)\sin(azimuth)$$

are attractive because they and all their derivatives are continuous and smooth through the zenith.

4. With large-format CCD cameras, it is often difficult to properly correct the science images for vignetting and other spatial variations in the throughput of the optical system. As mentioned above, when taking flat-field exposures of a target inside the dome, or of the bright twilight sky, if the telescope is not perfectly baffled it is possible for scattered light to reach the detector through the side of the open telescope tube without taking the normal path through the telescope optics. The resultant flat-field images can contain a gradient of extra light which represents an additive (and not multiplicative) component on top of the flux pattern produced by the telescope. When science exposures are divided by these flat-field images, the relative quantum efficiencies of adjacent pixels are effectively rectified, but a variable flux *scale* is introduced across the astronomical field. This manifests itself as a variation of the photometric zero point as a function of position within the image. One way to ameliorate this problem *ex post facto* is to include additional calibration terms incorporating the x and y coordinates of each star's location in the natural coordinate system of the detector.

5. In photometric bandpasses where the atmospheric opacity includes a significant contribution from molecular bands, as the near-infrared K band at 2.2 μm contains water-vapor bands, the molecular absorption features can saturate at moderate airmass; beyond that point the line-of-sight extinction increases less rapidly than linearly with airmass. To second order, in the calibration equation this can be represented by a quadratic term in airmass, with a negative coefficient.

[4]This is a suitable place to maintain some semblance of mathematical rigor by admitting that there are no physical laws requiring the effects discussed here to follow a polynomial as opposed to any other functional form. Low-order polynomial expansions have the advantages that they are single valued, continuous, and smooth. In these properties they resemble the physical situations that we typically encounter. Classical *δ-versus-ϵ* arguments can be used to show that a polynomial of some order can model the relevant physical situations to within any required tolerance. Polynomials have the further advantage that they are fast to compute. That said, there is no objection in principle to the use of other functional forms if they can be shown to produce improved fitting residuals without requiring additional free fitting parameters.

3.6 General Comments

The terrestrial atmosphere is certainly not static, and any variations in extinction on angular scales comparable to the separations between target and standard fields, or on time scales comparable to the duration of an oberving sequence, introduce noise that cannot be modeled out. As a result, corrections for atmospheric extinction are probabilistic rather than deterministic. I have seen students, when analyzing a series of CCD photometric measurements, correct the instrumental magnitudes from one exposure of the series for extinction by application of the standard formula, and then refer all the other exposures to the magnitude system of that one by the application of simple, empirical zero-point shifts. Then the results are averaged. This is quick and easy, but it is not the best practise: even on a good, photometric night the instantaneous extinction in a particular direction at a particular time can depart by easily ~1% or 2% from the prediction of the best extinction model for the night. Usable nights can have even larger erratic variations than that.

In many cases, stellar colors are of greater importance to the researcher than stellar magnitudes. In matching the observed principal sequences of a star cluster to theoretical models, for instance, a systematic 2% error in the apparent magnitude produces a 1% error in the inferred distance, which is probably of no consequence. A systematic 2% error in color, however, translates to an appreciable error in temperature, age, or chemical abundance, which may be important. With photoelectric photometry the strategy of rapidly, repeatedly cycling through the filters is very effective for obtaining highly reliable colors without necessarily getting magnitudes of comparable quality. This observing strategy is not readily available to the CCD observer. To obtain good colors with a standard CCD imaging camera, it is essential to first measure reliable *magnitudes*; only then can they be differenced to produce useful colors. To obtain magnitudes whose uncertainties are both small and well understood, it is important to repeat each measurement a number of times, at different times of night, if possible on different nights, even with different equipment, so that the unmodeled vagaries of a given direction at a particular moment can be beaten down. By correcting *one* observation for extinction, and then referring all subsequent observations to the same zero point, the aforementioned student incurs the full amount of whatever error in the extinction correction occurred for that one observation. By correcting *each* of N magnitude measurements independently for its own (imperfectly estimated) extinction value, and *then* averaging the results, the student could benefit from the usual $\sqrt{N}$ reduction in the net effect of the atmospheric circumstances on each of the observing occasions.

When designing a photometric observing program, it is probably not wise to plan to observe standard stars spanning a much wider range of color, airmass, or azimuth than the target stars. For instance, if the standards span a much broader range of color than the target stars, neglect of unrecognized higher-order terms in the color-based correction for bandpass mismatch can introduce a systematic error in the calibrated results for the target objects. It has also been known to happen that the observer measures a few standard stars at the zenith and a few standard stars at very high airmass in evening and morning twilight, and concludes that the "linear" extinction "law" for the night has been determined with a very high degree of precision. This is not necessarily the case, of course; the observations at high airmass, in particular, are especially subject to seeing and other irregularities in the atmosphere, and moreover it takes time to slew the telescope to large zenith distances (not to mention potential changes to the instrumentation caused by the unusual gravity vector). It is preferable to choose standards in the same general part of the sky as the targets, and to circulate frequently among the standard and target fields using short slews. Errors in the extinction correction scale as errors in the extinction coefficient *times* the difference in the airmass. By matching the standard airmasses to the target airmasses these systematic errors can be minimized more efficiently and reliably than by trying to reduce the uncertainties in the extinction coefficients alone.

3.7 Differential Photometry; Time-Domain Photometry

Differential photometry can be performed when the target stars and suitable standard stars appear in the same CCD images. In this case, it is usually possible to assume that all the stars in the exposure are subject to the same atmospheric extinction: if

$$X = \sec z = \frac{1}{\cos z},$$

then

$$\frac{dX}{dz} = \frac{\sin z}{\cos^2 z} = \sec z \tan z.$$

Consider a CCD of dimension 20 arcminutes = 0.006 radians pointed toward a field at airmass = 2.0, or $z = 60^\circ = 1.05$ radians. The difference in airmass between the top and bottom of the field is $(0.0058) \cdot (2) \cdot \tan 60^\circ \approx 0.02$; the top-to-bottom difference in the extinction is this times the extinction coefficient which is, say, ~0.14, and the root-mean-square variation in the extinction averaged over the entire field is smaller by a further factor of $1/\sqrt{12}$. So, the root-mean-square photometric uncertainty contributed by neglect of the exinction gradient in the comparison of one star to another in the same digital image is ~0.0008 mag. The differential extinction decreases very rapidly for still smaller airmasses.

Useful differential photometry can be performed on occasions when conditions are not sufficiently photometric (in the sense defined above) for obtaining absolute photometry. As seen through a telescope, terrestrial clouds are so far out of focus, and they move so rapidly across the line of sight, that to a very good approximation their effect is uniform across the field of view of a CCD. Algebraically, the equations describing this situation reduce to

$$v_{ij} \doteq V_i + \alpha_j + \beta\,(B\text{–}V)_i \tag{1.7}$$

$$b_{ik} \doteq B_i + \delta_k + \zeta\,(B\text{–}V)_i. \tag{1.8}$$

Here, v_{ij} is an instrumental magnitude measured for some star i in a CCD image j; b_{ik} is a magnitude measured for the same star in a different image k obtained in a different bandpass. Here, each individual CCD image j, k, etc., has its own photometric zero point, whose value depends upon the instantaneous transmission of the atmosphere in that direction and at that wavelength, which is not necessarily simply related to the transmision in any other direction, time, or wavelength. As long as each CCD image contains at least one star of known photometric properties, *and* as long as at least *some* images contain several standard stars spanning an appreciable range of B–V colors, this network of equations can still be solved for unique optimum values for the color coefficients β and ζ, and for all the different zero points α_j and δ_k.

Differential photometry in this sense cannot be obtained with a simple photoelectric photometer, because truly simultaneous observations of photometric standard and science target are not possible. However, special-purpose photometers have been built with multiple photomultipliers that can exploit the advantages of simultaneous observations. (1) In particular, variable stars can be observed in poor conditions by focusing the image of the target star on one photomultiplier, while another monitors the signal from a reference star a few arcminutes away (e.g., Bernacca et al. 1978; Piccioni et al. 1979), and perhaps a third photomultiplier monitors the instantaneous brightness of the sky. (2) In other photometers, beam-splitters and/or diffraction gratings have been used to direct different photometric bands for a single star to different photomultipliers, allowing the precise measurement of stellar colors under marginal observing conditions (e.g., Grønbech et al. 1976; Grønbech and Olsen 1977).

Of these two types of multi-channel photoelectric photometer, the former is less sensitive to the assumption that clouds are gray across the full spectral range, and therefore this method may still be used to obtain variable-star light curves under conditions where other types of photoelectric observation are impractical. For this reason, time-domain photometry often closely resembles differential photometry in its practical aspects. If the variable star that is the scientific target of the investigation can be observed simultaneously with one or more nearby stars of constant magnitude and similar color, a highly reliable light curve can be obtained whether the observing conditions are good or marginal.

4 Measuring the Instrumental Magnitudes

In the previous section, I have described how it is possible to relate observed, instrumental magnitudes to magnitudes on a scientifically useful, calibrated, standard photometric system. I glossed over the question of how those instrumental magnitudes are obtained.

4.1 Photoelectric Photometry

With a photoelectric photometer, the definition of instrumental magnitude is conceptually simple. Through a focal-plane aperture of rigidly fixed size, one measures the incident flux coming from the target of scientific interest, along with its attendant, unavoidable sample of emission from the foreground sky. A separate observation – using the same aperture – of a nearby field containing no perceptible astronomical objects establishes the current sky brightness, and subtraction of this from the accompanying observation (of star plus sky) represents the flux from the science target alone. By repeating these measurements several times in rapid succession, the astronomer can reduce the net effects of any unpredictable minor variations in either the transmission or the emission of the terrestrial sky. The instrumental magnitude is then equal to some convenient constant *minus* 2.5 log(detected flux from the target).

A straightforward error analysis leads to the most efficient division of the observing time between the target field and the blank sky field. Let us define the arrival rate of star photons as $\rho_* = N_*/t_*$, where N_* is the number of photoelectrons from the star detected in integration time t_*. Similarly, the arrival rate of photons from the terrestrial sky is $\rho_s = N_s/t_s$. The quantities actually observed are total counts from the star and sky together, N_{*+s} detected in integration time t_*, and from sky alone, N_s in time t_s; the desired quantity, ρ_* is to be inferred from these:

$$\begin{aligned}\rho_* &= \rho_{*+s} - \rho_s \\ &= \frac{N_{*+s}}{t_*} - \frac{N_s}{t_s}.\end{aligned}$$

Assuming that the only observational uncertainties affecting the individual measurements are the Poisson statistics of the photon arrivals, $\sigma^2(N_{*+s}) = N_{*+s}$ and $\sigma^2(N_s) = N_s$, then by standard propagation of errors

$$\begin{aligned}\sigma^2(\rho_*) &= \sigma^2(\rho_{*+s}) + \sigma^2(\rho_s)\\ &= \left(\frac{d\,\rho_{*+s}}{d\,N_{*+s}}\right)^2 \sigma^2(N_{*+s}) + \left(\frac{d^2\,\rho_s}{d\,N_s}\right)^2 \sigma^2(N_s)\\ &= \frac{N_{*+s}}{t_*^2} + \frac{N_s}{t_s^2}.\end{aligned}$$

The goal is to minimize $\sigma^2(\rho_*)$ subject to the constraint that the total observing time, $T = t_* + t_s$, be constant.

$$\begin{aligned}\sigma^2(\rho_*) &= \frac{N_{*+s}}{t_*^2} + \frac{N_s}{(T-t_*)^2}\\ \frac{d\,\sigma^2(\rho_*)}{d\,t_*} &= -2\frac{N_{*+s}}{t_*^3} + 2\frac{N_s}{(T-t_*)^3} \equiv 0.\end{aligned}$$

Since $T - t_* = t_s$ and $N/t = \rho$, this becomes

$$\begin{aligned}\frac{\rho_{*+s}}{t_*^2} &= \frac{\rho_s}{t_s^2},\\ t_s/t_* &= \sqrt{\rho_s/\rho_{*+s}}.\end{aligned}$$

The optimum ratio of time spent integrating on the empty sky patch to time spent integrating on the science target is equal to the square root of the ratio of the observed count rates in the two positions. If the observed (star plus sky) counting rate is 100 times greater than the (sky only) counting rate, 10 s should be spent observing the sky for every 100 s spent observing the target for the most precise possible results in a fixed amount of observing time. If the science target represents only a minor excess in flux above the diffuse sky brightness, equal time should be spent on the target and sky pointings.

4.2 Aperture Photometry with CCDs

Transferring the concept of fixed-aperture photometry to observations with CCDs is straightforward. Given a two-dimensional data array representing the quantitative distribution of flux as a function of position in the target field, one simply defines a fixed pattern of pixels to represent a synthetic observing aperture. A box of this shape is centered upon each of the target objects in the frame, and the sum of the electrons in these pixels becomes N_{*+s}. Data extracted from an identical box positioned in a star-free region of the frame become N_s, and the analysis proceeds from this point as for counting rates detected with a photomultiplier.[5]

However, CCD cameras offer several important advantages over the photoelectric photometer. First of all, as area detectors CCDs are capable of measuring star and blank sky simultaneously. In the simple case of a single, fixed synthetic aperture used for both star and sky measurements, this results in a straightforward savings of a factor ~2 in observing time even beyond the significant increase in the quantum efficiency of a CCD over a photocathode.

[5] Please note that for convenience, I am considering any dark current in the detector to be merely a part of the diffuse-sky contribution to the signal. For our immediate purposes, there are no practical differences between dark current and sky brightness.

Second, the aperture in a photoelectric photometer is a hole in a piece of metal, and it must be chosen at the beginning of the night before the seeing conditions are well known, and used consistently throughout. With a CCD the optimum aperture size can be chosen after examination of all the data from the night. Third, with a CCD the synthetic aperture can be centered on the target after the observation is complete; no allowance for possible errors in the telescope tracking is required. For both of the latter two reasons, the synthetic aperture employed for CCD data can be much smaller than the physical aperture used with a photoelectric photometer. As a result, contamination of the science target by uninteresting – but noise-contributing – diffuse sky light is greatly reduced: that faint star is now measured against a few square arcseconds of sky instead of a few hundred, reducing the Poisson noise of *both* the (star+sky) and (sky alone) measurements in the above equations. (Since astronomers typically want to observe the faintest possible targets, the target flux is often small compared to the sky flux, and it is appropriate that equal integration times be spent on target and sky; this, of course, is automatically the case with a CCD.)

More importantly, with digital images it is not strictly necessary to use the same aperture for sky measurements as for star measurements. Assuming that the images have been suitably pre-processed to yield data numbers truly proportional to flux in all pixels, the contribution of the diffuse sky emission to each pixel should be a constant value everywhere in the field. Any actual spatial variations in the skyglow will be out of focus, and will move rapidly across the field in any exposure of non-negligible duration. This means that it is possible to estimate a mean sky flux in units of counts per pixel over any convenient object-free area of the frame, scale this to the number of pixels in the synthetic aperture, and subtract it from the observed (star plus sky) counts. The error contribution from the uncertainty of the sky-alone flux can thus be reduced to insignificance compared to the Poisson statistics of the star+sky counts from inside the synthetic aperture. (This assertion is *not* valid for target stars superimposed on a structured astronomical source of diffuse luminosity – a nebula or a galaxy, for instance. Sky determination under such circumstances can be a difficult problem requiring customized treatment.)

Note that the strictly simultaneous observation of targets and sky renders CCD photometry relatively immune to sporadic fluctuations in the sky brightness. However, as mentioned above, the overhead incurred by the readout time and the reluctance to deal with large data volumes both tend to discourage symmetric observing sequences of many short exposures. As a result, CCD photometry tends to be *more* sensitive than photoelectric photometry to unmodeled variations in the atmospheric extinction, *especially* for the measurement of stellar colors, as already mentioned above.

In a CCD with non-negligible readout noise, this provides an additional contribution of uncertainty to each measurement even in the case of short exposures where the diffuse-sky contribution is negligible (as in, for instance, observations of bright photometric standard stars). In general, if the expected number of actual photoelectrons in a given pixel is N, then the uncertainty in the measurement has variance

$$\sigma^2(N) = (N + R^2), \tag{1.9}$$

where R is the root-mean-square readout noise of the detector expressed in units of electrons, and the N on the right-hand side of the equation represents the Poisson noise in the photoelectrons themselves.

In practical use, CCDs do not usually produce data directly in units of detected photoelectrons; instead, the output amplifiers are experimentally tuned for optimum operating characteristics, and their output voltages are then digitized into units which are related to electron counts through some gain factor, g, which is usually estimated in units of electrons per DN: $D = N/g$, where D is the signal in arbitrary data-number units from a pixel which contained N photoelectrons. In this notation, the variance of the measurement (➋ 1.9) becomes

$$\begin{aligned}\sigma^2(D) &= \left(\frac{dD}{dN}\right)^2 \sigma^2(N) \\ &= \frac{1}{g^2}\left(gD + R^2\right) \\ &= D/g + r^2,\end{aligned} \tag{1.10}$$

where $r = R/g$ is now the readout noise expressed in units of DN rather than electrons.

A reasonably accurate estimate of the readout noise, r (in units of data numbers), is easy to obtain at the telescope, by a suitable statistical analysis of a sequence of frames of zero exposure time. The average of many such frames is traditionally called a *bias frame*. As discussed above, this contains the repeatable fluctuations in the detector output signal from a single readout of the device (if there are any). When this mean bias pattern has been subtracted from an individual frame of zero exposure time, only non-repeatable signal fluctuations remain. In addition to the raw amplifier noise, these can include stray electromagnetic emissions from other equipment in the observing environment that have been picked up by the CCD camera circuitry. The residual root-mean-square dispersion of the data values in the individual pixels of a bias-subtracted zero-exposure image is a measure of r. This can often be estimated with a precision of order 0.1 DN, or better, from a suitable series of individual bias frames. An unexpectedly large value of r or a sudden change from one night to the next can be an indicator of equipment problems.

The gain g is harder to estimate at the telescope. Successive observations of an out-of-focus, diffusely illuminated target in front of the telescope inside the dome can be averaged together to produce a "dome flat" (or a "lamp flat"). As discussed above, the average dome flat contains information on the relative quantum efficiencies of the individual pixels, as well as some information on any fixed vignetting or illumination pattern in the optical system (as well as the possibility of some scattered light). When one of the individual exposures from a sequence of successive flat-field images is *divided* by the average flat-field image, the root-mean-square residuals provide information on the gain factor: a plot of $\sigma^2(D)$ versus D for images obtained with different exposure times should have slope $1/g$ and y-intercept r^2. In practice, it is often found that there is appreciable scatter of the individual points about a best-fitting line. This may be due to flickering of the artificial illumination on the target, variations in daylight leaking into the dome, or other unrecognized causes that are difficult to isolate outside a well-controlled laboratory. But whatever the cause, it is often found that g can only be estimated without a very high level of precision (uncertainties of many percent).

However, by defining the instrumental magnitude as *constant* $-2.5 \log D$ instead of *constant* $- 2.5 \log N$, the observer neatly avoids any indeterminacy of g. The difference between the two constants, which is just $2.5 \log g$, becomes simply one more contribution to the zero points (α or δ in (➋ 1.1) and (➋ 1.2)) which must be empirically estimated from standard stars in any case. It is true that a numerical value for g is needed for estimating the internal uncertainty of the magnitudes via (➋ 1.10). Still, an uncertainty of several percent in the estimate of an uncertainty is not very important, particularly since this represents only one component of a much more extensive error budget. It should be noted that sky flats – images of the twilight sky – while also extremely useful for removing the instrumental signature from the raw observations (see above) are less suitable for estimating the gain factor because they may contain images of astronomical objects that add pixel-to-pixel flux variations beyond those of Poisson photon statistics.

4.3 Concentric-Aperture Photometry with CCDs

In defining aperture magnitudes, it is not absolutely essential (nor generally possible) to use a measuring aperture large enough to measure *all* the incident radiation from each target object; it is sufficient merely to measure a *consistent fraction* of the light. King (1971) demonstrated that the radial profile of a star image is well described by a combination of several components. At very small radii, where the profile is dominated by the average seeing that prevailed during the observation, the star image is very nearly Gaussian in form:

$$S(a) \propto e^{-a^2/2a_0^2}$$

where I have used S to represent surface brightness in units of flux per unit solid angle, and a is the angular distance from the center of the star image. The quantity a_0 quantifies the profile width produced by the atmospheric seeing at the time of the observation; it typically has a value of order one arcsecond. In the opposite limit of very large radii, scattering in the atmosphere and the telescope optics produces a decline in surface brightness roughly obeying a power law of the form

$$S(a) \propto a^{-p}, \qquad p \sim 2, \;\; a \gtrsim \text{a few arcseconds.}$$

(We know that S cannot be strictly proportional to a^{-2}, because that would imply a divergent total flux when integrated to infinity. Here it is sufficient to assert that star images tend to have a perceptible fraction of their flux in a smooth component extending well beyond the range of easy measurement.) Between these limiting cases is a transition zone where the fixed aberrations in the telescope optics and perhaps the *range* of seeing that occurred during the observation contribute to the profile shape; this region is as well represented by an exponential formula as by any other:

$$S(a) \propto e^{-a/a_1}.$$

Among these profile features, the power-law scattering tail is the most likely to remain fixed from one observation to the next; the seeing-dominated core is the most likely to vary with time and direction. Provided that the measuring aperture is large enough to contain that range of radius where the profile changes from moment to moment and from target to target, it need not include radii where the stellar profile is constant: as long as a fixed fraction, f, of the total stellar flux is measured consistently, then a constant correction for the lost light, $2.5 \log f$, simply becomes subsumed into the zero points (α and δ in (➋ 1.1) and (➋ 1.2), or (➋ 1.7) and (➋ 1.8)). Furthermore, if the lost light varies systematically with zenith distance, then an empirical correction for the first-order variation of f with X becomes absorbed into the extinction coefficients δ and η; under these circumstances, it may no longer be true that one expects the empirical zero points α and δ to be strictly constant from one night to the next.

The integrated flux, F_A, measured for an azimuthally symmetric stellar profile in a circular aperture of fixed outer radius A is given by

$$F_A = \int_{a=0}^{A} [S(a) + 2\pi as]\, da, \\ = \int_{a=0}^{A} S(a)\, da + \pi A^2 s,$$

where s is the sky flux. If we presume S and s to be measured in units of CCD data numbers per pixel, then by analogy with (➋ 1.9) we can say

$$\sigma^2(F_A) = F_A/g + \pi(r^2 + s/g)A^2$$

where r^2 is the variance of the readout noise in one pixel, s/g is the variance of the diffuse sky flux per pixel due to Poisson statistics, and $\pi(r^2+s/g)A^2$ is the sum of these over the πA^2 pixels contained in the measuring aperture. I have assumed here that the mean surface brightness of the sky, s, has negligible uncertainty of its own because it has been estimated from a very large number of blank-sky pixels.

From this we wish to infer the impact of a particular choice of aperture radius, A, on the signal-to-noise ratio of the detected target flux. Obviously, when $A = 0$ the signal and the noise are both zero, so the signal-to-noise ratio is undefined. However, for very small values of a, $S(a)$ is – to first order – constant, so the signal out to aperture radius A grows as $F_A \sim 2\pi S(0)A^2$ while the noise $\sigma(F_A)$ grows as somewhere between $\sqrt{F_A/g}$ and $\sqrt{\pi r^2 + s/g}\ A$, depending upon whether the star is very bright or very faint compared to the combination of sky flux and readout noise. In either case, for a small aperture the signal grows $\sim A^2$ and the noise grows $\sim A$ as the aperture radius is increased, so the signal grows faster than the noise.

Conversely, if we choose a value for A that places the edge of the aperture in the power-law wing of the stellar profile, the increase in the signal occasioned by an incremental change in the aperture radius is

$$F_{A+dA} \sim F_A + (2\pi A)\, S(A)\, dA \\ \sim F_A + (2\pi A)\, A^{-p}\, dA \\ \sim F_A + 2\pi A^{1-p} dA. \tag{1.11}$$

For large A, the net increase in signal achieved by making the aperture still larger *decreases* as long as $p > 1$; if $p > 2$, F_A is asymptotically approaching a constant value. The combination of readout noise and Poisson noise in the sky, however, continues to grow as A for any aperture radius. It follows that eventually the signal-to-noise ratio will *decrease* with increasing A, and thus there must be some intermediate aperture radius where the signal-to-noise ratio is maximized.

However, from the above discussion it is clear that the optimum aperture radius is not a fixed number. In particular, it depends upon the ratio of the central surface brightness of the star, S(0), to the readout noise of the detector and the diffuse sky brightness in the frame. This means that a given aperture radius *cannot* be optimum for stars of all apparent magnitudes, even in a CCD frame where the form of the stellar profile is independent of the stellar brightness, and where the diffuse sky brightness is everywhere the same. This is one more case where the obviously simplest approach – a fixed synthetic aperture for all measurents – is demonstrably not the *best* approach. A less obvious, more elaborate approach to the problem can achieve significantly improved photometric results.

In order to extract aperture magnitudes having the best possible signal-to-noise ratios for stars spanning a range of apparent brightness, we can use the methodological device known as concentric aperture photometry: each star is measured through a series of synthetic apertures having the same center position but a range of limiting radii, A_i, $i = 1, \ldots, n$. It is to be hoped that at least some of these stars will be bright enough that the signal-to-noise ratio is maximized at a large radius where the stellar profile is effectively constant over time and direction. If we take $m_{i,j} = constant - 2.5 \log F_{i,j}$ to represent the instrumental magnitude corresponding to the measured flux out to radius A_i in star j, then for these bright stars we can determine a series of mean aperture corrections from the smaller ones to the largest:

$$\begin{aligned}
\Delta_{1,n} &= \left\langle m_{1,j} - m_{n,j} \right\rangle_j \\
\Delta_{2,n} &= \left\langle m_{2,j} - m_{n,j} \right\rangle_j \\
&\ldots \\
\Delta_{n-1,n} &= \left\langle m_{n-1,j} - m_{n,j} \right\rangle_j
\end{aligned}$$

where the notation "$\langle \ldots \rangle_j$" represents an average value taken over some suitable subset of stars j having good measurements in *all* apertures. The set of numbers Δ is often called a "growth curve" because it quantifies the rate at which the enclosed stellar flux grows with increasing aperture radius. With these corrections in place, we can choose for each individual target star the measurement through the particular aperture that most nearly maximizes that star's signal-to-noise ratio, and then correct that measurement to the instrumental magnitude that *would have been measured* in the largest aperture, n, by subtracting the appropriate value of Δ.

The above paragraph explains the basic principle of concentric-aperture photometry. In practical use, however, a slightly different approach is preferable. It makes more sense to define the raw aperture corrections not from aperture i to aperture n, but from aperture i to aperture $i+1$:

$$\begin{aligned}
\delta_{1,2} &= \left\langle m_{1,j} - m_{2,j} \right\rangle_j \\
\delta_{2,3} &= \left\langle m_{2,j} - m_{3,j} \right\rangle_j \\
&\ldots \\
\delta_{n-1,n} &= \left\langle m_{n-1,j} - m_{n,j} \right\rangle_j
\end{aligned}$$

Then,

$$\begin{aligned}
\Delta_{1,n} &= \sum_{i=1}^{n-1} \delta_{i,i+1} \\
\Delta_{2,n} &= \sum_{i=2}^{n-1} \delta_{i,i+1} \\
&\ldots \\
\Delta_{n-1,n} &= \delta_{n-1,n}
\end{aligned}$$

It is preferable to calculate the corrections Δ in this two-step fashion because the typical CCD image will usually contain relatively few stars that can be reliably measured in *all* the apertures from the smallest to the largest: noise considerations limit the maximum aperture radius for fainter stars, and even some bright stars may be unmeasurable in the larger apertures because of the proximity of neighbor stars, the edge of the frame, or detector blemishes. Thus there will be many more stars that can be profitably measured in apertures 1 and 2 (for instance) than there will be that can be measured in apertures 1 and n. The uncertainty of each of the individual small steps, δ, can therefore be beaten down by averaging over the largest possible subset of stars having valid measurements in the two adjacent apertures. As a result, the total uncertainty in the accumulated aperture corrections $\Delta = \sum \delta$ can be somewhat smaller than corrections based only on those stars that can be measured in *all* apertures.

Each weighted mean value $\delta_{i,i+1}$ will have some associated standard error of the mean associated with it, which can be estimated from the readout noise and sky brightness (➋ 1.10) but can also be validated by the star-to-star repeatability. These uncertainties will tend to be smallest for the δ values associated with the smallest apertures, both because the number of stars from which they can be estimated will be large and because the random scatter due to readout noise and sky-photon statistics will be small. Counting inward from $\Delta_{n-1,n}$ to $\Delta_{1,n}$, each successive value of Δ represents the sum of a larger number of components δ, each with its own uncertainty. The standard errors of the different Δ's, it is clear, increase monotonically from the outside in.

Since an instrumental magnitude is the logarithm of the flux, it follows that the uncertainty in the derived magnitude

$$\sigma(m) \sim \sigma(F)/F;$$

in other words, the magnitude uncertainty scales inversely as the signal-to-noise ratio of the flux. Since we have seen that, in general, the signal-to-noise ratio of the flux measurement for any given star j achieves a maximum for some aperture radius, it follows that the uncertainty of the instrumental aperture magnitude $m_{i,j}$ will be minimized in that same aperture. For very faint stars the most reliable measurement may be in the smallest synthetic aperture employed; for very bright stars the most reliable measurement may be in the largest aperture; for stars of intermediate brightness, some intermediate aperture will turn out to be best. However, the goal is to obtain the best possible estimate of the instrumental magnitude *that would have been measured for that star in the largest aperture*, where the stellar profile obtained with that equipment is constant over time and direction. In choosing the best aperture for any given star, therefore, one should consider not only the uncertainty of the raw instrumental magnitude, $\sigma(m_{i,j})$, but also the uncertainty of the correction to large aperture, $\sigma(\Delta_j)$. The best aperture for star i will be the one where the sum of the squares of these two quantities achieves its minimum value.

4.4 Profile-Fitting Photometry

At present, some form of concentric aperture photometry seems to be the most reliable way to obtain absolute, all-sky photometry with digital imaging cameras. For relative photometry within a particular CCD image, profile-fitting photometry offers some advantages. This involves a mathematical model representing the two-dimensional distribution of light within a recorded stellar profile, $S(x, y)$, where for simplicity of notation the orthogonal coordinates x and y are measured relative to the centroid of the star image. Here the surface brightness S is defined in a more general way than the strictly radial, azimuthally-averaged $S(a)$ that was used in the previous section because, when examined in enough detail, star images recorded by CCDs are not always round. I shall henceforth refer to the mathematical representation of a stellar profile, $S(x, y)$, as a "point-spread function," or "PSF."

For achieving the highest precision in photometry from ground-based observations, it is generally necessary to construct a model point-spread function for each individual CCD image.

It cannot be guaranteed that the seeing and tracking conditions of one exposure will duplicate those of another sufficiently well to allow a single model PSF to be used for both. Requisite conditions for a good definition of a model stellar profile are that the characteristic size of the star image (e.g., a_0 in ➋ Sect. 4.3) should not be small compared to the dimensions of a pixel; that the shape of the profile be independent of the brightness of the star; and that the shape of the profile be independent of the position of the star within the area of the CCD frame (or at worst mildly dependent on position in some easily modeled way). Given that those conditions are satisfied, in actual practise the PSF, $S(x, y)$, can be encoded as a continuous analytic function, as a two-dimensional table combined with an interpolation algorithm, or as a combination of the two. In general, an analytic PSF which is numerically integrated over the area of each pixel is most convenient for smooth, regular stellar profiles, particularly when they are undersampled by the detector pixel grid. A lookup table plus interpolation scheme is better for complex profiles (telescope aberrations, poor telescope tracking or focus) that have been oversampled.

In any case, for each CCD image a numerical model PSF should be constructed from the most isolated, best exposed stars in the image; in some cases it can be advantageous to use a first estimate of the PSF to model and subtract the profiles of fainter stars in the image, thus better isolating the images of brighter stars for the construction of an improved second-generation PSF. Some badly crowded CCD frames give the impression that there are *no* stars suitable for generating a model point-spread function. At these times it is reassuring to remember that, once the model PSF has been constructed from at least two of the best-exposed and most isolated stars in the image, however poor they may appear, from that point the PSF is already better defined than any target star to which it will be compared: the errors in fitting this model PSF to the images of individual stars will be dominated by the data for those stars, and not by uncertainties in the model PSF.

A numerical representation of a stellar profile having been defined, the actual observed data within some small patch of data frame around the star image, $D(x, y)$, are fitted by a statistical optimization technique to a scaled version of the model PSF:

$$D(x, y) \doteq \phi S(x, y) + s. \tag{1.12}$$

Again here, s is the contribution of the diffuse sky brightness to the signal derived from each pixel, which for simplicity is assumed to be the same for every pixel in the area of interest. Given a two-dimensional model for the PSF, $S(x, y)$, and the observed data, $D(x, y)$, we are left with only the flux-scaling factor, ϕ, as a free fitting parameter to be determined by a statistical estimation, such as the method of least squares. Since the form of the model profile, $S(x, y)$, is presumed to be known and fixed for every star in a given CCD frame, the volume of the profile – which is proportional to the total flux contained within it – scales simply as ϕ and

$$m = constant - 2.5 \log \phi$$

is a useful relative instrumental magnitude scale that is appropriate to that CCD frame (and not necessarily any other).

Profile-fitting photometry is better than aperture photometry in crowded fields where the profiles of different stars may overlap, so that synthetic apertures centered on a particular star of interest could contain significant flux contributions from other nearby stars. It is not necessary to use *all* the pixels in a star's profile to estimate a value for ϕ; given prior knowledge of the star's centroid position and the diffuse sky flux s, a least-squares estimate for ϕ can be determined from only a few pixels near the center of the star image where the signal-to-noise ratio is maximized (or, indeed, even from a single pixel, although in this case the problem is no longer overdetermined.) Thus, in the simplest case the instrumental magnitude of a star can be estimated from only those pixels that are minimally contaminated by the light of other, nearby stars. More generally and powerfully, multiple copies of the PSF can be fitted simultaneously to the different, mutually overlapping star images, thus obtaining individual values of ϕ for all the stars, each automatically corrected for the flux due to the others.

Profile-fitting magnitudes can also be derived for stars lying so close to frame boundaries or detector blemishes that simple aperture magnitudes cannot be obtained. Finally, profile-fitting magnitudes can often be obtained for stars whose images are mildly saturated, thus effectively extending the dynamic range of the photometric investigation; provided the effect of the saturation is simply to invalidate the signal from those pixels containing the highest flux levels (and not to redistribute electrons to other portions of the stellar profile), a flux-scaling factor ϕ can still be estimated by excluding the invalid pixels from the fits, relying instead on only the valid pixels in the flanks of the star image.

Most practical implementations of profile-fitting photometry do not require that the (X, Y) position of the stellar centroid (i.e., the star's position in the natural coordinate system of the digital CCD frame) be known with a high degree of accuracy. Given a reasonable initial guess at the location of the stellar centroid, incremental corrections can be included as additional fitting parameters in the least-squares adjustment, which now involves three fitting parameters for each star: ϕ, X_0, and Y_0. Since X-position, Y-position, and flux are all mutually orthogonal directions, to first order errors in one have no correlation with errors in another; the derived estimate for a star's flux, then, is little affected by uncertainties in its position, provided those uncertainties are small compared to the image size. If desired, the local sky brightness, s, can also be included as a fourth free parameter in the profile fits. In this case, however, there is a direct correlation between the estimate of the stellar flux and the estimate of the sky brightness; uncertain placement of the optimum sky level has a direct detrimental effect on the precision of the stellar photometry. For this reason it is often best to estimate the local sky brightness from a much larger area of the frame surrounding the image of the target star than is needed for fitting the model profile to the star image itself.

Profile-fitting photometry is effective and reliable for purposes of relative photometry, where target objects and photometric reference stars are contained in the same CCD frames. Here, absolute photometric zero points are not needed and the logarithm of the flux-scaling factor ϕ contains all the information necessary for establishing the magnitude differences among stars in the frame.

Profile-fitting photometry by itself is not as good for all-sky, absolute photometry. The reader might think that, once the model PSF $S(x, y)$ has been obtained for a particular CCD frame, it remains only to numerically integrate that model profile out to some suitably large radius, and then one knows the total flux represented by that PSF; the scaling factor ϕ then gives the total flux contained within the image of any star to which that profile has been fitted. That assertion is true, in principle, but it fails to take into account the measurement errors. Assume for the moment that a star with central surface brightness $S(0, 0) = 100$ DN per pixel sits atop a diffuse sky background. If for whatever reason the quantitative estimate of the diffuse sky brightness is incorrect by 1 DN, then that will produce a corresponding error of 1 DN in the estimate of the stellar flux in the central pixel: the inferred central height of the stellar profile will be incorrect by 1%. If a PSF constructed from this star were to be applied to other stars in the image, all of the scaling factors, ϕ, for those stars would be incorrect by that same 1%, which would not matter because it would be the same for all stars and would disappear in the magnitude differences. However, suppose that the total flux for that PSF is to be estimated by numerical integration of the profile out to a radius of, say, 10 pixels. The 1 DN per pixel error in the sky estimate now becomes a mistake of 314 DN in the integrated flux for the star, which might itself be only a few hundred DN (assuming the bulk of the star light is concentrated in the central few pixels). This level of precision is not good enough for all-sky photometry.

The best all-sky photometry, therefore, tends to rely on a combination of profile-fitting photometry (for precise relative instrumental magnitudes) with concentric aperture photometry (for the most reliable and repeatable total flux estimates). In each CCD frame total instrumental magnitudes are obtained for a number of bright, well-isolated stars lying in flat-sky areas of the field. These magnitudes are derived from apertures large enough to contain an invariant fraction of the flux, and/or have been corrected to such large apertures via growth curves, so they represent a measuring system that is consistent from one CCD frame to the next, regardless of seeing differences. For each frame, the mean difference between the total aperture magnitudes and the relative profile-fitting magnitudes for the same stars is computed, and is used as an additive magnitude correction to place the relative, internal, profile-fitting magnitude scale on the zero point of the absolute, external, aperture-magnitude measurements. These corrected magnitudes have the best achievable internal precision *and* external accuracy; they can be corrected for atmospheric extinction and bandpass mismatch just like any other all sky photometry.

This hybrid approach again tends to compartmentalize the error budget. The internally precise magnitudes derived from profile-fitting photometry provide the tightest possible observed principal sequences in the color-magnitude diagram for a star cluster, or the best-defined light curve for a variable star in the same field as one or more reference stars. The absolute placement of that cluster sequence or that light-curve in standard color- and magnitude-space relies on the all-sky capabilities of concentric aperture photometry, and is usually less well determined.

References

Allen, C. W. 1973, Astrophysical Quantities (3rd ed.; London: The Athlone Press)
Argue, A. N. 1963, MNRAS, 125, 557
Baade, D., et al., 1999, Messenger, 95, 15
Bernacca, P. L., Canton, G., Stagni, R., Leschiutta, S., & Sedmak, G. 1978, A&A, 70, 821
Bessell, M. S. 2005, ARAA, 43, 293
Burke, B., Jorden, P., & Vu, P. 2005, Exp Astron, 19, 69
Cousins, A. W. J. 1976, MmRAS, 81, 25
Forbes, M. C., et al. 1995, The Observatory, 115, 29
Giannì, G., Mazzitelli, I., & Natali, G. 1975, A&A, 44, 277
Grønbech, B., & Olsen, E. H. 1977, A&AS, 27, 443
Grønbech, B., Olsen, E. H., & Strömgren, B. 1976, A&AS, 26, 155
Grundahl, F., & Sørensen, A. N. 1996, A&A, 116, 367
Gunn, J. E., & Stryker, L. L. 1983, ApJS, 52, 121
Hamuy, M., Walker, A. R., Suntzeff, N. B., Grigoux, P., Heathcote, S. R., & Phillips, M. M. 1992, PASP, 104, 533
Hardie, R. H. 1962, in Astronomical Techniques (Chicago/London: University of Chicago Press), 178
Johnson, H. L. 1955, Ann Astrophys, 18, 292
Johnson, H. L. 1962, in Astronomical Techniques (Chicago/London: University of Chicago Press), 157
Johnson, H. L., & Harris, D. L., III. 1954, ApJ, 120, 196
Johnson, H. L., & Morgan, W. W. 1953, ApJ, 117, 313
Kaye, A. L. 1977, MNRAS, 180, 147
King, I. R. 1971, PASP, 83, 199
Kron, G. E., White, J. S., & Gascoigne, S. C. B. 1953, ApJ, 118, 502
Landolt, A. U. 1992, AJ, 104, 340
Latham, D. W. 1976, AAS Photo-Bulletin, No. 13
Leach, R. W., Schild, R. E., Gursky, H., Madejski, G. M., Schwartz, D. A. & Weeks, T. C. 1980, PASP, 92, 233
Massey, P., & Jacoby, G. H. 1992, in Astronomical CCD Observing and Reduction Techniques, ed. S. Howell, ASP Conf. Ser., 23 (San Francisco: Astronomical Society of the Pacific), 240
Massey, P., Strobel, K., Barnes, J. V., & Anderson, E. 1988, ApJ, 328, 315
Piccioni, A., Bartolini, C., Guarnieri, A., & Giovanelli, F. 1979, AcA, 29, 463

Shao, C.-Y., & Young, A. T. 1965, AJ, 70, 726
Smart, W. M. 1977, Textbook on Spherical Astronomy (6th ed.; Cambridge: Cambridge University Press)
Stetson, P. B. 1979, AJ, 84, 1056
Stetson, P. B. 1989a, Highlights of Astronomy, 8, 635
Stetson, P. B. 1989b, in Image and Data Processing; Interstellar Dust, V Advanced School of Astrphysics (São Paulo: Instituto Astronômico e Geofísico), 1
Sung, H., & Bessell, M. S. 2000, PASA, 17, 244
Suntzeff, N. B., & Walker, A. R. 1996, PASP, 108, 265
Young, A. T. 1974, Methods Exp Phys 12A, 123

2 Astronomical Spectroscopy

Philip Massey[1] · *Margaret M. Hanson*[2]
[1]Lowell Observatory, Flagstaff, AZ, USA
[2]Department of Physics, University of Cincinnati, Cincinnati, OH, USA

T.D. Oswalt, H.E. Bond (eds.), *Planets, Stars and Stellar Systems. Volume 2: Astronomical Techniques, Software, and Data*, DOI 10.1007/978-94-007-5618-2_2, © Springer Science+Business Media Dordrecht 2013

Abstract: Spectroscopy is one of the most important tools that an astronomer has for studying the universe. This chapter begins by discussing the basics, including the different types of optical spectrographs, with extension to the ultraviolet and the near-infrared. Emphasis is given to the fundamentals of how spectrographs are used, and the trade-offs involved in designing an observational experiment. It then covers observing and reduction techniques, noting that some of the standard practices of flat-fielding often actually degrade the quality of the data rather than improve it. Although the focus is on point sources, spatially resolved spectroscopy of extended sources is also briefly discussed. Discussion of differential extinction, the impact of crowding, multi-object techniques, optimal extractions, flat-fielding considerations, and determining radial velocities and velocity dispersions provide the spectroscopist with the fundamentals needed to obtain the best data. Finally the chapter combines the previous material by providing some examples of real-life observing experiences with several typical instruments.

Keywords: Instrumentation: spectrographs, Methods: observations, Methods: data analysis, Techniques: spectroscopic

1 Introduction

They're light years away, man, and that's pretty far
(lightspeed's the limit, the big speed limit)
But there's plenty we can learn from the light of a star
(split it with a prism, there's little lines in it)
–Doppler Shifting, Alan Smale (AstroCappella[1])

Spectroscopy is one of the fundamental tools at an astronomer's disposal, allowing one to determine the chemical compositions, physical properties, and radial velocities of astronomical sources. Spectroscopy is the means used to measure the dark matter content of galaxies, the masses of two stars in orbit about each other, the mass of a cluster of galaxies, the rate of expansion of the Universe, or to discover exoplanets around other stars, all using the Doppler shift. It makes it possible for the astronomer to determine the physical conditions in distant stars and nebulae, including the chemical composition and temperatures, by quantitative analysis of the strengths of spectral features, thus constraining models of chemical enrichment in galaxies and the evolution of the universe. As one well-known astronomer put it, "You can't do astrophysics just by taking pictures through little colored pieces of glass," contrasting the power of astronomical spectroscopy with that of broadband imaging.

Everyone who has seen a rainbow has seen the light of the sun dispersed into a spectrum, but it was Isaac Newton (1643–1727) who first showed that sunlight could be dispersed into a continuous series of colors using a prism. Joseph von Fraunhofer (1787–1826) extended this work by discovering and characterizing the dark bands evident in the sun's spectrum when sufficiently dispersed. The explanation of these dark bands was not understood until the work of Gustav Kirchhoff (1824–1887) and Robert Bunsen (1811–1899), who proposed that they were due to the selective absorption of a continuous spectrum produced by the hot interior of the sun by cooler gases at the surface. The spectra of stars were first observed visually by Fraunhofer and

[1] http://www.astrocappella.com/

Angelo Secchi (1818–1878), either of whom may be credited with having founded the science of astronomical spectroscopy.

This chapter will emphasize observing and reduction techniques primarily for optical spectroscopy obtained with charge coupled devices (CCDs) and the techniques needed for near-infrared (NIR) spectroscopy obtained with their more finicky arrays. Spectroscopy in the ultraviolet (UV) will also be briefly discussed. Very different techniques are required for gamma-ray, x-ray, and radio spectroscopy, and these topics will not be included here. Similarly, the emphasis here will be primarily on stellar (point) sources, but with some discussion of how to extend these techniques to extended sources.

The subject of astronomical spectroscopy has received a rich treatment in the literature. The volume on *Astronomical Techniques* in the original *Stars and Stellar Systems* series contains a number of seminal treatments of spectroscopy. In particular, the introduction to spectrographs by Bowen (1962) remains useful even 50 years later, as the fundamental physics remains the same even though photographic plates have given way to CCDs as detectors. The book on diffraction gratings by Loewen and Popov (1997) is also a valuable resource. Gray (1976) and Schroeder (1974) provide very accessible descriptions of astronomical spectrographs, while the "how to" guide by Wagner (1992) has also proven to be very useful. Similarly the monograph by Walker (1987) delves into the field of astronomical spectroscopy in a more comprehensive manner than is possible in a single chapter, and is recommended.

2 An Introduction to Astronomical Spectrographs

This section will concentrate on the hardware aspect of astronomical spectroscopy. The basics are discussed first. The following subsections then describe specific types of astronomical spectrographs, citing examples in current operation.

2.1 The Basics

When the first author was an undergraduate, his astronomical techniques professor, one Maarten Schmidt, drew a schematic diagram of a spectrograph on the blackboard, and said that all astronomical spectrographs contained these essential elements: a slit on to which the light from the telescope would be focused; a collimator, which would take the diverging light beam and turn it into parallel light; a disperser (usually a reflection grating); and a camera that would then focus the spectrum onto the detector. In the subsequent 35 years of doing astronomical spectroscopy for a living, the first author has yet to encounter a spectrograph that didn't meet this description, at least in functionality. In a multi-object fiber spectrometer, such as Hectospec on the MMT (Fabricant et al. 2005), the slit is replaced with a series of fibers. In the case of an echelle, such as MagE on the Clay 6.5 m telescope (Marshall et al. 2008), prisms are inserted into the beam after the diffraction grating to provide cross-dispersion. In the case of an objective-prism spectroscopy, the star itself acts as a slit "and the Universe for a collimator" (Newall 1910; see also Bidelman 1966). Nevertheless, this heuristic picture provides the reference for such variations, and a version is reproduced here in ➋ *Fig. 2-1* in the hopes that it will prove equally useful to the reader.

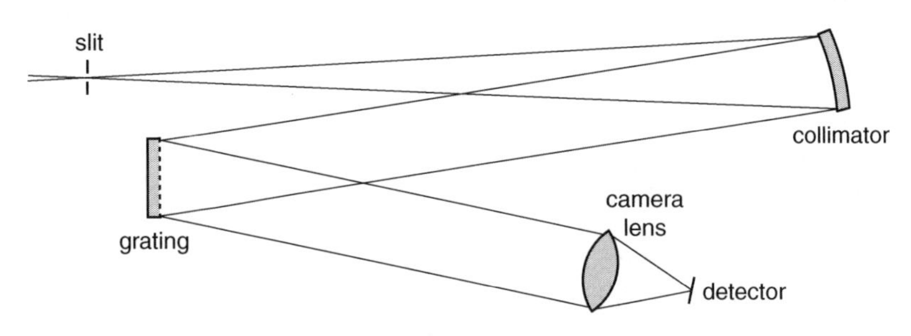

Fig. 2-1
The essential components of an astronomical spectrograph

The slit sits in the focal plane, and usually has an adjustable width w. The image of the star (or galaxy or other object of interest) is focused onto the slit. The diverging beam continues to the collimator, which has focal length L_{coll}. The *f-ratio* of the collimator (its focal length divided by its diameter) must match that of the telescope beam, and hence its diameter has to be larger the further away it is from the slit, as the light from a point source should just fill the collimator. The collimator is usually an off-axis paraboloid, so that it both turns the light parallel and redirects the light toward the disperser.

In most astronomical spectrographs the disperser is a grating, and is ruled with a certain number of grooves per mm, usually of order 100–1000. If one were to place one's eye near where the camera is shown in *Fig. 2-1* the wavelength λ of light seen would depend upon exactly what angle i the grating was set at relative to the incoming beam (the angle of incidence), and the angle θ the eye made with the normal to the grating (the angle of diffraction). How much one has to move one's head by in order to change wavelengths by a certain amount is called the dispersion, and generally speaking the greater the projected number of grooves/mm (i.e., as seen along the light path), the higher the dispersion, all other things being equal. The relationship governing all of this is called *the grating equation* and is given as

$$m\lambda = \sigma(\sin i + \sin \theta). \tag{2.1}$$

In the grating equation, m is an integer representing the *order* in which the grating is being used. Without moving one's head, and in the absence of any order blocking filters, one could see 8000 Å light from first order and 4000 Å light from second order at the same time.[2] An eye would also have to see further into the red and blue than human eyes can manage, but CCDs typically have sensitivity extending from 3000–10,000 Å, so this is a real issue. It is solved by inserting a *blocking filter* into the beam that excludes unwanted orders, usually right after the light has passed through the slit.

The angular spread (or dispersion[3]) of a given order m with wavelength can be found by differentiating the grating equation:

$$d\theta/d\lambda = m/(\sigma \cos \theta) \tag{2.2}$$

[2]This is because of the basics of interference: if the extra path length is any integer multiple of a given wavelength, constructive interference occurs.

[3]Although we derive the true dispersion here, the characteristics of a grating used in a particular spectrograph usually describe this quantity in terms of the "reciprocal dispersion," i.e., a certain number of Å per mm or Å per pixel. Confusingly, some refer to this as the dispersion rather than the reciprocal dispersion.

for a given angle of incidence i. Note, though, from (➋ 2.1) that $m/\sigma = (\sin i + \sin\theta)/\lambda$, so

$$d\theta/d\lambda = (\sin i + \sin\theta)/(\lambda\cos\theta) \tag{2.3}$$

The grating is most efficient when it is used in the Littrow configuration ($i = \theta$), in which case the angular dispersion $d\theta/d\lambda$ is given by:

$$d\theta/d\lambda = (2/\lambda)\tan\theta. \tag{2.4}$$

Consider a conventional grating spectrograph. These must be used in low order (m is typically 1 or 2) to avoid overlapping wavelengths from different orders, as discussed further below. These spectrographs are designed to be used with a small angle of incidence (i.e., the light comes into and leaves the grating almost normal to the grating), and the only way of achieving high dispersion is by using a large number of grooves per mm (i.e., σ is small in (➋ 2.2)). (A practical limit is roughly 1800 grooves per mm, as beyond this polarization effects limit the efficiency of the grating.) Note from the above that $m/\sigma = 2\sin\theta/\lambda$ in the Littrow condition. So, if the angle of incidence is very low, $\tan\theta \sim \sin\theta \sim \theta$, and the angular dispersion $d\theta/d\lambda \sim m/\sigma$. If m must be small to avoid overlapping orders, then the only way of increasing the dispersion is to decrease σ, i.e., use a larger number of grooves per mm. Alternatively, if the angle of incidence is very high, one can achieve high dispersion with a low number of groves per mm by operating in a high order. This is indeed how echelle spectrographs are designed to work, with usually $\tan\theta \sim 2$ or greater. A typical echelle grating might have ~80 grooves/mm, so, $\sigma \sim 25\lambda$ or so for visible light. The order m must be of order 50. Echelle spectrographs can get away with this because they cross-disperse the light (as discussed more below) and thus do not have to be operated in a particular low order to avoid overlap.

Gratings have a *blaze angle* that results in their having maximum efficiency for a particular value of mλ. Think of the grating as having little triangular facets, so that if one is looking at the grating perpendicular to the facets, each will act like a tiny mirror. It is easy to envision the efficiency being greater in this geometry. When speaking of the corresponding *blaze wavelength*, $m = 1$ is assumed. When the blaze wavelength is centered, the angle θ above is this blaze angle. The blaze wavelength is typically computed for the Littrow configuration, but that is seldom the case for real astronomical spectrographs due to design considerations, so the effective blaze wavelength is usually a bit different.

As one moves away from the blaze wavelength λ_b, gratings fall to 50% of their peak efficiency at a wavelength

$$\lambda = \lambda_b/m - \lambda_b/3m^2 \tag{2.5}$$

on the blue side and

$$\lambda = \lambda_b/m + \lambda_b/2m^2 \tag{2.6}$$

on the red side.[4] Thus the efficiency falls off faster to the blue than to the red, and the useful wavelength range is smaller for higher orders. Each spectrograph usually offers a variety of gratings from which to choose. The selected grating can then be tilted, adjusting the central wavelength.

[4]The actual efficiency is very complicated to calculate, as it depends upon blaze angle, polarization, and diffraction angle. See Miller and Friedman (2003) and references therein for more discussion. ➋ Equations 2.5 and ➋ 2.6 are a modified version of the "2/3–3/2 rule" used to describe the cut-off of a first-order grating as $2/3\lambda_b$ and $3/2\lambda_b$; see Al-Azzawi (2007).

The light then enters the camera, which has a focal length of L_{cam}. The camera takes the dispersed light, and focuses it on the CCD, which is assumed to have a pixel size p, usually 15 μm or so. The camera design often dominates in the overall efficiency of most spectrographs.

Consider the trade-off involved in designing a spectrograph. On the one hand, one would like to use a wide enough slit to include most of the light of a point source, i.e., be comparable or larger than the seeing disk. But the wider the slit, the poorer the spectral resolution, if all other components are held constant. Spectrographs are designed so that when the slit width is some reasonable match to the seeing (1-arsec, say) then the projected slit width on the detector corresponds to at least 2.0 pixels in order to satisfy the tenet of the Nyquist-Shannon sampling theorem. The magnification factor of the spectrograph is the ratio of the focal lengths of the camera and the collimator, i.e., L_{cam}/L_{coll}. This is a good approximation if all of the angles in the spectrograph are small, but if the collimator-to-camera angle is greater than about 15 degrees one should include a factor of r, the "grating anamorphic demagnification," where $r = \cos(t + \phi/2)/cos(t - \phi/2)$, where t is the grating tilt and ϕ is collimator-camera angle (Schweizer 1979).[5] Thus the projected size of the slit on the detector will be WrL_{cam}/L_{coll}, where W is the slit width. This projected size should be equal to at least 2 pixels, and preferably 3 pixels.

The spectral resolution is characterized as $R = \lambda/\Delta\lambda$, where $\Delta\lambda$ is the resolution element, the difference in wavelength between two equally strong (intrinsically skinny) spectral lines that can be resolved, corresponding to the projected slit width in wavelength units. Values of a few thousand are considered "moderate resolution," while values of several tens of thousands are described as "high resolution." For comparison, broadband filter imaging has a resolution in the single digits, while most interference-filter imaging has an $R \sim 100$.

The *free spectral range* $\delta\lambda$ is the difference between two wavelengths λ_m and $\lambda_{(m+1)}$ in successive orders for a given angle θ:

$$\delta\lambda = \lambda_m - \lambda_{m+1} = \lambda_{m+1}/m. \tag{2.7}$$

For conventional spectrographs that work in low order ($m = 1 - 3$) the free spectral range is large, and blocking filters are needed to restrict the observation to a particular order. For echelle spectrographs, m is large ($m \geq 5$) and the free spectral range is small, and the orders must be cross-dispersed to prevent overlap.

Real spectrographs do differ in some regards from the simple heuristic description here. For example, the collimator for a conventional long-slit spectrograph must have a diameter that is larger than would be needed just for the on-axis beam for a point source, because it has to efficiently accept the light from each end of the slit as well as the center. One would like the exit pupil of the telescope imaged onto the grating, so that small inconsistencies in guiding, etc., will minimize how much the beam "walks about" on the grating. An off-axis paraboloid can do this rather well, but only if the geometry of the rest of the system is a good match.

2.1.1 Selecting a Blocking Filter

There is often confusion over the use of order separation filters. ➋ *Figure 2-2* shows the potential problem. Imagine trying to observe from 6000 Å to 8000 Å in first order. At this particular angle, one will encounter overlapping light from 3000 Å to 4000 Å in second order, and, in principle, 2000 Å to 2666 Å in third order, etc.

[5]Note that some observing manuals give the reciprocal of r. As defined here, $r \leq 1$.

Since the atmosphere transmits very little light below 3000 Å, there is no need to worry about third or higher orders. However, light from 3000 Å to 4000 Å *does* have to be filtered out. There are usually a wide variety of blue cut-off filters to choose among; these cut off the light in the blue but pass all the light longer than a particular wavelength. In this example, any cut-off filter that passed 6000 Å and higher would be fine. The transmission curves of some typical order blocking filters are shown in *Fig. 2-3*. The reader will see that there are a number of good choices, and that either a GG455, GG475, GG495, OG530, or an OG570 filter could be used. The GG420 might work, but it looks as if it is still passing *some* light at 4000 Å, so why take the chance?

What if instead one wanted to observe from 4000 Å to 5000 Å in second order? Then there is an issue about first order red light contaminating the spectrum, from 8000 Å on. Third order light *might* not be a problem – at most there would be a little 3333 Å light at 5000 Å, but one could trust the source to be faint there and for the atmosphere to take its toll. So, a good choice for a blocking filter would seem be a $CuSO_4$ filter. However, one should be relatively cautious though in counting on the atmosphere to exclude light. Even though many astronomers would argue that the atmosphere doesn't transmit "much" in the near-UV, it is worth noting that actual extinction at 3200 Å is typically only about 1 magnitude per airmass, and is 0.7 mag/airmass at 3400 Å. So, in this example if one were using a very blue spectrophotometric standard to flux

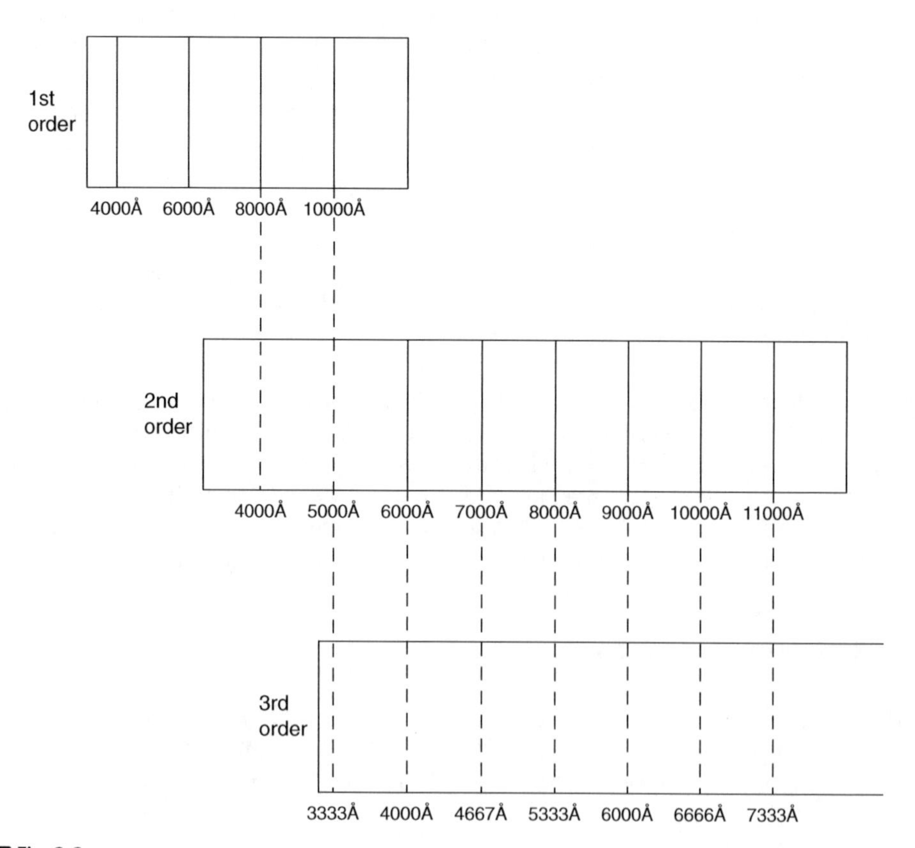

Fig. 2-2
The overlap of various orders is shown

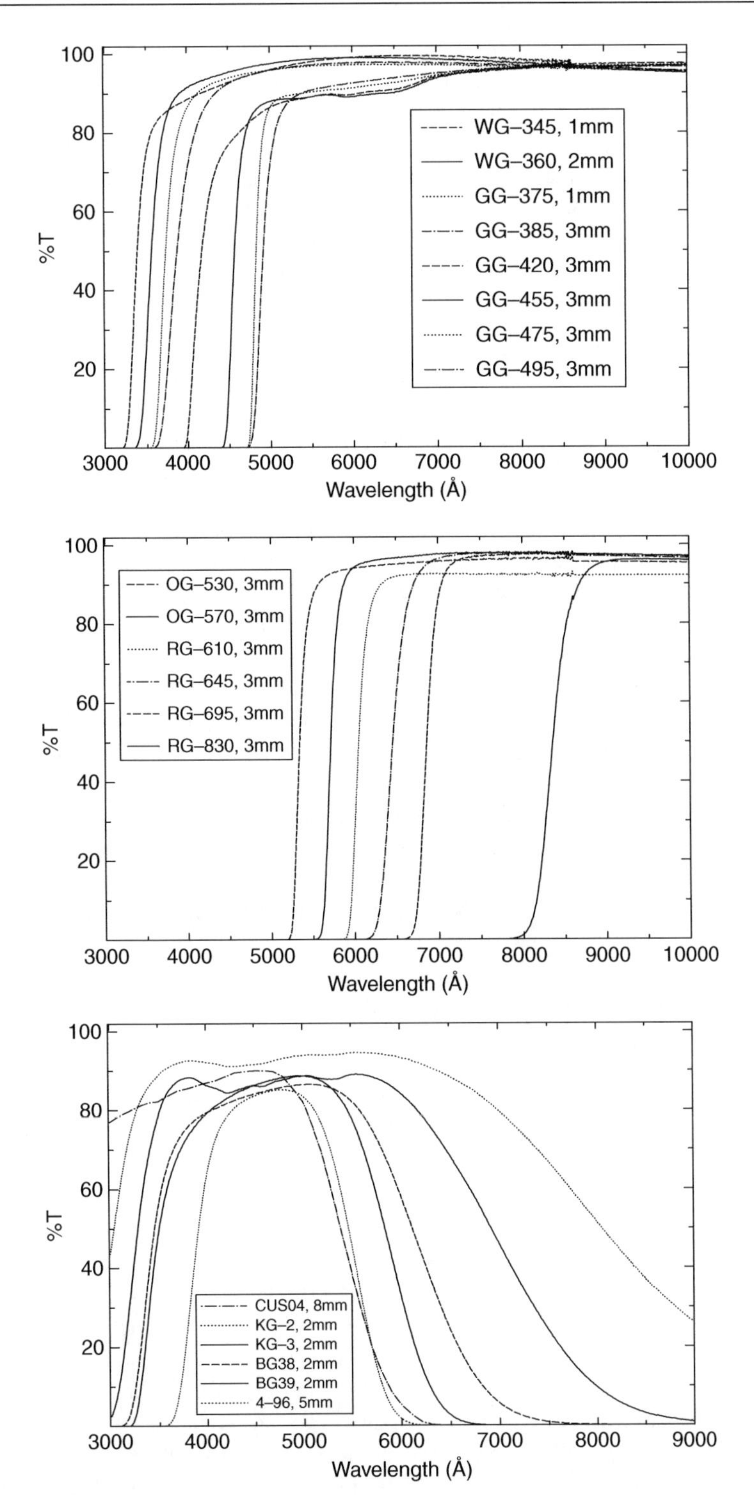

Fig. 2-3
Examples of the transmission curves of order blocking filters (Taken from Massey et al. 2000)

calibrate the data, one could only count on its second-order flux at 3333 Å being attenuated by a factor of 2 (from the atmosphere) and another factor of 1.5 (from the higher dispersion of third order). One might be better off using a BG-39 filter (➋ *Fig. 2-3*).

One can certainly find situations for which no available filter will do the job. If instead one had wanted to observe from 4000 Å to 6000 Å in second order, one would have such a problem: not only does the astronomer now have to worry about >8000 Å light from first order, but also about <4000 Å light from third order. And there simply is no good glass blocking filter that transmits well from 4000 Å to 6000 Å but also blocks below 4000 Å and long-wards of 8000 Å. One could buy a special (interference) filter that did this but these tend to be rather expensive and may not transmit as well as a long pass filter. The only good solution in this situation is to observe in first order with a suitably blazed grating.

2.1.2 Choosing a Grating

What drives the choice of one grating over another? There usually needs to be some minimal spectral resolution, and some minimal wavelength coverage. For a given detector these two may be in conflict, i.e., if there are only 2000 pixels and a minimum (3-pixel) resolution of 2 Å is needed, then no more than about 1300 Å can be covered in a single exposure. The larger the number of lines per mm, the higher the dispersion (and hence resolution) for a given order. Usually the observer also has in mind a specific wavelength region, e.g., 4000–5000 Å. There may still be various choices to be made. For instance, a 1200 line/mm grating blazed at 4000 Å and a 600 line/mm grating blazed at 8000 Å may be (almost) equally good for such a project, as the 600 line/mm could be used in second order and will then have the same dispersion and effective blaze as the 1200 line grating. The primary difference is that the efficiency will fall off much faster for the 600 line/mm grating used in second order. As stated above (➋ Equations 2.5 and ➋ 2.6), gratings fall off to 50% of their peak efficiency at roughly $\lambda_b/m - \lambda_b/3^2$ and $\lambda_b/m + \lambda_b/2m^2$, where λ_b is the first-order blaze wavelength and m is the order. So, the 4000 Å blazed 1200 line/mm grating used in first order will fall to 50% by roughly 6000 Å. However, the 8000 Å blazed 600 line/mm grating used in second order will fall to 50% by 5000 Å. Thus most likely the first order grating would be a better choice, although one should check the efficiency curves for the specific gratings (if such are available) to make sure one is making the right choice. Furthermore, it would be easy to block unwanted light if one were operating in second order in this example, but generally it is a lot easier to perform blocking when one is operating in first order, as described above.

2.2 Conventional Long-Slit Spectrographs

Most of what has been discussed so far corresponds to a conventional long-slit spectrograph, the simplest type of astronomical spectrograph, and in some ways the most versatile. The spectrograph can be used to take spectra of a bright star or a faint quasar, and the long-slit offers the capability of excellent sky subtraction. Alternatively the long-slit can be used to obtain spatially resolved spectra of extended sources, such as galaxies (enabling kinematic, abundance, and population studies) or HII regions. They are usually easy to use, with straightforward acquisition using a TV imaging the slit, although in some cases (e.g., IMACS on Magellan, discussed below in ➋ Sect. 2.4.1) the situation is more complicated.

Table 2-1
Some long slit optical spectrographs

Instrument	Telescope	Slit length	Slit scale (arcsec/pixel)	R	Comments
LRIS	Keck I	2.9′	0.14	500–3000	Also multi-slits
GMOS	Gemini-N,S	5.5′	0.07	300–3000	Also multi-slit masks
IMACS	Magellan I	27′	0.20	500–1200	f/2.5 camera, also multi-slit masks
		15′	0.11	300–5000	f/4 camera, also multi-slit masks
Goodman	SOAR 4.2 m	3.9′	0.15	700–3000	also multi-slit masks
RCSpec	KPNO 4 m	5.4′	0.7	300–3000	
RCSpec	CTIO 4 m	5.4′	0.5	300-3000	
STIS	HST	0.9′	0.05	500–17,500	
GoldCam	KPNO 2.1 m	5.2′	0.8	500–4000	
RCSpec	CTIO 1.5 m	7.5′	1.3	300–3000	

Table 2-1 provides characteristics for a number of commonly used long-slit spectrographs. Note that the resolutions are given for a 1-arcsec wide slit.

2.2.1 An Example: The Kitt Peak RC Spectrograph

Among the classic workhorse instruments of the Kitt Peak and Cerro Tololo Observatories have been the Ritchey–Chretien (RC) spectrographs on the Mayall and Blanco 4 m telescopes. Originally designed in the era of photographic plates, these instruments were subsequently outfitted with CCD cameras. The optical diagram for the Kitt Peak version is shown in *Fig. 2-4*. It is easy to relate this to the heuristic schematic of *Fig. 2-1*. There are a few additional features that make using the spectrograph practical. First, there is a TV mounted to view the slit jaws, which are highly reflective on the front side. This makes it easy to position an object onto the slit. The two filter bolts allow inserting either neutral density filters or order blocking filters into the beam. A shutter within the spectrograph controls the exposure length. The f/7.6 beam is turned into collimated (parallel) light by the collimator mirror before striking the grating. The dispersed light then enters the camera, which images the spectrum onto the CCD.

The "UV fast camera" used with the CCD has a focal length that provides an appropriate magnification factor. The magnification of the spectrograph $rL_{\mathrm{cam}}/L_{\mathrm{coll}}$ is $0.23r$, with r varying from 0.6 to 0.95, depending upon the grating. The CCD has 24 μm pixels and thus for 2.0 pixel resolution one can open the slit to 250 μm, corresponding to 1.6 arcsec, a good match to less than perfect seeing.

The spectrograph has 12 available gratings to choose among, and their properties are given in *Table 2-2*.

How does one choose from among all of these gratings? Imagine that a particular project required obtaining radial velocities at the Ca II triplet ($\lambda\lambda$8498, 8542, 8662) as well as MK classification spectra (3800–5000 Å) of the same objects. For the radial velocities, suppose that 3–5 km s^{-1} accuracy was needed, a pretty sensible limit to be achieved with a spectrograph

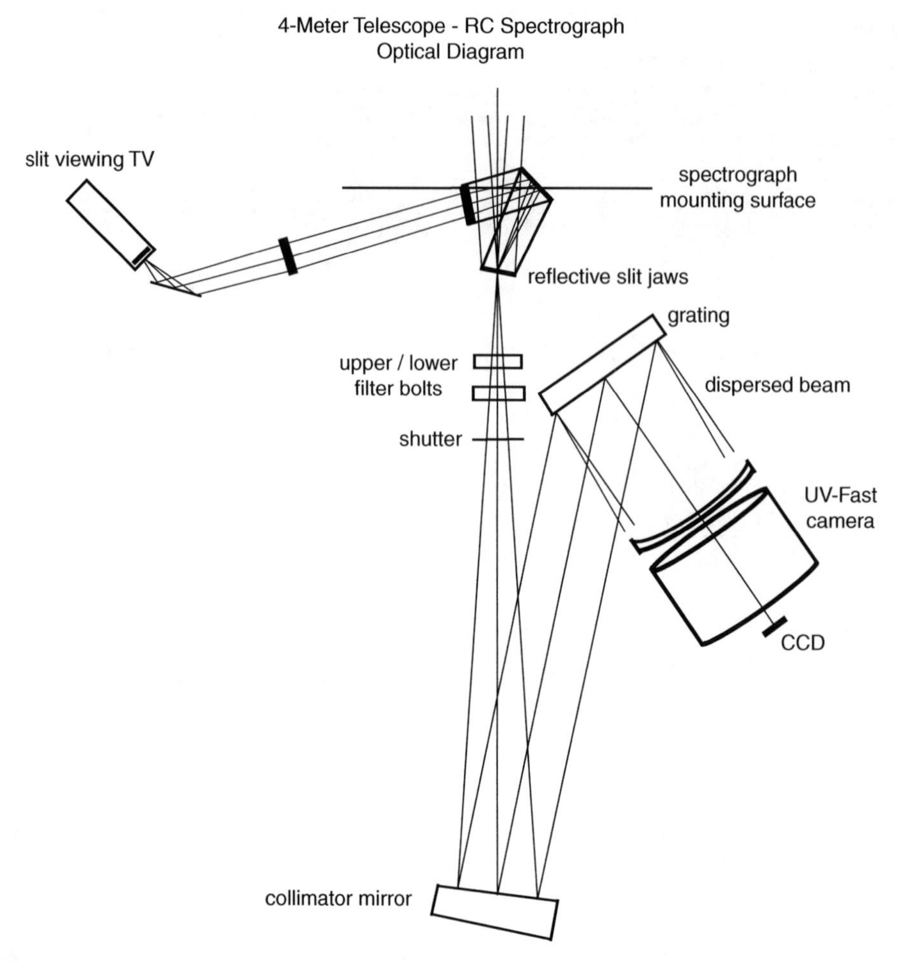

Fig. 2-4
The optical layout of the Kitt Peak 4 m RC Spectrograph (This figure is based upon an illustration from the Kitt Peak instrument manual by James DeVeny)

mounted on the back end of a telescope and the inherent flexure that comes with this. At the wavelength of the Ca II lines, 5 km s^{-1} corresponds to how many angstroms? A velocity v will just be $c\Delta\lambda/\lambda$ according to the Doppler formula. Thus, for an uncertainty of 5 km s^{-1} one would like to locate the center of a spectral line to 0.14 Å. In general, it is easy to centroid to 1/10th of a pixel, and so one needs a reciprocal dispersion smaller than about 1.4 Å/pixel. It is hard to observe in the red in second order so one probably wants to look at gratings blazed at the red. One could do well with KPC-22B (first order blaze at 8500 Å) with a reciprocal dispersion of 1.44 Å per pixel. One would have to employ some sort of blocking filter to block the blue second order light, with the choice dictated by exactly how the Ca II triplet was centered within the 2450 Å wavelength coverage that the grating would provide. The blue spectrum can then be obtained by just changing the blocking filter to block first order red while allowing in

Table 2-2
4 m RC spectrograph gratings

Name	l/mm	Order	Blaze (Å)	Coverage(Å) 1500 pixels	Coverage(Å) 1700 pixels	Reciprocal Dispersion (Å/pixel)	Resolution[a] (Å)
BL 250	158	1	4000	1 octave[b]		5.52	13.8
BL 400	158	1	7000	1 octave[b]		5.52	13.8
		2	3500	<4100[c]		2.76	6.9
KPC-10A	316	1	4000	4100	4700	2.75	6.9
BL 181	316	1	7500	4100	4700	2.78	7.0
		2	3750	<2000[c]		1.39	3.5
KPC-17B	527	1	5540	2500	2850	1.68	4.2
BL 420	600	1	7500	2300	2600	1.52	3.8
		2	3750	1150	1300	0.76	1.9
KPC-007	632	1	5200	2100	2350	1.39	3.5
KPC-22B	632	1	8500	2150	2450	1.44	3.6
		2	4250	1050	1200	0.72	1.8
BL 450	632	2	5500	1050	1200	0.70	1.8
		3	3666	690	780	0.46	1.2
KPC-18C	790	1	9500	1700	1900	1.14	2.9
		2	4750	850	970	0.57	1.4
KPC-24	860	1	10,800	1600	1820	1.07	2.7
		2	5400	800	900	0.53	1.3
BL 380	1200	1	9000	1100	1250	0.74	1.9
		2	4500	550	630	0.37	0.9

[a] Based on 2.5 pixels FHWM corresponding to 300 μm slit (2 arcsec) with no anamorphic factor
[b] Spectral coverage limited by overlapping orders
[c] Spectral coverage limited by grating efficiency and atmospheric cut-off

second order blue. The blue 1200 Å coverage would be just right for covering the MK classification region from 3800 Å to 5000 Å. By just changing the blocking filter, one would then obtain coverage in the red from 7600 Å to 1 μm, with the Ca II lines relatively well centered. An OG-530 blocking filter would be a good choice for the first order red observations. For the second order blue, either the BG-39 or $CuSO_4$ blocking filters would be a good choice as either would filter out light with a wavelength of >7600Å, as shown in *Fig. 2-3*.[6] The advantage to this setup would be that by just moving the filter bolt from one position to another one could observe in either wavelength region.

[6]The BG-38 also looks like it would do a good job, but careful inspection of the actual transmission curve reveals that it has a significant red leak at wavelengths >9000 Å. It's a good idea to check the actual numbers.

2.3 Echelle Spectrographs

In the above sections the issue of order separation for conventional slit spectrographs have been discussed extensively. Such spectrographs image a single order at a given time. On a large two-dimensional array most of the area is "wasted" with the spectrum of the night sky, unless one is observing an extended object, or unless the slit spectrograph is used with a multi-object slit mask, as described below.

Echelle spectrographs use a second dispersing element (either a grating or a prism) to *cross disperse* the various orders, spreading them across the detector. An example is shown in ➋ *Fig. 2-5*. The trade-off with designing echelles and selecting a cross-dispersing grating is to balance greater wavelength coverage, which would have adjacent orders crammed close together, with the desire to have a "long" slit to assure good sky subtraction, which would have adjacent orders more highly separated.[7]

Echelles are designed to work in higher orders (typically $m \geq 5$) and both i and θ in the grating equation (➋ Sect. 2.1) are large.[8] At the detector one obtains multiple orders side-by-side. Recall from above that the wavelength difference $\delta\lambda$ between successive orders at a given angle (the free spectral range) will scale inversely with the order number (➋ 2.7). Thus for low order numbers (large central wavelengths) the free spectral range will be larger.

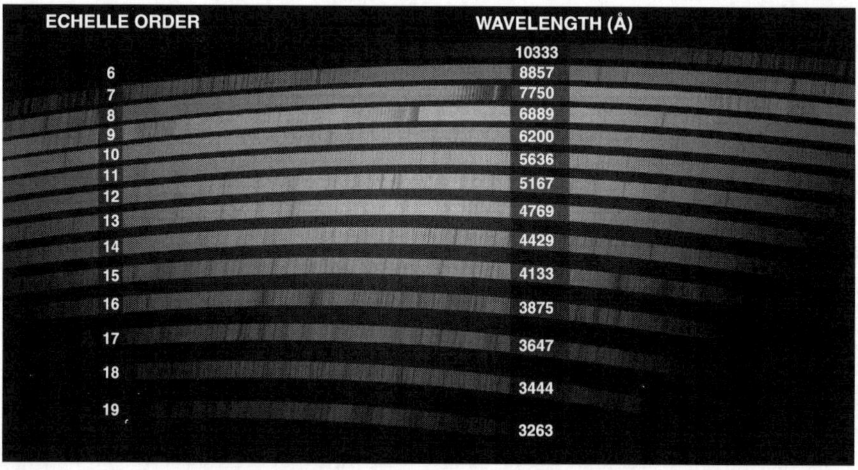

Fig. 2-5
The spectral format of MagE on its detector. The various orders are shown, along with the approximate central wavelength

[7]Note that some "conventional" near-IR spectrographs are cross dispersed in order to take advantage of the fact that the JHK bands are coincidently centered one with the other in orders 5, 4, and 3 respectively (i.e., 1.25 μm, 1.65 μm, and 2.2 μm).

[8]Throughout this section the term "echelle" is used to include the so-called echellette. Echellette gratings have smaller blaze angles ($\tan\theta \leq 0.5$) and are used in lower orders ($m = 5 - 20$) than classical echelles ($\tan\theta \geq 2$, $m = 20 - 100$). However, both are cross-dispersed and provide higher dispersions than conventional grating spectrographs.

The angular spread $\delta\theta$ of a single order will be $\delta\lambda d\theta/d\lambda$. Combining this with the equation for the angular dispersion (➋ 2.4) then yields:

$$\lambda/\sigma\cos\theta = \delta\lambda(2/\lambda)\tan\theta,$$

and hence the wavelength covered in a single order will be

$$\delta\lambda = \lambda^2/(2\sigma\sin\theta). \tag{2.8}$$

The angular spread of a single order will be

$$\Delta\theta = \lambda/(\sigma\cos\theta). \tag{2.9}$$

Thus the number of angstroms covered in a single order will increase by the *square* of the wavelength (➋ 2.8), while the length of each order increases only *linearly* with each order (➋ 2.9). This is apparent from ➋ *Fig. 2-5*, as the shorter wavelengths (higher orders) span less of the chip. At lower orders the wavelength coverage actually exceeds the length of the chip. Note that the same spectral feature may be found on adjacent orders, but usually the blaze function is so steep that good signal is obtained for a feature in one particular order. This can be seen for the very strong H and K Ca II lines apparent near the center of order 16 and to the far right in order 15 in ➋ *Fig. 2-5*.

If a grating is used as the cross-disperser, then the separation between orders should increase for lower order numbers (larger wavelengths) as gratings provide fairly linear dispersion, and the free spectral range is larger for lower order numbers. (There is more of a difference in the wavelengths between adjacent orders and hence the orders will be more spread out by a cross-dispersing grating.) However, ➋ *Fig. 2-5* shows that just the opposite is true for MagE: the separation between adjacent orders actually decreases toward lower order numbers. Why? MagE uses prisms for cross-dispersing, and (unlike a grating) the dispersion of a prism is greater in the blue than in the red. In the case of MagE the decrease in dispersion toward larger wavelength (lower orders) for the cross-dispersing prisms more than compensates for the increasing separation in wavelength between adjacent orders at longer wavelengths.

Some echelle spectrographs are listed in ➋ *Table 2-3*. HIRES, UVES, and the KPNO 4 m echelle have a variety of gratings and cross-dispersers available; most of the others provide a fixed format but give nearly full wavelength coverage in the optical in a single exposure.

Table 2-3
Some Echelle spectrographs

Instrument	Telescope	*R* (1 arcsec slit)	Coverage(Å)	Comments
HIRES	Keck I	39,000	Variable	
ESI	Keck II	4000	3900–11,000	Fixed format
UVES	VLT-UT2	40,000	Variable	Two arms
MAESTRO	MMT	28,000	3185–9850	Fixed format
MIKE	Magellan II	25,000	3350–9500	Two arms
MagE	Magellan II	4100	<3200–9850	Fixed format
Echelle	KPNO 4 m	~30,000	Variable	

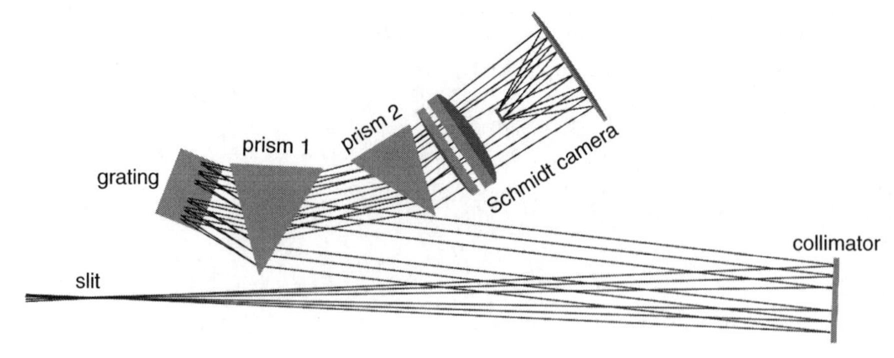

Fig. 2-6
The optical layout of MagE (Based upon Marshall et al. 2008)

2.3.1 An Example: MagE

The Magellan Echellette (MagE) was deployed on the Clay (Magellan II) telescope in November 2007, and provides full wavelength coverage from 3200 Å to 10,000 Å in a single exposure, with a resolution *R* of 4100 with a 1 arcsec slit. The instrument is described in detail by Marshall et al. (2008). The optical layout is shown in *Fig. 2-6*. Light from the telescope is focused onto a slit, and the diverging beam is then collimated by a mirror. Cross dispersion is provided by two prisms, the first of which is used in double pass mode, while the second has a single pass. The echelle grating has 175 lines/mm and is used in a quasi-Littrow configuration. The Echelle Spectrograph and Imager (ESI) used on Keck II has a similar design (Sheinis 2002). MagE has a fixed format and uses orders 6–20, with central wavelengths of 9700–3125 Å, respectively.

The spectrograph is remarkable for its extremely high throughput and ease of operation. The spectrograph was optimized for use in the blue, and the measured efficiency of the instrument alone is >30% at 4100 Å. (Including the telescope the efficiency is about 20%.) Even at the shortest wavelengths (3200 Å and below) the overall efficiency is 10%. The greatest challenge in using the instrument is the difficulties of flat-fielding over that large a wavelength range. This is typically done using a combination of in- and out-of-focus Xe lamps to provide sufficient flux in the near ultraviolet, and quartz lamps to provide good counts in the red. Some users have found that the chip is sufficiently uniform that they do better by not flat-fielding the data at all; in the case of very high signal-to-noise one can dither along the slit. (This is discussed in general in Sect. 3.2.6.) The slit length of MagE is 10 arcsec, allowing good sky subtraction for stellar sources, and still providing clean separation between orders even at long wavelengths (*Fig. 2-5*).

It is clear from an inspection of *Fig. 2-5* that there are significant challenges to the data reduction: the orders are curved on the detector (due to the anamorphic distortions of the prisms) and in addition the spectral features are also tilted, with a tilt that varies along each order. One spectroscopic pundit has likened echelles to space-saving storage travel bags: a lot of things are packed together very efficiently, but extracting the particular sweater one wants can be a real challenge.

2.3.2 Coude Spectrographs

Older telescopes have equatorial mounts, as it was not practical to utilize an altitude-azimuth (alt-az) design until modern computers were available. Although alt-az telescopes allow for a more compact design (and hence a significant cost savings in construction of the telescope enclosure), the equatorial systems provided the opportunity for a coude focus. By adding three additional mirrors, one could direct the light down the stationary polar axis of an equatorial system. From there the light could enter a large "coude room," holding a room-sized spectrograph that would be extremely stable. Coude spectrographs are still in use at Kitt Peak National Observatory (fed by an auxiliary 0.9 m telescope), McDonald Observatory (on the 2.7 m telescope), and at the Dominion Astrophysical Observatory (on a 1.2 m telescope), among other places. Although such spectrographs occupy an entire room, the basic principles are the same, and these instruments afford very high stability and high dispersion. To some extent, these functions are now provided by high resolution instruments mounted on the Nasmyth foci of large alt-az telescopes, although these platforms provide relatively cramped quarters to achieve the same sort of stability and dispersions offered by the classical coude spectrographs.

2.4 Multi-object Spectrometers

There are many instances where an astronomer would like to observe multiple objects in the same field of view, such as studies of the stellar content of a nearby, resolved galaxy, the members of a star cluster, or individual galaxies in a group. If the density of objects is relatively high (tens of objects per square arcminute) and the field of view small (several arcmins) then one often will use a *slit mask* containing not one but dozens or even hundreds of slits. If instead the density of objects is relatively low (less than 10 per square arcminute) but the field of view required is large (many arcmins) one can employ a multi-object fiber positioner feeding a bench-mounted spectrograph. Each kind of device is discussed below.

2.4.1 Multi-slit Spectrographs

Several of the "long slit" spectrographs described in ➋ Sect. 2.2 were really designed to be used with multi-slit masks. These masks allow one to observe many objects at a time by having small slitlets machined into a mask at specific locations. The design of these masks can be quite challenging, as the slits cannot overlap spatially on the mask. An example is shown in ➋ *Fig. 2-7*. Note that in addition to slitlet masks, there are also small alignment holes centered on modestly bright stars, in order to allow the rotation angle of the instrument and the position of the telescope to be set exactly.

In practice, each slitlet needs to be at least 5 arcsec in length in order to allow sky subtraction on either side of a point source. Allowing for some small gap between the slitlets, one can then take the field of view and divide by a typical slitlet length to estimate the maximum number of slitlets an instrument would accommodate. ➋ *Table 2-1* shows that an instrument such as GMOS on the Gemini telescopes has a maximum (single) slit length of 5.5 arcmin, or 330 arcsec. Thus at most, one might be able to cram in 50 slitlets, were the objects of interest properly aligned on the sky to permit this. An instrument with a larger field of view, such as

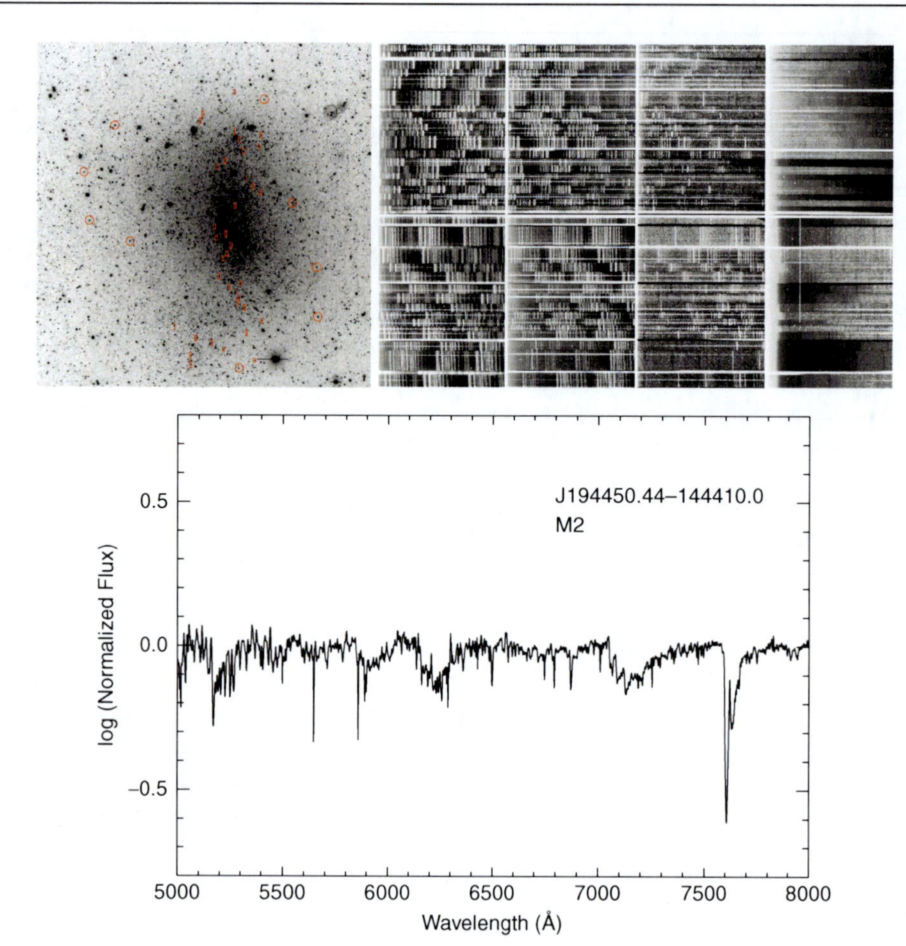

Fig. 2-7

Multi-object mask of red supergiant candidates in NGC 6822. The *upper left* figure shows an image of the Local Group galaxy NGC 6822 taken from the Local Group Galaxy Survey (Massey et al. 2007). The *red circles* correspond to "alignment" stars, and the small rectangles indicate the position of red supergiant candidates to be observed. The slit mask consists of a large metal plate machined with these holes and slits. The *upper right* figure is the mosaic of the eight chips of IMACS. The *vertical lines* are night-sky emission lines, while the spectra of individual stars are *horizontal narrow lines*. A sample of one such reduced spectrum, of an M2 I star, is shown in the lower figure. These data were obtained by Emily Levesque, who kindly provided parts of this figure

IMACS (described below) really excels in this game, as over a hundred slitlets can be machined onto a single mask.

Multi-slit masks offer a large multiplexing advantage, but there are some disadvantages as well. First, the masks typically need to be machined weeks ahead of time, so there is really no flexibility at the telescope other than to change exposure times. Second, the setup time for such masks is non-negligible, usually of order 15 or 20 min. This is not an issue when exposure times are long, but can be a problem if the objects are bright and the exposure times short. Third, and perhaps most significantly, the wavelength coverage will vary from slitlet to slitlet, depending upon location within the field. As shown in the example of *Fig. 2-7*, the mask field has been rotated so that the slits extend north and south, and indeed the body of the galaxy is mostly located north and south, minimizing this problem. The alignment holes are located well to the east and west, but one does not care about their wavelength coverage. In general, though, if one displaces a slit off center by X arcsec, then the central wavelength of the spectrum associated with that slit is going to shift by Dr(X/p), where p is the scale on the detector in terms of arcsec per pixel, r is the anamorphic demagnification factor associated with this particular grating and tilt (≤ 1), and D is the dispersion in Å per pixel.

Consider the case of the IMACS multi-object spectrograph. Its basic parameters are included in *Table 2-1*, and the instrument is described in more detail below. The field of view with the f/4 camera is 15 arcmins × 15 arcmins. A slit on the edge of the field of view will be displaced by 7.5 arcmin, or 450 arcsec. With a scale of 0.11 arcsec/pixel this corresponds to an offset of 4090 pixels ($X/p = 4090$). With a 1200 line/mm grating centered at 4500 Å for a slit on-axis, the wavelength coverage is 3700–5300 Å with a dispersion $D = 0.2$ Å/pixel. The anamorphic demagnification is 0.77. So, for a slit on the edge the wavelengths are shifted by 630 Å, and the spectrum is centered at 5130 Å and covers 4330 Å to 5930 Å. On the other edge the wavelengths will be shifted by −630 Å, and will cover 3070–4670 Å. The only wavelengths in common to slits covering the entire range in X is thus 4330–4670 Å, only 340 Å!

Example: IMACS

The Inamori-Magellan Areal Camera & Spectrograph (IMACS) is an innovative slit spectrograph attached to the Nasmyth focus of the Baade (Magellan I) 6.5 m telescope (Dressler et al. 2006). The instrument can be used either for imaging or for spectroscopy. Designed primarily for multi-object spectroscopy, the instrument is sometimes used with a mask cut with a single long (26-in. length!) slit. There are two cameras, and either is available to the observer at any time: an f/4 camera with a 15.4 arcmin coverage, or an f/2.5 camera with a 27.5 arcmin coverage.

The f/4 camera is usable with any of 7 gratings, of which 3 may be mounted at any time, and which provide resolutions of 300–5000 with a 1-arcsec wide slit. The delivered image quality is often better than that (0.6 arcsec is not unusual) and so one can use a narrower slit resulting in higher spectral resolution. The spectrograph is really designed to take advantage of such good seeing, as a 1-arcsec wide slit projects to nine unbinned pixels. Thus binning is commonly used. The f/2.5 camera is used with a grism,[9] providing a longer spatial coverage but lower dispersion. Up to two grisms can be inserted for use during a night.

The optical design of the spectrograph is shown in *Fig. 2-8*. Light from the f/11 focus of the Baade Magellan telescope focuses onto the slit plate, enters a field lens, and is collimated by transmission optics. The light is then either directed into the f/4 or f/2.5 camera. To direct the light into the f/4 camera, either a mirror is inserted into the beam (for imaging) or a diffraction grating is inserted (for spectroscopy). If the f/2.5 camera is used instead, either the light enters directly (in imaging mode) or a transmission "grism" is inserted. Each camera has its own mosaic of eight CCDs, providing 8192 × 8192 pixels. The f/4 camera provides a smaller field of view but higher dispersion and plate scale; see *Table 2-1*. Pre-drilled "long-slit" masks are available in a variety of slit widths. Up to six masks can be inserted for a night's observing, and selected by the instrument's software.

[9]A "grism" is a prism attached to a diffraction grating. The diffraction grating provides the dispersive power, while the (weak) prism is used to displace the first-order spectrum back to the straight-on position. The idea was introduced by Bowen and Vaughan (1973), and used successfully by Art Hoag at the prime focus of the Kitt Peak 4 m telescope (Hoag 1976).

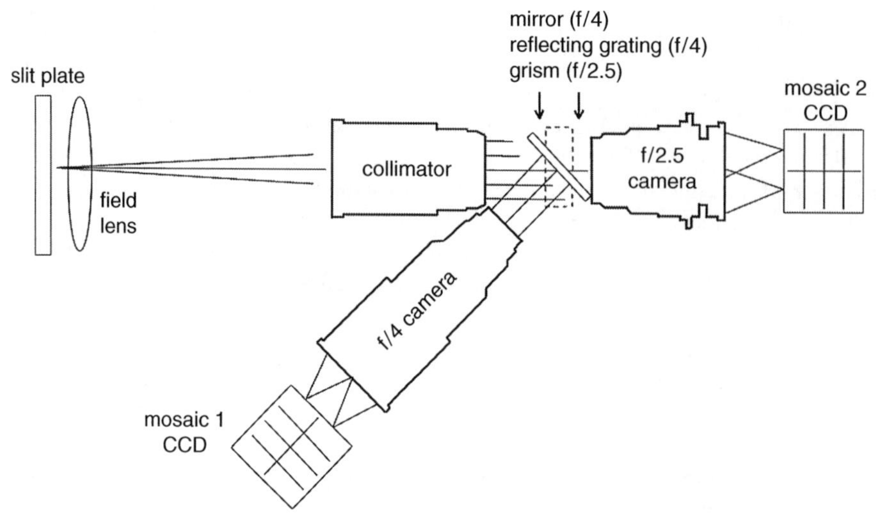

Fig. 2-8
Optical layout of Magellan's IMACS (This is based upon an illustration in the IMACS user manual)

2.4.2 Fiber-fed Bench-Mounted Spectrographs

As an alternative to multi-slit masks, a spectrograph can be fed by multiple optical fibers. The fibers can be arranged in the focal plane so that light from the objects of interest enter the fibers, while at the spectrograph end the fibers are arranged in a line, with the ends acting like the slit in the model of the basic spectrograph (*Fig. 2-1*). Fibers were first commonly used for multi-object spectroscopy in the 1980s, prior even to the advent of CCDs; for example, the Boller and Chivens spectrograph on the Las Campanas du Pont 100 in. telescope was used with a plug-board fiber system when the detector was an intensified Reticon system. Plug-boards are like multi-slit masks in that there are a number of holes pre-drilled at specific locations in which the fibers are then "plugged." For most modern fiber systems, the fibers are positioned robotically in the focal plane, although the Sloan Digital Sky Survey used a plug-board system. A major advantage of a fiber system is that the spectrograph can be mounted on a laboratory air-supported optical bench in a clean room, and thus not suffer flexure as the telescope is moved. This can result in high stability, needed for precision radial velocities. The fibers themselves provide additional "scrambling" of the light, also significantly improving the radial velocity precision, as otherwise the exact placement of a star on the slit may bias the measured velocities.

There are three down sides to fiber systems. First, the fibers themselves tend to have significant losses of light at the slit end, i.e., not all of the light falling on the entrance end of the fiber actually enters the fiber and makes it down to the spectrograph. These losses can be as high as a factor of 3 or more compared to a conventional slit spectrograph. Second, although typical fibers are 200–300 μm in diameter, and project to a few arcsec on the sky, each fiber must be surrounded by a protective sheath, resulting in a minimal spacing between fibers of 20–40 arcsec. Third, and most importantly, sky subtraction is never "local." Instead, fibers are assigned to blank sky locations just like objects, and the accuracy of the sky subtraction is dependent on how accurately one can remove the fiber-to-fiber transmission variations by flat-fielding.

An Example: Hectospec

Hectospec is a 300-fiber spectrometer on the MMT 6.5 m telescope on Mt Hopkins. The instrument is described in detail by Fabricant et al. (2005). The focal surface consists of a 0.6 m diameter stainless steel plate onto which the magnetic fiber buttons are placed by two positioning robots (➋ *Fig. 2-9*). The positioning speed of the robots is unique among such instruments and is achieved without sacrificing positioning accuracy (25 μm, or 0.15 arcsec). The field of view is a degree across. The fibers subtend 1.5 arsec on the sky, and can be positioned to within 20 arcsec of each other. Light from the fibers is then fed into a bench-mounted spectrograph, which uses either a 270 line/mm grating ($R \sim 1000$) or a 600 line/mm grating ($R \sim 2500$). The same fiber bundle can be used with a separate spectrograph, known as Hectochelle.

Another unique aspect of Hectospec is the "cooperative" queue manner in which the data are obtained, made possible in part because multi-object spectroscopy with fibers is not very flexible, and configurations are done well in advance of going to the telescope. Observers are awarded a certain number of nights and scheduled on the telescope in the classical way. The astronomers design their fiber configuration files in advance; these files contain the necessary positioning information for the instrument, as well as exposure times, grating setups, etc. All of the observations however become part of a collective pool. The astronomer goes to the telescope on the scheduled night, but a "queue manager" decides on a night-by-night basis which fields should be observed and when. The observer has discretion to select alternative fields and vary exposure times depending upon weather conditions, seeing, etc. The advantages of this over classical observing is that weather losses are spread among all of the programs in the scheduling period (4 months). The advantages over normal queue scheduled observations is that the astronomer is actually present for some of his/her own observations, and there is no additional cost involved in hiring queue observers.

Fig. 2-9
View of the focal plane of Hectospec (From Fabricant et al. 2005. Reproduced by permission)

2.5 Extension to the UV and NIR

The general principles involved in the design of optical spectrographs extend to those used to observe in the ultraviolet (UV) and near-infrared (NIR), with some modifications. CCDs have high efficiency in the visible region, but poor sensitivity at shorter (<3000 Å) and longer (>1 μm) wavelengths. At very short wavelengths (x-rays, gamma-rays) and very long wavelengths (mid-IR through radio and mm) special techniques are needed for spectroscopy, and are beyond the scope of this chapter.

Here we provide examples of two non-optical instruments, one whose domain is the ultraviolet (1150–3200 Å) and one whose domain is in the near-infrared (1–2 μm).

2.5.1 The Near Ultraviolet

For many years, astronomical ultraviolet spectroscopy was the purview of the privileged few, mainly instrument Principle Investigators (PIs) who flew their equipment on high-altitude balloons or short-lived rocket experiments. The Copernicus (Orbiting Astronomical Observatory 3) was a longer-lived mission (1972–1981), but the observations were still PI-driven. This all changed drastically due to the *International Ultraviolet Explorer (IUE)* satellite, which operated from 1978 to 1996. Suddenly any astronomer could apply for time and obtain fully reduced spectra in the ultraviolet. *IUE*'s primary was only 45 cm in diameter, and there was considerable demand for the community to have UV spectroscopic capability on the much larger (2.4 m) *Hubble Space Telescope (HST)*.

The Space Telescope Imaging Spectrograph (STIS) is the spectroscopic work-horse of *HST*, providing spectroscopy from the far-UV through the far-red part of the spectrum. Although a CCD is used for the optical and far-red, another type of detector (multi-anode microchannel array, or MAMA) is used for the UV. Yet, the demands are similar enough for optical and UV spectroscopy that the rest of the spectrograph is in common to both the UV and optical. The instrument is described in detail by Woodgate (1998), and the optical design is shown in ➲ *Fig. 2-10*.

In the UV, STIS provides resolutions of ~1000–10,000 with first-order gratings. With the echelle gratings, resolution as high as 114,000 can be achieved. No blocking filters are needed as the MAMA detectors are insensitive to longer wavelengths. From the point of view of the astronomer who is well versed in optical spectroscopy, the use of STIS for UV spectroscopy seems transparent.

With the success of the Servicing Mission 4 in May 2009, the Cosmic Origins Spectrograph (COS) was added to *HST's* suite of instruments. COS provides higher through-put than *STIS* (by factors of 10–30) in the far-UV, from 1100 Å to 1800 Å. In the near-UV (1700–3200 Å) STIS continues to win out for many applications.

2.5.2 Near-Infrared Spectroscopy and OSIRIS

Spectroscopy in the near-infrared (NIR) is complicated by the fact that the sky is much brighter than most sources, plus the need to remove the strong telluric bands in the spectra. In general, this is handled by moving a star along the slit on successive, short exposures (dithering), and subtracting adjacent frames, such that the sky obtained in the first exposure is subtracted from

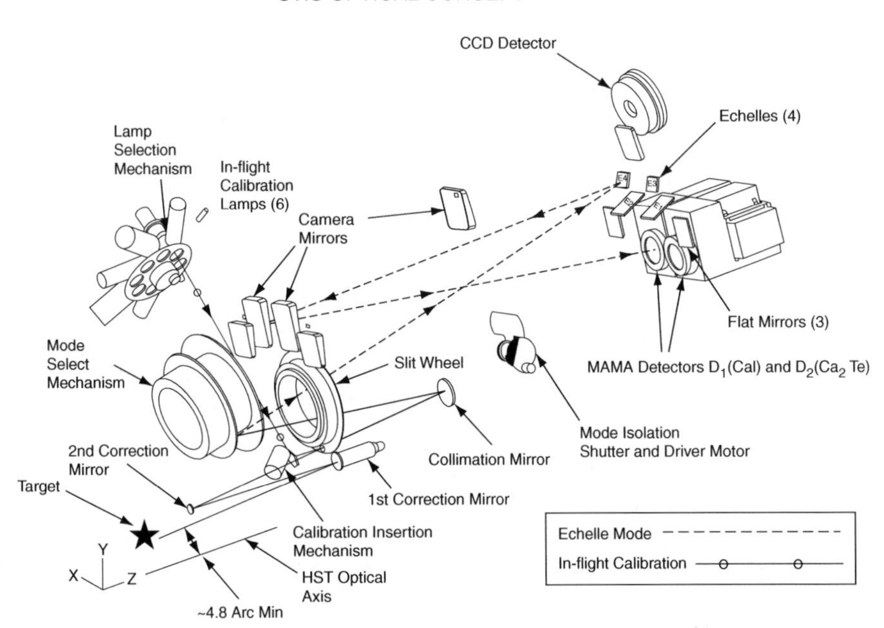

Fig. 2-10
Optical design of STIS (From Woodgate (1998). Reproduced by permission)

the source in the second exposure, and the sky in the second exposure is subtracted from the source in the first exposure. Nearly featureless stars are observed at identical airmasses to that of the program object in order to remove the strong telluric absorption bands. These issues will be discussed further in Sects. 3.1.2 and 3.3.3 below.

The differences in the basics of infrared arrays compared to optical CCDs also affect how NIR astronomers go about their business. CCDs came into use in optical astronomy in the 1980s because of their very high efficiency (≥50%, relative to photographic plates of a few percent) and high linearity (i.e., the counts above bias are proportional to the number of photons falling on their surface over a large dynamic range). CCDs work by exposing a thin wafer of silicon to light and to collect the resulting freed charge carriers under electrodes. By manipulating the voltages on those electrodes, the charge packets can be carried to a corner of the detector array where a single amplifier can read them out successively. (The architecture may also be used to feed multiple output amplifiers.) This allows for the creation of a single, homogenous silicon structure for an optical array (see Mackay 1986 for a review). For this and other reasons, optical CCDs are easily fabricated to remarkably large formats, several thousand pixels to a side.

Things are not so easy in the infrared. The band gap (binding energy of the electron) in silicon is simply too great to be dislodged by an infrared photon. For detection between 1 and 5 μm, either Mercury-Cadmium-Telluride (HgCdTe) or Indium-Antimonide (InSb) are typically used, while the read out circuitry still remains silicon-based. From 5 to 28 μm, silicon-based (extrinsic photoconductivity) detector technology is used, but they continue to use similar approaches to array construction as in the near-infrared. The two layers are joined electrically and mechanically with an array of Indium bumps. (Failures of this Indium bond lead to dead pixels in the array.) Such a two-layered device is called a *hybrid* array (Beckett 1995). For the silicon integrated circuitry, a CCD device could be (and was originally) used for NIR arrays, but the very cold temperatures required for the photon detection portion of the array produced high read noise. Instead, an entirely new structure that provides a dedicated readout amplifier for each pixel was developed (see Rieke 2007 for more details). These direct readout arrays are the standard for infrared instruments and allow for enormous flexibility in how one reads the array. For instance, the array can be set to read the charge on a specific, individual pixel without even removing the accumulated charge (non-destructive read). Meanwhile, reading through a CCD removes the accumulated charge on virtually every pixel on the array.

Infrared hybrid arrays have some disadvantages, too. Having the two components (detection and readout) made of different materials limits the size of the array that can be produced. This is due to the challenge of matching each detector to its readout circuitry to high precision when flatly pressed together over millions of unit cells. Even more challenging is the stress that develops from differential thermal contraction when the hybrid array is chilled down to very cold operating temperatures. However, improvements in technology now make it possible to fabricate 2K × 2K hybrid arrays, and it is expected that 4K × 4K will eventually be possible. Historically, well depths have been lower in the infrared arrays, though hybrid arrays can now be run with a higher gain. This allows for well depths approaching that available to CCD arrays (hundreds of thousands of electrons per pixel). All infrared hybrid arrays have a small degree of nonlinearity, of order a few percent, due to a slow reduction in response as signals increase (Rieke 2007). In contrast, CCDs are typically linear to a few tenths of a percent over five orders of magnitude. Finally, the infrared hybrid arrays are far more expensive to build than CCDs. This is because of the extra processing steps required in fabrication and their much smaller commercial market compared to CCDs.

The Ohio State Infrared Imager/Spectrometer (OSIRIS) provides an example of such an instrument, and how the field has evolved over the past two decades. OSIRIS is a multi-mode infrared imager and spectrometer designed and built by The Ohio State University (Atwood et al. 1992; Depoy et al. 1993). Despite being originally built in 1990, it is still in operation today, most recently spending several successful years at the Cerro Tololo Inter-American Observatory Blanco 4 m telescope. Presently, OSIRIS sits at the Nasmyth focus on the 4.1 m Southern Astrophysical Research (SOAR) Telescope on Cerro Pachón.

When built 20 years ago, the OSIRIS instrument was designed to illuminate the best and largest infrared-sensitive arrays available at the time, the 256 × 256 pixel NICMOS3 HgCdTe arrays, with 27 μm pixels. This small array has long since been upgraded as infrared detector technology has improved. The current array on OSIRIS is now 1024 × 1024 in size, with 18.5 μm pixels (NICMOS4, still HgCdTe). As no design modifications could be afforded to accommodate this upgrade, the larger array now used is not entirely illuminated due to vignetting in the optical path. This is seen as a fall off in illumination near the outer corners of the array.

OSIRIS provides two cameras, $f/2.8$ for lower resolution work ($R \sim 1200$ with a 3.2 arcmin long slit) and $f/7$ for higher resolution work ($R \sim 3000$, with a 1.2 arcminute long slit). One then uses broadband filters in the J (1.25 μm), H (1.65 μm), or K (2.20 μm) bands, to select the desired order, fifth, fourth, and third, respectively. The instrument grating tilt is set to simultaneously select the central regions of these three primary transmission bands of the atmosphere. However, one can change the tilt to optimize observations at wavelengths near the edges of these bands. OSIRIS does have a cross-dispersed mode, achieved by introducing a cross-dispersing grism in the filter wheel. A final filter, which effectively blocks light outside of the J, H, and K

bands, is needed for this mode. The cross-dispersed mode allows observing at low resolution ($R \sim 1200$) in all three bands simultaneously, albeit it with a relatively short slit (27 arcsecs).

Source acquisition in the infrared is not so straightforward. While many near-infrared objects have optical counterparts, many others do not or show rather different morphology or central position offsets between the optical and infrared. This means acquisition and alignment must be done in the infrared, too. OSIRIS, like most modern infrared spectrometers, can image its own slit onto the science detector when the grating is not deployed in the light path. This greatly facilitates placing objects on the slit (some NIR spectrometers have a dedicated slit viewing imager so that objects may be seen through the slit during an actual exposure). This quick-look imaging configuration is available with an imaging mask too, and deploys a flat mirror in place of the grating (without changing the grating tilt) thereby displaying an infrared image of the full field or slit with the current atmospheric filter. This change in configuration only takes a few seconds and allows one to align the target on the slit; then quickly returns the grating to begin observations. Even so, the mirror/grating flip mechanism will only repeat to a fraction of a pixel when being moved to change between acquisition and spectroscopy modes. The most accurate observations may then require new flat-fields and/or lamp spectra be taken before returning to imaging (acquisition) mode.

For precise imaging observations, OSIRIS can be run in "full" imaging mode which includes placing a cold mask in the light path to block out-of-beam background emission for the telescope primary and secondary. Deployment of the mask can take several minutes. This true imaging mode is important in the K-band where background emission becomes significant beyond 2 μm due to the warm telescope and sky.

There are fantastic new capabilities for NIR spectroscopy about to become available as modern multi-object spectrometers come on-line on large telescopes (LUCIFER on the Large Binocular Telescope, MOSFIRE on Keck, FLAMINGOS-2 on Gemini-South, and MMIRS on the Clay Magellan telescope). As with the optical, utilizing the multi-object capabilities of these instruments effectively requires proportionately greater observer preparation, with a significant increase in the complexity of obtaining the observations and performing the reductions. Such multi-object NIR observations are not yet routine, and as such details are not given here. One should perhaps master the "simple" NIR case first before tackling these more complicated situations.

2.6 Spatially Resolved Spectroscopy of Extended Sources: Fabry-Perots and Integral Field Spectroscopy

The instruments described above allow the astronomer to observe single or multiple point sources at a time. If instead one wanted to obtain spatially resolved spectroscopy of a galaxy or other extended source, one could place a long slit over the object at a particular location and obtain a one-dimensional, spatially resolved spectrum. If one wanted to map out the velocity structure of an HII region or galaxy, or measure how various spectral features changed across the face of the object, one would have to take multiple spectra with the slit rotated or moved across the object to build up a three-dimensional image "data cube": two-dimensional spatial location plus wavelength. Doing this is sometimes practical with a long slit: one might take spectra of a galaxy at four different position angles, aligning the slit along the major axis, the minor axis, and the two intermediate positions. These four spectra would probably give a pretty good indication of the kinematics of the galaxy. But, if the velocity field or ionization structure is complex, one would really want to build up a more complete data cube. Doing so by stepping the slit by its width over the face of an extended object would be one way, but clearly very costly in terms of telescope time.

An alternative would be to use a Fabry–Perot interferometer, basically a tunable filter. A series of images through a narrow-band filter is taken, with the filter tuned to a slightly different wavelength for each exposure. The resulting data are spatially resolved, with the spectral resolution dependent upon the step size between adjacent wavelength settings. (A value of 30 km s^{-1} is not atypical for a step size, i.e., a resolution of 10,000.) The wavelength changes slowly as a function of radial position within the focal plane, and thus a "phase-corrected" image cube is constructed which yields both an intensity map (such being a direct image) and radial velocity map for a particular spectral line (for instance, Hα).

This works fine in the special case where one is interested in only a few spectral features in an extended object. Otherwise, the issue of scanning spatially has simply been replaced with the need to scan spectrally.

Alternative approaches broadly fall under the heading of *integral field spectroscopy*, which simply means obtaining the full data cube in a single exposure. There are three methods of achieving this, following Allington-Smith et al. (1998).

Lenslet arrays. One method of obtaining integral field spectroscopy is to place a microlens array (MLA) at the focal plane. The MLA produces a series of images of the telescope pupil, which enter the spectrograph and are dispersed. By tilting the MLA, one can arrange it so that the spectra do not overlap with one another.

Fiber bundles. An array of optical fibers is placed in the focal plane, and the fibers then transmit the light to the spectrograph, where they are arranged in a line, acting as a slit. This is very similar to the use of multi-object fiber spectroscopy, except that the ends of the fibers in the focal plane are always in the same configuration, with the fibers bundled as close together as possible. There are of course gaps between the fibers, resulting in incomplete spatial coverage without dithering.
It is common to use both lenslets and fibers together, as for instance is done with the integral field unit of the FLAMES spectrograph on the VLT.

Image slicers. A series of mirrors can be used to break up the focal plane into congruent "slices" and arrange these slices along the slit, analogous to a classic Bowen image slicer.[10] One advantage of the image slicer technique for integral-field spectroscopy is that spatial information within the slice is preserved.

[10]When observing an astronomical object, a narrow slit is needed to maintain good spectral resolution, as detailed in ➋ Sect. 2.1. Yet, the size of the image may be much larger than the size of the slit, resulting in a significant loss of light, known as "slit losses." Bowen (1938) first described a novel device for reducing slit losses by changing the shape of the incoming image to match that of a long, narrow slit: a cylindrical lens is used to slice up the image into a series of strips with a width equal to that of the slit, and then to arrange them end to end along the slit. This needs to be accomplished without altering the focal ratio of the incoming beam. Richardson et al. (1971) describes a variation of the same principle, while Pierce (1965) provides detailed construction notes for such a device. The heyday of image slicers was in the photographic era, where sky subtraction was impractical and most astronomical spectroscopy was not background limited. Nevertheless, they have not completely fallen into disuse. For instance, GISMO (Gladders Image-Slicing Multislit Option) is an image slicing device available for use with multi-slit plates on Magellan's IMACS (➋ Sect. 2.4.1) to re-image the central 3.5 arcmin × 3.2 arcmin of the focal plane into the full field of view of the instrument, allowing an eightfold increase in the spatial density of slits.

3 Observing and Reduction Techniques

This section will begin with a basic outline of how spectroscopic CCD data are reduced, and then extend the treatment to the reduction of NIR data. Some occasionally overlooked details will be discussed. The section will conclude by placing these together by describing a few sample observing runs. It may seem a little backward to start with the data reduction rather than with the observing. But, only by understanding how to reduce data can one really understand how to best take data.

The basic premise throughout this section is that one should neither observe nor reduce data by rote. Simply subtracting biases because all of one's colleagues subtract biases is an inadequate reason for doing so. One needs to examine the particular data to see if doing so helps or harms. Similarly, unless one is prepared to do a little math, one might do more harm than good by flat-fielding. Software reduction packages, such as *IRAF*[11] or *ESO-MIDAS*[12] are extremely useful tools – in the right hands. But, one should never let the software provide a guide to reducing data. Rather, the astronomer should do the guiding. One should strive to understand the steps involved at the level that one could (in principle) reproduce the results with a hand calculator!

In addition, there are very specialized data reduction pipelines, such as *HST's* CALSTIS and IMACS's COSMOS, which may or may not do what is needed for a specific application. The only way to tell is to try them, and compare the results with what one obtains by more standard means. See ➲ Sect. 3.2.5 for an example where the former does not do very well.

3.1 Basic Optical Reductions

The simplest reduction example involves obtaining a spectrum of a star obtained from a long slit spectrograph. What calibration data does one need, and why?

- *The data frames themselves* doubtlessly contain an overscan strip along the side, or possibly along the top. As a CCD is read out, a "bias" is added to the data to assure that no values go below zero. Typically this bias is several hundred or even several thousand electrons (e^-).[13] The exact value is likely to change slightly from exposure to exposure, due to slight temperature variations of the CCD. The overscan value allows one to remove this offset. In some cases the overscan should be used to remove a one-dimensional bias structure, as demonstrated below.
- *Bias frames* allow one to remove any residual bias *structure*. Most modern CCDs, used with modern controllers, have very little (if any) bias structure, i.e., the bias levels are spatially uniform across the chip. So, it's not clear that one needs to use bias frames. If one takes enough bias frames (9 or more) and averages them correctly, one probably does very little damage to the data by using them. Still, in cases where read-noise dominates your program spectrum, subtracting a bias could increase the noise of the final spectrum.

---

[11] The Image Reduction and Analysis Facility is distributed by the National Optical Astronomy Observatory, which is operated by the Association of Universities for Research in Astronomy, under cooperative agreement with the National Science Foundation.

[12] The European Southern Observatory Munich Image Data Analysis System.

[13] We use "counts" to mean what gets recorded in the image; sometimes these are also known as analog-to-digital units (ADUs). We use "electrons" (e^-) to mean the things that behave like Poisson statistics, with the noise going as the square root of the number of electrons. The gain g is the number of e^- per count.

- *Bad pixel mask* data allows one to interpolate over bad columns and other non-linear pixels. These can be identified by comparing the average of a series of exposures of high counts with the average of a series of exposures of low counts.
- *Dark frames* are exposures of comparable length to the program objects but obtained with the shutter closed. In the olden days, some CCDs generated significant dark current due to glowing amplifiers and the like. Dark frames obtained with modern CCDs do little more than reveal light leaks if taken during daytime. Still, it is generally harmless to obtain them.
- *Featureless flats* (usually "dome flats") allow one to correct the pixel-to-pixel variations within the detector. There have to be sufficient counts to not significantly degrade the signal-to-noise ratio of the final spectrum. This will be discussed in more detail in ➋ Sect. 3.2.6.
- *Twilight flats* are useful for removing any residual spatial illumination mismatch between the featureless flat and the object exposure. This will be discussed in more detail in ➋ Sect. 3.2.6.
- *Comparison arcs* are needed to apply a wavelength scale to the data. These are usually short exposures of a combination of discharge tubes containing helium, neon, and argon (HeNeAr), or thorium and argon (ThAr), with the choices dictated by what lamps are available and the resolution. HeNeAr lamps are relatively sparse in lines and useful at low to moderate dispersion; ThAr lamps are rich in lines and useful at high dispersion, but have few unblended lines at low or moderate dispersions.
- *Spectrophotometric standard stars* are stars with smooth spectra with calibrated fluxes used to determine the instrument response. Examples of such stars can be found in Oke (1990), Stone (1977), Stone and Baldwin (1983), Stone (1996), Massey et al. (1988), Massey and Gronwall (1990), and particularly Hamuy et al. (1994). Exposures should be long enough to obtain at least 10,000 e^- integrated over 50 Å bandpass, i.e., at least 200 e^- per Å integrated over the spatial profile.

The basic reduction steps are described here. For convenience to the reader, reference is made to the relevant IRAF tasks. Nevertheless, the goal is to explain the steps, not identify what buttons need to be pushed.

1. *Fit and subtract overscan.* Typically the overscan occupies 20–40 columns on the right side (high column numbers) of the chip. The simplest thing one can do is to average the results of these columns on a row-by-row basis, and then fit these with a low-order function, possibly a constant.

   An example of such a nicely behaved overscan is shown in the left panel of ➋ *Fig. 2-11*. In other cases, there is a clear gradient in the bias section that should be removed by a higher-order fit. The IMACS chips, for instance, have an overscan at both the top and the right, and it is clear from inspection of the bias frames that the top overscan tracks a turn-up on the left side. Such an overscan is shown in the middle panel of ➋ *Fig. 2-11*. Occasionally the overscan allows one to correct for some problem that occurred during the readout of the chip. For instance, the GoldCam spectrometer chip suffers 10–20 e^- striping along rows if the controller electronics get a bit too hot. (This situation is sometimes referred to as "banding.") A similar problem may occur if there is radio transmission from a nearby walkie-talkie during readout. However, these stripes extend into the overscan, which can be used to eliminate them by subtracting the overscan values in a row-by-row manner rather than by subtracting a low- or high-order function. Such banding is shown in the right panel of ➋ *Fig. 2-11*. In IRAF, overscan subtraction is a task performed in the first pass through *ccdproc*.

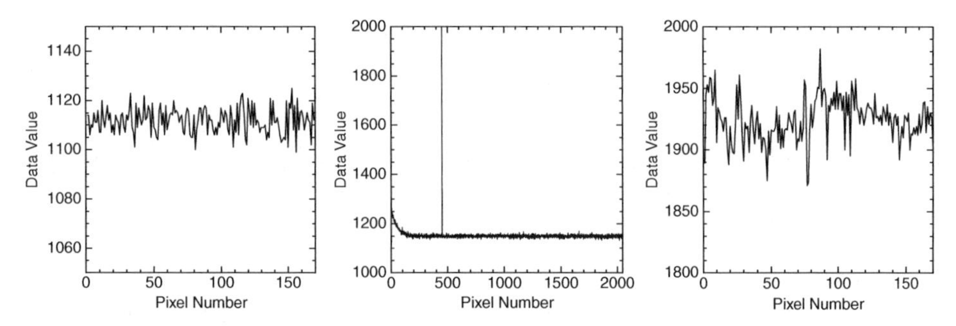

Fig. 2-11

Three examples of overscan structure. The overscan values have been averaged over the 20–40 widths of the overscan region, and are shown plotted against the read-axis. The overscan on the *left* is from the GoldCam spectrometer on the Kitt Peak 2.1 m and is best fit by a constant value. The overscan in the *middle* is from chip 3 of IMACS, and shows a pronounced upturn on the *left*, and is best fit with a low-order function. (Note too the bad column around pixel 420.) The overscan on the *right* is also from GoldCam, but during an electronics problem that introduced a number of bands with low values that were also in the data region. Subtracting the average values in the overscan row-by-row removed them to a good approximation

2. *Trim.* Most spectroscopic data do not extend over the entire face of the chip, and the blank areas will need to be trimmed off. Often the first or last couple of rows and columns are not well behaved and should be trimmed as well. At the very least, the overscan regions need to be removed. In IRAF this task is performed in the same step as the overscan subtraction by *ccdproc*.
3. *Dark subtraction.* In the highly unlikely event that there is significant dark current the dark frame should be subtracted from all of the images after scaling by the relevant exposure times.
4. *Interpolate over bad pixels.* This step is mostly cosmetic, but is necessary if one plans to flux-calibrate. Most CCDs have a few bad columns and non-linear pixels. To identify these one can use two series of exposures of featureless flats: one of high intensity level (but be careful not to saturate!) and one of low intensity. Enough of these low-intensity flats need to be taken in order for the average to have a reasonable signal-to-noise ratio. Typically three exposures of an internal quartz lamp (say) might be taken, along with 50 frames of similar exposure time with a 100× attenuating neutral density filter. The first series may contain 10,000 counts/pixel, while the latter only 100 counts/pixel. Averaging each series and dividing the results identifies the non-linear pixels. One can construct a bad pixel "mask" (an image that is only a few bits deep) to use to decide whether to interpolate along rows (in order to handle bad columns) or over rows and columns (to handle isolated bad pixels), etc. The relevant tasks in IRAF are *ccdmask* which generates a bad pixel mask, followed by using the mask in *ccdproc*.
5. *Construct (and use?) a master bias frame.* All the bias frames (which have been treated by steps 1–2 as well) are combined and examined to see if there is any spatial structure. If there is, then it is worth subtracting this master bias frame from all the program frames. The relevant IRAF tasks are *zerocombine* and *ccdproc*.

6. *Construct and apply a master normalized featureless flat.* All of the dome flat exposures are combined using a bad pixel rejection algorithm. The IRAF task for this combination is *flatcombine*. The question then is how to best normalize the flat?

   Whether the final step in the reduction process is to flux-calibrate the spectrum, or simply to normalize the spectrum to the continuum, the effort is going to be much easier if one makes the correct choice as to how to best normalize the flat. Physically the question comes down to whether or not the shape of the featureless flat in the wavelength direction is dominated by the same effects as the astronomical object one is trying to reduce, or if instead it is dominated by the calibration process. For instance, all flats are likely to show some bumps and wiggles. Are the bumps and wiggles due to the grating? If so, one is better off fitting the flat with a constant value (or a low order function) and using that as the flat's normalization. That way the same bumps and wiggles are present in both the program data as in the flat, and will divide out. If instead some of the bumps and wiggles are due to filters in front of the flat-field lamp, or the extreme difference in color temperature between the lamps and the object of interest, or due to wavelength-dependent variations in the reflectivity of the paint used for the flat-field screen, then one is better off fitting the bumps and wiggles by using a very high order function, and using the flat only to remove pixel-to-pixel variations. It usually isn't obvious which will be better *a priori*, and really the only thing to do is to select an object whose spectrum is expected to be smooth (such as a spectrophotometric standard) and reduce it through the entire process using each kind of flat. It will be easy to tell in the end. The IRAF task for handing the normalization of the flat is *response*, and the flat-field division is handled by *ccdproc*.

   An example is shown in *Fig. 2-12*. The flat-field has a very strong gradient in the wavelength direction, and a bit of a bend around pixel 825 (about 4020 Å). Will one do better by just normalizing this by a constant value, ascribing the effects to the grating and spectrograph? Or should one normalize the flat with a higher order function based on the assumption that these features are due to the flat-field system and very red lamps? (Blue is on the right and red is on the left.) The only way to answer the question is to reduce some data both ways and compare the results, as in *Fig. 2-12*. The flat fit by a constant value does a better job removing both the gradient and the bump. (The spectrum needs to be magnified to see that the bump at 4020 Å has been removed.) Thus the flat was a good reflection of what was going on in the spectrograph, and is not due to issues with the calibration system.

   Once one determines the correct normalized flat, it needs to be divided into all of the data. The IRAF task for handling this is again *ccdproc*.

7. *Construct and use illumination function correction.* The non-uniformity in the spatial direction is referred to as the "slit illumination function" and is a combination of any unevenness in the slit jaws, the vignetting within the focal plane or within the spectrograph, etc. Most of these issues should have been removed by dividing by the flat-field in Step 6. Is this good enough? If the remaining non-uniformity (relative to the night sky) is just a few percent, and has a linear gradient, this is probably fine for sky subtraction from a point source as one can linearly interpolate from the sky values on either side of the object. However, if the remaining non-uniformity is peaked or more complex, or if the goal is to measure the surface brightness of an extended source, then getting the sky really and truly flat is crucial, and worth some effort.

   The cheapest solution (in terms of telescope time) is to use exposures of the bright twilight sky. With the telescope tracking, one obtains 3–5 unsaturated exposures, moving the telescope slightly (perpendicular to the slit) between exposures. To use these, one can combine the exposures (scaling each by the average counts), averaging the spatial profile over all

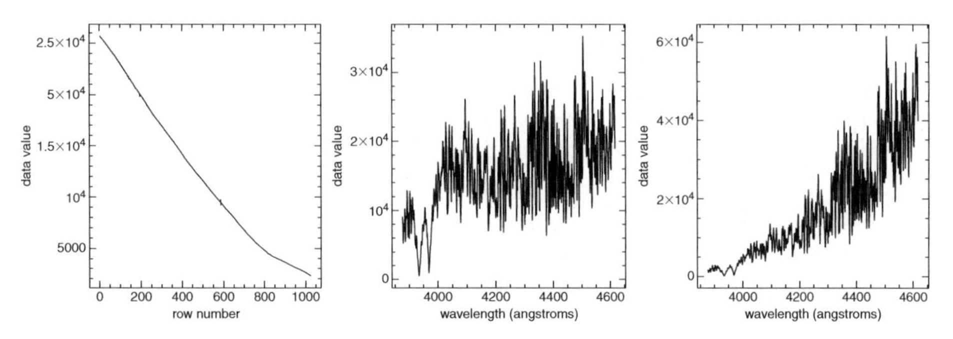

Fig. 2-12
Normalization of Flats. *Left.* The flat-field counts varies by a factor of 10 or more with wavelength, and there is a discontinuity in the derivative around pixel 820. *Middle.* Normalizing the flat by a constant yields a smooth stellar spectrum. *Right.* Normalizing the flat by a higher order function results in an artificially large gradient in the counts

wavelengths, and then fitting a smooth function to the spatial profile. The result is an image that is uniform along each row of the dispersion axis, and hopefully matches the night sky perfectly in the spatial direction. Sadly this is usually not the case, as during bright twilight there is light bouncing off of all sorts of surfaces in the dome. Thus one might need to make a further correction using exposures of the dark night sky itself. This process is illustrated in *Fig. 2-13*. The relevant IRAF tasks involved are *imcombine*, *illumination*, and *ccdproc*.

This concludes the basic CCD reductions, and one is now ready to move on to extracting and calibrating the spectrum itself.

8. *Identification of object and sky.* At this stage one needs to understand the location of the stellar spectrum on the detector, decide how wide an extraction window to use, and to select where the sky regions should be relative to the stellar spectrum. The location of the star on the detector will doubtless be a mild function of wavelength, due either to slight non-parallelism in how the grating is mounted or simply atmospheric differential refraction. This map of the location of the stellar spectrum with wavelength is often referred to as the "trace." Examples are shown in *Fig. 2-14*. In IRAF this is done either as part of *apall* or via *doslit*.
9. *Wavelength calibration.* Having established the aperture and tracc of the stellar spectrum, it behooves one to extract the comparison spectrum in exactly the same manner (i.e., using the same trace), and then identify the various lines and perform a smooth fit to the wavelength as a function of pixel number along the trace. *Figure 2-15* shows a HeNeAr comparison lamp. In IRAF this is done either by *identify* or as part of *doslit*.
10. *Extraction.* Given the location of the spectrum, and knowledge of the wavelength calibration, one needs to then extract the spectrum, by which one means adding up all of the data along the spatial profile, subtracting sky, and then applying the wavelength solution. Each of these steps requires some decision about how to best proceed.

    First, how does one establish an unbiased sky estimate? To guard against faint, barely resolved stars in the sky aperture, one often uses a very robust estimator of the sky, such as the median. To avoid against any remaining gradient in the spatial response, one might want to do this on either side of the profile an equal distance away. The assumption in this is that the wavelength is very close to falling along columns (or lines, depending upon the dispersion axis). If that is not the case, then one needs to geometrically transform the data.

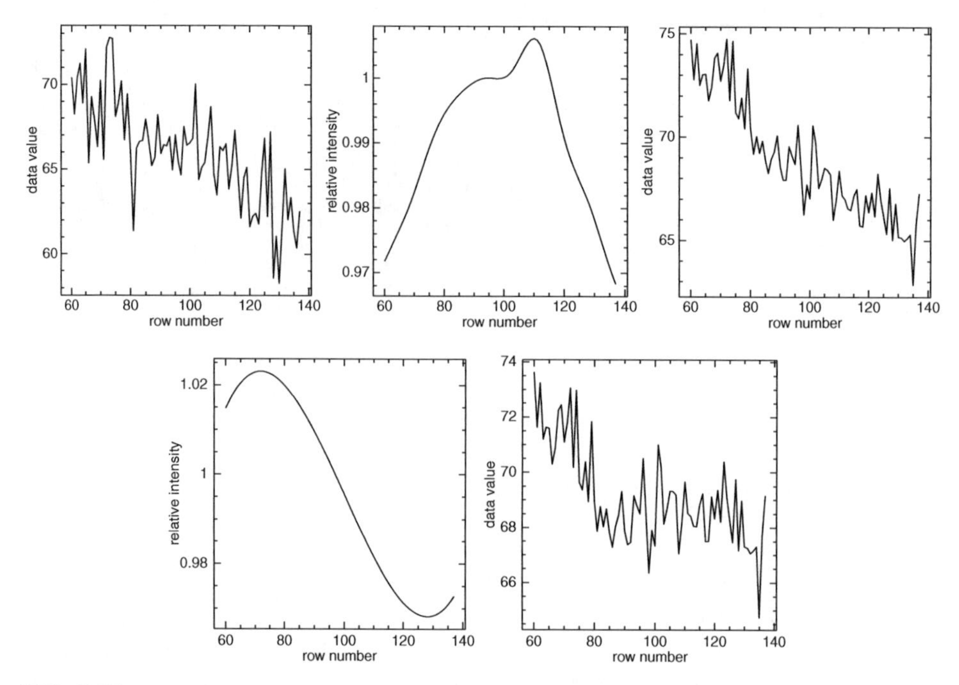

Fig. 2-13

The spatial profile after flattening by the dome flat (Step 6) is shown at the *upper left*. Clearly there is a gradient from *left to right*. The profile of the twilight sky is shown in the *middle*. It's not a particularly good match, and after dividing by it we see that the situation is not improved (*top right*). The profile of all of the sky exposures (after division by the twilight profile) is shown at *lower left*, and was used to re-correct all of the data. After this, the spatial profile was flatter, as shown at *lower right*, although there is now a modest kink

Second, how should one sum the spectrum along the stellar profile? One could simply subtract the sky values, and then sum over the spatial profile, weighting each point equally, but to do this would degrade the signal-to-noise ratio. Consider: the data near the sky add little to the signal, as the difference is basically zero, and so adding those data in with the data of higher signal will just add to the noise. Instead, one usually uses a weighted sum, with the statistics expected from the data (depending upon the gain and the read-noise). Even better would be to use the "optimal extraction" algorithm described below (➋ Sect. 3.2.5), which would also allow one to filter out cosmic rays.

Third, how should the wavelength solution be applied? From the efforts described in Step 9 above, one knows the wavelength as a function of pixel value in the extracted spectrum. Generally one has used a cubic spline or something similarly messy to characterize it. So, rather than try to put a description of this in the image headers, it is usually easier to "linearize" the spectrum, such that every pixel covers the same number of Angstroms. The interpolation needs to be carefully done in order not to harm the data.

In IRAF, these tasks are usually handled by *apsum* and *dispcor* or else as part of *doslit*.

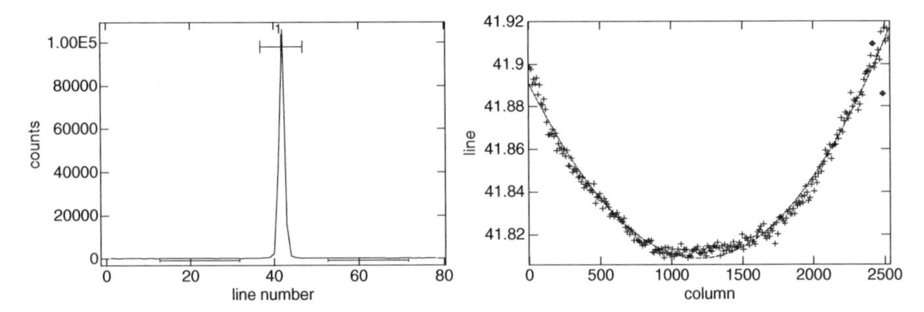

Fig. 2-14
Locating the spectrum. On the *left* is a cut along the spatial profile near the *middle* of the array. A generously large extraction aperture (shown by the numbered bar at the *top*) has been defined, and sky regions located nicely away from the star, as shown by the two bars near the *bottom*. On the *right* the trace of the spectrum along the detector is shown. Although the shape is well defined, the full variation is less than a tenth of a pixel

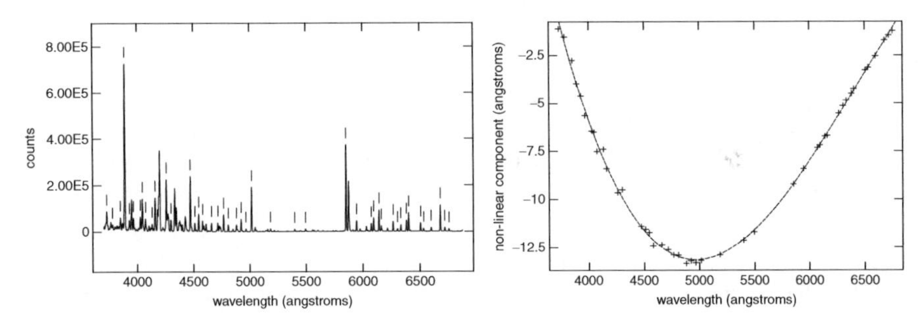

Fig. 2-15
Wavelength calibration. On the *left* is shown a HeNeAr spectrum, with prominent lines marked and identified as to wavelength. On the *right* is shown the non-linear portion of the fit of a first order cubic spline; the residuals from the fit were about 0.2 Å

11. *Finishing up: normalization or flux calibrating?* In some instances, one is primarily interested in modeling line profiles, measuring equivalent widths in order to perform abundance analysis, measuring radial velocities, or performing classical spectral type classification spectra. In this case, one wants to normalize the spectrum. In other cases one might want to model the spectral energy distribution, determine reddening, and so on, which requires one to "flux the data," i.e., determine the flux as a function of wavelength. *Figure 2-16* shows a raw extracted spectrum of Feige 34, along with a normalized version of the spectrum and a fluxed-calibrated version of the spectrum.

    Normalization is usually achieved by fitting a low order function to the spectrum. In order to exclude absorption lines, one might wish to eliminate from the fit any point more than (say) 2.5σ below the fit and iterate the fit a few times. At the same time, one might want to avoid any emission lines and reject points 3σ too high. The relevant IRAF task is *continuum*.

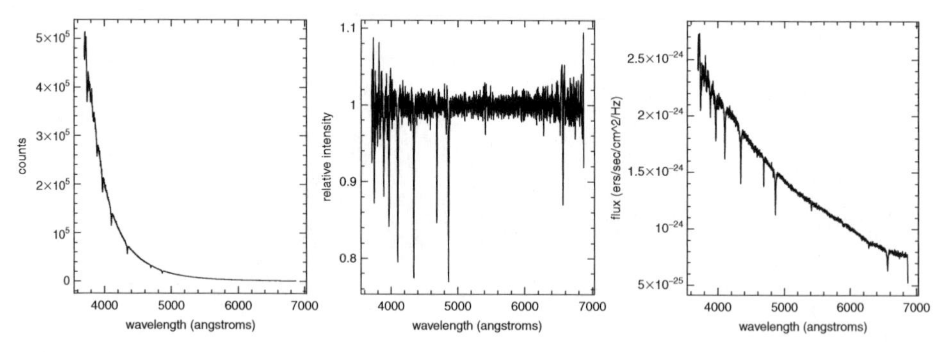

Fig. 2-16
Finishing the reductions. The *left panel* shows the extracted spectrum of the star Feige 34. There are far more counts in the blue than in the *red* owing to the normalization of the flat-field. The *middle panel* shows the normalized version of the same plot; a higher order function was fit to the extracted spectrum with low and high points being rejected from the fit iteratively. The *right panel* shows the fluxed version of the spectrum, with F_ν vs wavelength

Spectrophotometric standards are stars whose fluxes have been measured in various wavelength ranges, some of them at 50 Å intervals.[14] By observing the standard stars and summing the counts over the same bandpasses used to calibrate the standards, one can associate a count rate (as a function of wavelength) with the published fluxes. In practice, one uses a low order fit to the observed counts per second and the spectrophotometric magnitudes for all of the standard stars, applying a gray shift (if needed) to reduce the effects of slit losses. Optionally one might try to derive a wavelength-dependent extinction curve with such data, but experience suggests that the 3–5% errors in the calibration of the standards preclude much improvement over the mean extinction curve.[15] Finally one corrects the observed data by the mean extinction curve and applies the flux calibration to the data. The final spectrum is then in terms of either f_ν or f_λ vs wavelength.

The relevant IRAF tasks are *standard, sensfunc*, and *calib*.

The steps involved in this reduction are straightforward, and when observing stars using a long slit it is hard to imagine why an astronomer would not have fully reduced data at the end of the night. Carrying out these reductions in real time allows one to see that the signal-to-noise ratio is as expected, and avoids there being any unpleasant surprises after the observing run. The IRAF task *doslit* was specifically designed with this in mind, to allow a quick-look capability that was in fact done sufficiently correctly so that the data reduced at the telescope could serve as the final reduction.

[14]The fluxes are often expressed in terms of spectrophotometric "AB magnitudes," which are equal to $-2.5\log f_\nu - 48.60$, where f_ν is the flux in ergs cm^{-2} s^{-1} Hz^{-1}. The spectrophotometric magnitude is essentially equal to V for a wavelength of 5555 Å.

[15]Note that this is not equivalent to adopting a mean extinction; instead, that mean extinction term is absorbed within the zero-point of the fit. Rather, this statement just means that changes in the extinction tend to be independent of wavelength (gray).

3.1.1 Multi-object Techniques

Most of what has just been discussed will also apply to multi-object data, be it fibers or slitlets. There are a few differences which should be emphasized.

For fibers, sky subtraction is accomplished by having some fibers assigned to blank sky. However, how well sky subtraction works is entirely dependent upon how well the flat-fielding removes fiber-to-fiber sensitivity variations (which will be wavelength-dependent) *as well as* how well the flat-fields match the vignetting of a particular fiber configuration. Since the vignetting is likely to depend upon the exact placement of a fiber within the field, the best solution is to either model the vignetting (as is done in the Hectospec pipeline) or to make sure that flat-fields are observed with each configuration. For some telescopes and instruments (such as CTIO/Hydra) it is more time efficient to simply observe some blank sky, making a few dithered exposures near the observed field.

For multi-slits one of the complications is that the wavelength coverage of each slitlet will depend upon its placement in the field, as discussed above in Sect. 2.4.1. This results in some very challenging reduction issues, particularly if one plans to flux-calibrate. Consider again the IMACS instrument (Sect. 2.4.1). There are eight chips, each of which will have its own sensitivity curve, and even worse, the wavelength coverage reaching each chip will depend upon the location of each slitlet within the field. The perfect solution would be to observe a spectrophotometric standard down each of the slits, but that would hardly be practical. Flux calibrating multislit data is *hard*, and even obtaining normalized spectra can be a bit of a challenge.

3.1.2 NIR Techniques

When reducing near-infrared spectra, the basic philosophy outlined in Sect. 3.1 for the optical generally applies. But there are two significant differences between the infrared and optical. First the detectors are different, as discussed in Sect. 2.5.2. This leads to some minor modifications of the standard optical reduction. But, more importantly, observing in the near-infrared presents significant challenges that do not occur in the optical, in the form of significant background radiation and strong absorption from the Earth atmosphere. In order to correct for the sky background, which is often much brighter than the science target, a sky frame is obtained equal in exposure time to the object. For point sources this is often accomplished by moving the object along the slit (dithering). Alternate dithers can be subtracted to remove the sky. Or, the entire dither set might be median combined to form a sky frame.

Although this process works relatively well for subtracting the night-sky emission, it does nothing for removing the telluric absorption features. Instead, telluric standards (stars with nearly featureless intrinsic spectra) need to be observed at nearly *identical* airmasses as the program stars: differences of less than 0.1 airmasses is advisable. Why? The IR telluric absorption spectrum is due to a number of molecules (e.g., CO, CO_2, H_2O, CH_4, etc.), which can vary in relative concentration, especially H_2O. In addition, some of the absorption lines are saturated and thus do not scale linearly with airmass. Both of these factors can make it difficult or impossible to scale the telluric absorption with airmass.

Additional calibration observations and reduction steps will be required to address these challenges. Newcomers to spectroscopy might well want to reduce optical spectra before attempting to tackle the infrared.

Infrared hybrid arrays do not have an overscan strip along their side like in a CCD. However, bias structure is a very real problem and absolutely must be removed. Direct bias images are not typically used to do this. This is because in the infrared, one of the first stages in reduction is always subtracting one data frame from another, of the same integration time, but with differing sky locations or with differing sources on or off. Thus the bias is automatically removed as a result of always subtracting "background" frames, even from flat-field and comparison exposures.

During the day, one should take the following, although they may not always be needed in the reductions.

- *Dark frames* are used to subtract the background from twilight flats and comparison lamps, so it is important that the darks have the same exposure times as the twilight and comparison lamp exposures. Dark frames are also used to make a bad pixel mask.
- *A bad pixel mask* is constructed from (1) the dark frames, which identify hot pixels, and from (2) the dome flats, which are used to identify dead and flaky pixels.
- *Dome flats* are critical. And, they are done differently in the near-infrared. They must be obtained for *each* grating setting and filter combination you use during the night, the same as with optical data. Furthermore, they should be obtained with exposure levels that push to the linearity limits in order to maximize the signal-to-noise ratio. This linearity limit can be as low as 10,000 e^- for some IR detectors. However, *unlike the case for the optical*, one must also obtain an equivalent set (in terms of exposure time and number) of dome flats with the dome illumination turned off, as the dome flat itself is radiating in the NIR. One can demonstrate the necessity for this by a simple experiment: subtract an image obtained with the lamps turned off from an image with the lamps turned on, and see how many counts there really are over the entire wavelength range. One may be surprised how many counts were removed from the original exposure, particularly in the longer wavelength K band. But it is this *subtracted* dome flat that will be used to flatten the data frames, so there needs to be a lot of counts in it in order not to reduce the signal-to-noise! For the K band, where there is high background counts and because the dynamic range of the detector is small, one may need 20 or more flat-on, flat-off pairs to get sufficient real signal.
- *Quartz flats* will be taken during the night, but daytime exposures will test the exposure levels and reveal any structural differences between it and the dome and twilight flats. Of course, dark frames with matching integration times will be needed.
- *Comparison lamps* should be taken throughout the night, but obtaining a few during the day may also prove to be useful. Be sure to get dark frames to match the integration time.

During the night, one will take several additional calibration exposures.

- *Telluric standard stars* are (nearly) featureless stars that will be used to derive the spectrum of the telluric absorption bands for removal from the program data. These are not spectrophotometric stars. They must be chosen to lie at similar (within 0.1) airmasses as the target stars for the reasons explained above, and observed within an hour or two of the target (more often for observations beyond 3 μm).
- *Quartz lamps* may be useful or not, depending upon the specific instrument.
- *Comparison lamp* exposures are a good idea. In the infrared, one can almost always use the night sky emission lines as the wavelength reference source. But separate comparison lamp exposures may prove useful if there turns out not to be enough counts in the sky lines, or if some regions are too void of sky lines, such as the long-K region. It is also a good idea to

take these if you are moving the grating often, the program integration times are short, or if the dispersion is high.

- *Twilight flats* may wind up not being used, but they could be handy. Many NIR spectrographs will need some illumination correction (as in the optical) and the twilight flats are the easiest means to make such corrections. Be sure to get dark frames of the same integration time to remove dark current and bias levels later.
- *Stellar flats* are spectra of a very bright star which is moved by small amounts along the slit. The images can be co-added to create a very rough point-source flat, and used much like a twilight flat to check on illumination corrections or other irregularities which can uniquely appear in point-source observations.

The basic two-dimensional reduction steps for NIR spectroscopic data are presented here; one should review ➋ Sect. 3.1 to compare and contrast these with what is involved in optical reductions. The relevant IRAF tasks are mentioned for convenience, although again the goal here is not to identify what buttons to press but rather to be clear about the steps needed.

1. *Subtract one slit position frame from another.* This step removes the bias and all other uniform additive sources, such as evenly illuminated sky. (Lumpy, nebular emission will subtract horribly.) In long-slit spectroscopy one has to observe the star in two (or more) positions along the slit. This basic subtraction step needs to be done while one is observing, and it behooves the observer to *always look at the result.* There should be a zero-level background throughout, with two stars, one positive, one negative, running along the dispersion axis. The primary thing to check is to see if the OH night-sky emission lines are canceling. There may be residual positive and/or negative OH lines running perpendicular to the dispersion axis. This can happen just as a consequence of the temporal and spatial variations in the OH spectra, but certainly will happen if light clouds move through the field. As there will be many frames, try all combinations with unique slit positions to find the best OH removal. Nothing else should be there, and if there is, there are some serious problems. The only IRAF tool needed is *imarith.*
2. *Construct and apply a master, normalized, featureless flat.* The "lamp on" flats and the "lamp off" flats are median averaged using a rejection algorithm (IRAF: *imcombine*) and then the "lamps off" average is subtracted from the "lamps on" average to produce a master flat. As with the optical, one must use one's own judgment about how to normalize. This flat is then divided into each of the subtracted frames. The process is repeated for each grating/filter combination.
3. *Construct and use an illumination correction.* This step is identical to that of the optical, in that twilight flats can be used to correct any vignetting left over from the flat-field division. Alternately, telluric standards can be obtained at the same slit positions as science targets. Making a ratio of the object to telluric standard will also correct for vignetting along the slit.
4. *Make the bad pixel mask.* A good bad pixel mask can be constructed from a set of dark and lamps-on flat-field images. A histogram from a dark image reveals high values from "hot" pixels. Decide where the cut-off is. Copy the dark image, call it "hot," set all pixels below this level to zero, then set all pixels above zero to 1. Display a histogram of your flat and decide what low values are unacceptable. Copy the flat image calling it dead and set all values below this low value to 1. Set everything else to 0. Finally, take many identically observed flat (on) exposures, average them, and determine a sigma map (in IRAF, this is done using *imcomb* and entering an image name for "sigma"). Display a histogram, select your upper

limit for acceptable sigma, and set everything below that to zero in the sigma map. Then set everything above that limit to 1. Now average (no rejection) your three images: dead, hot, and sigma. All values above 0.25 get set to 1.0 and *voila!*, one has a good bad pixel mask. One should then examine it to see if one was too harsh or too lax with your acceptable limits.

5. *Trace and extract spectrum.* This step is nearly identical to what is done for the optical: one has to identify the location of the stellar spectrum on the array and map out its location as a function of position. Although the sky has already been subtracted to first order (Step 1), one might want to do sky subtraction again during this stage for a couple of reasons, namely, if one needs to remove astrophysical background (nebular emission or background stellar light), or if the previous sky subtraction left strong residuals due to temporal changes, particularly in the sky lines. Be sure *not* to use optimal extraction, as the previous sky subtraction has altered the noise characteristics. If you do subtract sky at this stage, make sure it is the median of many values, else one will add noise. (It may be worth reducing a sample spectrum with and without sky subtraction turned on in the extraction process to see which is better.) Bad pixels can be flagged at this stage using the bad pixel mask constructed in Step 4 and will disappear (one hopes!) when all of the many extracted frames are averaged below. In IRAF the relevant task is *apall*.
6. *Determine the wavelength scale.* One has a few choices for what to use for the wavelength calibration, and the right choice depends upon the data and goals. One can use the night-sky emission spectrum as a wavelength reference, and in fact at high dispersion at some grating settings these may be the only choice. (One needs to be at high dispersion though to do so as many of the OH emission lines are hyperfine doubles; see Osterbrock et al. 1996, 1997.) In this case, one needs to start with the raw frames and median average those with unique stellar positions with a rejection algorithm that will get rid of the stellar spectra. Alternatively, in other applications one may be able to use the comparison lamp exposures after a suitable dark has been removed. Whichever is used, one needs to extract a one-dimensional spectrum following the same location and trace as a particular stellar exposure. One can then mark the positions of the reference lines and fit a smooth function to determine the wavelength as a function of position along the trace. One should keep fit order reasonably low. A good calibration should result in a wavelength scale whose uncertainty is smaller than one tenth of the resolution element. In principle one can also use the telluric absorption lines for wavelength calibration, but this is only recommended if there is no other recourse, as it is very time-consuming. The comparisons will have to be extracted and calibrated for each stellar trace. The resulting fits can be applied to the stellar spectra as in the optical. The relevant IRAF tasks are *identify*, *reidentify*, and *dispcor*.
7. *Average the spectra.* If one took the data properly, there should be 6–12 spectra that can be averaged to increase the signal-to-noise and remove bad pixels. One needs to check that the spectra have the same shape. The spectra need to be averaged in wavelength, not pixel, space, with a rejection algorithm to remove the bad pixels and with appropriate scaling and weighting by the number of counts. In IRAF the relevant task is *scombine*.
8. *Remove telluric absorption.* How do the spectra look? Not so good? That's because there is still one last important step, namely left the removal of telluric absorption features. The telluric standards need to be treated with the same care and effort used for the target spectra. There is a bit of black magic that must occur before one can divide the final target spectrum by the final telluric spectrum in (wavelength space) to remove the Earth absorption lines, namely, that any stellar features intrinsic to the telluric standards have to be removed. A-type stars are often used as telluric standards as they have hardly any lines other than hydrogen. But the hydrogen lines are huge, deep, and complex: broad wings, deep cores and far from

Gaussian to fit, particularly if $v \sin i$ is low. Some prefer to use early G-type dwarfs, correcting the intrinsic lines in the G star by using a Solar spectrum, available from the National Solar Observatory. For lower resolution work or in a low signal-to-noise regime, this is probably fine and the differences in metallicity, $v \sin i$, temperature, and gravity between the telluric and the Sun may not be significant. Probably the best plan is to think ahead and observe at least one telluric star during the night for which the intrinsic spectrum is already known to high precision (see the appendix in Hanson et al. 2005). This gives a direct solution from which you can bootstrap additional telluric solutions throughout the night. More about this is discussed below in ➋ Sect. 3.3.3.

3.2 Further Details

The description of the reduction techniques above were intended as a short introduction to the subject. There are some further issues that involve both observation and reduction techniques that are worth discussing in some additional depth here; in addition, there are some often-neglected topics that may provide the spectroscopist with some useful tools.

3.2.1 Differential Refraction

Differential refraction has two meanings to the spectroscopist, both of them important. First, there is the issue of refraction as a function of wavelength. Light passing through the atmosphere is refracted, so that an object will appear to be higher in the sky than it really is. The amount of refraction is a function of the wavelength and zenith distance, with blue light being refracted the most, and the effect being the greatest at low elevation. If one looks at a star near the horizon with sufficient magnification one will notice that the atmosphere itself has dispersed the starlight, with the red end of the spectrum nearest the horizon, and the blue part of the spectrum further from the horizon. The second meaning has to do with the fact that the amount of refraction (at any wavelength) depends upon the object's zenith distance, and hence the amount of refraction will differ across the field in the direction toward or away from the zenith.

This wavelength dependence of refraction has important implications for the spectroscopist. If one is observing a star at low elevation with the slit oriented parallel to the horizon, the blue part of the star's light will be above the slit, and the red part of the star's light below the slit if one has centered on the slit visually. Thus much of the light is lost. Were one to instead rotate the slit so it was oriented perpendicular to the horizon then all of the star's light would enter the slit, albeit it at slightly different spatial locations along the slit. So, the spectrum would appear to be tilted on the detector, but the light would not have been selectively removed.

The position angle of the slit on the sky is called the *parallactic angle*, and so it is good practice to set the slit to this orientation if one wishes to observe very far from the zenith. How much does it matter? Filippenko (1982) computed the amount of refraction expected relative to 5000 Å for a variety of airmasses making realistic assumptions. Even at a modest airmass of 1.5, the image at 4000 Å is displaced upward (away from the horizon) by 0.71 arcsec compared to the image at 5000 Å. So, if one were observing with a 1 arcsec slit, the 4000 Å image would be shifted out of the slit! The degree of refraction scales as the tangent of the zenith distance, with the wavelength dependence a function of the ambient temperature, pressure, and amount of water vapor in the atmosphere (see Filippenko 1982 and references therein for details).

The parallactic angle η will be 0° or 180° for an object on the meridian (depending, respectively, if the object is north or south of the zenith); at other hour angles h and declinations δ the relationship is given by

$$\sin\eta = \sin h \cos\phi/[1-(\sin\phi\sin\delta+\cos\phi\cos\delta\cos h)^2]^{0.5},$$

where ϕ is the observer's latitude. Fortunately most telescope control systems compute the parallactic angle automatically for the telescope's current position.

When using fibers or multi-slit masks there is not much one can do, as the fiber entrance aperture is circular, and the multi-slit masks are designed to work at one particular orientation on the sky. Thus these instruments almost invariably employ an atmospheric dispersion corrector (ADC). (There are several ways of constructing an ADC, but most involve two counter-rotating prisms; for a more complete treatment, see Wynne 1993.) This is good for an additional reason having to do with the second meaning of "differential refraction," namely, the fact that objects on one side of the field will suffer slightly different refraction than objects on the other side of the field as the zenith distances are not quite the same. Imagine that the instrument has a 1° field of view, and that one is observing so that the "top" of the field (the one furthest from the horizon) is at a zenith distance of 45°. The lower part of the field will have a zenith distance of 46°, and the difference in tangent between these two angles is 0.036. At 5000 Å the typical amount of refraction at a zenith distance of 45° is roughly 1 arcmin, so the differential refraction across the field is 0.036 arcmin, or about 2 arcsec! Thus the separation between multi-slits or fibers would have to be adjusted by this amount in the absence of an ADC. Among the fiber positioners listed in *Table 2-4*, only Hydra on WIYN lacks an ADC. There one must employ rather large fibers and adjust the position of the fibers depending upon the proposed wavelength of observation. Nevertheless, ADCs have their drawbacks, and in particular their transmission in the near-UV may be very poor.

3.2.2 Determining Isolation

An interesting question arises when obtaining spectra of stars in crowded fields: how much light from neighboring objects is spilling over into the slit or fiber? In principle this can be answered given a complete catalog of sources. For a star centered in a slit with a width of $2a$, the relative contamination from a star a separation s away will depend upon the seeing. We can characterize

Table 2-4
Some fiber spectrographs

Instrument	Telescope	# fibers	Fiber size (μm)	Fiber size (")	Closest spacing (")	FOV (′)	Setup (mins)	*R*
Hectospec	MMT 6.5 m	300	250	1.5	20	60	5	1000–2500
Hectochelle	MMT 6.5 m	240	250	1.5	20	60	5	30,000
MIKE	Clay 6.5 m	256	175	1.4	14.5	23	40	15,000–19,000
AAOMega	AAT 4 m	392	140	2.1	35	120	65	1300–8000
Hydra-S	CTIO 4 m	138	300	2.0	25	40	20	1000–2000
Hydra (blue)	WIYN 3.5 m	83	310	3.1	37	60	20	1000–25,000
Hydra (red)	WIYN 3.5 m	90	200	2.0	37	50	20	1000–40,000

the latter by a Gaussian with a σ of $0.85f/2$, where the seeing full-width-at-half-maximum is f. Following (➋ 2.8) in Filippenko (1982), the relative contribution of a nearby star is

$$10^{(\Delta V/-2.5)} F(a,s,\sigma)/F(a,0,\sigma)$$

where the definition of F depends upon whether or not the second star is located partially in the slit or not. Let $a1 = a - s$ and $a2 = a + s$ if the star is in the slit ($s < a$). Then $F = 0.5(G(a1,\sigma) + G(a2,\sigma))$. (Use this for the denominator as well, with $a1 = a2 = a$.) If the star is located outside of the slit ($s > a$) then $F = 0.5(G(a1,\sigma) - G(a2,\sigma))$. $G(z,\sigma)$ is the standard Gaussian integral,

$$G(z,\sigma) = \frac{1}{\sqrt{2\pi\sigma}} \int_{-z}^{z} e^{-x^2/2\sigma^2} dx$$

The simplifying assumption in all of this is that the slit has been oriented perpendicular to a line between the two stars. But, this provides a mechanism in general for deciding in advance what stars in one's program may be too crowded to observe.

3.2.3 Assigning Fibers and Designing Multi-slit Masks

In order to design either a fiber configuration or a multi-slit mask, one invariably runs highly customized software which takes the celestial coordinates (right ascension and declination) of the objects of interest and computes optimal centers, rotation, etc. that allow the fibers to be assigned to the maximum number of objects, or the most slitlet masks to be machined without the spectral overlapping. However, a key point to remind the reader is that such instruments work only if there are alignment stars that are on the same coordinate system as the program objects. In other words, if one has produced coordinates by using catalog "X" to provide the reference frame, it would be good if the alignment stars were also drawn from the same catalog. This was much harder 10 years ago than today, thanks to the large number of stars in uniform astrometric catalogs such as the 2MASS survey or the various USNO publications. The most recent of the latter is the CCD Astrograph Catalogue Part 3 (UCAC3). One advantage of the proper motion catalogs such as the UCAC3 is one that can then assure that the relative proper motion between the alignment stars and the program objects are small.

This point bears repeating: there is a danger to mixing and matching coordinates determined from one catalog with coordinates from another. The coordinates need to be on the same system, or there is significant risk of being quite disappointed in the final throughput at the telescope.

3.2.4 Placing Two Stars on a Long Slit

If one is observing multiple objects whose separations are smaller than the slit length (such as stars within a nearby galaxy or within a star cluster), and one does not need to be at the parallactic angle (either because the instrument has an ADC or because one is observing either near the zenith or over a small wavelength range) one may want to multiplex by placing two stars on the slit by rotating the slit to the appropriate angle. To plan such an observation, one must first precess the coordinates of both stars to the current equinox, as precession itself introduces a rotation. Then, one must compute the "standard coordinates," i.e., de-project the spherical coordinates to the tangent plane.

Assume that one of the two stars is going to be centered in the slit, and that its precessed coordinates are α_1 and δ_1, converted to radians. Assume that the other star's coordinates are α_2 and δ_2, where again these are expressed in radians after precession. Then the standard coordinates ξ and η of star 2 will be

$$\xi = \cos\delta_2 \sin(\alpha_2 - \alpha_1)/F$$

$$\eta = (\sin\delta_2 \cos\delta_1 - \cos\delta_2 \sin\delta_1 \cos(\alpha_2 - \alpha_1))/F,$$

where

$$F = \sin\delta_2 \sin\delta_1 + \cos\delta_2 \cos\delta_1 \cos(\alpha_2 - \alpha_1).$$

The position angle from star 1 to star 2 will then simply be the arctangent of (ξ/η). If η is 0, then the position angle should be 90° or 270° depending upon whether ξ is positive or negative, respectively; if ξ is 0, then the position angle should be 0° or 180°, depending on whether η is positive or negative, respectively. The distance between the two objects will be $\sqrt{\xi^2 + \eta^2}$.

3.2.5 Optimal Extraction

With CCDs coming into common use as detectors, Horne (1986) pointed out that simply summing the data over the spatial profile of a point source did an injustice to the data, in that it degraded the signal-to-noise ratio. Consider the case of a faint source with lots of sky background. A large extraction aperture in which the data were summed in an unweighted manner would be far noisier than one in which the extraction aperture was small and excluded more sky: the sky adds only noise, but no new information. The mathematically correct way to extract the data is to construct the sum by weighting each point by the inverse of the variance. (Recall that the variance is the square of the expected standard deviation.)

The assumption in the Horne (1986) algorithm is that the spatial profile P(λ) varies slowly and smoothly with wavelength λ. At each point along the spatial direction i at a given wavelength λ, $C_i(\lambda)$ photons are measured. This value is the sum of the number of photons from the star $A_i(\lambda)$ and of the sky background $B(\lambda)$, where the latter is assumed to be a constant at a given λ. In the absence of optimal-weighting, one would determine the total sum $T(\lambda)$ of the sky-subtracted object by:

$$T(\lambda) = \sum_i A_i(\lambda) = \sum_i (C_i(\lambda) - B(\lambda))$$

as $A_i(\lambda) = C_i(\lambda) - B(\lambda)$. In practice the summation would be performed over some "sensible" range to include most of the spatial profile. For optimal weighting one would instead weight each point in the sum by $W_i(\lambda)$:

$$T(\lambda) = \sum_i W_i(\lambda)(C_i(\lambda) - B(\lambda)) / \sum_i W_i(\lambda)$$

The weighting function $W_i(\lambda)$ is taken to be $P^2(\lambda)/\sigma_i^2(\lambda)$, where $P(\lambda)$ is the intrinsic spatial profile of the spectrum at wavelength λ. P is usually forced to vary smoothly and slowly with wavelength by using a series of low order functions to represent the profile as a function of wavelength, and is normalized in such a way that the integral of $P(\lambda)$ at a particular λ is always unity. (For the more complicated situation that corresponds to echelle or other highly distorted spectra, see Marsh 1989.) The variance $\sigma_i^2(\lambda)$ is readily determined if one assumes that the statistical uncertainty of the data is described by simple Poisson statistics plus read-noise. The errors add in quadrature, and will include the read-noise R. The variance will also include the photon-noise due to the object $\sqrt{A_i(\lambda)}$ and the photon-noise due to the sky $\sqrt{B(\lambda)}$. Finally, the variance will also include a term that represents the uncertainty in the background determination. Typically one determines the sky value by averaging over a number N of pixels located far from the star, and hence the uncertainty in the sky determination is $\sqrt{B(\lambda)/(N-1)}$. Thus

$$\sigma_i^2(\lambda) = R^2 + A_i(\lambda) + B(\lambda) + B(\lambda)/(N-1).$$

One can eliminate cosmic-rays by substituting $T(\lambda)P_i(\lambda)$ for $A_i(\lambda)$:

$$\sigma_i^2(\lambda) = R^2 + T(\lambda)P_i(\lambda) + B(\lambda) + B(\lambda)/(N-1),$$

and continuing to solve for $T(\lambda)$ iteratively.

An example of the improvement obtained by optimal extraction and cleaning is shown in *Fig. 2-17*, where Massey et al. (2004) compare a standard pipeline reduction version of an *HST* STIS optical spectrum (upper spectrum) with one re-reduced using IRAF with the optimal extraction algorithm (lower spectrum).

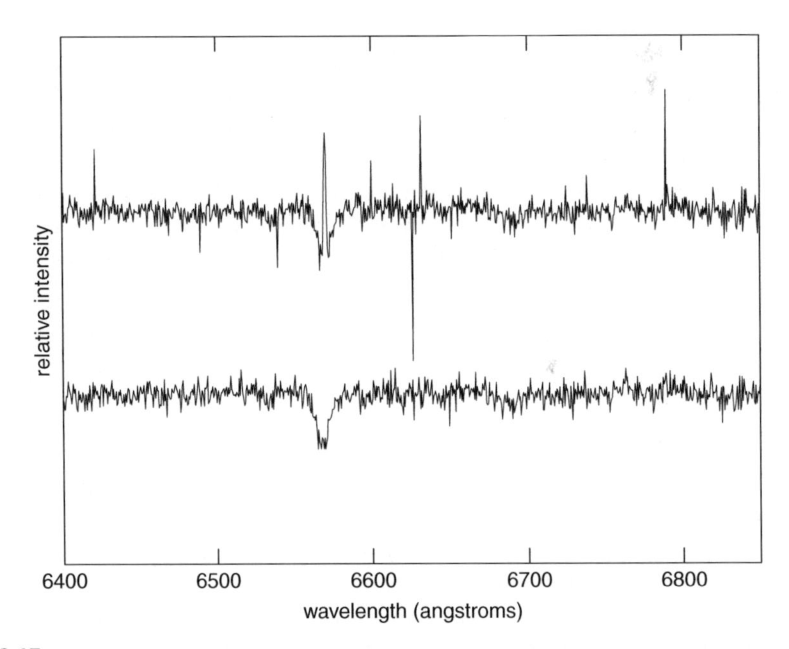

Fig. 2-17
Two reductions of a STIS spectrum obtained of an early O-type star in the LMC, LH101:W3-24. The *upper* spectrum is the standard reduction produced by the *HST* STIS "CALSTIS" pipeline, while the *lower* spectrum has been reduced using the CCD spectral reduction package *doslit* in IRAF using optimal extraction. The signal-to-noise ratio in the upper spectrum is 18 in a spike-free region, while that of the *lower* spectrum is 22. The many cosmic-ray spikes in the *upper* further degrade the signal-to-noise ratio. Note in particular the difference in the profile of the strong Hα absorption line at 6562 Å (From Massey et al. 2004. Reproduced by permission of the AAS)

3.2.6 Long-Slit Flat-Fielding Issues

With conventional long-slit spectroscopy, generally two kinds of flats are needed: (1) a series of exposures of a featureless continuum source illuminating the slit (such as a dome flat) and (2) something that removes any residual spatial mis-match between the flat-field and the night-sky. Typically exposures of bright twilight are used for (2), although dithered exposures of the night sky itself would be better, albeit costly in the sense of telescope time.

Featureless Flats

The featureless flats are primarily useful to remove pixel-to-pixel gain variations within the CCD: some pixels are a little more or a little less sensitive to light than their neighbors. Good flat-fielding is vital in the high signal-to-noise ratio regime. But, with modern chips it is a little less needed than many realize at modest and low signal-to-noise ratios. And poor flat-fielding is worse than no flat-fielding. How one evaluates what is needed is briefly discussed here.

First, one needs to consider the heretical question of what would be the effect of *not* flat-fielding one's data? The answer to this depends upon the signal-to-noise ratio regime one is attempting to achieve. If CCDs were perfectly uniform and required no flat-fielding, one would expect that the scatter σ (root-mean-square variation, or rms) in counts would be related to the number of counts n as

$$\sigma = \sqrt{ng}/g$$

where g is the gain in e^- per count. (For simplicity it is assumed here that read-noise R is inconsequential, i.e., $ng >> R^2$, and that the background counts are not significant.) Consider chip "3" in the 8-chip mosaic used for the f/4 IMACS camera. The gain is 0.90 e/ADU. According to the formula above, for a flat-field with 6055 counts, one would expect an rms of only 82 counts, but in reality the rms is 95 counts. The extra scatter is due to the "graininess" (non-uniformity) of a CCD. Let p be the average ratio of the gain of one pixel compared to another. Then the noise (in e^-) added to the signal will just be npg and the total rms σ_T (what one measures, in counts) is

$$\sigma_T^2 = n/g + n^2 p^2. \tag{2.10}$$

Solving for p in the above example, one finds that $p = 0.008$, i.e., about 0.8%. The signal-to-noise S will be

$$S = ng/\sqrt{ng + n^2 p^2 g^2}.$$

Thus if one had a stellar spectrum exposed to the same level, one would achieve a signal-to-noise of 63 rather than 73. An interesting implication of this is that if one is after a fairly small signal-to-noise ratio, 10 per pixel, say (possibly 30 when one integrates over the spectral resolution elements and spatially) then if one assumes perfect flat-fielding, $ng = 100$. Adding in the additional noise term with $p = 0.008$ leads to the fact that the signal-to-noise one would achieve would be 9.97 rather than 10. In other words, if one is in the low signal-to-noise regime flat-fielding is hardly necessary with modern detectors.

It should be further emphasized that were one to use bad flat-fields (counts comparable to those of the object) one actually can significantly *reduce* the signal-to-noise of the final spectra. How many counts does one need for a "good" flat field? The rule of thumb is that one does not want the data to be limited by the signal in the flat field. Again let n be the number of counts per pixel in the spectrum. Let m be the number of counts per pixel in the flat-field. Consider this purely from a propagation of errors point of view. If σ_f is the error in the flattened data, and

σ_n and σ_m are the rms-scatter due to photon-statistics in the spectrum and flat-field, respectively, then

$$\frac{\sigma_f^2}{(ng)^2} = \frac{\sigma_n^2}{(ng)^2} + \frac{\sigma_m^2}{(mg)^2}.$$

(The assumption here is that the flat has been normalized to unity.) But each of those quantities is really just the inverse of the signal-to-noise squared of the respective quantities; i.e., the inverse of the final signal-to-noise squared will be the sum of the inverses of the signal-to-noise squared of the raw spectrum and the inverse of the signal-to-noise squared of the flat-field. To put another way:

$$1/S^2 = 1/ng + 1/mg,$$

where S is the final signal-to-noise. The quantity $1/ng$ needs to be much greater than $1/mg$. If one has just the same number of counts in one's flat-field as one has in one's spectrum, one has degraded the signal-to-noise by $1/\sqrt{2}$.

Consider that one would like to achieve a signal-to-noise ratio per spectral resolution element of 500, say. That basically requires a signal-to-noise ratio per pixel of 200 if there are seven pixels in the product of the spatial profile with the number of pixels in a spatial resolution element. To achieve this one expects to need something like $200^2 = 40{,}000\ e^-$ in the stellar spectrum. If one were not to flat-field *that*, one would find that the signal-to-noise ratio was limited to about 100 per pixel, not 200. If one were to obtain only 40,000 counts in the flat-field, one would achieve a signal-to-noise ratio of 140. To achieve a signal-to-noise ratio of 190 (pretty close to what was wanted) one would need about $400{,}000\ e^-$ (accumulated) in all of the flat-field exposures – in other words, about 10 times what was in the program exposure. In general, obtaining 10 times more counts in the flat than needed for the program objects will mean that the signal-to-noise ratio will only be degraded by 5% over the "perfect" flat-fielding case. (This admittedly applies only to the peak counts in the stellar profile, so is perhaps conservative by a factor of two or so.)

What if one cannot obtain enough counts in the flat-field to achieve this? For instance, if one is trying to achieve a high signal-to-noise ratio in the far blue (4000 Å, say) it may be very hard to obtain enough counts with standard calibration lamps and dome spots. One solution is to simply "dither" along the slit: move the star to several different slit positions, reducing the effect of pixel-to-pixel variations.

A remaining issue is that there is always some question about exactly how to normalize the featureless dome flat. If the illumination from a dome spot (say) had the same color as that of the objects or (in the case of background-limited work) the night sky, then simply normalizing the flat by a single number would probably work best, as it would remove much of the grating and detector response. But, in most instances the featureless flat is much redder than the actual object (or the sky), and one does better by fitting the flat with a function in the wavelength direction, and normalizing the flat by that function. In general, the only way to tell which will give a flatter response in the end is to try it both ways on a source expected to have a smooth spectrum, such as a spectrophotometric standard star. This is demonstrated above in Step 6 ➲ Sect. 3.1.

Illumination Correction Flats

The second kind of flat-fielding that is useful is one that improves the slit illumination function. As described above, one usually uses a featureless flat from observations of the dome spot to remove the pixel-to-pixel variations. To a first approximation, this also usually corrects for the slit illumination function, i.e., vignetting along the slit, and correction for minor irregularities in the slit jaws. Still, for accurate sky subtraction, or in the cases where one has observed an extended source and wants accurate surface brightness estimates, one may wish to correct the illumination function to a better degree than that.

A popular, albeit not always successful, way to do this is to use exposures of the bright twilight sky. Shortly after the sun has set, one takes exposures of the sky offsetting the telescope slightly between exposures to guard against any bright stars that might fall on the slit. The spectrum appears to be that of a G2 V star. How then can this be used to flatten the data? In this case one is interested only in correcting the data along the spatial axis. So, the data can be collapsed along the wavelength axis and examined to see if there is any residual non-uniformity in the spatial direction after the featureless flat has been applied. If so, one might want to fit the spatial profile with a low order function and divide that function into the data to correct it.

The twilight sky exposures come for free, as the sky is too bright to do anything else. But, in some critical applications, the twilights may not match the illumination of the dark night sky well enough. The only solution then is to use some telescope time to observe relatively blank night sky, dithering between exposures in order to remove any faint resolved stars. Since these illumination-correction data are going to be collapsed in the wavelength direction, one does not need very many counts per pixel in order to achieve high signal-to-noise ratio in the spatial profile, but excellent bias correction is needed in this case.

Summary

When using a CCD for spectroscopy for the first time, it is worthwhile to understand how "grainy" the pixel-to-pixel variations are in the first place in order to better understand one's flat-fielding needs. One can do this by obtaining a flat-field exposure and comparing the rms of a fairly uniform section with that expected from photon statistics. If n are the number of counts, g is the gain, and σ_T is the total rms (also in counts), then the intrinsic pixel-to-pixel variation p can be found by solving (➜ 2.10):

$$p = \sqrt{\sigma_T^2/n^2 - 1/ng}.$$

It is only in the case that the precision needed is comparable to this quantity p that one really has to worry about flat-fielding. In other words, if $p = 0.01$ (1%) as is typical, and one wants a signal-to-noise ratio of 100 (1%), flat-fielding is needed. If one wants a signal-to-noise ratio of 20 (5%), flat-fielding is not going to improve the data. In addition, one needs to obtain a final flat-field that has enough counts in order not to degrade one's data. In general one wants the flat-field to have about 10 times the number of counts (per pixel) than the most heavily exposed pixel in the object spectrum in order not to damage the signal-to-noise significantly. A good rule of thumb is to take the desired signal-to-noise per spectral resolution element, square it, and aim for that number of counts (per pixel) in the flat field. This assumes that the number of pixels in the spatial profile times the number of pixels in the resolution element is of order 10.

Slit illumination corrections can be obtained (poorly but cheaply) from bright twilight flats, or (well but expensive) by observing blank sky at night. However, one needs only enough counts integrated over the entire wavelength range to achieve 0.5% precision (40,000 e^- per spatial pixel integrated over all wavelengths), i.e., if there are 2000 pixels in the wavelength direction one doesn't need more than about 20e^- per pixel. At these low count levels though accurate bias removal is essential.

3.2.7 Radial Velocities and Velocity Dispersions

Often the goal of the observations is to obtain radial velocities of the observed object. For this, one needs to obtain sufficient calibration to make sure that any flexure in the instrument is removed. Even bench-mounted instruments may "flex" as the liquid N_2 in the CCD dewar evaporates during the night. The safe approach is to make sure that the wavelength calibration spectra are observed both before and after a series of integrations, with the wavelength scale then interpolated.

To obtain radial velocities themselves, the usual technique is to observe several radial velocity standard stars of spectral type similar to the object for which one wants the velocity, and then cross-correlate the spectrum of each standard with the spectrum of the object itself, averaging the result.

Spectra are best prepared for this by normalizing and then subtracting 1 so that the continuum is zero. This way the continuum provides zero "signal" and the lines provide the greatest contrast in the cross-correlation. IRAF routines such as *fxcor* and *rvsao* will do this, as well as (in principle) preparing the spectra by normalizing and subtracting the continuum.

One way of thinking of cross-correlation is to imagine the spectrum of the standard and the program object each containing a single spectral line. One then starts with an arbitrary offset in wavelength and sums the two spectra. If the lines don't match up, then the sum is going to be zero. One then shifts the velocity of one star slightly relative to the other and recomputes the sum. When the lines begin to line up, the cross-correlation will be non-zero, and when they are best aligned the cross-correlation is at a maximum. In practice such cross-correlation is done using Fourier transforms. The definitive reference to this technique can be found in Tonry and Davis (1979).

The Earth is both rotating and revolving around the sun. Thus the Doppler shift of an object has not only the object's motion relative to the sun, but also whatever the radial component is of those two motions. This motion is known as the heliocentric correction. The rotation component is at most $\pm 0.5\,\text{km s}^{-1}$, while the orbital motion is at most $\pm 29.8\,\text{km s}^{-1}$. Clearly the heliocentric correction will depend both on latitude, date, time of day, and the coordinates of the object. When one cross-correlates an observation of a radial velocity standard star against the spectrum of a program object, one has to subtract the standard star's heliocentric correction from its cataloged value, and then add the program object's heliocentric correction to the obtained relative velocity.

It should be noted that cross-correlation is not always the most accurate method for measuring radial velocities. Very early-type stars (such as O-type stars) have so few lines that it is often better to simply measure the centers of individual spectral lines. Broad-lined objects, such as Wolf–Rayet stars, also require some thought and care as to how to best obtain radial velocities.

The velocity dispersion of a galaxy is measured using a similar method in order to obtain the width of the cross-correlation function. The intrinsic line widths of the radial velocity standards must be removed in quadrature. Thus sharp-lined radial velocity standards work better than those with wide lines.

Precision Radial Velocities

The desire to detect exoplanets has changed the meaning of "precision" radial velocities from the 1–2 km s^{-1} regime to the 1–100 m s^{-1} regime. Several methods have been developed to achieve this. The traditional method of wavelength calibration does not achieve the needed precision as the comparison arcs and the starlight are not collimated in exactly the same method. Early work by Campbell and Walker (1979) achieved a precision of 10 m s^{-1} using a hydrogen fluoride cell inserted into the beam that provided an evenly spaced absorption spectrum. More modern techniques achieve 1 m s^{-1} precision. One example is the High-Accuracy Radial Velocity Planetary Searcher (HARPS) deployed on an ESO's 3.6 m telescope on La Silla. It uses a conventional ThAr discharge lamp for wavelength reference. The ThAr source and the star's light each enter its own fiber, which then feeds a bench-mounted, extremely stable spectrograph. Both the ThAr reference source and the star are observed simultaneously (see Pepe et al. 2004). Another innovate instrument is the CHIRON high resolution spectrometer, which is achieving unpreciented precision by using and iodine cell, fiber scrambling, and high cadance observations. The goal is to search for earth-like planets around alpha Centuri, and hence they are not hurting for photons. Each observation achieves 3 m s^{-1} precision, and by coadding many spectra, 30 cm s^{-1} precision in a night. The instrument is deployed at the CTIO 1.5-m telescope (Schwab et al. 2010)

The recent emphasis on planet detection among late-type (cool) stars has driven some of this work into the NIR, where ammonia gas cells provide a stable reference (Bean et al. 2010; Valdivielso et al. 2010).

Laboratory Wavelengths

The traditional source of wavelengths for astronomers has been Moore (1972), a reprint of her original 1945 table. More up-to-date line lists can be found on the Web, allowing one to search the National Institute of Standards and Technology Atomic Spectra Database. The official site is http://www.nist.gov/physlab/data/asd.cfm, but a very useful search interface can be found at http://www.pa.uky.edu/~peter/atomic/. One must make sure that one is using the "air" wavelengths rather than "vacuum" wavelengths when observing from the ground.

One problem is that many lines in stellar (and reference sources!) are actually blends of lines. So, tables of "effective wavelengths" can be found for stellar lines for stars of different spectral types. These are generally scattered throughout the literature; see, for example, Conti et al. (1977) for a line list for O-type stars.

A good general line list (identifying what lines may be found in what type of stars) is the revised version of the "Identification List of Lines in Stellar Spectra (ILLSS) Catalog" of Coluzzi (1993), which can be obtained from http://cdsarc.u-strasbg.fr/cgi-bin/Cat?VI/71. Also useful is the Meinel et al. (1968) *Catalog of Emission Lines in Astrophysical Objects* and the Tokunaga (2000) list of spectral features in the NIR region. Lists of wavelengths of the night sky OH lines can be found in both Osterbrock et al. (1996, 1997) and Oliva and Origlia (1992).

3.2.8 Some Useful Spectral Atlases

The spectroscopist is often confronted by the "What Is It?" question. Of course the answer may be a quasar or other extragalactic object, but if the answer is some sort of star, the following resources may be useful in identifying what kind.

- Jacoby et al. (1984) *A Library of Stellar Spectra* provides moderate resolution (4.5 Å) dereddened spectrophotometry from 3500 Å to 7400 Å for normal stars of various spectral types and luminosity classes. Intended primarily for population synthesis, one deficiency of this work is the lack of identification of any spectral features. The digital version of these spectra may be found through VizieR.

- Turnshek et al. (1985) *An Atlas of Digital Spectra of Cool Stars* provides spectra of mid-to-late type stars (G-M, S and C) along with line identifications.
- Walborn and Fitzpatrick (1990) *Contemporary Optical Spectral Classification of the OB Stars – A Digital Atlas* provides moderate resolution normalized spectra from 3800Å to 5000 Å of early-type stars, along with line identification. The spectral atlas only extends to the early B-type stars for dwarfs, and to B8 for supergiants. An updated atlas with higher signal-to-noise can be found in Sota et al. (2011).
- Hanson et al. (2005) *A Medium Resolution Near-Infrared Spectral Atlas of O and Early-B Stars* provides moderately high resolution (R ~8000–12,000) spectra of O and early B-type stars in the H- and K-bands. The lower resolution spectra shown in Hanson et al. (1996) also remains very useful for spectroscopists working at $R \leq 1000$. Ivanov et al. (2004) provides *A Medium Resolution Near-infrared Spectral Library of Late-Type Stars*, which includes a nice listing of other useful work.
- Hinkle et al. (2003) describe in details various *High Resolution Infrared, Visible, and Ultraviolet Spectral Atlases of the Sun and Arcturus.*

3.3 Observing Techniques: What Happens at Night

One of the goals of this chapter has been to provide observing tips, and possibly the best way of doing this is to provide some examples of what some typical nights are actually like. Included here are examples of observing with a long-slit spectrograph, observing with a fiber spectrograph, and some advice on what to do when observing with a NIR spectrometer.

A common theme that emerges from these (mostly true) stories is that the observers spend a lot of time thinking through the calibration needs of their programs. For the optical this is mainly an issue of getting the flat-fields "right" (or at least good enough), while there are more subtle issues involved in NIR spectroscopy. Throughout these the same philosophy holds: obtaining useful spectra involves a lot more than just gathering photons at the right wavelength.

3.3.1 Observing with a Long-Slit Spectrograph

The GoldCam spectrometer on the Kitt Peak 2.1 m provides an interesting example of a classical long-slit instrument. The observing program described here was aimed at obtaining good (<5%) spectrophotometry of a sample of roughly 50 northern Galactic red supergiant stars whose V-band magnitude ranged from 6 to 11. The goal was to match both the spectral features and continuum shapes to a set of stellar atmosphere models in order to determine basic physical properties of the stars, especially the effective temperatures. Broad wavelength coverage was needed (4000–9000 Å), but only modest spectral resolution (4–6 Å) as the spectra of these late-type stars are dominated by broad molecular bands. An observing program was carried out for southern stars using a similar spectrograph on the Cerro Tololo 1.5 m telescope.

At the stage of writing the observing proposal, the choice of the telescopes was pretty well set by the fact that these stars were bright and the exposure times on even 2 m class telescopes would be short. The number of nights needed would be dominated by the observing overhead rather than integration time. The gratings and blocking filters that would meet the requirements were identified, based upon the needed spectral resolution and wavelength coverage (➋ Sect. 2.1). It was not possible to obtain the needed spectral resolution with a single grating setting over the entire wavelength range, and so the northern (2.1 m) sample would have to be observed twice, once with a 600 line/mm grating that would provide wavelength coverage from 3200 Å to 6000 Å, and a 400 line/mm grating that would provide coverage from 5000 Å to 9000 Å. Both gratings would be used in first order. The blue grating would thus require no blocking filter, as inspection of *Fig. 2-2* reveals: at 6000 Å the only overlap would be with 3000 Å from the second order. Given the poor atmospheric transparency at that wavelength, and the extreme red color of such stars, it is safe to assume that there would be negligible contamination. For the red setting (5000–9000 Å) there would be overlap with second order light <4500 Å, and so a GG495 blocking filter was chosen. As shown by *Fig. 2-3*, this filter would begin to have good transmission for wavelengths greater than 5000 Å, and no transmission below 4750 Å, a good match.

The 2.1 m GoldCam does not have any sort of ADC, and so the major challenge would be to observe in such a way that good spectrophotometry was maintained, i.e., that differential refraction (Sect. 3.2.1) would not reduce the accuracy of the fluxed spectra. The GoldCam spectrograph is mounted on a rotator with a mechanical encoder, and would have to be turned by hand (at zenith) to the anticipated parallactic angle for each observation. How much "slop" in the parallactic angle could be tolerated would be a function of time during the night for any given star, and in advance of the observing these values were computed. It was clear that the slit would have to be kept open as wide as possible without sacrificing the needed spectral resolution, and this was one of the factors that entered into the grating choice. With the higher dispersion blue grating one could have a slit width of 3 arcsec (250 μm) and still have 4 Å spectral resolution, while with the lower dispersion red grating one needed a skinnier 2 arcsec slit (170 μm) in order to maintain adequate resolution (6 Å), but this was made possible by the fact that differential refraction was less significant in the red. Without doing the math (using the equations given in Sect. 3.2.1 and adapting the equations in Sect. 3.2.2), all of this would have been a guess.

The first afternoon of the run was spent performing the basic setup for both gratings. First the grating was inserted, the blocking filter (if any) was put in the beam. The grating tilt was set to some nominal value based on the instrument manual and the desired central wavelength setting. The HeNeAr comparison source was turned on, and an exposure made to see if the grating tilt was a little too red or blue. This required some finagling, as bad columns dominate the first 300 columns of the 3000 pixel detector, and the focus gets soft in the final 500 columns. Thus 2200 columns are useful, but the center of these is around column 1600 and not 1500. Several exposures of the comparison arc were needed, with small tweaks (0.02–0.1 degrees) of the grating tilt used to get things just so.

Next, the spectrograph was focused. The HeNeAr comparison source was left on, and the slit set to a skinny 100 μm. Typically the collimator should be at a certain distance (the "autocollimate" position) from the slit so that it is just filled by the diverging beam from the slit (*Fig. 2-1*). But this luxury is achieved only in spectrographs in which the camera lens can be moved relative to the detector to obtain a good focus of the spectrum. In many instances – and in fact, one might have to say *most* instances – that is not the case, and the camera focus is fixed or at least difficult to adjust. Instead, the collimator is moved slightly in order to achieve a good focus. Each of the two gratings would require a different collimator setting as one had a blocking filter and the other did not; inserting such a filter after the slit changes the optical path length slightly.

The observers decided to start with the red grating, since this was their first night and the red observations should be a little less demanding. It was not practical to change gratings during the night and so there would be specific nights devoted to either the red or blue observations. The grating tilt and focus were adjusted to the values found earlier. The slit was opened to the needed amount, and the optical path was carefully checked. Was the appropriate blocking filter in place? Was the collimator set to the right focus?

The detector covers 512 (spatial) rows but each star would cover only a few of them. Good sky subtraction was important particularly in the blue (the observations would be made with considerable moonlight) but even so, this was unnecessarily excessive. The chip was reformatted to read out only the central 250 rows. The observers could have spatially binned the data by 3 (say) and reduced the readout time, but to do so would slightly compromise the ability to reject cosmic rays by having a nicely sampled spatial profile (➋ Sect. 3.2.5). Besides, it was clear that the observing overhead would be dominated by the need to move the telescope to zenith and manually rotate to the parallactic angle, not the readout time of the chip. The telescope could be slewed during the readouts. (Telescopes are usually tracking during readout.)

With the CCD and spectrograph set, the observers next proceeded to take some calibration data. Since there was still plenty of time before dinner, they decided to make a bad pixel mask. The calibration source housed two lamps: the HeNeAr source and a quartz lamp. The quartz lamp could provide a "featureless flat" but its illumination of the slit was sufficiently different that it provides a very poor match to the night sky compared to the dome flat. It does, however, have the advantage that it can be run in place. For just hitting the detector with enough light to identify bad pixels it would be plenty good enough. A 5 s exposure had 30,000 counts, plenty of counts, and well below the expected saturation of the detector. Two more were obtained just to protect against a single exposure having been hit by a strong cosmic ray. In order to obtain frames with a scant number of counts, a 5-magnitude (100× attenuation) "neutral density" filter was placed in front of the slit. This resulted in a 5 s exposure having about 300 counts. A series of 50 of these would take about an hour to run, given the readout time, but the average would then have good statistics. During the break, the observers attempted to get the music system connected to their iPods, and discussed some unanticipated flat-fielding issues.

Long experience at the 2.1 m had taught the observers that the dome flat exposures do a far better job at matching the illumination of the night sky in the spatial direction than do the internal quartz exposures. But, even superficial inspection of the bad pixel mask data revealed that there were significant, 10–20% fringes in the red (>7000 Å) region. How to remove these? If the instrument were absolutely stable (no flexure) then the fringes should divide out in the flat-fielding process. The internal lamp offered an additional option: it could be used in place without moving to the dome spot. Thus for safety the observers decided to take the standard dome flat exposures but also planned to take some internal quartz lamp exposures during the night at various positions and see how much (if any) the fringes moved. The blue data would be straightforward, and just require long exposures as the dome spot has poor reflectivity in the far blue.

During the course of the run the observers discovered that the fringes moved significantly during the first half hour after the nightly fill of the CCD dewar but were quite stable after that. So, in the end they wound up combining the quartz lamp exposures taken throughout the night and using that as the featureless flat in the red, and using dome flats as the featureless flat for the blue.

The mirror cover was next opened and the telescope moved to the dome flat position. The illumination lamps were turned on, and the comparison optics (HeNeAr/quartz) were removed from the beam. A short test exposure was run. Much surprise and consternation was expressed when a nearly blank exposure read out. What was the problem? Generations of astronomers have answered this in the same way: think about the light path from the one end to the other and at each point consider what could be blocking the light. The lamps were on. The telescope was pointed in the right position, as confirmed by visual inspection. The mirror covers were open. The comparison optics were out, at least according to the control unit. Wait! The filter wheel above the slit was still set to the 5-magnitude neutral density filter. Setting this back in the clear position solved the problem. A series of five dome flats were quickly obtained, and the telescope was slewed back to zenith and the mirror cover closed. The observers went to dinner, leaving a series of 15 biases running.

Shortly before the sun set, the observers filled the CCD dewar, opened the telescope dome, and brought the telescope fully on line with tracking turned on. After watching the sunset, they hurried back inside, where they took a series of exposures of the twilight sky. These would be used to correct for the mismatch (a few percent) between the projector flats and the night sky illumination along the slit, improving sky subtraction. They slewed to a bright star nearly overhead, and checked that the pointing was good. The slit was visible on the TV camera, with the reflective metal to either side showing the sky.

Next they moved the telescope back to zenith so they could manually adjust the rotator to the parallactic angle planned for the first observation, which would be of a spectrophotometric standard. The star would be relatively near the zenith, and so knowing exactly when they could get started was not critical, as the allowed tolerance on the rotator angle is very large for good spectrophotometry. They moved the platform out of the way, and slewed the telescope to the star. When the sky was judged to be sufficiently dark, they carefully centered it in the slit and began a 5-min exposure. Since all of the exposures would be short, they decided not to bother with the considerable overhead of setting up the guider, but would hand guide for all of the exposures, using the hand paddle to tweak the star's position on the slit if it seemed to be slightly off-center. They observed two more spectrophotometric standards, each time first moving the telescope to zenith, unstowing the platform, rotating to the parallactic angle, restowing the platform, and slewing to the next target. There was no need to measure radial velocities (a difficult undertaking with broad-lined stars) and so a single HeNeAr comparison would be used to reduce all of the data during the night.

After each exposure read down, the data were examined by running IRAF's *splot* plotting routine in a somewhat unconventional manner. The dispersion axis runs along rows, and normally one would plan to first extract the spectrum before using *splot*. Instead, the astronomers used this as a quick method for measuring the integrated counts across the spatial profile subtracting off the bias and sky level, by specifying *splot image[1600,*]* to make a cut across the middle of the spectrum. The "e" keystroke which is usually used to determine an equivalent width is then run on the stellar profile to determine the number of counts integrated under the profile, listed as the "flux." This neatly removes any bias level from the counts and integrates across the spatial profile. This could be checked at several different columns (500, 1600, 2200) to make sure there are good counts everywhere. After things settled down for the night, the observers were in a routine, and used *ccdproc* to trim the data and remove the overscan. Flat-fielding would be left until they had thought more about the fringes. Nevertheless, *doslit* could be used to extract the spectrum with a wavelength calibration.

The observers began observing their red supergiant sample. The *splot* trick proved essential to make sure they were obtaining adequate counts on the blue side, given the extreme cool temperatures of these stars. Every few hours they would take a break from the red supergiants to observe spectrophotometric standards, two or three in a row. By the end of the first night they had observed 28 of their program objects, and 11 spectrophotometric standards, not bad

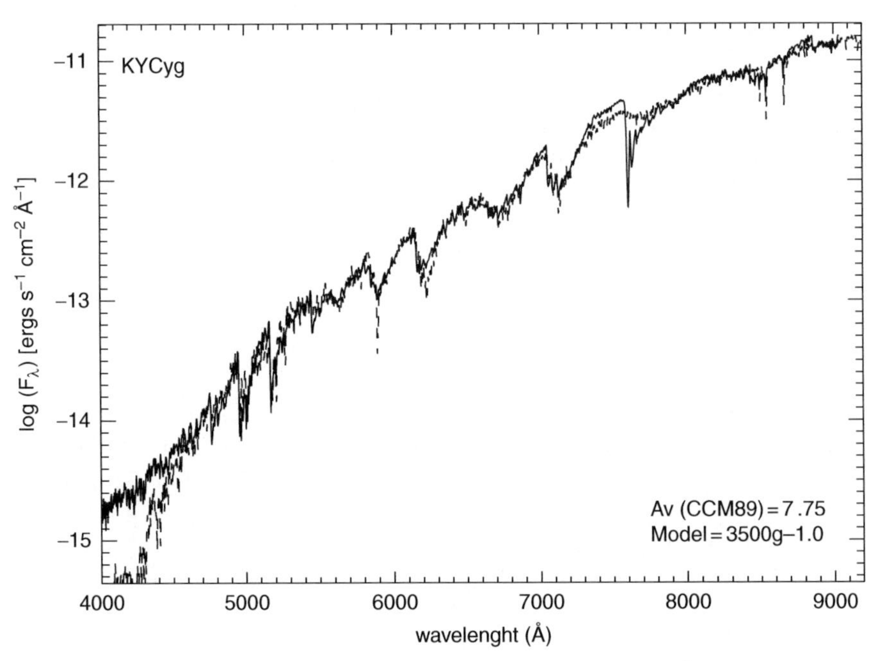

Fig. 2-18

Model fitting KY Cyg. The solid black shows the fluxed spectrum obtained as part of the 2.1 m GoldCam observing described here. The *dotted line* shows the best fit model, with an effective temperature of 3500 K and a log g=-1.0 [cgs]. The study by Levesque et al. (2005) showed that this was one of the largest stars known. The extra flux in the star in the far *blue* is due to scattering by circumstellar dust

considering the gymnastics involved in going to the parallactic angle. Some older telescopes have rotators that are accessible remotely (CTIO and KPNO 4 m) while all alt-az telescopes have rotators that can be controlled remotely by necessity.

The analogous observations at CTIO were obtained similarly. Since the detector there was smaller, three gratings were needed to obtain full wavelength coverage with similar dispersion. Going to the parallactic angle was even less convenient since the control room was located downstairs from the telescope. Fortunately the slit width at the CTIO telescope could be controlled remotely, and therefore the observations were all made with two slit settings, a narrow one for good resolution, and a really wide one to define the continuum shape.

In the end the data were all fluxed and combined after several weeks of work. A few stars had been observed both from CTIO and KPNO and their fluxed spectra agreed very well. The comparison of these spectra with model atmospheres began. A sample spectrum, and model fit are shown in *Fig. 2-18*. The work (Levesque et al. 2005) established the first modern effective temperature scale for red supergiants, removing the discrepancy between evolutionary theory and the "observed" locations of red supergiants in the H-R diagram, discovered circumstellar reddening due to these stars dust production, and identified the three "largest stars known," not bad for a few nights of hard labor hand guiding!

3.3.2 Observing with a Multi-fiber Spectrometer

The CTIO 4 m Hydra fiber spectrometer was commissioned in late 1998, and was the third version of this instrument, with earlier versions having been deployed on the KPNO 4 m and at the WIYN 3.5 m telescope (Barden and Ingerson 1998). It consists of 138 fibers each with 2.0 arcsec (300 μm) diameter, and is located at the Ritchey–Chretien focus covering a 40 arcmin diameter field. It is used with an ADC, removing the need to worry about differential refraction when placing the fibers. The observing program described here was carried out in order to identify yellow supergiants in the Small Magellanic Cloud based on their radial velocities measured using the strong Ca II triplet lines ($\lambda\lambda$ 8498, 8542, 8662) in the far red. In addition, the observations would test the use of the OI λ7774 line as a luminosity indicator at the relatively low metallicity of the SMC.

The astronomers needed to observe about 700 stars, only a small fraction of which were expected to be bona fide supergiants. The rest would be foreground stars in the Milky Way. This list of 700 stars had been selected on the basis of color and magnitude, and had already been culled from a much larger sample based on having negligible proper motions. The stars were relatively bright ($V < 14$) and the major limitation would be the overhead associated with configuring the Hydra instrument, which requires about 8 s per fiber, or 20 min in total. Because the SMC is large compared even to Hydra's field of view, several dozen fields would be needed to cover most of the targets.

The situation was further complicated by the desire to not only have spectra in the far red (including at least 7770–8700 Å) but also in the blue to obtain some of the classical luminosity indicators used for Galactic yellow supergiants. To observe each fiber configuration twice, on "blue nights" and "red nights" would require twice as many observing nights, given the large overhead in each fiber configuration.

The fibers feed a bench spectrograph mounted in a dark room a floor below the telescope. Changing the grating tilt might require refocusing the spectrograph (a manual operation, impractical at night) but simply changing blocking filters could be done remotely. If the filters were of similar thickness, and if the camera's focus was fairly achromatic (which would be expected of a Schmidt camera) then one could configure the fibers, observe in the red, and simply by changing blocking filters, observe in the blue. No one was quite sure if this would work, as no one could remember the spectrograph having been used this way, but it would be easy enough to check on the first afternoon of the run. A 790 line/mm grating blazed at 8500 Å in first order was available, and would yield 2.6 Å resolution in the red in first order, and 1.3 Å resolution in the blue in second order, providing wavelength coverage of 7300–9050 Å in the red and 3650–4525 Å in the blue. Obviously the red observations would require a blocking filter that removed light <4525 Å, while the blue observations needed a red cut-off filter that removed any light >7300 Å. Among the available filters, an OG515 did an excellent job in the first case, and a BG39 did a good job in the second case while still transmitting well over the region of interest (see *Fig. 2-3*). (The same argument was presented above in Sect. 2.2.1.)

Prior to the observing run, 30 fiber configuration fields had been designed in order to obtain as many of the target stars as possible. Since bad weather is always a possibility (even at Cerro Tololo) the fields were designed in a particular order, with field centers chosen to include the maximum number of stars that had not been previously assigned. Although the fiber configuration program is flexible in providing various weighting schemes for targets, it was found necessary to slightly rewrite the code to allow for stars that had been previously assigned to be added "for free," i.e., without displacing any not-yet assigned star. (It helped that the first author had written the original version of the code some years back.) The process took a week or more to refine the code, but the assignments themselves then were straightforward.

The first afternoon at the telescope, the astronomers arrived to find that everything appeared to be in good shape. Instrument support personnel had inserted the grating and blocking filter, checked the grating tilt, and had focused the spectrograph, substantiating the fact that the focus was unchanged between the red and the blue setups. A comparison arc had been used to focus the spectrograph, and examination confirmed the expectation that at the best focus the spectral resolution covered about seven pixels. The observers decided thus to bin by a factor of 2 in the wavelength direction. Even though radial velocities were desired, there was no advantage in having that many pixels in a spectral resolution element: 3 would be plenty, and 3.5 generous, according to the Nyquist–Shannon criterion (➋ Sect. 2.1). No binning was applied to the spatial direction as clean separation of one fiber from another is desirable.

The fibers could be positioned only with the telescope at zenith. This is quite typical for fiber instruments. Fibers are not allowed to "cross" or get so close to another fiber to disturb it, and thus in order to have reliable operations the fibers are configured only at a certain location. In mid-afternoon then the astronomers had the telescope moved to the zenith and configured the fibers into a circle for observing the dome spot.

With fiber instruments, sky subtraction is never "local," as it is with a long slit. Some fibers are pointing at objects, while other fibers have been assigned to clean sky positions. In order to subtract the sky spectrum from the object spectrum, flat-fielding must remove the fiber-to-fiber sensitivity, which itself is wavelength dependent. In addition, it must compensate for the different illumination that a fiber will receive from the sky when it is placed in the middle as opposed to somewhere near the edge. In other words, under a perfectly clear sky, the same fiber would have somewhat different counts looking at blank sky depending upon its location.

In addition, the pixel-to-pixel gain variations need to be removed, just as in long slit observations. But, the profiles of the fibers output are quite peaked, and they may shift slightly during the night as the liquid nitrogen in the dewar turns to gas and the weight changes.

On Hydra CTIO there are four possible flats one can take: an instrument support person can place a diffuser glass in back of the fibers, providing a somewhat uniform illumination of the CCD when the fibers are pointed at a bright light source. This is called a "milk flat," and would be suitable for removing the pixel-to-pixel gain variations. A second flat is the dome flat, with the fibers configured to some standard (large circle) configuration. This would also work for pixel-to-pixel gain variations as long as the output location of the fibers were stable on the detector. A third flat involves putting in a calibration screen and illuminating it with a lamp. This can be done in place with the fibers in the same position for the actual observations (unlike the dome flats) but the illumination by the lamp is very non-uniform, and thus has little advantage over the dome flat in terms of removing the vignetting. The fourth possibility is to observe blank sky with the same configuration. Since the SMC F/G supergiants were bright, and sky subtraction not critical, it was easy to eliminate the fourth possibility.

Exposures of the dome flat quickly revealed that although there was plenty of light for the red setting, obtaining a proper flat (one that would not degrade the observations; see ➋ Sect. 3.2.6) in the blue would take on the order of days, not minutes.

Given this, an arguable decision was reached, namely, that the observing would (provisionally) rely upon the calibration screen flats obtained at each field during the night. There were several arguments in favor of using the calibration screen flats. First, by having the rest of the fibers stowed, and only the fibers in use deployed, the flat-field would be useful in unambiguously identifying which fibers mapped to which slit positions on the detector. Second, and more importantly, it provided a real-time mapping of the trace of each fiber on the array. Third, it would cost little in overhead, as the radial velocities already required observing the HeNeAr calibration lamps with the screen in place, and that most of the overhead in the calibration itself would be moving the calibration screen in and out of the beam. The definitive argument, however, was that the stars were very bright compared to the sky, and so even if there was no sky subtraction, the science data would not be much compromised. Had the objects been comparable to the sky values, the best alternative would have been to do blank sky exposures, despite the use of extra telescope time. In any event, dome flats in the red were run each afternoon as it provided a good chance to exercise the instrument during the afternoon and ascertain that everything remained copacetic.

Prior to dinner, the observers configured the instrument to their first field. A problem was immediately revealed: one of the assigned fibers did not deploy. Why? Although there are 138 fibers, several fibers have become broken over time or have very low through-put and they are "locked" into the park position. A "concentricities" file is provided with the software used to assign fibers and test the configurations, and after a little probing it became clear that the concentricities file used in the assignments had been out of date. Therefore, assignments for the entire 30 fields would have to be recomputed. Fortunately most of the preparation work was simply in getting the software system set up, and before dinner the observers had managed to get the first few fields recomputed, enough to get them going, and the remainder were easily recomputed during the night. The new configuration files were transferred from the observer's laptop to the instrument computer, and the first configuration was again configured, this time without incident. The observers began a series of biases running and left for dinner.

Shortly before sunset the instrument assistant opened the dome to allow any heat in the dome to escape. The first actual target would be a bright radial velocity standard star (➋ Sect. 3.2.7). Without disturbing the other fibers, the astronomers moved an unused fiber to the center of the field, and deployed an unused alignment fiber to the location of another bright star near the radial velocity standard. As discussed in ➋ Sect. 3.2.3 there has to be some way to align (and guide!) fiber instruments. In the case of Hydra, these functions are accomplished by means of any of 12 "field orientation probes" (FOPs). These are each bundles of five fibers around a central sixth fiber. These are deployed like regular fibers, but the other ends of these fiber bundles are connected to a TV rather than feeding the spectrograph. Thus an image of six dots of light are seen for each FOP. When the telescope is in good focus, the centering is good, and the seeing is excellent, all of the light may be concentrated in the central fiber of the six. The telescope is guided by trying to maximize the amount of light in each of the central FOP fibers. In principle, a single FOP should be sufficient for alignment and guiding, since the only degrees of freedom are motions in right ascension and declination, and not rotation. But in practice a minimum of 3 is recommended. The assigned SMC fields had 3–5 each, but for the bright radial velocity standard a single FOP was judged sufficient as the exposure would be a few seconds long at most and no guiding would be needed.

Once the two new fibers were in place, the focal plane plate was "warped," i.e., bent into the curved focal surface using a vacuum. (The fibers had to be deployed onto the plate when it was flat.) The telescope was slewed to the position of the radial velocity standard, and the "gripper" – the part of the instrument which moves the fibers, was inserted into the field. The gripper has a TV camera mounted on it in such a way that it can view the reflection of the sky. Thus by positioning the gripper over a deployed fiber (such as a FOP) one can also see superimposed on the image any stars near that position. In this case, the gripper was placed in the center of the field of view, and the bright radial velocity standard carefully centered. The gripper was then moved to the single "extra" FOP and the presence of a bright star near that position was also confirmed. As the gripper was removed from the field of view, the light from the single FOP was visible. The telescope was next focused trying to maximize the light in the central fiber of the bundle.

While this was going on, the observers carefully checked the spectrograph configuration using the spectrograph GUI. Was the correct blocking filter in place for the red? Were the grating tilt and other parameters still set to what they were in the afternoon? When the operator announced that the telescope was focused, the observers then took a 5 s exposure. At the end, the CCD read down, and light from a single fiber was obvious. A cut across the spectrum showed that there were plenty of counts. The voltage on the TV was then turned down to protect the sensitive photocathode, the calibration screen was moved into place, and both a projector flat and HeNeAr comparison arc exposure were made. The first observation was complete!

Rather than waste time removing the two extra fibers (which would have required going back to the zenith), the telescope was slewed to the SMC field for which the fibers had been configured. The gripper was moved into the field, and sent to one of the deployed FOPs. A bright star was seen just to the upper left. The gripper was then moved to a second FOP. Again a bright star was seen just to the upper left. These must be the alignment stars. The operator then moved the telescope to center the reflection of the star image on the FOP. Going to a third FOP confirmed that there was now a bright star superimposed on that FOP. The gripper was moved out of the field, and the images of 5 illuminated FOPs appeared on the guider TV. The guider was activated, and after a short struggle the telescope motion seemed to be stable. "Okay," the operator announced. The astronomers took three exposures of 5 min each. Guiding was stopped, the voltage was turned down on the TV, the calibration screen reinserted into the beam, and a short projector flat and HeNeAr exposure were made. Then the blocking filter was changed from the red (OG515) to the blue (BG39), and new projector flats and HeNeAr exposures were made. The calibration screen was removed. Examination of the FOP guide TV showed that the telescope had drifted only slightly, and guiding was again initiated. The observers took three-10 min exposures for blue spectra of the same stars. Then the telescope was moved to the zenith, the plate flattened, and the next field was configured. The process was repeated throughout the night, interrupted from time to time to observe new radial velocity standards in the red.

Throughout the night the observers made cuts through the spectra, but the first efforts to reduce the data to "final" spectra failed as there was an ambiguity in how the slit positions were numbered. The assignment files assigned fiber 103 to a specific star. But, where did fiber 103 map to on the detector? The concentricities file was supposed to provide the mapping between fiber number and slit position, but the image headers also contained a mapping. These agreed for the first couple of dozen fibers but after that there was an offset of one. After a few dozen more fibers they differed by several. The problem appeared to be that there were gaps in the output slit. The concentricities file numbered the slit positions consecutively in providing the mapping, while the header information was derived apparently assuming there was more or less even spacing. The problem this introduced was not just being sure which object was which, but which spectra were that of sky in order to sky subtract. ➋ *Figure 2-19* shows the problem.

Fortunately (?) the following night was cloudy, and the astronomers spent the evening with the telescope pointed at zenith, creating a mapping between fiber number, slit position in header, and pixel number on the detector. In the case of any uncertainty, a fiber could be moved from the outer region to the central region and exposed to the calibration screen, and the position on the detector measured unambiguously. The third night was clear, and by then data could be reduced correctly in real time using the mapping and the IRAF task *dohydra*.

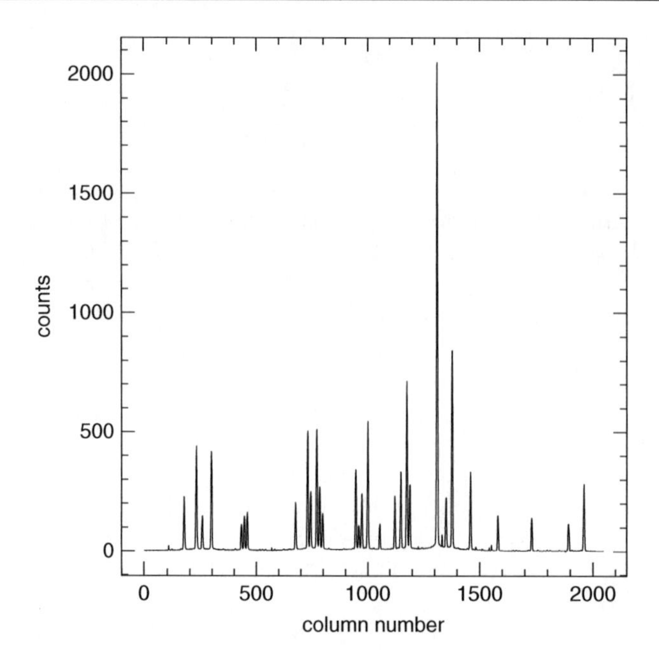

Fig. 2-19
Spatial cut across Hydra data. The spectra of many stars have been obtained in a single exposure, but which star is which?

How well did the project succeed? Radial velocities were obtained for approximately 500 stars. ➋ *Figure 2-20* shows the Tonry and Davis (1979) r parameter (a measure of how well the cross-correlation worked) versus the radial velocity of each star. There are clearly two distributions, one centered around a velocity of zero (expected for foreground stars) and one centered around 160 km s^{-1}, the radial velocity of the SMC. All together 176 certain and 16 possible SMC supergiants were found, and their numbers in the H-R diagram were used to show that the current generation of stellar evolutionary models greatly over estimate the duration of this evolutionary phase (Neugent et al. 2010). Since a similar finding had been made in the higher metallicity galaxy M31 (using radial velocities from Hectospec; Drout et al. 2009) the SMC study established that uncertainties in the mass-loss rates on the main-sequence must not be to blame. The OI λ7774 line proved to be a useful luminosity indicator even at relatively low metallicities.

3.3.3 Observing with a NIR Spectrometer

With near-infrared observations, virtually all the steps taken in preparing and undertaking a run in the optical are included and will not be repeated here. What will be done here is a review of the few additional steps required for near-infrared spectroscopy. These steps center around the need to remove OH sky emission lines (numerous and strong, particularly in the H-band) and to correct for the absorption lines from the Earth's atmosphere. It will not be possible to do a very thorough reduction during the night like one can with the optical, but one can still

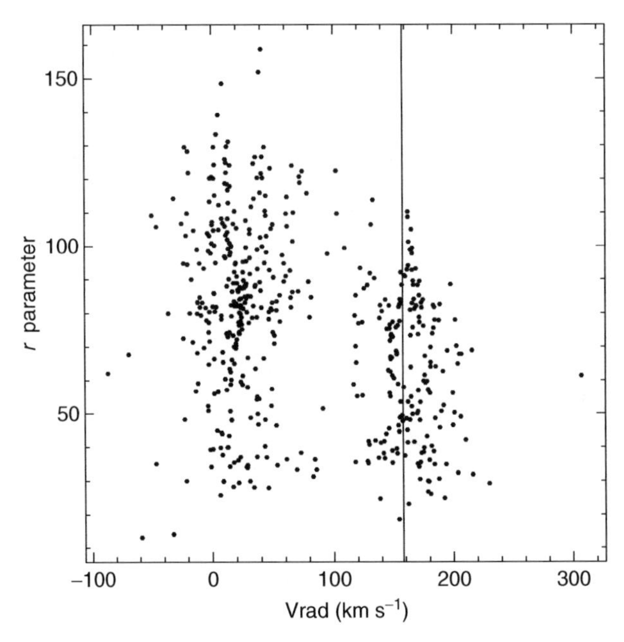

Fig. 2-20
The radial velocities of SMC F/G supergiant candidates. The Tonry and Davis (1979) *r* parameter is plotted against the radial velocity of approximately 500 stars observed with Hydra at the CTIO 4 m telescope. The stars with radial velocities <100 km s^{-1} are Milky Way foreground stars. The line at 158 km s^{-1} denotes the systematic radial velocity of the SMC (Based upon Neugent et al. (2010))

perform various checks to ensure the data will have sufficient signal-to-noise and check if most of the sky and thermal emission is being removed.

The near-infrared spectroscopist is far more obsessed with airmass than the optical spectroscopist. It is advised that the observer plot out the airmass of all of the objects (targets and standards) well in advance for a run. This can be done using online software, http://catserver.ing.iac.es/staralt/index.php. The output of this program is given in *Fig. 2-21*. For this run, the authors were extraordinarily lucky that two of the telluric standards from the Hanson et al. (2005) catalog were well suited to be observed during a recent SOAR run: HD 146624 during the first half, and HD 171149 during the second half. Looking at this diagram, one can make the best choices about when to observe the telluric standard relative to any observations made of a target object. If the target observations take about 30 min (this includes total real time, such as acquisition and integration) then observing Telluric Object 1 just before the program target during the early part of the night will mean the target star will pass through the exact same airmass during its observations, optimizing a telluric match. Later in the night, Object 3 can also be used, though observed after the target. As hour angle increases, airmass increases quickly. Note the non-linear values of the ordinate on the right of *Fig. 2-21*. One must be ready to move quickly to a telluric standard, or the final spectra may be quite disappointing. This observer has been known to trace in red pen in real time on such a diagram, the sources being observed as time progresses, to know when it is time to move between object and telluric

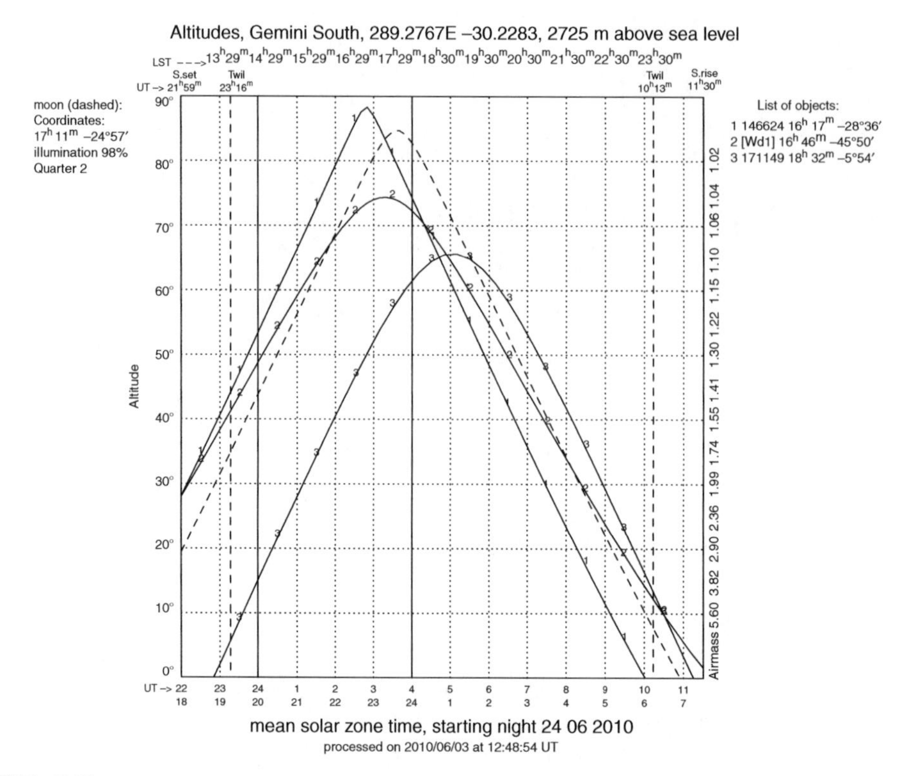

Fig. 2-21

Output from the Staralt program. Here the date, observatory location, and objects for the run have been entered uniquely. Object 2 is the science target, Westerlund 1. Object 1 and Object 3 are HD numbers for telluric standards (Taken from Hanson et al. (2005))

and vice versa. The goal should be to observe the telluric standard when its airmass is within 0.1 of that of the observation of the program object.

How to select telluric standards? As was mentioned in Sect. 2.5.2, early-A dwarfs or solar analogues are typically used. Ideally, one should seek telluric standards which are bright (for shorter integration times), have normal spectral types (no anomalies), and are not binaries (visible or spectroscopic). But also, location in the sky is important. Referring back to *Fig. 2-21* again, stars passing close to zenith at meridian have a different functional form to their airmass curves than do stars that remain low even during transit. This can make it hard to catch both target and telluric at the same airmass if their curves are very different. So, attempt to select a telluric standard that has a similar declination to the target object, but that transits 30–60 min before or after the target object.

Background emission is the second serious concern for the infrared spectroscopist. Even optical astronomers are aware of the increase in night sky brightness with increasing wavelength, with U and B brightness of typically >22 mag arcsec^{-2}, V around 21.5, $R \sim 21$, and $I \sim 20$ at the best sites. But this is nothing compared to what the infrared observer must endure. A very nice review of infrared astronomy is given by Tokunaga (2000) that all new (and seasoned!) infrared astronomers should read. He lists the sky brightness in mag arcsec^{-2} as 15.9, 13.4, and

14.1 at J, H, and K_S. For the L and M bands, the sky is 4.9 and around 0 mag arcsec^{-2}, respectively! In the latter bands, this is dominated by thermal emission, while in the J, H, and K_s, it is dominated by OH airglow. This background emission will dominate one's spectrum if not removed.

Removal of background emission is done by stepping the object along the slit between exposures or periodically offsetting the telescope to a blank field. Since the background emission is ubiquitous, offsetting a compact target along the slit allows one to measure the background spectrum at the first target position in the second exposure, and vice versa. In the near-infrared, where the sky background intensity is modest, one may step the target to several positions in the slit, observing the target all of the time, while simultaneously observing the sky background in the rest of the slit. However, if the field is densely populated with stars, or the target itself is extended or surrounded by nebulosity, it is necessary to offset to a nearby patch of blank sky periodically to obtain the sky spectrum for background subtraction. At mid-infrared wavelengths, the sky background is significantly larger and even small temporal variations can overwhelm the signal from the science target, so it is necessary to carry out sky subtraction on a much shorter time scale. This is often done by taking very short exposures (to avoid saturation) and chopping the target on and off the slit at a few Hz, typically using a square-wave tip/tilt motion of the telescope secondary mirror. The chopping and data taking sequences are synchronized, so that the on- and off-source data can be stored in separate data buffers and the sky subtraction carried out in real time.

How often to step along the slit? This depends on a few things. One always wants to maximize the counts for any single step integration, letting the exposure time be determined by the limits of the detector. Remember that you must stay within the linear regime *while including* background emission in any single frame! If the integration is too long in the H-band and the OH airglow lines are saturated, they won't properly subtract. Always check that the counts are not too high before any subtraction is done. The number of steps should be at least four, to remove bad pixels and six is a more typical minimum number. Build up signal-to-noise through multiple sets of optimized offsets, returning to a telluric standard as needed between sets. Finally, always check, as the data is coming in, that when you do subtract one slit (or sky offset) position from another that you do get zero counts outside of the star. What can go wrong here, as was mentioned in ➋ Sect. 2.5.2, is that the strengths of the OH bands vary with time. This is particularly true if clouds move in or the seeing changes between slit positions, with the result that the OH lines will not cancel entirely and reduction will be much more difficult. If this occurs, one is forced to use a shorter integration per step to find a time frame over which the sky emission is sufficiently stable.

Note that even if one does have "perfect" sky subtraction of the OH spectra, the large signal in the OH lines invariably adds noise to the final reduced spectra. This is unavoidable, and makes the entire concept of defining the signal-to-noise-ratio in the NIR tricky to define, as it is bound to vary depending upon the OH spectra within a particular region.

What could possibly go wrong? A lot. The second author has never worked on an infrared spectrometer that didn't offer additional challenges which fell outside the standard operation as listed above. To keep this brief, four of the most common problems that occur will now be discussed. They include: flexure, fringing, wavelength calibration problems, and poor telluric matching. Each of these can lead to greatly reduced signal-to-noise in the spectra, far below what would be predicted by counts alone.

- *Flexure.* No spectrograph is absolutely rigid! Flexure can be a larger issue for infrared spectrographs because the optics must be cooled to reduce the thermal background, and it is challenging to minimize the thermal conduction to the internal spectrograph bench while also minimizing the mechanical flexure between the bench and outer structure of the instrument. Depending on how the instrument is mounted, when observing to the east, versus looking to the west, for instance, or as the telescope passes meridian, the internal light path will shift due to structure shifts. The amount of flexure depends on the instrument and telescope setup. Previous users or the support astronomer for the instrument can help the new user decide if this needs to be considered and how to mitigate the effects. This is best addressed during the observing, keeping telluric standards, lamps and objects on the same side of the meridian, for instance.

- *Fringing.* Optical CCDs fringe in the red wavelength regions due to interference within the surface layers of the detector; for IR spectrographs, fringing can occur due to parallel optical surfaces, such as blocking filters, in the optical path. While in principle, this should cancel out with the dome flats, due to flexure in the system and light path differences, they typically do not cancel well. Common remedies include obtaining quartz lamps at the exact same sky location as the observations. This might work and should be included in the observing plan. However, the fringes are often not similar between the quartz lamp and point sources. The second author has instead turned to Fourier filtering methods to simply remove fringes outright. Software exists within IRAF in the STSDAS package to lock in on the fringe and remove it. This is easily done with flat-fields, and virtually impossible for point source images. Stellar flats (➋ Sect. 3.1.2) can be used to create a very crude two-dimensional illumination of a point source. If the fringe pattern is relatively strong, the Fourier filtering packages should be able to lock in on the overarching pattern and create a fringe correction which can be applied to all your frames before extraction.
- *Wavelength calibration.* For many long slit spectrometers, the wavelength solution is a function of position on the slit. It was already suggested that wavelength calibration should be applied using comparison lamp solutions which were extracted at the exact same location as the star was extracted. If this is not done, then there will be slight variations in solution with slit position and the resolution will be inadvertently reduced by co-adding such data. Moreover, this can lead to even more serious problems later on when applying telluric corrections.
- *Telluric correction.* The telluric standards need to have been observed at a very similar airmass and hopefully fairly close in time to the program object. When observing, one needs to be sure that the telluric standard observations extend to the same range of airmasses as the program objects (both minimum and maximum). If the match isn't great, one can interpolate as needed by using hybrid spectra of two telluric stars to get a better airmass correction. Also, IRAF has very useful software which actually uses the telluric lines with a statistical minimization routine, to make the best match. Finally, if one is working in an area of very strong telluric lines, there may be a good match with the features *only* when the target and standard were observed in the *exact same location in the slit*. This requires keeping all spectra separate until telluric removal, then combining the final set of spectra as the last step.

Possibly the strongest recommendation is that one needs to talk to a previous user of the instrument, preferably one who actually knows what they are doing. Make sure that the answers make sense, though. (Better still would be to talk to several such previous users, and average their responses, possibly using some strong rejection algorithm.) Maybe they will even be willing to share some of their data before the observing run, so that one can really get a sense of what things will look like.

Wrap-Up and Acknowledgments

The authors hope that the reader will have gained something useful from this chapter. They had a lot of fun in writing it. If there was any simple conclusion to offer it would be that astronomical spectroscopy is a stool that must sit on three legs for it to be useful: simply taking data with sufficient counts and resolution is not enough. One has to carefully think through the calibration requirements of the data as needed by one's program, and one must perform the reductions in such a way that honors the data. Quantitative quick-look at the telescope is essential for the process.

The authors' knowledge of spectroscopy is due to contact with many individuals over the years, and in particular the authors want to acknowledge the influence of Peter S. Conti, Bruce Bohannan, James DeVeny, Dianne Harmer, Nidia Morrell, Virpi Niemela, Vera Rubin, and Daryl Willmarth. Constructive comments were made on parts or the whole of a draft of this chapter by Travis Barman, Bob Blum, Howard Bond, Dianne Harmer, Padraig Houlahan, Deidre Hunter, Dick Joyce, Emily Levesque, Stephen Levin, Nidia Morrell, Kathryn Neugent, George Rieke, Sumner Starrfield, Mark Wagner, and Daryl Willmarth. Neugent was responsible for constructing many of the figures used here.

References

Al-Azzawi, A. 2007, Physical Optics: Principles and Practices (Boca Raton: CRC Press), 48

Allington-Smith, J., Content, R., & Haynes, R. 1998, SPIE, 3355, 196

Atwood, B., Byard, P., Depoy, D.L., Frogel, J., & O'brien, T. 1992, European Southern Observatory Conference and Workshop Proceedings, 42, 693

Barden, S. C., & Ingerson, T. E. 1988, In ASCP Conf. Ser. 152, Fiber Optics in Astronomy III, ed. S. Arribas, E., Mediavilla, & F. Watson (San Francisco: ASP), 60

Bean, J. L., Seifahrt, A., Hartman, H., Nisson, H., Wiedemann, G., Reiners, A., Dreizler, S., & Henry, T. J. 2010, ApJ, 713, 410

Beckett, M. G. 1995, Ph.D. thesis, Univ. of Cambridge, Cambridge

Bidelman, W. P. 1966, Vistas in Astronomy, 8, 53

Bowen, I. S. 1938, ApJ, 88, 113

Bowen, I. S. 1962, in Astronomical Techniques, ed. W. A. Hiltner (Chicago, IL: Univ. Chicago Press), 34

Bowen, I. S., & Vaughan, A. H., Jr. 1973, PASP, 85, 174

Campbell, B., & Walker, G. A. H. 1979, PASP, 91, 540

Coluzzi, R. 1993, Bull. Inf. CDS 43, 7

Conti, P. S., Leep, E. M., & Lorre, J. J. 1977, ApJ, 214, 759

Depoy, D. L., Atwood, B., Byard, P. L., Frogel, J., O'Brien, T. P. 1993, SPIE, 1946, 667

Dressler, A., Hare, T., Bigelow, B. C., & Osip, D. J. 2006, SPIE, 6269, 62690F

Drout, M. R., Massey, P., Meynet, G., Tokarz, S., & Caldwell, N. 2009, ApJ, 703, 441

Fabricant, D. et al. 2005, PASP, 117, 1411

Filippenko, A. V. 1982, PASP, 94, 715

Gray, D. F. 1976, The Observation and Analysis of Stellar Photospheres (Cambridge: Cambridge University Press), 40

Hamuy, M., Suntzeff, N. B., Heathcote, S. R., Walker, A. R., Gigoux, P., & Phillips, M. M. 1994, PASP, 106, 566

Hanson, M. M., Conti, P. S., & Rieke, M. J. 1996, ApJS, 107, 281

Hanson, M. M., Kudritzki, R.-P., Kenworthy, M. A., Puls, J., & Tokunaga, A. T. 2005, ApJS, 161, 154

Hinkle, K., Wallace, L., Livingston, W., Ayres, T., Harmer, D., & Valenti, J. 2003, in The Future of Cool-Star Astrophysics, ed. A. Brown, G. M. Harper & T. R. Ayres (Boulder: Univ. Colorado), 851

Hoag, A. 1976, PASP, 88, 860

Horne, K. 1986, PASP, 98, 609

Ivanov, V. D., Rieke, M. J., Engelbracht, C. W., Alonso-Herrero, A., Rieke, G. H., & Luhman, K. L. 2004, ApJS, 151, 3871

Jacoby, G. H., Hunter, D. A., & Christian, C. A. 1984, ApJS, 56, 257

Levesque, E. M., Massey, P., Olsen, K. A. G., Plez, B., Josselin, E., Maeder, A., & Meynet, G. 2005, ApJ, 628, 973

Loewen, E. G., & Popov, E. 1997, Diffraction Gratings and Applications (New York: Marcel Dekker)

Mackay, C. D. 1986, ARA&A, 24, 255

Marsh, T. R. 1989, PASP, 101, 1032

Marshall, J. L. et al. 2008, Proc. SPIE, 7014, 701454

Massey, P., Bresolin, F., Kudritzki, R. P., Puls, J., & Pauldrach, A. W. A. 2004, ApJ, 608, 1001

Massey, P., DeVeny, J., Jannuzi, B., & Carder, E. 2000, Low-to-Moderate Resolution Optical Spectroscopy Manual for Kitt Peak (Tucson: NOAO)

Massey, P., Olsen, K. A. G., Hodge, P. W., Jacopy, G. H., McNeill, R. T., Smith, R. C., & Strong, S. 2007, AJ, 133, 2393

Massey, P., Strobel, K., Barnes, J. V., & Anderson, E. 1988, ApJ, 328, 315

Massey, P., & Gronwall, C. 1990, ApJ, 358, 344

Meinel, A. B., Aveni, A. F., & Stockton, M. W. 1968, A Catalog of Emission Lines in Astrophysical Objects (Tucson: Optical Sciences Center and Steward Observatory)

Miller, J. L., & Friedman, E. 2003, Optical Communications Rules of Thumb (New York: McGraw-Hill Professional), 368.

Moore, C. E. 1972, A Multiplet Table of Astrophysical Interest (Washington: National Bureau of Standards)

Neugent, K. F., Massey, P., Skiff, B., Drout, M. R., Meynet, G., & Olsen, K. A. G. 2010, ApJ, 719, 1784

Newall, H. F. 1910, The Spectroscope and Its Work (London: Society for Promoting Christian Knowledge), 97 (http://books.google.com/books?id=YCs4AAAAMAAJ)

Oke, J. B. 1990, AJ, 99, 1621

Oliva, E., & Origlia, L. 1992, A&A, 254, 466

Osterbrock, D. E., Fulbright, J. P., Martel, A. R., Keane, M. J., Trager, S. C., & Basri, G. 1996, PASP, 108, 277

Osterbrock, D. E., Fulbright, J. P., & Bida, T. A. 1997, PASP, 109, 614

Pepe, F. et al. 2004, A&A, 423, 385

Pierce, A. K. 1965, PASP, 77, 216

Richardson, E. H., Brealey, G. A., & Dancey, R. 1971, Publ. Dominion Astrophysical Observatory, 14, 1

Rieke, G. H. 2007, ARA&A, 45, 77

Schroeder, D. J. 1974, in Methods of Experimental Physics, Astrophysics: Optical and Infrared, Vol. 12, Part A, ed. N. Carleton (New York: Academic), 463

Schwab, C., Spronck, J. F. P., Tokovinin, A., & Fisher, D. A. 2010, in SPIE, 7735, 77354G

Schweizer, F. 1979, PASP, 91, 149

Sheinis, A. I. et al. 2002, PASP, 114, 851

Sota, A., Maiz Apellaniz, J., Walborn, N. R., Alfaro, E. J., Barba, R. H., Morrell, N. I., Garmen, R. C., & Arias, J. I. 2011, ApJS, 193, 24

Stone, R. P. S. 1996, ApJS, 107, 423

Stone, R. P. S. 1977, ApJ, 218, 767

Stone, R. P. S., & Baldwin, J. A. 1983, MNRAS, 204, 347

Tokunaga, A. T. 2000, Allen's Astrophysical Quantities, ed. A.N. Cox (4th ed.; Berlin: Springer), 143

Tonry, J., & Davis, M. 1979, AJ, 84, 1511

Turnshek, D. E., Turnshek, D. A., Craine, E. R., & Boeshaar, P. C. 1985, An Atlas of Digitial Spectra of Cool Stars (Types G, K, M, S, and C) (Tucson: Western Research Company)

Valdivielso, L., Esparza, P., Martin, E. L., Maukonen, D., & Peale, R. E. 2010, ApJ, 715, 1366

Wagner, R. M. 1992, in ASP Conf. Ser. 23, Astronomical CCD Observing and Reductions Techniques, ed. S. B. Howell (San Francisco: ASP), 160

Walborn, N. R., & Fitzpatrick, E. L. 1990, PASP, 102, 379

Walker, G. 1987, Astronomical Observations: An Optical Perspective (Cambridge, MA: Press Syndicate of the University of Cambridge)

Woodgate, B. E. et al. 1998, PASP, 110, 1183

Wynne, C. G. 1993, MNRAS, 262, 741

3 Infrared Astronomy Fundamentals

Alan T. Tokunaga[1] · *William D. Vacca*[2] · *Erick T. Young*[2]
[1]Institute for Astronomy, University of Hawaii, Honolulu, HI, USA
[2]SOFIA, NASA Ames Research Center, Moffett Field, CA, USA

T.D. Oswalt, H.E. Bond (eds.), *Planets, Stars and Stellar Systems. Volume 2: Astronomical Techniques, Software, and Data*, DOI 10.1007/978-94-007-5618-2_3, © Springer Science+Business Media Dordrecht 2013

Abstract: This chapter provides basic information on infrared astronomy as practiced from the ground, in the air, and in space. The focus in this chapter is on atmospheric and background limitations, basic data reduction techniques, absolute calibration, and photometry.

1 Introduction

This chapter provides a basic introduction to infrared (IR) astronomy. As the scope of IR astronomy is vast, key references are provided to allow the reader to pursue subjects in greater detail. This work does not attempt to cover all aspects of IR astronomy; it focuses on basic background information, observing techniques, photometry, and absolute calibration. Instrumentation and adaptive optics are covered in other volumes in this series. Heterodyne spectroscopy, submillimeter astronomy, and polarimetry are left to other accounts of these subjects.

The foundation for the IR detectors used today was established in the 1960s and 1970s, and this led to the growth of ground-based IR astronomy as a mature field. Efforts to get above the atmosphere with suborbital rockets started about 1965, and these led to a series of increasingly sophisticated space-based observatories. See ➋ Sect. 6.5 for a brief summary and Price (2009) and Rieke (2009) for detailed historical accounts.

IR astronomy now spans the entire 1–1,000 μm spectral range with a suite of ground-based and spacebased observatories. One of the great advantages of IR observations is that interstellar extinction is much lower than it is at visible wavelengths. This is dramatically illustrated in ➋ *Fig. 3-1* with the near-IR image of our galaxy obtained by 2MASS, a near-IR all sky survey (➋ Sect. 7.2; Skrutskie et al. 2006; http://www.ipac.caltech.edu/2mass/).

Many astronomical sources emit most of their luminosity in the infrared. A few examples are shown in ➋ *Fig. 3-2a, b*. The entire wavelength range of 1–1,000 μm has been accessible in

Fig. 3-1
Near-IR image of the sky obtained by the 2MASS near-IR sky survey. Note that the central bulge of the Milky Way is readily apparent due to the relatively low extinction in the near infrared compared with that at visible wavelengths. See Mellinger (2009) to compare this image to a similar visible image in which the obscuration from interstellar dust is very prominent

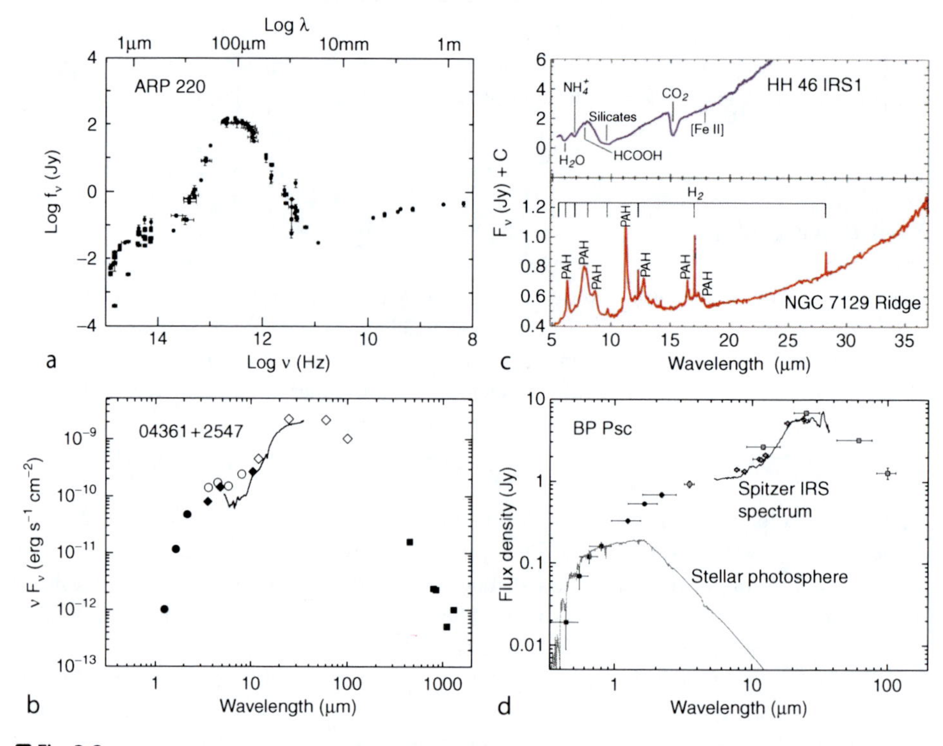

Fig. 3-2

Examples of spectra and spectral energy distributions in the infrared. (a) Ultraluminous IR galaxy Arp 220. Data from the NASA/IPAC Extragalactic Database (NED). (b) Class I young stellar object (Furlan et al. 2008). (c) Ice absorption bands and polycyclic aromatic hydrocarbon emission bands in starforming regions (From Werner et al. 2006; see also van Dishoeck 2004 and Peeters 2011). (d) Star with a cool dust envelope (Melis et al. 2010)

the past few decades with ground-based and space-based observatories. Note the range in flux density of nearly 10^5 and the spectral range of 10^4 in the examples shown in *Fig. 3-2a–d.*

The IR spectral range is also important because many molecules, dust grains, and ice grains in the interstellar medium and circumstellar disks can only be observed in the infrared (see *Fig. 3-2c, d*). The infrared is indispensable for the study of the composition of the interstellar medium as well as comets, planetary atmospheres, and cool stellar atmospheres due to the strong molecular bands at IR wavelengths. In addition to compositional information, the abundance and distribution of interstellar dust can be studied only in the far infrared. An example of the IR "cirrus" is shown in *Fig. 3-3.*

The power of adaptive optics (AO) coupled with a 10-m class telescope is shown in *Fig. 3-4.* An image of an exoplanet taken with an adaptive optics system is shown in *Fig. 3-4* (left). The sharpened images with the AO system coupled with a coronagraph allow extremely faint objects near to the primary star to be detected. Note that adaptive optics operates in the near infrared because the wavefront sensors operate at visible wavelengths and the wavefront corrections are easier to accommodate in the infrared than at visible wavelengths. Direct detection of very cool objects such as exoplanets can only be achieved in the infrared.

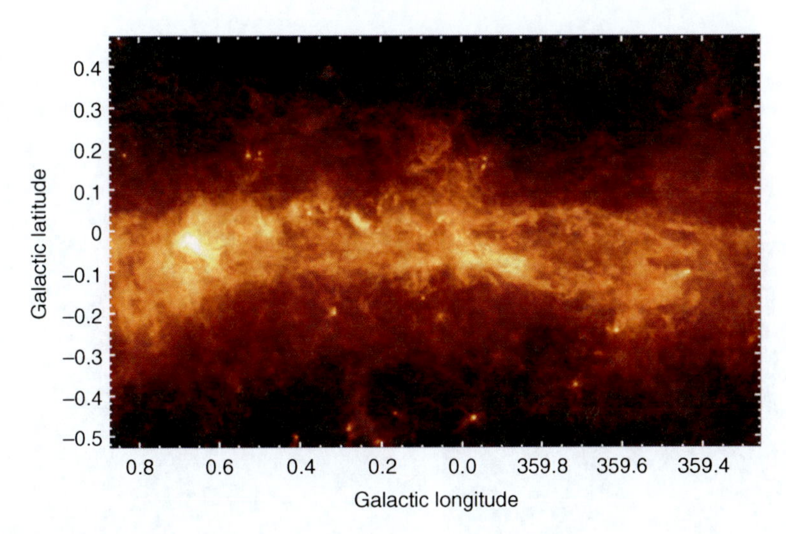

Fig. 3-3
Herschel observation of the Galactic Center region at 250 μm showing the IR cirrus emission from cold dust grains (Molinari et al. 2011)

Figure 3-4 (right) shows the stars near the Galactic Center and their motion around the black hole. The amount of information contained in the $1'' \times 1''$ image is impressive, and it demonstrates the impact on high angular resolution imaging that the next generation of large telescopes will have.

Much of modern cosmology is based on infrared observations of objects and phenomena that range from high redshift galaxies to the cosmic background. See for example Soifer et al. (2008). For this reason the *James Webb Space Telescope* emphasizes observations in the near infrared (Gardner et al. 2006). A striking example of a high redshift object is shown in *Fig. 3-5*. Most importantly, the cosmic background radiation provides critical information on the large-scale structure of the universe and high precision constraints on cosmological parameters (Jarosik et al. 2011).

1.1 Terminology and Units

The following definitions are used for convenience: near infrared (1–5 μm), mid infrared (5–25 μm), far infrared (25–200 μm), and submillimeter (submm) (200–1,000 μm). Broadband measurement of radiation from celestial sources are referred to as "photometry" although the term "radiometry" is technically the proper term in fields other than astronomy.

Because terminology for units is sometimes confused, they are summarized in *Table 3-1*.

It is best to plot in consistent units and use SI units. *Figure 3-2* shows typical ways that data are plotted. *Figure 3-2a* shows a log–log plot with units of Jy on the y-axis and frequency on the x-axis. This plot could also be shown as F_λ on the y-axis and λ on the x-axis. *Figure 3-2b* is also correct since $\nu F_\nu = \lambda F_\lambda$ and the equivalent of W m^{-2} is plotted on the y-axis with λ on the x-axis. *Figure 3-2c, d* show inconsistent units on the y-axis

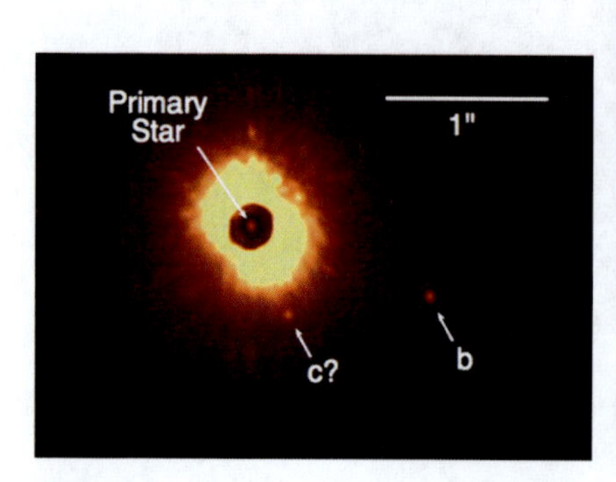

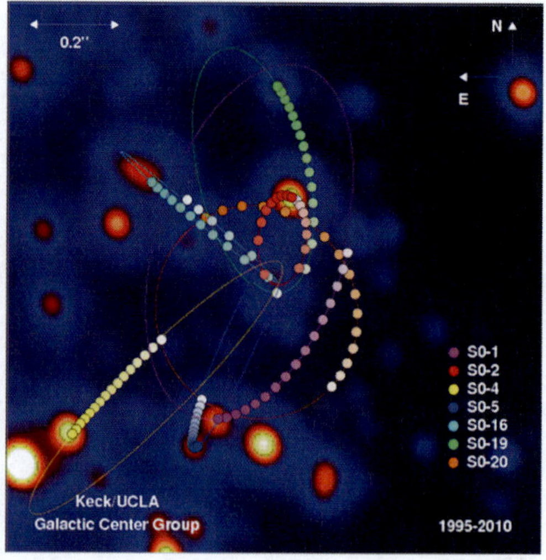

Fig. 3-4

Examples of adaptive optics imaging. *Left*: A young star with one confirmed planetary-mass companion (b) with an estimated mass of 5 M_{Jup} and a possible interior candidate of similar brightness (c?), as imaged with Keck-II/NIRC2 with natural guide star adaptive optics. The system was observed at K′ with a 300 milliarcsec coronagraph that attenuates the light of the primary star by 7.2 mag. Both faint sources are 6.5 mag fainter than the primary star at *K*′ and have projected separations of 1.1″ and 0.55″ (A. Kraus, private communication). *Right*: The orbits of stars within the central 1.0″×1.0″ of our Galaxy obtained with the Keck Observatory AO system. In the background, the central portion of a diffraction-limited image taken in 2010 is displayed. The motions of seven stars are shown. The annual average positions for these seven stars are plotted as *colored dots*, which have increasing color saturation with time. Also plotted are the best-fitting simultaneous orbital solutions. These orbits provide evidence for a *black hole* at the center of the Galaxy with a mass of 4×10^{6} M_{sun} (A. Ghez, UCLA Galactic Center Group; http://www.astro.ucla.edu/~ghezgroup/gc/pictures/index.shtml)

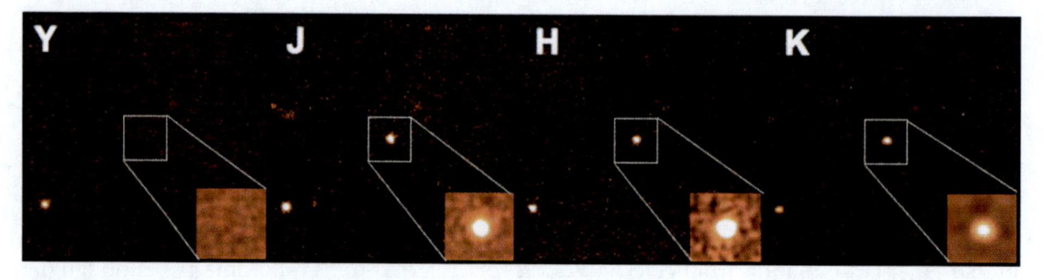

Fig. 3-5

Multiband images at Y (1.02 μm), J (1.26 μm), H (1.65 μm), and K (2.15 μm) of the afterglow of gamma ray burster GRB 090423 obtained with the UKIRT Wide Field Camera and the Gemini-North Near Infrared Imager. The images shown are approximately 40 arcsec on a side, with *North up* and *East left*. The lack of a *Y*-band image implies a redshift greater than 7.8. A spectrum obtained by the Very Large Telescope (VLT) shows the Lyman-alpha break and a redshift determination of 8.26 (Tanvir et al. 2009)

Table 3-1
Units

Units	Radiometric name	Astronomical name
W	Flux	Luminosity
$W m^{-2}$	Irradiance; radiant exitance	Flux
$W sr^{-1}$	Intensity	–
$W m^{-2} sr^{-1}$	Radiance	Intensity
$W m^{-2} \mu m^{-1}$; $W m^{-2} Hz^{-1}$	Spectral irradiance	Flux density
$W m^{-2} \mu m^{-1} sr^{-1}$; $W m^{-2} Hz^{-1} sr^{-1}$	Spectral radiance	Surface brightness; specific intensity
$10^{-26} W m^{-2} Hz^{-1}$	–	Jansky (Jy)

From Tokunaga (2000); see also Rieke (2003)

and x-axis. Although it is common to see such mixed units in the literature, it should be avoided. To use Jy on the y-axis and show a wavelength scale, follow the example of *Fig. 3-2a*.

Proper comparison to physical models requires consistent units. See Soffer and Lynch (1999) for a discussion of how the choice of units affects our interpretation and perception of data.

1.2 Blackbody Radiation and Emissivity

Every object in the universe emits thermal radiation. For a perfect blackbody, the emissivity $(\varepsilon) = 1.0$ and the spectral radiance is given by the Planck function:

$$\begin{aligned} B_\lambda(T) &= 2hc^2\lambda^{-5}/(e^{hc/k\lambda T} - 1) \\ &= 1.1910 \times 10^8 \lambda_{\mu m}^{-5}/(e^{14387.7/\lambda_{\mu m} T} - 1)\ \mathrm{W\,m^{-2}\,\mu m^{-1} sr^{-1}}, \end{aligned}$$

where $\lambda_{\mu m}$ is the wavelength in micrometers.

In frequency units (ν in Hz),

$$\begin{aligned} B_\nu(T) &= 2h\nu^3 c^{-2}/(e^{h\nu/kT} - 1) \\ &= 1.4745 \times 10^{-50} \nu^3/(e^{4.79922\times 10^{-11}\nu/T} - 1)\ \mathrm{W\,m^{-2} Hz^{-1} sr^{-1}} \end{aligned}$$

Figure 3-6 shows spectral radiance of blackbodies for a range of temperatures. Real objects have an emissivity that is less than unity. For example, an aluminized mirror has a reflectivity of about 95–97% and by Kirchhoff's law, it therefore has an emissivity of about 3–5%. Thus, the spectral radiance of an aluminized mirror at room temperature is approximated by $0.04 B_\lambda(300\,\mathrm{K})$.

The Wien wavelength displacement law gives the wavelength of the maximum blackbody emission for a given temperature:

$$\lambda_{max}(\mu m) = 2,897.8/T(\mathrm{K})$$

From this equation the wavelength for the peak blackbody emission from a telescope at a temperature of 0°C is 10.6 μm. This has important consequences for observing in the mid infrared (see Sects. 3.2 and 6.4.2).

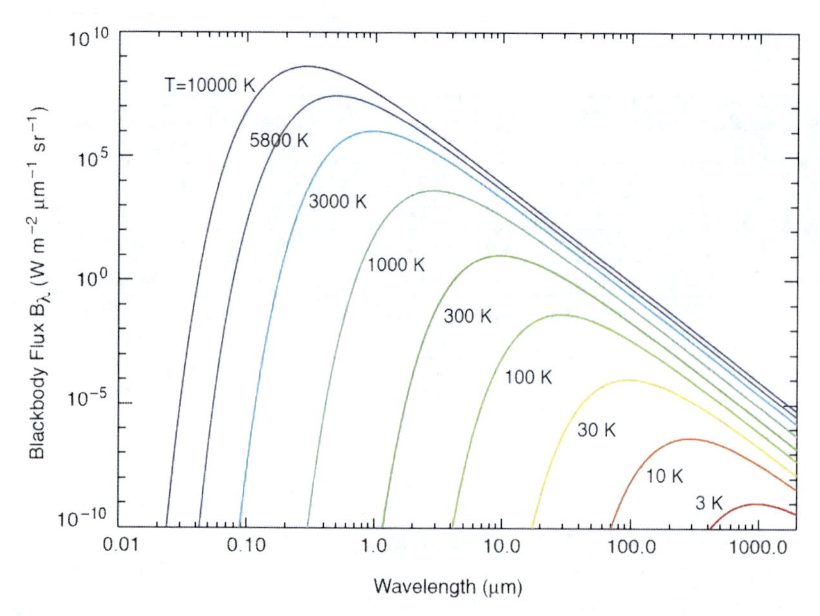

Fig. 3-6
Blackbody emission curves. The peak of the blackbody curve indicates the temperature of the object. Note that room-temperature objects peak near 10 μm. In *Fig. 3-2a*, **the dust emission in Arp 220 peaks near 100 μm, and therefore, the typical temperature of the dust is about 20 K**

It is frequently necessary to convert from F_λ to F_ν or vice versa, where F_λ and F_ν are the spectral irradiance (flux density). The conversion equations are (Ω is the solid angle in steradians):

$$F_\lambda\left(\mathrm{W\,m^{-2}\mu m^{-1}}\right)=\Omega\,B_\lambda,\quad F_\nu\left(\mathrm{W\,m^{-2}Hz^{-1}}\right)=\Omega\,B_\nu,\quad F_\lambda=3.0\times10^{14}F_\nu/\lambda^2_{\mu\mathrm{m}}.$$

Key References

Glass, I. S. 1999, Handbook of Infrared Astronomy (Cambridge: Cambridge University Press)
McLean, I. 2008, Electronic Imaging in Astronomy: Detectors and Instrumentation (2nd ed.; New York: Springer)
Rieke, G. H. 2009, History of infrared telescopes and astronomy. Exp. Astron., 25, 125
Tokunaga, A. T. 2000, Infrared astronomy, in Allen's Astrophysical Quantities, ed. A. N. Cox (4th ed.; New York: Springer), 143

2 Overview of Atmospheric Transmission from 1 to 1,000 μm

IR astronomy from the ground is largely constrained by the absorption and emission of Earth's atmosphere and the thermal emission from the telescope. Many of the techniques employed in IR observations are aimed at mitigating the effects of these limitations.

Major atmospheric absorbers and central wavelengths of absorption bands are as follows: H_2O (0.94, 1.12, 1.37, 1.87, 2.7, 3.2, 6.3, $\lambda > 25 \mu m$); CO_2 (2.0, 2.7, 4.3, 15 μm); CO (4.7μm); CH_4 (3.3, 6.5, 7.7 μm); and O_3 (4.7, 9.6 μm). See Crisp (2000) and Killinger et al. (1995) for more information.

Water vapor accounts for much of the telluric absorption. This is a major problem because the water vapor content in the line of sight to the celestial object is constantly changing. At mid-IR wavelengths the sky emission is highly variable on short time scales (fraction of a second) and background subtraction is needed on time scales of 0.3–0.5 s or shorter for imaging.

A low-resolution spectrum of Earth's atmospheric transmission at 1–20 μm is shown in ❯ *Fig. 3-7.*

High-resolution model transmission spectra of Earth's atmosphere computed at 0.9–26 μm for various airmasses, water vapor columns, and altitudes are provided at the Gemini Observatory website: http://www.gemini.edu/?q=node/10789.

The transmission of Earth's atmosphere from 1 to 1,000 μm at different altitudes is shown in ❯ *Fig. 3-8.* This figure shows the advantage of going to higher altitudes, especially to airborne (14 km) and balloon altitudes (>28 km), and of course in going completely outside of the atmosphere.

A higher-resolution calculated spectrum of the atmospheric transmission at a high ground-based site and airborne telescope is shown in ❯ *Fig. 3-9.*

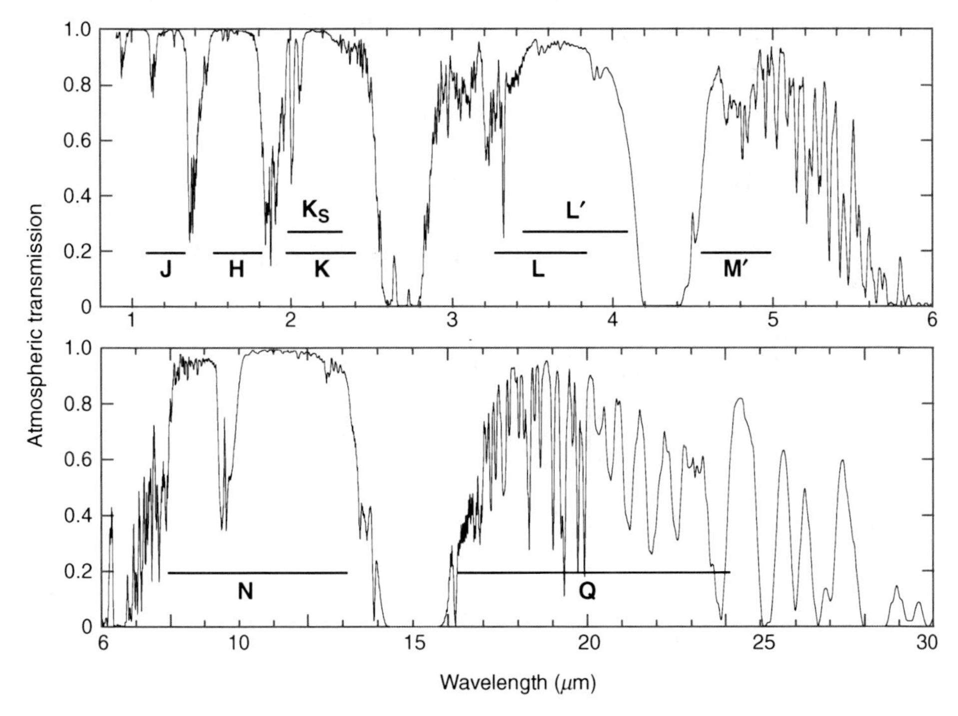

❏ Fig. 3-7

Model atmospheric transmission at 1–20 μm at Mauna Kea (altitude = 4.2 km, airmass = 1.15, precipitable water vapor = 1 mm). The passbands of commonly used filters are shown as *horizontal bars* (Tokunaga 2000). For a discussion of the effects of the atmosphere on the filter effective wavelengths, see Manduca and Bell (1979)

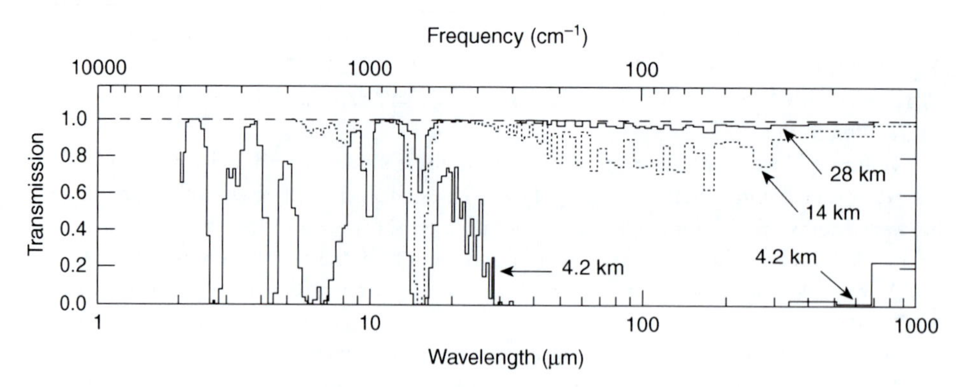

Fig. 3-8
Model atmospheric transmission from 1 to 1,000 μm showing the average transmission at a high ground-based site (4.2 km), airborne telescope (14 km), and balloon-borne telescope (28 km) (Figure adapted from Traub and Stier (1976). See this reference for high-spectral resolution plots of the atmospheric transmission)

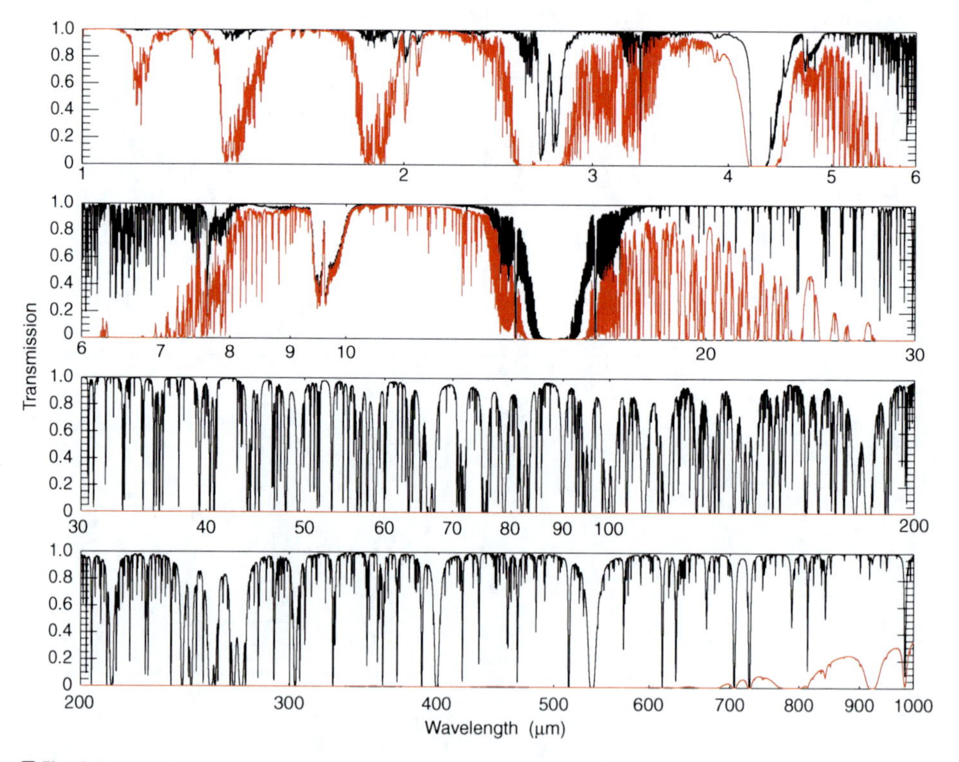

Fig. 3-9
Comparison of the atmospheric transmission at Mauna Kea (*red curve*) and at the altitude the *Stratospheric Observatory for Far-Infrared Astronomy* (*SOFIA; black curve*). Assumptions: For Mauna Kea, altitude = 4.2 km, precipitable water vapor (PWV) = 3.4 mm, airmass = 1.4. For *SOFIA*, altitude = 12.5 km, PWV = 7.3 μm, airmass = 1.4. With PWV of about 1 mm or less, observations at 350 μm are possible from Mauna Kea. Calculated using ATRAN (Lord 1992)

The strong telluric absorption over much of the 1–1,000-μm spectral range has driven the quest to build IR telescopes at high sites in Hawaii and Chile. A comparison of the major sites in Hawaii, Chile, and Mexico that were considered for the Thirty Meter Telescope is given by Schöck et al. (2009) and subsequent papers in this series. The site in Hawaii was selected for the Thirty Meter Telescope. A similar study of sites for the European Extremely Large Telescope (E-ELT) was conducted by Vernin et al. (2011) and Kerber et al. (2010) and the Cerro Armazones site in Chile near the Cerro Paranal site of ESO's VLT was selected. The site for the Giant Magellan Telescope is Cerro Las Campanas; the site characteristics were summarized by Thomas-Osip et al. (2010). For similar reasons, the Atacama Large Millimeter Array (ALMA) is located at a 5,000-m altitude site. In addition a very high-altitude site at Llano de Chajnantor in the Atacama Desert of northern Chile is being considered for a large millimeter and submillimeter telescope (Sebring 2010). Characteristics of this site are discussed by Giovanelli et al. (2001a, b).

➋ *Figures 3-8* and ➋ *3-9* show that the 350–1,000-μm spectral region is accessible from the ground. This allowed pioneering observations to be made at wavelengths longer than 350 μm from high altitude sites in Hawaii, Chile, and the South Pole. As an example, ➋ *Fig. 3-10* shows the computed atmospheric transmission from Llano de Chajnantor. At such a high site, observations at 30 μm are possible (Miyata et al. 2008).

The advantages of the South Pole for IR astronomy are well known (Burton 2010) and sites at the South Pole have much lower water vapor than the high sites in Chile. The much lower atmospheric attenuation due to water vapor and very low thermal background are the most critical advantages for infrared and submillimeter astronomy. In addition there is a minimum in the combination of airglow and thermal emission at ~2.4 μm – there are no airglow lines and the thermal emission is very low. At this wavelength, the measured sky surface brightness at this wavelength is 20–50 times lower than Siding Spring (Lawrence et al. 2002; Phillips et al. 1999). At 3–5 μm, the measured sky surface brightness is 10–100 times lower than at Siding Spring in Australia and Mauna Kea in Hawaii (Phillips et al. 1999). These measurements refer to the South Pole station. Other sites being developed at Dome C and Dome A are superior to the South Pole station except for the turbulence in the free atmosphere (Saunders et al. 2009).

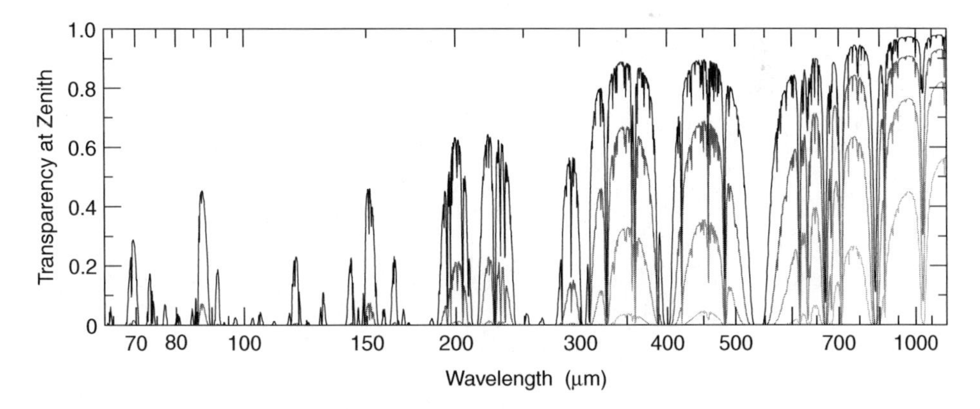

Fig. 3-10

Computed atmospheric transparency at zenith between 70 and 1,000 μm for a site at an altitude of 5,000 m (Llano de Chajnantor) and different H_2O column densities: The tracings, *top* to *bottom*, correspond to 0.1, 0.4, 1.0, and 3.0 mm of PWV (Giovanelli et al. 2001b). The typical PWV at this site is 1.2 mm

In addition to going into space, airborne and balloon-borne telescopes have been developed. *SOFIA* became operational in 2010. It allows frequent access to nearly the entire infrared spectral range, 1–1,000 μm (see ➋ Sect. 6.5.1). Balloon-borne telescopes carry out experiments at even higher altitudes, up to approximately 40 km.

2.1 Atmospheric Extinction

The atmospheric extinction consists of Rayleigh scattering by molecules, molecular absorption, and aerosol scattering by particulates, as shown in ➋ *Fig. 3-11*. (Hayes and Latham 1975; Killinger et al. 1995). These are highly dependent on the site, seasonal weather patterns, and natural events such as dust storms and volcanic eruptions. Long-term extinction trends show these effects clearly as discussed by Lombardi et al. (2011) and Burki et al. (1995) for ESO and Frogel (1998) for CTIO. As emphasized by Frogel, extinction coefficients in the infrared vary considerably because of the effects of water vapor absorption; thus, the extinction coefficient should be measured throughout the night for accurate photometry.

Infrared photometric observations are reduced to an airmass of unity and are not extrapolated to zero airmass. The extrapolation to zero airmass is nonlinear because the atmospheric absorption bands in the infrared are saturated. Calculations of this are shown explicitly by Manduca and Bell (1979), Volk et al. (1989), and Tokunaga and Vacca (2007). Thus, caution is advised if extrapolation to above the atmosphere is required when using broadband filters. Narrowband filters in regions clear of atmospheric absorption are used when it is necessary to extrapolate to above the atmosphere as in absolute calibration experiments comparing a

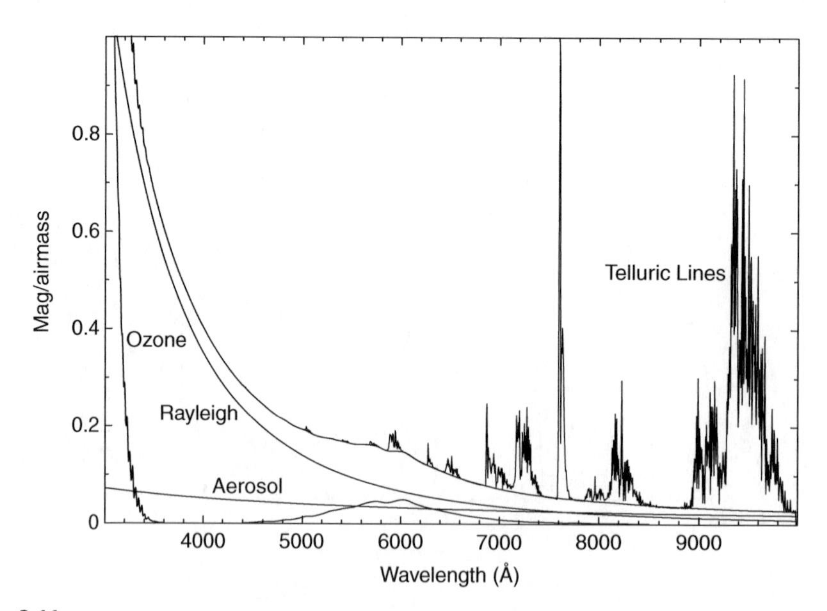

Fig. 3-11

Simulation of typical extinction by Earth's atmosphere, showing the various components. Note that in the near infrared all sources of extinction are very small compared with those in the visible (Smalley et al. 2007)

calibrated blackbody to a stellar source. See for example the absolute calibration work by Blackwell et al. (1983) and the use of narrowband filters proposed by Milone and Young (2007).

2.2 Atmospheric Refraction

There are practical difficulties introduced by atmospheric refraction (atmospheric dispersion) over a wide wavelength range. First of all, guiding is typically done with a CCD (i.e., at optical wavelengths), while the observations are conducted in the infrared. Thus, the IR object will drift in the field as the airmass changes. This limits the integration time unless steps are taken to compensate for the atmospheric differential refraction. Roe (2002) discusses the calculation of the effect and the implications for observations with adaptive optics. The effects are especially severe for the next generation of large ground-based telescopes, where corrections at the milliarcsecond level are critical (Kendrew et al. 2008; Phillips et al. 2010; Zuther et al. 2010).

The problem of differential refraction is most severe for spectroscopy, where the light loss will be a function of wavelength due to atmospheric refraction unless the slit is oriented along the direction of the atmospheric refraction. It is also a problem for adaptive optics (AO) observations where the AO system typically measures the wavefront distortions with a visible light sensor, while the observations are conducted in the near infrared. Secondly, for wide-bandpass imaging the images could be smeared out due to atmospheric refraction. One example of the atmospheric refraction effects is shown in ➋ *Fig. 3-12* (Phillips et al. 2010).

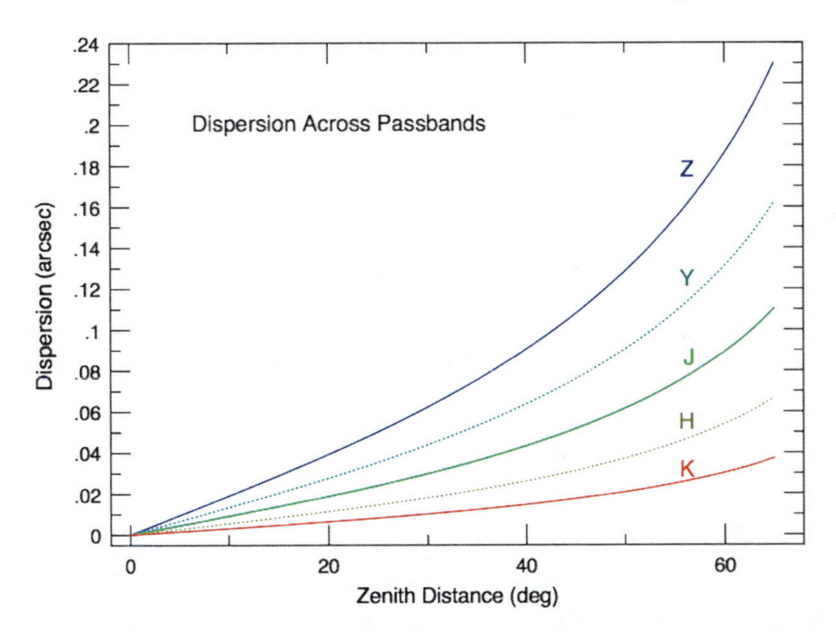

◘ Fig. 3-12
Atmospheric dispersion within the various passbands as a function of zenith distance. This figure shows the dispersion from the short to long wavelength transmitted by the filter. Thus, a point source observed through the *J* filter would be spread out by about 30 milliarcsec at an angle of 30° from the zenith. The dispersion has been calculated for the Thirty Meter Telescope site on Mauna Kea (Phillips et al. 2010)

For a spectrograph the slit should be oriented parallel to the atmospheric refraction if observing only a single object in the slit. This angle on the sky is perpendicular to the horizon and is known as the parallactic angle. Computation of this angle is discussed by Filippenko (1982) and Stone (1996).

Mid-IR observations typically rely on visible guide cameras while observing at 10 μm. The atmospheric dispersion effects need to be compensated for but the calculation of the effect is hampered by a lack of suitable data. See Livengood et al. (1999) and Skemer et al. (2009) for an empirical measurement of the atmospheric differential refraction in the mid infrared.

Key References

Hayes, D. S., & Latham, D. W. 1975, A rediscussion of the atmospheric extinction and the absolute spectral-energy distribution of Vega. ApJ, 197, 593. Good discussion of atmospheric extinction

Killinger, D. K., Churnside, J. H., & Rothman, L. S. 1995, Atmospheric Optics, in Handbook of Optics, ed. M. Bass, E. W. V. Stryland, D. R. Williams, & W. L. Wolfe (New York: McGraw-Hill), 44.1 Discussion of properties of the atmosphere

Lombardi, G., Mason, E., Lidman, C., Jaunsen, A. O., & Smette, A. 2011, A study of near-IR atmospheric properties at Paranal Observatory. A&A, 528, 43. Quantitative analysis of factors affecting the near-IR extinction

3 Background Emission from the Ground

3.1 Near-IR Airglow

Strong emission from the hydroxyl radical OH in the visible and up to 2.3 μm is a source of background that limits the sensitivity of near-IR observations. At high-spectral resolution the OH emission is a series of very narrow lines as shown in *Fig. 3-13*.

OH emission displays spatial and 5–10% temporal variability on time scales of minutes (see, e.g., Ramsay et al. 1992; Frey et al. 2000; High et al. 2010). Thus sky subtraction at a frequency of 1 or 2 min is required for spectroscopy and imaging to have good subtraction of the OH lines as well as to avoid saturating on the OH lines. Movies of the OH emission by J. Adams and M. Strutskie are shown at http://astsun.astro.virginia.edu/~mfs4n/2mass/airglow/airglow.html. They find the airglow can vary by up to a factor of 3 during the night and it is strongest at twilight and decreases by about 0.6 mag during the night. Patat (2008) discusses the time variability of the OH emission lines at visible wavelengths and provides a movie of the variability in the OH emission lines at http://www.eso.org/~fpatat/science/skybright/. This movie vividly demonstrates the necessity for frequent sky subtraction.

The OH lines are not as closely packed as *Fig. 3-13* suggests. Spectra of the OH emission have been published by Oliva and Origlia (1992) and Rousselot et al. (2000; see *Fig. 3-14*). The latter shows that at high spectral resolution the intervals between the lines are very dark. This has led to instruments and concepts to reduce the OH emission by various means to get to the level of sensitivity limited by the background between the OH lines.

The near-IR sky surface brightness at major observatories is given in *Table 3-2*.

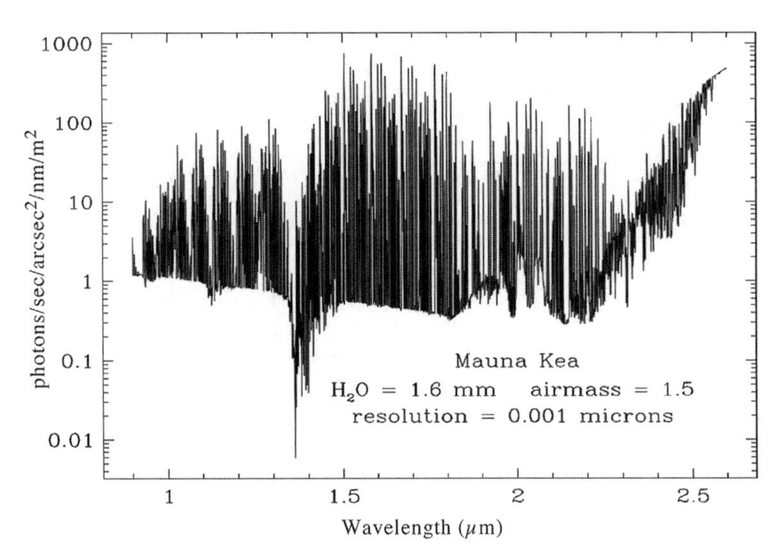

Fig. 3-13
High-resolution model of the sky emission, showing the OH lines, the zodiacal continuum emission (approximated by a 5,800-K black body), and thermal emission from the atmosphere (assumed to be a 273-K black body for Mauna Kea) (From: http://www.gemini.edu/node/10787)

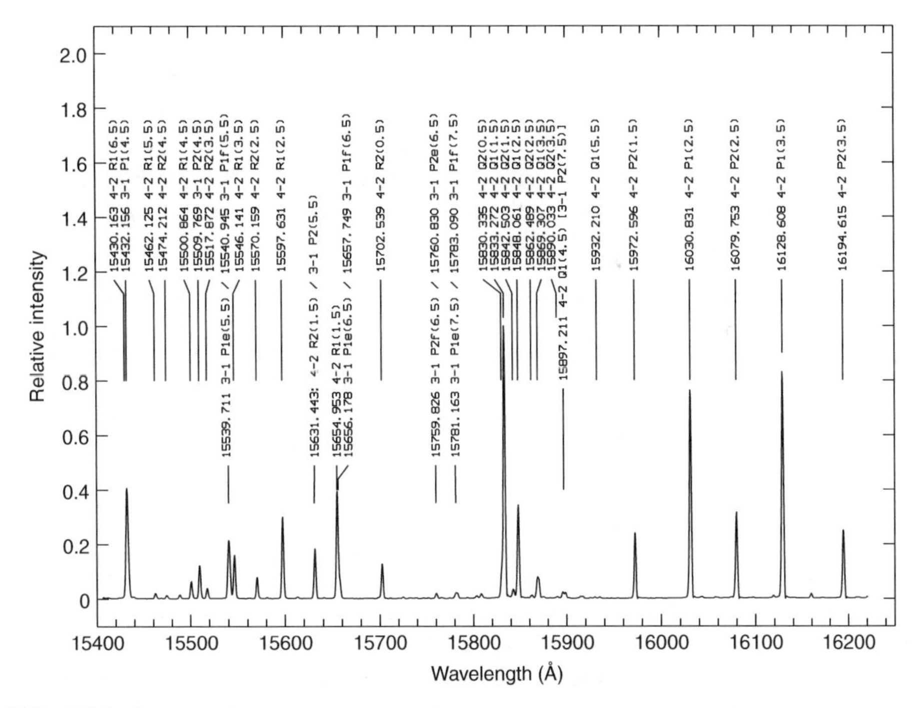

Fig. 3-14
Atlas of the OH emission spectrum at a resolving power of 8,000 intended for wavelength calibration (Rousselot et al. 2000). Laboratory data of OH line positions are given by Maillard et al. (1976) and Abrams et al. (1994)

Table 3-2
The night sky surface brightness (mag/arcsec^2)

Descriptio	Altitude (m)	Type	*J*	*H*	K_s or *K*
La Palma	2,500	Average	15.5	14.0	12.6
Paranal	2,635	Darkest	16.5	14.4	13.0
		Darkest	–	–	13.5
Cerro Pachon	2,200	–	16.0	13.9	13.5
Mt. Graham	1,926	–	–	–	13.5
Mauna Kea	4,200	Average	15.6	14.0	13.4
		Darkest	16.75	14.75	14.75
Mt. Hamilton	1,283	–	16.0	14.0	13.0
Kitt Peak	2,096	–	15.7	13.9	13.1
Anglo-Australian Obs	1,164	–	15.7	14.1	13.5
South Pole	2,800	–	16.4	14.7	15.3
		Darkest	16.8	15.2	15.8

Note: Sky brightness due to OH is highly variable. Data from Sánchez et al. (2008), Table 4. South Pole values from Phillips et al. (1999). Kenyon and Storey (2006) expect the South Pole sky brightness to be comparable to other sites such as Mauna Kea. The cold temperature at the South Pole reduces thermal emission in the K band, so the apparent sky brightness is lower than at other sites

3.2 Thermal Emission from Sky and Telescope

To visualize the thermal emission from the sky and the telescope, imagine putting your eye at the focus of a telescope. Looking upward at the sky, you will see the secondary mirror, which will reflect an image of the spiders, the hole in the primary mirror, and the sky. Each element (including the telescope mirrors) will emit thermal radiation toward the detector. Each item can be approximated by the product of the Planck function and an emissivity factor. For the telescope mirrors and spiders,

$$\text{Telescope Intensity} = \varepsilon_{\text{tel}} B_\lambda(T_{\text{tel}}) A_{\text{tel}} + B_\lambda(T_{\text{tel}}) A_{\text{spiders}} + B_\lambda(T_{\text{tel}}) A_{\text{hole}} \ \ \text{Wm}^{-2}\text{sr}^{-1},$$

where ε_{tel} is the total emissivity of the telescope mirrors, B_λ is the Planck function, T_{tel} is the telescope temperature, A_{tel} is the telescope area, A_{spiders} is the area of the secondary spiders, and A_{hole} is the area of the hole in the primary mirror For an IR-optimized telescope like the NASA Infrared Telescope Facility the second and third terms are much smaller than the first term. For the sky emission,

$$\text{Sky Intensity} = (1 - e^{-\tau}) B_\lambda(T_{\text{sky}}) A_{\text{tel}},$$

where τ is the optical depth of the atmosphere and T_{sky} is the characteristic sky temperature. A good approximation of the atmospheric characteristic temperature at Mauna Kea is 250 K. The optical depth is typically calculated using a line-by-line atmospheric transmission program such at ATRAN or HITRAN. Note that $(1 - e^{-\tau})$ is the atmospheric transmission. This is a single slab model for the atmosphere. In order to have a more realistic accounting of the sky emission, a line-by-line calculation and a realistic model of the sky temperature profile is required.

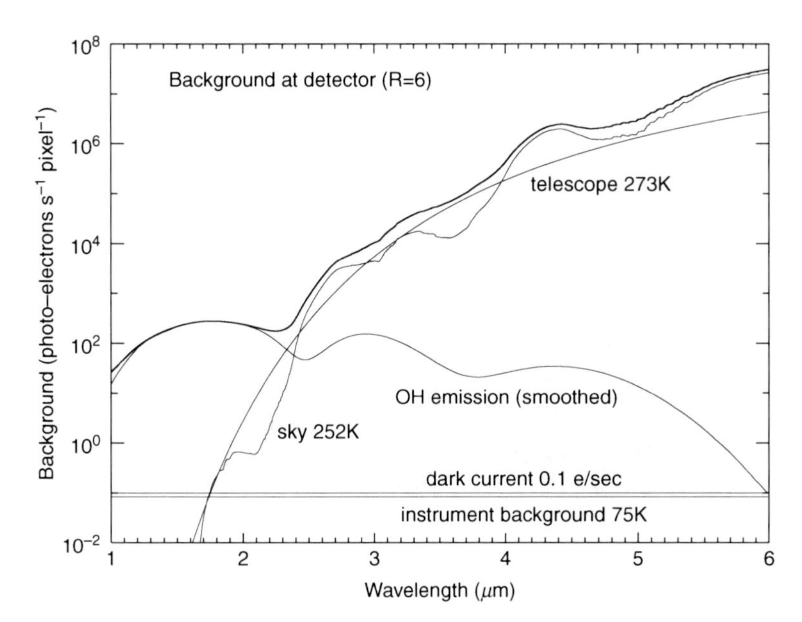

Fig. 3-15
Calculated background from the sky, instrument, and telescope at a resolving power of 6 calculated by J. Rayner. A 3.0-m telescope is assumed, with 0.1 emissivity, telescope temperature of 273 K, sky effective temperature of 252 K, total throughput (telescope, instrument, detector quantum efficiency) of 0.1, pixel size of 0.125″ and a slit width of 3 pixels. The emissivity of the sky is estimated as 1 – atmospheric transmission, where the atmospheric transmission was computed from the ATRAN software (Lord 1992) using an airmass of 1.5 and a precipitable water vapor value of 2.0 mm. The instrument is assumed to be cooled to 75 K so that the background integrated over all wavelengths from the instrument is below that of the dark current. Note that the required instrument temperature also depends on the long wavelength cutoff of the detector (see, e.g., Yasui et al. 2008)

Examples of the sky background at imaging and high spectral resolution are shown in *Figs. 3-15* and *3-16* The sky background at mid-infrared wavelengths is shown in *Fig. 3-17*.

The rise in the background starting at about 1.6 μm is a result of the thermal emission from the telescope. For broadband imaging, the thermal emission from the telescope and instrument becomes larger than the detector dark current at wavelengths longer than about 1.6 μm (in between the OH lines). This is the wavelength where the telescope thermal emission curve crosses the dark current level in *Fig. 3-14*. Thus, cooling of the instrument structure and optics is not required for wavelengths less than 1.6 μm. For high-resolution spectroscopy (R = 70,000), the thermal emission from the sky and telescope is greater than the detector dark current for wavelengths greater than 2.5 μm (see *Fig. 3-16*).

At wavelengths longer than 3 μm the sky and telescope background emission limits the signal-to-noise ratio that can be achieved from the ground. The large increase in the background emission is shown in *Figs. 3-3* and *3-4*. The fluctuations in the sky background due to the atmosphere are the major problem at thermal wavelengths. Thus, techniques discussed in Sect. 6.4 are employed to reduce the effects of a changing background radiance.

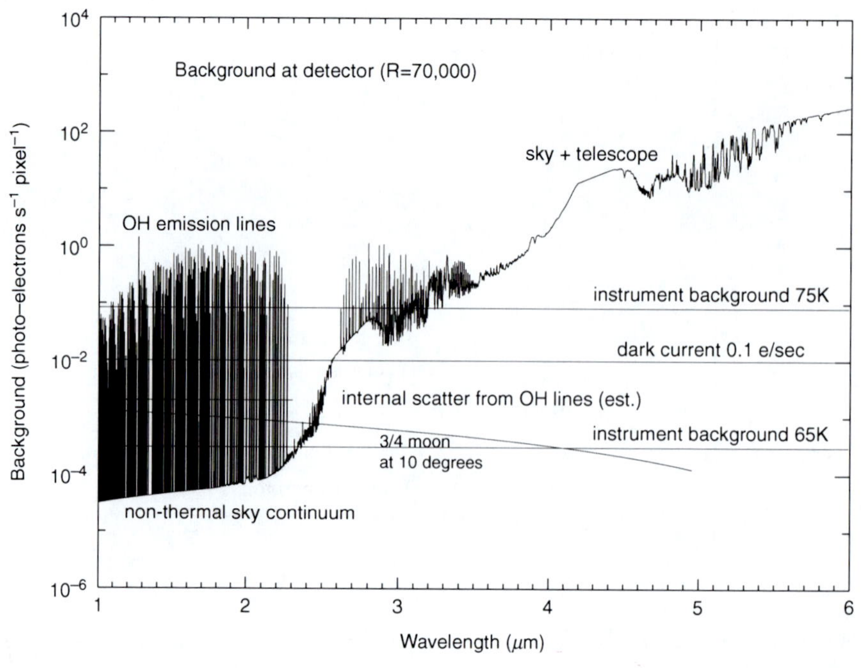

Fig. 3-16

Calculated background from the sky, instrument, and telescope at a resolving power of 70,000 calculated by J. Rayner. Same parameters as in ***Fig. 3-15*****. The OH lines are from McCaughrean (1988). The nonthermal continuum emission between the OH lines is from Maihara et al. (1993). An instrument model was used to estimate the internal scattered light from OH. The instrument is assumed to be cooled to 65 K so that the background integrated over all wavelengths from the instrument is below that of other sources**

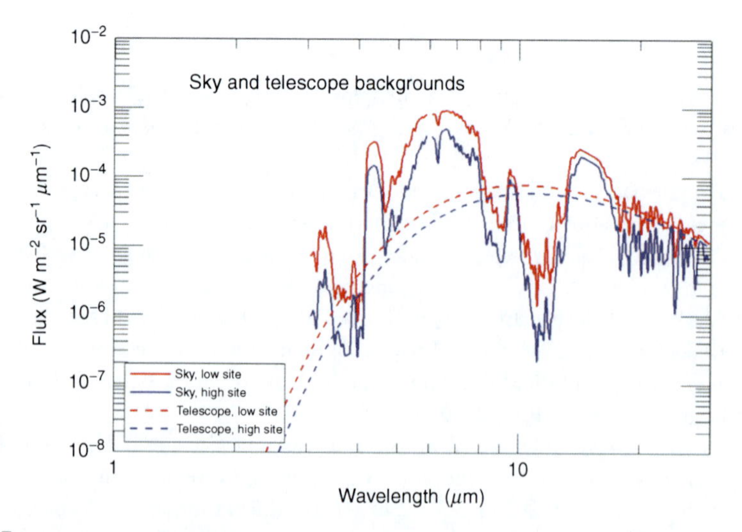

Fig. 3-17

Radiance from telescope (*dashed lines*) and sky (*solid lines*) for Paranal (*red*) and a higher and drier (*blue*) observing site at 5,000 m (Kendrew et al. 2010)

In ➊ *Fig. 3-16* the Moon brightness was scaled from Krisciunas and Schaefer (1991) assuming the color of the Moon is that of a G2V star. This is a rough estimate and it is well below the level of the dark current and so is negligible.

Key References

Leinert, C., Bowyer, S., Haikala, L. K., et al. 1998, The 1997 reference of diffuse night sky brightness. A&AS, 127, 1. Comprehensive discussion of the night sky brightness and backgrounds in space from the ultraviolet to the far infrared

4 Background Emission from Space

The infrared background as seen from space consists of a number of distinct components. Although numerous attempts to accurately measure this background were conducted from balloons and rockets, accurate full sky images of the far-IR background were not produced until the launching of the *Infrared Astronomical Satellite* (*IRAS*) in 1983 (Neugebauer et al. 1984). Even though *IRAS* did not have a cold shutter to establish a reliable zero point, variations in detector response due to ionizing radiation were used to infer this zero point (Beichman et al. 1988). The *IRAS* data showed the structure of the far-IR emission of dust in interplanetary space, the contribution of dust from the Milky Way galaxy, and the possibility of an extragalactic component at 100 μm, the longest wavelength observed by *IRAS* (Rowan-Robinson et al. 1990).

The *Cosmic Background Explorer* (*COBE*) satellite, launched in 1989, was specifically designed to measure the absolute background from the near infrared to millimeter wavelengths (Boggess et al. 1992). *COBE* had three instruments. The Diffuse Infrared Background Experiment (DIRBE) was designed to measure the background between 1.25 and 240 μm in 10 bands with a $0.7^\circ \times 0.7^\circ$ beam. The Far Infrared Absolute Spectrometer (FIRAS) was a polarizing Michelson interferometer that produced low-resolution spectra between 125 μm and 10 mm. FIRAS had a 7° beam. The Diffuse Microwave Radiometer (DMR) was designed to measure primeval fluctuations in the cosmic microwave background (CMB) at 31.5, 53, and 90 GHz.

➊ *Figure 3-18* shows the annually averaged maps from DIRBE for four wavelengths, 3.5, 25, 100, and 240 μm, plotted in Galactic coordinates for a solar elongation angle of 90°. At 3.5 μm, the background is primarily starlight in the Galactic plane and thermal emission from zodiacal dust in the solar system, which traces out the "S"-shaped pattern in this projection. At 25 μm, the zodiacal dust is dominant, with a much more modest contribution coming from stars and HII regions. At 100 μm, the relative zodiacal contribution becomes small, and at 240 μm, the emission is dominated by dust in the Galaxy. The annual average data smooth out the variations in the emission due to Earth's slightly eccentric orbital motion.

The zodiacal emission provides valuable insight into the dust processes in the inner solar system. It has been modeled by a number of workers including Reach et al. (1996) and Fixsen and Dwek (2002). Because the interplanetary dust is subject to drag from the Poynting-Robertson effect, this dust must be replenished via collisions in the asteroid belt, deposition from comets, etc. Fixsen and Dwek find that $\sim 10^{11}$ kg year^{-1} of dust needs to be generated to maintain the zodiacal cloud. At the shortest wavelengths, the zodiacal light is due to scattered sunlight rather than thermal emission from the dust. A detailed summary of the zodiacal observations including temporal and spatial variations can be found in Leinert et al. (1998).

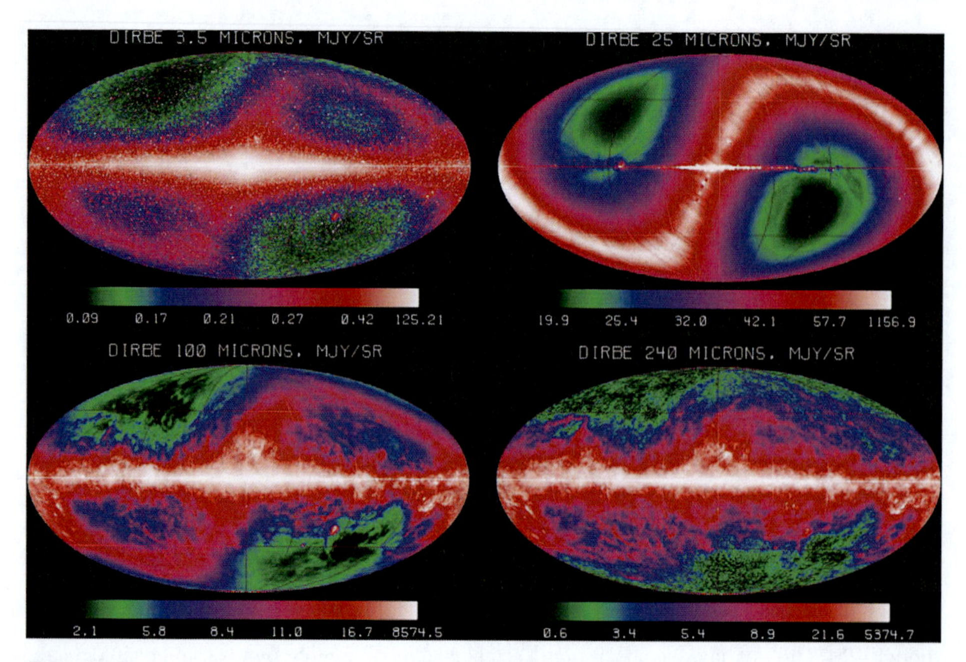

Fig. 3-18
DIRBE annual average maps at 3.5, 25, 100, and 240 μm. Galactic coordinate Mollweide projection maps of the entire sky at four wavelengths showing emission from stars and dust in the Galactic plane (*horizontal* feature) and *light* scattered and emitted by dust in the solar system (S-shape) (COBE Slide Set, Legacy Archive for Microwave Background Data Analysis, NASA Goddard Space Flight Center)

At wavelengths longer than ~50 μm, thermal emission from dust in our Galaxy becomes dominant. This emission comes from dust in both clouds and the more diffuse, filamentary IR "cirrus" (Low et al. 1984; see *Figs. 3-3* and *3-19*). In recent years, images of unprecedented detail have been obtained with the *Herschel Space Observatory*, enabling enhanced studies of the structure of the interstellar medium (Martin et al. 2010; Miville-Deschênes et al. 2010). *Figure 3-19* is an image of the Rosette Nebula that illustrates the interaction of the cloud with the young stars being formed in the nebula and shows both emission from the molecular cloud and the cirrus.

Besides these "local" components, there is a contribution from extragalactic sources. The extragalactic infrared background provides important observational constraints on the population of infrared galaxies and the early formation of galaxies in the universe. In practice, the determination of this cosmic infrared background is challenging because it is difficult to remove the foreground emission. For example, the zodiacal emission at 25 μm is roughly three orders of magnitude larger than the extragalactic component, making foreground modeling the largest uncertainty. Observational evidence for this infrared background was first reported by Puget et al. (1996) and Hauser et al. (1998). The situation regarding this extragalactic background in the infrared was reviewed by Hauser and Dwek (2001), who attribute the bulk of this background to infrared galaxies. The study of the cosmic infrared background light is an active area of research, and it has been revolutionized with the advent of the datasets from *Herschel* and

Fig. 3-19
Image of the Rosette Nebula obtained with the *Herschel Space Telescope*. This image is a three-color composite showing infrared wavelengths of 70 μm (*blue*), 160 μm (*green*), and 250 μm (*red*). It was made with observations from *Herschel*'s Photodetector Array Camera and Spectrometer (PACS) and the Spectral and Photometric Imaging REceiver (SPIRE) instruments (ESA and the PACS, SPIRE and HSC Consortia)

Planck (a European Space Agency satellite studying fluctuations in the CMB) (e.g., Béthermin and Dole 2011; Planck Collaboration et al. 2011b, c).

Figure 3-20 illustrates a simple decomposition of the infrared background in the direction of the ecliptic pole. Scattered sunlight is the dominant background at wavelengths shorter than 3 μm. Between 3 and 100 μm, thermal emission from the zodiacal dust is most important. It has long been noted that there is a local minimum in the background near 3.5 μm. This local minimum has been exploited in conducting very deep surveys with the IRAC camera on the *Spitzer Space Telescope* and will also be an important wavelength band for the *James Webb Space Telescope* (*JWST*).

The much colder far-infrared background (FIRB) is roughly half due to background infrared galaxies and half due to Galactic dust. Moreover, the FIRB spectrum is much more complicated than this simple blackbody estimate due to emission bands in the IR cirrus, as shown in *Fig. 3-21* (Compiègne et al. 2011).

Key References

Hauser, M. G., & Dwek, E. 2001, The cosmic infrared background: measurements and implications. ARA&A, 39, 249

Leinert, C., Bowyer, S., Haikala, L. K., et al. 1998, The 1997 reference of diffuse night sky brightness. A&AS, 127, 1

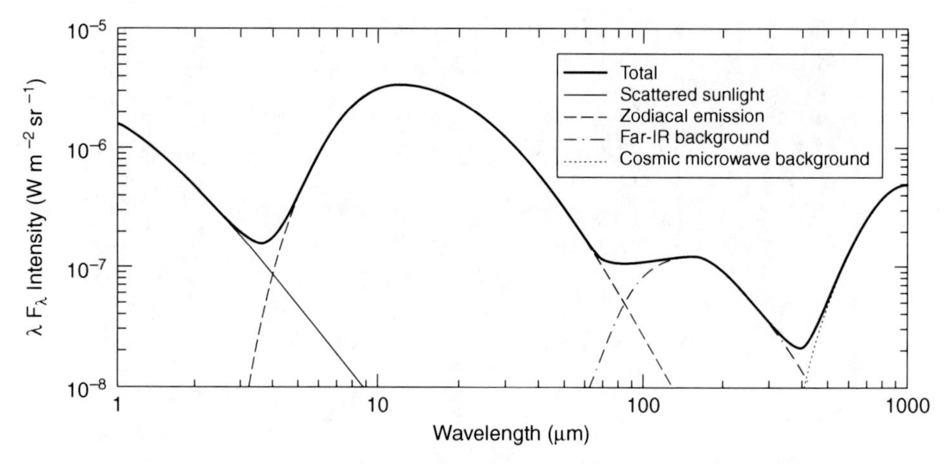

Fig. 3-20
Decomposition of the infrared background to illustrate the key components seen at the ecliptic pole 1 AU from the Sun. The scattered sunlight assumes gray scattering with an optical depth of 1.7×10^{-13} for the zodiacal dust particles. The zodiacal thermal emission uses a temperature of 290 K and a λ^{-1} emissivity beyond 40 μm. The far-infrared background has a temperature of 17 K and a λ^{-2} emissivity law beyond 100 μm. The parameters were adjusted to fit the DIRBE data

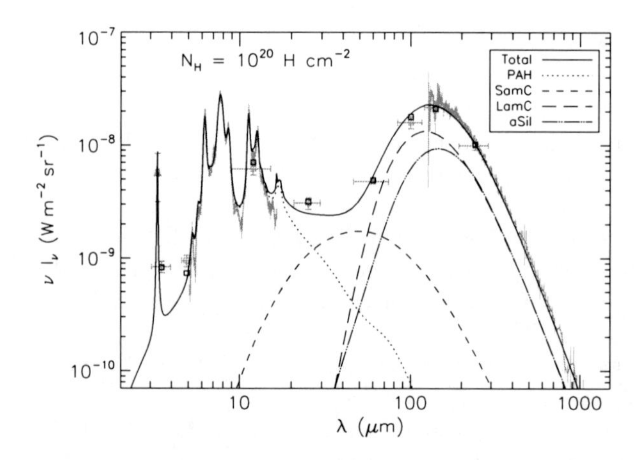

Fig. 3-21
Spectrum of the diffuse interstellar medium at high Galactic latitude. The mid-IR (~5–15 μm) and far-IR (~100–1,000 μm) spectra are from ISOCAM/CVF on the Infrared Space Observatory (ISO) and FIRAS (COBE), respectively. The *triangle* at 3.3 μm is a narrowband measurement from the AROME balloon experiment. *Squares* are the photometric measurements from DIRBE (COBE). *Black lines* are the model output and *black squares* the modeled DIRBE points, taking into account instrumental transmission and color corrections. *PAH* polycyclic aromatic hydrocarbon, *SamC* very small carbon grains, *LamC* large carbon grains, *aSil* silicate grains (Compiègne et al. 2011)

5 Detectors Used in Infrared Astronomy

The IR spectral range covers over the three orders of magnitude in wavelength and therefore a wide range of different physical phenomena have been employed to detect IR radiation. The topic of IR detectors is large, and a number of excellent reviews of the technology are available. In particular, Rieke (2003, 2007) and McLean (2008) provide comprehensive coverage of recent developments. The chapters on detectors in this book series also provide more detailed coverage of the subject.

5.1 Thermal Detectors

The earliest IR detectors were thermal detectors. The IR radiation produces a change in the temperature of the detector, and this temperature change is measured. Depending on the sophistication of the temperature sensor, the sensitivity can be very low (for the case of William Herschel's thermometer) to exquisitely high (for modern superconducting sensors.) The most important form of thermal detector is the bolometer. All bolometers share common characteristics, as depicted in ➋ *Fig. 3-22*. First, there is a sensing element that intercepts the IR radiation. It is isolated from a cold sink by a link of thermal conductance (G). When radiation hits the absorber, its temperature rises and this is sensed by the thermometer, for example by a change in resistance or change in volume.

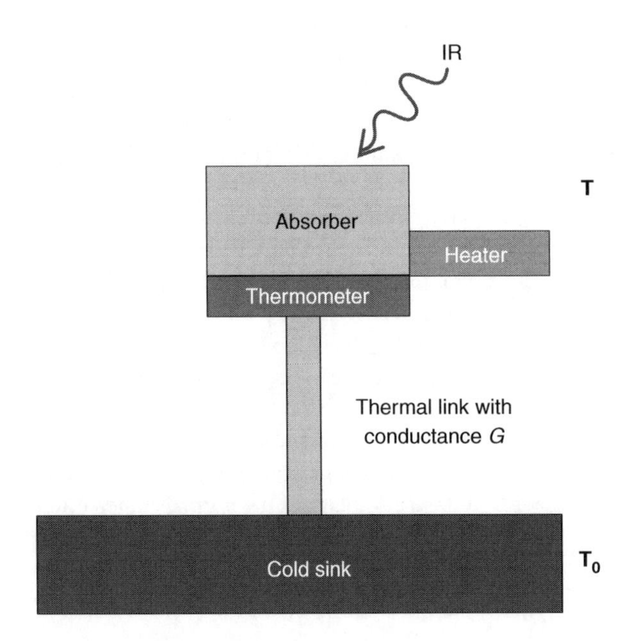

Fig. 3-22

Schematic representation of a bolometer. The absorber receives infrared energy and the rise in temperature is sensed by the thermometer. The absorber is connected to the cold sink which is at temperature T_0, by a thermal link of conductance *G*. In some designs, power in the heater is modulated by a feedback signal from the thermometer to maintain a constant absorber temperature *T*

The fundamental limit on the performance of a bolometer is from the thermal fluctuations in the thermal link between the cold sink and the absorber. Hence, all high performance bolometer systems are cooled to very low temperatures, often into the sub-Kelvin regime. Most of the design effort in bolometers has been in producing the most sensitive thermometers, the smallest thermal links, and the smallest thermal masses of the absorber and thermometer. The key figure of merit for a bolometer is the power of the infrared signal that is equal to the noise of the bolometer. The noise equivalent power (NEP) of a bolometer with no background is given by

$$\mathrm{NEP} = (4kT^2G)^{0.5}/\eta,$$

where k is Boltzmann's constant, T is the temperature, G is the thermal conductance, and η is the quantum efficiency (See Mather (1982) for a more detailed discussion).

Various types of thermometers have been used in the astronomical bolometer. The first highly sensitive bolometers were developed by Low (1961) and utilized gallium-doped germanium as the thermometer. When cooled to liquid helium temperatures, this material exhibits large changes in conductivity with temperature. More recently, doped-silicon bolometers have been used on a number of instruments, most notably, the PACS instrument on *Herschel* and the High Frequency Instrument (HFI) on the *Planck* mission (Planck HFI Core Team et al. 2011).

Superconducting thin films have also been used in astronomical bolometers. These systems take advantage of the very steep dependence of the film resistance with temperature near the superconducting transition. By sensing the changes in current flow with changes in illumination, these transition edge sensor (TES) bolometers make exceedingly sensitive thermal detectors. The most advanced TES detectors utilize superconducting quantum interference devices (SQUIDs) to measure the current flowing in the circuit (e.g., Benford et al. 2004). Recent examples of TES detectors are the Goddard IRAM Superconducting 2-mm Observer (GISMO) (Staguhn et al. 2008) and the Submillimetre Common-User Bolometer Array (SCUBA-2) instrument (Bintley et al. 2010).

The desire to maximize sensitivity by making the thermal conductance G as small as possible is complicated by corresponding increase in the thermal time constant $\tau = C/G$ of the bolometer, where C is the heat capacity of the bolometer element. An important improvement in the response speed of bolometers was realized by making the bolometer part of a thermal feedback circuit, as illustrated in the ➋ *Fig. 3-22*. The bolometric element is heated slightly above the bath temperature using a heater. Any required temperature changes due to the absorption of infrared radiation are sensed by the thermometer and a corresponding reduction in the heater power is applied by the feedback circuit. This electrothermal feedback reduces the effect of the heat capacity of the bolometer since the temperature is no longer required to change. In the TES bolometers, it is possible to take advantage of the normal–superconducting transition to provide this feedback since the power dissipation in the normal part of the sensing element acts as the heater.

5.2 Photon Detectors

Photoconductors are semiconducting materials that produce free charge carriers when photons of sufficient energy hit the material. The minimum energy is known as the band gap

(E_g) of the semiconductor and corresponds to the energy between the conduction and valence bands of the material. The longest wavelength that can be detected by a photoconductor is related to the band gap by

$$\lambda_c = hc/E_g = 1.24\ \mu\text{m}/E_g\ (\text{eV})$$

where h = Planck's constant, c = speed of light and E_g is the band gap in electron volts. For silicon, the band gap is 1.1 eV, meaning that the longest infrared wavelength that can be detected is just over 1.1 μm. Fortunately, other semiconductor materials have smaller band gaps. The most important materials for infrared astronomy have been indium antimonide, with a 5.3-μm cutoff, and mercury cadmium telluride. For mercury cadmium telluride, the band gap can be adjusted by varying the fraction x of the cadmium vs. mercury in the material ($Hg_{(1-x)}Cd_xTe$). In this way, high performance detectors have been built with cutoff wavelengths from as short as 1 μm to beyond 10 μm. *Table 3-3* lists the cutoff wavelengths of some important intrinsic semiconductors.

In addition to utilizing the intrinsic band gap of semiconductor materials, detectors have employed the energy levels associated with dopants in semiconductors. *Figure 3-23* shows the band gap diagram for silicon doped with arsenic. The impurity energy levels of arsenic are only 0.05 eV from the conduction band, meaning that a photon of wavelength less than 23 μm would be energetic enough to ionize a charge carrier into the conduction band. This process

Table 3-3
Cutoff wavelengths for some important intrinsic photoconductors (From Bratt 1977)

Intrinsic photoconductors	
Material	Cutoff wavelength (μm)
HgCdTe	0.8 to >20
Si	1.1
Ge	1.6
InSb	5.5

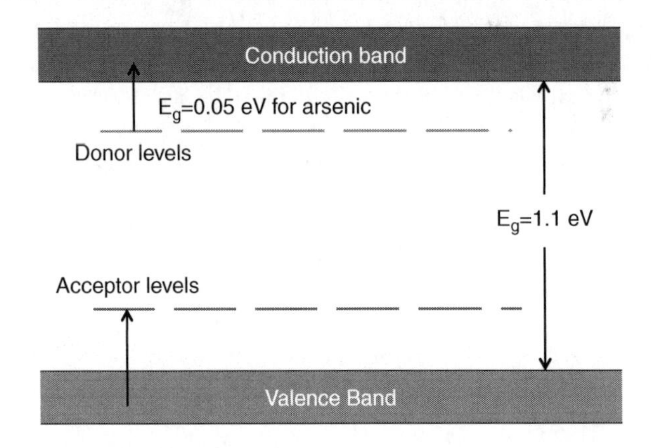

Fig. 3-23
Band gap diagram of silicon showing extrinsic impurity levels of arsenic. The cutoff wavelength of an extrinsic impurity is set by the energy between the extrinsic level and either the conduction band or valence band in the semiconductor

Table 3-4

Cutoff wavelengths for some important extrinsic dopants in silicon and germanium (Bratt 1977)

Extrinsic photoconductors			
Material	Dopant	Type	Cutoff wavelength (μm)
Si	Ga	p	17
Si	As	n	23
Si	B	p	28
Si	Sb	n	29
Ge	Be	p	52
Ge	Ga	p	115
Ge	Sb	n	129

is called extrinsic photoconductivity. Extrinsic photoconductors with dopants in both silicon and germanium have been successfully used at wavelengths as long as 200 μm in astronomy, notably on the *IRAS*, *ISO*, *Akari* (a Japanese infrared satellite), *Spitzer*, and *Herschel* missions.

While in principle it is possible to make sensitive infrared detectors using simple photoconductors, practical modern direct detectors use more complicated structures. For the intrinsic photoconductors, the detectors are almost always made as photodiodes, where n-doped material forms a junction with p-doped material. By forming a p-n junction, a charge-free region, the depletion region, is formed, and any photoexcited charge carriers formed in this region are swept to the contacts by the electric field in the region. Photodiodes using InSb and HgCdTe are the most important intrinsic photodiodes in the near-infrared regime. High performance photodiodes have been formed in HgCdTe for wavelengths as long as 10 μm (Bacon et al. 2004).

From ~10 μm to nearly 40 μm, the most sensitive detectors have used the impurity band conduction (IBC) structure. Also known as the blocked impurity band (BIB) detector, these devices overcome many of the fundamental limitations of classical bulk photoconductors. The photon absorption efficiency of a photoconductor is set by the product of the photoionization cross section of the dopant and the doping concentration of the desired impurity. The latter is limited by excessively high dark currents when the concentration gets too high. Real values for photoionization cross sections and maximum concentrations force conventional bulk photoconductors to be physically large, typically with millimeter-scale absorption path lengths. IBC detectors, by incorporating an undoped "blocking" layer to block the dark current, allow much higher doping concentrations in the IR absorption layer. An additional advantage to the high doping in the IR absorption layer is an extension of the cutoff wavelength beyond the normal extrinsic limit. For Si:As IBC detectors, the cutoff is 28 μm, while for Si:Sb IBC detectors, the cutoff is ~36 μm. IBC detectors have been successfully used on *ISO*, *Akari*, *Spitzer*, and the *Wide Field Infrared Survey Explorer* (*WISE*) and will be flown on *JWST*.

5.3 Detector Arrays

Arguably one of the most important technical developments in infrared astronomy has been the transition from single detectors to arrays. Increases in overall information gathering power of many orders of magnitude have been realized over the few decades. Following the great

Fig. 3-24
***IRAS* focal plane array consisting of 62 discrete photoconductors**

advances in semiconductor fabrication technology, infrared detector arrays have seen an exponential growth in array format mimicking Moore's Law (Moore 1965). A good illustration of the advances in just three decades is to compare the *IRAS* focal plane array (*Fig. 3-24*), which had 62 detectors, with one of the focal plane arrays for the *JWST* NIRCam instrument (*Fig. 3-25*), which has 16 million pixels. The entire NIRCam instrument has 40 million pixels.

The key to this continuing advance has been the utilization of many technologies from the semiconductor industry. *Figure 3-26* illustrates the principal architecture used on large format detector arrays. Using integrated circuit technology, a readout integrated circuit (ROIC) is fabricated out of silicon. This circuit typically has a unit cell amplifier for each pixel and all the multiplexing circuitry needed to send the signals down a manageable number of output lines. The inputs to all these amplifiers are arranged in a grid on the top of the integrated circuit. The detector array is constructed out of the appropriate material for infrared absorption and carrier generation such as an array of photodiodes fabricated out of HgCdTe or an array of IBC detectors on a silicon substrate. The need to connect the outputs of the detectors to the inputs of the amplifiers is accomplished with indium bump bonds. These exactly matched microscopic indium bumps are deposited on both the detector array and the ROIC. When the two sets of bumps are aligned and pressed together, a robust cold weld is formed in the indium. This technology has reached the level of a fine art, with as many as sixteen million interconnections being made with very high success rate (>99%) and high reliability.

Fig. 3-25

One of two 1–2.5-μm focal plane arrays that will be used on the NIRCam instrument on *JWST*. It is a 2 × 2 mosaic of 2048 × 2048 pixel HgCdTe arrays. The complete instrument has a pair of these mosaics in addition to two 2048 × 2048 pixel 2- to 5-μm focal planes, for a total of 40 million pixels. See Rieke (2007) for an illustration of the construction of this focal plane

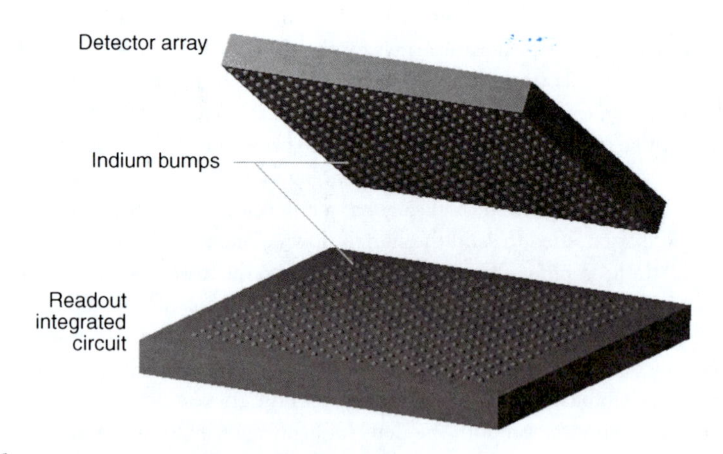

Fig. 3-26

Indium bump bonding of an infrared detector array to a silicon readout integrated circuit is the principal architecture for detectors in the 1- to 40-μm range

Table 3-5
Some representative focal plane arrays used in astronomy

Array	Material	Format	Wavelength (μm)	Operating temp (K)	Notable example	References
Teledyne H2RG (JWST)	HgCdTe	2048 × 2048	0.6–5.3	38	JWST, NIRSpec, and NIRCam	Smith et al. (2009)
Teledyne H4RG	HgCdTe	4096 × 4096	0.6–5.3	38		Blank et al. (2011)
Raytheon Orion	InSb	2048 × 2048	0.6–5.5	32	NEWFIRM	Hoffman et al. (2004)
DRS MEGAMIR	Si:As Si:Sb	1024 × 1024	5–28 5–38	7	SOFIA FORCAST	Mainzer et al. (2005)
Raytheon JWST MIRI	Si:As	1024 × 1024	5–28	7	JWST MIRI	Ressler et al. (2010)
CEA Si bolometer	Si	64 × 32 32 × 16	60–85 85–130 130–210	0.3	Herschel PACS	Billot et al. (2009)
SCUBA-2 bolometer	TES	4 × 32 × 40	850 450	0.3	SCUBA-2	Bintley et al. (2010)

This simplistic description, of course, bypasses many of the extremely difficult engineering developments that have been needed to make the whole process work. In particular, low temperature operation of the readout, accommodation of the thermal properties of materials, and the development of good infrared detector materials have required the work of many dedicated scientists and engineers. *Table 3-5* lists the characteristics of some of the important infrared detector arrays in astronomy.

5.4 Microwave Kinetic Induction Detectors

While a superconductor film has a zero D.C. impedance, the Cooper pairs in the superconductor present a surface inductance to A.C. signals due to the inertia of the Cooper pairs. The breaking of the Cooper pairs by infrared photons creates unpaired electrons or quasiparticles resulting in a change in the kinetic inductance of the film. Microwave kinetic induction detectors (MKIDs) utilize this creation of quasiparticles, in a superconducting film to detect photons (Day et al. 2003). To sense these quasiparticles, the superconductor is made part of a highly tuned resonant microwave circuit that is excited by an external oscillator, typically in the 1–10-GHz range. The change in inductance causes a change in the tuning of the circuit that can be sensed by observing either amplitude or phase changes in the output.

The promise of MKIDs is that it is possible to frequency multiplex many detectors on a single output line by having each pixel tuned to a slightly different microwave frequency. This tuning is accomplished geometrically in the fabrication using standard microlithographic techniques. In operation, each pixel is excited by a tone matched to its resonant frequency. Hence, each pixel is associated with single microwave frequency. Because the quality factor (Q) of this circuit can be as high as 10^5, highly specific frequency measurements are possible, and many channels can be fit into a given frequency range. MKID arrays are an active area of detector work, and a number of cameras for ground-based telescopes, including the Caltech Submillimeter Observatory (Maloney et al. 2009), APEX (Heyminck et al. 2010), and IRAM (Monfardini et al. 2011), have been developed. A comprehensive review of the physics of MKID detectors can be found in Zmuidzinas (2012).

Key References

McLean, I. 2008, Electronic Imaging in Astronomy: Detectors and Instrumentation (2nd ed.; New York: Springer)
Rieke, G. H. 2003, Detection of Light: From the Ultraviolet to the Submillimeter (Cambridge: Cambridge University Press)
Rieke, G. H. 2007, Infrared detector arrays for astronomy. ARA&A, 45, 77

6 Optimizing the Signal-to-Noise Ratio

The high IR background and the characteristics of IR detector arrays require techniques that differ from CCD imaging and spectroscopy to optimize the signal-to-noise ratio (SNR). The basic concepts are discussed in this section. Section 6.1 begins with the signal-to-noise equation and emphasizes the influence of the background on the SNR. Section 6.2 discusses techniques to reduce the background in ground-based telescopes, while Sect. 6.3 covers IR-optimized telescopes. Section 6.4 discusses data taking techniques in the presence of high background. Finally, Sect. 6.5 discusses key airborne and space observatories that have been developed or are planned to avoid the ground-based limitations arising from the high background.

6.1 Signal-to-Noise Equation

The discussion in this section applies to photon detectors such as those described in Sect. 5.2 and it does not apply to thermal detectors such as bolometers.

The SNR is simply the number of electrons from the science target (the signal measurement) divided by the estimate of the noise, or uncertainty, in the measurement. Noise arises from multiple sources, including the science target itself, as well as background contributors. Noise terms add in quadrature, and all astronomical sources of noise are generally assumed to be Poissonian. The measured SNR on a single image of a source is given by Merline and Howell (1995):

$$\mathrm{SNR} = N_s / \sqrt{N_s + n_p \left(1 + \frac{n_p}{n_B}\right) (N_B + N_D + N_{\mathrm{RN}}^2)},$$

where N_s is the number of electrons from the source, N_B is the number of electrons per pixel from the background (sky and telescope), N_D is the number of electrons per pixel arising from the dark current, N_{RN}^2 is the number of electrons per pixel associated with reading out the array (read noise), n_p is the number of pixels the source subtends, and n_B is the number of pixels used to estimate the background level. Usually, $n_B \gg n_p$, in which case the usual S/N equation is obtained (see below).

We can make the dependence on the exposure time more explicit by expressing the number of electrons detected from the source, the background, and the dark current as the products of a rate and the time so that

$$S/N = St/\sqrt{St + n_p(Bt + Dt + N_{\mathrm{RN}}^2)},$$

where S is the number electrons per second from the science target, t is the exposure time in seconds, B is the number electrons from all background contributors per second per pixel, D is the number of electrons per second arising from the dark current at each pixel, and $n_B \gg n_p$ is assumed.

The SNR can now be predicted for an observation with a given exposure time. If the science target has an intrinsic spectrum f_λ, then the number of electrons generated per second by a telescope of area A is given by

$$S = A \int \lambda f_\lambda(\lambda) Q(\lambda) T_a(\lambda) d\lambda / hc,$$

where Q is the overall throughput of the telescope, instrument, detector, and spectral element (e.g., filter) in electrons/photon and T_a is the atmospheric transmission. The integral is performed over the wavelength range of the filter passband where the throughput is >0. For spectroscopy, a slit is usually used, and the above equation needs to account for the fraction of the light transmitted by the slit.

Similarly, the number of electrons measured per second in each pixel from a thermal background contributor at a temperature T is given by

$$B = A\Omega \int \lambda \varepsilon(\lambda) B_\lambda(\lambda, T) Q(\lambda) d\lambda / hc,$$

where $B_\lambda(\lambda, T)$ is the blackbody function, $\varepsilon(\lambda)$ is the emissivity, and Ω is the solid angle of a pixel. The sky background is given by $\varepsilon_s(\lambda) B_\lambda(\lambda, T)$, where $\varepsilon_s(\lambda) = 1 - T_a(\lambda) = 1 - e^{-\tau(\lambda)}$ and $\tau(\lambda)$ is the optical depth of the telluric lines. The telescope background is given by $\varepsilon_t(\lambda) B_\lambda(T_{\mathrm{tel}})$, where $\varepsilon_t(\lambda)$ is the emissivity of the telescope. In addition, OH emission from the sky is a bright nonthermal source of background electrons.

These equations transfer directly to spectroscopic observations, for which the integral over $d\lambda$ can usually be taken as multiplication by the spectral resolution element = λ/R, where R is the spectral resolution.

The signal-to-noise equation demonstrates that as long as the observations are not in the read noise-limited or dark current-limited regime (i.e., $Bt \gg N_{\mathrm{RN}}^2$ and $B \gg D$), then

$$\mathrm{SNR} = S\sqrt{t}/\sqrt{S + n_p B},$$

and the SNR increases only as the square root of the telescope area or the exposure time. For most mid- and far-IR observations (especially from the ground), we have $B \gg S$, which is referred to as the background-limited regime. In this case, SNR = $S\sqrt{t}/\sqrt{n_p B}$ and the SNR increases linearly with the source flux (as opposed to the situation often encountered in the optical, where $S \gg B$ and the SNR increases as the square root of the source flux).

Inspection of ➋ *Fig. 3-15* shows that the background emission from the sky and telescope completely dominates the detector dark current for imaging and low resolution spectroscopic observations. To maximize the SNR, the array is exposed long enough so that shot noise from the background is larger than the read noise. Current near-IR arrays have a read noise of about 20 electrons, so integration times of a few seconds are sufficient to be background limited with broadband imaging.

The situation is very different for high-resolution spectroscopy. In the example shown in *Fig. 3-16*, the background levels are extremely low in between the OH emission lines. Here one would set the exposure for as long as possible, without driving the OH lines into saturation or the nonlinearity regime, before reading out the array. If the array performance is very stable, long integrations of 600–1,200 s are possible at 1–2.5 μm.

6.2 Ground-Based Observations in the Infrared

Unlike observations from space, ground-based IR observations must account for the sky, which acts as both an absorber and an emitter. As shown in Sects. 2 and 3, the magnitude of the absorption varies strongly with wavelength. Because the atmospheric constituents vary with altitude, the transmission is also dependent on observatory altitude as well as local atmospheric conditions (water vapor). The sky emission is dependent on the effective temperature of the atmosphere.

In addition, the telescope and instrument also emit thermal radiation and generate background flux. For most of the IR wavelength regime, the background from the sky, telescope, and instrument completely dominates the flux from almost all astronomical sources. For example, at 10 μm, the photon flux from a 1 Jy source is approximately 1.5×10^6 photons $m^{-2}\ s^{-1} \mu m^{-1}$, while the sky background alone can produce 3.6×10^9 photons $m^{-2}\ s^{-1} \mu m^{-1}$ in 1 $arcsec^2$, a factor of over 2,000 times larger.

Although its level can be so much larger than that detected from an astronomical source, the variability of the background poses the real difficulty for ground-based IR observations. While the telescope and instrument generally have large thermal masses, and so the temperatures are fairly constant, the sky emission exhibits significant variations on very short time scales. As shown by Kaeufl et al. (1991), the sky noise at 10 μm requires sky subtraction at a rate of 5–8 Hz in order for the sky noise to be below the detector noise. Note that this is dependent on the site, the detector used, and local sky conditions during the observations. In another study of sky noise, Miyata et al. (2000) have found that the sky subtraction frequency should be greater than 0.5 Hz through narrowband filters around 10 μm at Mauna Kea (see *Fig. 3-33*). The optimum sky subtraction frequency needs to be determined empirically for each site. This necessity to subtract the sky rapidly has led to the adoption of chopping and nodding techniques described in Sect. 6.4.2. At shorter wavelengths such as 1–2.5 μm, the sky observations can be taken every 1–2 min or longer by nodding the telescope.

6.3 IR-Optimized Telescopes

This section contains a brief discussion of methods used to reduce the thermal background at ground-based and airborne telescopes. It is instructive to imagine putting your eye at the focus of a telescope. Looking up at the secondary mirror, you would see the secondary mirror itself, the supporting structure behind the secondary mirror, the secondary mirror spiders, the sky behind the secondary mirror, and the telescope top ring holding the secondary mirror in place (see *Fig. 3-27*). An instrument working at thermal IR wavelengths ($\lambda > 3\,\mu m$) would see

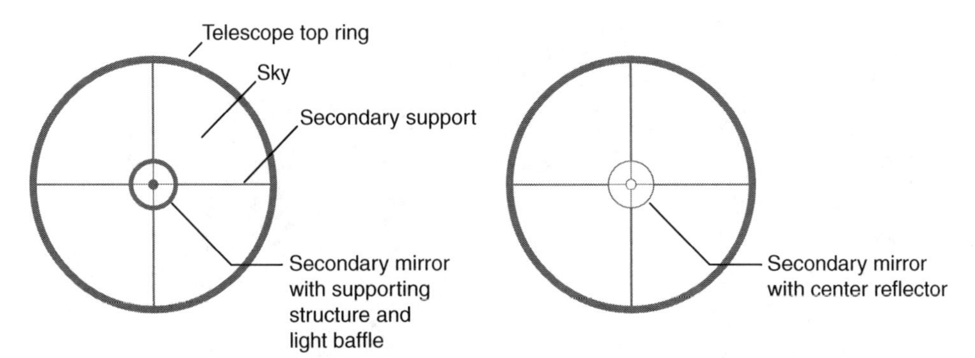

Fig. 3-27

(*Left*) View of secondary mirror in a conventional telescope, showing the secondary mirror, spiders, sky, and telescope *top ring*. High emissivity regions are indicated as a *dark color* and can be seen by the detector unless precautions are taken. (*Right*) View of the secondary mirror of an IR-optimized telescope. The center of the secondary mirror has a reflector or a hole to prevent thermal emission from the center of the primary mirror reaching the detector. The secondary mirror is made as small as possible (high f/no) and it is undersized so that thermal emission from the primary mirror cell and the telescope *top ring* does not reach the detector

thermal emission from all of these surfaces. Even the primary and secondary mirrors would be emitting surfaces. The sky would also be radiating thermally. Light baffles would be a major source of thermal emission as well.

An IR-optimized telescope is designed to keep all radiating surfaces to a minimum, as shown in *Fig. 3-27* (right). The secondary mirror supporting structure is constructed to be completely behind the secondary mirror. The secondary mirror is undersized so that highly emissive parts of the mirror cell are not seen. There are no light baffles; stray light control is designed into the instrument. There is a tilted flat mirror at the center of the secondary mirror to block thermal emission arising from the central hole in the primary mirror (see *Fig. 3-28*). Alternatively, the secondary mirror can be designed with a clear hole in the center of the secondary mirror to prevent emission from the hole in primary reaching the detector. Low emissivity coatings such as overcoated silver are used on the primary and secondary mirrors. Some or all of these features are designed into telescopes such as the 3.8-m United Kingdom Infrared Telescope, the 3.0-m NASA Infrared Telescope Facility (IRTF), and the 8.0-m Gemini North telescope. The IRTF has an f/38 secondary mirror, which reduces the central obscuration of the primary mirror and this minimizes the thermal emission from the telescope. The Gemini North telescope employs overcoated silver (Boccas et al. 2004; Vucina et al. 2006) so that the emissivity is about 2% per mirror surface in the near infrared and about 1.4% in the mid infrared. This can be compared to >25% for the total emissivity for a conventional telescope.

Figure 3-28 shows a thermal image of the IRTF secondary mirror. The thermal emission from the primary mirror, secondary mirror, and spiders are emitting toward the detector. The supporting structure for the secondary mirror is completely behind the secondary mirror. There are no light baffles in the telescope. All IR instruments are designed to have a cold pupil stop within the instrument. At this stop there is an image of the secondary mirror, and if, necessary, a cold mask can be located here to block the thermal emission from the spiders, the hole in

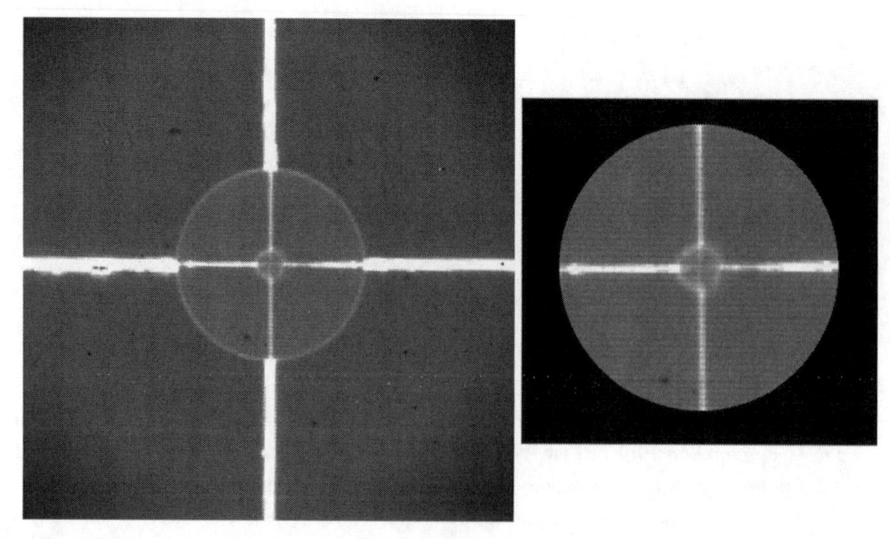

Fig. 3-28
(*Left*) Thermal image of the IRTF secondary mirror taken at a wavelength of 3.3 μm. The dome is opened to the sky. The secondary mirror spiders are thermally radiating and are very bright; in fact, the spiders are saturating the camera. The sky and the reflection of the sky in the secondary mirror are much colder than the spiders and appear darker. The thermal emission from the hole in the primary mirror is reflected away by a small tilted mirror in the center of the secondary mirror. The *dark* and *white spots* are bad pixels in the array. (*Right*) Pupil image formed inside of the instrument at a cold stop. This shows the thermal emission that the detector sees. *Light* from the sky is prevented from going directly to the detector. The background emission from the environment is limited to the sky emission, thermal emission from the spiders and telescope mirrors, and instrument window. Thermal emission from the telescope mirrors can be reduced by using low emissivity coatings (such as overcoated silver) and by keeping the mirrors free of dust particles. The thermal emission from the spiders and the central hole in the primary mirror can also be blocked by a cold mask located at the pupil image

the primary mirror, and any warm light baffles. Such measures are often taken with optical telescopes that have high emissivity.

Telescopes at the South Pole take advantage of the much lower temperatures of both the sky and the telescope (Burton 2010). *SOFIA* accomplishes this to a much greater degree by going to an altitude of 14 km (see Sect. 6.5.1), although the SOFIA telescope is not as cold as the South Pole.

6.4 Data Taking in the Presence of High Background Emission

Although every attempt is made to reduce the sky, telescope, and instrument background, the background levels are still very high compared with most astronomical objects. Therefore, data-taking techniques have been developed to maximize the signal-to-noise ratio, as described in the following subsections.

6.4.1 Near-infrared Imaging

The basic elements of data taking and data reduction are discussed in this section. In practice, special handling of data is always required due to problems arising from array defects and array readout characteristics, fringing effects arising from the array, instrumental effects (scattered light, nonuniform illumination), and variable weather. The discussion by Joyce (1992) and the book by McLean (2008) provide a good introduction to data taking with IR arrays.

Unlike in the optical, where an observer can integrate on the astronomical source for many hours, time variability of the IR background means that one cannot observe for extended periods without obtaining a measure of the background (telescope and sky). Depending on the site, the telescope, and the wavelength, imaging exposure times can be as short as a few minutes due to the highly variable OH emission in the near IR (see Sect. 3.1). At the J band longer exposure times are feasible. At longer wavelengths, however, the variability of the telluric emission lines, as well as quick filling of the array wells by the huge number of photons detected, implies that much shorter integrations are necessary. At 10 μm, for example, it is not uncommon for exposure times of individual frames to be on the order of tens of milliseconds.

To understand the steps employed in the reduction of IR data, it is useful to begin by writing the equation for the number of counts detected at a given pixel i in an array:

$$I_i^{\mathrm{obs}} = F_i \times (S_i + B_i) + D_i,$$

where S_i is the signal from the source, B_i is the background, D_i is the dark current, and F_i is the relative response of the pixel (the "flat field"), which is related to the value of the gain at each pixel.

When observing in the infrared, the background flux can be orders of magnitude larger than the flux from astronomical sources, and therefore, accurate subtraction of the background is crucial. For the near-IR regime, in which the thermal background is relatively stable, a background frame is often generated as part of the observations as follows: If the region around the targeted source is not extremely crowded and does not contain extended emission, and if the source itself is compact, this background frame can be generated from the observations of the target itself, typically from "dithered" observations. These consist of multiple observations of the source, with the telescope moved slightly between each observation. The amount of each move, or dither, depends on the expected extent of the sources as well as the need to avoid bad regions of the array. The latter may consist of clusters of bad pixels or regions of low quantum efficiency (QE).

To construct a background frame, the individual images can then be averaged or median combined using various algorithms to exclude outliers. That is, at each pixel in the array the counts in that pixel from each frame are used to generate a mean background value in the pixel and a standard deviation. It is best to mask out bright objects in the frame before combining. Because the background is time variable, it is usually necessary to offset each frame to the median level of all the frames before combining. The resulting background frame should have a high signal-to-noise ratio, if a suitable number of frames were used in the combining process.

If the source is extended or the surrounding area has extended emission, or is crowded, it is usually necessary to move the telescope off the source to obtain a background image. These frames should have the same exposure time as the source frames and should also be dithered to remove any objects present. (As a general principle, whatever is done to the source frames should be done to the background frames.) These frames can then be combined, as described above, to generate a mean or median background image with high SNR.

At this point, the counts in the background image at each pixel i are given by

$$I_i^{\mathrm{bk}} = F_i \times B_i + D_i.$$

It can be seen that subtracting the background image from a source image removes the dark current immediately, and the source signal can be obtained by dividing the result by the flat field,

$$S_i = (I_i^{\mathrm{obs}} - I_i^{\mathrm{bk}})/F_i.$$

Again, because the background is time variable, it is usually necessary to offset the background image by a constant factor equal to the difference in the median values between the source frame and the background image before subtraction. The result should yield a "background-subtracted" image with a median level around zero.

The flat-field F_i can be obtained from the background frames if the count levels in the sky background are large enough. The background frames are normalized by the mean, median, or modal values in each frame and then median combining them. Dithered observations of blank sky at twilight can also be used to generate flat fields. In both cases, the practice of using actual observations to produce flat fields assumes that the background is uniform across the array, and therefore the division by the mean value in the frame yields the relative response of each pixel.

In principle, an estimate of the dark current at each pixel should be subtracted from these background frames before they are normalized and combined to produce the flat field. The dark current image can be generated by combining (averaging) several frames taken (usually during the afternoon before observing or in the morning after observing) with the dome closed and all light sources turned off. These frames should have the same exposure times as the background frames used to generate the flat fields. However, for most near-IR arrays, the contribution to the detected counts due to dark current is negligible compared with that from other background sources. In addition, many arrays behave nonlinearly at low count levels. These effects imply that dark subtraction can be skipped in most cases. If dark subtraction is performed, the observer must be careful to assure that the process is not simply increasing the noise.

Naylor (1998) has pointed out that when generating sky and flat-field images from the sources frames, the statistically correct method requires the exclusion of the frame to be processed from the stack combined to make the background and flat images. The implication is that each source frame must be processed with its own unique background and flat images, generated from all the other source frames. As stated by Naylor (1998, p. 342), "Each frame must be flat-fielded with the median of all the frames in the group excluding itself."

If the thermal background (e.g., from the telescope) is not uniform, then flat fields generated from the observations of the sky background will not yield photometrically flat photometry (i.e., the photometric measurements of an object will vary with position on the array). In this case, a flat field should be generated from an internal lamp or from the uniform illumination of a spot on the interior of the dome (dome flat). By taking sets of exposures with the lights on and off and subtracting the two sets, the telescope thermal background (including the dark current) is automatically removed and the resulting images can be normalized and combined to produce a flat field.

Each background-subtracted and flat-fielded frame provides one image of the source. Each of the source frames can be shifted to align the astronomical sources and then combined ("shift and add"). Since the original observations of the source were dithered, any dead or "hot" pixels (which are fixed in the array but appear to move relative to the source) will be removed from the final combined image by setting the reduction software to reject outlying values pixel by pixel. Bad pixels can also be eliminated by including their locations on the array in a "bad-pixel

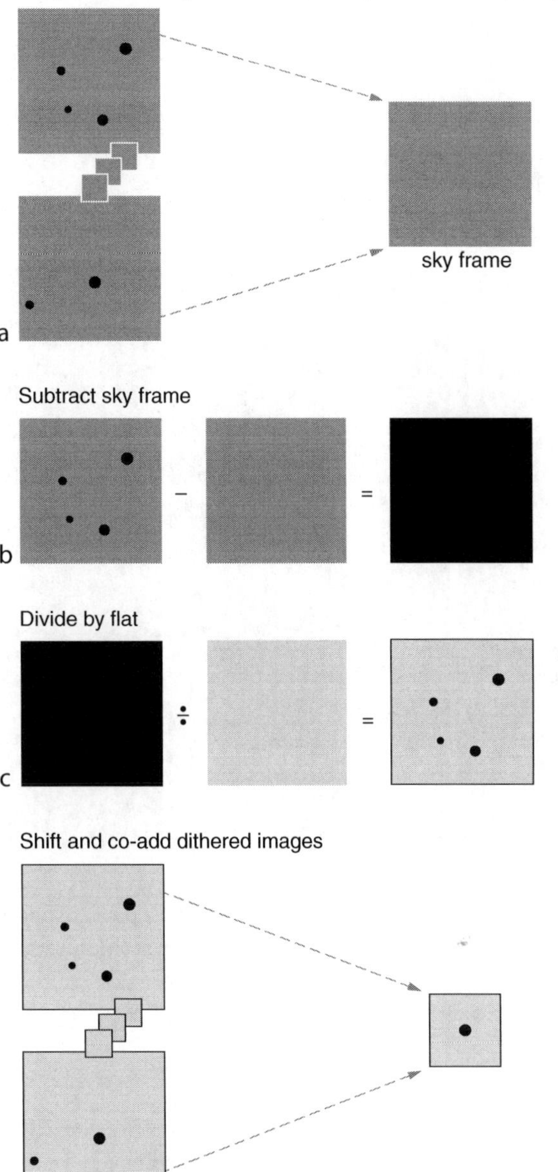

Fig. 3-29

Schematic of making a sky-subtracted image from a set of dithered images. In making the median filtered sky frame bright stars should be masked out

mask" and employing the mask at each step in the reduction process to identify and exclude them from the arithmetic calculations.

Figure 3-29 shows a schematic of the process of removing the sky. *Figures 3-30* and *3-31* show actual data and the removal of the sky background.

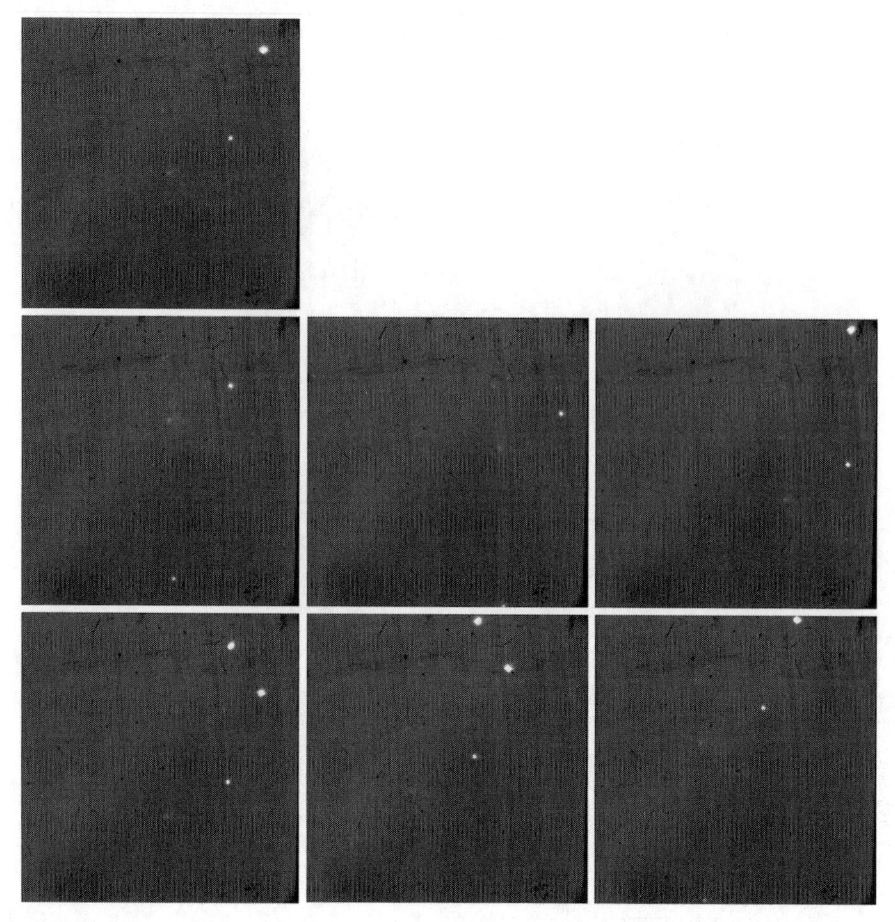

Fig. 3-30
Seven-point dither images. Dither pattern is in the shape of a hexagon with one image in the center of the hexagon (*top row*). The corner to corner distance across the hexagon was 20″. Image was taken with the slit viewer of SpeX on the NASA IRTF (Images from M. Connelley, private communication)

We summarize in *Table 3-6* the various problems associated with IR arrays and techniques used to achieve a good image. Many of these effects can be reduced to a tolerable level by dithering and by taking the object and calibration exposures with the same settings.

Near-IR arrays are inherently nonlinear because of the method used to sense the number of photoelectrons generated by the detector. The nonlinearity for each pixel can be measured and a correction factor applied (McLean 2008; Vacca et al. 2004).

Bright sources create latent images in near-IR arrays that arise from "charge trapping" in the readout amplifier (Solomon et al. 1993). The trapped charge decays exponentially, but it can take many hours to be reduced to a negligible level. Some reduction can be achieved by "flushing" the array by reading out the array with short exposure times. Since latent images cannot be entirely suppressed, this effect should be considered in how the data are taken and reduced.

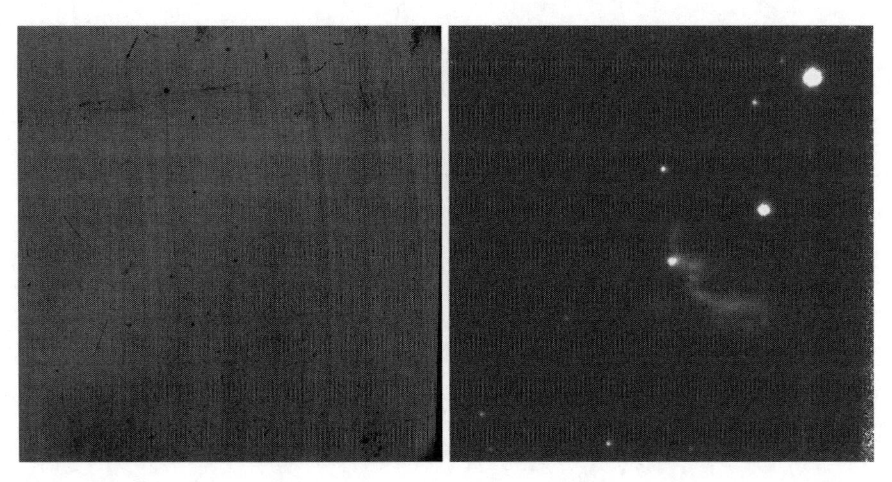

Fig. 3-31
(*Left*) Sky image generated from a median coaddition of the dithered images. (*Right*) Final image obtained by subtracting the sky image and realigning and coadding the dithered images

Table 3-6
Typical problems with IR arrays

Problem	Remedy
Nonuniform quantum efficiency	Divide by flat field
Bad pixels; cosmic-ray hits	Mask or filter out by dithering
Dark current and hot pixels	Remove by sky subtraction
Nonlinearity	Stay within linear range or apply linearity correction
Latent images	Avoid bright sources or saturation before exposing on faint sources
High background	Dither to get sky frame

6.4.2 Mid-infrared Imaging

In an early primer on IR photometry, Low and Rieke (1974, p. 444) stated, "Observing at 10 μm with a ground-based telescope has been likened to observing visually through a telescope lined with luminescent panels and surrounded by flickering lights as though the telescope dome were on fire." As shown in *Fig. 3-6*, a room-temperature blackbody peaks near 10 μm so that every warm surface is bright in the infrared. The sky emission is up to 10^6 brighter than the celestial background (see *Fig. 3-35*) and is highly variable due to variations in the atmospheric constituents, especially water vapor.

Low and Rieke (1974) introduced the technique of temporally modulating the sky emission on the detector by using a chopping secondary mirror (see *Fig. 3-32*). This allows removal of the sky emission at a fraction of a second rates (typically 1–10 Hz). "Sky chopping," as it is now known, is used to reduce the low-frequency noise by a factor of 10^3–10^4. The optimum frequency and amplitude of the sky chopping depends on the sky conditions and must be determined empirically (Allen and Barton 1981; Papoular 1983; Kaeufl et al. 1991; Miyata et al. 2000;

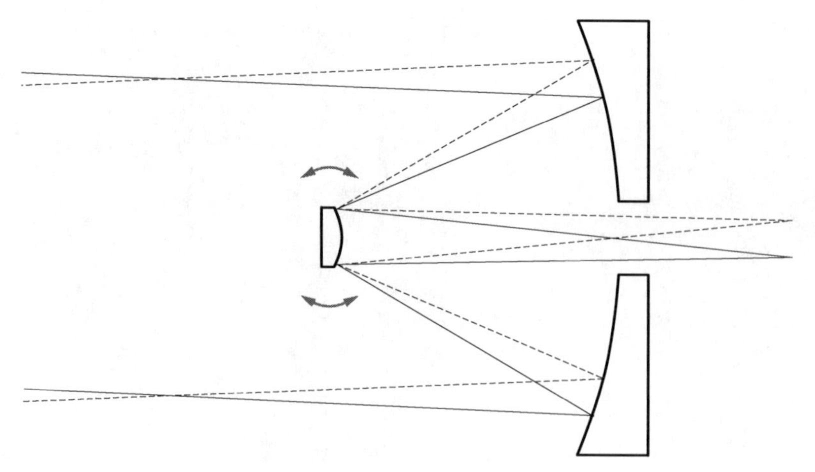

Fig. 3-32
Schematic of chopping secondary. The secondary is oscillated, producing two images in the focal plane. Subtraction of these images eliminates most of the low frequency noise components. Note that the secondary mirror must be undersized so that the edge of the beam on the primary mirror does not reach warm surfaces of the mirror cell

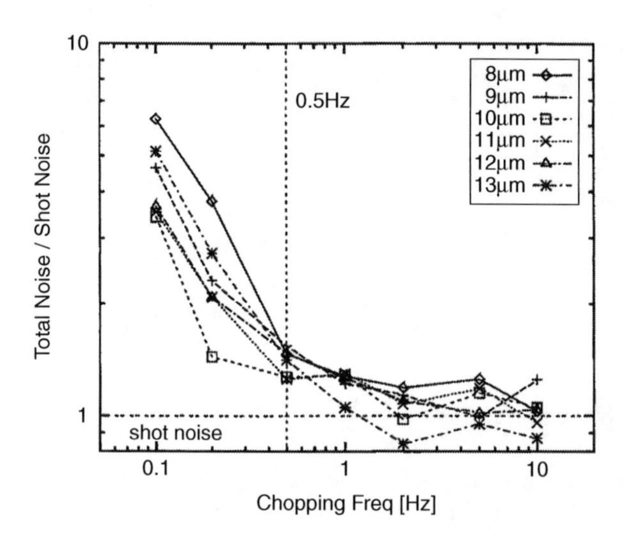

Fig. 3-33
Example of sky noise showing frequency and wavelength dependence of sky noise. "1/f" noise dominates at low frequencies and arises from sky emission fluctuations and detector noise. The sky noise depends on the wavelength, with higher noise at wavelengths where thermal emission from water vapor is high. This data taken at Mauna Kea shows that the sky subtraction should be done at a frequency ≥0.5 Hz (From Miyata et al. (2000))

Mason et al. 2008; see *Fig. 3-33*). For a point source, a chop frequency of 10 Hz and amplitude of 0.5–1 arcmin is typical, but the chop amplitude is larger at far-IR wavelengths due to the larger beam size. The smallest chop amplitude possible should be used so that the same path through the atmosphere is observed by the detector.

Although sky chopping eliminates most of the variable sky emission, there is an offset signal that arises from small temperature imbalances on the telescope optics and structure. To remove this offset signal, the telescope is moved or "nodded" to a different location on the sky. The difference between the first and second positions removes the unwanted offset signal.

For point sources, chopping and nodding amplitudes are usually small enough such that the object remains on the detector during both chopping and nodding. For extended sources, however, it is often necessary to chop off the object to a sky region and to nod the telescope to another sky region.

Mid-IR images are reduced using techniques somewhat different from those employed in the reduction of near-IR images. A set of images usually comprise four frames, two frames in one telescope nod position (usually called nod A) and two frames in the second nod position (nod B). The two frames at each nod position correspond to the two positions of the chopping secondary. Each individual raw frame is completely dominated by the background emission, and usually no astronomical sources can be seen (unless they are extremely bright).

An example is shown in ➋ *Fig. 3-34*. The first step in the reduction process is to subtract the pair of "chopped" frames at each nod position. This immediately removes a large fraction

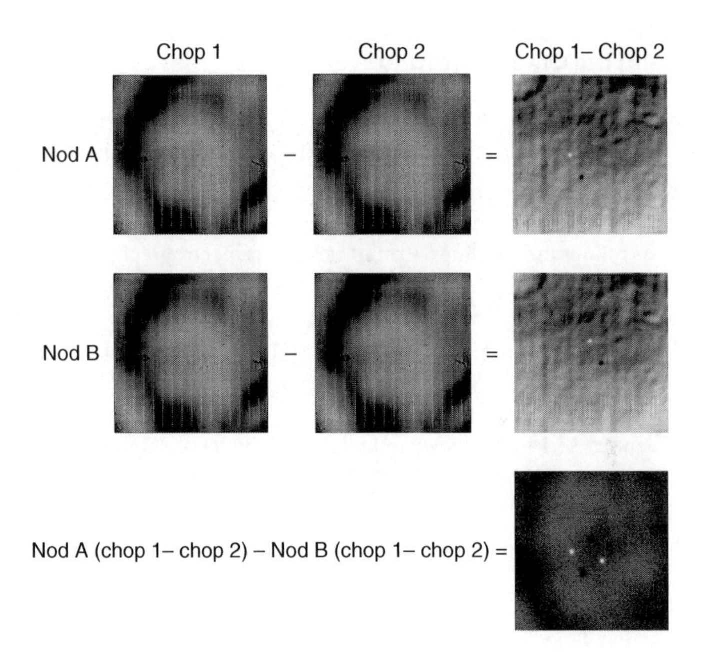

Fig. 3-34

Example of reduction of mid-IR imaging data. The four images on the *left* are raw frames of a standard star (γDra) obtained at 24 μm with the FORCAST instrument (Herter et al. 2012) on *SOFIA*. Each frame represents the sum of ~6,000 integrations, each ~2 ms, for a total integration time of ~12 s at one of the two chop positions (1 and 2) and one of the two nod positions (A and B). The chop throw was 30 arcsec, and the chop frequency was ~4 Hz. The final image at the *lower right* is the result of subtracting the chopped images at each nod position (chop subtraction) and then subtracting the two chop-subtracted images at the two nod positions (nod subtraction). This double subtraction removes the thermal background levels as well as the offset levels due to the chopping. No flat fielding has been applied

of the total background due to the sky, telescope, optics, etc. However, the subtraction does not completely remove all of the background. The residual (or offset) background is removed by further subtracting the chop-subtracted frame obtained at nod position B. Since the frames obtained at the second nod position should have been obtained in exactly the same manner as those at nod position A, the residual background should be the same as well. This double subtraction technique will then yield an image of the source.

If the secondary chop angle was small enough (relative to the size of the detector array) such that the astronomical source remained on the array in both chop positions, then two images (a positive and negative) of the source will appear. If the telescope nod angle was also small relative to the size of the array, four instances of the source will appear in the final chop-nod-subtracted image. The separate instances of the object can then be shifted and aligned with one another (by making copies of the reduced image) and then combined (averaged or median combined). It can be shown that the final SNR achieved with this technique is approximately the same as would be measured on a single frame if the source could be seen in that frame. Flat fielding of mid-IR image data may not be possible due to the difficulty of obtaining a stable flatfield image. In this case, averaging many images is necessary to average out fluctuations in the background.

6.4.3 Reducing the OH Background

As *Figs. 3-13* and *3-14* show, the strong OH lines are widely spaced. This has led to attempts to suppress the strongest OH lines. Content (1996) contains a general discussion of the gains that are possible if the OH lines could be suppressed. One approach is to image the OH spectrum onto a mask in which only the light in between the strong OH lines is reflected to form a second spectrum without the OH lines. Examples of such spectrographs are described by Iwamuro et al. (2001) and Kimura et al. (2010). Sensitivity gains of about 1.2 mag for low-resolution spectroscopy are possible.

In another approach, Bland-Hawthorn et al. (2011) have pioneered a method of using a cluster of fiber Bragg gratings to selectively reject the strongest of the OH emission lines. This is still in the experimental stage but shows promise of reducing the OH background by a factor of 30–60 in the *H* band. A series of narrowband filters with high transmission between the strong OH lines has been proposed (Offer and Bland-Hawthorn 1998), but not yet tried.

Antarctic sites have very dark skies at 2.35–2.45 μm due to a combination of low OH emission and thermal background emission at these wavelengths (Ashley et al. 1996). Compared to Siding Spring Observatory, they find the sky brightness is about 100 times smaller at the South Pole. See also Phillips et al. (1999) for a detailed discussion of the near-IR sky brightness at the South Pole, including the 3–5-μm region.

6.5 Airborne and Space Infrared Missions

As noted in Sect. 2 and this section, observations from ground-based facilities suffer from both limited atmospheric transmission and the extremely high thermal emission from the warm telescope and atmosphere. *Figure 3-35* illustrates the background difference between one of the best terrestrial sites and representative airborne and space missions (see also Sect. 2 for plots of atmospheric transmission). Airborne, balloon-borne, and space-based observatories

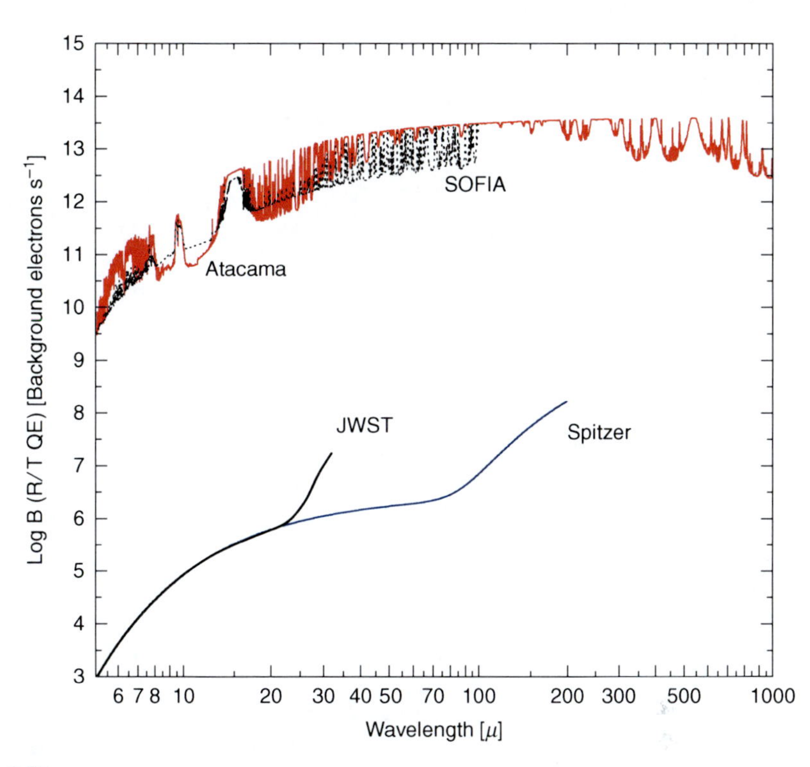

Fig. 3-35

The total backgrounds for the very best Atacama conditions (precipitable water vapor 0.2 mm and seeing of 0.25″) compared with *SOFIA*, *JWST*, and *Spitzer*. Assumed temperatures of Atacama, *SOFIA*, *JWST*, and *Spitzer* are 260, 240, 40, and 5.5 K, respectively. *SOFIA* flies above 12-km altitude. For the *JWST*, the background is set by zodiacal emission except beyond 20 μm where the telescope emission becomes important. Note the calculation for *SOFIA* does not go beyond 100 μm (Adapted from Giovanelli et al. (2001b))

enable measurements that are infeasible from the ground. For example, the 6 orders of magnitude difference in background means that even modest-sized space-based observatories like *IRAS*, *Spitzer*, and *WISE* can be up to 10^3 times more sensitive than ground-based facilities.

6.5.1 Airborne Astronomy

The history of astronomy from aircraft dates back to the 1920s (Dolci 1997), but the first efforts in the infrared began with the work of Kuiper et al. (1967), who measured the infrared spectrum of Venus from the NASA Convair 990 aircraft. They were the first to demonstrate that the atmosphere of Venus was dry, contrary to predictions at the time.

The principal motivation for infrared observing from aircraft is the opening up of the atmospheric transmission between 20 and 1,000 μm. Most of the absorption is due to water vapor in the troposphere, and flights above the tropopause result in dramatic reductions in this

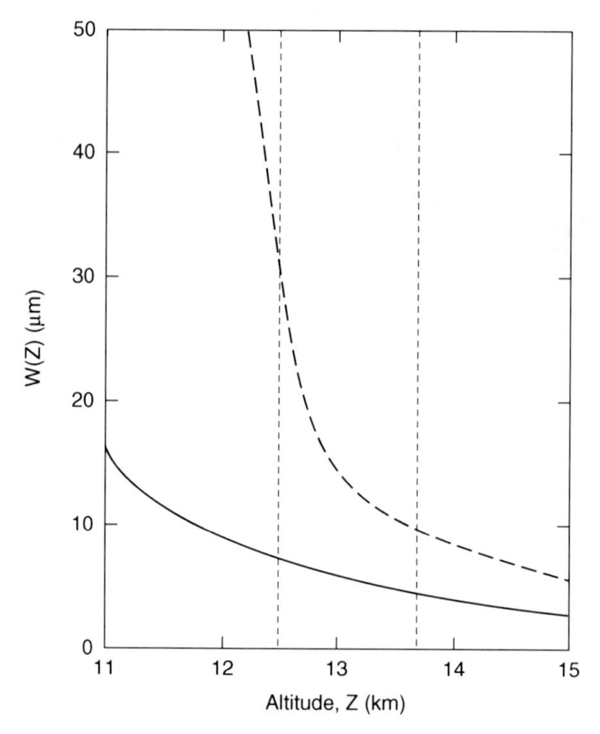

Fig. 3-36

Precipitable column depth of water above altitude Z. Typical *SOFIA* operating altitudes are between 12.5 and 13.7 km (41,000–45,000 feet), shown as *vertical dotted lines*. The *solid curve* is for northern midlatitudes and has 4.2 precipitable microns of water above 13.7-km altitude. The *dashed curve* is the *solid curve* shifted by +2 km, corresponding to a higher tropopause with 10 μm of precipitable water overhead at 13.7-km altitude (Erickson 1998)

absorption. *Figure 3-36* illustrates dependence of precipitable water vapor on altitude and the sudden reduction in water vapor overburden at the tropopause.

Frank Low began a series of pioneering flights with the NASA Learjet to measure the emission from a range of astronomical objects in the late 1960s. Using a 30-cm telescope and a helium-cooled bolometer system, Aumann and Low (1970), for example, made some of the first far-infrared measurements of extragalactic sources. These efforts have been documented by Low et al. (2007). Among the other key discoveries from the Learjet were the observation of the dipole in the cosmic microwave background (Smoot et al. 1977; Smoot and Lubin 1980) and the first detection of the [CII] fine structure line in the ISM (Russell et al. 1980).

Following the Learjet, NASA developed the Kuiper Airborne Observatory (KAO). Originally a Lockheed C-141A cargo aircraft, the airplane was modified to accept a 91-cm diameter telescope (*Fig. 3-37*). The KAO was operated out of Ames Research Center and flew research missions from mid-1974 to fall 1995. The history of the KAO has been summarized by Larson (1992) and Erickson and Meyer (2010). In its 21 years of operation, more than 600 investigators used the facility, resulting in more than a 1,000 publications.

Motivated by the need for a larger, more-capable observatory, NASA, in collaboration with the German Aerospace Center DLR, developed *SOFIA*, a highly modified Boeing 747SP aircraft with a 2.7-m diameter telescope. A cutaway schematic of *SOFIA* is shown in *Fig. 3-38*.

Fig. 3-37
The NASA Kuiper airborne observatory

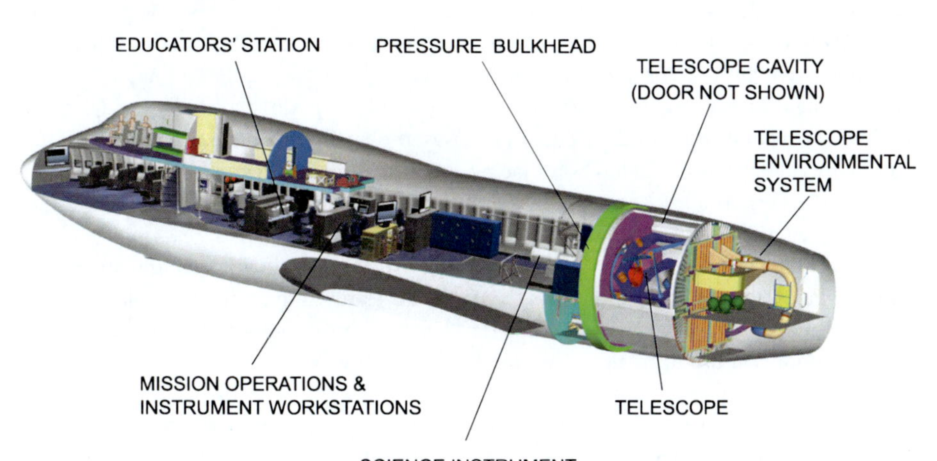

Fig. 3-38
Cutaway view of *SOFIA*. The facility is a highly modified Boeing 747SP with a 2.7-m-diameter telescope (Young et al. 2012)

The telescope resides in an open cavity at the rear of the airplane and is supported on a 1.2-m-diameter spherical bearing. It has an unvignetted elevation range of 23°–58°. The telescope is a conventional Cassegrain system, with the beam sent down the Nasmyth tube that goes through the bearing. This configuration allows the scientific instrument to remain in the pressurized cabin.

SOFIA is based out of the Dryden Aircraft Operations Facility in Palmdale, California, with the *SOFIA* Science Center located at NASA Ames Research Center. Most of the *SOFIA* flights originate and end in Palmdale, but the program also conducts deployments to other airfields to accommodate scientific needs. Two examples of this deployment capability are flights to observe occultations of stars by solar system objects and deployments to the Southern Hemisphere to observe targets not visible on Palmdale-based missions.

SOFIA began science operations in 2010 and has an anticipated 20-year lifetime. SOFIA has a range of instruments including photometers, cameras, grating spectrometers, and heterodyne spectrometers (Young et al. 2012).

6.5.2 Space Infrared Missions

Because of the immense advantages of observing from space, there has been much activity in this area. Initial rocket-borne observations were conducted by Price and collaborators at the Air Force Geophysics Laboratory beginning in the late 1970s (Price et al. 1981). Despite the very limited observing time, these early experiments illustrated the potential power of space infrared observations.

Table 3-7 lists a number of the important astronomical missions in the ensuing years, with references to the papers describing the missions.

IRAS

The *Infrared Astronomical Satellite* (*IRAS*) was launched in 1983 and was the first sensitive all-sky survey in the infrared. The satellite had a 60-cm beryllium telescope and a focal plane that consisted of 62 discrete photoconductor detectors (see *Fig. 3-39*). *IRAS* was launched into a 900-km Sun-synchronous orbit around Earth aligned with the day-night terminator. The inclination of the orbit was such that the orbital precession (approximately $\sim 1^{\circ}$ per day) exactly matched Earth's motion around the Sun. In this way, the satellite maintained a very stable thermal environment with one side always facing the Sun and the other side always in shadow (*Fig. 3-40*). This orbit has been used by a number of other missions, notably *COBE* (Boggess et al. 1992), *Akari* (Murakami et al. 2007), and *WISE* (Wright et al. 2010).

The focal plane array on *IRAS* marked the first use of extrinsic photoconductors in orbit by the civilian astronomical community. These detectors proved to be highly sensitive, but significant complications associated with radiation effects in bulk photoconductors needed to be solved to produce a useful catalog. The correction for these effects is described in Beichman et al. (1988).

The wavelengths observed were 12, 25, 60, and 100 μm. In addition to a point source catalog of more than a quarter of a million sources, *IRAS* also produced images of the whole sky at those wavelengths (Beichman et al. 1988). *IRAS* proved to be an extremely influential mission as the only all-sky catalog in the mid- and far infrared for almost 25 years.

Spitzer Space Telescope

Spitzer (originally named the *Space Infrared Telescope Facility*) was the infrared component of the "Great Observatories" program (Harwit 2009). Spitzer was launched in 2003 and covers the wavelengths 3.6–160 μm. With its 85-cm mirror, this observatory has a relatively modest aperture, but the cold telescope and highly sensitive detectors enable Spitzer to impact all areas of astrophysics. There are three instruments on this observatory. The Infrared Array Camera (IRAC) has three 256×256 pixel cameras operating at 3.6, 4.5, 5.8, and 8.0 μm (Fazio et al. 2004).

Table 3-7
Selected Space Infrared Missions

Mission	Launch	Aperture (m)	Instruments	Function	Wavelength (μm)	Reference
IRAS	1983	0.6	62 detectors	All-sky survey	12, 25, 60, 100	Neugebauer et al. (1984)
			DAX	spectrometer	7.5–23	
COBE	1989	0.6	DIRBE	IR radiometer	1.25–240	Boggess et al. (1992)
			FIRAS	FTS spectrometer	0.1–10 mm	
			DMR	Microwave radiometer	3.3, 5.6, 9.3 mm	
ISO	1995	0.6	SWS	Spectrometer	2.4–45	Kessler et al. (1996)
			LWS	Spectrometer	45–197	
			ISOCAM	Camera	2.5–17	
			ISOPHOT	Photometer	2.5–240	
MSX	1996	0.35	SPIRIT III	6-channel radiometer	4.3–21.3	Egan et al. (1999)
HST	2002	2.4	NICMOS w/cryocooler	Camera	0.8–2.5	Viana et al. (2009)
Spitzer	2003	0.85	IRAC	Camera	3.6, 4.5, 5.8, 8.0	Werner et al. (2004)
			MIPS	Imaging photometer	24, 70, 160	
			IRS	Spectrometer	5.2–38	
Akari	2006	0.68	IRC	Infrared camera	2.4–24	Murakami et al. (2007)
			FIS	Far IR surveyor	65, 90, 140, 160	
HST	2009	2.4	WFC3/ IR	Camera	0.8–1.7	Dressel (2011)
Herschel	2009	3.5	PACS	Photometer/spectrometer	60–210; 51–220	Pilbratt et al. (2010)
			SPIRE	Photometer/spectrometer	250, 350, 500;	
			HIFI	Heterodyne spectrometer	157–212; 240–625	
Planck	2009	1.9 × 1.5	LFI	Radiometer	4.3–10 mm	Planck Collaboration et al. (2011a)
			HFI	Radiometer	0.35–3 mm	
WISE	2009	0.4	4 Arrays	All-sky survey	3.4, 4.6, 12, 22	Wright et al. (2010)
JWST	2018	6.5	NIRCam	Camera	0.6–5	Gardner et al. (2006) Clampin (2011)
			NIRSpec	spectrometer	0.6–5	
			MIRI	Camera/spectrometer	5–28	

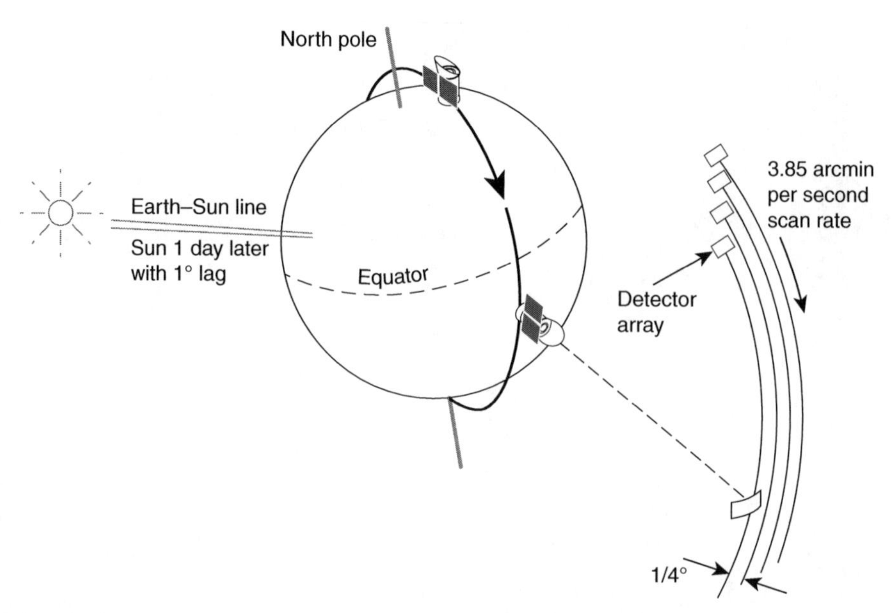

Fig. 3-39
***IRAS* orbit. The satellite orbit lies at the day-night terminator. Orbital precession moves this orbital plane in step with the motion of Earth around the Sun, maintaining this favorable geometry (Beichman et al. 1988)**

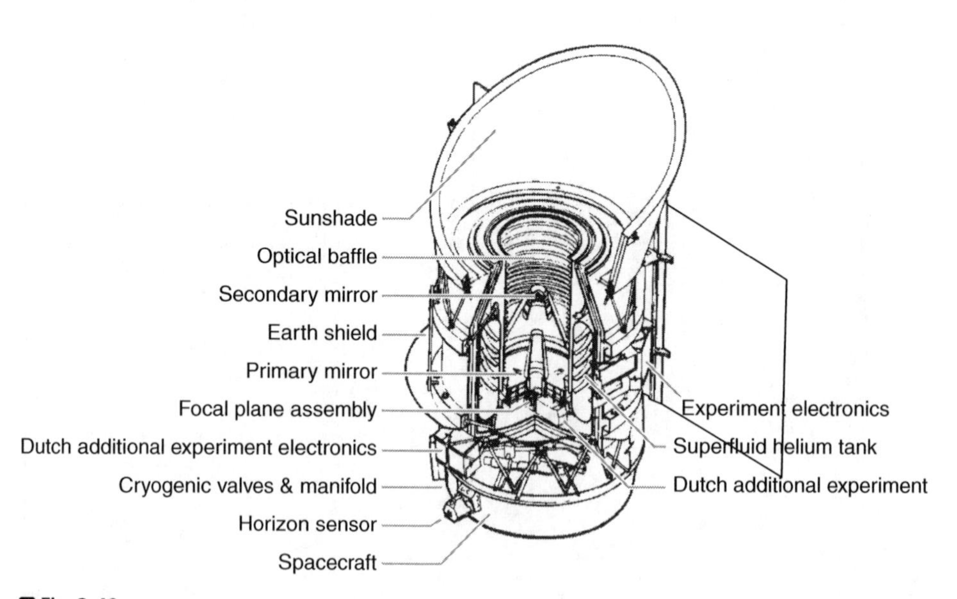

Fig. 3-40
Cutaway schematic of the *IRAS* satellite showing the 60-cm telescope enclosed in the cryostat (Beichman et al. 1988)

The Multiband Imaging Photometer for Spitzer (MIPS) had three imagers operating at 24, 70, and 160 μm (Rieke et al. 2004). The infrared spectrometer (IRS) provided low- and medium-resolution (R ~ 60–600) spectra between 5 and 37 μm (Houck et al. 2004).

Spitzer pioneered several key innovations that have been adopted by subsequent missions. *Spitzer* demonstrated the "warm launch" concept for the first time. Previous missions like *IRAS* and *ISO* had enclosed the telescope in the cryostat for launch. While simplifying the testing and verification phase of development, enclosing the telescope clearly had limits as larger and larger telescopes were required. Instead, *Spitzer* enclosed only the instrument package in the cryostat, greatly reducing the total cold volume. The telescope was cooled after launch with a combination of radiative cooling and cold vapor from the helium boil off. By decoupling the size of the telescope from the size of the instrument package, the warm launch of *Spitzer* validated the warm launch concept, thereby clearing the path for much larger missions such as *Herschel* and *JWST*. ➲ *Figure 3-41* shows a cutaway view of the *Spitzer* observatory, illustrating the separation of the telescope from the cryostat.

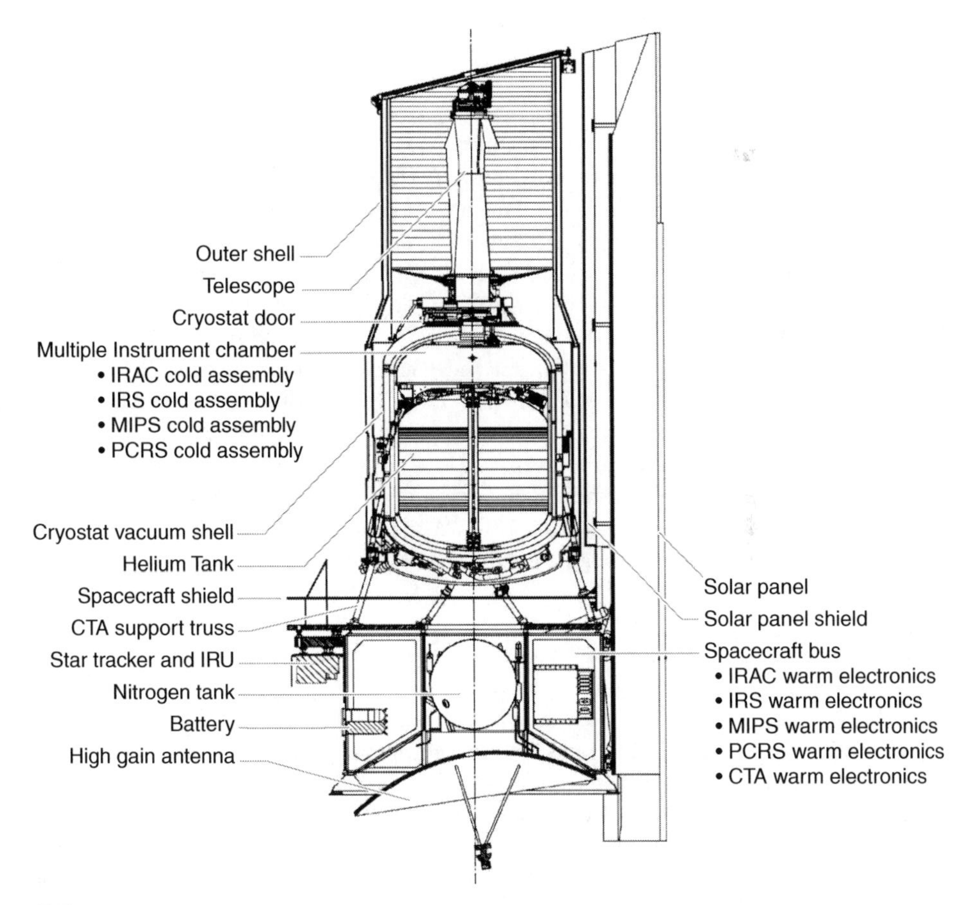

Fig. 3-41

Cutaway view of Spitzer space telescope. The observatory is approximately 4.5 m tall and 2.1 m in diameter (Werner et al. 2004)

Spitzer's second key innovation was demonstrating the importance of radiative cooling in infrared missions. Spitzer was launched into an Earth-trailing drift-away orbit around the Sun, which has many significant benefits (Kwok 1993). By leaving the Earth's vicinity, the largest heat load on the observatory after the Sun was eliminated. Rather than having half the sky filled with a 300-K blackbody, which is the case for near-Earth-orbiting satellites, the solid angle subtended by Earth quickly became insignificant thermally. By careful design of the solar shielding, Spitzer radiatively cooled to an outer shell temperature of 34–34.5 K (Werner et al. 2004). This low outer shell temperature greatly reduced parasitic thermal loads on the superfluid helium, and Spitzer was able to attain a cryogenic lifetime of more than 5 years with only 337 liters of helium. In comparison, *IRAS* had a load of 500 liters and lasted 11 months. Radiative cooling of the telescope in an orbit far from Earth is an important part of the mission design for *Herschel* and *JWST*. Those two missions, however, utilize halo orbits at the L2 Lagrangian point of the Earth-Sun system rather than the Earth-trailing orbit of *Spitzer*.

The ability to radiatively cool the telescope and detectors has allowed *Spitzer* to remain operation even after the depletion of the helium. The focal plane arrays equilibrated to a temperature of ~30 K, cold enough to allow operation of the two shortest IRAC bands. As of this writing, Spitzer continues to produce unique data, nearly 8 years after launch and 3 years after helium depletion.

Herschel Space Telescope

The Herschel Space Telescope (Pilbratt et al. 2010), one of the Cornerstone ESA missions, was launched on 2009 May 14 into an L2 halo orbit. With a 3.5-m silicon carbide primary mirror, it is the largest infrared telescope ever launched, a key consideration for the far-infrared and submillimeter wavelengths used by Herschel. The observatory has three instruments. The Photodetector Array Camera and Spectrometer (PACS) has a camera and an integral field spectrometer for the wavelength range 60 – 210 μm (Poglitsch et al. 2010). The Spectral and Photometric Imaging REceiver (SPIRE) is a submillimeter camera and spectrometer for Herschel. It is a three-band imaging photometer and Fourier transform imaging spectrometer covering the wavelength range of 194 – 671 μm (Griffin et al. 2010). The Heterodyne Instrument for the Far-Infrared (HIFI) consists of seven single pixel heterodyne receivers covering the range 480 – 1250 GHz (625 – 240 μm) and 1410 - 1910 GHz (212 – 157 μm). The spectrometers associated with HIFI provide extremely high spectral resolution, and provide velocity resolution of up to 0.1 km s^{-1} (de Graauw et al. 2010). As a general purpose observatory with an anticipated cryogenic mission through 2013 March, Herschel has made dramatic impacts on all areas of astronomy.

Wide Field Infrared Survey Experiment

The Wide Field Infrared Survey Experiment (WISE) mapped the entire sky at 3.4, 4.6, 12, and 22 μm. WISE was launched into a sun-synchronous on 2009 December 14 and covered the sky in the four bands until 2010 August 6. The optical system consists of an afocal 40 cm diameter telescope, scan mechanism, and camera cooled with solid hydrogen (Wright et al. 2010). By taking advantage of modern detector technology, WISE provides more than 100 times the sensitivity of IRAS at 12 μm, and is the only all-sky survey at 3.4 and 4.6 μm except for the COBE catalog which only had 0.7-degree angular resolution (Smith et al. 2004). The WISE All-Sky Data Release is available at the WISE web site (http://wise2.ipac.caltech.edu/docs/release/allsky/). The data release provides photometric and positional information for more than 563 million objects as well as images in the four bands. After the depletion of cryogens on 2010, operations were continued until 2011 Feb in the two shortest wavelengths. An enhancement in the data processing of the WISE dataset named NEOWISE allows the detection of moving objects, with particular emphasis on discovering Near Earth Objects (NEO) (Mainzer et al. 2011).

James Webb Space Telescope

The *James Webb Space Telescope* (JWST; see ➋ *Fig. 3-42*) represents the transition from monolithic telescopes to ones that must be deployed after launch. With an effective diameter of 6.5 m, no available rocket would be able launch the telescope if it were built as a single monolithic mirror. Instead, the telescope is made of 18 hexagonal beryllium mirrors that will be deployed after launch to produce a collecting area of more than 25 m^2. *JWST* will have four scientific instruments. The Near-Infrared Camera (NIRCam) will cover 0.6–5 μm using large-format focal plane arrays. The Near Infrared Spectrograph (NIRSpec) will produce multiobject spectra between 1 and 5 μm. The Mid-Infrared Instrument (MIRI) produces both images and spectra from 5 to 27 μm. The Fine Guidance Sensor includes a Tunable Filter Instrument (TFI) that will permit narrowband imaging between 1 and 5 μm.

JWST is scheduled for launch aboard an Ariane 5 rocket in 2018 and is designed to orbit about the L2 Lagrangian point of the Earth-Sun system. This particular orbit is well suited to infrared missions. With a nearly constant Earth-Sun geometry, *JWST* will have a very stable thermal environment, far removed from the heat of Earth. A number of other spacecraft have also used this orbital configuration, notably *Wilkinson Microwave Anisotropy Probe* (*WMAP*), *Herschel*, and *Planck*. To take full advantage of the L2 orbit, *JWST* incorporates a highly effective five-layer sunshade to minimize thermal loading from the Sun (Clampin 2011).

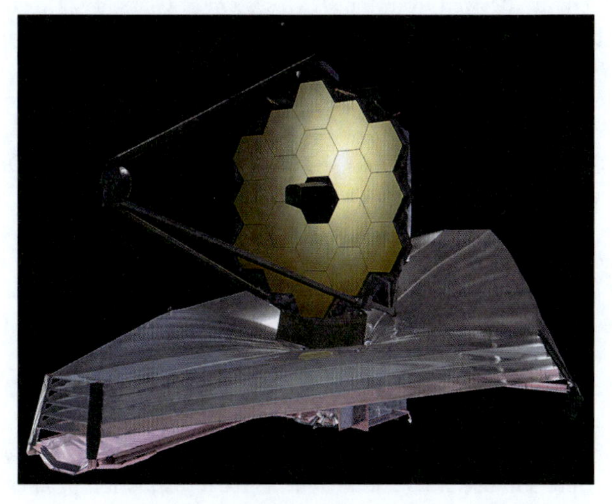

◘ Fig. 3-42
September 2009 artist conception of JWST (NASA)

Key References

Howell, S. B. 2006, Handbook of CCD Astronomy (Cambridge: Cambridge University Press)
McLean, I. 2008, Electronic Imaging in Astronomy: Detectors and Instrumentation (2nd ed.; New York: Springer). Introduction to all aspects of IR astronomy
Merline, W. J., & Howell, S. B. 1995, A realistic model for point-sources imaged on array detectors: the model and initial results. Exp. Astron. 6, 163. Comprehensive discussion of detector noise including the effects of digitization noise on the signal-to-noise
Papoular, R. 1983, The processing of infrared sky noise by chopping, nodding and filtering. A&A, 117, 46. Basic concepts of how to suppress sky noise
Price, S. D. 2009, Infrared sky surveys. Space Sci. Rev., 142, 233. Comprehensive history of IR astronomy from space

7 Infrared Standards and Absolute Calibration

This chapter describes IR photometric systems in use and the absolute calibration to allow conversion of magnitudes to flux density units. Note that we are not able to cover all the nuances of the various photometric systems, and this subject is continuing to evolve. The reader should always check the primary references.

7.1 Ground-Based Photometry

The words of Bouchet et al. (1991, p. 409) should be kept in mind when practicing IR photometry: "It is usually believed that there are as many *JHKLM* photometric systems as observatories. Although all of them are derived from the *JHKL* system of Glass (1974), which followed the pioneering work by Johnson (the 'Arizona' system defined by Johnson (1965) and Johnson et al. (1966)), and use the 'same' InSb detectors, the filters used can be significantly different, and the passbands are affected by the atmospheric transparency which varies from one site to the other. The detectors can also be different."

This section is an overview of infrared photometry as practiced today. Detailed comparison of one photometric system to another requires keen attention to details of the photometric zero points (what defines 0 mag in the system), filter profiles, adopted filter wavelength definition, color transformations, and adopted absolute calibration for 0 mag. A comprehensive list of optical and IR photometric systems is given in the Asiago Database on Photometric Systems (Moro and Munari 2000). As of 2004, there were 226 systems listed on their website, http://ulisse.pd.astro.it/Astro/ADPS/.

As noted above, there is no standard infrared photometric system in use. The set of filters in use for ground-based infrared astronomy is matched to the regions of high atmospheric transmission (see *Fig. 3-7*). Unfortunately, the fabrication of such filters has evolved over time with no standard set of filters in use. Some of the older filter transmission profiles were very wide, and as a result, the effective width of the filters was set by the atmospheric absorption. In this case, changes in the water vapor content in the atmosphere can affect the photometric accuracy. See, for example, Stephens and Leggett (2004) and Cohen et al. (2003b).

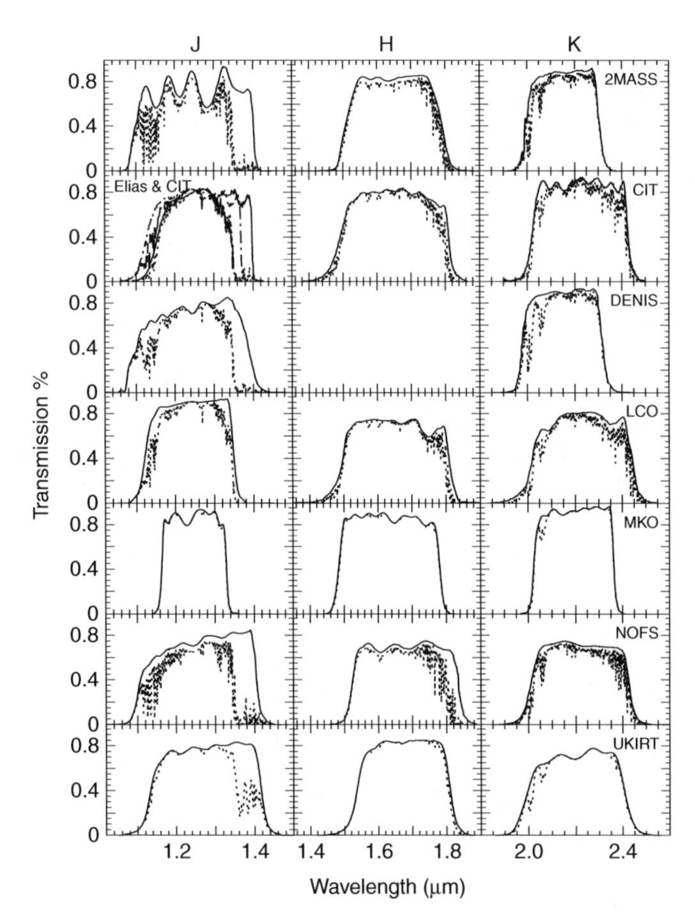

Fig. 3-43
Filter profiles (*solid lines*) and atmospheric transmission (*dashed lines*) from Stephens and Leggett (2004). Note that in some cases, the atmospheric absorption defines the effective width of the filter. This is the case for the 2MASS, CIT, DENIS, NOFS, and UKIRT *J* filters, for example. The MKO filter set was designed to avoid this problem

A discussion of the filter profiles, the effect of the atmospheric absorption, and color transformations between photometric systems is given by Stephens and Leggett (2004). They also show the different filter profiles in use (*Fig. 3-43*), and this demonstrates clearly why there is a problem in comparing results from observatories that use different filters. Limitations of the infrared filter sets are discussed by Milone and Young (2007) and Tokunaga and Vacca (2007). Milone and Young argue for the use of narrowband filters to allow proper transformation of colors from one system to another. This has not been adopted thus far due to the lower system throughput (and, hence, signal-to-noise) compared with using wideband filters.

Simons and Tokunaga (2002) and Tokunaga et al. (2002) proposed a set of near-infrared filters ($JHKL'M'$) that minimizes the effects of telluric absorption, the so-called Mauna Kea Observatories (MKO) filter set. These filters have been adopted at many observatories. A summary of the characteristics of these filters is given in Tokunaga et al. (2002).

Table 3-8
Selected photometric systems

Name	Bands	References	Notes
Johnson (Arizona)	*JHKLM NQ*	1,2,3,4,5	Earliest near-IR system established
CIT/CTIO (Elias)	*JHK*	6,7	Northern and southern hemispheres
Persson (LCO)	$JHKK_s$	8	Set of faint standards based on CIT/CTIO
ESO	*JHKL′MNQ*	9,10,11	
UKIRT (MKO)	*JHKL′M′*	12,13,14	UKIRT and MKO standards
SAAO (Carter)	*JHKL*	15,16	
Bessell and Brett	*JHKL*	17	Stellar colors on Johnson-Glass system
Tenerife (Kidger)	*JHK*	18,19	Supports Cohen et al. (1999)
2MASS	$JHKK_s$	20,21	Based on Persson standards
UKIDSS	*ZYJHK*	22,23	Based on 2MASS point source catalog

Notes: Only key references are given here. It is necessary to check these references for how the system was developed, the list of standard stars, and color transformation between the various systems
References: (1) Johnson (1965) (2) Johnson et al. (1966) (3) Campins et al. (1985) (4) Rieke et al. (1985) (5) Rieke et al. (2008) (6) Elias et al. (1982) (7) Elias et al. (1983) (8) Persson et al. (1998) (9) Bouchet et al. (1991) (10) Bersanelli et al. (1991) (11) van der Bliek et al. (1996) (12) Hawarden et al. (2001) (13) Leggett et al. (2003) (14) Leggett et al. (2006) (15) Carter (1990) (16) Carter and Meadows (2001) (17) Bessell and Brett (1988) (18) Kidger and Martín-Luis (2003) (19) Cohen et al. (1999) (20) Skrutskie et al. (2006) (21) Cohen et al. (2003b) (22) Hewett et al. (2006) (23) Hodgkin et al. (2009)
Other standard star list: Hunt et al. (1998)

Infrared photometric systems have been established at major observatories, and a selected set is shown in *Table 3-8*. Glass (1999) and Bessell (2005) discuss these and other photometric systems.

In some photometric systems, the zero point of the photometric system is determined by assuming that the magnitude of Vega is zero. This is the case, for example, in those defined by Elias et al. (1982). Since the magnitude of Vega at visible wavelengths is usually taken to be 0.02 or 0.03, some care is needed in comparing magnitudes at visible and infrared wavelengths.

Lists of near-IR standard stars are given in the references in *Table 3-8*. For the mid infrared (8–25 μm), the choices for standard stars are much more limited since only the brightest stars in the sky are suitable due to the high background emission. Lists of bright mid-IR standards suitable for 3–6-m telescopes can be found in Hanner et al. (1984), Tokunaga (1984), Rieke et al. (1985), Bouchet et al. (1989), Hammersley et al. (1998), Cohen et al. (1999) and Ishihara et al. (2006). Note that the late-type stars (spectral type M) are variable and should be avoided where possible, or they should be used only as secondary standards. A list of standard stars derived from Cohen et al. (1992, 1995) is given by http://www.gemini.edu/sciops/instruments/michelle/calibration?q=node/10188. A fainter set of mid-IR standards for large telescopes can be found at http://www.eso.org/sci/facilities/lasilla/instruments/timmi/html/stand.html.

Bessell and Brett (1988) give the colors of main-sequence stars in the Johnson-Glass system, and Leggett et al. (2002) give the colors for late-M, L, and T dwarfs. Color transformation from one system to another is given in the references provided in *Table 3-8*. Color transformation from 2MASS to other systems is given by Carpenter (2001; see also an update at http://www.astro.caltech.edu/~jmc/2mass/v3/transformations/. The color transformation for L and T dwarfs is given by Stephens and Leggett (2004).

7.2 Near-IR Sky Surveys

Between 1997 and 2001, the entire sky was surveyed at *JH*, and K_s by the Two Micron All-Sky Survey (2MASS) project (Skrutskie et al. 2006). This was the first all-sky near-IR survey undertaken with an infrared array, and about 500 million point sources and 1.6 million extended sources were cataloged. Because of the accessibility, uniformity, and quality of its data, the 2MASS database has had a substantial impact on research by connecting visible and near-IR photometry and by providing a photometric foundation for the large-area mapping projects that followed. In a more practical application, the availability of near-IR images of the entire sky has allowed looking in regions that are totally obscured by interstellar extinction in the visible and making finding charts.

The 2MASS near-IR images are available through the Image Services at the NASA/IPAC Infrared Science Archive (IRSA): http://irsa.ipac.caltech.edu/index.html. This site also provides access to an extensive set of databases, including *IRAS*, *ISO*, *Akari*, *Midcourse Space Experiment* (*MSX*), *Spitzer*, *WISE*, *Herschel*, and *Planck*.

Deeper large-scale imaging of the sky in the near infrared has been undertaken with larger telescopes and larger focal plane arrays. The UKIRT Infrared Deep Sky Survey (UKIDSS) started in 2005 (Hewett et al. 2006) and is conducted in five filters: *Z*, *Y*, *J*, *H*, and *K*. The *JHK* filters adopted were identical to the MKO near-IR filter set. The *Y* filter was defined in a way to allow cool brown dwarfs to be separated from high-redshift objects. The photometric system for this survey was set up to be consistent with 2MASS (Hodgkin et al. 2009). At approximately the same time, the CFHT WIRCam came online to conduct deep surveys at *J*, *H*, and K_s (http://www.cfht.hawaii.edu/Instruments/Imaging/WIRCam/).

In the Southern Hemisphere, the Visible and Infrared Survey Telescope for Astronomy (VISTA) is conducting six survey programs. VISTA consists of a wide-field telescope with the largest IR camera currently in use with 67 million pixels. It has a set of filters similar to that used by UKIDSS. The survey programs started in 2010 and will continue for 5 years (Emerson and Sutherland 2010).

7.3 Space Infrared Calibration

The situation for calibration of space infrared data is somewhat different from that for ground-based observations. In general, each facility is unique, and the various missions have produced prodigious amounts of data that is internally self-consistent. In particular, the stability of the space environment (except for instrumentally induced effects) has made possible the pursuit of calibration at precision levels and accuracy levels unattainable for ground-based observations, which have to contend with highly variable atmospheric transmission and backgrounds. The goal of calibration for space missions has been to tie each facility to absolute physical units and to be consistent with other facilities. The attainment of precise absolute calibration has enabled key areas of research such as debris disks and the cosmic infrared background. This section discusses the calibration strategy for key IR space observatories.

IRAS

As the first all-sky infrared survey, many of the approaches for the calibration of space infrared observations were first used on *IRAS*. In particular, the use of stellar atmospheric models at the shorter wavelengths and the extrapolation and linkage to asteroid thermal models was

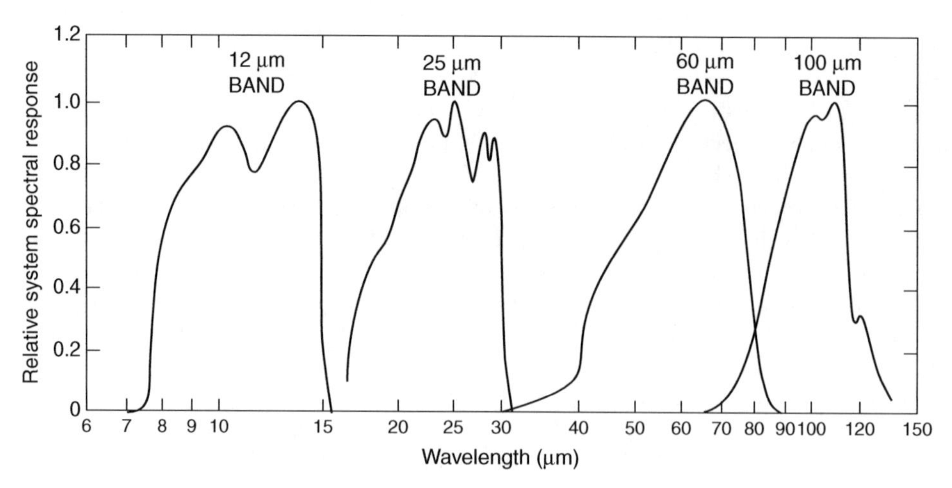

Fig. 3-44
The relative spectral response of the *IRAS* optics and detectors (Form Neugebauer et al. (1984))

pioneered on *IRAS*. *IRAS* had detectors at 12, 25, 60, and 100 μm, and the filter profiles are shown in *Fig. 3-44*.

The *IRAS* calibration is based on the absolute calibration of Rieke et al. (1985) at 10.0 μm, extrapolated to the *IRAS* 12 μm wavelength using stellar models. The primary standard for *IRAS* is α Tau, assuming that the color-corrected flux density at 12 μm is 448 Jy, based on the system of Rieke et al. (1985). Extrapolation to the 25- and 60-μm bands was made for selected calibration stars based on several stellar atmospheric models. The resulting flux densities for α Boo were 448, 103.8, and 18.01 Jy at 12, 25, and 60 μm, respectively. α Lyr (Vega) was one of the fundamental *IRAS* calibration stars, but it was found to have an infrared excess due to a debris disk around the star (Aumann et al. 1984).

For the 100-μm band, the extrapolation from shorter wavelengths was too uncertain to be useful, and observations of the asteroids Hygiea, Europa and Bamberga were compared to the asteroid standard model of Morrison (1973) and Jones and Morrison (1974); see Beichman et al. (1988).

IRAS popularized the use of color corrections in the reporting of flux densities. Since flux densities are specified at a particular wavelength, a correction often needs to be made to accommodate different input spectral energy distributions that are integrated over the broad spectral filters (see Sect. 7.6). The *IRAS* catalog specified fluxes assuming a "flat" input spectrum, i.e., with constant flux per octave. Such a source would have $f_\nu \propto \nu^{-1}$. Color corrections were supplied that allowed adjustments in the flux density so that it would be reported correctly at the standard wavelengths.

MSX

The *Midcourse Space Experiment* (*MSX*) was a US Department of Defense satellite that carried a variety of experiments covering the ultraviolet into the infrared. One of the experiments, the Spatial Infrared Imaging Telescope III (SPIRIT III), had six spectral bands that spanned the mid infrared: 8.28, 4.29, 4.35, 12.13, 14.65, and 21.34 μm, and was particularly important for calibration work in the infrared.

The unique aspect of *MSX* was that it carried emissive reference spheres that were occasionally ejected and measured with the SPIRIT III instrument. These 2-cm-diameter spheres were manufactured out of 6061T6 aluminum alloy and coated with a Martin Black finish (a special coating designed to be very black at IR wavelengths). These spheres had well-determined thermal properties and provided an independent, nonstellar reference source. In all, the irradiance of seven reference stars was determined with 1.4% accuracy (Price 2004).

Spitzer

Spitzer has three instruments spanning 3–160 μm. The Infrared Array Camera (IRAC) is a four-channel camera with dedicated focal plane arrays operating at 3.6, 4.5, 5.8, and 8.0 μm (see ➋ *Fig. 3-45*). With the depletion of helium on *Spitzer*, only the two shortest wavelength channels are operational. The principal calibration strategy on IRAC was to observe a set of selected stars that had been characterized prior to launch. The standards were either A dwarfs or K giants that had been characterized using the methodology of Cohen et al. (2003a). The absolute flux calibration is estimated to be better than 3% in all bands, with the uncertainties primarily dominated by modeling errors (Reach et al. 2005).

IRAC has also proven to be remarkably stable during the more than 7 years of Spitzer operation. Photometry is stable to better than 1.5% over a year (Reach et al. 2005) and better than 0.01% over periods of hours for carefully crafted observations of transiting planets

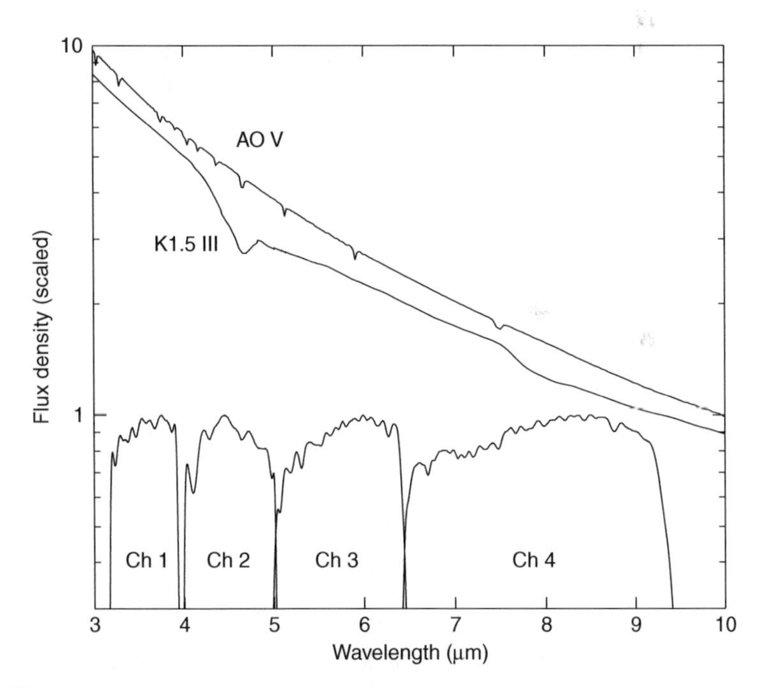

Fig. 3-45
Predicted spectra of two primary calibrators, an A dwarf and a K giant, together with the relative spectra responses of the four IRAC channels. Both the spectra (flux density units Jy) and relative spectral response (in electrons per photon) were scaled arbitrarily to unity to fit on this plot (Reach et al. 2005)

(Charbonneau et al. 2005). This stability has proven to be particularly critical in exoplanet transit observations, where the amplitude of events is typically much less than 1%.

The Multiband Imaging Photometer for Spitzer (MIPS) had three camera modes and one low-resolution spectrometer mode (Rieke et al. 2004). The MIPS detectors were a Si:As impurity band conduction array at 24 μm, a Ge:Ga photoconductor array at 70 μm, and a stressed Ge:Ga array at 160 μm. Because of the significant differences in the physics of these devices, a variety of methods were needed to account for the many instrumental effects. The calibration for the MIPS 24-μm band is based on A-dwarf and solar-type stellar standards developed by Rieke et al. (2008), has been described by Engelbracht et al. (2007), and has an estimated accuracy of 2%. The repeatability of the 24-μm observations of a star in the constant-viewing zone is excellent at 0.4%.

The calibration of the 70-μm channel on MIPS has been described by Gordon et al. (2007). Because this band utilized extrinsic Ge:Ga photoconductors, a number of severe detector-related effects had to be accounted for in the calibration of observations. In particular, the changes in detector responsivity with ionizing radiation and the long, multiple response time constants of the detectors needed to be corrected in the calibration analysis. These corrections in the data pipeline have been described in Gordon et al. (2005). Like the 24-μm channel, the absolute calibration of the MIPS 70-μm band is based on stellar photosphere models and has an estimated accuracy of 5%. The calibration of the spectral energy distribution mode has been documented separately by Lu et al. (2008).

The calibration of the 160-μm channel on MIPS was complicated by a near-infrared (1–1.6 μm) leak in the filter stack. Since the stellar calibrators put out prodigious numbers of photons at the shorter wavelengths relative to 160 μm, the use of stars was precluded since the leak would dominate any signal. Additionally, the 160-μm channel utilized bulk photoconductors that exhibited characteristic nonlinear behavior. To address the spectral leak, Stansberry et al. (2007) developed a calibration using asteroid observations. Observations using the 24-, 70-, and 160-μm bands were taken of calibrator asteroids close in time, and the 160-μm flux density was extrapolated from the two shorter wavelengths using the standard thermal model (Lebofsky and Spencer 1989). Including the uncertainties in the 24- and 70-μm absolute calibrations, Stansberry et al. estimate an uncertainty of 12% for the MIPS 160-μm band.

WISE

WISE is calibrated on the same absolute basis as *Spitzer*. For the wavelengths of *WISE*, a network of calibration stars used by IRAC and MIPS provided the ability for cross calibration. The spectral energy distributions of the calibration stars were constructed by Cohen et al. (2003a) to tie directly to the absolute mid-infrared calibrations by *MSX* (Price 2004). *Figure 3-46* shows the *WISE* spectral bands.

The *WISE* team and *Spitzer* teams collaborated on a survey of the constant-viewing ecliptic pole regions with the object being the verification of the cross calibration of the two missions. The achieved photometric accuracy of the *WISE* calibration, relative to the *Spitzer* and *MSX* systems, was 2.4%, 2.8%, 4.5%, and 5.7% for W1 (3.4 μm), W2 (4.6 μm), W3 (12 μm), and W4 (22 μm), respectively (see *Fig. 3-47*). The *WISE* photometry was internally stable to better than 0.1% over the cryogenic lifetime of the mission (Jarrett et al. 2011). As the largest unbiased catalog of infrared sources in the mid infrared, the *WISE* observations will likely be a basis for future calibration efforts. A comparison of asteroid diameters computed using the standard thermal model and *IRAS* observations has shown good agreement with the *WISE* derivations (Mainzer et al. 2011).

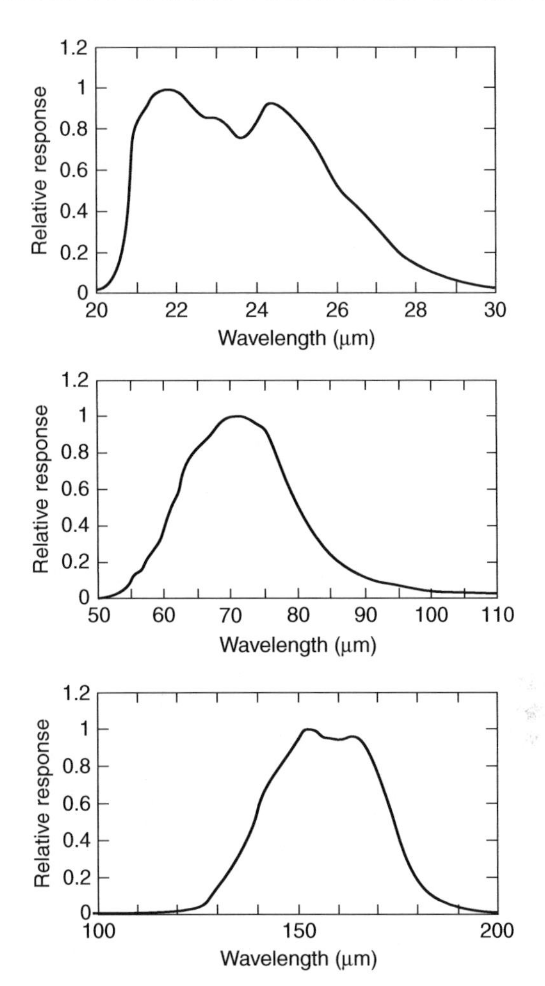

Fig. 3-46

Relative spectral response for the three MIPS bands 24 μm (*top*), 70 μm (*middle*), and 160 μm (*bottom*) (MIPS Instrument and MIPS Instrument Support Teams 2011)

Hubble Space Telescope and JWST

All *Hubble Space Telescope* flux calibrations can be traced to three primary white dwarf standards, G191B2B, GD71, and GD153. The absolute flux calibration is tied to the flux of Vega, determined by Mégessier (1995) to be 3.46×10^{-11} W m^{-2} nm^{-1} at 555.6 nm, and the slopes of the white dwarf spectral energy distributions for extrapolation to longer wavelengths are determined from model calculations (Bohlin et al. 2011). This calibration system is expected to be used for the *JWST* instruments.

Herschel Space Observatory

Flux calibration of *Herschel* data is based on observations and models of a few bright stars, asteroids, and planets (e.g., Griffin et al. 2010; Swinyard et al. 2010; Poglitsch et al. 2010). A discussion

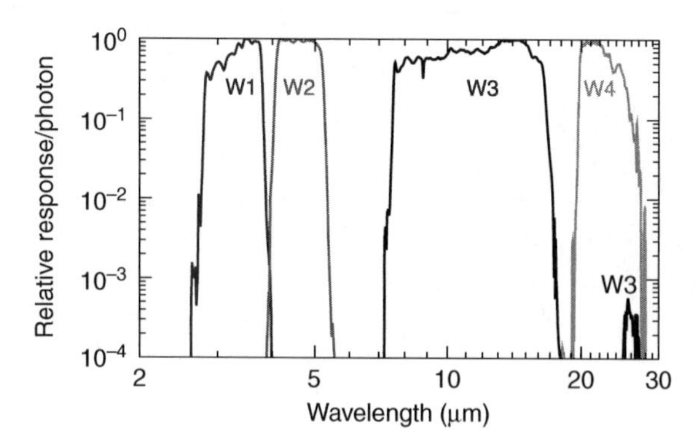

Fig. 3-47
The QE-based (response per photon) relative system response (RSR) curves normalized to a peak value of unity, on a logarithmic scale (Cutri et al. 2011)

of the stellar models used for the calibration can be found in Dehaes et al. (2011). The thermophysical models of the asteroids have been generated by Müller and Lagerros (1998, 2002). Models of various planets have been generated by Moreno (1998). The stellar models are scaled to match the observed K band magnitudes using a zero point derived from the Kurucz theoretical spectrum of Vega, the same one used by Cohen et al. (1992), so these models should be on the same absolute scale as those in the Cohen et al. series of papers.

The calibration for the mid-IR instruments aboard *SOFIA* is based on the same stellar and asteroid models used by *Herschel* (Dehaes et al. 2011; Decin and Eriksson 2007).

7.4 Absolute Calibration

The conversion of magnitudes to flux density (e.g., $\mathrm{W\,m^{-2}\,\mu m^{-1}}$) requires knowledge of the flux density of a 0 mag source above the atmosphere. This section describes the current state of knowledge regarding the absolute calibration in the infrared. This is a complex subject, and we can only touch on the major points in this section. Price (2004) gives a detailed account of various methods used for absolute calibration in the infrared, and he gives references that allow a much deeper study of this subject.

The flux density of Vega (HR 7001) was established as the primary standard by a series of experiments during the 1970s and 1980s that tied the flux from this star to calibrated blackbody sources. See Cohen et al. (1992) and Mégessier (1995) and for references to these papers. As will be seen below, Vega is no longer considered a viable primary flux standard in the infrared. Instead, the definition of the 0 mag flux density is tied to a synthetic stellar spectrum that is close to Vega's spectral type and flux density.

The following is a very brief synopsis of the advances since 1992 in our understanding of the absolute calibration for 1–30 μm:

1. Cohen et al. (1992) used the Hayes et al. (1985) summary as the value of the absolute flux density for Vega at 0.5556 μm, $3.44 \pm 0.05 \times 10^{-8}\ \mathrm{W\,m^{-2}\,\mu m^{-1}}$. Using a carefully chosen atmospheric model for Vega, they derived the absolute flux density from Vega out to 20 μm. Cohen et al. rejected the use of near-IR absolute calibration measurements, citing

reservations about the quality of the observations. Since Vega has cool dust around it that affects the infrared continuum, Cohen et al. recommended that Sirius be adopted as the primary calibrator.

2. Mégessier (1995) critically reviewed the existing set of absolute calibrations measurements at 0.5556 μm and derived a value of $3.46 \pm 0.025 \times 10^{-8}$ W m^{-2}μm^{-1}. Mégessier also used near-IR absolute calibration observations to derive the near-IR absolute flux density from Vega. This method is independent of an atmospheric model.
3. Bohlin and his collaborators established a set of white dwarf and solar-type stars for the spectrophotometric calibration of *HST*. Bohlin and Gilliland (2004) established Vega as an absolute calibrator for *HST* in conjunction with the absolute calibration of Mégessier (1995) and a best-fitting atmospheric model for Vega. This work allowed the fainter white dwarf stars to be absolutely calibrated to Vega.
4. More evidence accumulated regarding the variability of Vega and the fact that it is a pole-on star, thus negating the validity of existing model atmospheres for Vega (see, e.g., Gray 2007). A recent summary of the difficulties with using Vega as a standard is given by Engelke et al. (2010), who recommend 109 Vir and Sirius to be the primary standards in the visible and infrared, respectively.
5. Price et al. (2004) described the results of the *MSX*, which used emissive spheres as calibration sources to obtain absolute calibration of bright mid-IR standards, including Vega and Sirius. Overall accuracy of ~1% was achieved at 4.29, 4.35, 8.28, 12.13, 14.65, and 21.34 μm. This was a significant observational result since it provided a completely independent absolute calibration in the mid infrared that did not depend on stellar models.
6. Cohen and collaborators built upon the absolute calibration determination of Cohen et al. (1992) to construct a network of near-IR and mid-IR standards on a common system through a formidable series of papers during the years 1992–2003. Price (2004) and Price et al. (2004) give a summary of the results of these papers and their connection to *MSX* absolute calibration work. Although the atmospheric model for Vega used by Cohen et al. (1992) is not appropriate for the pole-on case, various methods comparing the Cohen et al. (1992) calibration to direct absolute calibration measurements in the infrared have shown it to be valid at the level of about 2% (Engelke et al. 2010; Price et al. 2004). Tokunaga and Vacca (2005) showed that the absolute calibration of Cohen et al. (1992) and Mégessier (1995) is consistent to about 1% in the near infrared. Price et al. (2004) found the Cohen et al. (1992) is consistent with the highly accurate space-based *MSX* measurements made in the mid infrared to about 1%.
7. Rieke et al. (2008) combined ground-based and *MSX* absolute calibration measurements to construct a flux density spectrum from 0.2 to 30 μm of an A star (with 0 mag) and the Sun. Two methods were used to establish the 0 mag absolute calibration – comparisons of black bodies to stellar sources and the comparison of solar-type stars to the Sun. The overall accuracy is 2% at 1–25 μm. Recommended corrections for the *Spitzer* IRAC, 2MASS, and *IRAS* calibrations are provided, thus allowing for a unified absolute calibration for both ground-based and space-based observations.
8. Engelke et al. (2010) used a different approach to derive a 0 mag flux density spectrum of an A star from 0.2 to 30 μm. This spectrum overlaps closely with the Rieke et al. (2008) 0 mag spectrum at wavelengths <2.5 μm, but it is systematically up to 2.6% lower at longer wavelengths. The Engelke et al. (2010) mid-IR 0 mag flux densities are in agreement with the *MSX* absolute calibration, while Rieke et al. (2008) derived an absolute calibration at 10.6 μm that is based on ground-based, solar analog stars, and the MSX results.

9. Bohlin et al. (2011) provided absolute calibration of the *Spitzer* IRAC instrument based on *HST* white dwarf standard stars, which in turn have been calibrated to Vega at 0.5556 μm. These fainter white dwarf stars will be the basis for providing the absolute flux calibration of *JWST*.

We suggest the use of the Engelke et al. (2010) absolute calibration for the ultraviolet to 30 μm. This absolute calibration is consistent with Cohen et al. (1992) and Rieke et al. (2008) to about 2%. For longer wavelengths, one must resort to methods used for the *Herschel* absolute calibration discussed in Sect. 7.3.

7.5 Definition of a Filter Wavelength

One might naively assume that it is a simple matter to determine the effective wavelength of a filter. However, there is no unique way to define the wavelength of a filter that is suitable for all types of objects that might be observed. The crux of the problem is that when we plot data to show the spectral energy distribution (SED) of a source, we assume that the flux density of the source can be presented at single wavelength as a monochromatic flux density. In the case of a broad filter, the SED of a hot star weights the flux density toward shorter wavelengths, while that of a cool star weights it toward longer wavelengths. Therefore, a single wavelength would not accurately represent the monochromatic flux density for both objects. However, a correction factor can be determined so that the use of a single wavelength is possible (see Sect. 7.6).

This discussion covers a few of the common filter wavelength definitions. All filter wavelength definitions have advantages and disadvantages, and in the context of IR filters, this subject has been recently discussed by Tokunaga and Vacca (2005), Rieke et al. (2008), Bohlin et al. (2011), and Bessell and Murphy (2012).

Following Tokunaga and Vacca (2005), we define the isophotal wavelength of the filter. We start with an expression for the number of photoelectrons detected per second from a source with a flux density as a function of wavelength given by $F_\lambda(\lambda)$. This quantity is

$$\begin{aligned} N_p &= \int F_\lambda(\lambda) S(\lambda)/h\nu d\lambda \\ &= \frac{1}{hc} \int \lambda F_\lambda(\lambda) S(\lambda) d\lambda. \end{aligned}$$

$S(\lambda)$ is the total system response given by

$$S(\lambda) = T(\lambda) Q(\lambda) R(\lambda) A_{\text{tel}}$$

where $T(\lambda)$ is the atmospheric transmission; $Q(\lambda)$ is the product of the throughput of the telescope, instrument, and quantum efficiency of the detector; $R(\lambda)$ is the filter response function; and A_{tel} is the telescope collecting area. From the equation for N_p and the mean value theorem for integration, there exists an isophotal wavelength, λ_{iso}, such that

$$F_\lambda(\lambda_{\text{iso}}) \int \lambda S(\lambda) d\lambda = \int \lambda F_\lambda(\lambda) S(\lambda) d\lambda.$$

Note that this formulation is valid only for photon-counting detectors such as those used in the optical, near infrared, and mid infrared. However, for bolometers, the equations should be cast in terms of energy, not photons. See the discussion by Bessell and Murphy (2012) for the appropriate equations for detectors that are not photon counting.

Rearranging this equation, we obtain

$$F_\lambda(\lambda_{\rm iso}) = \langle F_\lambda \rangle = \frac{\int \lambda F_\lambda(\lambda) S(\lambda) d\lambda}{\int \lambda S(\lambda) d\lambda}.$$

$F_\lambda(\lambda_{\rm iso})$ is therefore the mean value of the intrinsic flux density above the atmosphere over the passband of the filter. In other words, $\lambda_{\rm iso}$ is the wavelength at which the monochromatic flux density, $F_\lambda(\lambda)$, equals the mean flux in the passband. In practice, if there is an absorption line at $\lambda_{\rm iso}$, then one must interpolate over the absorption line to determine $F_\lambda(\lambda_{\rm iso})$. Golay (1974) discusses the isophotal wavelength in greater detail.

Since the isophotal wavelength definition requires knowing the SED of the source, there has been criticism of this definition. Rieke et al. (2008) use the mean wavelength defined by

$$\lambda_0 = \frac{\int \lambda S(\lambda) d\lambda}{\int S(\lambda) d\lambda},$$

Bessell and Murphy (2012) recommend the use of only the isophotal wavelength and the pivot wavelength, with the latter defined as

$$\lambda_p = \sqrt{\frac{\int \lambda S(\lambda) d\lambda}{\int \lambda^{-1} S(\lambda) d\lambda}}.$$

They suggest, "The pivot wavelength should be used as part of a description of the filter system, while the isophotal wavelength should be used to plot the fluxes as broadband magnitudes against wavelength" (p. 136). The advantages of this are that the pivot wavelength depends only on the filter profile, and the mean flux density conversion from wavelength to frequency units is given by the simple relationship

$$\langle F_\nu \rangle = \langle F_\lambda \rangle \lambda_p^2 / c$$

On the other hand, the isophotal wavelength incorporates the SED of the source and gives the mean flux density at a single wavelength. Thus, it provides the best approximation to use when plotting an SED.

7.6 Correction to a Monochromatic Wavelength

Calibration of an observation of an object generally entails converting the quantity that the detector system actually counts (e.g., electrons for most near-IR and mid-IR arrays, energy units for bolometers in the far infrared) for the object into astrophysical flux units. Usually this is done by observing another object (a "standard") for which the intrinsic SED (or flux density) is assumed to be known. For a photon-counting device, the ratio of the number of photoelectrons generated during the observations of the object to those generated during the observations of the standard gives the ratio of the fluxes. That is,

$$\frac{\int \lambda F_\lambda^{\rm obj} S(\lambda) d\lambda}{\int \lambda F_\lambda^{\rm std} S(\lambda) d\lambda} = \frac{N_p^{\rm obj}}{N_p^{\rm std}},$$

and therefore, using the above equations, we have

$$F_\lambda^{\rm obj}(\lambda_{\rm iso}^{\rm obj}) = \langle F_\lambda^{\rm obj} \rangle = \frac{N_p^{\rm obj}}{N_p^{\rm std}} \langle F_\lambda^{\rm std} \rangle = \frac{N_p^{\rm obj}}{N_p^{\rm std}} F_\lambda^{\rm std}(\lambda_{\rm iso}^{\rm std}).$$

From this equation, we can see that the difficulty in interpreting the result of this process is not so much the determination of the mean flux density but rather the wavelength at which the mean flux density value is valid. As stated above, the difficulty with the use of the isophotal wavelength lies in the fact that its determination requires knowledge of the SED of the object observed, a property that is usually the goal of the observations in the first place. Furthermore, it clearly varies from object to object, and therefore, the isophotal wavelength for an observation of a cool star or a galaxy will be different from that for a hot star (e.g., a standard star). To account for these differences, it is common to simply adopt a reference wavelength for a filter (often the isophotal wavelength of a standard star) and make corrections to the observed fluxes of other objects based on their assumed SEDs. These corrections, known as color corrections or K-corrections, can be computed for a range of SEDs, such as blackbodies of various temperatures, or power laws with various spectral slopes. The corrections, K, are defined such that

$$F_\lambda^{\rm obj}(\lambda_{\rm iso}^{\rm std}) = K\langle F_\lambda^{\rm obj}\rangle = KF_\lambda^{\rm obj}(\lambda_{\rm iso}^{\rm obj}),$$

which gives the corrected flux density of the object at the reference wavelength (the isophotal wavelength of the standard) and $\langle F_\lambda^{\rm obj}\rangle$ is the measured flux density of the object. The values of K can be computed for any assumed SED shape from the above equations, and a value can then be applied to the observed flux density of an object. Of course, this does assume that the observer has some general knowledge of what the intrinsic SED of the source should be. However, unless the SED for the object is extreme, the K-corrections for most flux distributions are well behaved.

Many observatories (e.g., *IRAS*, *Spitzer*) provide the fluxes of objects at a reference wavelength under the assumption that they exhibit a nominal spectral shape. It is usually assumed that $\nu F_\nu^{\rm nom} = \lambda F_\lambda^{\rm nom} = C$ where C is a constant. With this assumption, we have

$$F_\lambda^{\rm obj} = F_\lambda^{\rm nom}\left(\frac{\lambda_{\rm ref}}{\lambda}\right),$$

where $\lambda_{\rm ref}$ is a reference wavelength and $F_\lambda^{\rm nom}$ is the value of the flux density at the reference wavelength. If we take $\lambda_{\rm ref} = \lambda_{\rm iso}^{\rm std}$, then it is straightforward to show that

$$\langle F_\lambda^{\rm obj}\rangle = F_\lambda^{\rm obj}(\lambda_{\rm iso}^{\rm obj}) = F_\lambda^{\rm nom}(\lambda_{\rm iso}^{\rm std})\frac{\lambda_{\rm iso}^{\rm std}}{\lambda_0}.$$

Since

$$\langle F_\lambda^{\rm obj}\rangle = \frac{hcN_p^{\rm obj}}{\int \lambda S(\lambda)d\lambda},$$

the nominal flux at the isophotal wavelength of the standard $F_\lambda^{\rm nom}(\lambda_{\rm iso}^{\rm std})$ can be computed from the measured flux of the object.

See also other discussions of computing the correction factor by Hanner et al. (1984), Glass (1999), and Bersanelli et al. (1991).

Key References

Glass, I. S. 1999, Handbook of Infrared Astronomy (Cambridge: Cambridge University Press)

Bessell, M. S. 2005, Standard photometric systems. ARA&A, 43, 293. Review of visible and near-IR photometry

Bessell, M. S., & Murphy, S. J. 2012, Spectrophotometric libraries, revised photonic passbands and zero-points for UBVRI, Hipparcos and Tycho photometry. PASP, 124, 140. Good discussion of filter wavelength definitions
Golay, M. 1974, Astrophysics and Space Science Library, Vol. 41, Introduction to Astronomical Photometry (Dordrecht: Reidel)
Price, S. D. 2004, Infrared irradiance calibration. Space Sci. Rev., 113, 409. Comprehensive review of infrared absolute calibration techniques

8 Infrared Spectroscopy

8.1 Spectroscopic Standards (Spectral Libraries)

Although perhaps not yet at the level of detail and accuracy found in the optical, spectral classification in the near infrared has made considerable progress in the last decade or so. Numerous spectral libraries of O through M stars covering various regions of the near-IR wavelength regime have appeared in the literature, as summarized in Table 1 of Rayner et al. (2009). Many diagnostics for estimating the spectral type of an object based solely on its near-IR spectrum have been explored in the literature. These diagnostics are usually established by examining the equivalent widths, or the ratios of equivalent widths, of various atomic and molecular absorption features as a function of spectral type for a sample of stars with well-defined spectral classifications. (Except for the coolest brown dwarfs, these spectral classifications are generally based on optical spectra and definitions.) Near-IR spectral features frequently used to classify cool stars include CO, H_2O, FeH, and various atomic metal lines (e.g., Lançon and Rocca-Volmerange 1992; Meyer et al. 1998; Burgasser et al. 2002; Ivanov et al. 2004; Cushing et al. 2005; Rayner et al. 2009). The identification and wavelengths of a large number of atomic and molecular features present in the near-IR spectra of cool stars (F through T subtypes) can be found in Cushing et al. (2005) and Rayner et al. (2009). The classification of hot stars typically relies on the relative strengths of H and He lines (Hanson et al. 1996; Blum et al. 1997; Hanson et al. 1998; Lenorzer et al. 2002).

8.2 Taking Near-IR Spectra: An Example

The acquisition and reduction of near-IR spectra employ techniques similar to those outlined above for near-IR imaging observations. Typically, sequential observations are obtained with the object located on one of two positions on the slit, labeled A and B. The frames acquired with the object at these two positions are subtracted to remove the dark current and thermal background from the sky as well as the telescope. Dithering along the slit, similar to the standard practice for near-IR imaging, is possible, but not usually done.

➋ *Figure 3-48* shows a raw exposure taken with a 0.8- to 2.4-μm cross-dispersed spectrograph of an M6 V star (GL 406) obtained with the SpeX instrument (Rayner et al. 2003) at the IRTF. The wavelength range is covered in six orders with an Aladdin 3 1024 × 1024 pixel InSb array. Several features seen in this raw image are worth noting. The width of the stellar spectrum (a point source) is determined by the seeing and is about 1 arcsec FWHM in this case (about 7 pixels). The spectra themselves are not continuous; they show the effects of telluric absorption,

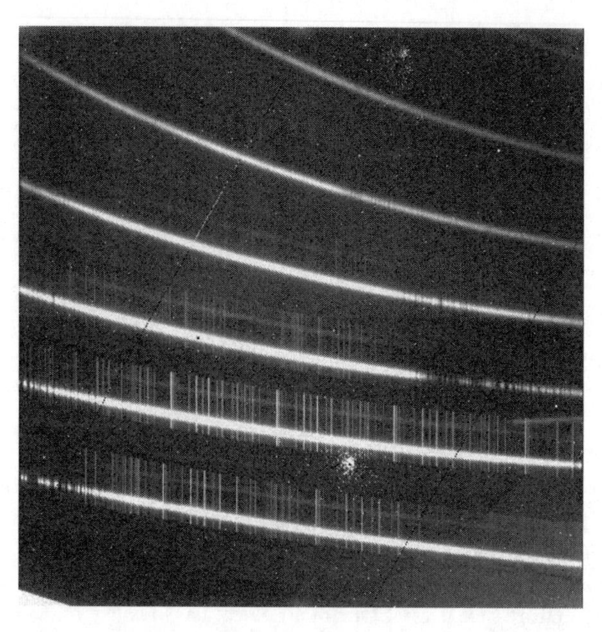

Fig. 3-48

A raw exposure taken with a 0.8- to 5.5-μm cross-dispersed spectrograph (Rayner et al. 2003) of an M6 V star (GL 406) with a resolving power of 2,000. The continuum of the star stretches across the orders and atmospheric absorption lines can be seen. Bright emission lines from the sky (mostly OH) bands are resolved and fill the length of the slit

particularly at ~1.8–2.0 μm (top order) and ~1.35–1.5 μm (second and third orders from the top). Bright emission lines from the sky (mostly OH) are apparent in the *J*, *H*, and *K* bands. These lines are quite well resolved at this resolving power (R = 2,000) and fill the length of the slit. The array itself is not perfect, and "bad" pixels, appearing as both dark (not sensitive to light) spots and bright (very high, unstable dark current) regions, can be seen. Bad pixels occupy ~1% of the array. These are entered into a bad pixel mask and either are ignored in the reduction process or are replaced by interpolation. An area of higher than average dark current is visible in the bottom-right corner of the array. Also visible is a residual image (about 1%) of the spectrum from a previous exposure (see below).

In this example, the various orders are curved on the array, while the 15-arcsec slit (as traced by the emission lines from the sky) is nearly vertical and coincident with the array columns. In many cases, the slit may appear to be tilted or even curved on the array. Although this introduces a number of complications in the data reduction procedure, a tilted slit provides increased spectral sampling of the background, enabling a high-fidelity empirical model of the two-dimensional background emission (or the residual emission after pair subtraction) to be generated using B-splines and removed from the data (e.g., Kelson 2003). Creation of such a background image can be important in the spectroscopic reduction and analysis of very faint objects.

It should be noted that because of differential atmospheric refraction the location of the spectrum along the slit (i.e., along the y-axis of the order in this case) will vary with wavelength. That is, the spectrum will not necessarily be parallel to the edges of the orders.

The usual observing technique involves moving the telescope between two nod positions, which moves the object between two positions on the slit. This results in the spectrum of the object moving between two positions on the array. Although it is possible to dither the object along the slit, in a manner similar to that used in imaging, the more common observing technique is to nod the telescope between only two positions. The frames acquired at these two positions form a pair and can be subtracted to immediately remove the background emission, including any dark current. For this technique to be successful, the array must not be dominated by bad, hot, or dead pixels, and the integration times at the two positions must be equal.

Figure 3-49 shows the result of one A-B image pair in which the star spectrum is positive in the A position at the top of the slit and negative in the B position at the bottom of the slit. Exposure times at each nod position are typically on the order of a few minutes, short enough that the sky emission level does not change substantially and the OH emission lines do not saturate. In this case, the residuals from the subtraction, which are due to the differences in the mean level of the sky emission between the two exposures, should be small. The subtraction also removes any diffuse scattered light. For the example shown, the slit is long enough (15 arcsec) that a point source can be moved between the two nod positions without the spatial point spread functions at the two positions overlapping substantially and with sufficient space on either side

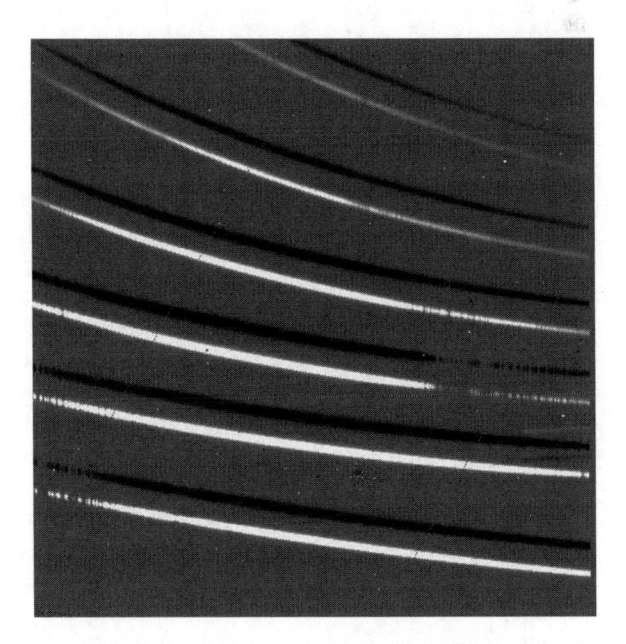

Fig. 3-49

Result of one A-B image pair in which the star spectrum is positive in the A beam (*top* of slit) and is negative in the B beam (*bottom* of slit). The exposures are timed such that the time between the A and B beams is short enough (typically 1 or 2 min) for the sky to be subtracted without any significant residual image. Subtraction also removes diffuse scattered light features

to measure (and fit) the sky residuals as a function of position along the slit at each wavelength. For substantially shorter slits, the two nod positions are located on the source (A beam) and off the source, at some region of blank sky (B beam). The subtraction will then effectively remove the background but yield only a single source spectrum. As many AB pairs are obtained as necessary to build up the desired signal-to-noise ratio.

Flat fields for near-IR spectra are usually generated from internal lamps, by subtracting a lamp-off frame from a lamp-on frame and median combining several of these pairs. The subtraction should remove the dark current and the thermal background produced by the telescope. To remove the spectral signature of the lamp from these frames and differences between the lamp and sky illumination of the slit, the spectra can be smoothed and normalized by fitting them with smooth functions in both the spectral and spatial dimensions. Care must be taken to ensure that real pixel-to-pixel variations are not included in the fits, as these are what the flat fields are designed to remove from the science data. To account for the effects of flexure, the flat-field exposures should be taken at the same telescope elevation angle as that for the object observations.

A full description of the general data reduction procedures for the data shown in *Figs. 3-48* and *3-49* can be found in Cushing et al. (2004). A brief overview is given here.

After pair subtraction, the image is divided by the flat field, the positions of the object spectra are determined at each position on the array (traced), the extraction apertures for the object and the background are determined, and residual background levels at each wavelength are subtracted. The spectra are extracted at each wavelength point by either summing the background-subtracted signal in the spatial direction (along the slit) at each wavelength or by using an optimal extraction technique. The latter technique involves generating the spatial profile of the spectrum at each wavelength and computing the weighted mean of the flux estimates provided by each spatial pixel at each wavelength point. The weights are given by the square of the profile divided by the variance (noise) of the data values at each point. The result of this procedure is a one-dimensional spectrum.

Wavelength calibration of the extracted spectrum can be achieved by observing internal arc lamps whose emission features have well-known wavelengths or from the wavelengths of the observed telluric emission and absorption features. Again, to avoid wavelength discrepancies resulting from flexure, it is advisable that internal arc exposures be obtained at the same telescope elevation angle as those for the object. These wavelength observations should be traced and extracted at the same detector position as the object spectrum. The result provides a conversion between pixel location and wavelength that can be applied to the object spectrum.

Telluric absorption and the instrument response as a function of wavelength can be achieved by observing a spectral standard star whose intrinsic spectrum (in physical units) is known a priori (see Sect. 8.3). The standard star should be observed and reduced in the same manner as the science object. The ratio of the reduced and wavelength calibrated spectrum of the standard star to the known spectrum yields a correction spectrum that can then be applied to a science object (provided that the observations of the science object and the standard star occur at similar airmasses and near in time).

The extracted and fully reduced spectrum of GL 406 is shown in *Fig. 3-50* (Rayner et al. 2009), with the addition of a separate 2.2- to 5.0-μm spectrum. A telluric transmission spectrum is overplotted for comparison. The signal-to-noise in the spectrum is over 100 except in regions of low (less than ~20%) telluric transmission. Further examples of spectra obtained in the mode discussed above are given in *Figs. 3-51* and *3-52*, showing spectral sequences of cool dwarfs at about 1.0 and 2.0 μm (R = 2,000).

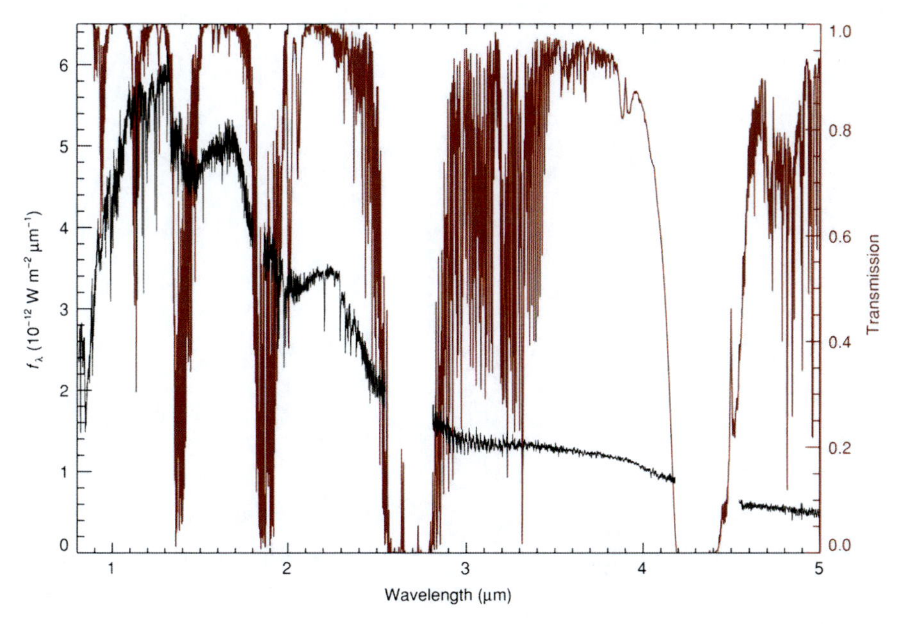

Fig. 3-50
Atmospheric transmission for Mauna Kea (4,200 m, airmass 1.15, precipitable water vapor 2 mm; *red line*) overplotted with the spectrum of Gl 406 (M6V; *black line*) (Rayner et al. 2009)

8.3 Telluric Correction

As discussed in Sects. 2 and 3, near-IR spectra acquired from the ground suffer from several effects introduced by the atmosphere, including the presence of OH emission lines and deep atmospheric absorption features. As long as the OH emission lines are not saturated in the spectra, they can be removed by various methods, depending on the time scale. These include simple subtraction as well as more complicated methods, for example, the use B-splines (Kelson 2003; Bolton and Burles 2007; Bochanski et al. 2009).

The intrinsic source spectrum can never be recovered in completely opaque wavelength regions of the atmosphere. However, the observed spectrum in regions of nonzero absorption can be corrected to yield the source spectrum. The most common method relies on knowledge of the intrinsic spectrum of another object. The observed spectrum of this object can be divided by the known intrinsic spectrum (usually, a model spectrum for the object) to derive the atmospheric transmission, which can then be divided into the observed spectrum of the science target to yield its intrinsic spectrum. Such an object with a "known" intrinsic spectrum is often referred to as a "telluric standard."

Given the spatial and temporal variations in the atmospheric transmission, the success of this method depends on how close the observations of the telluric standard are in time to those of the scientific object of interest as well as how near the two objects are on the sky. However, an all-sky network of objects whose spectra are known and modeled does not yet exist. It is common therefore to choose objects as telluric standards whose spectra are known reasonably well without requiring detailed modeling.

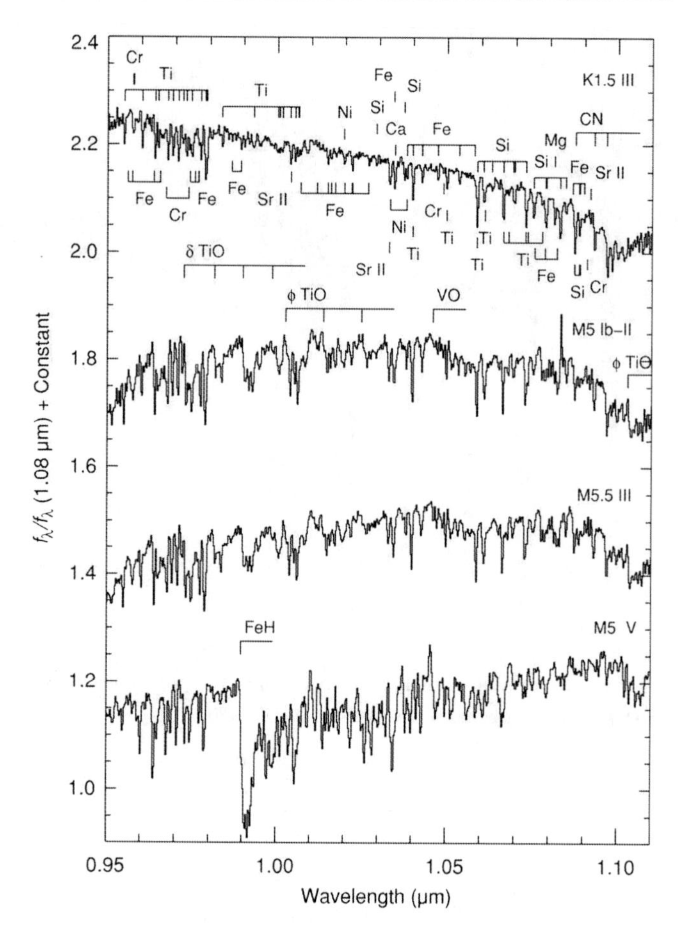

Fig. 3-51
Spectra of spectroscopic standards in the *Y* band (Rayner et al. 2009)

Because the intrinsic spectrum of the sun is considered to be reasonably well known, solar analogs have often been used (e.g., Maiolino et al. 1996). However, the spectra of late-type stars contain numerous weak metal lines whose strengths vary depending on the details of the metal abundances. Therefore, accurate telluric corrections can be achieved only with observations of stars nearly identical to the sun and may also require modifications to the models to match the stellar parameters.

White dwarfs can be used but are fairly faint in the IR. Early type (usually B- and A-type) stars are often selected because their spectra do not contain large numbers of metal lines (whose strength would be dependent on the detailed abundances) or molecular features. Furthermore, the metal lines in these stars are fairly weak. Of course, both of these classes exhibit strong and rotationally broadened hydrogen absorption lines.

Any differences in the strengths or widths of the lines between the observed stellar spectra and the models must be taken into account in the correction process. It is also necessary to smooth the model to observed resolution and bin it to the observed sampling. In addition, it is usually necessary to account for small shifts in the wavelength solutions between the spectra

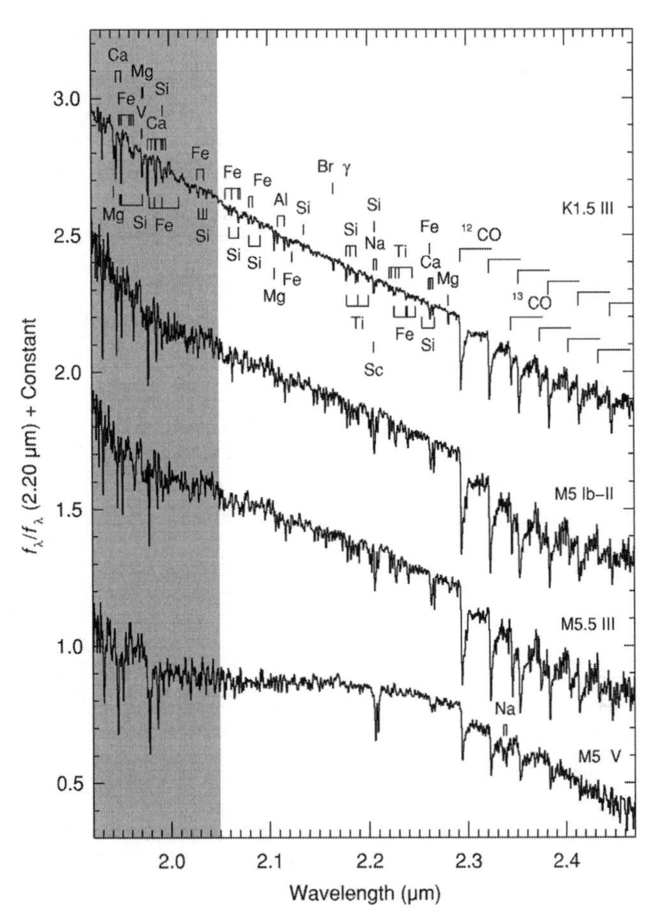

Fig. 3-52
Spectra of standard stars in the *K* band (Rayner et al. 2009)

of the object and the telluric standard that arise as a result of instrument flexure. Differences in the wavelength solution amounting to even a few tenths of a pixel can produce large residual features in the telluric corrected spectrum.

Vacca et al. (2003) describe a software package, originally developed for the SpeX near-infrared spectrograph at the IRTF, that relies on a theoretical spectrum of an A0 V star to generate a telluric correction spectrum from an observation of a star with a similar spectral classification. This software package incorporates all of the aforementioned effects and has been adapted for use with other near-infrared spectrographs. No matter which empirical technique is used, good telluric corrections require very high S/N observations of telluric standards close in time and airmass to those of the science objects and a relatively stable atmosphere (on the time scale of the duration of the object and standard observations). If the strength of the atmospheric features varies wildly on a time shorter than that between the observations of the object and the standard, it will be extremely difficult to derive spectra free of strong residual telluric absorption of emission features. Of course, no telluric correction technique can recover an object's intrinsic spectrum in those regions where the atmosphere is completely opaque.

It is also possible to generate telluric corrections by directly modeling the atmospheric transmission spectrum. Computer codes like ATRAN (Lord 1992) and MODTRAN make use of data for molecular transitions and models of the structure of the Earth's atmosphere (e.g., temperature and pressure profiles) to generate atmospheric transmission spectra at any desired resolution, sampling, airmass, and observatory altitude. These can then be used to correct observed spectra and derive intrinsic spectra. Accurate telluric corrections require knowledge of the precipitable water vapor content and possibly other atmospheric constituents at the time of the observations.

Acknowledgments

J. Rayner and M. Connelley provided some of the content of this chapter.

We are grateful for the assistance of K. Teramura and Louise Good in the preparation of the figures and editing of the text as well as to T. Geballe, R. Knacke, and N. Kobayashi for helpful comments.

References

Abrams, M. C., Davis, S. P., Rao, M. L. P., Engleman, R., Jr., & Brault, J. W. 1994, ApJS, 93, 351

Allen, D. A., & Barton, J. R. 1981, PASP, 93, 381

Ashley, M. C. B., Burton, M. G., Storey, J. W. V., et al. 1996, PASP, 108, 721

Aumann, H. H., & Low, F. J. 1970, ApJ, 162, L79

Aumann, H. H., Beichman, C. A., Gillett, F. C., et al. 1984, ApJ, 278, L23

Bacon, C. M., McMurtry, C. W., Pipher, J. L., et al. 2004, Proc. SPIE, 5167, 313

Beichman, C. A., Neugebauer, G., Habing, H. J., et al. 1988, Infrared Astronomical Satellite (IRAS) Catalogs and Atlases. Volume 1: Explanatory Supplement. http://irsa.ipac.caltech.edu/IRASdocs/exp.sup/credits.html

Benford, D. J., Staguhn, J., Stacey, G., et al. 2004, Proc. SPIE, 5498, 647

Bersanelli, M., Bouchet, P., & Falomo, R. 1991, A&A, 252, 854

Bessell, M. S. 2005, ARA&A, 43, 293

Bessell, M. S., & Brett, J. M. 1988, PASP, 100, 1134

Bessell, M., & Murphy, S. 2012, PASP, 124, 140

Béthermin, M., & Dole, H. 2011, Proceedings Cosmic Radiation Fields: Sources in the Early Universe – CRF2010, Desy, 9–12 Nov 2010

Billot, N., Rodriguez, L., Okumura, K., Sauvage, M., & Agnèse, P. 2009, in EAS Pub. Ser., 37, Astrophysics Detector Workshop 2008, ed. P. Kern, (Cambridge University Press) 119

Bintley, D., MacIntosh, M. J., Holland, W.S., et al. 2010, Proc. SPIE, 741, 7741106

Blackwell, D. E., Leggett, S. K., Petford, A. D., et al. 1983, MNRAS, 205, 897

Bland-Hawthorn, J., Ellis, S. C., Leon-Saval, S. G., et al. 2011, Nat. Commun., 2, 581

Blank, R., Anglin, S., Beletic, J. W., et al. 2011, in ASP Conf. Ser. 437, Solar Polarization, Vol. 6, ed. J. R. Kuhn et al. (San Francisco: ASP), 383

Blum, R. D., Ramond, T. M., Conti, P. S., et al. 1997, AJ, 113, 1855

Boccas, M., Vucina, T., Araya, C., et al. 2004, Proc. SPIE 5494, 239

Bochanski, J. J., Hennawi, J. F., Simcoe, R. A., et al. 2009, PASP, 121, 1409

Boggess, N. W., Mather, J. C., Weiss, R., et al. 1992, ApJ, 397, 420

Bohlin, R. C., & Gilliland, R. L. 2004, AJ, 127, 3508

Bohlin, R. C., Gordon, K. D., Rieke, G. H., et al. 2011, AJ, 141, 173

Bolton, A. S., & Burles, S. 2007, New J. Phys., 9, 443

Bouchet, P., Slezak, E., Le Bertre, T., et al. 1989, A&AS, 80, 379

Bouchet, P., Schmider, F. X., & Manfroid, J. 1991, A&AS, 91, 409

Bratt, P. R., 1977, in Semiconductors and Semimetals, Vol. 12, ed. R. K. Willardson & A. C. Beer (New York: Academic), 39

Burgasser, A. J., Kirkpatrick, J. D., Brown, M. E., et al. 2002, ApJ, 564, 421

Burki, G., Rufener, F., Burnet, M., et al. 1995, A&AS, 112, 383
Burton, M. G. 2010, A&AR, 18, 417
Campins, H., Rieke, G. H., & Lebofsky, M. J. 1985, AJ, 90, 896
Carpenter, J. M. 2001, AJ, 121, 2851
Carter, B. S. 1990, MNRAS, 242, 1
Carter, B. S., & Meadows, V. S. 1995, MNRAS, 276, 734
Charbonneau, D., Allen, L. E., Megeath, S. T., et al. 2005, ApJ, 626, 523
Clampin, M. 2011, Proc. SPIE, 8146, 814605
Cohen, M., Walker, R. G., Barlow, M. J., & Deacon, J. R. 1992, AJ, 104, 1650
Cohen, M., Witteborn, F. C., Walker, R. G., et al. 1995, AJ, 110, 275
Cohen, M., Walker, R. G., Carter, B., et al. 1999, AJ, 117, 1864
Cohen, M., Megeath, S. T., Hammersley, P. L., Martín-Luis, F., & Stauffer, J. 2003a, AJ, 125, 2645
Cohen, M., Wheaton, W. A., & Megeath, S. T. 2003b, AJ, 126, 1090
Compiègne, M., Verstraete, L., Jones, A., et al. 2011, A&A, 525, A103
Content, R. 1996, ApJ, 464, 412
Crisp, D. 2000, in Allen's Astrophysical Quantities, ed. A. N. Cox (New York: Springer), 268
Cushing, M. C., Vacca, W. D., & Rayner, J. T. 2004, PASP, 116, 362
Cushing, M. C., Rayner, J. T. & Vacca, W. D. 2005, ApJ, 623, 1115
Cutri, R. M., Wright, E. W., Conrow, T., et al. 2011, Explanatory Supplement to the WISE Preliminary Data Release Products. http://wise2.ipac.caltech.edu/docs/release/prelim/expsup/wise_prelrel_toc.html
Day, P., Leduc, H., Mazin, B., Vayonakis, A., & Zmuidzinas, J. 2003, Nature, 425, 817
Decin, L., & Eriksson, K. 2007, A&A, 472, 1041
de Graauw, Th., Helmich, F. P., Phillips, T. G., et al. 2010, A&A, 518, L6
Dehaes, S., Bauwens, E., Decin, L., et al. 2011, A&A, 533, 107
Dolci, W. W. 1997, AIAA, Reprint 97-5609
Dressel, L. 2011, Wide Field Camera 3 Instrument Handbook, Version 4.0 (Baltimore: STScI)
Egan, M. P., Price, S. D., Moshir, M. M., Cohen, M., et al. 1999, The Midcourse Space Experiment Point Source Catalog Version 1.2 Explanatory Guide, Air Force Research Laboratory Technical Report AFRL-VS-TR-1999-1522 (Lincoln: Hanscom AFB, Air Force Research Laboratory)
Elias, J. H., Frogel, J. A., Matthews, K., & Neugebauer, G. 1982, AJ, 87, 1029; erratum: AJ, 87, 1893
Elias, J. H., Frogel, J. A., Hyland, A. R., & Jones, T. J. 1983, AJ, 88, 1027
Emerson, J.P., & Sutherland, W.J. 2010, Proc. SPIE, 7733, 773306
Engelbracht, C. W., Blaylock, M., Su, K. Y. L., et al. 2007, PASP, 119, 994
Engelke, C. W., Price, S. D., & Kraemer, K. E. 2010, AJ, 140, 1919
Erickson, E. F. 1998, PASP, 110, 1098
Erickson, E. F., & Meyer, A. W. 2010, NASA SP-Y-2009-571
Fazio, G. G., Ashby, M. L. N., Barmby, P., Hora, J. L., et al. 2004, ApJS, 154, 10
Furlan, E., McClure, M., Calvet, N., et al. 2008, ApJS, 176, 184
Filippenko, A. V. 1982, PASP, 94, 715
Fixsen, D. J., & Dwek, E., 2002, ApJ, 578, 1009
Frey, H. U., Mende, S. B., Arens, J. F., McCullough, P. R., & Swenson, G. R. 2000, Geophys. Res. Lett., 27, 41
Frogel, J. A. 1998, PASP, 110, 200
Gardner, J. P., Mather, J.C., Clampin, M., et al. 2006, Space Sci. Rev., 123, 485
Giovanelli, R., Darling, J., Sarazin, M., et al. 2001a, PASP, 113, 789
Giovanelli, R., Darling, J., Henderson, C., et al. 2001b, PASP, 113, 803
Glass, I. S. 1974, MNASSA, 33, 53
Glass, I. S. 1999, Handbook of Infrared Astronomy (Cambridge: Cambridge University Press)
Golay, M. 1974, Introduction to Astronomical Photometry, Astrophysics and Space Science Library, Vol. 41 (Dordrecht: D. Reidel)
Gordon, K. D., Rieke, G. H., Engelbracht, C. W., et al. 2005, PASP, 117, 503
Gordon, K. D. Engelbracht, C. W. Fadda, D., et al. 2007, PASP, 119, 1019
Gray, R. O. 2007, in ASP Conf. Ser. 364, The Future of Photometric, Spectrophotometric and Polarimetric Standardization, ed. C. Sterken (San Francisco: ASP), 305
Griffin, M. J., Abergel, A., Abreu, A., et al. 2010, A&A, 518, L3
Hammersley, P. L., Jourdain de Muizon, M., Kessler, M. F., et al. 1998, A&AS, 128, 207
Hanner, M. S., Tokunaga, A. T., Veeder, G. J., & Ahearn, M. F. 1984, AJ, 89, 162
Hanson, M. M., Conti, P. S., & Rieke, M. J. 1996, ApJS, 107, 281
Hanson, M. M., Rieke, G. H., & Luhman, K. L. 1998, AJ, 116, 1915
Harwit, M., 2009, Exp. Astron., 26, 163
Hauser, M. G., & Dwek, E. 2001, ARA&A, 39, 249
Hauser, M. G., Arendt, R. G., Kelsall, T., et al. 1998, ApJ, 508, 25

Hawarden, T. G., Leggett, S. K., Letawsky, M. B., et al. 2001, MNRAS, 325, 563
Hayes, D. S., & Latham, D. W. 1975, ApJ, 197, 593
Hayes, D. S., Pasinetti, L. E., & Philip, A. G. D. 1985, in Calibration of Fundamental Stellar Quantities, ed. D. S. Hayes, L. E. Pasinetti, & A. G. D. Philip (Dordrecht: D. Reidel), 225
Herter, T. L., Adams, J. D., De Buizer, J. M., et al. 2012, ArXiv e-prints, 1202.5021
Hewett, P. C., Warren, S. J., Leggett, S. K., & Hodgkin, S. T. 2006, MNRAS, 367, 454
Heyminck, S., Klein, B., Güsten, R., et al. 2010, Twenty-First International Symposium on Space Terahertz Technology (Oxford Univ.), 262
High, F. W., Stubbs, C. W., Stalder, B., Gilmore, D. K., & Tonry, J. L. 2010, PASP, 122, 722
Hunt, L. K., Mannucci, F., Testi, L., et al. 1998, AJ, 115, 2594 (tied to the UKIRT system)
Hodgkin, S. T., Irwin, M. J., Hewett, P. C., & Warren, S. J. 2009, MNRAS, 394, 675
Hoffman, A. W., Corrales, E., Love, P. J., et al. 2004, Proc. SPIE, 5499, 59
Houck, J. R., Roellig, T. L. van Cleve, J., et al. 2004, ApJS, 154, 18
Ishihara, D., Onaka, T., Kataza, H., et al. 2006, AJ, 131, 1074
Ivanov, V. D., Rieke, M. J., Engelbracht, C. W., et al. 2004, ApJS, 151, 387
Iwamuro, F., Motohara, K., Maihara, T., Hata, R., & Harashima, T. 2001, PASJ, 53, 355
Jarosik, N., Bennett, C. L., Dunkley, J., et al. 2011, ApJS, 192, 14
Jarrett, T. H., Cohen, M., Masci, F., et al. 2011, ApJ, 735, 112
Johnson, H. L. 1965, Commun. Lunar Planet. Lab., 3, 73
Johnson, H. L., Iriarte, B., Mitchell, R. I., & Wisniewskj, W. Z. 1966, Commun. Lunar Planet. Lab., 4, 99
Jones, T. J., & Morrison, D. 1974, AJ, 79, 892
Joyce, R. R. 1992, in ASP Conf. Ser. 23, Astronomical CCD Observing and Reduction Techniques, ed. S. B. Howell (San Francisco: ASP), 25
Kaeufl, H. U., Bouchet, P., van Dijsseldonk, A., & Weilenmann, U. 1991, Exp. Astron., 2, 115
Kelson, D. D. 2003, PASP, 115, 688
Kerber, F., Querel, R. R., Hanuschik, R. W., et al. 2010, Proc. SPIE, 7733, 77331M
Kendrew, S., Jolissaint, L., Mathar, R. J., et al. 2008, SPIE, 7015, 70155T
Kendrew, S., Jolissaint, L., Brandl, B., et al. 2010, SPIE, 7735, 77355F
Kenyon, S. L., & Storey, J. W. V. 2006, PASP, 118, 489
Kessler, M. F., Steinz, J. A., Anderegg, M. E., et al. 1996, A&A, 315, L27
Kidger, M. R., & Martín-Luis, F. 2003, AJ, 125, 3311
Killinger, D. K., Churnside, J. H., & Rothman, L. S. 1995, in Handbook of Optics, ed. M. Bass, E. W. V. Stryland, D. R. Williams, & W. L. Wolfe (New York: McGraw-Hill), 44.1
Kimura, M., Maihara, T., Iwamuro, F., et al. 2010, PASJ, 62, 1135
Krisciunas, K., & Schaefer, B. E. 1991, PASP, 103, 1033
Kuiper, G. P., Forbes, F. E., & Johnson, H. L. 1967, Commun. Lunar Planet. Lab., 6, 155
Kwok, J. H. 1993, in Adv. Astronaut Sci. 85, AAS/AIAA Astrodynamics Conference, 16–19 Aug 1993, ed. V. J. Modi et al. (San Diego: Univelt/AAS), 1401
Lançon, A., & Rocca-Volmerange, B. 1992, A&AS, 96, 593
Larson, H. P. 1992, PASP, 104, 146
Lawrence, J. S., Ashley, M. C. B., Burton, M. G., et al. 2002, PASA, 19, 328
Lebofsky, L. A., & Spencer, J. R. 1989, in Asteroids II, ed. R. P. Binzel, T. Gehrels, & M. S. Matthews (Tucson: The University of Arizona Press), 128
Leggett, S. K., Golimowski, D. A., Fan, X., et al. 2002, ApJ, 564, 452
Leggett, S. K., Hawarden, T. G., Currie, M. J., et al. 2003, MNRAS, 345, 144
Leggett, S. K., Currie, M. J., Varricatt, W. P., et al. 2006, MNRAS, 373, 781
Leinert, C., Bowyer, S., Haikala, L. K., et al. 1998, A&AS, 127, 1
Lenorzer, A., Vandenbussche, B., Morris, P., et al. 2002, A&A, 384, 473
Livengood, T. A., Fast, K. E., Kostiuk, T., et al. 1999, PASP, 111, 512
Lombardi, G., Mason, E., Lidman, C., et al. 2011, A&A, 528, 43
Lord, S. D. 1992, NASA Technical Memorandum 103957 (Washington D.C: NASA)
Low, F. J. 1961, JOSA, 51, 1300
Low, F. J. 1984, in NASA Conf. Publ., Vol. 2353, Airborne Astronomy Symposium, ed. H. A. Thronson & E. F. Erickson, (Washington D.C: NASA) 1
Low, F. J., & Rieke, G. H. 1974, in Methods of Experimental Physics: Optical and Infrared, Vol. 12A, ed. N. P. Carleton (New York: Academic), 415
Low, F. J., Young, E., Beintama, D., et al. 1984, ApJ, 278, L19
Low, F. J., Rieke, G. H., & Gehrz, R. D. 2007, ARA&A, 45, 43
Lu, N., Smith, P. S., Engelbracht, C. W., et al. 2008, PASP, 120, 328
Maihara, T., Iwamuro, F., Yamashita, T., et al. 1993, PASP, 105, 940

Maillard, J. P., Chauville, J., & Mantz, A. W. 1976, J. Mol. Spec., 63, 120
Mainzer, A., Bauer, J., Grav, T., et al. 2011, ApJ, 731, 53
Mainzer, A. K., Eisenhardt, P., Wright, E. L., et al. 2005, Proc. SPIE, 5881, 253
Mainzer, A. K., Grav, T., Masiero, J., et al. 2011, ApJL, 737, L9
Maiolino, R., Rieke, G. H., & Rieke, M. J. 1996, AJ, 111, 537
Maloney, P. R., Czakon, N. G., Day, P. K. et al. 2009, AIPC, 1185, 176
Manduca, A., & Bell, R. A. 1979, PASP, 91, 848
Martin, P. G., Miville-Deschênes, M., Roy, A., et al. 2010, A&A, 518, L105
Mason, R., Wong, A., Geballe, T., et al. 2008, Proc. SPIE, 7016, 70161Y
Mather, J.C. 1982, Applied Optics, 21, 1125
McCaughrean, M. J. 1988, Ph.D. Thesis, University of Edinburgh
McLean, I. 2008, Electronic Imaging in Astronomy: Detectors and Instrumentation (2nd ed.; New York: Springer)
Mégessier, C. 1995, A&A, 296, 771
Melis, C., Gielen, C., Chen, C. H., et al. 2010, ApJ, 724, 470
Mellinger, A. 2009, PASP, 121, 1180
Merline, W. J., & Howell, S. B. 1995, Exp. Astron., 6, 163
Meyer, M. R., Edwards, S., Hinkle, K. H., & Strom, S. E. 1998, ApJ, 508, 397
Milone, E. F., & Young, A. T. 2007, in ASP Conf. Ser. 364, The Future of Photometric, Spectrophotometric and Polarimetric Standardization, ed. C. Sterken (San Francisco: ASP), 387
MIPS Instrument and MIPS Instrument Support Teams 2011, MIPS Instrument Handbook, Version 3.0. http://irsa.ipac.caltech.edu/data/SPITZER/docs/mips/mipsinstrumenthandbook/MIPS_Instrument_Handbook.pdf
Miville-Deschênes, M., Martin, P. G., Abergel, A., et al. 2010, A&A, 518, L104
Miyata, T., Kataza, H., Okamoto, Y., et al. 2000, iProc. SPIE, 4008, 842
Miyata, T., Sako, S., Nakamura, T., et al. 2008, Proc. SPIE, 7014, 701428
Molinari, S., Bally, J., Noriega-Crespo, A., et al. 2011, ApJL, 735, L33
Monfardini, A., Swenson, L. J., Bideaud, A., et al. 2011, A&A, 521, A29
Moore, G. 1965, Electronics, 38, 114
Moreno, R. 1998, Ph.D. Thesis, Université de Paris VI
Moro, D., & Munari, U. 2000, A&AS, 147, 361
Morrison, D. 1973, AJ, 194, 203
Müller, T. G., & Lagerros, J. S. V. 1998, A&A, 338, 340
Müller, T. G., & Lagerros, J. S. V. 2002, A&A, 381, 324
Murakami, H., Baba, H., Barthel, P., et al. 2007, PASJ, 59, 369
Naylor, T. 1998, MNRAS, 296, 339
Neugebauer, G., Habing, H. J., Van Duinen, R., et al. 1984, ApJ, 278, L1
Offer, A. R., & Bland-Hawthorn, J. 1998, MNRAS, 299, 176
Oliva, E., & Origlia, L. 1992, A&A, 254, 466
Papoular, R. 1983, A&A, 117, 46
Patat, F. 2008, A&A, 481, 575
Peeters, E. 2011, in IAU Symp. 280, The Molecular Universe, ed. J. Cernicharo & R. Bachiller (Cambridge: Cambridge University Press), 149
Persson, S. E., Murphy, D. C., Krzeminski, W., et al. 1998, AJ, 116, 2475
Phillips, A., Burton, M. G., Ashley, M. C. B., et al. 1999, ApJ, 527, 1009
Phillips, A. C., Bauman, B. J., Larkin, J. E., et al. 2010, Proc. SPIE, 7735, 189
Pilbratt, G. L., Riedinger, J. R., Passvogel, T., et al. 2010, A&A, 518, L1
Planck Collaboration, et al. 2011a, A&A, 536, A1
Planck Collaboration, et al. 2011b, A&A, 536, A18
Planck Collaboration, et al. 2011c, A&A, 536, A24
Planck HFI Core Team, et al. 2011, A&A, 536, A4
Poglitsch, A., Waelkens, C., Geis, N., et al. 2010, A&A, 518, L2
Price, S. D. 2004, Space Sci. Rev., 113, 409
Price, S. D. 2009, Space Sci. Rev., 142, 233
Price, S. D., Murdock, T. L., & Shivanandan, K. 1981, Proc. SPIE, 280, 33
Price, S. D., Paxson, C., Engelke, C., & Murdock, T. L. 2004, AJ, 128, 889
Puget, J. L., Abergel, A., Benard, J.-P., et al. 1996, A&A, 308, 5
Ramsay, S. K., Mountain, C. M., & Geballe, T. R. 1992, MNRAS, 259, 751
Rayner, J. T., Toomey, D. W., Onaka, P. M., et al. 2003, PASP, 115, 362
Rayner, J. T., Cushing, M. C., & Vacca, W. D. 2009, ApJS, 185, 289
Reach W. T., Franz B. A., Kelsall T., & Weiland J. L. 1996, in AIP Conf. Proc., Vol. 348, Unveiling the Cosmic Infrared Background, ed. E. Dwek (Woodbury: AIP), 37
Reach, W. T., et al. 2005, PASP, 117, 978
Ressler, M. E., Cho, H., Lee, R. A. M., et al. 2010, Proc. SPIE, 7021, 7021O
Rieke, G. H. 2003, Detection of Light from the Ultraviolet to the Submillimeter (Cambridge: Cambridge University Press)
Rieke, G. H. 2007, ARA&A, 45, 77
Rieke, G. H. 2009, Exp. Astron., 25, 125
Rieke, G. H., Lebofsky, M. J., & Low, F. J. 1985, AJ, 90, 900

Rieke, G. H., Young, E. T., Engelbracht, C., et al. 2004, ApJS, 154, 25
Rieke, G. H., Blaylock, M., Decin, L., et al. 2008, AJ, 135, 2245
Roe, H. G. 2002, PASP, 114, 450
Rowan-Robinson, M., Hughes, J., Vedi, K., & Walker, D. W. 1990, MNRAS, 246, 273
Rousselot, P., Lidman, C., Cuby, J.-G., Moreels, G., & Monnet, G. 2000, A&A, 354, 1134
Russell, R. W., Melnick, G., Gull, G. E., & Harwit, M. 1980, ApJ, 240, L99
Sánchez, S. F., Thiele, U., Aceituno, J., Cristobal, D., Perea, J., & Alves, J. 2008, PASP, 120, 1244
Saunders, W., Lawrence, J. S., Storey, J. W. V., et al. 2009, PASP, 121, 976
Schöck, M., Els, S., Riddle, R., et al. 2009, PASP, 121, 384
Sebring, T. 2010, Proc. SPIE, 7733, 77331X
Stephens, D. C., & Leggett, S. K. 2004, PASP, 116, 9
Simons, D. A., & Tokunaga, A. 2002, PASP, 114, 169
Skemer, A. J., Hinz, P. M., Hoffmann, W. F., et al. 2009, PASP, 121, 897
Skrutskie, M. F., Cutri, R. M., Stiening, R., et al. 2006, AJ, 131, 1163
Smith, E. C., Rauscher, B. J., Alexander, D., et al. 2009, Proc. SPIE, 7419, 741907
Smalley, B., Gulliver, A. F., & Adelman, S. J. 2007, in ASP Conf. Ser. 364, The Future of Photometric, Spectrophotometric and Polarimetric Standardization, ed. C. Sterken (San Francisco: ASP), 265
Smith, B. J., Price, S. D., & Baker, R. I. 2004, ApJS, 154, 673
Smoot, G. L., & Lubin, P. 1980, ApJ, 234, L83
Smoot, G. F., Gorenstein, M. V., & Muller, R. 1977, Phys. Rev. Lett., 39, 898
Soffer, B. H., & Lynch, D. K. 1999, Am. J. Phys., 67, 946
Soifer, B. T., Helou, G., & Werner, M. 2008, ARA&A, 46, 201
Solomon, S. L., Garnett, J. D., & Chen, H. 1993, Proc. SPIE, 1946, 33
Staguhn, J. G., Benford, D. J., Allen, C. A., et al. 2008, Proc. SPIE, 7020, 702004
Stansberry, J. A., Gordon, K. D., Bhattacharya, B., et al. 2007, PASP, 119, 1038
Stone, R. C. 1996, PASP, 108, 1051
Swinyard, B. M., Ade, P., Baluteau, J.-P., et al. 2010, A&A, 518, L4
Tanvir, N. R., Fox, D. B., Levan, A. J., et al. 2009, Nature, 461, 1254
Thomas-Osip, J. E., McCarthy, P., Prieto, G., Phillips, M. M., & Johns, M. 2010, Proc. SPIE, 7733, 47
Tokunaga, A. T. 1984, AJ, 89, 172
Tokunaga, A. T. 2000, in Allen's Astrophysical Quantities, ed. A. N. Cox (4th ed.; New York: Springer), 143
Tokunaga, A. T., & Vacca, W. D. 2005, PASP, 117, 421
Tokunaga, A. T., & Vacca, W. D. 2007, in ASP Conf. Ser. 364, The Future of Photometric, Spectrophotometric and Polarimetric Standardization, ed. C. Sterken (San Francisco: ASP), 409
Tokunaga, A. T., Simons, D. A., & Vacca, W. D. 2002, PASP, 114, 180
Traub, W. A., & Stier, M. T. 1976, Appl. Opt., 15, 364
Vacca, W. D., Cushing, M. C., & Rayner, J. T. 2003, PASP, 115, 389
Vacca, W. D., Cushing, M. C., & Rayner, J. T. 2004, PASP, 116, 352
van Dishoeck, E. F. 2004, ARA&A, 42, 119
van der Bliek, N. S., Manfroid, J., & Bouchet, P. 1996, A&AS, 119, 547
Vernin, J., Muñoz-Tuñón, C., Sarazin, M., et al. 2011, PASP, 123, 1334
Viana, A., Wiklind, T., Koekemoer, D., et al. 2009, NICMOS Instrument Handbook, Version 11.0 (Baltimore: STScI)
Volk, K., Clark, T. A., & Milone, E. F. 1989, in Lecture Notes in Physics, Vol. 341, Infrared Extinction and Standardization, ed. E. F. Milone (New York: Springer), 15
Vucina, T., Boccas, M., Araya, C., & Ahhee, C. 2006, Proc. SPIE, 6273, 62730W
Werner, M. J., Roellig, T. L., Low, F. J., et al. 2004, ApJS, 154, 1
Werner, M., Fazio, G., Rieke, G., et al. 2006, ARA&A, 44, 269
Wright, E. L., Eisenhardt, P. R. M., Mainzer, A. K., et al. 2010, AJ, 140, 1868
Yasui, C., Kondo, S., Ikeda, Y., et al. 2008, Proc. SPIE, 7014, 701433
Young, E. T., Becklin, E. E., Marcum, P. M., et al. 2012, ApJL, 749, L17
Zmuidzinas, J. 2012, Ann. Rev. Condens. Matter Physics, 3, 169
Zuther, J., Eckart, A., Bertram, T., et al. 2010, Proc. SPIE, 7734, 773448

4 Astronomical Polarimetry: Polarized Views of Stars and Planets

Frans Snik · Christoph U. Keller
Sterrewacht Leiden, Universiteit Leiden, Leiden, The Netherlands

T.D. Oswalt, H.E. Bond (eds.), *Planets, Stars and Stellar Systems. Volume 2: Astronomical Techniques, Software, and Data*, DOI 10.1007/978-94-007-5618-2_4, © Springer Science+Business Media Dordrecht 2013

Abstract: Polarization is a fundamental property of light from astronomical objects, and measuring that polarization often yields crucial information, which is unobtainable otherwise. This chapter reviews the useful formalisms for describing polarization in the optical regime, the mechanisms for the creation of such polarization, and methods for measuring it. Particular emphasis is given on how to implement a polarimeter within an astronomical facility, and on how to deal with systematic effects that often limit the polarimetric performance.

1 Introduction

Light from any astronomical source is only completely described as a vector quantity: its polarization **S** as a function of position $[x, y]$ on the sky, wavelength λ, and time t. As the light from any astronomical source is polarized to some degree, measuring its polarization (i.e., *polarimetry*) by its very nature yields additional information compared to merely measuring its scalar properties. In fact, polarimetry often provides crucial information on the physical state of the medium that produced the light captured by our telescopes. For instance, scattering processes and magnetic fields leave uniquely discernible imprints on polarization, and polarimetry is the only method to unequivocally characterize scattering media and measure the direction and strength of magnetic fields anywhere in the universe in the optical domain of the electromagnetic spectrum. Astronomical polarimetry is therefore capable of addressing challenging science cases for a wide range of targets: from solar system objects to exoplanetary systems, and from our Sun to other stars and stellar systems.

In general, polarization is created and/or modified at any location of interaction with light that involves some kind of breaking of symmetries. And it is such asymmetries that allow us to enhance our understanding of astrophysics beyond simple, symmetric models. For instance, magnetic fields were frequently swept under the rug in many astrophysical models in the past but are now often identified as a crucial, missing ingredient. Jokingly, many astronomers hypothesize that all astronomical phenomena that are not understood are somehow caused by magnetic fields. Magnetic fields are indeed ubiquitous in astronomical contexts at all scales, but their exact influence on many physical processes is often poorly understood. The only way to make progress on the observational side therefore comes from measuring magnetic fields through polarimetry.

Moreover, polarimetry can often derive physical properties at scales well below the spatial resolution attainable with the largest telescopes currently available. Using polarization measurements, one can diagnose magnetic fields on the Sun with subpixel scales, and one can obtain magnetic field maps of spotted, unresolved stars. Furthermore, one can derive the presence of stellar asymmetries and clumps in stellar winds. By measuring the polarization of starlight that has been scattered on micron-sized dust particles, one can not only derive their sizes but also their shapes and material properties. Often, polarimetry can even make the invisible visible. One can, for instance, peek "around the corner" into an obscured source. One can also reduce the unpolarized "glare" from a bright central source such as to image its faint, polarized circumstellar structure that would otherwise be completely swamped in the photons from the central star. This makes polarimetry not only a powerful technique to detect exoplanets through direct imaging, but as the polarization spectrum from an exoplanet contains crucial information on its atmospheric constituents (e.g., water clouds, oxygen, ozone) and surface properties (e.g., oceans, forests, deserts), polarimetry is also uniquely suited to characterize exoplanets and, ultimately, detect (the conditions for) extraterrestrial life.

In the light of its scientific potential, polarimetry is an underexploited technique, and it is often considered difficult to implement and to interpret. On the instrumental side, this view is mostly caused by the fact that polarization optics have completely different functionalities than all of the other optics in astronomical instrumentation, which is often primarily geared toward imaging and/or spectroscopy. The introduction of polarization optics is then often seen as a nuisance because it reduces the overall transmission and optical performance, and adds cost and complexity. Vice versa, the imaging optics often degrade the potential polarimetric performance, particularly when the polarimeter is designed as an add-on to the imaging or spectroscopy common-mode as is very often the case. For instance, a well-performing adaptive optics (AO) system may deliver a good image quality, but its many mirrors may have modified the source polarization and added polarization of its own (instrumental polarization) in the process. This apparent mutual exclusivity of polarization optics and imaging optics is wrong as can be seen when a clear, systematic view on implementing polarimetry within astronomical instrumentation is employed in the design process. Moreover, polarimetry, being a differential technique, is usually fraught with all kinds of systematic errors. Spurious polarization signals can be created by, e.g., varying atmospheric properties or by imperfect knowledge of the optics or the detector properties. Also, polarimetry is a photon-hungry technique requiring large-aperture telescopes and efficient optics. The implementation of polarimetry in an astronomical instrument therefore always requires a careful instrument design.

Polarimetry is a technique that is still maturing, and it can profit significantly from technological developments in various commercial sectors[1]. Polarimetry being an underexploited technique only strengthens the drive to develop better astronomical polarimeters and data-reduction techniques, as well as to develop better design and systems engineering methods for polarimetric instrumentation. Only then can the full potential of polarimetric observations be achieved.

This review gives an overview of opportunities and challenges related to the implementation of astronomical polarimeters. In particular, this chapter focuses on methods to deal with systematic errors that usually plague polarimetric observations. First, the mathematical framework to describe polarization in the optical regime (UV to mid-IR) is introduced in ➋ Sect. 2. The most common mechanisms for the creation (and/or modification) of polarization in an astrophysical context are outlined in ➋ Sect. 3, which implies a myriad of science that polarimetry can address, but in no way intends to provide a complete listing of science cases. We then follow the flux of photons from the astronomical sources through our telescopes and describe how their polarization properties can be measured. In ➋ Sect. 4, the optical components are presented that are typically used in polarimeters as well as their limitations. ➋ Sect. 5 describes various methods for the measurement of polarization, and discusses the sources of error in this measurement in the context of seeing and instrumental polarization due to the telescope and the instrument optics. Various solutions and calibration methods are listed to deal with these systematic effects. Finally, ➋ Sect. 6 describes several successful astronomical polarimeters and presents a brief outlook on future telescopes and instruments.

This review is restricted to the optical regime (~200 nm–20 μm, which comprises UV, visible, and infrared light), and does not cover X-ray or submillimeter/radio polarimetry, however interesting these techniques are, both scientifically and technically. The focus is on techniques and components that are frequently used these days or are particularly promising for the near

[1]The profit may even be bidirectional as some developments for astronomical polarimetry have already found their spin-off in other applications, like remote sensing (Tyo et al. 2006) and biomedical applications.

future. Lengthy mathematical and technical details have been omitted whenever this is possible without confounding the reader's understanding of the big picture. Many such details can be found in the books by Tinbergen (1996) and Clarke (2009) and in the review by Keller (2002a).

2 What is Polarization?

Polarization is a fundamental property of light and it describes the evolution of the electric field vector of the electro-magnetic (EM) wave upon propagation. Several formalisms exist to describe polarization. The Jones formalism describes single EM waves, which, by definition, are 100% polarized and contain phase information. In the context of astronomical polarimetry, this formalism applies to the submillimeter/radio regime, where the detectors are antennae that can directly measure the E-vector. Therefore this formalism is only briefly described. The Stokes formalism applies to intensity measurements of photon fluxes as performed by detectors in the optical regime. It can describe partial polarization, which is particularly useful for astronomical polarimetry because most sources have a degree of polarization of at most a few percent. Interference phenomena can, however, not be described with the Stokes formalism.

2.1 Jones Formalism

A monochromatic electromagnetic wave can be described with the Jones formalism. As the electric and magnetic fields in such a wave are linked through Maxwell's equations, it is sufficient to describe the wave's *polarization* as the spatial and temporal properties of its electric vector. The polarization of a monochromatic EM wave is described by the (complex) Jones vector $\vec{E}$:

$$\vec{E} = \begin{pmatrix} E_x \\ E_y \end{pmatrix} = \begin{pmatrix} \hat{E}_x e^{i(kz-\omega t+\phi_x)} \\ \hat{E}_y e^{i(kz-\omega t+\phi_y)} \end{pmatrix} = \begin{pmatrix} \hat{E}_x \\ \hat{E}_y e^{i\Delta\phi_{x,y}} \end{pmatrix}, \tag{4.1}$$

for absolute and relative phases, respectively. In general, the E-vector outlines an ellipse upon propagation. Linear polarization, where the E-vectors vibrates in one direction only, and circular polarization, where the E-vector is described by perfect left- or right-handed corkscrew are special cases of this. Since Jones vectors describe a fully polarized EM wave, depolarization cannot be taken into account by the 2×2 complex Jones matrices that are used to propagate Jones vectors.

2.2 Stokes Formalism

All astronomical sources are partially polarized, and therefore the creation and measurement of polarization in astronomical contexts is best described with the Stokes–Mueller formalism.

The *Stokes vector* **S** is defined as (Stokes 1852; Chandrasekhar 1946):

$$\mathbf{S} = \begin{pmatrix} I \\ Q \\ U \\ V \end{pmatrix} = \begin{pmatrix} \langle E_x^* E_x + E_y^* E_y \rangle \\ \langle E_x^* E_x - E_y^* E_y \rangle \\ \langle E_x^* E_y + E_y^* E_x \rangle \\ \langle i(E_y^* E_x - E_x^* E_y) \rangle \end{pmatrix} = \begin{pmatrix} I'_{0^\circ} + I'_{90^\circ} \text{ or } I'_{45^\circ} + I'_{-45^\circ} \text{ or } I'_{RHC} + I'_{LHC} \\ I'_{0^\circ} - I'_{90^\circ} \\ I'_{45^\circ} - I'_{-45^\circ} \\ I'_{RHC} - I'_{LHC} \end{pmatrix}. \quad (4.2)$$

The first definition is in terms of ensemble averages of the correlations of the Jones vector components. The second definition gives an operational description in terms of photon flux (I') measurements through polarizers at certain angles and after filtering for right- and left-handed circular polarization (RHC, LHC)[2]. I denotes the intensity regardless of polarization. Q and U describe the (two-dimensional) state of linear polarization, and V represents circular polarization.

The Stokes formalism is complete in the sense that it describes all partially polarized polarization states. Through the incoherent summing in the first definition, the phase information is lost, and indeed the Stokes formalism cannot be used to describe interference phenomena.

The corollary of ➋ Eq. 4.2 gives the linear combination of Stokes parameters for a single intensity measurement after filtering for linear or circular polarization:

$$\begin{aligned} I'_{0^\circ} &= \frac{1}{2}(I + Q); \\ I'_{90^\circ} &= \frac{1}{2}(I - Q); \\ I'_{45^\circ} &= \frac{1}{2}(I + U); \\ I'_{-45^\circ} &= \frac{1}{2}(I - U); \\ I'_{RHC} &= \frac{1}{2}(I + V); \\ I'_{LHC} &= \frac{1}{2}(I - V). \end{aligned} \quad (4.3)$$

From ➋ Eqs. 4.2 and ➋ 4.3 it becomes clear that polarimetry is a differential technique: to verify that light is linearly polarized in the vertical direction, one needs to measure the intensity through a vertical polarizer, but also through a horizontal polarizer to compare with. The polarized Stokes parameters Q, U, and V are all obtained through differences of such photon flux measurements.

In many cases the absolute values for Q, U, and V are not very relevant, and the Stokes vector is normalized by the intensity I. This makes the normalized Stokes parameters independent of

[2]The sign of Q depends on the choice of the observer. It is usually parallel or perpendicular to some convenient direction, such as the coordinate system used on the sky, or the optical table. More fundamental issues are related to the handedness, i.e., the direction from $+Q$ to $+U$, and the sign of V. Particularly the latter has generated a lot of confusion in the literature (see Tinbergen (1996) and Clarke (2009) for extensive discussions on this matter), because the rotation direction of the E-vector changes orientation when looking at the propagating beam from the source's or from the instrument's point of view, and also when considering a co-moving or a fixed reference plane within which the E-vector is evaluated. Several (conflicting) conventions exist for positive handedness, but the best one can hope for is that the signs of Q, U, and V are unambiguously defined for each individual publication.

the absolute photometric accuracy of an instrument. The degree of polarization P is then:

$$P = \frac{\sqrt{Q^2 + U^2 + V^2}}{I} \leq 1. \tag{4.4}$$

It is also useful to define the degree of linear polarization (P_L) and the angle of linear polarization (ϕ_L):

$$P_L = \frac{\sqrt{Q^2 + U^2}}{I}, \tag{4.5}$$

$$\phi_L = \frac{1}{2} \arctan \frac{U}{Q}. \tag{4.6}$$

Finally, also the fractional polarizations Q/I, U/I, and V/I are often used.

2.3 Mueller Matrices

The interaction of Stokes vectors with matter is described by 4 × 4 real *Mueller matrices*:

$$\mathbf{S}_{\text{out}} = \mathsf{M} \cdot \mathbf{S}_{\text{in}}, \tag{4.7}$$

with the general structure of a Mueller matrix given by:

$$\mathsf{M} = \begin{pmatrix} I \to I & Q \to I & U \to I & V \to I \\ I \to Q & Q \to Q & U \to Q & V \to Q \\ I \to U & Q \to U & U \to U & V \to U \\ I \to V & Q \to V & U \to V & V \to V \end{pmatrix}. \tag{4.8}$$

The [1,1] element merely describes the transmission of unpolarized light. When using normalized Stokes vectors, the Mueller matrices can be normalized with the [1,1] element. Mueller matrices for a train of n optical elements are multiplicative, but are not commutative:

$$\mathbf{S}_{\text{out}} = \mathsf{M}_n \mathsf{M}_{n-1} \dots \mathsf{M}_2 \mathsf{M}_1 \cdot \mathbf{S}_{\text{in}} = \mathsf{M}_{\text{total}} \cdot \mathbf{S}_{\text{in}}. \tag{4.9}$$

The Mueller matrix for an ideal *polarizer* with its transmitting axis in the $\pm Q$ direction is given by:

$$\mathsf{M}_{\text{pol}} = \frac{1}{2} \begin{pmatrix} 1 & \pm 1 & 0 & 0 \\ \pm 1 & 1 & 0 & 0 \\ 0 & 0 & 0 & 0 \\ 0 & 0 & 0 & 0 \end{pmatrix}. \tag{4.10}$$

It fully polarizes unpolarized light ($\mathbf{S}_{\text{in}} = (1, 0, 0, 0)^T$) in the $\pm Q$ direction and reduces the light level by half ($\mathbf{S}_{\text{out}} = (\frac{1}{2}, \pm\frac{1}{2}, 0, 0)^T$). Light that is fully linearly polarized in the perpendicular direction ($\mp Q$) is completely blocked.

The Mueller matrix for an ideal retarder (or wave plate; see below) with retardance δ and its fast axis aligned with $+Q$ is given by:

$$\mathsf{M}_{\mathrm{ret}}(\delta) = \begin{pmatrix} 1 & 0 & 0 & 0 \\ 0 & 1 & 0 & 0 \\ 0 & 0 & \cos\delta & \sin\delta \\ 0 & 0 & -\sin\delta & \cos\delta \end{pmatrix}. \tag{4.11}$$

The *retardance* is a phase delay between the components of the propagating E-vector along the *fast axis* and the axis perpendicular to that, the *slow axis* (see *Fig. 4-1*). If the phase delay is zero (or any other multiple of 2π), the retarder does not affect the polarization. For $\delta=\pi/2$, the retarder induces a phase change of a quarter of a wave between the projected components of the input polarization on the retarder's axes (at $\pm Q$ in this case). This *quarter-wave plate* (QWP) therefore transforms U into V and vice versa. For $\delta=\pi$, the retarder is a *half-wave plate* (HWP) and it "mirrors" linear polarization around its fast axis. In this case it flips the directions

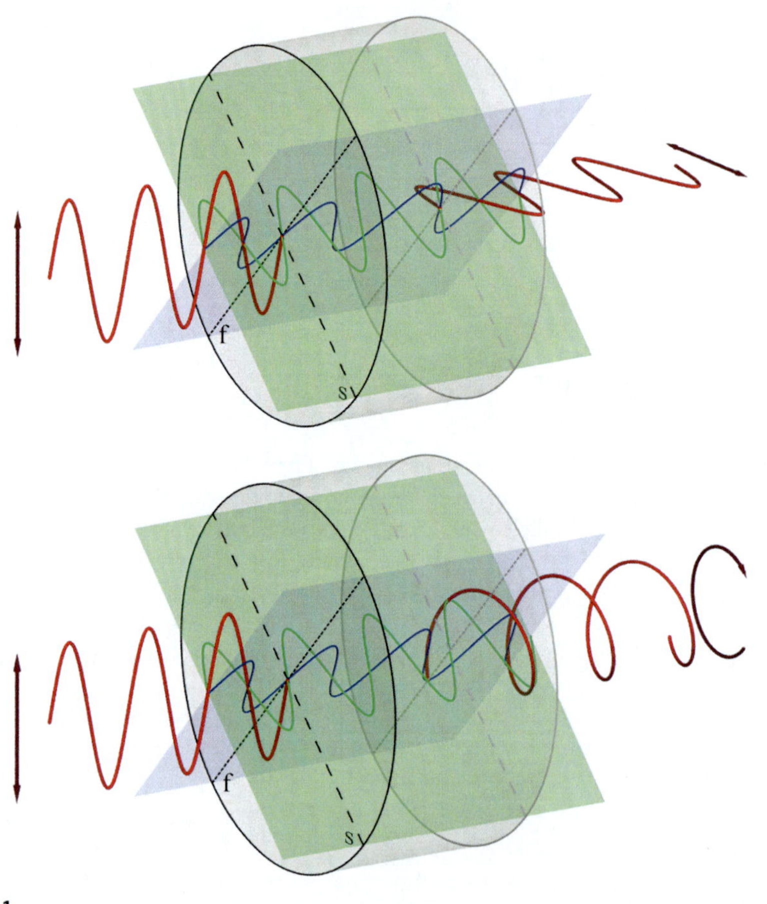

Fig. 4-1
Sketch of the principle of retardance. *above*: the case of a quarter-wave plate; *below*: the case of a half-wave plate

for $\pm U$ and also changes the sign of V. Light that is fully polarized in the $\pm Q$ direction is an "eigenvector" of this retarder, as it is unaffected by it.

➋ Section 4 discusses practical implementations of polarizers and retarders.

The Mueller matrix for a rotation of the Stokes coordinate system by an angle α (measured from $+Q$ to $+U$) around the optical axis is given by:

$$\mathsf{M}_{\mathrm{rot}}(\alpha) = \begin{pmatrix} 1 & 0 & 0 & 0 \\ 0 & \cos(2\alpha) & \sin(2\alpha) & 0 \\ 0 & -\sin(2\alpha) & \cos(2\alpha) & 0 \\ 0 & 0 & 0 & 1 \end{pmatrix}. \tag{4.12}$$

For a freely rotating element with Mueller matrix M, the rotation is described as[3]:

$$\mathsf{M}'(\alpha) = \mathsf{M}_{\mathrm{rot}}(-\alpha) \cdot \mathsf{M} \cdot \mathsf{M}_{\mathrm{rot}}(\alpha). \tag{4.13}$$

Any physical process that creates or modifies polarization, be it in an astrophysical medium or an optical component, can be described by a Mueller matrix. Note that not every 4×4 matrix is a Mueller matrix; it has to obey certain conditions (Landi Degl'Innocenti and Del Toro Iniesta 1998), and can be decomposed into elementary Mueller matrices describing (de-)polarizing and retardance action (Lu and Chipman 1996).

With the three listed elements (polarizer, retarder, and rotation), most manipulations of the Stokes parameters Q, U, and V can be described: the creation of all (fully polarized) polarization states and the measurement of all polarization states. The creation of partially polarized light can also be described with Mueller matrices, e.g., the action of a partial (imperfect) polarizer or a depolarizer. The latter has no practical counterpart as an optical component, but in real physical environments or instruments depolarization does occur, for instance, due to scattering. The incoherent superposition of differently polarized waves can also depolarize, and this is the principle behind pseudo-depolarizers.

2.4 The Poincaré sphere

In a three-dimensional coordinate system formed by $[Q/I, U/I, V/I]$, the normalized Stokes vector for fully polarized light is located on a sphere with radius 1 (➋ Eq. 4.4): the *Poincaré sphere*, see ➋ *Fig. 4-2*. The center of the sphere represents completely unpolarized light. Partially polarized light is located inside the sphere. The equator plane of the sphere $[Q/I, U/I]$ contains linearly polarized light (V/I=0) with the two horizontal axes described by $-1 \leq Q/I \leq 1$ and $-1 \leq U/I \leq 1$. The axis connecting $\pm Q$ then corresponds to all partial polarization states in the Q-direction. The two poles represent pure right- and left-handed circular polarization (V/I=±1). Note that whereas the linear polarization directions $\pm Q$ (and also $\pm U$) are orthogonal in real space, they are located on opposite sides of the Poincaré sphere.

The Poincaré sphere is indeed useful to visually represent polarization and the influence of polarization optics, whereas Mueller matrices are largely a numerical tool. For example, the action of a retarder on some input polarization state is represented by a rotation of δ radians around an axis in the Poincaré sphere corresponding to the orientation of the fast axis of the

[3]Note that this is not valid if the coordinate system is modified by M itself, for instance, if it contains an odd number of mirrors.

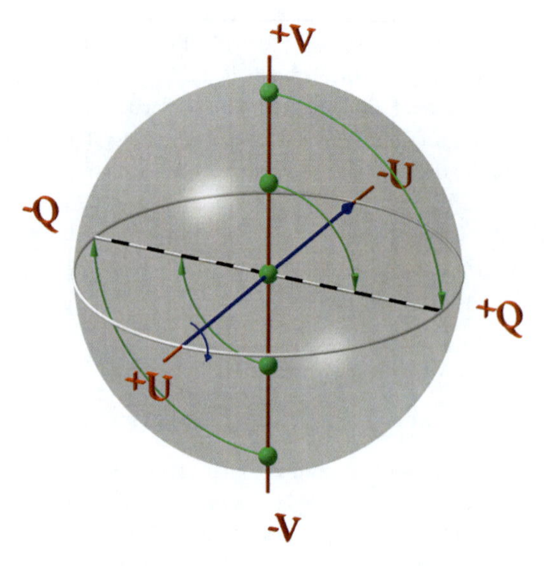

Fig. 4-2
The Poincaré sphere. The actions of a QWP with its fast axis in the *U*-direction and a polarizer in the +*Q*-direction are indicated

retarder. This can be an axis in the $[Q/I, U/I]$ equatorial plane for linear retarders, but also circular retarders (due to *optical activity* or the Faraday effect) exist, and in general a retarder has elliptical axes and eigenvectors. With this visualization in the Poincaré sphere, it is easily seen that a QWP with its fast axis aligned with $+U$ can turn $+Q$ into $+V$ and vice versa, and that a HWP with the same orientation merely changes the signs of Q and V. When the input polarization is aligned with the eigenvector of the retarder, it is not modified because it rotates around itself. The effect of more complex combinations of retarders can thus be elegantly visualized using the Poincaré sphere.

The effect of a linear polarizer is equivalent to the projection of the location of the input polarization on or inside the sphere onto the axis of the polarizer. For a polarizer in the $+Q$ direction, incoming light corresponding to this point on the sphere corresponds to a normalized transmitted intensity of 1, whereas the $-Q$ direction is fully blocked, i.e., an intensity of 0. A linear scale (the "ruler" in the figure) describes the intensity for input of any other polarization state. For instance, any point in the $[U/I, V/I]$ plane of the sphere yields an emerging intensity of $\frac{1}{2}$ times the maximum intensity. The physical reason for this is that both U and V have equal amounts of photon fluxes in the $+Q$ and the $-Q$ directions.

3 Polarizing Mechanisms

To obtain some insight into the kind of polarization one needs to measure to obtain valuable information on a wide range of astronomical sources, a quick overview of polarizing mechanisms in astrophysical contexts is provided. A much more extensive review is provided by Clarke (2009) and Hough (2007). In general, polarization is created (and/or modified) whenever the symmetry around the propagation direction of the light is broken. This is the case for anisotropic optics such as polarizers and retarders (and unfortunately also non-polarimetric optics), but this breaking of the symmetry can also be due to the change of direction of the light itself in a scattering (or reflection) process or due to the presence of magnetic fields. These two physical effects (scattering and magnetic fields) are the main theme of most of the science cases for polarimetric instruments. However, as will become apparent, the types of signal due to the various polarizing mechanisms are very different indeed. This justifies the wide range of polarimetric instruments that are (and will be) available to the astronomical community. A distinction is made between broadband polarization and spectral line polarization that require very different spectral resolution and therefore very different types of instruments. The single common aspect of astronomical polarization signals is their generally very small degree of polarization ($P \lesssim 10\%$ all the way down to $\sim 10^{-7}$), which explains the significant technical challenges in designing polarimetric instruments.

3.1 Broadband Polarization

3.1.1 Scattering and Reflection

Scattering is the most well-known mechanism for creating polarization since it is the reason that the blue sky is polarized and that Polaroid sunglasses are so effective at blocking scattered sunlight. In fact, the measurements of the polarization of the blue sky and of the moon by Arago in the early 19th century are generally regarded as the first implementation of polarimetry as an astronomical technique. The (continuum) scattering polarization is largest for a scattering angle of 90° (see ➋ *Fig. 4-3*) and the degree of polarization can exceed 50%. The observation of scattered and reflected sunlight from solar system bodies

Fig. 4-3
Schematic representation of 90° scattering for unpolarized incoming light

(planets, comets, etc.) has long been a major driver for astronomical polarimetry, and it still is (see Hovenier and Muños 2009). In fact, Hansen and Travis (1974) have shown that the microphysical properties (i.e., size, shape and chemical composition) of small particles like dust grains and atmospheric aerosols can only be unambiguously determined using spectropolarimetry[4]. As an example, the properties of the small sulphuric acid cloud droplets in Venus' atmosphere could only be established through phase-resolved polarimetry at different wavelengths, see Hovenier and Muños (2009). Polarimetry is a particularly useful diagnostic for imaging and characterizing stellar ejecta and circumstellar material, such as (protoplanetary and debris) disks and exoplanets. The presence of scattering blobs, disks, jets, or other conglomerates of circumstellar material (gas and/or dust) can directly be confirmed with the measurement of the polarization of the integrated light, assuming that there is a geometrical asymmetry in their distribution. The orientation of linear polarization due to single scattering is always azimuthal with respect to the central star, and the polarization averages out if the circumstellar structure is fully symmetric and unresolved. Even if circumstellar disks and exoplanets can, in principle, be spatially distinguished from the central star, they are still drowned in the light of the star that is diffracted and scattered within the instrument itself. But, because the light of the circumstellar material is heavily polarized and the direct starlight is, to a large extent, unpolarized, polarimetry constitutes a powerful technique to suppress the direct starlight and enable the direct imaging of these disks (e.g., Kuhn et al. 2001) and exoplanets. Moreover, the polarization spectrum contains information on the scattering medium, e.g., due to the $1/\lambda^4$ behavior of Rayleigh scattering (or $1/\lambda^2$ due to Mie scattering), the specific scattering phase behavior of different (cloud) particles and surface features, depolarization in spectral bands, and the albedo. Various models have been constructed to predict the polarization behavior of Jupiter-like exoplanets (Seager et al. 2000) and Earth-like exoplanets (Stam 2008), such that (future) observations can be guided and interpreted. Therefore, polarimetry of scattered sun- or starlight is a very powerful tool for characterizing circumstellar dust features and exoplanetary atmospheres[5]. In fact, polarimetry is unique in its prospect to detecting clouds or oceans of water, which are considered prerequisites for the development of life, and the presence of biomarker atmospheric gasses.

Asymmetries are thought to play a key role in the physics of supernovae and gamma-ray bursts. Indeed, their light and afterglows are found to be polarized (Wang and Wheeler 2008; Vink 2010; Steele et al. 2009), which strongly constrains models of massive stars at the end of their lives. Also "light echos" from nova outbursts are heavily polarized (up to 50%), and they can be used to directly establish the distance to the star (Sparks et al. 2008).

Further out in the universe, polarimetry has played a key role in the establishment of a unified model of active galactic nuclei (AGN). For a long time it was thought that there were two distinct classes of AGNs (Seyfert 1 and 2), which differ significantly in their spectral properties. It turns out that the difference is merely due to the inclination angle of the dust torus, which in the case of Seyfert 1 galaxies obscures the nucleus. By observing linearly polarized light, which is light from the nucleus that did find its way toward us by scattering around the torus, it was discovered that the spectrum is identical to that of

[4]Determining the properties of anthropogenic aerosols that pose health hazards and can significantly influence global climate is therefore a main science case for the implementation of polarimetry on Earth-observing platforms.

[5]The expected polarization from an exo-Jupiter is generally much larger than from our own Jupiter since the latter can never be observed at large scattering angles. The integrated polarization signals of the outer planets are therefore small, although specific polarized structures appear at large spatial resolution.

Seyfert 2 galaxies (Antonucci and Miller 1985). In such cases, polarimetry can be used as a periscope to indirectly observe obscured galactic (and also stellar) sources, and, by analyzing the scattering polarization geometry, one can establish the position of the obscured source.

3.1.2 Polarization-Dependent Absorption and Emission

Since dust grains are generally not spherical, they can act as partial polarizers and as retarders. However, there is only a net polarization effect if dust grains are aligned, i.e., not randomly oriented. Such *grain alignment* can be induced by a magnetic field. Although the exact mechanism for this is still unknown (see Hough 2007), the smallest dimension of nonspherical dust grains are always aligned with the direction of the local magnetic field. Absorption of background light by interstellar dust that is aligned with the galactic magnetic field thus creates *interstellar polarization*, which can be a nuisance when trying to measure intrinsic, broad-band stellar polarization, but is a gold mine for mapping and consequently understanding the galactic magnetic field structure. In the mid-IR this polarization-dependent absorption turns into polarization-dependent emission. The corresponding sign change of the measured Stokes parameters can be used to characterize this transition. Medium-resolution polarization spectra can be used for mineralogical studies of the dust grains.

When aligned dust grains are illuminated anisotropically and scatter this light, they can also produce circular polarization, and not just linear polarization (see Hough 2007; Hovenier and Muños 2009). This can be due to multiple scattering, where the linear polarization is produced in the first scattering effect, and converted into circular by the retardance action of the nonspherical particles in consecutive scattering events. Also, magnetic alignment of the dust grains in a direction that breaks the symmetry once more can cause circular polarization after a single-scattering event. It can also be that the complex dust grains themselves have a preferred handedness (i.e., circular retardance or optical activity). In any case, broadband circular polarization is observed, but the signals are much lower than the typical linear polarization signals due to single scattering. It is hypothesized e.g., Hough 2007 that this preference for a certain handedness in regions of star and planet formation is the cause of homochirality in complex molecules that constitute the basic building blocks of life. In fact, measuring circular polarization turns out to be a unique remote-sensing technique for the detection of (extraterrestrial) life, as complex molecules (e.g., chlorophyll) produce particular circular polarization spectra at the 10^{-4} level (Sparks et al. 2009).

3.1.3 Synchrotron and Cyclotron Radiation

In more extreme environments, magnetic fields produce linear and circular polarization through synchrotron emission. When a relativistic electron gyrates around a strong magnetic field, the outward radiation is strongly linearly polarized throughout all wavelength regimes, with a direction perpendicular to the magnetic field. When the magnetic field has a net component along the line of sight, also circular polarization is generated (*synchrotron radiation*). These effects are used to study magnetic fields in high-energy objects such as AGN jets and neutron stars (including pulsars). Indeed, the study of the polarization of the Crab nebula (e.g., Oort and Walraven 1956; Slowikowska et al. 2009) is one of the astronomical success stories of polarimetry, as it allows for a detailed description of the magnetic field and emission structure at the pulsar and in the wind nebula. To study the polarization properties of the pulsar itself, a polarimeter with very high time resolution is required.

3.2 Spectral Line Polarization

3.2.1 The Zeeman Effect

The *Zeeman effect* (Zeeman 1897; Stenflo 1994; Landi Degl'Innocenti and Landolfi 2004; Del Toro Iniesta 2003) is the most famous and most utilized spectral line polarization effect and describes the creation of spectral line polarization by a magnetic field. A magnetic field modifies the energy of the magnetic sublevels of an atom, which makes the spectral lines appear to be split into different components. These components have different polarization properties, depending on the quantum numbers of the corresponding sublevels and the orientation of the magnetic field with respect to the line of sight.

The impact of a magnetic field on the emergent line polarization through the Zeeman effect is explained with the cartoon model of ➋ *Fig. 4-4*. Here, the transition is represented by a three-dimensional oscillator. The magnetic field couples the linear oscillators perpendicular to it into two circular oscillators and makes them precess around it. As the right-handed precession around the magnetic field is energetically favored, it is blue-shifted, whereas the left-handed precession is red-shifted. The splitting in energy is proportional to the magnetic field strength. The third oscillator component, which is parallel to the magnetic field, stays at the same energy and therefore at the same wavelength.

In quantum mechanical terms, the circular oscillators correspond to the σ components of the transition, i.e., the ones with $\Delta m_J = \pm 1$, and therefore carry angular momentum in

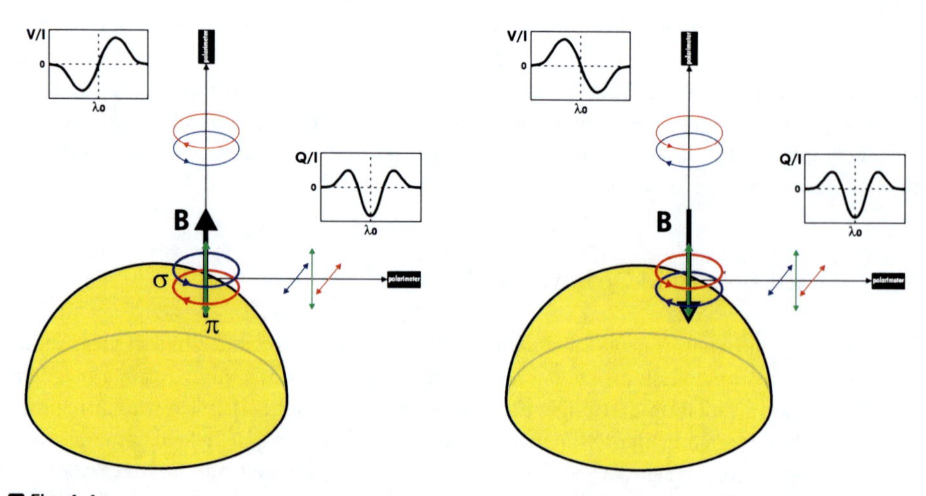

Fig. 4-4
Classical representation for the emergent line polarization caused by a magnetic field due to the Zeeman effect

the direction of the magnetic field vector. The linear oscillator corresponds to the π component ($\Delta m_J = 0$). The connection with polarization is obvious: the σ components are circularly polarized with opposite handedness when looking along the magnetic field direction. The π component then disappears. When the magnetic field is observed at an angle of 90°, the σ-component is linearly polarized, and perpendicular to the π-component.

With zero magnetic field, the σ and π components overlap, and no line polarization is created. The "sources" and "sinks" of circular and linear polarization become increasingly apparent when the components are split by the magnetic field. The spectral line pattern due to the Zeeman effect in Stokes V appears antisymmetric, as indicated in the figure. The pattern in linear polarization (Stokes Q or U) is typically symmetric. Note that the Stokes V pattern changes sign when the magnetic field direction is flipped, whereas the pattern in Stokes Q remains the same. This so-called 180° ambiguity is the only fundamental limitation to the full vector measurement of magnetic fields with the Zeeman effect.

To first order, the spectral line polarization due to the Zeeman effect is given by (assuming weak fields, no velocity or magnetic field gradients along the line of sight and within the area of formation of the spectral line, and the absence of magneto-optical effects):

$$V(\lambda) = -\frac{\mu_B}{hc} g_L \lambda^2 \frac{\partial I(\lambda)}{\partial \lambda} B \cos\gamma; \tag{4.14}$$

$$Q(\lambda) = -\left(\frac{\mu_B}{2hc} g_L \lambda^2\right)^2 \frac{\partial^2 I(\lambda)}{\partial \lambda^2} B^2 \sin^2\gamma \cos 2\chi; \tag{4.15}$$

$$U(\lambda) = -\left(\frac{\mu_B}{2hc} g_L \lambda^2\right)^2 \frac{\partial^2 I(\lambda)}{\partial \lambda^2} B^2 \sin^2\gamma \sin 2\chi, \tag{4.16}$$

with μ_B the Bohr magneton and g_L the Landé factor of the particular line. γ and χ represent the magnetic field inclination and azimuth, respectively. The Stokes-V pattern due to the longitudinal Zeeman effect is linear in (apparent) magnetic field strength B and can reach values for V/I of ~10% for kilogauss fields. The transverse Zeeman effect (in Q and U) is only a second-order effect and therefore yields polarization degrees of an order of magnitude lower for weak fields. It is clear from the equations that spectral lines with a large Landé factor, a large line wing gradient, and/or at a longer wavelength are more effective for measuring weak magnetic fields. Since the Zeeman effect scales with λ^2 whereas the Doppler broadening scales with λ, the effectiveness of the Zeeman effect is proportional to the wavelength.

The magnetic field strength B is measured in gauss, although technically it would be better to refer to the measured magnetic field as magnetic flux density [Mx/cm^2]. From the cartoon in ➋ *Fig. 4-4* it becomes obvious that if a resolution element consists of two equal but opposite magnetic field concentrations, the line polarizations in Stokes V cancel. The same holds for magnetic fields crossed by 90° in linear polarization. This property directly indicates the large weakness of the Zeeman effect: it is "blind" to isotropic magnetic field topologies within a resolution element.

The first application of the Zeeman effect, readily after its postulation, was the polarimetric measurement of Hale (1908) to confirm that sunspots harbor strong magnetic fields. Ever since, polarimetry using the Zeeman effect has been a major workhorse technique for solar observations (see De Wijn et al. 2009). With the advent of sensitive stellar polarimeters, large-scale magnetic fields have also been detected on other stars (Donati and Landstreet 2009; Kochukhov and Piskunov 2009). These fields can even be mapped on the stellar surface (and then extrapolated to the stellar surroundings) if the star rotates relatively fast, by interpreting time series of polarimetric measurements of spectral lines exhibiting the Zeeman effect (Zeeman–Doppler imaging). Note that this technique particularly applies to unresolved stars. This allows us to put solar magnetism into perspective and apply lessons learned to other stars, whose magnetic fields may have been generated in completely different ways.

3.2.2 The Hanle Effect and Other Line Polarization Effects

Similar to broadband polarization, spectral line polarization can be created by scattering of anisotropic radiation. For spectral lines, the induced atomic polarization depends largely on the quantum numbers of the transition. Scattering polarization is largest for scattering angles close to 90° and, indeed, in the case of the solar atmosphere, spectral line polarization is observed close to the solar limb. This linear polarization is oriented parallel to the limb as shown in ➲ *Fig. 4-5a.*

This spectral line scattering polarization is modified by magnetic fields through the *Hanle effect* (Hanle 1924; Trujillo Bueno 2009). This effect is highly complementary to the Zeeman effect, which creates polarization with increasing magnetic field strengths. On the other hand, the Hanle effect due to weak magnetic fields induces a rotation (and alignment) and in some cases destruction of existing spectral line polarization.

The scattering polarization is not modified by a vertical magnetic field as indicated in ➲ *Fig. 4-5b*. However, horizontal magnetic fields as sketched in ➲ *Fig. 4-5c, d*, do affect the emergent line polarization. Thus, the Hanle effect mostly depolarizes the line, although rotation of linear polarization may be observed for a magnetic field parallel to the line of sight (➲ *Fig. 4-5d*) and a transition ("oscillator") with a short lifetime.

The magnetic field strengths to which the Hanle effect is sensitive can be computed by comparing the Larmor precession period with the lifetime of the transition (or, equivalently, the magnetic splitting with the natural line broadening):

$$\frac{e}{2m_e} g_L B = 1/t_{\mathrm{life}}. \tag{4.17}$$

Typical values for magnetic field strengths that depolarize the scattering polarization of the line are up to ~10–100 G. Furthermore, mixed-polarity magnetic field topologies depolarize the line, which gives the Hanle effect additional complementarity to the Zeeman effect, which is essentially blind to such fields. For magnetic field strengths larger than the critical value (which may still be very small in the case of forbidden lines that have a large life-time), the saturated Hanle effect orients the linear spectral line polarization parallel or perpendicular to the local magnetic field direction (depending on the scattering angle). Unfortunately, the signals of the scattering polarization are very small (~1% down to 10^{-5}, depending on the anisotropy of the radiation field), which drives the need for very sensitive polarimetry to observe the polarization variations due to the Hanle effect.

Indirect observations of weak, turbulent magnetic fields in the solar atmosphere using the Hanle effect are now a standard tool in solar physics (see Trujillo Bueno 2009). Furthermore, it is now also pursued as a technique for diagnosing magnetic fields in stellar winds (Ignace 1999). Finally, absorbing clumps in extended stellar atmospheres are investigated with spectral line polarization created by optical pumping (Kuhn et al. 2007).

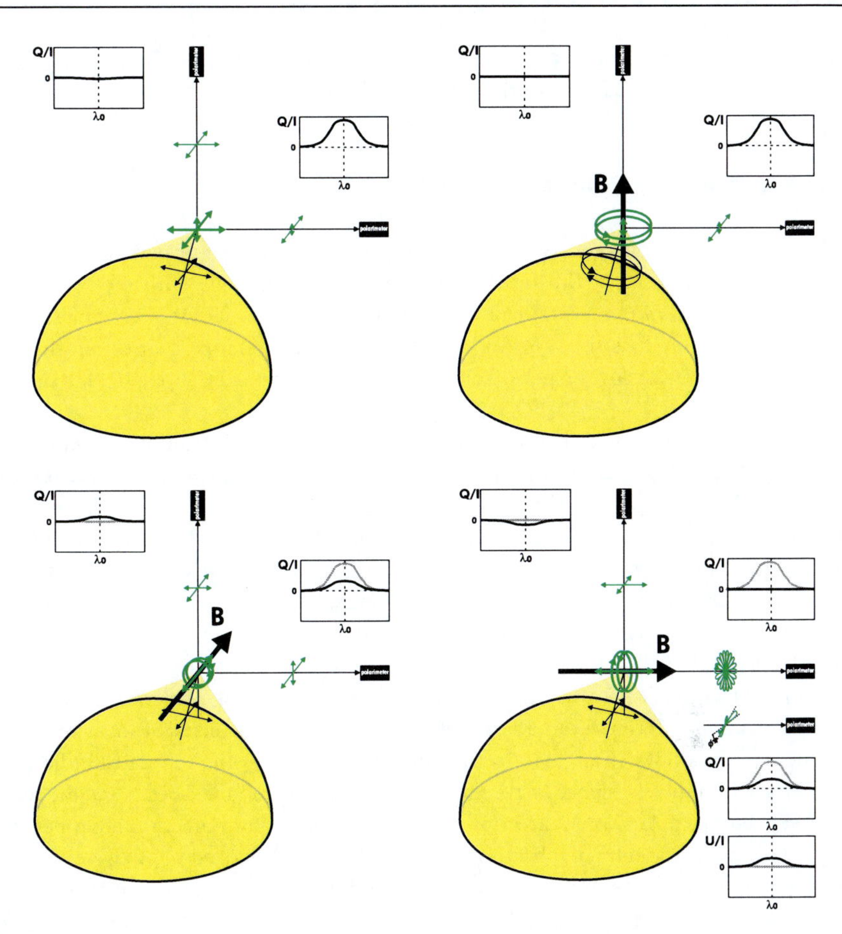

Fig. 4-5
Classical representation for the emergent line polarization caused by (a) scattering, (b) a vertical magnetic field, (c) a horizontal magnetic field perpendicular to the line of sight, and (d) a horizontal magnetic field along the line of sight

4 The Polarimetrist's Toolkit

A polarimeter requires optical components that are very different from the imaging optics used in non-polarimetric instrumentation. An overview is provided here of frequently used polarization optics and their limitations. Many more technical details are given by Keller (2002a) and Clarke (2009).

4.1 Polarizers

To measure polarization, one has to filter photon fluxes according to polarization states (cf. Eqs. 4.2 and 4.3). This filtering is, in most cases, performed by a linear polarizer (that only transmits one polarization direction) or a polarizing beam-splitter. The properties of the most commonly used polarizers are presented in Table 4-1.

4.1.1 Sheet or Plate Polarizers

Certain dichroic[6] materials consisting of aligned anisotropic crystals or polymers absorb light with one linear polarization direction and largely transmit the light with the perpendicular direction. Polaroid film is a prominent example of such a material for sheet polarizers. The *extinction ratio*, defined as the transmitted intensity for the polarizing direction divided by the transmitted intensity for the perpendicular direction, is typically about 10^2 for such sheet polarizers.

A similar polarizing effect is achieved with a pattern of conducting wires with spacings smaller than the wavelength. This wire-grid technique is often used in the submillimeter regime, but is now finding its way into the optical regime thanks to the increasingly small length scales achieved by lithographic processes. The cartoon picture of only vertical waves being able to slip through this "fence" is wrong, as the electric fields that describe this vertical polarization are suppressed by conduction in the wires. Wire-grid polarizers have fairly large extinction ratios and are applicable over large wavelength ranges. The non-transmitted beam is often reflected with high efficiency.

A simple piece of glass acts as a polarizer for light at non-normal incidence, as described by the Fresnel equations. It can become a complete polarizer for the reflected beam at an incidence angle equal to the Brewster angle $\theta_B = \arctan(n_2/n_1)$, with $n_{1,2}$ the refractive indices of the media involved. At this incidence angle, the angle between the refracted beam and the reflected beam is 90°, and the polarization direction parallel to the reflection plane cannot propagate into the reflected beam. The refracted beam is still only partially polarized. A Brewster window is the only polarizer of choice in the far UV regime, and the extinction ratio mostly depends on the incidence angles of the incoming beam.

4.1.2 Polarizing Beam-Splitters

Often, one would like to keep all the light that enters a polarizer as output, if only for efficient use of all the precious astronomical photons. In this case, the polarizer needs to act as a beam-splitter that discriminates between light at perpendicular (linear) polarization directions.

A cube beam-splitter can be turned into a polarizing beam-splitter by designing an appropriate multilayer coating for the internal hypothenuse plane of the cube. Fairly large extinction ratios are obtained when there are multiple internal reflections within the thin film stack at the Brewster angle.

Most other polarizing beam-splitters are based on *birefringent* crystals. As the name suggests, these crystals have different refractive indices for light propagating in the direction of the crystal axes. In the case of uniaxial crystals, one can describe their optical properties with the extraordinary index n_e, which is valid for electric field vectors (or linear polarization) parallel

[6]The term "dichroic" to describe the polarizing action of a material is particularly confusing since an identical term is also used to describe the wavelength dependence of coating reflectivities. The polarization effect was found to be related to color effects when it was discovered, and the terminology has stuck.

Table 4-1
Properties of commonly used polarizers. Numbers between parentheses for quartz; otherwise for calcite

Polarizer type	Splitting geometry	Extinction ratio	Wavelength range [nm]	Bandpass [nm]	Acceptance angle	Max clear aperture	Location
Sheet polarizer		10^2–10^4	310–2000	200	> 20°	1000 mm	Anywhere
Wire-grid polarizer		> 10^2	> 250	full	> 20°	200 mm	Anywhere
Brewster window		> 10^3	> 200	~ 700 nm	< 1°	~ 200 mm	Anywhere
Cube beam-splitter		> $5 \cdot 10^2$	400–1600	200–400	~ 10°	75 mm	Anywhere
Wollaston prism		> 10^6	(200) 300–2300	full	~ 20°	~ 150 mm	Pupil plane
Foster prism		10^6, 10^5	(200) 300–2300	full	~ 6°	> 12 mm	Depends on other beam-folding optics
Savart plate		> 10^4	(200) 300–2300	full	see text	> 50 mm	Close to image plane

to the crystal's *optic axis*. The ordinary index n_o applies to electric field vectors in the plane perpendicular to the crystal's optic axis. The birefringence of a material is then defined as $n_e - n_o$. Such a polarization-dependent refractive index can be exploited in several ways to create polarizing beam-splitters as sketched in *Table 4-1*. Common materials for such prisms are calcite (with $n_e - n_o \approx -0.17$) and quartz (with $n_e - n_o \approx +0.009$).

The *Wollaston prism* is often used in astronomical polarimeters as it splits the beam by an angle that is adjustable by design. Locating such a prism in a pupil plane results in a split field of view in the image plane. The amount of splitting in the image plane is therefore a design parameter. The angular splitting is enforced at the interface between the two birefringent crystal sections that have their optic axes crossed. At the interface, one polarization direction goes from a smaller to larger refractive index and gets refracted inward, and vice versa for the other polarization direction, in a symmetrical fashion. The image quality decreases with increasing splitting angle, although a combination of three crystal sections alleviates this (see, e.g., Hough et al. 2006). A wedged double Wollaston was designed by Oliva (1997) to split the incoming beam into four polarized beams according to $\pm Q$ and $\pm U$.

The *Foster prism* (the extension of a Glan–Thompson prism) consists of crystal sections that have their optic axes aligned, but are separated by a thin cement layer of smaller refractive index. For one polarization direction, the light (at normal incidence to the prism's entrance face) undergoes a total internal reflection and exits the prism at some angle (~135° for a calcite-based Foster prism). The other beam just goes straight through.

A *Savart plate* induces a physical splitting of the incident light beam without any angular separation. It is therefore best used close to a focal plane. Such a beam displacer makes use of the most basic property of a birefringent crystal, which is the creation of a double image when the optic axis is at some angle with the entrance surface. The splitting is maximum for an angle of 42° for calcite in the visible part of the spectrum, and for calcite the amount of splitting is about 0.08 times the thickness of the plate (~0.004 for quartz). Since the splitting due to a single calcite plate introduces a path length (and therefore focus) difference between the two beams, a Savart plate usually consists of two plates that are rotated by 90° with respect to each other. In a converging beam, birefringent crystals introduce astigmatism (see Semel 1987). This can be easily understood in terms of the effective refractive index differences experienced by rays at an angle with the chief ray. This astigmatism can be corrected by two separate cylindrical lenses for the two beams out of the Savart plate (which have opposite astigmatism) or by a single cylindrical lens when a HWP is placed between the two calcite plates (a modified Savart plate).

Furthermore, both the Savart plate as well as the Wollaston prism suffer from lateral chromatism. The nonideal effects of the various polarizers are listed in *Table 4-1*.

4.2 Retarders

According to Eq. 4.11 and *Fig. 4-1*, retarders can be used to manipulate polarization by converting circular into linear polarization and by rotating the angle of linear polarization. Retardance action is often achieved using another property of birefringent materials. When a crystal is cut with its optic axis parallel to the entrance surface, light under normal incidence propagates with different velocities dependent on the input linear polarization. When the linear polarization is parallel to the axis that has the smallest refractive index (either n_e or n_o), it travels faster than light polarized in the perpendicular direction. A phase retardance can thus be created between the two polarization components within the wave plate with thickness d:

$$\delta = \frac{2\pi \cdot d}{\lambda}(n_e - n_o). \tag{4.18}$$

Depending on the sign of the difference $(n_e - n_o)$, the optic axis is either the slow axis or the fast axis. The special cases of wave plate (quarter-wave plate and half-wave plate) are obtained after polishing the plate to the correct thickness d such that $\delta = (\pi/2 \bmod \pi)$ and $\delta = (\pi \bmod 2\pi)$, respectively. A true zero-order wave plate usually has more benign nonideal properties than multiple-order retarders. The properties of some common retarders are listed in ➋ *Table 4-2*.

4.2.1 Fixed Linear Retarders

Most fixed retarders are made of birefringent crystals (e.g., quartz, MgF_2, sapphire, or CdSe) or birefringent polymer sheets like stretched polyvinyl chloride, polycarbonate, and polymethylmethacrylat (PMMA; Samoylov et al. 2004). The latter are true zero-order retarders, whereas even weakly birefringent crystals usually have to be crossed to obtain a quasi zero-order retarder as otherwise the plates would become too thin to be polished. The fundamental issue with all birefringent wave plates is that they are inherently chromatic. Technically, a quarter-wave plate or a half-wave plate is only perfect at one wavelength. However, combinations of two different crystals that partially compensate each other's chromaticity (Beckers 1971), or stacks of three or five identical plates at different angles (Pancharatnam 1955), yield "*achromatic*" wave plates and even "*superachromatic*" wave plates when both tricks are used simultaneously (Serkowski 1974).

The exact retardance of wave plates in general is often a steep function of the incidence and azimuth angle of the light as well as temperature (see Keller 2002a). The acceptance angle can be maximized by using appropriate combinations of crystals with opposites signs of the birefringence. An athermal wave plate can be made by combining crystals with different thermal properties in a subtractive combination (Hale and Day 1988; Guimond and Elmore 2004). Wave plates are weak, partial polarizers (~1%), because the two polarization directions corresponding to the fast/slow axis have slightly different transmissions. Furthermore, if multiple internal reflections occur within the wave plate, the emergent spectral fringes are polarized at the ~1% level because the interference effects depend on the refractive indices corresponding to the two polarization directions along the fast and slow axes (Weenink et al. 2011). Such effects can be minimized by applying antireflection coatings, and by matching the refractive indices across each interface. This is best achieved using true zero-order birefringent polymers for which the polarized fringes are very broad in wavelength, if they are detectable at all.

There are two other ways of creating retarders, for which the retardance is independent of wavelength (within some wavelength range). The first method relies on a phase difference created by total internal reflection (TIR, which has reflectivities of 100% for both polarization directions). This retardance is derived from the Fresnel equations:

$$\delta_{\mathrm{TIR}} = 2\arctan\left(-\frac{\cos\theta\sqrt{(n(\lambda))^2\sin^2\theta - 1}}{n(\lambda)\sin^2\theta}\right). \tag{4.19}$$

For regular glasses it takes two TIRs to obtain quarter-wave retardance, as is implemented in a *Fresnel rhomb*. An achromatic half-wave retarder is obtained by combining two Fresnel rhombs,

Table 4-2
Properties of commonly used retarders

retarder type	Wavelength range [nm]	Bandpass [nm]	Acceptance angle	Max clear aperture	Type of modulation
Multiple-order wave plate	180–2700	< 20	~ 3°	~ 100 mm	Rotation
Quasi zero-order wave plate	180–2700	~ 100	~ 3°	~ 100 mm	Rotation
Achromatic wave plate	180–2700	$\sim 0.4 \cdot \lambda_0$	~ 3°	~ 100 mm	Rotation
Superachromatic wave plate	180–2700	$\sim 1.2 \cdot \lambda_0$	~ 3°	~ 100 mm	Rotation
Pancharatnam polymer wave plate	340–1800	$\sim 0.8 \cdot \lambda_0$	> 10°	60 mm	Rotation (beam displacement for QWP)
Fresnel rhomb	> 200	~ 700	~ 1°, 10°	> 25 mm	Rotation
Liquid crystal variable retarder	450–1800	~ 100	> 10°	40 mm	Electric tuning of retardance
Ferroelectric liquid crystal	400–1800	~ 100	> 10°	45 mm	Electric tuning of fast axis orientation
Photo-elastic modulator	> 200	~ 100	> 10°	88 mm	Variation of retardance through (resonant mechanical) stress

which has the additional benefit of no beam deviation. Also prisms with three or more TIRs (e.g., a K-prism) can be designed to yield a specified total retardance. The deviation of the retardance from the design value is only limited by variations of the refractive index with wavelength of the glass from which the rhomb is made of. The actual retardance value can be tweaked with a dielectric coating on the TIR surface(s) (King 1966). The variation of retardance with incidence angle is slow for the direction perpendicular to the plane of the TIRs, but it is very steep in the other direction.

A second, novel method for creating achromatic retarders is through *form birefringence.* This is basically a subwavelength grating, i.e., a ruled pattern with spacings smaller than the wavelength. The effective refractive index is then different for the polarization directions parallel and perpendicular to these rulings. The effective birefringence can be tuned with the relative widths of the rulings, and through the refractive index of the substrate(s). By designing the wavelength dependence of the form birefringence such that it compensates for the inherent chromaticity of retardance of a regular wave plate, achromatic retarders can be obtained (see Kikuta et al. 1997).

4.2.2 Variable Retarders

Liquid crystals are another birefringent material, with the main distinction that their orientations can be manipulated on a microscopic level with the application of external electric fields. There are two major types of liquid crystal components that are frequently used: *liquid crystal variable retarders* (LCVRs) and *ferroelectric liquid crystals* (FLCs). The main difference is that one can tune the retardance of an LCVR device with increasing applied voltage, whereas the fast axis orientation of FLCs is flipped upon changing the polarity of the voltage. In both cases, a variable polarization manipulation can be achieved without physically rotating the retarder.

In LCVRs, the liquid crystal molecules are initially all aligned in some direction parallel to the surrounding substrates' surfaces, and the maximum value of retardance is obtained. These substrates are coated with a transparent electrode (e.g., indium tin oxide), such that an electric field can be applied that is perpendicular to the surfaces. An increasing electric field rotates the birefringent liquid crystal molecules until, for large voltages ($\gtrsim$10 V), they are all aligned with the electric field, and the device is effectively not a retarder anymore. For intermediate voltages, a stable retardance smaller than the maximum value is obtained. In other words: changing the applied voltage results in a change of retardance. The exact, highly nonlinear correspondence between retardance and voltage depends on the temperature; therefore LCVRs need to be frequently calibrated and/or stabilized in temperature. The switching times of LCVRs are relatively fast ($\sim$10 ms), and even faster components are being developed.

In FLCs, the orientation of the birefringent liquid crystals' optic axes has only two stable states, both perpendicular to the entrance surface normal. Most FLCs are engineered to provide a fast axis rotation of 45°, although the exact value of this angle is slightly dependent on voltage and temperature. FLCs therefore have a constant retardance, but their real strength lies in their extremely fast switching times of $\sim$100 μs, enabling kHz modulation rates.

Another frequently used variable retarder is the *photo-elastic modulator* (PEM; see Kemp et al. 1987). It relies on the fact that regular glass becomes birefringent when an external stress is exerted upon it. This effect can be used to create a fixed retarder, but when the stress is applied at the mechanical resonance frequency, it becomes a variable retarder. In fact, typical resonance frequencies are $\sim$50 kHz, so PEMs are the prime choice for ultrafast modulation.

4.3 Novel Components

Many useful novel polarization components have recently been developed, mostly owing to developments in the field of nanophotonics and by the liquid crystal display and telecommunications industries. For instance, true circular polarizers based on chiral nanostructures are being developed (Gansel et al. 2009), as well as linear polarizers that are 100% efficient by transmitting one polarization direction and rotating the other one by 90°. Passive devices based on twisted nematic liquid crystals that locally rotate the direction of linear polarization can be used to induce a coordinate system transformation on the sky. For instance, many astronomical objects exhibit a centrosymmetric linear polarization pattern around a central source. When it is properly aligned in a focal place, such a passive liquid crystal device (a "theta cell") aligns this pattern in one direction so that only one Stokes parameter (say, Q') has to be measured in the instrument's coordinate frame (Snik 2009). This makes the data acquisition and reduction easier and furthermore facilitates averaging over the pattern as otherwise the vector average of a centrosymmetric pattern yields zero by definition.

Another spin-off from liquid crystal manufacturing processes is the development of a "polarization grating" (Packham et al. 2010). Such a grating not only acts as a dispersion element, but as a polarizing beam-splitter as well, by diffracting one circular polarization direction $(+V)$ into order +1 and $-V$ into order −1. By using an additional QWP, the device splits linear polarization states. For spectropolarimeters, this polarization grating may then replace the combination of a grating (or prism or grism) and a Wollaston prism, with the added benefit that the splitting angle between polarizations may be much larger than what Wollaston prisms can offer.

Increasingly, fibers and fiber bundles are used to transport light from a focal plane to a spectrograph's slit plane. With polarization-maintaining (birefringent) fibers, one can now design polarimetric versions of these fiber-based integral field units (Lin and Versteegh 2006). And from there the next step is to implement polarization manipulation in integrated optics modules for astronomical instrumentation, as the telecom industry already has experience with. In such an approach, all light from a telescope is injected into fibers and fed into miniaturized photonic components that perform all desired manipulations, like dispersion, beam-combination and detection. This provides a cheap, scalable framework for astronomical instrumentation that will allow for a quantum leap in observational capabilities.

4.4 Detectors

The detector requirements for many polarimetric instruments are very similar as for other astronomical instruments: large quantum efficiency, low read-out noise, low dark current, high linearity, high dynamic range, small gain variations, etc. For polarimetry, the noise considerations are the most important. Eventually, the polarimetric noise levels should be dominated by photon noise and not read-out noise. Therefore, for every frame the read-out noise should be smaller than the photon noise. By using an electron-multiplying CCD (EMCCD), one can easily reduce the influence of read-out noise by cranking up the detector gain. Nonideal detector properties can also create unwanted polarization signals, as shown by Keller (1996) for dark drift and uncorrected nonlinearity.

The other main requirement on the detector and its associated electronics concerns its read-out speed. As we will see in the next section, polarimetry often requires fast read-out rates, sometimes >1 kHz. Single-pixel detectors like photomultiplying tubes (PMTs) and avalanche photodiodes (APDs), or arrays thereof, can cope with this speed as they have an analogue output and do not require a read-out cycle. Such detectors are most useful for aperture polarimetry of the integrated light of an astronomical source. CCD and other two-dimensional detectors are usually too slow to keep up with a kHz modulation rate[7]. Quasi-one-dimensional line-scan detectors[8] do approach such speeds and are particularly useful for short-slit spectropolarimetry and for applications that allow for scanning. Dedicated two-dimensional detectors for polarimetry have been developed as well. The current state of the art is represented by the ZIMPOL detectors (Povel et al. 1994) that consist of standard CCDs with an additional mask that covers alternate pixel rows. The masked pixels allow for the storage of photoelectrons according to two or more polarization directions. By shifting charges back and forth in synchrony with the polarization modulation, without reading out the CCD, interlaced images are built up at modulation rates limited only by the charge transfer time of one pixel row (and not of the entire array). When sufficient photoelectrons are collected after many modulation cycles, the entire CCD is read out. The masks decrease the filling factor of the pixels, which may be counteracted with a microlens array. The many charge transfers and the asymmetries of charge transfer efficiency somewhat limit the polarimetric performance of such ZIMPOL detectors. An ideal polarimetric imaging detector would consist of multiple CMOS read-out capacitors for each pixel, as proposed by Keller (2004).

5 Polarimeter Implementation: How to Deal with Systematic Effects

5.1 Some Definitions

The performance of a polarimeter is limited by systematic effects and/or noise. Before the design options for a polarimetric instrument and the methods to deal with systematic errors are described, an introduction of definitions and terminology to describe polarimetric performance is apropos. As there is considerable confusion and vagueness concerning this terminology in the current literature, the definitions are as concise as possible.

The *polarimetric sensitivity* is defined as the smallest polarization signal that a polarimeter can detect. It is therefore related to the final noise levels in Q/I, U/I, and V/I and is expressed as a fraction of the intensity I that should be $\ll 1$. However, it is not only photon and read-out noise that limit the polarimetric sensitivity. Other random effects like seeing-introduced noise-like *spurious polarization signals* limit the detectability of a certain signal. In general, the polarimetric sensitivity is determined by errors that are not "real" polarization effects.

Once a signal larger than the polarimetric sensitivity level is detected, its magnitude should be quantified. The metric for how well this can be performed is the *polarimetric accuracy*. It states the accuracy of the measured polarization as compared to the actual Stokes parameter(s) of the light incident on the telescope, in the absence of noise and other spurious polarization signals. This correspondence is parametrized by the position of the zero point for the measurement of Q, U, and V (often called the *absolute polarimetric accuracy*) and the measurement scale

[7]Fast, low-noise two-dimensional detectors are currently being developed mostly for application in wavefront sensors.

[8]These are commercially available for machine vision applications.

(the *relative polarimetric accuracy*). It is best to define the polarimetric accuracy using the 4×4 matrix X that relates the measured Stokes parameters to the real ones:

$$\mathbf{S}_{\text{out}} = \mathsf{X} \cdot \mathbf{S}_{\text{in}}. \tag{4.20}$$

Note that X is not a Mueller matrix, as it includes the effects of data reduction and calibration. The accuracy is then described by the matrix $\Delta\mathsf{X}$, which is defined as (see, e.g., Ichimoto et al. 2008):

$$\mathbf{S}_{\text{out}} = (\mathbb{I} + \Delta\mathsf{X}) \cdot \mathbf{S}_{\text{in}}, \tag{4.21}$$

with $\mathbb{I}$ as the 4×4 identity matrix. The diagonal elements of $\Delta\mathsf{X}$ describe the uncertainties in the scaling factors $Q \rightarrow Q$, $U \rightarrow U$, and $V \rightarrow V$. These scaling factors describe how much of the incident polarized Stokes parameters is actually transmitted by the system. They are often called the *polarization efficiencies*, also in the context of Mueller matrices. But as we will see in Sect. 5.2, it is more convenient to define the *polarimetric efficiencies* in terms of noise propagation in the Stokes parameters. Therefore it is preferable to just call the diagonal elements of a Mueller matrix or X what they are: the diagonal elements.

The off-diagonal elements in these matrices also have particular roles. The elements in the first column ($I \rightarrow Q, U, V$) describe the generation of polarization by the instrument, i.e., *instrumental polarization*. Often, the total combination of polarization effects induced by the instrument (as described by the instrumental Mueller matrix $M_{\text{instr.}}$ or X) is called instrumental polarization, but it is better to be more specific. The off-diagonal elements $Q, U, V \rightarrow U, V, Q$ are called *cross talk*, as they describe mixing between the polarized Stokes parameters. The two elements $Q \leftrightarrow U$ describe rotation of the direction of linear polarization. Cross talk is always accompanied by a decrease of the pertinent diagonal elements. Pure cross talk keeps the degree of polarization P constant. But in the case of *depolarization*, some of the Q, U, and/or V signals are lost and reappear in the unpolarized fraction of I. Finally, the elements on the first row of X describe how the intensity measurement I depends on the input polarization. But, because in most astronomical situations $P \ll 1$, these terms are not very important[9].

5.2 Modulation and Demodulation

Figure 4-6 presents a sketch of the general setup of an astronomical polarimeter. The core of any polarimeter is some kind of linear polarizer or polarizing beam-splitter. In other words, one has to filter the light for one polarization direction (by polarizing it) to measure its polarization, and this component is called the *analyzer*. The analyzer transforms the measurable polarization into an intensity signal. Obviously, one needs a detector to measure the photon fluxes after the analyzer. More often than not, a polarimeter contains a *modulator* that determines which polarization state is analyzed by the polarizer, as described below. The grey blocks in *Fig. 4-6* can contain all kinds of optical components that have polarization properties of their own. These components include the telescope itself (usually somewhere before the modulator) including all the rotations involved in telescope pointing. Other optical components form an image of the source that the telescope is pointing at, filtered for some particular wavelength range. In this case, the setup is an *imaging polarimeter*, and the data will consist of polarization as a function of $[x, y]$ on the detector. The basic instrument can also be a spectrograph and contain a slit and

[9]The elements of the first row of a Mueller matrix are, however, very important in the description of polarimetric modulation, see the next section.

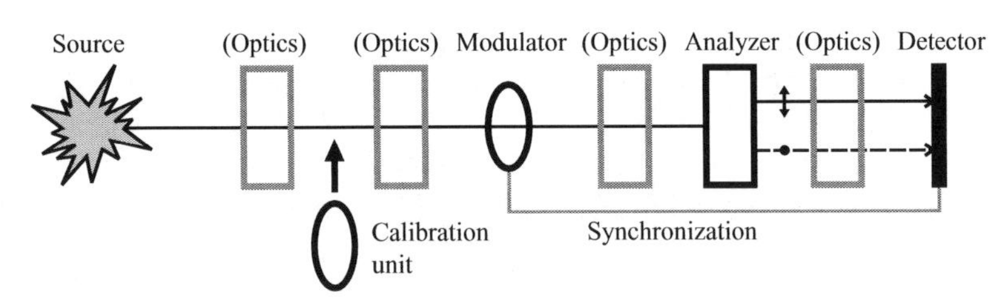

Fig. 4-6
General setup of an astronomical polarimeter. The grey blocks can contain any number of optical components (including none) and, e.g., rotations that relate to telescope pointing

a dispersing element (e.g., a prism or grating), such that the *spectropolarimetric* mode yields polarization data as a function of $[\lambda, y]$ (y along the slit).

Recalling the definition of the Stokes parameters in Eqs. 4.2 and 4.3, one requires at least two photon flux measurements to retrieve I and Q, U, or V. The modulation of the polarimeter describes how the various measurements yield (a part of) the Stokes vector. Classically, one can discern two types of modulation: *spatial modulation* and *temporal modulation*. With spatial modulation, the incident beam is somehow split up into two or more beams, whose intensities are measured by several detectors or by different parts of a single detector. It is important that these beams have been filtered for different polarization directions. As an example, a polarizing beam-splitter delivers two beams that have intensities $(I + Q)/2$ and $(I - Q)/2$ where $\pm Q$ are defined as the linear polarization directions of the beam-splitter. The *demodulation* process then involves a subtraction of these two intensity measurements to obtain a measure of Q_{out} and an addition to obtain I_{out}. Usually, Q_{out} is normalized by dividing by I_{out}. In principle, the complete Stokes vector can be measured using four beams that are appropriately split according to Q, U, and V.

With temporal modulation, the intensity measurements in particular polarization states are obtained sequentially. In principle, for measuring Q and U this could be implemented by rotating a polarizer, but as we will see in Sect. 5.5, this makes the measurement dependent on the polarization properties of the optics after the analyzer as well as before the analyzer. Having a fixed analyzer makes the measurement independent on the optics after it, and therefore the polarization properties of these elements are irrelevant inasmuch they do not act as a strong polarizer that is crossed with the (close to) 100% polarization that the analyzer delivers such that the transmission of the instrument is strongly reduced. It is often necessary to introduce additional polarization optics that act as a modulator. Its job is to convert the measurable polarization into the polarization that the analyzer actually analyzes (i.e., $+Q$). As we have seen in the previous sections, retarders have the property that they can convert U, V, and also $-Q$ into $+Q$. Let us consider a retarder with retardance δ that can be rotated by an angle α. The emergent intensity signal I' after the polarizer is then given by:

$$I'(\delta,\alpha) = \frac{1}{2}\bigg(I + \frac{Q}{2}((1+\cos\delta) + (1-\cos\delta)\cos 4\alpha) + \frac{U}{2}(1-\cos\delta)\sin 4\alpha - V\sin\delta\sin 2\alpha\bigg). \tag{4.22}$$

The temporal modulation is described by the terms containing α. It is clear that Stokes Q and U are modulated at four times the rotation angle, whereas V is modulated at twice the speed.

When the rotating retarder is a half-wave plate (HWP; $\delta = \pi$), the term with V becomes zero and the polarimeter only modulates linear polarization. Four measurements with the retarder at $\alpha = 0°, 45°$ and $\alpha = 22.5°, 67.5°$ then yield $(I \pm Q)/2$ and $(I \pm U)/2$, respectively. It speaks for itself that the readout of the detector needs to be accurately synchronized to the modulation sequencing. The modulation signals for Q and U are 90° out of phase. If the retarder is a quarter-wave plate (QWP; $\delta = \pi/2$) and measurements are performed at $\alpha = 45°, 135°$, Stokes V is measured optimally. In fact, measuring Stokes V always requires the use of a QWP to convert circular polarization into linear, since circular polarizers are not commonly available. If one wants to measure all Stokes parameters, a retarder with $\delta \approx 127°$ yields equal modulation amplitudes for Q, U, and V.

Such a temporal modulation scheme can be generalized with the use of (fixed) liquid crystals. For instance, a combination of two LCVRs with their fast axes at 0° and 45°, respectively, with respect to the polarizer's axis can be used to convert any linear combination of Q, U, and V into the linear polarization analyzed by the polarizer ($+Q$) by choosing appropriate retardances of the two LCVRs through the applied voltages. One could choose a sequence according to the Stokes definition scheme (cf. ➋ Eq. 4.3) such that the emergent intensities are $(I \pm Q)/2$, $(I \pm U)/2$, and $(I \pm V)/2$. In general, any sequence of four or more measurements of linear combinations of I, Q, U, and V is sufficient, as long as this sequence is nonredundant. Such a modulation scheme with n states is best described as an $n \times 4$ *modulation matrix* O:

$$\mathbf{I}' = \mathsf{O} \cdot \mathbf{S}_{\text{in}}. \tag{4.23}$$

The column vector $\mathbf{I}'$ contains the n intensity measurements according to the particular modulation state i. The n rows of the modulation matrix O are in fact the first rows of the complete instrumental Mueller matrix for modulation state i. These rows describe the conversion of the incident Stokes vector $\mathbf{S}_{\text{in}}$ to the intensity I'_i that is measured by the detector. If we let M_i denote the instrumental Mueller matrix (including the effects of the modulator in state i) up to the analyzer, and describe the analyzer as an ideal polarizer at $\pm Q$, then:

$$\mathsf{O}_\pm = \frac{1}{2}\begin{pmatrix} \mathsf{M}_{i=1,11} \pm \mathsf{M}_{i=1,21} & \mathsf{M}_{i=1,12} \pm \mathsf{M}_{i=1,22} & \mathsf{M}_{i=1,13} \pm \mathsf{M}_{i=1,23} & \mathsf{M}_{i=1,14} \pm \mathsf{M}_{i=1,24} \\ \mathsf{M}_{i=2,11} \pm \mathsf{M}_{i=2,21} & \mathsf{M}_{i=2,12} \pm \mathsf{M}_{i=2,22} & \mathsf{M}_{i=2,13} \pm \mathsf{M}_{i=2,23} & \mathsf{M}_{i=2,14} \pm \mathsf{M}_{i=2,24} \\ \vdots & \vdots & \vdots & \vdots \\ \mathsf{M}_{i=n,11} \pm \mathsf{M}_{i=n,21} & \mathsf{M}_{i=n,12} \pm \mathsf{M}_{i=n,22} & \mathsf{M}_{i=n,13} \pm \mathsf{M}_{i=n,23} & \mathsf{M}_{i=n,14} \pm \mathsf{M}_{i=n,24} \end{pmatrix}. \tag{4.24}$$

If the instrument in front of the polarimeter is only very weakly polarizing, the modulation matrix can be approximated by:

$$\mathsf{O}_\pm \approx \frac{1}{2}\begin{pmatrix} \mathsf{M}_{i=1,11} & \pm\mathsf{M}_{i=1,22} & \pm\mathsf{M}_{i=1,23} & \pm\mathsf{M}_{i=1,24} \\ \mathsf{M}_{i=2,11} & \pm\mathsf{M}_{i=2,22} & \pm\mathsf{M}_{i=2,23} & \pm\mathsf{M}_{i=2,24} \\ \vdots & \vdots & \vdots & \vdots \\ \mathsf{M}_{i=n,11} & \pm\mathsf{M}_{i=n,22} & \pm\mathsf{M}_{i=n,23} & \pm\mathsf{M}_{i=n,24} \end{pmatrix}. \tag{4.25}$$

Note that even non-polarimetric instruments are only fully described by ➋ Eq. 4.23. In that case, the 1×4 modulation matrix reads:

$$\tilde{\mathsf{O}} = \begin{pmatrix} \mathsf{M}_{11} & \mathsf{M}_{12} & \mathsf{M}_{13} & \mathsf{M}_{14} \end{pmatrix}, \tag{4.26}$$

where M describes the polarization properties of the instrument. The fact that the instrument does not intend to measure polarization does not mean that this matrix is fully diagonal and that the $Q, U, V \rightarrow I$ elements are zero. As we will see in ➋ Sect. 5.5, many optical components act as partial polarizers and weak retarders, even if they are not designed to do that. Hence, the intensity I' as measured by this non-polarimetric instrument depends on the input polarization (cf. Eq. 4.23). And since this polarization is not measured, this effect introduces uncalibratable photometric errors.

The measured intensities $\mathbf{I}'$ must be transformed into the Stokes parameters using

$$\mathbf{S}_{\text{out}} = \mathsf{D} \cdot \mathbf{I}'. \tag{4.27}$$

Obviously, for the obtained Stokes vector to be equal to the input Stokes vector (in the absence of noise), the *demodulation matrix* D should obey:

$$\mathsf{D} \cdot \mathsf{O} = \mathbb{I}. \tag{4.28}$$

For a four-stage temporal modulation, D is obtained after a regular matrix inversion of the 4×4 matrix O. But for $n > 4$ there is an infinite number of possibilities for D to satisfy the above equation. Del Toro Iniesta and Collados (2000) have shown that the optimal demodulation matrix[10] is the pseudo-inverse, namely:

$$\mathsf{D} = (\mathsf{O}^T \mathsf{O})^{-1} \mathsf{O}^T. \tag{4.29}$$

With this demodulation matrix, one optimizes the *polarimetric efficiency* of the system, defined as the 4-vector ϵ with components

$$\epsilon_k = \left(n \sum_{l=1}^{n} \mathsf{D}_{kl}^2 \right)^{-\frac{1}{2}}. \tag{4.30}$$

The four values of ϵ describe how efficiently the Stokes parameters $\mathbf{S}_k = (I, Q, U, V)^T$ are obtained using a certain demodulation matrix. Note that the efficiency is normalized with the number of modulation states n so that different modulation approaches can be compared. The efficiency in this context is related to the propagation of random noise that is present in the intensity measurements (mostly photon and readout noise):

$$\sigma_{\mathbf{S}_k}^2 = \frac{\sigma_{I'_k}^2}{\epsilon_k^2}. \tag{4.31}$$

The efficiencies obey:

$$\epsilon_1 \leq 1, \qquad \sum_{k=2}^{4} \epsilon_k^2 \leq 1. \tag{4.32}$$

For an *optimum polarimeter* the equal signs in the above equation are valid. It turns out (see Keller and Snik 2009) that an optimum in polarimetric efficiency is reached when the locations of the polarization state as sequentially selected by the modulator form the vertices of a platonic solid inscribed within the Poincaré sphere. For instance, the Stokes definition scheme has (optimal) efficiencies $\epsilon_{2,3,4} = 1/\sqrt{3}$ and the locations of its modulation states $((I \pm Q)/2, (I \pm U)/2,$ and $(I \pm V)/2)$ describe an octahedron on the Poincaré sphere. If one is only interested in linear polarization, then the locations of the optimal modulation states are described by an equilateral triangle, a square, etc. on the equator of the Poincaré sphere. As the minimum amount of intensity measurements to determine the complete Stokes vector is four, the optimal modulation sequence corresponds to a tetrahedron on the Poincaré sphere. This has identical efficiencies

[10] This description assumes that the entire Stokes vector is modulated. For a linear polarimeter or a circular polarimeter, the dimension of these matrices needs to be reduced to avoid singularities in the matrix inversion.

to an octahedron scheme, but the duty cycle of the modulation is faster. Sabatke et al. (2000) have shown that a rotating retarder with $\delta \approx 132°$ is capable of generating such an optimal four-state modulation[11]. In general, it is easier to design a modulator based on liquid crystals, possibly in combination with fixed retarders, that is able to generate modulation sequences with optimum efficiencies. In fact, Tomczyk et al. (2010) propose to use only this metric of polarimetric efficiency to design modulators that cover a very broad wavelength range. This is driven by the fact that the retardance of the wave plates used in a modulator (e.g., a QWP or a HWP) is often very chromatic, which drastically limits the operational wavelength range for such a modulator. By taking the (known) wavelength dependence of retardance of (liquid) crystals for granted, one can design "polychromatic" modulators based on stacks of these (liquid) crystals with particular thickness and orientations that have optimum polarimetric efficiencies over a certain wavelength range (which can be as large as several octaves). This means that the modulation and demodulation matrices for such modulators vary strongly with wavelength. Such an approach is only feasible for narrowband or spectroscopic measurements and is not suitable for broadband polarimetry.

5.3 Boosting Polarimetric Sensitivity

In the previous section, perfect behavior of the detector and the optics behind the analyzer has been assumed, as well as the absence of seeing and other time-dependent effects. As polarimetry is inherently a differential technique, differential effects can severely limit the polarimetric performance by creating spurious polarization signals. For spatial modulation, differential effects include (uncalibrated) transmission differences and differential aberrations between the beams, and the limited flat-fielding accuracy for the different pixels where the signal corresponding to the same location on the sky is measured. To simplify the equations, all these effects are now described as due to the two different transmissions (and gains) $g_{L,R}$ for two output beams of a polarizing beam-splitter that analyzes $\pm Q$. As an example, consider a polarimeter that merely consists of this beam-splitter. The emergent intensities are then:

$$I'_L = \frac{1}{2} g_L (I_{\rm in} + Q_{\rm in});$$
$$I'_R = \frac{1}{2} g_R (I_{\rm in} - Q_{\rm in}), \tag{4.33}$$

such that the demodulated result is:

$$I_{\rm out} = I'_L + I'_R = (g_L + g_R) I_{\rm in} + (g_L - g_R) Q_{\rm in};$$
$$Q_{\rm out} = I'_L - I'_R = (g_L + g_R) Q_{\rm in} + (g_L - g_R) I_{\rm in}. \tag{4.34}$$

Particularly the last term ($I \rightarrow Q$) is a nuisance, since often $Q \ll I$. And if the differences $(g_L - g_R)$ are dominated by detector gain errors, then this effect introduces a fixed noise-like structure in the Q images. Since flat-fielding accuracy is usually limited to no better than ~0.1%, this effect limits the polarimetric sensitivity for a polarimeter based merely on spatial modulation to a few tenths of a percent.

In the case of temporal modulation, the differential effects are variable in time. For astronomical instrumentation these effects are dominated by seeing, but also vibrations, pointing

[11]Note that this value is very close to the $\delta \approx 127°$ required for equal modulation amplitudes in Q, U, and V.

and guiding instabilities, and beam wobble due to rotating elements (including the modulator itself) play a role. In any case, these effects are modeled with a parameter t_i that has an average of 1 and takes on slightly different values during the modulation sequences[12]. The main consequence of seeing and other variable effects is a small, random displacement $\mathbf{r}_i$ of the intensity structure $I(x, y)$ on the detector (which can be an image or a spectrum)[13]. This is then approximated as (Lites 1987):

$$t_i = 1 + \mathbf{r}_i \cdot \frac{\nabla I(x, y)}{I(x, y)}. \tag{4.35}$$

Note that the largest degrading effects due to seeing take place at locations of large gradients in the intensity image, which, incidentally, is also the case for differential aberrations in the case of spatial modulation. Considering a similar example as for the spatial modulation, i.e., a single beam polarimeter measuring Stokes Q (or, equivalently, U or V) with a rotating HWP for the temporal modulation, the results are very similar:

$$\begin{aligned} I_1' &= \frac{1}{2} t_1 (I_{\text{in}} + Q_{\text{in}}) ; \\ I_2' &= \frac{1}{2} t_2 (I_{\text{in}} - Q_{\text{in}}), \end{aligned} \tag{4.36}$$

$$\begin{aligned} I_{\text{out}} &= I_1' + I_2' = (t_1 + t_2) I_{\text{in}} + (t_1 - t_2) Q_{\text{in}} ; \\ Q_{\text{out}} &= I_1' - I_2' = (t_1 + t_2) Q_{\text{in}} + (t_1 - t_2) I_{\text{in}}. \end{aligned} \tag{4.37}$$

Again, the last term limits the polarimetric sensitivity, but this time there is a way to reduce the influence of the differential effects. If the modulation cycle is much faster than the typical timescales at which t varies (~1 ms for natural seeing), then the real value of Q/I is approached with these measurements. Indeed, the record for polarimetric sensitivity was achieved with a fast temporal modulation approach. Kemp et al. (1987) found integrated broadband sunlight to be linearly unpolarized at a level of $3 \cdot 10^{-7}$ using an aperture polarimeter based on PEM modulation at more than 30 kHz[14]. Similar approaches using fast modulators like PEMs or FLCs capable of polarimetry with a sensitivity of 10^{-5}–10^{-6} have been applied to aperture polarimetry of stars (Hough et al. 2006) and for imaging or spectropolarimetry of the Sun, although the latter requires a dedicated detector that is capable of demodulating at kHz rates (Povel et al. 1994).

Polarimetric sensitivity at the ~10^{-5} level can also be attained with very slow modulation, but only when the spatial modulation and the temporal modulation approaches are combined. This *dual-beam technique* is most useful in situations when photons come at a premium (which is all too often the case in astronomy[15]: It allows for long exposures and does not throw away half of the light as a single-beam system does. To show how such a dual-beam technique can significantly decrease the amount of spurious polarization signals created by differential effects,

[12]Fortunately, the Earth's atmosphere is not polarizing nor birefringent, although the presence of dust in the atmosphere does create small polarization signals (Hough 2007). Scattered (sun or moon) light can create a polarized sky background.

[13]In the IR range, variable atmospheric transmission and background also play a role in this term t_i.

[14]This is an interesting finding by itself: this means that asymmetries in the solar shape, due to sunspots, and in the medium between the Sun and the Earth are very small indeed. This implies that the same is the case for similar stars and that detected polarization is solely due to interstellar absorption or circumstellar objects like exoplanets.

[15]Even for solar observations at high spatial and spectral resolution!

let us expand our example Stokes Q polarimeter. We now have four intensity measurements:

$$I'_{L,1} = \frac{1}{2} g_L t_1 (I_{\text{in}} + Q_{\text{in}}) \qquad I'_{R,1} = \frac{1}{2} g_R t_1 (I_{\text{in}} - Q_{\text{in}});$$
$$I'_{L,2} = \frac{1}{2} g_L t_2 (I_{\text{in}} - Q_{\text{in}}) \qquad I'_{R,2} = \frac{1}{2} g_R t_2 (I_{\text{in}} + Q_{\text{in}}). \tag{4.38}$$

The additional redundancy in the third and fourth measurements allows for a demodulation that strongly reduces the influence of the differential effects parameterized by $g_{L,R}$ and t_i. There are two methods for obtaining Q/I. The first one is through a *double difference*:

$$\frac{Q_{\text{out}}}{I_{\text{out}}} \approx \frac{1}{2}\left[\frac{I'_{L,1} - I'_{R,1}}{I'_{L,1} + I'_{R,1}} - \frac{I'_{L,2} - I'_{R,2}}{I'_{L,2} + I'_{R,2}}\right] \approx \frac{1}{2}\left[\frac{I'_{L,1} - I'_{L,2}}{I'_{L,1} + I'_{L,2}} - \frac{I'_{R,1} - I'_{R,2}}{I'_{R,1} + I'_{R,2}}\right]. \tag{4.39}$$

It is easily verified that for $Q \ll I$ all the terms with differences of $g_{L,R}$ and t_i disappear. The second demodulation method for such a dual-beam system is through a *double ratio*:

$$\frac{Q_{\text{out}}}{I_{\text{out}}} \approx \frac{\sqrt{R} - 1}{\sqrt{R} + 1}, \qquad \text{with} \qquad R = \frac{I'_{L,1}}{I'_{R,1}} \frac{I'_{R,2}}{I'_{L,2}}. \tag{4.40}$$

In this case, all factors $g_{L,R}$ and t_i automatically divide out. For small degrees of polarization, these methods yield equivalent results, and all differential effects are eliminated to first order (Semel et al. 1993; Tinbergen 1996; Bagnulo et al. 2009). However, for larger degrees of polarization ($P \gtrsim 10\%$), the ratio method is more generally applicable. And for observations with low light levels, the ratios are sometimes better avoided, and a difference to obtain the pure Stokes parameters (without dividing by I) is the best solution. If four additional redundant measurements are obtained, for instance, by changing the orientation of the rotating HWP or QWP by two additional incremental steps, one can also obtain a null measurement (see Bagnulo et al. 2009). Such a null measurement can then be used to assess the inherent noise level in the observations and the presence of residual spurious polarization signals.

In case that only one polarized Stokes parameter (Q, U, or V) is obtained through four intensity measurements of a dual-beam system, this is called a *beam exchange*, for obvious reasons. For the general case of some modulation matrix O, the equations read:

$$t = \text{diag}(t_1, t_2, \ldots, t_n); \tag{4.41}$$

$$\mathbf{S}_{\text{out},L} = (\mathsf{O}_+^T \mathsf{O}_+)^{-1} \mathsf{O}_+^T \cdot g_L \cdot t \cdot \mathsf{O}_+ \cdot \mathbf{S}_{\text{in}};$$
$$\mathbf{S}_{\text{out},R} = (\mathsf{O}_-^T \mathsf{O}_-)^{-1} \mathsf{O}_-^T \cdot g_R \cdot t \cdot \mathsf{O}_- \cdot \mathbf{S}_{\text{in}}. \tag{4.42}$$

Obviously, the ratio method does not generally work, unless beam-exchange pairs are present in the modulation sequence. This is only the case when the solid body connecting modulation states in the Poincaré sphere has mirror symmetry (e.g., an octahedron or a cube). In all other cases, a dual-beam polarimeter should be demodulated with a difference method:

$$\frac{\mathbf{S}_{\text{out}}}{I_{\text{out}}} = \frac{1}{2}\left[\frac{\mathbf{S}_{\text{out},L}}{I_{\text{out},L}} - \frac{\mathbf{S}_{\text{out},R}}{I_{\text{out},R}}\right]. \tag{4.43}$$

Also in this case, the seeing-induced spurious polarization signals and cross-talk are eliminated to first order, under the assumption of a weakly polarizing telescope (see Eq. 4.25). Furthermore, since the effect of seeing as modeled by Eq. 4.35 has a noise-like appearance ($I' \cdot t_i \approx I' + \sigma_{I',i}$) its propagation is minimized in the first place by optimizing the polarimetric efficiency. The effects of seeing are eliminated to an absolute minimum when the dual-beam technique is combined with very fast modulation.

Only once systematic effects like the ones described here are eliminated, is the polarimetric efficiency limited by photon noise. In that case the noise level decreases with the increasing number of detected photons. Because photon noise is described by Poisson statistics, the noise level in the measurement of a normalized Stokes parameter (here again Q) scales like

$$\sigma_{Q/I} \propto \frac{\sqrt{N}}{N} = \frac{1}{\sqrt{N}}, \tag{4.44}$$

with N the amount of detected photoelectrons, which is proportional to the exposure time, or the amount of pixels that are binned[16]. To reach a polarimetric sensitivity of 10^{-5}, one needs to capture at least 10^{10} photons. And to boost the sensitivity of some polarimetric observation from 10^{-4} to 10^{-5}, one needs to capture 100× more photons by increasing the exposure time and/or (cleverly) binning detector pixels. Polarimetry is therefore a particularly photon-hungry technique, requiring large-aperture telescopes and high-transmission instrumentation. Furthermore, the local degree of polarization of an extended astrophysical source generally increases with increasing spatial resolution as structures are being resolved at their fundamental length scales[17]. The push for telescopes with increasingly large apertures is therefore particularly promising for polarimetry.

5.4 Spectral Modulation

In some cases, the implementation of temporal and/or spatial modulation is impossible. For instance, in space applications an active modulator is undesirable since it consumes power and can fail. Furthermore, splitting beams according to some spatial modulation may yield an intolerably large and heavy instrument. If the instrument to be designed is a spectropolarimeter, and the temporal and spatial dimensions cannot be modulated, then a modulation in the spectral domain can be devised. Such a *spectral modulation* consists of one or several sinusoidal patterns that are superimposed on the intensity spectrum.

The full-Stokes version of spectral modulation was introduced by Nordsieck (1974) and reinvented by Oka and Kato (1999). The modulator consists of two thick retarder plates, oriented at 0° and 45° from a polarizer. The modulated intensity spectrum is then described by:

$$\begin{aligned} I'(\lambda,\delta_1,\delta_2) = \frac{1}{2}\Big[& I_{\text{in}}(\lambda) \\ & \pm \big[Q_{\text{in}}(\lambda)\cos(\delta_2(\lambda)) \\ & + U_{\text{in}}(\lambda)\sin(\delta_1(\lambda))\sin(\delta_2(\lambda)) \\ & - V_{\text{in}}(\lambda)\cos(\delta_1(\lambda))\sin(\delta_2(\lambda))\big]\Big], \end{aligned} \tag{4.45}$$

with $\delta_{1,2}(\lambda)$ the wavelength-dependent retardances ($\delta_i = 2\pi d_i(n_e - n_o)/\lambda$) of the two crystal plates with thicknesses $d_{1,2}$, and the ± depending on the orientation of the polarizer. After some trigonometry it becomes clear that there are three modulation "frequencies" (one determined by δ_2 and two determined by the sums and differences of $\delta_{1,2}$) that carry the information on the polarization, which may be retrieved using Fourier transform methods. Note that the modulations are periodic in $1/\lambda$. It is clear that the spectral resolution of the spectrometer needs

[16]For a description of the propagation of Gaussian noise to derived observables like the degree or angle of linear polarization, see Sparks and Axon (1999), Patat and Romaniello (2006) and chapter 5 of Clarke (2009).
[17]The same is true for an increase in spectral resolution.

to be significantly increased to record the additional polarization information compared to the spectral resolution that is required for the data-product.

In case one only wants to perform linear spectropolarimetry with medium spectral resolution with Q and U often being the only broadband observables, one can reduce the number of modulation frequencies to one, for which the amplitude scales with the degree of linear polarization P_L and the phase with the angle of linear polarization ϕ_L. Such a spectral modulation for linear spectropolarimetry is obtained by combining an achromatic QWP with a single multiple-order retarder with value $\delta(\lambda)$ (Snik et al. 2009). In this case, the modulated intensity spectrum is described by:

$$I'(\lambda,\delta) = \frac{1}{2} I_{\text{in}}(\lambda) \cdot \left[1 \pm P_L(\lambda) \cdot \cos(\delta(\lambda) + 2 \cdot \phi_L(\lambda))\right]. \tag{4.46}$$

In this case, the increase of spectral resolution is minimal, and the demodulation for P_L and ϕ_L is relatively easy. In both implementations of spectral modulation, the intensity spectrum at full spectral resolution is trivially obtained by using a polarizing beam-splitter and adding the two beams. The pure modulation envelopes are then obtained through division of each signal by the intensity spectrum. This envelope can then be used to perform the demodulation on, without confusion from spectral features in the intensity spectrum. The thus obtained polarization spectra will have a significantly reduced spectral resolution compared to that of the spectrometer. It is therefore only applicable to broadband polarization. If there are many spectral lines that have similar polarization behavior, the (normalized) signals in these lines may be phase-folded and demodulated to derive one common polarization signal, which is different from the continuum polarization.

The major advantage of such spectral modulation approaches is that all information is contained within a single measurement and that it is therefore not susceptible to differential effects. The modulator itself is fully passive, and can be inserted in front of any spectrometer to turn it into a spectropolarimeter. The polarimetric sensitivity is only limited by the effectiveness of the demodulation algorithm, and by noise. The sensitivity limits of this technique have not yet been explored, but it is expected that it will be below the $\sim 10^{-3}$ level. The inherent polarimetric accuracy is mostly limited by alignment and by variations of the multiple-order retardance with temperature. Snik et al. (2009) therefore implemented an athermal combination of crystals, although it will only be perfectly athermal at one wavelength. However, to first order, a variation of the retardance only influences the measurement of ϕ_L in ➋ Eq. 4.46, and the measurement of the degree of linear polarization (often the prime observable) is unaffected.

5.5 Instrumental Polarization Effects

To measure a small asymmetry (i.e., a slight imbalance between polarization directions) within the light collected by a telescope, it is crucial to maximize the symmetry of the measurement setup and procedure. This is the case for the modulation approaches described in the previous sections, but it is even more pertinent to the instrumental setup as a whole. Only an instrument that is perfectly rotationally symmetric can be considered "polarization free."[18] Any oblique

[18] It is often sufficient to just have 90° point symmetry such that the polarizing and/or retardance properties of a part of some optical component is compensated with another, identical, crossed part of the same component. This can be the case for, e.g., a square aperture.

reflection or refraction creates and/or modifies polarization, as determined by the Fresnel equations. Indeed, the Mueller matrix of a telescope measured at the center of the diffraction-limited point spread function (PSF) of a Cassegrain or Gregorian focus is fully diagonal (Sánchez Almeida and Martínez Pillet 1992). Only some depolarization occurs. However, the mere breaking of the symmetry by moving off-axis introduces instrumental polarization effects. Sánchez Almeida and Martínez Pillet (1992) show that already the diffraction rings of the PSF are polarized at the 10^{-4}–10^{-3} level (with the average over the PSF being zero again). Furthermore, seeing and variations of mirror coating properties across the aperture introduce polarization, even on-axis. So, for all practical purposes, when one operates an astronomical polarimeter, one has to assume that some level of instrumental polarization and cross talk is present.

Fold mirrors are a very common source of instrumental polarization. For instance, at a Nasmyth focal station, the light has been reflected off a 45° mirror. For such a mirror consisting of a pure aluminum layer, the normalized Mueller matrix at 500 nm is given by (from measurements and models by Van Harten et al. 2009):

$$\mathsf{M}_{\text{alu}}(500\ \text{nm}) = \begin{pmatrix} 1.000 & 0.028 & 0.000 & 0.000 \\ 0.028 & 1.000 & 0.000 & 0.000 \\ 0.000 & 0.000 & -0.974 & 0.225 \\ 0.000 & 0.000 & -0.225 & -0.974 \end{pmatrix}. \tag{4.47}$$

These numbers are fully determined by the refractive index of aluminum, which is a complex number that takes into account the strong absorption within the metal (which boosts the reflectivity). This Mueller matrix shows ~3% of instrumental polarization and a cross talk of ~23% from Stokes U into Stokes V, and vice versa. In many cases, these numbers are much larger than the requirements for polarimetric accuracy, and the problem has to be dealt with in the instrument design or through calibration (see below). This issue becomes even more problematic if the Mueller matrix of the mirror is variable, for instance because it is rotating with respect to the polarimeter, which is the case for a Nasmyth mirror (see Harrington and Kuhn 2008 for a more extreme example). Furthermore, an aluminum mirror in air always has a natural, dielectric aluminum oxide layer, which modifies its polarization properties. Measurements by Van Harten et al. (2009) show that this layer has a thickness of ~4 nm and yields the following total Mueller matrix:

$$\mathsf{M}_{\text{alu+d}}(500\ \text{nm}) = \begin{pmatrix} 1.000 & 0.030 & 0.000 & 0.000 \\ 0.030 & 1.000 & 0.000 & 0.000 \\ 0.000 & 0.000 & -0.961 & 0.276 \\ 0.000 & 0.000 & -0.276 & -0.961 \end{pmatrix}. \tag{4.48}$$

It is clear that the instrumental polarization as well as the cross talk terms are significantly modified. Fortunately, the thickness of the oxide layer is found to be stable in time within a few days after coating the mirror. However, during that time also layers of, e.g., oil or dust may be formed if one is not careful enough. An extreme layer of dust that results in a 25% loss of reflectivity yields the following Mueller matrix (Snik et al. 2011):

$$\mathsf{M}_{\text{alu+d+dust}}(500\ \text{nm}) = \begin{pmatrix} 1.000 & 0.026 & 0.000 & 0.000 \\ 0.026 & 1.000 & 0.000 & 0.000 \\ 0.000 & 0.000 & -0.959 & 0.281 \\ 0.000 & 0.000 & -0.281 & -0.959 \end{pmatrix}. \tag{4.49}$$

The dust creates an additional nuisance as it scatters slightly polarized light, which is a particular problem for performing polarimetry in regions next to a bright source (Keller 2002a).

Other optical components can also have nonzero polarization properties. For instance, any piece of glass causes some retardance action, even if it is used at normal incidence. Although most glasses are isotropic in theory, birefringence will be present at some level because of internal stresses as a result of the production process, and because of unidirectional external stress applied by the mount of the component. Common glasses have an inherent stress birefringence of at least ~5 nm per cm of glass, with the notable exception of fused silica, which can be manufactured with less than 1 nm of birefringence per cm of glass. However, fused silica has a sizable stress-optic constant, which means that it can become very birefringent again upon external stress (hence its applicability within PEMs). This is the reason why the objective lenses of refracting telescopes and entrance windows always need to be carefully mounted, particularly when the glass elements seal off a vacuum system.

In spectrographs, both the slit (see Keller 2002a) and the grating act as partial linear polarizers. A low-order grating can in fact act as a nearly perfect polarizer for certain wavelengths (Wood's anomaly). This can be particularly problematic when the analyzer is located before the grating and its emergent polarization is crossed with the transmission axis of the grating. Also in non-polarimetric spectrometers, the grating can make the instrument very polarization sensitive.

These instrumental polarization effects are often unpredictable and they interact in surprising ways. For instance, Keller (1996) has shown that spurious polarization signals can be created through the interaction of instrumental polarization (or polarized fringes) with the detector nonlinearity and/or dark drift and stray light. An instrumental polarization of 1% and an uncorrected nonlinearity of 1% then cause a print-through of I in, e.g., Q/I at the 10^{-4} level.

For many reasons it is therefore crucial to deal with instrumental polarization effects already during the design of the polarimeter. The main solution is to locate the polarimeter as far upstream as possible. Ideally, the polarization analysis is completed before the first optical component of the telescope, which is indeed an option for small instruments. For larger telescopes, if at all possible, the polarimeter should be built at a symmetric focus, i.e., at a Cassegrain or Gregorian station. For AO-assisted polarimetry, this is only possible with an adaptive secondary mirror, see Packham and Jones (2008) for an example of this. With a fixed analyzer, all polarization effects behind it merely translate into transmission losses. In the case of a dual-beam polarimeter, having two beams to be propagated through all subsequent optics may be very difficult or simply impossible. One can then still decide to keep the modulator in an upstream location and move the analyzer further down (see, e.g., Ichimoto et al. 2008). The polarization properties of the optics in between then still need to be controlled, but the requirements are not as strict as for optics in front of the modulator. The reason for this is that the analyzer has already converted the measurable polarization into the linear polarization that is analyzed, say Stokes Q'. It is then only important that this Stokes Q' is transferred by the optics in between with a large efficiency. If, for instance, these would act as a QWP at 45°, the analyzer could not anymore distinguish between the polarization states that the modulator intends to measure. Therefore, it is important that all optics with retarding properties have their fast or slow axes aligned with the polarization direction(s) of the analyzer, such that this direction becomes the "eigenvector" of the system. In the case of mirrors, this means that all s and p directions[19] need to

[19]These are the directions perpendicular ("*s*enkrecht" in German) and *p*arallel to the reflection plane.

coincide with this eigenvector. Since the s and p directions generally have different reflectivities (in other words, the mirror is also a weak polarizer, see Eq. 4.47), this causes transmission differences between the two beams in a dual-beam polarimeter. If possible, these imbalances need to be minimized to maximize the symmetry of the system, which always boosts the polarimetric performance.

One way to deal with this instrumental polarization is to turn the upstream modulator into a *polarization switch* (Tinbergen 2007), and build a complete polarimeter downstream. The function of this switch is to change the sign of the source polarization with respect to the instrumental polarization, which is potentially much larger. So with a first complete measurement one obtains the sum of these two signals and with a second measurement the difference. The source polarization is then obtained through the difference of these two measurements, a "triple difference" in case of a dual-beam system.

There are various tricks to reduce the instrumental polarization effects of optical components. For a mirror, it is clear from Eqs. 4.47– 4.49 that its polarization properties are modified by the dielectric layer on top. This fact can be used to one's advantage, because the Mueller matrix elements $U \leftrightarrow V$ can be made to disappear when a dielectric layer with the correct thickness is overcoated. For an aluminum mirror, an aluminum oxide layer of ~100 nm is required for this (see Keller 2002a), although the exact thickness depends on the complex value of the refractive index of aluminum, which seems to vary significantly from mirror to mirror (see, e.g., Harrington and Kuhn, 2008). Furthermore, this correction is only valid for a single wavelength. A more general method for correcting all polarization properties of a mirror is by crossing it with an identical mirror such that the s and p directions are exchanged. The combined Mueller matrix is then necessarily diagonal, provided the two mirrors are perfectly aligned and are truly identical in their polarization properties.

The most general way of *polarization compensation* is by inserting components into the beam that are able to (partially) correct for instrumental polarization. A tiltable and rotatable glass plate can compensate for (flat) linear polarization introduced by the instrument of up to ~10%. The required amount of tilt can be computed in advance using the Fresnel equations, but the tilt and rotation angle of the compensating plate can also be controlled in real time by feeding back the measured polarization at a location of the image or spectrum that should have zero polarization. An instrumentally induced circular polarization can be corrected using retarders. A combination of a rotatable QWP and HWP, or two QWPs can bring any state of polarization back to the desired Stokes Q' direction. Another option for a polarimeter consisting of two LCVRs consists of applying offset retardances to its modulation scheme, such that the overall polarimetric efficiencies are maximized again.

5.6 Calibration

Even after minimizing instrumental polarization, calibration is required in most circumstances, even if only to check that it is still minimal and non-varying. Such *polarimetric calibration* is performed by injecting known polarization states into the polarimetric instrument. If at all possible, a calibration procedure is enabled by temporarily inserting calibration optics into the beam, as much upstream as possible. Such polarization calibration optics may consist of a rotatable polarizer, or a rotatable polarizer in combination with a rotatable QWP, which is able to generate all fully polarized Stokes vectors. If the calibration unit generates m different Stokes vectors, then the measured intensities at the polarimeter's detector for every modulation state i and calibration input j are:

$$I'_{i,j} = (\mathsf{O}_{i,1}, \mathsf{O}_{i,2}, \mathsf{O}_{i,3}, \mathsf{O}_{i,4}) \cdot \mathbf{S}_{\mathrm{cal},j}. \tag{4.50}$$

Rearranging this equation yields:

$$\begin{pmatrix} I'_{i,1} \\ I'_{i,2} \\ \vdots \\ I'_{i,m} \end{pmatrix} = \begin{pmatrix} I_1 & Q_1 & U_1 & V_1 \\ I_2 & Q_2 & U_2 & V_2 \\ \vdots & \vdots & \vdots & \vdots \\ I_m & Q_m & U_m & V_m \end{pmatrix} \cdot \begin{pmatrix} \mathsf{O}_{i,1} \\ \mathsf{O}_{i,2} \\ \mathsf{O}_{i,3} \\ \mathsf{O}_{i,4} \end{pmatrix}, \tag{4.51}$$

$$\mathbf{I}'_i = \mathsf{C} \cdot \mathsf{O}_i^T. \tag{4.52}$$

Again, the least-squares solution for the row of the modulation matrix O is obtained after pseudo-inversion:

$$\mathsf{O}^T_{\mathrm{cal},i} = (\mathsf{C}^T\mathsf{C})^{-1}\mathsf{C}^T \cdot \mathbf{I}'_i. \tag{4.53}$$

The calibration optics themselves also have errors associated with them, for instance an offset to the rotation angles, or to the retardance of the QWP. If the calibration procedure is overdetermined (i.e., $m > 4$), most of these unknown errors can be determined from the calibration data as well. One then sets up a Mueller matrix model for the generation of the calibration states $\mathbf{S}_{\mathrm{cal},m}$, that contains the known dependencies on the parameters with unknown values. Since the pseudo-inverse operation is equivalent to a least-squares fit, one can find the values of these parameters once the χ^2 of the modeled intensities is minimal when compared with the actual measurements. This fitting procedure can also be adopted for the instrumental Mueller matrix if it is known to depend on a limited number of free parameters (Skumanich et al. 1997). This is a particularly good method for setting up a complete instrument model that includes Mueller matrices that depend on, e.g., the telescope pointing. The polarimetric accuracy as described by $\Delta\mathsf{X}$ in ➋ Eq. 4.21 after rigorous calibration is typically ~0.01 for the diagonal elements and $\sim 10^{-3}$ for the instrumental polarization and the cross talk (see, e.g., Skumanich et al. 1997; Ichimoto et al. 2008). These numbers are usually not limited by the accuracy of the calibration itself, but by long-term stability issues between the times of calibration and observations due to certain unknown parameters that depend on, e.g., temperature or telescope pointing.

Often, the calibration unit does not calibrate the polarization properties of the telescope. These are either assumed to be small up to the location of the calibration unit, or it has to be calibrated in other ways. For small telescopes ($D \lesssim 1$ m), one can mount a large, rotatable polarizer sheet in front of the aperture. In all other cases one needs to observe standard polarized or unpolarized stars (Whittet et al. 1992; Heiles 2000; Clarke 2009) that yield input light which polarization is known with some accuracy. The instrumentally induced polarization is trivially obtained from the observation of unpolarized stars or other sources that (on average) can be assumed to be unpolarized, like the center of the solar disk in the absence of active regions. When the instrumental polarization and the observed astronomical polarization are both small, the instrumental polarization can just be subtracted from the observations, once it has been determined. The polarimetric efficiency can be determined by observing stars with a known nonzero polarization, either in the continuum or in spectral lines. There are quite many stars with continuum linear polarization, either due to the presence of a circumstellar disk or stellar asymmetry, or due to interstellar absorption by magnetically aligned dust grains. Often, such linearly polarized standard stars are also used to fix the orientation of the $[Q, U]$ coordinate system of a certain polarimetric instrument (or the position angle ϕ_L) on the sky. Unfortunately, there are not many stars with sizable circular polarization in their continuum, with the exception of some very magnetic white dwarfs (e.g., Kemp et al. 1970). Such observations of polarized standard stars can also be used to calibrate polarization cross talk. For instance, stars with a strong, stable, and geometrically simple magnetic field should only exhibit antisymmetric spectral line patterns in Stokes V, and symmetric ones in Q and U. Any offset from this likely indicates cross talk problems which are somehow caused by the instrument.

5.7 Performance Prediction

In the ideal case, a polarimeter's performance should be predicted before it is built. For optical designs and AO performance predictions, it is common practice to fully assess the influence of potential wavefront errors and assign tolerances to each optical component before the instrument is built. These error budgeting models are usually fairly straightforward as one can assume that errors are scalar, random, small, and independent of each other. For polarimetric instruments, errors are included in Mueller matrices, and, moreover, they are often systematic, large, correlated, and even unexpected. The first step for polarimetric performance prediction is to construct a complete Mueller matrix model, which contains values that are known from measurements, modeling, or experience. One can then assess the influence of residual errors after calibration on the final observables (see, e.g., Perrin et al. 2010). The next step is to create a formal model that includes all potential error contributions to the individual Mueller matrices, including their known dependencies on physical parameters like wavelength, temperature, time, field, and pupil position. Linearizing those errors, one can then define separate error matrices (Keller and Snik 2009), like:

$$\mathsf{M}\left(p_1 + \delta p_1, p_2 + \delta p_2\right) \approx \mathsf{M}\left(p_1, p_2\right) + \delta p_1 \cdot \mathsf{m}_1 + \delta p_2 \cdot \mathsf{m}_2, \tag{4.54}$$

These error matrices $\mathsf{m}_{i,j}$ belonging to Mueller matrix M_i and physical parameter j can be merely the first-order Taylor expansion, or it can be a statistical error term derived from some kind of assumed distribution (flat for systematic errors, see Keller and Snik 2009). The error propagation for any error induced by a physical parameter $p_{i,j}$ in Mueller matrix M_i is then to be computed as:

$$\begin{aligned} \mathsf{M}_n \cdot \mathsf{M}_{n-1} \cdot \mathsf{M}_{n-2} \ldots \mathsf{M}_{i+1} \cdot \mathsf{M}_i(p_j + \delta p_j) \cdot \mathsf{M}_{i-1} \ldots \mathsf{M}_3 \cdot \mathsf{M}_2 \cdot \mathsf{M}_1 \approx \\ \mathsf{M}_n \cdot \mathsf{M}_{n-1} \cdot \mathsf{M}_{n-2} \ldots \mathsf{M}_3 \cdot \mathsf{M}_2 \cdot \mathsf{M}_1 + \\ \delta p_{i,j} \cdot \left(\mathsf{M}_n \cdot \mathsf{M}_{n-1} \cdot \mathsf{M}_{n-2} \ldots \mathsf{M}_{i+1} \cdot \mathsf{m}_{i,j} \cdot \mathsf{M}_{i-1} \ldots \mathsf{M}_3 \cdot \mathsf{M}_2 \cdot \mathsf{M}_1\right). \end{aligned} \tag{4.55}$$

Hence the matrix product of the last line describes the separate error propagation of $p_{i,j}$ to the total Mueller matrix of the system in the second line. In these equations, the formalism only considers Mueller matrices, and does not take into account the measurement of the polarization nor the calibration of the polarimeter. Fortunately, most of these processes are described by matrices and linear operators, to which errors can be added in a very similar fashion. With the assumption that the errors can be linearized, the sensitivity analysis to individual errors can be performed rather easily in this way. By assuming certain error distributions, one can assign tolerances to each parameter by comparing the (quadratic) sum of them with the overall accuracy requirement ΔX. In the end, only a full-blown Monte Carlo analysis can give a complete analysis of the polarimetric performance, but this is very computationally intensive, and, moreover, does not yield any insight into how certain errors influence the polarimetric sensitivity and accuracy.

A notorious modeling issue for polarimetric instruments concerns polarized diffraction effects. By definition (➋ Eq. 4.2), the Stokes parameters are obtained after an incoherent sum, and cannot describe coherence effects. The Jones formalism is able to deal with diffraction, but only traces 100% polarized rays. However, by first using Jones matrices, Sánchez Almeida and Martínez Pillet (1992) derive Mueller matrix properties as a function of the diffraction-limited PSF pattern of a telescope. This approach is extended by Keller (2002b) to enable polarized ray-tracing within the optical design program ZEMAX. However, the accuracy of such modeling in situations with strong diffraction effects needs to be verified with real measurements.

6 Modern Polarimeters

6.1 Requirements

The exact implementation of any polarimeter depends largely on its requirements, which in turn depend on the scientific objective(s). The required polarimetric sensitivity and accuracy are usually the most stringent factors and constrain the design of the overall set-up. Furthermore, the required wavelength range, field-of-view, duty cycle, etc. can also pose significant design challenges. The best approach for designing a polarimeter is when it is considered simultaneously with the overall optical design from the very beginning. Frequently, the requirements and practical constraints lead to conflicts within the overall design, but then at least a fair trade-off between certain nonideal solutions can be made.

In the following sections, an overview is provided of current and future astronomical polarimeters (both imaging polarimeters and spectropolarimeters). This list is by no means complete, but it represents the current state of the art at astronomical facilities. Some polarimetric instruments and systems have already been mentioned in the text. It is clear that all these polarimeters are very different, because of their very different requirements and practical constraints. It is remarkable but not so surprising that most currently operating polarimeters cover only the visible or near-IR regimes. The UV and, to a lesser extent, the mid-IR therefore constitute blind spots to polarimetry, although developments in those regimes will no doubt be scientifically very valuable, particularly at high spectral resolution.

6.2 Dual-Beam Polarimeters

Most successful polarimeters at nighttime telescopes are of the dual-beam type. Moreover, the most frequently used polarimeters are those mounted at a Cassegrain focus, as calibration is often regarded as too time-consuming. One successful Cassegrain polarimeter operates at FORS at the 8-m VLT (see Patat and Romaniello 2006), and it has both an imaging as well as a medium-resolution spectropolarimetry mode. The polarimeter consists of a Wollaston prism that is inserted into the collimated section of the focal reduction optics, and a rotating superachromatic QWP or HWP as a modulator. For imaging polarimetry, a stripe mask is inserted in the first (i.e., Cassegrain) focal plane such that the Wollaston creates two interlaced, half-images. To obtain complete information on the field of view, one needs to dither the image. For spectropolarimetry, the focal mask contains a slit or slitlets (a slitless mode can also be employed for point sources), and an additional grism is inserted close to the reimaged pupil plane. The focal reduction optics create considerable off-axis instrumental polarization at the 1% level, which is symmetric around the center to a large degree, and, moreover, stable and therefore (partially) correctable. Typical polarimetric sensitivities attainable by FORS are $\sim 10^{-4}$. In the infrared range, several dual-beam imaging polarimeters are available to the community (see, e.g., Kuhn et al. 2001), and more will come online soon (e.g., Packham and Jones 2008).

High-resolution spectropolarimetry is also best performed at a Cassegrain focus. The most prominent instruments of this type are the twins ESPaDOnS at the 3.6-m CFHT and NARVAL at the 2-m TBL (Donati et al. 2006). In both instruments, the light is transported by two fibers to a spectrograph, after the polarimetric analysis has been performed. These two fibers are fed after splitting the beam by a Wollaston prism. To prevent spectral fringes and to cover a large wavelength range, the modulation is performed by a combination of five Fresnel rhombs in a collimated beam, of which the two half-wave pairs are rotatable. The collimating optics introduce some cross talk due to stress birefringence. After combining the many magnetically sensitive lines in its wavelength range (370–1050 nm) through least-squares deconvolution (LSD; Donati et al. 1997), these instruments can reach a polarimetric sensitivity of $\sim 10^{-5}$. A comparable capability has recently been commissioned for the famous HARPS spectrograph at ESO 3.6-m telescope (Snik et al. 2008). In HARPS' main mode, the two fibers are fed by starlight and calibration light at the Cassegrain focus, and in the polarimetry mode they transport the light from the two beams of the dual-beam polarimeter. Beam-splitting by a Foster prism (in combination with a cylindrical lens and another channeling prism) to deal with the large fiber separation delivers an optically achromatic system. The two modulators are a Pancharatnam-type QWP and HWP consisting of a stack of five true zero-order PMMA layers that enable a large wavelength range with minimal spectral fringing and allow for operation in a converging beam. The spectral resolution of HARPS ($R = 110,000$) is larger than for ESPaDONs ($R = 65,000$) whereas the spectral range is more limited (380–690 nm), and both instruments are very complementary in sky-coverage, as the former is in the southern hemisphere and the latter in the northern. An example of HARPSpol data is presented in ➋ *Fig. 4-7* and shows its polarimetric sensitivity of $\sim 10^{-5}$ after LSD.

6.3 Polarizer-Only Polarimeters

Several successful polarimeters are in operation that just use polarizers to measure linear polarization. Important examples are the polarimetric modes of ACS and NICMOS on the Hubble Space Telescope. Both consist of three different sheet polarizers at (ideally) 0°, 60°, and 120° within a filter wheel. However, this is technically a polarimeter with temporal modulation by exchanging the polarizers, and the absence of seeing in space makes that mostly the differences in transmission properties between the polarizers and their individual angle offsets limit the polarimetric performance (as with spatial modulation). Even after careful calibration (see Batcheldor et al. 2009) the NICMOS and ACS instruments are still limited in polarimetric sensitivity to $\sim 0.5\%$.

RINGO (Steele et al. 2009) is a very dedicated linear polarimeter for time-resolved detection of linear polarization in bright transient point sources like GRB afterglows. It consists

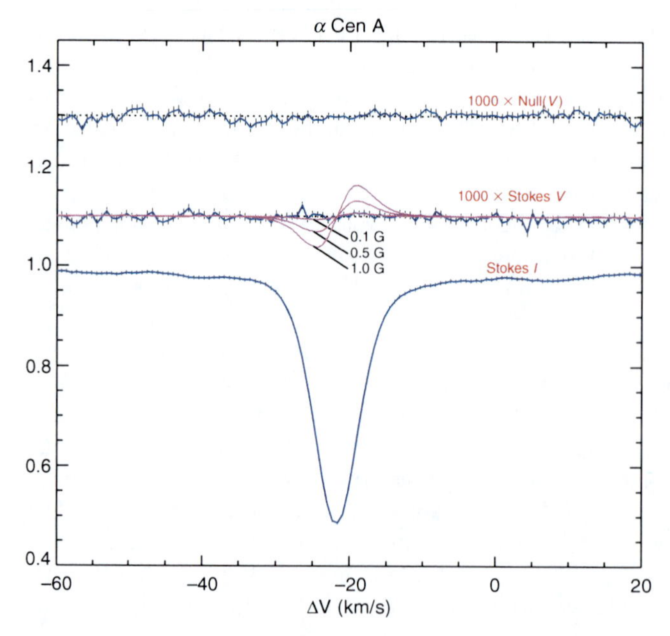

Fig. 4-7
HARPSpol observations of α Cen A. The Stokes *I* and *V* (and null) profiles are obtained after combining most spectral lines using LSD. The final polarimetric sensitivity of ~10^{-5} puts an upper limit on the star's global magnetic field of <0.1 G. In contrast, an identical twin of our Sun would yield a signal corresponding to ~0.5 G (Figure courtesy: Oleg Kochukhov)

of a rapidly rotating polarizer, in combination with a corotating wedge prism, such that for exposures much longer than the rotation period of the polarizer, each point source turns into a ring, whose intensity structure contains information on the degree and angle of linear polarization.

6.4 Solar Polarimetry

Polarimetry has been a well-established technique for solar observations for a long time, and significant technical progress has been made in that field. The current flag-ship for solar polarimetry is the diffraction-limited Solar Optical Telescope onboard the Japanese *Hinode* satellite (see Ichimoto et al. 2008). The continuously rotating athermal modulator is located after the Gregorian focus of the 0.5-m telescope. It serves several instruments, both filter imagers and spectropolarimeters operating in the visible regime, that each have their own polarizer or polarizing beam-splitter. The modulator is not achromatic, and it is optimized for high, full-Stokes polarimetric efficiencies at certain wavelengths, like for the Fe I lines at 630 nm, which are used by both the spectropolarimeter as the filter polarimeter to measure magnetic fields on the solar surface (Guimond and Elmore 2004). The optics between modulator and analyzer are folded such that the linear polarization for the analyzers are eigenvectors of the system.

The SOLIS Vector-Spectromagnetograph (VSM, Keller et al. 2003) is an example of a ground-based solar telescope optimized for polarimetric observations of the full solar disk. It has a compact 55-cm diameter telescope with a high-throughput vector polarimeter using ferroelectric liquid crystal (FLC) polarization modulators and two high-speed imaging detectors. The VSM records vector and deep longitudinal magnetic field maps in two Fe I lines around 630.2 nm, longitudinal magnetograms in Ca II 854.2 nm, and intensity in He I 1083.0 nm. Each wavelength range has its own polarization modulator and polarizing beam-splitter to provide optimum performance at the chosen wavelengths. Each of the three modulator packages consists of one or two half-wave FLC modulators and fixed polymer quarter-wave true zero-order retarders. The modulation scheme for vector polarimetry is chosen such that Stokes Q and U have the same noise characteristics and have slightly better S/N than Stokes V. The latter makes sense as circular polarization signals are almost always significantly larger than linear polarization signals. To achieve the theoretically expected performance with the real FLC modulators, the measured retardance and fast axis direction of all elements were entered into an optimizing code to determine the appropriate rotation angle of each element. The polarizing beam splitters are located just in front of the detectors to produce two orthogonal, linearly polarized spectral images. This minimizes the amount of differential geometric distortion between orthogonal polarization states. The images are offset in the spectral direction with two calcite plates and a half-wave plate in between (a modified Savart plate) so that both ordinary and extraordinary beams travel equal optical path lengths and experience the same amount of crystal astigmatism, which is compensated for by a weak cylinder lens. The spectrograph and the associated optics (slit, grating) are arranged in such a way as to minimize the instrumental polarization between the modulators and the polarizing beam splitters. The advantage of this approach is that there are no moving parts for the polarization analysis, that the switching of the polarization states can occur as rapidly as the polarization modulators and the detectors allow (~100 Hz), and that both polarization states are detected simultaneously after having passed through the same optics. The only optical elements whose polarization properties cannot be calibrated regularly are the entrance window and the primary and secondary mirrors. All the other optical elements are located after a polarization calibration unit that can insert a wire-grid polarizer and a quarter-wave plate. The entrance window is athermally "floating" in RTV silicone such that gravity does not introduce any non-radial forces. The static birefringence of the window is therefore due to remaining stress from the glass manufacturing. While the telescope design is "polarization free," the axial symmetry only holds for the center of the field of view. The Mueller matrix for a field point of 0.25° corresponding to the solar limb produces a Stokes I to Q cross talk of $4 \cdot 10^{-5}$ and a Stokes V to Q cross talk of $8 \cdot 10^{-5}$, both of which can be safely neglected.

Sensitive and accurate polarimetry is often achieved with heavily polarizing solar telescopes, after rigorous calibration. For the next generation of solar telescopes, such as the 4-m Advanced Technology Solar Telescope (ATST) and the similar-sized European Solar Telescope (EST)[20], polarimetry will be fully integrated at facility level, with minimal instrumental polarization.

[20]Note that solar telescopes are generally an order of magnitude smaller than nighttime telescopes, because of the enormous heat load in the prime focus

6.5 Exoplanet Detection and Characterization

A major driver for new polarimetric instrumentation is the direct detection and, consecutively, characterization of exoplanets, which are generally polarized at visible wavelengths. A first attempt to detect the polarized light reflected off an exoplanet within the integrated starlight was performed by (Hough et al. 2006). Their PlanetPol instrument was designed to achieve maximum polarimetric sensitivity by combining the dual-beam approach with ultrafast modulation by a PEM, in combination with PMT and APD single-pixel detectors. The final sensitivity after correction for the minute instrumental polarization due to the telescope (the 4-m WHT) at the symmetric Cassegrain focus was better than 10^{-6}. Unfortunately, hot Jupiters prove to have small albedos and no signals from them were detected by this instrument. This drives the need to physically resolve exoplanets from their parent star, and this is the goal of a number of future instruments, like SPHERE at the VLT and GPI at one of the 8-m Gemini telescopes. The main goals for these instrument is to confine the starlight to only the first few diffraction zones (λ/D) of the PSF using an extreme AO system, suppress it by using coronagraphy, and suppress its halo even further by using polarimetry as it is unpolarized whereas the circumstellar structure generally is not. To pave the road for these polarimetric exoplanet imagers, Rodenhuis et al. (2008) developed the Extreme Polarimeter (ExPo), which is currently used at the Nasmyth platform of the WHT. ExPo does not (yet) contain an AO system and is meant to push the boundaries of polarimetry in the context of high-contrast imaging. It is a dual-beam polarimeter based on ~35 Hz FLC modulation in combination with a cube beam-splitter and folding prisms that yield two images on an EMCCD detector. To deal with the variable instrumental polarization due to the

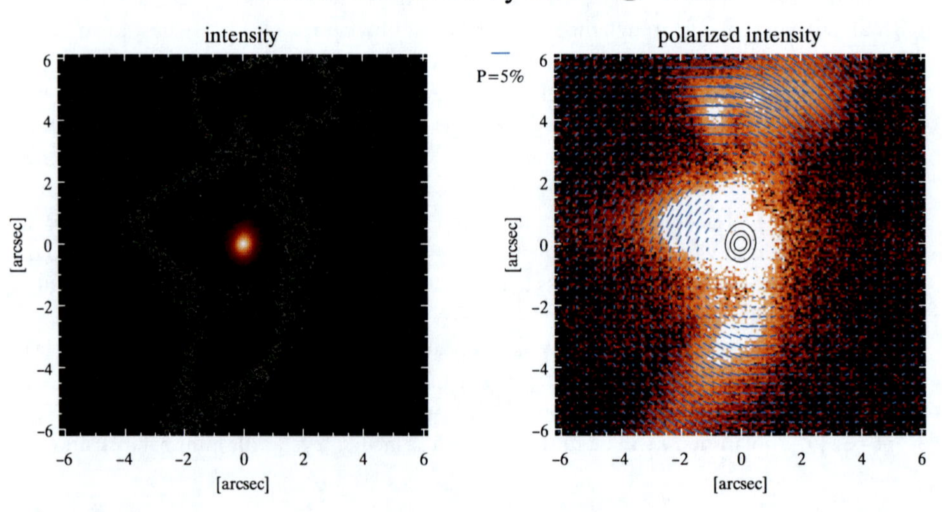

Fig. 4-8
ExPo observations of T Tauri. *Left*: intensity image, *right*: polarized intensity after calibration and stray light correction. The contours in one image correspond to the structure in the other. The blue polarization vectors are scaled with the degree of linear polarization. Data courtesy: the ExPo team (Héctor Canovas Cabrera, Michiel Rodenhuis, Sandra Jeffers, Christoph Keller)

45° Nasmyth mirror of the telescope, an active polarization compensator consisting of a tilting and rotating glass plate is employed. An example of ExPo data of circumstellar structure, and of the power of suppressing the light from a bright central source by polarimetry is presented in ➋ *Fig. 4-8*.

High contrast polarimetric instruments are now coming online on 8-m. class telescopes. The polarimetric mode of HiCIAO at the Subaru telescope is a classical dual-beam set-up consisting of a rotating HWP and a Wollaston prism, and is already producing some spectacular results on circumstellar disks (e.g., Hashimoto et al., 2011).

The polarimetric arm of SPHERE (Thalmann et al. 2008) will consist of two ZIMPOL detectors that run at kHz rates, with synchronous modulation by an FLC. This rapid modulation minimizes the influence of seeing and limited AO performance. The differential aberrations (which are a major limiting factor for this type of high-contrast imaging) due to the FLC switching have been minimized in its production process. A cube beam-splitter acts as the analyzer for both individual detectors. The polarization of the Nasmyth mirror is compensated with an achromatic HWP and a crossed mirror. Another HWP acts as a polarization switch to deal with the instrumental polarization created by the AO system, which does have linear eigenvectors. The polarimetric mode for the Gemini Planet Imager (Perrin et al. 2010) has a rotating achromatic HWP as a modulator at an intermediate position in the optical train. All GPI modes use a microlens array in the penultimate focal plane to sample the image, and, afterward, to effectuate spectral dispersion or polarizing beam-splitting using a Wollaston prism. The latter creates two images on the detector that are interleaved with a checkerboard pattern, and have minimal differential aberrations between them as they have already been identically spatially sampled by the microlenses. The instrumental polarization for GPI is considerable and will be modeled and calibrated.

The first instrument that will be able to directly detect rocky exoplanets, and characterize many Jupiter-like exoplanets through low-resolution spectropolarimetry, is EPICS at the future ~40-m E-ELT. Its design heritage stems from SPHERE-ZIMPOL and ExPo, and it combines the rapid modulation/demodulation with a true beam-exchange, which performance models show reduces the effects of differential aberrations by another order of magnitude (Keller et al. 2010).

It is concluded that the construction of large-aperture telescopes in combination with the ongoing development of new polarimetric techniques, components, and design approaches promises a very bright future for astronomical polarimetry.

Acknowledgments

The authors thank Gerard van Harten, Bill Sparks and Daphne Stam for their valuable input to this manuscript.

References

Antonucci, R. R. J., & Miller, J. S. 1985, ApJ, 297, 621

Bagnulo, S., Landolfi, M., Landstreet, J. D., Landi Degl'Innocenti, E., Fossati, L., & Sterzik, M. 2009, PASP, 121, 993

Batcheldor, D., Schneider, G,; Hines, D. C., Schmidt, G. D., Axon, D. J., Robinson, A., Sparks, W., & Tadhunter, C. 2009, PASP, 121, 153

Beckers, J. 1971, Appl Optics, 10, 973

Chandrasekhar, S. 1946, ApJ, 104, 110

Clarke, D. 2009, Stellar Polarimetry (Germany: Wiley-VCH)

del Toro Iniesta, J. C. 2003, Introduction to Spectropolarimetry (cambridge: Cambridge University Press)

del Toro Iniesta, J. C., & Collados, M. 2000, Appl Optics, 39, 1637

de Wijn, A. G., Stenflo, J. O., Solanki, S. K., & Tsuneta, S. 2009, Space Sci Rev, 144, 275

Donati, J.-F., Catala, C., Landstreet, J. D., & Petit, P. 2006, in ASP Conf Ser 358, Solar Polarization 4, ed. R. Casini, & B. W. Lites (San Francisco: ASP), 362

Donati, J.-F., Semel, M., Carter, B. D., Rees, D. E., & Collier Cameron, A. 1997, MNRAS, 291, 658

Donati, J.-F., and Landstreet, J. D. 2009, ARA&A, 47, 333

Gansel, J. K., Thiel, M., Rill, M. S., Decker, M., Bade, K., Saile, V., von Freymann, G., Linden, S., & Wegener, M. 2009, Science, 325, 1513

Guimond, S., & Elmore, D. 2004, OE Magazine, May edition

Hale, G. E. 1908, ApJ, 28, 315

Hale, P. D., & Day, G. W. 1988, Appl Optics, 27, 5146

Hanle, W. 1924, Z Phys, 30, 93

Hansen, J. E., & Travis, L. D. 1974, Space Sci Rev, 16, 527

Harrington, D. M., & Kuhn, J. R. 2008, PASP, 120, 89

Hashimoto, J., Tamura, M., Muto, T., Kudo, T., Fukagawa, M., Fukue, T., Goto, M., Grady, C. A., Henning, T., Hodapp, K., Honda, M., Inutsuka, S., Kokubo, E., Knapp, G., McElwain, M. W., Momose, M., Ohashi, N., Okamoto, Y. K., Takami, M., Turner, E. L., Wisniewski, J., Janson, M., Abe, L., Brandner, W., Carson, J., Egner, S., Feldt, M., Golota, T., Guyon, O., Hayano, Y., Hayashi, M., Hayashi, S., Ishii, M., Kandori, R., Kusakabe, N., Matsuo, T., Mayama, S., Miyama, S., Morino, J.I., Moro-Martin, A., Nishimura, T., Pyo, T.S., Suto, H., Suzuki, R., Takato, N., Terada, H., Thalmann, C., Tomono, D., Watanabe, M., Yamada, T., Takami, H., & Usuda, T. 2011, ApJL, 729, L17

Heiles, C. 2000, ApJ, 119, 923

Hough, J. H. 2007, JQRST, 106, 122

Hough, J. H., Lucas, P. W., Bailey, J. A., Tamura, M., Hirst, E., Harrison, D., & Bartholomew-Biggs, M. 2006, PASP, 118, 1302

Hovenier. J. W., and Muños, O. 2009, JQSRT, 110, 1280

Ichimoto, K., Lites, B., Elmore, D., Suematsu, Y., Tsuneta, S., Katsukawa, Y., Shimizu, T., Shine, R., Tarbell, T., Title, A., Kiyohara, J., Shinoda, K., Card, G., Lecinski, A., Streander, K., Nakagiri, M., Miyashita, M., Noguchi, M., Hoffmann, C., & Cruz, T. 2008, Solar Phys, 249, 233

Ignace, R., Cassinelli, J. P., & Nordsieck, K. H. 1999, ApJ, 520, 335

Keller, C. U. 1996, Solar Phys, 164, 243

Keller, C. U. 2002a, in Astrophysical spectropolarimetry, ed. J. Trujillo-Bueno, F. Moreno-Insertis, & F. SâŁ¡nchez (Cambridge: Cambridge University Press), 303

Keller, C. U. 2002b, ATST Technical Report #0007

Keller, C. U., Harvey, J. W., & the SOLIS Team 2003, in ASP Conf Proc 307, Solar Polarization, ed. J. Trujillo-Bueno, & J. Sanchez Almeida (San Francisco: ASP), 13

Keller, C. U. 2004, Proc SPIE, 5171, 239

Keller, C. U., & Snik, F. 2009, in ASP Conf Ser 405, Solar Polarization 5, ed. S. V. Berdyugina, K. N. Nagendra, & R. Ramelli (San Francisco: ASP), 371

Keller, C. U., Schmidt, H. M., Venema, L. B., Canovas, H., Hanenburg, H. H., Jager, R., Jeffers, S. V., Kasper, M. E., Martinez, P., Min, M., Rigal, F., Rodenhuis, M., Roelfsema, R., Snik, F., Stam, D. M., Verinaud, C., & Yaitskova, N. 2010, Proc SPIE, 7735, 239

Kemp, J. C., Swedlund, J. B., Landstreet, J. D., & Angel, J. R. P. 1970, ApJ 161, L77 (ApJL Homepage)

Kemp, J. C., Henson, G. D., Steiner, C. T., & Powell, E. R. 1987, Nature, 326, 270

Kikuta, H., Ohira, Y., & Iwata, K. 1997, Appl Optics, 36, 1566

King, R. J. 1966, J Sci Instrum, 43, 617

Kochukhov, O., & Piskunov, N. 2009, in Proc IAU 259, Cosmic Magnetic Fields: From Planets, to Stars and Galaxies, 653

Kuhn, J. R., Potter, D., & Parise, B. 2001, ApJ, 553, L189

Kuhn, J. R., Berdyugina, S. V., Fluri, D. M., Harrington, D. M., & Stenflo, J. O. 2007, ApJ, 668, L63

Landi Degl'Innocenti, E., and del Toro Iniesta, J. C. 1998, JOSA A, 15, 533

Landi Degli'Innocenti, E., and Landolfi, M. 2004, Polarization in Spectral Lines (Dordrecht: Kluwer)

Lites, B. W. 1987, Appl Optics, 26, 3838

Lin, H., & Versteegh, A. 2006, Proc. SPIE, 6269, 62690K

Lu, S.-Y., and Chipman, R. A. 1996, JOSA A, 13, 1106

Nordsieck, K. H. 1974, PASP, 86, 324

Oka, K., & Kato, T. 1999, Optics Lett, 24, 1475

Oliva, E. 1997, A&AS, 123, 589

Oort, J. H., & Walraven, Th. 1956, Bull Astronomi Instit Netherlands, 12, 285
Packham, C., & Jones, T. J. 2008, Proc. SPIE, 7014, 70145F
Packham, C., Escuti, M., Ginn, J., Oh, C., Quijano, I., & Boreman, G. 2010, PASP, 122, 1471
Pancharatnam, S. 1955, Proc Indian Acad Sci, Sect A, 42, 24
Patat, F., & Romaniello, M. 2006, PASP, 118, 146
Perrin, M. D., Graham, J. R., Larkin, J. E., Wiktorowicz, S., Maire, J., Thibault, S., Doyon, R., Macintosh, B. A., Gavel, D. T., Oppenheimer, B. R., Palmer, D. W., Saddlemeyer, L., & Wallace, J. K. 2010, Proc SPIE, 7736, 218
Povel, H. P., Keller, C. U., & Yadigaroglu, I.-A. 1994, Appl Optics, 33, 4254
Rodenhuis, M., Canovas, H., Jeffers, S. V., & Keller, C. U. 2008, Proc SPIE, 7014, 70146T
Sabatke, D. S., Descour, M. R., Dereniak, E. L., Sweatt, W. C., Kemme, S. A., & Phipps, G. S. 2000, Optics Lett, 25, 802
Samoylov, A. V., Samoylov, V. S., Vidmachenko, A. P., & Perekhod, A. V. 2004, JQSRT, 88, 319
Sánchez Almeida, J., & Martínez Pillet, V. 1992, A&A, 260, 543
Seager, S., Whitney, B. A., & Sasselov, D. D. 2000, ApJ, 540, 504
Serkowski, K. 1974, Polarization Techniques Ãł Chapter 8 from Methods of Experimental Physics, 12, Astrophysics: Part A Ãł Optical & Infrared, ed. M. L. Meeks, & N. P. Carleton, (New York: Academic Press)
Semel, M., Donati, J.-F., & Rees, D. E. 1993, A&A, 278, 231
Semel, M. 1987, A&A, 178, 257
Skumanich, A., Lites, B. W., Mart/'inez Pillet, V., & Seagraves, P. 1997, ApJ Suppl, 110, 357
Slowikowska, A., Kanbach, G., Kramer, M., & Stefanescu, A. 2009, MNRAS, 397, 103
Snik, F. 2010, in ASP Conf. Ser. (in press), Astronomical Polarimetry 2008 - science from small to large telescopes, Ed. P. Bastien, (San Francisco: ASP)
Snik, F. 2010, in ASP Conf Ser (in press), Astronomical Polarimetry 2008 - Science from Small to Large Telescopes, ed. P. Bastien, (San Francisco: ASP)
Snik, F., Jeffers, S. V., Keller, C. U., Piskunov, N., Kochukhov, O., Valenti, J., & Johns-Krull, C. 2008, Proc. SPIE,7014, 70140O
Snik, F. 2009, in ASP Conf. Ser. (in press), Astronomical Polarimetry 2008 - science from small to large telescopes, Ed. P. Bastien, (San Francisco: ASP)
Snik, F., Karalidi, T., & Keller, C. U. 2009, Appl Optics, 48, 1337
Snik, F., van Harten, G, & Keller, C. U. 2011, PASP (submitted)
Sparks, W. B., & Axon, D. J. 1999, PASP, 111, 1298
Sparks, W. B., Bond, H. E.; Cracraft, M., Levay, Z., Crause, L. A., Dopita, M. A., Henden, A. A., Munari, U., Panagia, N., Starrfield, S. G., Sugerman, B. E., Wagner, R. M., & White, R. L. 2008, AJ, 135, 605
Sparks, W. B., Hough, J. H., Kolokolova, L., Germer, T. A., Chen, F., DasSarma, S., DasSarma, P., Robb, F. T., Manset, N., Reid, I. N., Macchetto, F. D., & Martin, W. 2009, JQSRT, 110, 1771
Stam, D. M. 2008, A&A, 482, 989
Steele, I. A., Mundell, C. G., Smith, R. J., Kobayashi, S., & Guidorzi, C. 2009, Nature, 462, 767
Stenflo, J. O. 1994, Solar Magnetic Fields: Polarized Radiation Diagnostics (Dordrecht:Kluwer)
Stokes, G. G. 1852, Trans Cambridge Phil Soc, 9, 399
Thalmann, C., Schmid, H. M., Boccaletti, A., Mouillet, D., Dohlen, K., Roelfsema, R., Carbillet, M., Gisler, D., Beuzit, J.-L., Feldt, M., Gratton, R., Joos, F., Keller, C. U., Kragt, J., Pragt, J. H., Puget, P., Rigal, F., Snik, F., Waters, R., & Wildi, F. 2008, Proc SPIE, 7014, 70143F-70143F-12
Tinbergen. J. 1996, Astronomical Polarimetry (Cambridge: Cambridge University Press)
Tinbergen, J. 2007, PASP, 119, 1371
Tomczyk, S., Casini, R., de Wijn, A. G., & Nelson, P. G. 2010, Appl Optics, 49, 3580
Trujillo Bueno, J. 2009, in ASP Conf Ser 405, Solar Polarization 5, ed. S. V. Berdyugina, K. N. Nagendra, & R. Ramelli (San Francisco: ASP), 65
Tyo, J. S., Goldstein, D. L., Chenault, D. B., & Shaw, J. A. 2006, Appl Optics, 45, 5453
van Harten, G., Snik, F., & Keller, C. U. 2009, PASP, 12, 377
Vink, J. 2010, Messenger, 140, 46
Wang, L., and Wheeler, J. C. 2008, ARA&A, 46, 433
Weenink, J. G., Snik, F., & Keller, C. U. 2011, A&A (submitted)
Whittet, D. C. B., Martin, P. G., Hough, J. H., Rouse, M. F., Bailey, J. A., & Axon, D. A. 1992, ApJ, 386, 562
Zeeman, P. 1897, Nature, 55, 347

5 Sky Surveys

S. George Djorgovski · Ashish Mahabal · Andrew Drake · Matthew Graham · Ciro Donalek
California Institute of Technology, Pasadena, CA, USA

We dedicate this chapter to the memory of three pioneers of sky surveys, Fritz Zwicky (1898–1974), Bogdan Paczynski (1940–2007), and John Huchra (1948–2010).

T.D. Oswalt, H.E. Bond (eds.), *Planets, Stars and Stellar Systems. Volume 2: Astronomical Techniques, Software, and Data*, DOI 10.1007/978-94-007-5618-2_5, © Springer Science+Business Media Dordrecht 2013

Abstract: Sky surveys represent a fundamental data basis for astronomy. We use them to map in a systematic way the universe and its constituents and to discover new types of objects or phenomena. We review the subject, with an emphasis on the wide-field, imaging surveys, placing them in a broader scientific and historical context. Surveys are now the largest data generators in astronomy, propelled by the advances in information and computation technology, and have transformed the ways in which astronomy is done. This trend is bound to continue, especially with the new generation of synoptic sky surveys that cover wide areas of the sky repeatedly and open a new time domain of discovery. We describe the variety and the general properties of surveys, illustrated by a number of examples, the ways in which they may be quantified and compared, and offer some figures of merit that can be used to compare their scientific discovery potential. Surveys enable a very wide range of science, and that is perhaps their key unifying characteristic. As new domains of the observable parameter space open up thanks to the advances in technology, surveys are often the initial step in their exploration. Some science can be done with the survey data alone (or a combination of data from different surveys), and some require a targeted follow-up of potentially interesting sources selected from surveys. Surveys can be used to generate large, statistical samples of objects that can be studied as populations or as tracers of larger structures to which they belong. They can be also used to discover or generate samples of rare or unusual objects and may lead to discoveries of some previously unknown types. We discuss a general framework of parameter spaces that can be used for an assessment and comparison of different surveys and the strategies for their scientific exploration. As we are moving into the Petascale regime and beyond, an effective processing and scientific exploitation of such large data sets and data streams pose many challenges, some of which are specific to any given survey and some of which may be addressed in the framework of Virtual Observatory and Astroinformatics. The exponential growth of data volumes and complexity makes a broader application of data mining and knowledge discovery technologies critical in order to take a full advantage of this wealth of information. Finally, we discuss some outstanding challenges and prospects for the future.

Keywords: Archives, Asteroid surveys, Astroinformatics, Astronomical photography, Catalogs, Classification, Data mining, Data processing pipelines, Deep surveys, Digital imaging, Figures of merit for sky surveys, History of astronomy, Measurement parameter space (MPS), Microlensing surveys, Multi-wavelength astronomy, Observable parameter space (OPS), Physical parameter space (PPS), Sky surveys, Software, Space-based astronomy, Statistical studies, Supernova surveys, Systematic exploration, Technology, Time domain, Virtual observatory (VO), Wide-field surveys

List of Abbreviations: *AAVSO*, American Association of Variable Star Observers, http://www.aavso.org/; *CCD*, Charge Coupled Device; *CfA*, Harvard-Smithsonian Center for Astrophysics, http://cfa.harvard.edu; *CFHT*, Canada-France-Hawaii Telescope, http://www.cfht.hawaii.edu; *CMBR*, Cosmic Microwave Background Radiation; *EB*, Exabyte (10^{18} bytes); *ESO*, European Southern Observatory, http://eso.org; *FITS*, Flexible Image Transport System, http://heasarc.nasa.gov/docs/heasarc/fits.html; *FoM*, Figure of Merit; *FOV*, Field of view; *FWHM*, Full Width at Half Maximum; *GB*, Gigabyte (10^9 bytes); *HST*, Hubble Space Telescope, http://www.stsci.edu/hst; *ICT*, Information and Computing Technology; *LSS*, Large-Scale Structure; *MB*, Megabyte (10^6 bytes); *MJD*, Modified Julian Date; *MPS*, Measurement Parameter Space; *NOAO*, National Optical Astronomy Observatory, http://noao.edu; *NRAO*, National Radio Astronomy Observatory, http://nrao.edu; *OPS*, Observable Parameter Space; *PB*, Petabyte (10^{15} bytes);

PPS, Physical Parameter Space; *SETI* Search for Extraterrestrial Intelligence, http://www.seti.org; *SN* Supernova; *TB*, Terabyte (10^{12} bytes); *USNO*, United States Naval Observatory, http://www.usno.navy.mil/; *VLA*, NRAO Very Large Array, http://www.vla.nrao.edu; *VO*, Virtual Observatory, http://www.ivoa.net; *WWT*, WorldWide Telescope, http://www.worldwidetelescope.org
Additional abbreviations for the various sky surveys and catalogs are listed in the Appendix.

1 Introduction

1.1 Definitions and Caveats

Sky surveys are at the historical core of astronomy. Charting and monitoring the sky gave rise to our science, and today large digital sky surveys are transforming the ways astronomy is done. In this chapter we review some of the general issues related to the strategic goals, planning, and execution of modern sky surveys and describe some of the currently popular ones, at least as of this writing (late 2011). This is a rapidly evolving field, and the reader should consult the usual sources of information about the more recent work.

Some caveats are in order: The very term "sky surveys" is perhaps too broad and loosely used, encompassing a very wide range of the types of studies and methods. Thus, we focus here largely on the wide-field, panoramic sky surveys, as opposed, for example, to specific studies of deep fields or to heavily specialized surveys of particular objects, types of measurements, etc. We also have a bias toward the visible regime, reflecting, at least partly, the authors' expertise, but also a deeper structure of astronomy: most of the science often requires a presence of a visible counterpart regardless of the wavelength coverage of the original detection. We also focus mainly on the imaging surveys, with a nod to the spectroscopic ones. However, many general features of surveys in terms of the methods, challenges, strategies, and so on are generally applicable across the wavelengths and types of observations. There are often no sharp boundaries between different kinds of surveys, and the divisions can be somewhat arbitrary. Finally, while we outline very briefly the kinds of science that are done with sky surveys, we do not go into any depth for any particular kind of studies or objects, as those are covered elsewhere in these volumes.

It is tempting to offer a working definition of a survey in this context. By a *wide-field survey*, we mean a large data set obtained over areas of the sky that may be at least of the order of ~1% of the entire sky (admittedly an arbitrary choice) that can support a variety of scientific studies even if the survey is devised with a very specific scientific goal in mind. However, there is often a balance between the depth and the area coverage. A *deep survey* (e.g., studies of various deep fields) may cover only a small area and contain a relatively modest number of sources by the standards of wide-field surveys, and yet it can represent a survey in its own right, feeding a multitude of scientific studies. A wide-field coverage by itself is not a defining characteristic: for example, all-sky studies of the CMBR may be better characterized as focused experiments rather than as surveys, although that boundary is also getting fuzzier.

We also understand that "a large data set" is a very relative and rapidly changing concept, since data rates increase exponentially, following Moore's law, so perhaps one should always bear in mind a conditional qualifier "at that time." Some types of surveys are better characterized by a "large" number of sources (the same time-dependent conditional applies), which is also a heavily wavelength-dependent measure; for example, nowadays a thousand is still a large number of γ-ray sources, but a trivial number of visible ones.

Perhaps the one unifying characteristic is that surveys tend to support a broad variety of studies, many of which haven't been thought of by the survey's originators. Another unifying characteristic is the exploratory nature of surveys, which we address in more detail in ➋ Sect. 3 below. Both approaches can improve our knowledge of a particular scientific domain and can lead to surprising new discoveries.

The meaning of the word "survey" in the astronomical context has also changed over the years. It used to refer to what we would now call a sky atlas (initially hand-drawn sky charts and later photographic images), whereas catalogs of sources in them were more of a subsidiary or derived data product. Nowadays the word largely denotes catalogs of sources and their properties (positions, fluxes, morphology, etc.), with the original images provided almost as a subsidiary information, but with an understanding that sometimes they need to be reprocessed for a particular purpose. Also, as the complexity of data increased, we see a growing emphasis on carefully documented metadata ("data about the data") that are essential for the understanding of the coverage, quality, and limitations of the primary survey data.

1.2 The Types and Goals of Sky Surveys

We may classify surveys in regard to their scientific motivation and strategy, their wavelength regime, ground-based vs. space-based, the type of observations (e.g., imaging, spectroscopy, polarimetry), their area coverage and depth, their temporal character (one-time vs. multi-epoch), as panoramic (covering a given area of the sky with all sources therein) or targeted (observing a defined list of sources), and can have any combination of these characteristics. For example, radio surveys generally produce data cubes, with two spatial and one frequency dimension, and are thus both imaging and spectroscopic and often include the polarization as well. X-ray and γ-ray images generally also provide some energy resolution. Slitless spectroscopy surveys (images taken through an objective prism, grating, or a grism) provide wavelength-dispersed images of individual sources. Surveys can be also distinguished by their angular, temporal, or energy resolution.

Surveys may be scientifically motivated by a census of particular type of sources, for example, stars, galaxies, or quasars, that may be used for statistical studies such as the Galactic structure or the large-scale structure (LSS) in the universe. They may be aimed to discover significant numbers of a particular type of objects, often relatively rare ones, for the follow-up studies, for example, supernovae (SNe), high-redshift galaxies or quasars, and brown dwarfs. When a new domain of an observable parameter space opens up, for example, a previously unexplored wavelength regime, it usually starts with a panoramic survey to see what kinds of objects or phenomena populate it.

Therein lies perhaps the key scientific distinction between surveys and the traditional, targeted astronomical observations: surveys aim to map and characterize the astrophysical contents of the sky or of the populations of objects of particular kinds in a systematic manner, whereas the traditional observations focus on detailed properties of individual sources or relatively small numbers of them. Surveys are often the ways to find such targets for detailed studies.

The first type of survey science – use of large, statistical samples of objects of some kind (stars, galaxies, etc.) as probes of some collective properties (e.g., Galactic structure or LSS) – may be done with the survey data alone or may be supplemented by additional data from other sources. The other two types of survey science – as a discovery mechanism for rare, unusual, or new types of objects or phenomena or as a pure initial exploration of some new domain of the observable parameter space – require targeted follow-up observations. Thus surveys become a backbone of much of astronomical research today, forming a fundamental data infrastructure of astronomy. This may make them seem less glamorous than the successful targeted observations that may be enabled by surveys, but it does not diminish their scientific value.

Imaging surveys are commonly transformed into catalogs of detected sources and their properties, but in some cases, images themselves represent a significant scientific resource, for example, if they contain extended structures of diverse morphologies; like images of star-forming regions or stellar bubbles and SN remnants in Hα images.

The process of detection and characterization of discrete sources in imaging surveys involves many challenges and inevitably introduces biases, since these processes always assume that the sources have certain characteristics in terms of a spatial extent, morphology, and so on. We discuss these issues further in ➋ Sect. 5.

Like most astronomical observations, surveys are often enabled by new technologies and push them to their limits. Improved detector and telescope technologies can open new wavelength regimes or more sensitivity or resolution, thus providing some qualitatively new view of the sky. A more recent phenomenon is that information and computation technologies (ICT) dramatically increased our ability to gather and process large quantities of data and that quantitative change has led to some interesting qualitative changes in the ways we study the universe.

A direct manifestation of this is the advent of large synoptic sky surveys that cover large areas of the sky repeatedly and often, thus opening the time domain as new arena for exploration and discovery. They are sometimes described as a transition from a panoramic cosmic photography to a panoramic cosmic cinematography. We describe some examples below.

Spectroscopic surveys, other than the data cubes generated in radio astronomy, typically target lists of objects selected from imaging surveys. In case of extragalactic surveys, the primary goal is typically to obtain redshifts, as well as to determine some physical properties of the targets, for example, star formation rates, or presence and classification of active galactic nuclei (AGN), if any. If the targets are observed with long slit or multi-slit mask spectrographs, or integral field units (IFU), information can be obtained about the kinematics of resolved structures in galaxies, typically emission-line gas. In case of Galactic survey, the goals are typically to measure radial velocities and sometimes also the chemical abundances of stars.

Spectroscopic surveys depend critically on the quality of the input catalogs from which the targets are selected, inheriting any biases that may be present. Their observing strategies in terms of the depth, source density, spectroscopic resolution, etc., are determined by the scientific goals. Since spectroscopy is far more expensive than imaging in terms of the observing time, some redshift surveys have adopted a sparse-sampling strategy, for example, by observing every Nth (where N = 2, or 10, or…) source in a sorted list of targets, thus covering a larger area, but with a corresponding loss of information.

Our observations of the sky are no longer confined to the electromagnetic window. Increasingly, sky is being monitored in high-energy cosmic rays (Kotera and Olinto 2011), neutrinos (Halzen and Klein 2010), and even gravitational waves (Centrella 2010). So far, these information channels have been characterized by a paucity of identified sources, largely due to the lack of a directional accuracy, with the exceptions of the Sun and SN 1987A in neutrinos, but they will likely play a significant role in the future.

Finally, as numerical simulations become ever larger and more complex and theory is expressed as data (the output of simulations), we may start to see surveys of simulations as means of characterizing and quantifying them. These surveys of theoretical universes would have to be compared to the measurements obtained in the surveys of the actual universe. New knowledge often arises as theories are confronted with data, and in the survey regime, we will be doing that on a large scale.

1.3 The Data Explosion

In 1990s, astronomy transitioned from a relatively data-poor science to an immensely data-rich one, and the principal agent of change were large digital sky surveys. They, in turn, were enabled by the rapid advances in ICT. Sky surveys became the dominant data sources in astronomy, and this trend continues (Brunner et al. 2001c). The data volume in astronomy doubles at Moore's law pace, every year to a year and a half (Szalay and Gray 2001; Gray and Szalay 2006), reflecting the growth of the technology that produces the data. (Obviously, the sheer size of data sets by itself does not imply a large scientific value; for example, very deep images from space-based observatories may have a modest size in bits but an immense scientific value.)

In the past, surveys and their derived catalogs could be published as printed papers or small sets of volumes that can be looked up "by hand" (this is still true in some regimes, e.g., the γ-ray astronomy, or other nascent fields). But as the data volumes entered the Terascale regime in the 1990s, and the catalogs of sources started containing millions of objects, there was an inevitable transition to a purely electronic publication and dissemination, for example, in the form of the web-accessible archives, that also provide access to the necessary metadata and other documentation. Databases, data mining, web services, and other computational tools and techniques became a standard part of astronomy's tool chest, although the community is still gradually gaining their familiarity with them. This is an aspect of an inevitable culture change as we enter the era of a data-rich, data-intensive science.

The growth of data quantity, coupled with an improved data homogeneity, enabled a new generation of statistical or population studies: with samples of hundreds of millions of sources, the Poissonian errors were no longer important, and one could look for subtle effects simply not accessible with the more limited data sets. Equally important was the growth of data quality and data complexity. The increased information content of the modern sky surveys enabled a profitable data mining: the data could be used for a much broader variety of studies than it was possible in the past.

For these reasons, survey-enabled astronomy became both popular and respectable. But it was obvious that data fusion across different surveys (e.g., over different wavelengths) has an even higher scientific potential, as it can reveal knowledge that is present in the combined data, but cannot be recognized in any individual data set, no matter how large. Historical examples from multi-wavelength cross-correlations abound, for example, the discoveries of quasars, ultraluminous starbursts, and interpretation of γ-ray bursts. The new, data-rich astronomy promised to open this discovery arena wholesale.

There are many nontrivial challenges posed by the handling of large, complex data sets and knowledge discovery in them: how to process and calibrate the raw data; how to store, combine, and access them using modern computing hardware and networks; and how to visualize, explore, and analyze these great data sets quickly and efficiently. This is a rapidly developing field, increasingly entailing collaborative efforts between astronomers and computer scientists.

The rise of data centers was the response to dealing with *individual* large data sets, surveys, or data collections. However, their fusion and the scientific synthesis required more than just their interoperability. This prompted the rise of the *Virtual Observatory* (VO) concept as a general, distributed research environment for astronomy with large and complex data sets (Brunner et al. 2001a; Hanisch 2001, 2010; Djorgovski and Williams 2005). Today, sky surveys are naturally included in an evolving worldwide ecosystem of astronomical data resources and services. The reader is directed to the VO-related websites or their future equivalents for an up-to-date description of the data assets and services and access to them.

Astronomy was not alone in facing the challenges and the opportunities of an exponential data growth. Virtual scientific organizations with similar mandates emerged in many other fields and continue to do so. This entire arena of a computationally enabled, data-driven science is sometimes referred to as cyber-infrastructure, or e-Science, unified by the common challenges and new scientific methodologies (Atkins et al. 2003; Hey and Trefethen 2003, 2005; Djorgovski 2005; Hey et al. 2009; Bell et al. 2009). Nowadays we also see the blossoming of "science informatics," for example, Astroinformatics (by analogy with its bio-, geo-, etc., counterparts). These are broader concepts of scientific and methodological environments and communities of interest that seek to develop and apply new tools for the data-rich science in the twenty-first century.

2 A (Very) Brief History of Sky Surveys

We cannot do justice to this subject in the limited space here, and we just mention some of the more important highlights in order to put the subject in a historical context. The interested reader may wish to start with a number of relevant, excellent Wikipedia articles and other web resources for additional information and references.

Historically, surveying of the sky started with a naked eye (that may be looking through a telescope), and we could consider Charles *Messier*'s catalog from the middle of the eighteenth century as a forerunner of the grander things to come. In the prephotography era, the most notable sky surveying was done by the Herschel family (brother William, sister Caroline, and son John), starting in the late eighteenth century. Among the many notable achievements, the Herschels also introduced stellar statistics – counting of stars per unit area – that can be used to learn more about our place in the universe. Their work was continued by many others, leading to the publication of the first modern catalogs in the late nineteenth century, for example, the still-used *New General Catalogue* (NGC) and *Index Catalogue* (IC) by John Dreyer (Dreyer 1888, 1895).

Visually compiled star catalogs culminated in the mid/late nineteenth century, including the *Bonner Durchmusterung* (BD) in the North and its Southern equivalent, the *Cordoba Durchmusterung* (CD), that contained nearly 900,000 stars.

The field was transformed by the advent of a new technology – photography. Surveys that may be recognized as such in the modern sense of the term started with the first photographic efforts that covered systematically large areas of the sky at the end of the nineteenth century. Perhaps the most notable of those is the *Harvard Plate Collection* that spans over a century of sky coverage and that is currently being scientifically rejuvenated through digitization by the *Digital Access to Sky Century at Harvard* project (DASCH; http://hea-www.harvard.edu/DASCH; Grindlay et al. 2009). A roughly contemporaneous, international effort, *Carte du Ciel*, led to the production of the *Astrographic Catalogue* (AC), reaching to $m \sim 11$ mag, that served as the basis of the more modern astrometric catalogs.

Another important innovation introduced at the Harvard College Observatory at the turn of the nineteenth century was systematic monitoring of selected areas on the sky, a precursor of the modern synoptic sky surveys. Photographic monitoring of the Magellanic Clouds enabled Henrietta Leavitt (Leavitt and Pickering 1912) to discover the period-luminosity relations for Cepheids, thus laying the groundwork for the cosmological distance scale and the discovery of the expanding universe by Edwin Hubble and others in the 1920s.

In the early decades of the twentieth century, Edward Pickering, Annie Jump Cannon, and their collaborators at Harvard produced the *Henry Draper Catalogue* (HD), named after the donor, that was eventually extended to ~360,000 stars, giving spectral types based on objective prism plates. Around the same time, Harlow Shapley and collaborators catalogued for the first time tens of thousands of galaxies in the Southern sky and noted the first signs of the large-scale structure in the universe.

Around the same time, roughly the first third of the twentieth century, the Dutch school (Jakobus Kapteyn, Pieter van Rhijn, Jan Oort, Bart Bok, and their students and collaborators) started systematic mapping of the Milky Way using star counts and laid foundations for the modern studies of Galactic structure. Kapteyn also introduced *Selected Areas* (SA), a strategically chosen set of directions where star counts can yield information about the Galactic structure, without the need to survey the entire sky. In 1966, the heavily used *Smithsonian Astrophysical Observatory Catalog* was published that contained positions, proper motions, magnitudes, and (usually) spectral types for over 250,000 stars.

A key figure emerging in the 1930s was Fritz Zwicky, who, among many other ideas and discoveries, pioneered systematic sky surveys in the ways that shaped much of the subsequent work. He built the first telescope on Mt. Palomar, the 18-in. Schmidt, then a novel design for a wide-field instrument. With Walter Baade, he used it to search for supernovae (a term he and Baade introduced), leading to the follow-up studies and the physical understanding of this phenomenon (Baade and Zwicky 1927). Zwicky's systematic, panoramic mapping of the sky led to many other discoveries, including novel types of compact dwarf galaxies, and more evidence for the LSS. In 1960s, he and his collaborators E. Herzog, P. Wild, C. Kowal, and M. Karpowitz published the *Catalogue of Galaxies and of Clusters of Galaxies* (CGCG; 1961–1968) that served as an input for many redshift surveys and other studies in the late twentieth century. All this established a concept of using wide-field surveys to discover rare or interesting objects to be followed up by larger instruments.

The potential of Schmidt telescopes as sky mapping engines was noted, and a 48-in. Schmidt was built at Palomar Mountain, largely as a means of finding lots of good targets for the newly built 200-in., for a while the largest telescope in the world. A major milestone was the first *Palomar Observatory Sky Survey* (POSS-I), conducted from 1949 to 1958 and spearheaded mainly by Edwin Hubble, Milton Humason, Walter Baade, Ira Bowen, and Rudolph Minkowski and was funded mainly by the National Geographic Society (Minkowski and Abel 1963).

POSS-I mapped about 2/3 of the entire sky, observable from Palomar Mountain, initially from the North Celestial Pole down to Dec $\sim -30°$, and later extended to Dec $\sim -42°$ (the Whiteoak extension). The survey used 14-in.-wide photographic plates, covering roughly $6.5° \times 6.5°$ fields of view (FOV) each, but with a useful, unvignetted FOV of $\sim 6° \times 6°$, with some overlaps, with a total of 936 fields in each of the two bandpasses, one using the blue-sensitive Kodak 103a-O plates and one using the red-sensitive Kodak 103a-F emulsion. Its limiting magnitudes vary across the survey but are generally close to $m_{lim} \sim 21$ mag. Reproduced as glass copies of the original plates and as paper prints, POSS-I served as a fundamental resource, effectively a roadmap for astronomy for several decades, and its digital form is still used today.

Its cataloguing was initially done by eye, and some notable examples include the *Uppsala General Catalogue* (UGC) of ~13,000 galaxies with apparent angular diameters >1 arcmin (Nilson 1973), *Morphological Catalog of Galaxies* (MCG) containing ~30,000 galaxies down to $m \sim 15$ mag (Vorontsov-Velyaminov and Arkhipova 1974), Abell's (1958) catalog of ~2,700 clusters of galaxies, and many others.

The *Second Palomar Observatory Sky Survey* (POSS-II), conducted about four decades later, was the last of the major photographic sky surveys (Reid et al. 1991). Using an improved telescope optics and improved photographic emulsions, it covered the entire Northern sky with ~900 partly overlapping 6.5° fields spaced by 5°, in 3 bandpasses corresponding to Kodak IIIa-J (blue), IIIa-F (red), and IV-N (far red) emulsions.

A number of other surveys have been conducted at the Palomar 48-in. Schmidt (renamed to *Samuel Oschin Telescope* in honor of the eponymous benefactor) in the intervening years, including Willem Luyten's measurements of stellar proper motions, the original *Hubble Space Telescope Guide Star Catalog* (GSC), and a few others.

Southern sky equivalents of the POSS surveys in terms of the coverage, depth, and distribution were conducted at the European Southern Observatory's 1.0-m Schmidt telescope and the 1.2-m UK Schmidt at the Anglo-Australian Observatory in the 1970s and 1990s, jointly called the *ESO/SERC Southern Sky Survey* Andris Lauberts (1982) produced a Southern sky equivalent of the UGC catalog.

Both POSS and ESO/SERC surveys have been digitized independently by several groups in the 1990s. Scans produced at the Space Telescope Science Institute were used to produce both the second-generation HST Guide Star Catalog (GSC-2; Lasker et al. 2008) and the *Digital Palomar Observatory Sky Survey* (DPOSS; Djorgovski et al. 1997a, 1999); the images are distributed through several *Digitized Sky Survey* (DSS) servers worldwide. The US Naval Observatory produced the astrometric *USNO-A* and *USNO-B* catalogs that also include proper motions (Monet et al. 2003). These digital versions of photographic sky surveys are described in more detail below. The surveys were also scanned by the *Automated Plate Measuring* facility (http://www.ast.cam.ac.uk/~mike/casu; APM) in Cambridge, the *SuperCOSMOS* group (http://www-wfau.roe.ac.uk/sss; Hambly et al. 2001), and the *Automated Plate Scanner* group (APS; http://aps.umn.edu; Cabanela et al. 2003). Scans of the ESO/SERC Southern Sky Survey plates resulted in the *APM Galaxy Survey* (Maddox et al. 1990a, b). DSS scans and other surveys can be also accessed through *SkyView* (http://skyview.gsfc.nasa.gov; McGlynn et al. 1997).

The middle of the twentieth century also saw an appearance of a plethora of sky surveys at other wavelengths as the new regimes opened up (IR, radio, X-ray, γ-ray, etc.). Enabled by the new technologies, for example, electronics, access to space, computers, etc., these new, panchromatic views of the sky led to the discoveries of many previously unknown types of objects and phenomena, e.g., quasars, pulsars, various X-ray sources, protostars, or γ-ray bursts, to name but a few.

A good account of the history of radio astronomy is by Sullivan (2009). Following the pioneering work by Karl Jansky and Grote Reber, radio astronomy started to blossom in the 1950s, fueled in part by the surplus radar equipment from the World War II. A key new development was the technique of aperture synthesis and radio interferometers, pioneered by Martin Ryle, Anthony Hewish, and collaborators. This led to the first catalogs of radio sources, the most enduring of which were the Third Cambridge (3C) and Fourth Cambridge (4C) catalogs, followed by the plethora of others. Optical identifications of radio sources by Walter Baade, Rudolph Minkowski, Maarten Schmidt, and many others in the 1950s and 1960s opened whole new areas of research that are still thriving today. Among the very many surveys for AGN and/or objects with a UV excess, we can mention the classic Palomar Green (PG; Green 1976) survey, and an extensive body of work by Benjamin Markarian at Byurakan Observatory in Armenia, from 1960's to 1980's.

While the infrared (IR) light was discovered already by William Herschel circa 1800, the IR astronomy started in earnest with the advent of first efficient IR detectors in the 1960s. A pioneering 2 μm IR sky survey (TMASS) was done by Neugebauer and Leighton 1969, resulting in a catalog of ~5,600 sources. Its modern successor some three decades later, the *Two Micron All-Sky Survey* (2MASS), catalogued over 300 million in three bandpasses (*JHK*); it is described in more detail below. Deeper surveys followed, notably the *United Kingdom Infrared Telescope (UKIRT) Infrared Deep Sky Survey* (UKIDSS), and more are forthcoming.

A wholesale exploration of the mid/far-IR regime required a space-borne platform, and the *Infrared Astronomy Satellite* (IRAS), launched in 1983, opened a huge new area of research, continued by a number of space IR missions since. Some of the milestones include the DIRBE and FIRAS experiments on the *Cosmic Background Explorer satellite* (COBE), launched in 1989, and more recently the *Wide-field Infrared Survey Explorer* (WISE), launched in 2009.

Astronomy at higher energies, i.e., UV beyond the atmospheric cutoff limit, X-rays, and γ-rays, required access to space that was a beneficent product of the space race and the cold war starting shortly after the World War II.

The UV (i.e., with $\lambda < 320$ nm or so) astronomy from space started with a number of targeted missions starting from the 1960s, the first, shallow all-sky survey was done with the TD-1A satellite in 1972, followed by the *Extreme UV Explorer* (EUVE, 1992–2001). The next major milestone was the *Galaxy Evolution Explorer* (GALEX, 2003–2011) that surveyed most of the sky in two broad bands, down to the UV equivalent of the optical POSS surveys, and a much smaller area about 2–3 mag deeper.

Following the birth of the X-ray astronomy with rocket-borne experiments in the 1960s, the first X-ray all-sky surveys started with the SAS-A *Uhuru* (1970–1973) satellite, HEAO-1 mission (1977–1979), and HEAO-2 *Einstein* (1978–1981), followed by many others, including *Rosat* (1990–1999). Many other missions over the past three decades followed the survey ones with pointed observations of selected targets. They led to the discoveries of many new aspects of the previously known types of sources, for example, accreting binaries, active galactic nuclei, and clusters of galaxies, and provided key new insights into their nature.

Gamma-ray astronomy is still mostly done with space-borne instruments that cover most or all of the sky. This is largely because γ-rays are hard to focus and shield from, so the instruments tend to look at all directions at once. While the first cosmic γ-ray emission was detected in 1961 by the Explorer-11 satellite, with a total of about 100 photons collected, γ-rays surveys really started with the SAS-2 (1972) and the COS-B (1975–1982) satellites. A major milestone was the *Compton Gamma-Ray Observatory* (CGRO, 1991–2000), followed by the *Fermi Gamma-Ray Space Telescope* (FGST), launched in 2008. These and other missions uncovered many important phenomena in the high-energy universe, but perhaps the most spectacular are the cosmic γ-ray bursts (GRBs), discovered in 1967 by the *Vela* satellites and finally understood in 1997 thanks to the *BeppoSAX* mission.

Descriptions of all of these fascinating discoveries are beyond the scope of this chapter, and can be found elsewhere in these volumes.

In the 1990s, the era of fully digital sky surveys began in earnest. Aside from the digitized versions of the photographic sky surveys, a major milestone was the *Sloan Digital Sky Survey* (SDSS), which really helped change the culture of astronomy. Another key contribution at that time was the *Two Micron All-Sky Survey* (2MASS). Radio astronomy contributed the *New VLA*

Sky Survey (NVSS) and the *Faint Images of Radio Sky at Twenty-centimeter* (FIRST). Significant new surveys appeared at every wavelength and continue to do so. We discuss a number of them in more detail in ➲ Sect. 4.

The early redshift observations by Vesto Melvin Slipher in the 1920s led to the discovery of the expanding universe by Edwin Hubble. Galaxy redshift surveys in a more modern sense can be dated from the pioneering work by Humason et al. (1956).

The early 1980s brought the first extensive redshift surveys, driven largely by the studies of the large-scale structure (LSS). An important milestone was the first Center for Astrophysics (CfA) redshift survey, conducted by Marc Davis, David Latham, John Huchra, John Tonry (Davis et al. 1982; Huchra et al. 1983), and their collaborators based on the optical spectroscopy of ~2,300 galaxies down to $m_B \approx 14.5$ mag, selected from the Zwicky and Nilson catalogs, obtained at Mt. Hopkins, Arizona. Arecibo redshift survey, conducted by Riccardo Giovanelli, Martha Haynes, and collaborators, used the eponymous radio telescope to measure ~2,700 galaxy redshifts through their H I 21-cm line. These surveys gave us the first significant glimpses of the LSS in the nearby universe. They were followed in a short order by the second CfA redshift survey, led by John Huchra and Margaret Geller (Geller and Huchra 1989), that covered galaxies down to $m_B \approx 15.5$ mag; it was later combined with other survey for a total of ~18,000 redshifts, compiled in the *Updated Zwicky Catalog* (UZC; Falco et al. 1999). Additional H I surveys from Arecibo added up to ~8,000 galaxies. Several redshift surveys used target selection on the basis of FIR sources detected by the IRAS satellite in order to minimize the selection effects due to the interstellar extinction (Strauss et al. 1992, Fisher et al. 1995, Saunders et al. 2000). Many other redshift surveys of galaxies, obtained one redshift at a time, followed.

Together, these surveys added a few tens of thousands of galaxy redshifts, mainly used to study the LSS. Good reviews include, Giovanelli and Haynes (1991), Salzer and Haynes (1996), Lahav and Suto (2004), and Geller et al. (2011). As of this writing, John Huchra's ZCAT website is still maintained at https://www.cfa.harvard.edu/~dfabricant/huchra/zcat, and a good listing of redshift surveys is currently available at http://www.astro.ljmu.ac.uk/~ikb/research/galaxy-redshift-surveys.html.

Development of highly multiplexed (multi-fiber or multi-slit) spectrographs in the late 1980s and 1990s brought a new generation of massive redshift surveys, starting with the Las Campanas Redshift Survey, led by Steve Schectman, Gus Oemler, Bob Kirshner, and their collaborators (Shectman et al. 1996), that produced ~26,400 redshifts in sky strips covering ~700 deg^2. The field was changed dramatically in the late 1990s and 2000s by two massive redshift surveys, 2dF and SDSS, described in more detail below. Together, they contributed more than a million galaxy redshifts, and changed the field.

This blossoming of sky surveys created the exponential data explosion discussed in ➲ Sect. 1 and transformed the way astronomy is done, both scientifically and technologically. The state of the art at the transition from the photography era to the digital era is encapsulated well in the IAU Symposia 161 (MacGillivray et al. 1994) and 179 (McLean et al. 1998).

3 A Systematic Exploration of the Sky

Surveys are our main path toward a systematic exploration of the sky. They often follow an introduction of some new technology that allows us to observe the sky in some new way, for example, in a previously unexplored wavelength regime, or to do so more efficiently or more quickly. It is worth examining this process in a general sense. The discussion below follows in part the ideas first discussed by Harwit (1975); see also Harwit (2003). Harwit pointed out that fundamentally new discoveries are often made by opening of the new portions of the observable parameter space and the key role of technology in enabling such advances. An interesting, complementary discussion was provided by Kurtz (1995).

Here we propose a framework for the representation of sky surveys (or, indeed, any astronomical observations), derived catalogs of measurements, and data exploration and analysis. The purpose is to have a way of quantitatively describing and comparing these large and complex data sets and the process for their systematic exploration. We distinguish three types of data parameter spaces, for the observations, measured properties of detected sources in these observations, and physical (rather than apparent) properties of these sources.

In the context of exploration of large digital sky surveys and Virtual Observatory, some of these ideas have been proposed by Djorgovski et al. (2001a, b, c, 2002). The present treatment develops further the concept of observational or data parameter space.

3.1 The Role of Technology

Progress in astronomy has always been driven by technology. From the viewpoint of surveys, the milestone technologies include the development of astrophotography (late 1800s), Schmidt telescopes (1930s), radio electronics (1940s to the present), access to space (from 1960s onward), ubiquitous and inexpensive computing, and digital detectors in every wavelength regime, notably the CCDs (1980s to the present).

A question naturally arises, what is the next enabling technology? The same technology that gave us computers with capabilities increasing according to Moore's law and digital detectors like the CCDs, namely, VLSI, also enabled the rise of the modern information and computation technology (ICT) in both hardware and software forms, the internet and the web. Most of the computing now is not about number crunching and solving a lot of differential equations quickly (although of course we still do that), but it is about manipulation, processing, and searching of information from the raw data to the high-level knowledge products. This is sometimes described as the Third (computation-intensive) and the Fourth (data-intensive) Paradigms of science, the First and the Second being the experiment and the analytical theory (Hey et al. 2009). Science progresses using all four of them.

The realm of key enabling technologies thus evolved historically from telescopes to detectors, and now to information and computing technologies (ICT). In astronomy, the principal research environments in which this process unfolds incorporate the Virtual Observatory framework and, more broadly, Astroinformatics. These are reviewed elsewhere in these volumes. Equivalent situations exist in many other fields of science.

ICT is therefore the enabling technology behind the modern sky surveys and their scientific exploitation. Moore's law underlies both the exponential data growth and the new methods needed to do science with these data. Surveys are already generating or enabling a lot of good science and will dominate the data-rich astronomy at least in the first part of the twenty-first century.

3.2 Data Parameter Spaces

Every astronomical observation, surveys included, covers some finite portion of the *Observable Parameter Space* (OPS), whose axes correspond to the observable quantities, for example, flux wavelength and sky coverage (see below). Every astronomical observation or a set thereof, surveys included, subtends a multidimensional volume (hypervolume) in this multidimensional parameter space.

The OPS can be divided for convenience into four domains, roughly corresponding to the type of observations, although each of the four domains is really a limited projection of the OPS:

- The spectrophotometric domain, whose axes include the wavelength λ (or the photon energy), flux F_λ (that may be integrated over some finite bandpass), the spectroscopic resolution $R = \lambda/\Delta\lambda$, the inverse flux precision $F/\Delta F$ (so that the higher is better), and a polarization subdomain that could consist, for example, of the inverse precision of the measurements of Stokes parameters. One could divide it into spectroscopic and photometric subdomains, but photometry can be seen as an extremely low-resolution spectroscopy, and as the bandpasses shrink, the distinctions blur.
- The astrometric domain, whose axes include the pairs of coordinates (Equatorial, Ecliptic, Galactic...) and the astrometric accuracy $\Delta\theta$. The two coordinates can be collapsed into a net area coverage Ω if it doesn't matter where on the sky the observations are done; in this context, we use the word "area" to mean "solid angle," following the common, if somewhat misleading, usage.
- The morphological domain, whose axes include the surface brightness μ and the angular resolution $\Delta\alpha$, that represents a "beam size" and should not be confused with the astrometric accuracy; given a sufficient S/N, it is possible to have $\Delta\theta \ll \Delta\alpha$.
- The time domain, whose axes include the time (say, the MJD or the UT) if it matters when the data are obtained, time baselines or sampling Δt, and the number of epochs (samples) N_{exp} obtained at each Δt. These reflect the cadence and determine the window function of a survey.

Obviously, the four domains can share some of the axes and seldom make sense in isolation. An obvious example would be a single image: it covers a finite field of view, in some finite bandpass, with some limiting magnitude and a saturation level, and a limited angular resolution determined by the pixel size and/or the PSF. An imaging survey represents a sum of the OPS hypervolumes of the individual images. Another example is the "cosmic haystack" of SETI searches, whose axes are the area, depth, and frequency (see, e.g., Tarter 2001).

We are also starting to add the nonelectromagnetic information channels: neutrinos, gravity waves, and cosmic rays; they add their own dimensions to the general OPS.

The dimensionality of the OPS is given by the number of characteristics that can be defined for a given type of observations, although some of them may not be especially useful and could be ignored in a particular situation. For example, time-domain axes make little sense for the observations taken in a single epoch. Along some axes, the coverage may be intrinsically discrete rather than continuous. An observation can be just a single point along some axis, but have a finite extent in others. In some cases, polar coordinates may be more appropriate than the purely Cartesian ones.

Some parts of the OPS may be excluded naturally, for example, due to quantum limits, diffraction limits and opacity and turbulence of the Earth's atmosphere or the Galactic ISM on some wavelengths; see Harwit (1975, 2003). Others are simply not accessible in practice due to limitations of the available technology, observing time, and funding.

We can thus, in principle, measure a huge amount of information arriving from the universe, and so far we have sampled well only a relatively limited set of subvolumes of the OPS in general much better along some axes than others: we have fairly good coverage in the visible, NIR, and radio; more limited X-ray and FIR regimes; and very poor at higher energies still. The discrimination becomes sharper if we also consider their angular resolution, etc.

The coverage and the dimensionality of the OPS determine what *can* be detected or measured; it fully describes the scope and the limitations of our *observations*. It is in principle enormous, and as our observing technologies improve, we cover an ever greater portion of it. Selection effects due to the survey limitations are also more apparent in such a representation. Examining the coverage of the OPS can yield insights for the optimal strategies for future explorations (➋ *Fig. 5-1*).

Alternatively, knowing how given types of objects populate the OPS, one can optimize the coverage strategy to find them; for example, supernovae, or brown dwarfs. We note that the inverse of the area coverage represents a limiting source density: if a survey covers the area Ω, in order to detect N sources of a given species, their surface density on the sky must be greater than N/Ω. For example, down to $r < 20$ mag, the surface density of quasars at redshifts $z > 4$ is $\sim 4.1 \times 10^{-2}$ deg^{-2}; thus, in a survey with such limiting magnitude, in order to find ~100 such objects for subsequent studies, one has to cover at least ~2,500 deg^2.

Two relatively poorly explored domains of the OPS may be especially interesting: the low-surface-brightness universe and the time domain, at any wavelength (Djorgovski et al. 2001a, b; Brunner et al. 2001b; Diercks 2001). For example, the subject of possible missing large populations of low-surface-brightness galaxies has been debated extensively in the literature (see, e.g., Impey and Bothun 1997), but a systematic exploration of the low-surface-brightness universe is just starting. The time domain is currently undergoing a vigorous development, thanks to

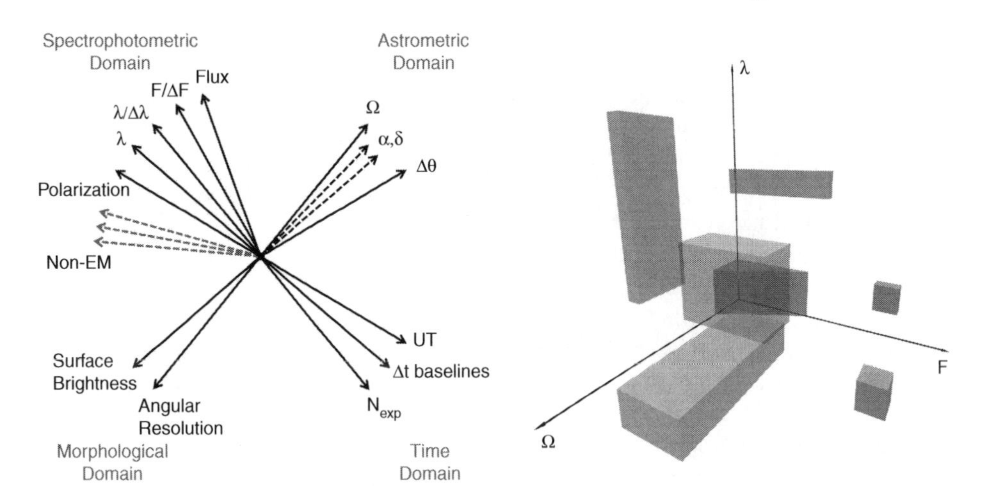

Fig. 5-1

A schematic illustration of the observable parameter space (OPS). All axes of the OPS corresponding to independent measurements are mutually orthogonal. Every astronomical observation, surveys included, carves out a finite hypervolume in this parameter space. *Left:* Principal axes of the OPS, grouped into four domains; representing such high-dimensionality parameter spaces on a 2D paper is difficult. *Right:* A schematic representation of a particular 3D representation of the OPS. Each survey covers some solid angle (Ω), over some wavelength range (λ), and with some dynamical range of fluxes (F). Note that these regions need not have orthogonal or even planar boundaries

the advent of the modern synoptic sky surveys (Paczynski 2000; Djorgovski et al. 2012, and the volume edited by Griffin et al. 2012); we review some of them below.

As catalogs of sources and their measured properties are derived from imaging surveys, they can be represented as points (or vectors) in the *Measurement Parameter Space* (MPS). Every measured quantity for the individual sources has a corresponding axis in the MPS. Some could be derived from the primary measured quantities; for example, if the data are obtained in multiple bandpasses, we can form axes of flux ratios or colors; a difference of magnitudes in different apertures forms a concentration index; surface brightness profiles of individual objects can be constructed and parametrized, for example, with the Sersic index; and so on. Some parameters may not even be representable as numbers but rather as labels; for example, morphological types of galaxies or a star versus a galaxy classification.

While OPS represents the scope and the limitations of *observations*, MPS is populated by the detected *sources and their measured properties*. It describes completely the content of catalogs derived from the surveys.

Some of the axes of the OPS also pertain to the MPS, for example, fluxes or magnitudes. In those cases, one can map objects back to the appropriate projections of the OPS. However, although there is only one flux axis in a given photometric bandpass in the OPS, there may be several corresponding derived axes in the MPS, for example, magnitudes measured in different apertures (obviously, such sets of measurement axes are correlated and would not be mutually orthogonal). Some axes of the OPS would not have meaningful counterparts in the MPS, for example, the overall area coverage. There may also be axes of the MPS that are produced by measurements of images that are meaningless in the OPS, for example, parameters describing the morphology of objects, like concentration indices, ellipticities, etc. For all of these reasons, it makes sense to separate the OPS from the MPS.

Each detected source is then fully represented as a feature vector in the MPS ("features" is a commonly used computer-science term for what we call measured parameters here). Modern imaging surveys may measure hundreds of parameters for each object, with a corresponding dimensionality of the MPS; however, many axes can be highly correlated (e.g., magnitudes in a series of apertures). It is thus a good idea to reduce the dimensionality of the MPS to a minimum set of independent axes before proceeding to further data analysis.

The MPS in itself can be a useful research tool. For example, morphological classification of objects (stars vs. galaxies, galaxies of different types, etc.), as well as a detection and removal of measurement artifacts, which often appear as outliers from the bulk of the measurements, can be accomplished very effectively in the MPS. Sometimes, a judiciously chosen subset of parameters can be used for such tasks. Statistical and machine learning tools like the Kohonen Self-Organizing Maps (SOM; Kohonen 1989) can be used to find the most discriminating parameters for a given problem.

Observed properties of detected sources can then be translated through the process of data reductions and analysis into their physical properties, usually requiring some additional or interpretative knowledge, for example, distances, and/or assuming some data model. These form the *physical parameter space* (PPS), where typically the scientific discoveries are made. The MPS describes *observations* of individual sources found in a catalog; the PPS is populated by astronomical *objects*, again quantified as feature vectors. For example, an observed color-magnitude diagram for stars in some field is a 2-dimensional projection of the MPS; the corresponding temperature-luminosity diagram is a projection of the PPS. A magnitude belongs to the MPS; a luminosity is a feature of the PPS. There is of course no reason to stop at 2

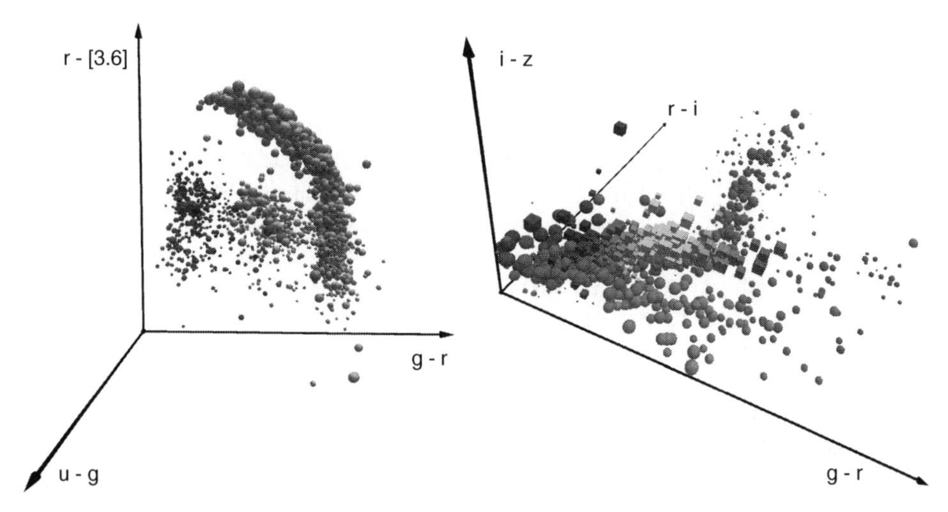

◘ Fig. 5-2

Examples of 3D projections of the measurement parameter space (MPS), specifically colors. Only a couple of thousand of objects are plotted, for clarity. *Left:* Colors of quasars in the SDSS $u'g'r'$ and the *Spitzer* 3.6 μm band, from Richards et al. (2009). *Right:* Colors of stars, quasars, and galaxies in the SDSS $u'g'r'i'z'$ color space. Objects of different types or subtypes form clusters and sequences in the MPS that may be apparent in some of its projections but not in others. In these examples, distributions in the space of observable parameters also carry a physical meaning and can be used to separate the classes or to search for rare types of objects, for example, high-redshift quasars. Thus, visual inspection and exploration of data represented in the MPS can lead directly to new insights and discoveries. In addition to the three spatial axes, one can also use the sizes, shapes, and colors of the data points in order to encode additional dimensions. A more rigorous approach would be to use various data mining techniques, such as the clustering, classification, anomaly or outlier searches, correlation searches, and similar tasks

dimensions, and in principle, the PPS can have at least as many axes as the MPS from which it is derived, although often many can be deliberately ignored in a given analysis. Also, the PPS can have additional axes of derived parameters, for example, masses, chemical abundances (derived from the observed equivalent widths in the MPS), and so on.

The MPS and the PPS may partly overlap, typically with the axes that represent distance-independent quantities, for example, colors or some other measures of the spectral energy distributions, such as the spectral indices, or hardness ratios; surface brightness; and measures of source morphology, such as concentration indices; or various measures of the variability, such as periods, if any, decay times. (In a cosmological context, relativistic corrections have to be applied to the surface brightness and any time intervals.)

Surfaces delimit the hypervolumes covered by surveys in the OPS map into the selection effects in the MPS and its projections to lower dimensionality parameter spaces. These, in turn, map directly into the selection effects in the PPS. The combination of these parameter spaces thus represents a quantitative framework for data analysis and exploration.

3.3 Exploring the Parameter Spaces

An early vision of such a systematic exploration of the observable universe was promoted by Zwicky (1957), who was as usual far ahead of his time, dubbing it the "Morphological Box Approach." That was perhaps an unfortunate wording; reading Zwicky's writings, it is clear that what he meant is very much like the OPS and PPS defined above. Zwicky was convinced that a systematic, orderly exploration of these parameter spaces may lead to the discoveries of previously unknown phenomena. While Zwicky was limited by the observational technology available to him at the time, today we can bring these ideas to fruition, thanks to the thorough and systematic coverage of the sky in so many different ways by sky surveys.

Astronomical objects and phenomena are characterized by a variety of physical parameters, for example, luminosities, spectral energy distributions, morphologies, or variability patterns. Sometimes, correlations among them exist, and their interpretation leads to a deeper physical understanding. Objects of different kinds, for example, stars, quasars, and galaxies of different kinds form clusters in the PPS, reflecting their physical distinctions, and that in itself represents some knowledge and understanding. Conversely, identifying such clusters (or outliers form them) in the space of physical properties can lead to discoveries of new types or subtypes of objects or phenomena.

It is important to note that the finite, specific hypervolumes of the OPS in any given survey are effectively "window functions" that determine what *can* be discovered: if a particular type of objects or phenomena avoids the probed region, then none will be found, and we do not know what we are missing in the portions of the OPS that were not covered. This argues for an OPS coverage, spread among a number of surveys, that covers as much of the accessible OPS hypervolume as possible and affordable.

The OPS and the PPS may overlap in some of the axes, and new discoveries may be made in the OPS alone, typically on the basis of the shapes of the spectral energy distributions. This generally happens when a new wavelength window opens up, for example, some blue "stars" turned out to be a new phenomenon of nature, quasars, when they were detected as radio sources, and similar examples can be found in every other wavelength regime. The unexpectedness of such discoveries typically comes from some hidden assumptions, for example, that all star-like objects will have a thermal emission in a certain range. When astronomical sources fail to meet our expectations, discoveries are made. Improvements in angular resolution or depth can also yield discoveries, simply by inspection. Of course, noticing something that appears new or unusual is just the first step, and follow-up observations and interpretative analysis are needed.

Sometimes we expect to find objects in some region of the parameter space. For example, most quasars (especially the high-redshift ones) and brown dwarfs are found in particular regions of the color space on the basis of an assumed (correct) model of their spectral energy distributions. Thus, PPS approach can thus be put to a good use for a *directed discovery* to form samples of known, interesting types of objects by introduction of a prior knowledge. However, it also enables discoveries that were unanticipated, and approach we may call an *organized serendipity*.

The PPS is *not* uniformly populated: astronomical objects cluster and correlate in particular regions, for good physical reasons, and are likewise absent from most of the PPS hypervolume. For example, only a few narrow, well-defined sequences are populated in the HR diagram, reflecting the physics of stellar structure and evolution. Some of these clusters correspond to the known classes of objects, but in principle, other, as yet unknown clusters may exist in the

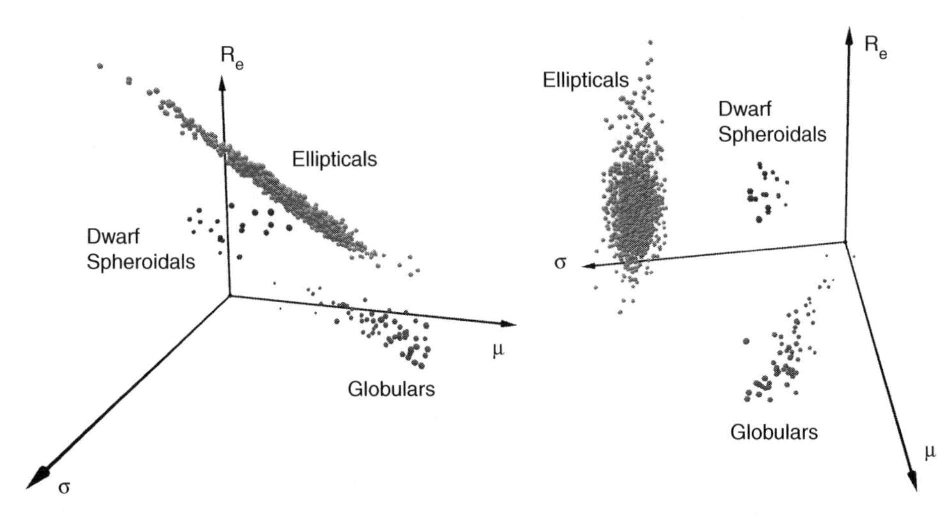

Fig. 5-3

A simple illustration of a physical parameter space (PPS) viewed from two different angles. Families of dynamically hot stellar systems are shown in the parameter space whose axes are logs of effective radii, mean surface brightness, and the central velocity dispersion in the case of ellipticals, and the maximum rotational speed in the case of spirals. In the *panel on the left*, we see an edge-on view of the fundamental plane (FP) for elliptical galaxies (Djorgovski and Davis 1987). On the *right* we see a clear separation of different families of objects, as first noted by Kormendy (1985). They form distinct clusters, some of which represent correlations of their physical parameters that define them as families in an objective way. This galaxy parameter space (Djorgovski 1992a, b) is a powerful framework for the exploration of such correlations and their implications for galaxy evolution. Data for ellipticals are from (La Barbera et al. 2009), and for dwarf spheroidals and globulars are compiled from the literature

unexplored corners of the PPS. Both the existence of clusters in the OPS, and the empty regions thereof in principle contain valuable information about the cosmic phenomena. Therein lies the knowledge discovery potential of this approach.

Various data mining (DM) techniques, such as supervised or unsupervised clustering analysis, outlier and anomaly searches, and multivariate analysis, can be used to map out the PPS and thus the kinds of the astronomical objects that populate it (e.g., de Carvalho et al. 1995; Yoo et al. 1996). If applied in the catalog domain, the data can be viewed as a set of N points or vectors in a Ddimensional parameter space, where N can be in the range of many millions or even billions and D in the range of a few tens to many hundreds. The data may be clustered in $k$statistically distinct classes that may be uncovered through the DM process. This is computationally and algorithmically a highly nontrivial problem that is now being addressed in the Astroinformatics and related disciplines. An example of a data exploration system along these lines is *Data Mining and Exploration* (DAME; Brescia et al. 2010, 2012; http://dame.dsf.unina.it). A caveat is in order: most of the commonly used DM techniques implicitly or explicitly assume that the data are complete, error-free, and free of artifacts, none of which is true in reality. Thus, they cannot be applied blindly, and scientists should understand their limitations when using them.

Effective visualization techniques can go a long way toward an exploration and intuitive understanding of all of these parameter spaces. Representation of highly dimensional parameter spaces in 3D projections (which in themselves tend to be projected to 2D surfaces, such as the paper or a screen) poses many challenges of its own, as we are biologically limited to see in 3 dimensions. Various tricks can be used to encode up to a dozen dimensions in a pseudo-3D plot, but there is no reason why the nature would limit itself to interesting or meaningful structures in a small number of dimensions.

However, not all parameters may be equally interesting or discriminating, and lowering the dimensionality of the PPS to some more appropriate, judiciously chosen subset of parameters may be necessary. This is where the scientists' skill comes into the play.

Discovery of particularly rare types of objects, whether known or unknown, can be done as a search for outliers in a judiciously chosen projection of the PPS. Such things might have been missed so far either because they are rare or because they would require a novel combination of data or a new way of looking at the data. A thorough, large-scale, unbiased, multi-wavelength census of the universe will uncover them if they do exist (and surely we have not yet found all there is out there). Surveys that detect large numbers of sources are the only practical way to find extremely rare ones; for example, down to the ~20 mag, approximately one in a million starlike objects at high Galactic latitude is actually a quasar at $z > 4$, and that is not even a particularly rare type of an object these days.

Sometimes new phenomena are found (or rediscovered) independently in different regions of the OPS. Whether such repeated findings imply that there is a finite (and even a relatively modest) number of fundamentally different astronomical phenomena in the universe is an interesting question (Harwit and Hildebrand 1986).

Most of the approaches described so far involve searches in some parameter or feature space, i.e., catalogs derived from survey images. However, we can also contemplate a direct exploration of sky surveys in the image (pixel) domain. Automated pattern recognition and classification tools can be used to discover sources with a particular image morphology (e.g., galaxies of a certain type). An even more interesting approach would be to employ machine learning and artificial intelligence techniques to search through panoramic images (perhaps matched from multiple wavelengths) for unusual image patterns; known example may be the gravitationally lensed arcs around rich clusters or supernova remnants.

Finally, an unsupervised classification search for unusual patterns or signals in astronomical data represents a natural generalization of the SETI problem (Djorgovski 2000). Manifestations of ETI may be found as "anomalous" signals and patterns in some portion of the OPS that has not been well explored previously.

4 Characteristics and Examples of Sky Surveys

4.1 A Sampler of Panoramic Sky Surveys and Catalogs

Sky surveys, large and small, are already too numerous to cover extensively here, and more are produced steadily. In the appendix, we provide a listing, undoubtedly incomplete, of a number of them, with the web links that may be a good starting point for a further exploration. Here we describe some of the more popular panoramic (wide-field, single-pass) imaging surveys, largely for the illustrative purposes, sorted in wavelength regimes, and in a roughly chronological order.

However, our primary aim is to illustrate the remarkable richness and diversity of the field and a dramatic growth in the data volumes, quality, and homogeneity. Surveys are indeed the backbone of contemporary astronomy.

➋ *Table 5-1* lists basic parameters for a number of popular surveys, at least as of this writing. It is meant as illustrative rather than complete in any way. In the appendix, we list a much larger number of the existing surveys and their websites, from which the equivalent information can be obtained. We also refrain from talking too much about the surveys that are still in the planning stages; the future will tell how good they will be.

4.1.1 Surveys and Catalogs in the Visible Regime

The POSS-II photographic survey plate scans obtained at the Space Telescope Science Institute (STScI) led to the*Digitized Palomar Observatory Sky Survey* (DPOSS; Djorgovski et al. 1994, 1997a, 1999; Weir et al. 1995a; http://www.astro.caltech.edu/~george/dposs), a digital survey of the entire Northern sky in three photographic bands, calibrated by CCD observations to SDSS-like *gri*. The plates are scanned with 1.0 arcsec pixels, in rasters of 23,040 square, with 16 bits per pixel, producing about 1 GB per plate or about 3 TB of pixel data in total. They were processed independently at STScI for the purposes of constructing a new guide star catalog for the HST (Lasker et al. 2008) and at Caltech for the DPOSS project. Catalogs of all the detected objects on each plate were generated, down to the flux limit of the plates, which roughly corresponds to the equivalent limiting magnitude of $m_B \sim 22$ mag. A specially developed software package, *Sky Image Cataloging and Analysis Tool* (SKICAT; Weir et al. 1995c), was used to process and analyze the data. Machine learning tools (decision trees and neural nets) were used for star-galaxy classification (Weir et al. 1995b, Odewahn et al. 2004), accurate to 90% down to ~1 mag above the detection limit. An extensive program of CCD calibrations was done at the Palomar 60-in. telescope (Gal et al. 2005). The resulting object catalogs were combined and stored in a relational database The DPOSS catalogs contain ~50 million galaxies and ~500 million stars, with over 100 attributes measured for each object. They generated a number of studies, for example, a modern, objectively generated analog of the old Abell cluster catalog (Gal et al. 2009).

The *United States Naval Observatory Astrometric Catalog* (USNO-A2 and an improved version, USNO-B1, http://www.nofs.navy.mil/data/FchPix) was derived from the POSS-I, POSS-II, and ESO/SERC Southern sky survey plate scans, done using the precision measuring machine (PMM) built and operated by the United States Naval Observatory Flagstaff Station (Monet et al. 2003). They contain about a billion sources over the entire sky down to a limiting magnitude of $m_B \sim 20$ mag, and they are now commonly used to derive astrometric solutions for optical and NIR images. A combined astrometric data set from several major modern catalogs is currently being served by the USNO's *Naval Observatory Merged Astrometric Dataset* (NOMAD; http://www.usno.navy.mil/USNO/astrometry/optical-IR-prod/nomad).

The *Sloan Digital Sky Survey* (SDSS; Fukugita et al. 1996; Gunn et al. 1998, 2006; York et al. 2000; http://sdss.org) is a large international astronomical collaboration focused on constructing the first CCD photometric survey high Galactic latitudes, mostly in the Northern sky. The survey and its extensions (SDSS-II and SDSS-III) eventually covered ~14,500 deg^2, i.e., more than a third of the entire sky. A dedicated 2.5 m telescope at Apache Point, New Mexico, was specially designed to take wide-field ($3^\circ \times 3^\circ$) images using a mosaic of thirty 2,048 × 2,048-pixel CCDs, with 0.396 arcsec pixels and with additional devices used for astrometric calibration.

Table 5-1
Basic properties of some of the popular wide-field surveys. The quoted limiting magnitudes are on the Vega zero-point system for DSS, 2MASS, and UKIDSS; on the AB_ν system for GALEX; and SDSS zero-points are close to the AB_ν system. See the text for more details and websites

Survey	Type	Duration	Bandpasses	Lim. flux	Area coverage	$N_{sources}$	Notes
DSS scans	Visible	1950s–1990s	*B* (~450 nm) *R* (~650 nm) *I* (~800 nm)	21–22 mag 20–21 mag 19–20 mag	Full sky	~10^9	Scans of plates from the POSS and ESO/SERC surveys
SDSS-I SDSS-II SDSS-III	Visible	2000–2005 2005–2008 2009–2014	u(~800 nm) g(~800 nm) r(~800 nm) i(~800 nm) z(~800 nm)	22.0 mag 22.2 mag 22.2 mag 21.3 mag 20.5 mag	14,500 deg^2	4.7×10^8	Numbers quoted for DR8 (2011). In addition, spectra of ~1.6 million objects
2MASS	Near IR	1997–2001	J(~1.25 μm) H(~1.65 μm) K_s(~2.15 μm)	15.8 mag 15.1 mag 14.3 mag	Full sky	4.7×10^8	
UKIDSS	Near IR	2005–2012	Y(~1.05 μm) J(~1.25 μm) H(~1.65 μm) *K*(~2.2 μm)	20.5 mag 20.0 mag 18.8 mag 18.4 mag	7,500 deg^2	~10^9	Estim. final numbers quoted for the LAS; deeper surveys over smaller areas also done
IRAS	Mid/Far-IR (space)	1983–1986	12 μm 25 μm 60 μm 100 μm	0.5 Jy 0.5 Jy 0.5 Jy 1.5 Jy	Full sky	1.7×10^5	
NVSS	Radio	1993–1996	1.4 GHz	2.5 mJy	32,800 deg^2	1.8×10^6	Beam ~45 arcsec
FIRST	Radio	1993–2004	1.4 GHz	1 mJy	10,000 deg^2	8.2×10^5	Beam ~5 arcsec
PMN	Radio	1990	4.85 GHz	~30 mJy	16,400 deg^2	1.3×10^4	Combines several surveys
GALEX	UV (space)	2003–2012	135–175 nm 175–275 nm	20.5 mag AIS 23 mag MIS	AIS 29,000 MIS 3,900	6.5×10^7 1.3×10^7	As of GR6 (2011); also some deeper data
Rosat	X-ray (space)	1990–1999	0.04–2 keV	~10^{-14} erg cm^{-2} s^{-1}	Full sky	1.5×10^5	Deeper, small area surveys also done
Fermi LAT	γ-ray (space)	2008–?	20 MeV to 30 GeV	4×10^{-9} erg cm^{-2} s^{-1}	Full sky	~2×10^3	LAT instrument only; in addition, GRBM

The imaging data were taken in the drift scanning mode along great circles on the sky. The imaging survey uses five passbands ($u'g'r'i'z'$), with limiting magnitudes of 22.0, 22.2, 22.2, 21.3, and 20.5 mag in the $u'g'r'i'z'$ bands respectively. The spectroscopic component of the survey is described below. The total raw data volume is several tens of TB, with an at least comparable amount of derived data products.

The initial survey (SDSS-I, 2000–2005) covered ~8,000 deg^2. SDSS-II (2005–2008) consisted of *Sloan Legacy Survey* that expanded the area coverage to ~8,400 deg^2 and catalogued 230 million objects, the *Sloan Extension for Galactic Understanding and Exploration* (SEGUE) that obtained almost a quarter million spectra over ~3,500 deg^2, largely for the purposes of studies of the Galactic structure, and the *Sloan Supernova Survey* which confirmed 500 Type Ia SNe. The latest survey extension, SDSS-III (2008–2014), is still ongoing at this time.

The data have been made available to the public through a series of well-documented releases, the last one (as of this writing), DR8, occurred in 2011. In all, by 2011, photometric measurements for ~470 million unique objects have been obtained and ~1.6 million spectra.

More than perhaps any other survey, SDSS deserves the credit for transforming the culture of astronomy in regards to sky surveys: they are now regarded as fundamental data sources with a rich information content. SDSS enabled – and continues to do so – a vast range of science, by not just the members of the SDSS team (of which there are many), but also the rest of the astronomical community, spanning topics from asteroids to cosmology. Due to its public data releases, SDSS may be the most productive astronomical project so far.

The *Panoramic Survey Telescope and Rapid Response System* (Pan-STARRS; http://pan-starrs.ifa.hawaii.edu; Kaiser et al. 2002) is a wide-field panoramic survey developed at the University of Hawaii's Institute for Astronomy and operated by an international consortium of institutions. It is envisioned as a set of four 1.8 m telescopes with a 3° field of view each, observing the same region of the sky simultaneously, each equipped with a ~1.4-Gigapixel CCD camera with 0.3 arcsec pixels. As of this writing, only one of the telescopes (PS1) is operational, at Haleakala Observatory on Maui, Hawaii; it started science operations in 2010. The survey may be transferred to the Mauna Kea Observatory if additional telescopes are built. The CCD cameras, consisting of an array of 64, 600 × 600-pixel CCDs, incorporate a novel technology, orthogonal transfer CCDs, that can be used to compensate the atmospheric tip-tilt distortions (Tonry et al. 1997), although that feature has not yet been used for the regular data taking. PS1 can cover up to 6,000 deg^2 per night and generate up to several TB of data per night, but not all images are saved, and the expected final output is estimated to be ~1 PB per year.

The primary goal of the PS1 is to survey ~3/3 of the entire sky (the "3π" survey) with 12 epochs in each of the five bands (*grizy*). The coadded images should reach considerably deeper than SDSS. A dozen key projects, some requiring additional observations, are also under way. The survey also has a significant time-domain component. The data access is restricted to the consortium members until the completion of the "3π" survey.

Its forthcoming Southern sky counterpart is *SkyMapper* (http://msowww.anu.edu.au/skymapper), developed by Brian Schmidt and collaborators at the Australian National University. It is undergoing the final commissioning steps as of this writing, and it will undoubtedly be a significant contributor when it starts full operations. The fully automated 1.35 m wide-angle telescope is located at Siding Spring Observatory, Australia. It has a ~270-Megapixel CCD mosaic camera that covers ~5.7 deg^2 in a single exposure, with 0.5 arcsec pixels. *SkyMapper* will cover the entire Southern sky 36 times in six filters (SDSS $g'r'i'z'$, a Strömgren system-like u, and a unique narrow v band near 4,000 Å). It will generate ~100 MB of data per second during every clear night, totaling about 500 TB of data at the end of the survey. A distilled version of the survey data will be made publicly available. Its scientific goals include discovering dwarf planets in the outer solar system, tracking asteroids, creating a comprehensive catalog of the stars in our Galaxy, studying dark matter, and so on.

Another Southern sky facility is the *Very Large Telescope (VLT) Survey Telescope* (VST; http://www.eso.org/public/teles-instr/surveytelescopes/vst.html) that is just starting operations as of this writing (2011). This 2.6 m telescope covers a ~1 deg^2 field, with a ~270-Megapixel CCD mosaic consisting of 32, 2,048 × 4,096-pixel CCDs. The VST ATLAS survey will cover 4,500 deg^2 of the Southern sky in five filters (*ugriz*) to depths comparable to those of the SDSS. A deeper survey, KIDS, will reach ~2–3 mag deeper over 1,500 deg^2 in *ugri* bands. An additional survey, VPHAS+, will cover 1,800 deg^2 along the Galactic plane in *ugri* and Hα bands to depths comparable to the ATLAS survey. These and other VST surveys will also be complemented by near-infrared data from the VISTA survey, described below. They will provide an imaging base for spectroscopic surveys by the VLT. Scientific goals include studies of dark matter and dark energy, gravitational lensing, quasar searches, structure of our Galaxy, etc.

An ambitious new *Dark Energy Survey* (DES; http://www.darkenergysurvey.org) should start in 2012 using the purpose-built *Dark Energy Camera* (DECam) on the Blanco 4 m telescope at CTIO. It is a large international collaboration, involving 23 institutions. DECam is a 570-Megapixel camera with 74 CCDs, with a 2.2 deg^2 FOV. The primary goal is to catalog 5,000 deg^2 in the Southern hemisphere, in *grizy* bands, in an area that overlaps with several other surveys for a maximum leveraging. It is expected to see over 300 million galaxies during the 525 nights and find and measure ~3,000 SNe and use several observational probes of dark energy in order to constrain its physical nature.

The GAIA mission (http://gaia.esa.int) is expected to be launched in 2012, expected to revolutionize astrometry and provide fundamental data for a better understanding of the structure of our Galaxy, distance scale, stellar evolution, and other topics. It will measure positions with a 20-μarcsec accuracy for about a billion stars (i.e., at the depth roughly matching the POSS surveys), from which parallaxes and proper motions will be derived. In addition, it will obtain low-resolution spectra for many objects. It is expected to cover the sky ~70 times over a projected 5-year mission and thus discover numerous variable and transient sources.

Many other modern surveys in the visible regime invariably have a strong synoptic (time domain) component, reflecting some of their scientific goals. We describe some of them below.

4.1.2 Surveys in the Infrared

The *Two Micron All-Sky Survey* (2MASS; 1997–2002; http://www.ipac.caltech.edu/2mass; Skrutskie et al. 2006) is a near-IR (J, H, and K_S bands) all-sky survey done as a collaboration between the University of Massachusetts, which constructed the observatory facilities and operated the survey, and the Infrared Processing and Analysis Center (IPAC) at Caltech, which is responsible for all data processing and archives. It utilized two highly automated 1.3-m telescopes, one at Mt. Hopkins, AZ, and one at CTIO, Chile. Each telescope was equipped with a three-channel camera with HgCdTe detectors, and was capable of observing the sky simultaneously at J, H, and K_S with 2 arcsec pixels. It remains as the only modern, ground-based sky survey that covered the entire sky. The survey generated ~12 TB of imaging data, a point source catalog of ~300 million objects, with a 10-σ limit of $K_S < 14.3$ mag (Vega zero point), and a catalog of ~500,000 extended sources. All of the data are publicly available via a web interface to a powerful database system at IPAC.

The *United Kingdom Infrared Telescope (UKIRT) Infrared Deep Sky Survey* (UKIDSS, 2005–2012, Lawrence et al. 2007, http://www.ukidss.org) is a NIR sky survey. Its wide-field component covers about 7,500 deg^2 of the Northern sky, extending over both high and low Galactic latitudes, in the *YJHK* bands down to the limiting magnitude of K = 18.3 mag (Vega zero point). UKIDSS is made up of five surveys and includes two deep extragalactic elements, one covering 35 deg^2 down to K = 21 mag and the other reaching K = 23 mag over 0.77 deg^2. The survey instrument is WFCAM on the UK Infrared Telescope (UKIRT) in Hawaii. It has four 2,048 × 2,048 IR arrays; the pixel scale of 0.4 arcsec gives an exposed solid angle of 0.21 deg^2. Four of the principal scientific goals of UKIDSS are: finding the coolest and the nearest brown dwarfs and studying high-redshift dusty starburst galaxies, elliptical galaxies and galaxy clusters at redshifts $1 < z < 2$, and the highest-redshift quasars, at $z \sim 7$. The survey succeeded on all counts. The data were made available to the entire ESO community immediately after they are entered into the archive, followed by a worldwide release 18 months later.

The *Visible and Infrared Survey Telescope for Astronomy* (VISTA; http://www.vista.ac.uk) is a 4.1 m telescope at the ESO's Paranal Observatory. Its 67-Megapixel IR camera uses 16 IR detectors, covers an FOV ~1.65° with 0.34 arcsec pixels, in $ZYJHK_s$ bands, and a 1.18 μm narrow band. It will conduct a number of distinct surveys, the widest being the *VISTA Hemisphere Survey* (VHS), covering the entire Southern sky down to the limiting magnitudes ~20–21 mag in $YJHK_s$ bands, and the deepest being the Ultra-VISTA that will cover a 0.73 deg^2 field down to the limiting magnitudes ~26 mag in the $YJHK_s$ bands and the 1.18 μm narrow band. VISTA is expected to produce data rates exceeding 100 TB/year, with the publicly available data distributed through the ESO archives.

A modern, space-based all-sky survey is *Wide-field Infrared Survey Explorer* (WISE; 2009–2011; Wright et al. 2010; http://wise.ssl.berkeley.edu). It mapped the sky at 3.4, 4.6, 12, and 22 μm bands, with an angular resolution of 6.1, 6.4, 6.5, and 12.0 arcsec in the four bands, the number of passes depending on the coordinates. It achieved 5-σ point source sensitivities of about 0.08, 0.11, 1 and 6 mJy, respectively, in unconfused regions. It detected over 250 million objects, with over two billion individual measurements, ranging from near-Earth asteroids to brown dwarfs, ultraluminous starbursts, and other interesting types of objects. Given its uniform, all-sky coverage, this survey is likely to play an important role in the future. The data are now publicly available.

Two examples of modern, space-based IR surveys over the limited areas include the *Spitzer Wide-area InfraRed Extragalactic* survey (SWIRE; http://swire.ipac.caltech.edu; Lonsdale et al. 2003) that covered 50 deg^2 in 7 mid-/far-IR bandpasses, ranging from 3.6 to 160 μm, and the *Spitzer Galactic Legacy Infrared Mid-Plane Survey Extraordinaire* (GLIMPSE; http://www.astro.wisc.edu/sirtf/) that covered ~240 deg^2 of the Galactic plane in 4 mid-IR bandpasses, ranging from 3.6 to 8.0 μm. Both were followed up at a range of other wavelengths and provided new insights into obscured star formation in the universe.

4.1.3 Surveys and Catalogs in the Radio

The *Parkes Radio Sources* (PKS, http://archive.eso.org/starcat/astrocat/pks.html) is a radio catalog compiled from major radio surveys taken with the Parkes 64 m radio telescope at 408 and 2,700 MHz over a period of almost 20 years. It consists of 8,264 objects covering all the sky south of Dec = +27° except for the Galactic plane and Magellanic Clouds. Discrete sources were selected with flux densities in excess of ~50 mJy. It has since been supplemented (PKSCAT90) with improved positions, optical identifications, and redshifts as well as flux densities at other frequencies to give a frequency coverage of 80 MHz–22 GHz.

Numerous other continuum surveys have been performed. An example is *Parkes-MIT-NRAO* (PMN, http://www.parkes.atnf.csiro.au/observing/databases/pmn/pmn.html) radio continuum surveys made with the NRAO 4.85 GHz seven-beam receiver on the Parkes telescope during June and November 1990. Maps covering four declination bands were constructed from the survey scans with the total survey coverage over -87.5° < Dec < $+10^\circ$. A catalog of 50,814 discrete sources with angular sizes approximately less than 15 arcmin and a flux density greater than about 25 mJy (the flux limit varies for each band but is typically this value) was derived from these maps. The positional accuracy is close to 10 arcsec in each coordinate. The 4.85 GHz weighted source counts between 30 mJy and 10 Jy agree well with previous results.

The *National Radio Astronomical Observatory* (NRAO) *Very Large Array* (VLA) *Sky Survey* (NVSS; Condon et al. 1998; http://www.cv.nrao.edu/nvss) is a publicly available, radio continuum survey at 1.4 GHz, covering the sky north of Dec = -40° declination, i.e., ~32,800 deg^2. The principal NVSS data products are a set of 2,326 continuum map data cubes, each covering an area of $4^\circ \times 4^\circ$ with three planes containing the Stokes I, Q, and U images and a catalog of discrete sources in them. Every large image was constructed from more than 100 of the smaller, original snapshot images (over 200,000 of them in total). The survey catalog contains over 1.8 million discrete sources with total intensity and linear polarization image measurements (Stokes I, Q, and U), with a resolution of 45 arcsec and a completeness limit of about 2.5 mJy. The NVSS survey was performed as a community service, with the principal data products being released to the public by anonymous FTP as soon as they were produced and verified.

The *Faint Images of the Radio Sky at Twenty-centimeter* (FIRST; http://sundog.stsci.edu; Becker et al. 1995) is a publicly available radio snapshot survey performed at the NRAO VLA facility in a configuration that provides higher spatial resolution than the NVSS, at the expense of a smaller survey footprint. It covers ~10,000 deg^2 of the North and South Galactic Caps with a ~5 arcsec resolution. A final atlas of radio image maps with 1.8 arcsec pixels is produced by coadding the twelve images adjacent to each pointing center. The survey catalog contains around one million sources with a resolution of better than 1 arcsec. A source catalog including peak and integrated flux densities and sizes derived from fitting a two-dimensional Gaussian to each source is generated from the atlas. Approximately 15% of the sources have optical counterparts at the limit of the POSS-I plates; unambiguous optical identifications (<5% false detection rates) are achievable to V ~ 24 mag. The survey area has been chosen to coincide with that of the SDSS. At the magnitude limit of the SDSS, approximately 50% of the FIRST sources have optical counterparts.

Two examples of modern H I 21 cm line surveys are the *HI Parkes All Sky Survey* (HIPASS; http://aus-vo.org/projects/hipass, http://www.atnf.csiro.au/research/multibeam/release/) and the Arecibo Legacy Fast ALFA survey (ALFALFA; http://egg.astro.cornell.edu; Giovanelli et al. 2005). These surveys cover the redshifts up to $z \sim 0.05 - 0.06$, detecing tens of thousands H I selected galaxies, high-velocity clouds, etc.

The new generation of ground-based radio sky surveys is intrinsically synoptic in nature; we discuss some of them below. A number of experiments are also addressing the epoch of reionization. The currently ongoing *Planck* mission (http://www.rssd.esa.int/Planck), in addition to its primary scientific goals of CMB-based cosmology, will also represent an excellent radio survey at high frequencies.

4.1.4 Surveys at Higher Energies

Numerous space missions have surveyed the sky at wavelengths ranging from UV to γ-rays. Some of the more modern, panoramic ones include:

Galaxy Evolution Explorer (GALEX; Martin et al. 2005; http://www.galex.caltech.edu), launched in 2003, is the first (and so far the only) nearly all-sky survey UV mission. Data are obtained in two bands (1,350–1,750 and 1,750–2,750 Å) using microchannel plate detectors that provide time-resolved imaging. The all-sky imaging survey (AIS; it excludes some fields containing bright stars) reaches the depth comparable to the POSS surveys ($m_{AB} \simeq 20.5$ mag); a medium depth survey covers ~3,900 deg^2 down to $m_{AB} \simeq 23$ mag, and a deep imaging survey covers ~350 deg^2 down to $m_{AB} \simeq 25$ mag. In addition, a survey of several hundred nearby galaxies and a low-resolution (R = 100–200) slitless grism spectroscopic survey was performed over a limited area. The data are released in roughly yearly intervals, with the final one (GR7) expected in 2012.

ROSAT (Röntgensatellit; 1990–1999; http://www.dlr.de/dlr/en/desktopdefault.aspx/tabid-10424/) was a German Aerospace Center-led satellite X-ray telescope, with instruments built by Germany, the UK, and the United States. It carried an imaging X-ray telescope (XRT) with three focal plane instruments: two position-sensitive proportional counters (PSPC), a high-resolution imager (HRI), and a wide-field camera (WFC) extreme ultraviolet telescope coaligned with the XRT that covered the wave band between 6 Å(0.042–0.21 keV). The X-ray mirror assembly was a grazing incidence fourfold nested Wolter I type telescope with an 84 cm diameter aperture and 240 cm focal length. The angular resolution was less than 5 arcsec at half energy width. The XRT assembly was sensitive to X-rays between 0.1 and 2 keV.

The *ROSAT All Sky Survey* (Voges et al. 1999) was publicly released in 2000, and it is currently available from the MPE archive and the HEASARC archive at GSFC. It consists of 1,378 distinct fields, with the bulk of all sky survey obtained by PSPC-C in scanning mode. The telescope detected more than 150,000 X-ray sources, ~20 times more than were previously known. The survey enabled a detailed morphology of supernova remnants and clusters of galaxies, detection of shadowing of diffuse X-ray emission by molecular clouds, detection of isolated neutron stars, discovery of X-ray emission from comets, and many other results.

4.2 Synoptic Sky Surveys and Exploration of the Time Domain

A systematic exploration of the time domain in astronomy is now one of the most exciting and rapidly growing areas of astrophysics, and a vibrant new observational frontier (Paczynski 2000). A number of important astrophysical phenomena can be discovered and studied *only* in the time-domain, ranging from exploration of the solar system to cosmology and extreme relativistic phenomena. In addition to the studies of known, interesting time domain phenomena, for example, supernovae and other types of cosmic explosions, there is a real and exciting possibility of discovery of new types of objects and phenomena. As already noted, opening new domains of the observable parameter space often leads to new and unexpected discoveries.

The field has been fueled by the advent of the new generation of digital synoptic sky surveys, which cover the sky many times, as well as the ability to respond rapidly to transient events using robotic telescopes. This new growth area of astrophysics has been enabled by information technology, continuing evolution from large panoramic digital sky surveys to panoramic digital cinematography of the sky, opening the great new discovery space (Djorgovski et al. 2012).

Numerous surveys, studies, and experiments have been conducted in this arena, are in progress now, or are in planning stages, indicating the growing interest in time-domain astronomy, leading to the Large Synoptic Survey Telescope (LSST; Tyson et al. 2002, Ivezic et al. 2008). Today's surveys are opening the time-domain frontier, yielding exciting results, and can be used to refine the scientific and methodological strategies for the future.

The key to progress in this emerging arena of astrophysics is the availability of substantial event data streams generated by panoramic digital synoptic sky surveys, coupled with a rapid follow-up of potentially interesting events (photometric, spectroscopic, and multi-wavelength). This, in turn, requires a rapid, real-time processing of massive sky survey data streams, with event detection, filtering, characterization, and rapid dissemination to the astronomical community. Over the past few years, we have been developing both synoptic sky survey data streams and cyber-infrastructure needed for the rapid distribution and characterization of transient events and phenomena.

As in the previous subsection, here we describe some of the recent and ongoing synoptic sky surveys. This list is meant as illustrative rather than complete, and the interested reader can follow additional links listed in the Appendix.

4.2.1 General Synoptic Surveys in the Visible Regime

Many current wide-field transient surveys are targeted toward the discovery and characterization of GRB afterglows. One example is the *Robotic Optical Transient Search Experiment* (ROTSE-III; Akerlof et al. 2003; http://www.rotse.net), which uses four telescopes with an FOV = 3.4 deg^2 to survey 1,260 deg^2 to R ~17.5 mag (e.g., Rykoff et al. 2005; Yost et al. 2007). ROTSE searches for astrophysical optical transients on time scales of a fraction of a second to a few hours. While the primary incentive of this experiment has been to find the GRB afterglows, additional variability studies have been enabled, for example, searches for RR Lyrae.

The *Palomar-Quest* (PQ; Djorgovski et al. 2008; http://palquest.org) survey was a collaborative project between groups at Yale University and Caltech, with an extended network of collaborations with other groups worldwide. The data were obtained at the Palomar Observatory's Samuel Oschin telescope (the 48-in. Schmidt) using the QUEST-2 112-CCD, 161 Megapixel camera (Baltay et al. 2007). Approximately 45% of the telescope time was used for the PQ survey, with the rest being used by the NEAT survey for near-Earth asteroids, and miscellaneous other projects. The survey started in the late summer of 2003, and ended in September 2008.

In the first phase (2003–2006), data were obtained in the drift scan mode in 4.6° wide strips of a constant Dec, in the range $-25° < \mathrm{Dec} < +25°$, excluding the Galactic plane. The total area coverage is ~15,000 deg^2, with multiple passes, typically 5–10, but up to 25, with time baselines ranging from hours to years. Typical area coverage rate was up to ~500 deg^2/night in four filters. The raw data rate was on average ~70 GB per clear night. About 25 TB of usable data have been collected in the drift scan mode. Two filter sets were used, Johnson *UBRI* and SDSS *griz*. Effective exposures ~150/(cos δ) sec per pass, with limiting magnitudes are $r \sim 21.5$, $i \sim 20.5$, $z \sim 19.5$, $R \sim 22$, and $I \sim 21$ mag, depending on the seeing, lunar phase, etc. Photometric calibrations were done by matching to the SDSS. In the second phase of the survey (2007–2008), data were obtained mostly in the traditional point-and-track mode, in a single, wide-band red filter (RG610). The coverage and the cadence were largely optimized for the nearby supernova search, in collaboration with the LBNL NSNF group, and a search for dwarf planets, in collaboration with M. Brown at Caltech.

Data were processed with several different pipelines, optimized for different scientific goals: the Yale pipeline (Andrews et al. 2008), designed for a search for gravitationally lensed quasars; the Caltech real-time pipeline, used for real-time detections of transient events; and the LBNL NSNF pipeline, based on image subtraction and designed for detection of nearby SNe. The data are being publicly released. Coadded images (the *DeepSky* project, Nugent et al., in prep.; http://supernova.lbl.gov/~nugent/deepsky.html) reach to R ~ 23 mag over ~20,000 deg^2.

The *Nearby Supernova Factory* (NSNF; http://snfactory.lbl.gov; Aldering et al. 2002), operated by the Lawrence Berkeley National Laboratory (LBNL), searches for type Ia SNe in the redshift range $0.03 < z < 0.08$ in order to establish the low-redshift anchor of the SN Ia Hubble diagram. This experiment ran from 2003 to 2008 using data obtained by the NEAT and PQ surveys, resulting in several hundred spectroscopically confirmed SNe.

The *Palomar Transient Factory* (PTF; http://www.astro.caltech.edu/ptf; Rau et al. 2009; Law et al. 2009) operates in the way that is very similar to the PQ survey, on the same telescope, but with a much better camera that was previously used on the CFHT. The data are taken in a point-and-stare mode, with 2 exposures per field in a given night, mostly in the broad R and Gbands. PTF reaches a comparable depth to its predecessor, PQ, and covers a few hundred deg^2/night. The overall area coverage is ~½ of the entire sky. The survey mostly uses a cadence optimized for a SN discovery. The principal data processing pipeline is an updated version of the LBNL NSNF image subtraction pipeline. The survey has been very productive, but the data are proprietary to the PTF consortium, at least as of this writing.

The *Catalina Real-Time Transient Survey* (CRTS; Drake et al. 2009, 2012; Djorgovski et al. 2011a; Mahabal et al. 2011; http://crts.caltech.edu) leverages existing synoptic telescopes and image data resources from the Catalina Sky Survey (CSS) for near-Earth objects and potential planetary hazard asteroids (NEO/PHA), conducted by Edward Beshore, Steve Larson, and their collaborators at the University of Arizona. The solar system objects remain the domain of the CSS survey, while CRTS aims to detect astrophysical transient and variable objects, using the same data stream. Real-time processing, characterization, and distribution of these events, as well as follow-up observations of selected events, constitute the CRTS.

CRTS utilizes three wide-field telescopes: the 0.68-m Schmidt at Catalina Station, AZ (CSS); the 0.5-m Uppsala Schmidt (Siding Spring Survey, SSS) at Siding Spring Observatory, NSW, Australia; and the Mt. Lemmon Survey (MLS), a 1.5-m reflector located on Mt. Lemmon, AZ. Each telescope employs a camera with a single, cooled, 4,096 × 4,096-pixel, back-illuminated, unfiltered CCD. They are operated for 23 nights per lunation, centered on new moon. Most of the observable sky is covered up to four times per lunation. The total area coverage is ~30,000 deg^2, as it excludes the Galactic plane within $|b| < 10° - 15°$. In a given night, four images of the same field are taken, separated by ~10 min, for a total time baseline of ~30 min between first and last images. The combined data streams cover up to 2,500 deg^2/night to a limiting magnitude of V ~ 19–20 mag, and up to 275 deg^2/night to V ~ 21.5 mag. On time scales of a few weeks, the combined survey covers most of the available sky from Arizona and Australia, with a few sets of exposures.

Optical transients (OTs) and highly variable objects (HVOs) are detected as sources that display significant changes in brightness, as compared to the baseline comparison images and catalogs with significantly fainter limiting magnitudes; these are obtained by median stacking of >20 previous observations and reach ~2 mag deeper. The search is performed in the catalog domain, but an image subtraction pipeline is also being developed. Data cover time baselines from 10 min to several years.

CRTS practices an *open data* philosophy: detected transients are published electronically in real time, using a variety of mechanisms, so that they can be followed up by anyone. Archival light curves for ~500 million objects and all of the images are also being made public.

The *All Sky Automated Survey* (ASAS; http://www.astrouw.edu.pl/asas/; Pojmanski 1997) covers the entire sky using a set of small telescopes (telephoto lenses) at Las Campanas Observatory in Chile and Haleakala, Maui, in V and I bands. The limiting magnitude are $V \sim 14$ mag and $I \sim 13$ mag, with ~14 arcsec/pixel sampling . The *ASAS-3 Photometric V-band Catalogue* contains over 15 million light curves; a derived product is the online *ASAS Catalog of Variable Stars* (ACVS), with ~50,000 selected objects.

4.2.2 Supernova Surveys

Supernova (SN) searches have been providing science drivers for a number of surveys. We recall that the original Zwicky survey from Palomar was motivated by the need to find more SNe for the follow-up studies. In addition to those described in Sect. 4.2.1, some of the dedicated SN surveys include:

The *Calan/Tololo Supernova Search* (Hamuy et al. 1993) operated from 1989 to 1993 and provided some of the key foundations for the subsequent use of type Ia SNe as standard candle, which eventually led to the evidence for dark energy.

The *Supernova Cosmology Project* (SCP, Perlmutter et al. 1998, 1999) was designed to measure the expansion of the universe using high-redshift type Ia SNe as standard candles. Simultaneously, the *High-z Supernova Team* (Filippenko and Riess 1998) was engaged in the same pursuit. While the initial SCP results were inconclusive (Perlmutter et al. 1997), in 1998 both experiments discovered that the expansions of the universe was accelerating (Perlmutter et al. 1998; Riess et al. 1998), thus providing a key piece of evidence for the existence of the dark energy (DE). These discoveries are documented and discussed very extensively in the literature.

The *Equation of State: SupErNovae trace Cosmic Expansion* project (ESSENCE; http://www.ctio.noao.edu/essence) was a SN survey optimized to constrain the equation of state (EoS) of the DE, using type Ia SNe at redshifts $0.2 < z < 0.8$. The goal of the project was to determine the value of the EoS parameter to within 10%. The survey used 30 half-nights per year with the CTIO Blanco 4 m telescope and the MOSAIC camera between 2003 and 2007 and discovered 228 type Ia SNe. Observations were made in the *VRI* passbands with a cadence of a few days. SNe were discovered using an image subtraction pipeline, with candidates inspected and ranked for follow-up by eye.

Similarly, the *Supernova Legacy Survey* (SNLS; http://www.cfht.hawaii.edu/SNLS) was designed to precisely measure several hundred type Ia SNe at redshifts $0.3 < z < 1$ or so (Astier et al. 2006). Imaging was done at the Canada-France-Hawaii Telescope (CFHT) with the *Megaprime* 340 Megapixel camera, providing a 1 deg^2 FOV, in the course of 450 nights, between 2003 and 2008. About 1,000 SNe were discovered and followed up. Spectroscopic follow-up was done using the 8–10 m class telescopes, resulting in ~500 confirmed SNe.

The *Lick Observatory Supernova Search* (LOSS; Filippenko et al. 2001; http://astro.berkeley.edu/bait/public_html/kait.html) is an ongoing search for low-z ($z < 0.1$) SNe that began in 1998. The experiment makes use of the 0.76 m Katzman Automatic Imaging Telescope (KAIT). Observations are taken with a 500×500 pixel unfiltered CCD, with an FOV of 6.7×6.7 arcmin^2, reaching R ~ 19 mag in 25 s (Li et al. 1999). As of 2011, the survey made 2.3 million observations and discovered a total of 874 SNe by repeatedly monitoring ~15,000 large nearby galaxies

(Leaman et al. 2011). The project is designed to tie high-z SNe to their more easily observed local counterparts and also to rapidly respond to gamma-ray burst (GRB) triggers and monitor their light curves.

The *SDSS-II Supernova Search* was designed to detect type Ia SNe at intermediate redshifts ($0.05 < z < 0.4$). This experiment ran 3-month campaigns from September to November during 2005 to 2007 by doing repeated scans of the SDSS Stripe 82 (Sako et al. 2008). Detection were made using image subtraction, and candidate type Ia SNe were selected using the SDSS five-bandpass imaging. The experiment resulted in the discovery and spectroscopic confirmation of ~500 type Ia SNe and 80 core-collapse SNe. Although the survey was designed to discover SNe, the data have been made public, enabling the study of numerous types of variables objects including variables star and AGN (Sesar et al. 2007; Bhatti et al. 2010).

Many other surveys and experiments have contributed to this vibrant field. Many amateur astronomers have also contributed; particularly noteworthy is the Puckett Observatory World Supernova Search (http://www.cometwatch.com/search.html) that involves several telescopes and B. Monard's Supernova search.

4.2.3 Synoptic Surveys for Minor Bodies in the Solar System

The discovery that NEO events were potentially hazardous to human life motivated programs to discover the extent and nature of these objects, in particular the largest ones which could yield a significant destruction, the potentially hazardous asteroids (PHAs). A number of NEO surveys began working with NASA in order to fulfill the congressional mandate to discover all NEOs larger than 1 km in diameter. The number of NEOs discovered each year continues to increase, as surveys increase in sensitivity, with the current rate of discovery of ~1,000 NEOs per year.

The search for NEOs (near-earth objects) with CCDs largely began with the discovery of the asteroid 1989 UP by the *Spacewatch* survey (http://spacewatch.lpl.arizona.edu). They also made the first automated NEO detection of the asteroid 1990 SS. The survey uses the Steward Observatory 0.9 m telescope and a camera covering 2.9 deg2. A 1.8 m telescope with an FOV or 0.8 deg2 was commissioned in 2001. Until 1997, Spacewatch was the major discoverer of NEOs, in addition to a number of trans-Neptunian objects (TNO).

The *Near-Earth Asteroid Tracking* (NEAT; http://neat.jpl.nasa.gov) began in 1996 and ran until 2006. Initially, the survey used a three-CCD camera mounted on the Samuel Oschin 48-in. Schmidt telescope on Palomar Mountain, until it was replaced by the PQ survey's 160 Megapixel camera in 2003. In addition, in 2000 the project began using data from the Maui Space Surveillance Site 1.2 m telescope. A considerable number of NEOs was found by this survey. The work was combined with Mike Brown's search for dwarf planets and resulted in a number of such discoveries, also causing the demotion of Pluto to a dwarf planet status. The data were also used by the PQ survey team for a SN search.

The *Lincoln Near-Earth Asteroid Research* survey, conducted by the MIT Lincoln Laboratory (LINEAR; http://www.ll.mit.edu/mission/space/linear), began in 1997 using four 1 m telescopes at the White Sands Missile Range in New Mexico (Viggh et al. 1997). The survey has discovered ~2,400 NEOs to date, among 230,000 other asteroids and comet discoveries.

The *Lowell Observatory Near-Earth-Object Search* (LONEOS; http://asteroid.lowell.edu/asteroid/loneos/loneos.html) ran from 1993 to 2008. The survey used a 0.6-m Schmidt telescope with an 8.3-deg FOV, located near Flagstaff, Arizona. It discovered 289 NEOs.

The *Catalina Sky Survey* (CSS; http://www.lpl.arizona.edu/css) began in 1999 and was extensively upgraded in 2003. It uses the telescopes and produces the data stream that is also used by the CRTS survey, described above. It became the most successful NEO discovery survey in 2004, and it has so far discovered ~70% of all known NEOs. On October 6, 2008, a small NEO, 2008 TC3, was discovered by their 1.5-m telescope. The object was predicted to impact the Earth within 20 h of discovery. The asteroid disintegrated in the upper atmosphere over the Sudan desert; 600 fragments were recovered on the ground, amounting to "an extremely low cost sample return mission." This was the first time that an asteroid impact on Earth has been accurately predicted.

The Pan-STARRS survey, discussed above, also has a significant NEO search component. Another interesting concept was proposed by Tonry (2011).

Recently, the *NEOWISE* program (a part of the WISE mission) discovered 130 NEOs, among a much larger number of asteroids.

One major lesson of the PQ survey was that a joint asteroid/transient analysis is necessary, since slowly moving asteroids represented a major "contaminant" in a search for astrophysical transients. A major lesson of the Catalina surveys is that the same data streams can be shared very effectively between the NEO and transient surveys, thus greatly increasing the utility of the shared data.

4.2.4 Microlensing Surveys

The proposition of detecting and measuring dark matter in the form of *Massive Compact Halo Objects* (MACHOs) using gravitational microlensing was first proposed by Paczynski (1986). The advent of large format CCDs made the measurements of millions of stars required to detect gravitation lensing a possibility in the crowded stellar fields toward the Galactic bulge and the Magellanic Clouds. Three surveys began searching for gravitation microlensing, and the MACHO and EROS (*Expèrience pour la Recherche d'Objets Sombres*) groups simultaneously announced the discovery of events toward the LMC in 1993 (Alcock et al. 1993; Aubourg et al. 1993). Meanwhile, the *Optical Gravitational Lensing Experiment* (OGLE) collaboration searched and found the first microlensing toward the Galactic bulge (Udalski et al. 1993). All three surveys continued to monitor ~100 deg^2 in fields toward the Galactic bulge, and discovered hundreds of microlensing events. A similar area was covered toward the Large Magellanic Cloud, and a dozen events were discovered. This result limited the contribution of MACHO to halo dark matter to 20% (Alcock et al. 2000).

It was also predicted that the signal of planets could be detected during microlensing events (Mao and Paczynski 1991). Searches for such events were undertaken by the *Global Microlensing Alert Network* (GMAN) and PLANET (*Probing Lensing Anomalies NETwork*) collaborations beginning in 1995. The first detection of planetary lensing was in 2003, when the microlensing event OGLE-2003-BLG-235 was found to harbor a Jovian mass planet (Bond et al. 2004). Udalski et al. (2002) used the OGLE-III camera and telescope to perform a dedicated search for transits and discovered the first large sample of transiting planets.

Additional planetary microlensing event surveys such as *Microlensing Follow-Up Network* (MicroFun) and *RoboNet-I,II* (Mottram and Fraser 2008; Tsapras et al. 2009) continue to find planets by following microlensing events discovered by OGLE-IV and MOA (*Microlensing Observations in Astrophysics*; Yock et al. 2000) surveys, which discover hundreds of microlensing events toward Galactic bulge each year. Additional searches to further quantify the amount of matter in MACHOs have been undertaken in microlensing by AGAPE (Ansari et al. 1997) toward M31 and the LMC by the SuperMACHO project (http://www.ctio.noao.edu/supermacho/lightechos/SM/sm.html).

In addition to the microlensing events, the surveys revealed the presence of tens of thousands of variable stars and others sources, as also predicted by Paczynski (1986). The data collected by these surveys represent a gold mine of information on variable stars of all kinds.

4.2.5 Radio Synoptic Surveys

There has been a considerable interest recently in the exploration of the time domain in the radio regime. Ofek et al. (2011) reviewed the transient survey literature in the radio regime. The new generation of radio surveys and facilities is enabled largely by the advances in the radio electronics (themselves driven by the mobile computing and cell phone demands), as well as the ICT. These are reflected in better receivers and correlators, among other elements, and since much of this technology is off the shelf, the relatively low cost of individual units enables production of large new radio facilities that collect vast quantities of data. Some of them include:

The *Allen Telescope Array* (ATA; http://www.seti.org/ata) consists of forty-two 6 m dishes covering a frequency range of 0.5–11 GHz. It has a maximum baseline of 300 m and a field of view of 2.5 deg at 1.4 GHz. A number of surveys have been conducted with the ATA. The *Fly's Eye Transient* search (Siemion et al. 2012) was a 500-h survey over 198 square degrees at 209 MHz with all dishes pointing in different directions, looking for powerful millisecond timescale transients. The *ATA Twenty-Centimeter Survey* (ATATS; Croft et al. 2010, 2011) was a survey at 1.4 GHz constructed from 12 epoch visits over an area of ~700 deg^2 for rare bright transients and to prove the wide-field capabilities of the ATA. Its catalog contains 4,984 sources above 20 mJy (>90% complete to ~40 mJy) with positional accuracies better than 20 arcsec and is comparable to the NVSS. Finally the *Pi GHz Sky Survey* (PiGSS; Bower et al. 2010, 2011) is a 3.1-GHz radio continuum survey of ~250,000 radio sources in the 10,000 deg^2 region of sky with b >30° down to ~1 mJy with each source being observed multiple times. A subregion of ~11 deg^2 will also be repeatedly observed to characterize variability on timescales of days to years.

The Low Frequency Radio Array (LOFAR; http://www.lofar.org/) comprises 48 stations, each consisting of two types of dipole antenna: 48/96 LBA covering a frequency range of 10–90 MHz and 48/96 4 × 4 HBA tiles covering 110–240 MHz. There is a dense core of 6 stations, 34 more distributed throughout the Netherlands and 8 elsewhere in Europe (UK, France, Germany). LOFAR can simultaneously operate up to 244 independent beams and observe sources at Dec >−30. A recent observation reached 100 μJy (L-band) over a region of ~60 deg^2 comprising ~10,000 sources. A number of surveys are planned including all-sky surveys at 15, 30, 60, and 120 MHz and ~1,000 deg^2 at 200 MHz to study star, galaxy, and large-scale structure formation in the early universe and intercluster magnetic fields and explore new regions of parameter space for serendipitous discovery, probing the epoch of reionization and zenith and Galactic plane transient monitoring programs at 30 and 120 MHz.

The field is largely driven toward the next-generation radio facility, the Square Kilometer Array (SKA), described below. Two prototype facilities are currently being developed:

One of the SKA precursors is the *Australian Square Kilometer Array Pathfinder* (ASKAP; http://www.atnf.csiro.au/projects/askap), comprising of thirty-six 12 m antennae covering a frequency range of 700–1,800 MHz. It has an FOV of 30° and a maximum baseline of 6 km.

In 1 h it will be able to reach 30 μJy per beam with a resolution of 7.5 arcsec. An initial construction of six antennae (BETA) is due for completion in 2012. The primary science goals of ASKAP include galaxy formation in the nearby universe through extragalactic HI surveys (WALLABY, DINGO); formation and evolution of galaxies across cosmic time with high-resolution, confusion-limited continuum surveys (EMU, FLASH); characterization of the radio transient sky through detection and monitoring of transient and variable sources (CRAFT, VAST); and the evolution of magnetic fields in galaxies over cosmic time through polarization surveys (POSSUM).

MeerKAT (http://www.ska.ac.za/meerkat/index.php) is the South African SKA pathfinder due for completion in 2016. It consists of 64 13.5 m dishes operating at ~1 GHz and 8–14.5 GHz with a baseline between 29 m and 20 km. It complements ASKAP with a larger frequency range and greater sensitivity but a smaller field of view. It has enhanced surface brightness sensitivity with its shorter and longer baselines and will also have some degree of phased element array capability. The primary science drivers cover the same type of SKA precursor science area as ASKAP, but MeerKAT will focus on those areas where its has unique capabilities – these include extremely sensitive studies of neutral hydrogen in emission to z ~ 1.4 and highly sensitive continuum surveys to μJy levels at frequencies as low as 580 MHz. An initial test bed of seven dishes (KAT-7) have been constructed and is now being used as an engineering and science prototype.

4.2.6 Other Wavelength Regimes

Given that the variability is common to most point sources at high energies and the fact that the detectors in this regime tend to be photon counting with a good time resolution, surveys in that regime are almost by definition synoptic in character.

An excellent example of a modern high-energy survey is the *Fermi Gamma-Ray Space Telescope* (FGST; http://www.nasa.gov/fermi), launched in 2008, that is continuously monitoring the γ-ray sky. Its Large Area Telescope (LAT) is a pair-production telescope that detects individual photons with a peak effective area of ~8,000 cm^2 and a field of view of ~2.4 sr. The observations can be binned into different time resolutions. The stable response of LAT and a combination of deep and fairly uniform exposure produces an excellent all-sky survey in the 100 MeV–100 GeV range to study different types of γ-ray sources. This has resulted in a series of ongoing source catalog releases (Abdo et al. 2009, 2010, and more to come). The current source count approaches 2,000, with ~60% of them identified on other wavelengths.

4.3 Toward the Petascale Data Streams and Beyond

The next generation of sky surveys will be largely synoptic in nature and will move us firmly to the Petascale regime. We are already entering it with the facilities like PanSTARRS, ASKAP, and LOFAR. These new surveys and instruments not only depend critically on the ICT but also push it to a new performance regime.

The *Large Synoptic Sky Survey* (LSST; Tyson et al. 2002; Ivezic et al. 2009; http://lsst.org) is a wide-field telescope that will be located at Cerro Paranal in Chile. The primary mirror will be 8.4 m in diameter, but because of the 3.4 m secondary, the collecting area is equivalent to that of a ~6.7 m telescope. While its development is still in the relatively early stages as of this writing, the project has a very strong community support, reflected in its top ranking in the 2010 Astronomy and Astrophysics Decadal Survey produced by the US National Academy of Sciences, and it may become operational before this decade is out. The LSST is planned to produce a 6-bandpass (0.3–1.1 μm) wide-field deep astronomical survey of over 20,000 deg^2 of the Southern sky, with many epochs per field. The camera will have a ~3.2 Gigapixel detector array covering ~9.6 deg^2 in individual exposures, with 0.2 arcsec pixels.

LSST will take more than 800 panoramic images each night, with 2 exposures per field, covering the accessible sky twice each week. The data (images, catalogs, alerts) will be continuously generated and updated every observing night. In addition, calibration and coadded images, and the resulting catalogs, will be generated on a slower cadence and used for data quality assessments. The final source catalog is expected to have more than 20 billion rows, comprising 30 TB of data per night, for a total of 60 PB over the envisioned duration of the survey. Its scientific goals and strategies are described in detail in the *LSST Science Book* (Ivezic et al. 2009). Processing and analysis of this huge data stream pose a number of challenges in the arena of real-time data processing, distribution, archiving, and analysis.

The currently ongoing renaissance in the continuum radio astronomy at cm and m scale wavelengths is leading toward the facilities that will surpass even the data rates expected for the LSST, and that will move us to the Exascale regime.

The *Square Kilometer Array* (SKA; http://skatelescope.org) will be the world's largest radio telescope, hoped to be operational in the mid-2020s. It is envisioned to have a total collecting area of approximately one million m^2 (thus the name). It will provide continuous frequency coverage from 70 MHz to 30 GHz, employing phased arrays of dipole antennas (low frequency), tiles (mid frequency), and dishes (high frequency) arranged over a region extending out to ~3,000 km. Its key science projects will focus on studying pulsars as extreme tests of general relativity, mapping a billion galaxies to the highest redshifts by their 21-cm emission to understand galaxy formation and evolution and dark matter, observing the primordial distribution of gas to probe the epoch of reionization, investigating cosmic magnetism, surveying all types of transient radio phenomena, and so on.

The data processing for the SKA poses significant challenges, even if we extrapolate Moore's law to its projected operations. The data will stream from the detectors into the correlator at a rate of ~4.2 PB/s, and then from the correlator to the visibility processors at rates between 1 and 500 TB/s depending on the observing mode, which will require processing capabilities of ~200 Pflops – 2.5 Eflops. Subsequent image formation needs ~10 Pflops to create data products (~0.5–10 PB/day), which would be available for science analysis and archiving, the total computational costs of which could easily exceed those of the pipeline. Of course, this is not just a matter of hardware provision, even if it is built for a special purpose, but also high computational complexity algorithms for wide-field imaging techniques, deconvolution, Bayesian source finding, and other tasks. Each operation will also place different constraints on the computational infrastructure with some being memory bound and some CPU bound that will need to be optimally balanced for maximum throughput. Finally, the power required for all this processing will also need to be addressed – assuming the current trends, the SKA data processing will consume energy at a rate of ~1 GW. These are highly nontrivial hardware and infrastructure challenges.

The real job of science, data analysis, and knowledge discovery starts after all this processing and delivery of processed data to the archives. Effective, scalable software and methodology needed for these tasks does not exist yet, at least in the public domain.

4.4 Deep Field Surveys

A counterpart to the relatively shallow, wide-field surveys discussed above are various deep fields that cover much smaller, selected areas, generally based on an initial deep set of observations by one of the major space-based observatories. Their scientific goals are almost always in the arena of galaxy formation and evolution: the depth is required in order to obtain adequate measurements of very distant ones, and the areas surveyed are too small to use for much of the Galactic structure work. This is a huge and vibrant field of research, and we cannot do it justice in the limited space here; we describe very briefly a few of the more popular deep surveys. More can be found, for example, in the reviews by Ferguson et al. (2000), Bowyer et al. (2000), Brandt and Hassinger (2005), and the volume edited by Cristiani et al. (2002).

The prototype of these is the *Hubble Deep Field* (HDF; Williams et al. 1996; http://www.stsci.edu/ftp/science/hdf/hdf.html). It was imaged with the HST in 1995 using the Director's Discretionary Time, with 150 consecutive HST orbits and the WFPC2 instrument, in four filters ranging from near-UV to near-IR, F300W, F450W, F606W, and F814W. The ~2.5 arcmin field is located at RA = 12 : 36 : 49.4, Dec = +62 : 12 : 58 (J2000); a corresponding HDF-South, centered at RA = 22 : 32 : 59.2, Dec = −60 : 33 : 02.7 (J2000) was observed in 1998. An even deeper *Hubble Ultra Deep Field* (HUDF; Beckwith et al. 2006), centered at RA = 03 : 32 : 39.0, Dec = −27 : 47 : 29 (J2000) and covering 11 $arcmin^2$, was observed with 400 orbits in 2003–2004, using the ACS instrument in 4 bands: F435W, F606W, F775W, and F850LP, and with 192 orbits in 2009 with the WFC3 instrument in 3 near-IR filters, F105W, F125W, and F160W. Depths equivalent to V ~30 mag were reached. There has been a large number of follow-up studies of these deep fields, involving many ground-based and space-based observatories, mainly on the subjects of galaxy formation and evolution at high redshifts.

The *Groth Survey Strip* field is a 127 $arcmin^2$ region that has been observed with the HST in both broad V and I bands and a variety of ground-based and space-based observatories. It was expanded to the *Extended Groth Strip* in 2004–2005, covering 700 $arcmin^2$ using 500 exposures with the ACS instrument on the HST. The *All-Wavelength Extended Groth Strip International Survey* (AEGIS; Davis et al. 2007) continues to study this at multiple wavelengths.

The *Chandra Deep Field South* (CDF-S; http://www.eso.org/~vmainier/cdfs_pub/; Giacconi et al. 2001) was obtained by the *Chandra* X-ray telescope looking at the same patch of the sky for 11 consecutive days in 1999–2000, for a total of one million seconds. The field covers 0.11 deg^2 centered at RA = 03 : 32 : 28.0, Dec = −27 : 48 : 30 (J2000). These X-ray observations were followed up extensively by many other observatories.

The *Great Observatories Origins Deep Survey* (GOODS; http://www.stsci.edu/science/goods/; Dickinson et al. 2003; Giavalisco et al. 2004) builds on the HDF-N and CDF-S by targeting fields centered on those areas and covers approximately 0.09 deg^2 using a number of NASA and ESA space observatories: *Spitzer, Hubble, Chandra, Herschel, XMM-Newton*, etc., as well as a number of deep ground-based studies and redshift surveys. The main goal is to study the distant universe to faintest fluxes across the electromagnetic spectrum, with a focus on the formation and evolution of galaxies.

The *Subaru Deep Field* (http://www.naoj.org/; Kashikawa et al. 2004) was observed initially over 30 nights at the 8.2 m Subaru telescope on Mauna Kea, using the *SupremeCam* instrument. The field, centered at RA = 13 : 24 : 38.9, Dec = +27 : 29 : 26 (J2000), covers a patch of 34 by 27 $arcmin^2$. It was imaged in the *BVRi'z'* bands and narrow bands centered at 8,150 and 9,196 Å. One of the main aims was to catalog Lyman-break galaxies out to large redshifts and get samples of Lyα emitters (LAE) as probes of the very early galaxy evolution. Over 200,000 galaxies were detected, yielding samples of hundreds of high-redshift galaxies. These were followed by extensive spectroscopy and imaging on other wavelengths.

The *Cosmological Evolution Survey* (COSMOS; http://cosmos.astro.caltech.edu; Scoville et al. 2007) was initiated as an HST Treasury Project, but it expanded to include deep observation from a range of facilities, both ground-based and space-based. The 2 deg^2 field is centered at RA = 10 : 00 : 28.6, Dec = +02 : 12 : 21.0 (J2000). The HST observations were carried out in 2004–2005. These were followed by a large number of ground-based and space-based observatories. The main goals were to detect two million objects down to the limiting magnitude $I_{AB} > 27$ mag, including >35,000 Lyman-break galaxies and extremely red galaxies out to $z \sim 5$. The follow-up included an extensive spectroscopic and imaging on other wavelengths.

The Cosmic Assembly Near-IR Deep Extragalactic Legacy Survey (CANDELS; 2010–2013; http://candels.ucolick.org/; Koekemoer et al. 2011) uses deep HST imaging of over 250,000 galaxies with WFC3/IR and ACS instruments to study galactic evolution at $1.5 < z < 8$. The surveys is at 2 depths: moderate (2 orbits) over 0.2 deg^2 and deep (12 orbits) over 0.04 deg^2. An additional goal is to refine the constraints on time variation of cosmic-equation of state parameter leading to a better understanding of the dark energy.

A few ground-based deep surveys (in addition to the follow-up of those listed above) are also worthy of note. They include:

The *Deep Lens Survey* (DLS; 2001–2006; Wittman et al. 2002; http://dls.physics.ucdavis.edu) covered ~20 deg^2 spread over five fields, using CCD mosaics at KPNO and CTIO 4 m telescopes, with 0.257 arcsec/pixel, in *BVRz'* bands. The survey also had a significant synoptic component (Becker et al. 2004).

The *NOAO Deep Wide Field Survey* (NDWFS; http://www.noao.edu/noao/noaodeep; Jannuzi and Dey 1999) covered ~18 deg^2 spread over two fields, also using CCD mosaics at KPNO and CTIO 4 m telescopes, in $B_W RI$ bands, reaching point source limiting magnitudes of 26.6, 25.8, and 25.5 mag, respectively. These fields have been followed up extensively at other wavelengths.

CFHT Legacy Survey (CFHTLS; 2003–2009; http://www.cfht.hawaii.edu/Science/CFHLS) is a community-oriented service project that covered 4 deg^2 spread over four fields, using the MegaCam imager at the CFHT 3.6 m telescope, over an equivalent of 450 nights, in $u^* g' r' i' z'$ bands. The survey served a number of different projects.

Deep fields have had a transformative effect on the studies of galaxy formation and evolution. They also illustrated a power of multi-wavelength studies, as observations in each regime leveraged each other. Finally, they also illustrated a power of the combination of deep space-based imaging, followed by deep spectroscopy with large ground-based telescopes.

Many additional deep or semideep fields have been covered from the ground in a similar manner. We cannot do justice to them or to the entire subject in the limited space available here, and we direct the reader to a very extensive literature that resulted from these studies.

4.5 Spectroscopic Surveys

While imaging tends to be the first step in many astronomical ventures, physical understanding often comes from spectroscopy. Thus, spectroscopic surveys naturally follow the imaging ones. However, they tend to be much more focused and limited scientifically: there are seldom surprising new uses for the spectra beyond the original scientific motivation.

Two massive, wide-field redshift surveys that have dramatically changed the field used highly multiplexed (multi-fiber or multi-slit) spectrographs; they are the 2dF and SDSS. A few other significant previous redshift surveys were mentioned in ➋ Sect. 2.

Several surveys were done using the 2-deg field (2dF) multi-fiber spectrograph at the Anglo-Australian Telescope (AAT) at Siding Spring, Australia. The *2dF Galaxy Redshift Survey* (2dFGRS; 1997–2002; Colless 1999; Colless et al. 2001) observed ~250,000 galaxies down to $m_B \approx 19.5$ mag. The *2dF Quasar Redshift Survey* (2QZ; 1997–2002; Croom et al. 2004) observed ~25,000 quasars down to $m_B \approx 21$ mag. The spectroscopic component of the SDSS eventually produced redshifts and other spectroscopic measurements for ~930,000 galaxies and ~120,000 quasars, and nearly 500,000 stars and other objects. These massive spectroscopic surveys greatly expanded our knowledge of the LSS and the evolution of quasars and led to numerous other studies.

Deep imaging surveys tend to be followed by the deep spectroscopic ones, since redshifts are essential for their interpretation and scientific uses. Some of the notable examples include:

The *Deep Extragalactic Evolution Probe* (DEEP; Vogt et al. 2005; Davis et al. 2003) survey is a two-phase project using the 10 m Keck telescopes to obtain spectra of faint galaxies over ~3.5 deg^2 to study the evolution of their properties and evolution of clustering out to $z \sim 1.5$. Most of the data were obtained using the DEIMOS multi-slit spectrograph, with targets preselected using photometric redshifts. Spectra were obtained for ~50,000 galaxies down to ~24 mag, in the redshift range $z \sim 0.7$–1.55 with candidates, with ~80% yielding redshifts.

The *VIRMOS-VLT Deep Survey* (VVDS; Le Fevre et al. 2005) is an ongoing comprehensive imaging and redshift survey of the deep universe complementary to the Keck DEEP survey. An area of ~16 deg^2 in four separate fields is covered using the VIRMOS multi-object spectrograph on the 8.2 m VLT telescopes. With 10 arcsec slits, spectra can be measured for 600 objects simultaneously. Targets were selected from a *UBRI* photometric survey (limiting magnitude $I_{AB} = 25.3$ mag) carried out with the CFHT. A total of over 150,000 redshifts will be obtained (~100,000 to $I_{AB} = 22.5$ mag, ~50,000 to $I_{AB} = 24$ mag, and ~1,000 to $I_{AB} = 26$ mag) providing insight into galaxy and structure formation over a very broad redshift range, $0 < z < 5$.

zCOSMOS (Lilly et al. 2007) is a similar survey on the VLT using VIRMOS but only covering the 1.7 deg^2 COSMOS ACS field down to a magnitude limit of $I_{AB} < 22.5$ mag. Approximately 20,000 galaxies were measured over the redshift range $0.1 < z < 1.2$, comparable in survey parameters to the 2dFGRS, with a further 10,000 within the central 1 deg^2, selected with photometric redshifts in the range $1.4 < z < 3.0$.

The currently ongoing *Baryon Oscillation Spectroscopic Survey* (BOSS), due to be completed in early 2014 as part of SDSS-III, will cover an area of ~10,000 deg^2 and map the spatial distribution of ~1.5 million luminous red galaxies to $z = 0.7$ and absorption lines in the spectra of ~160,000 quasars to $z \sim 3.5$. Using the characteristic scale imprinted by the baryon acoustic oscillation in the early universe in these distributions as a standard ruler, it will be able to determine the angular diameter distance and the cosmic expansion rate with a precision of ~1 to 2% in order to constrain theoretical models of dark energy.

Its successor, the *BigBOSS* survey (http://bigboss.lbl.gov) will utilize a purpose-built 5,000 fiber muti-object spectrograph to be mounted at the prime focus of the KPNO 4 m Mayall telescope, covering a 3° FOV. The project will conduct a series of redshift surveys over ~500 nights spread over 5 years, with a primary goal of constraining models of dark energy, using different observational tracers: clustering of galaxies out to $z \sim 1.7$ and Lyα forest lines in the spectra of quasars at $2.2 < z < 3.5$. The survey plans to obtain spectra of ~20 million galaxies and ~600,000 quasars over a 14,000 deg^2 area in order to reach these goals.

Another interesting, incipient surveys is the *Large Sky Area Multi-Object Fiber Spectroscopic Telescope* (LAMOST; http://www.lamost.org/website/en), one of the National Major Scientific Projects undertaken by the Chinese Academy of Science. A custom-built 4 m telescope has a 5° FOV, accommodating 4,000 optical fibers for a highly multiplexed spectroscopy.

A different approach uses panoramic, slitless spectroscopy surveys, where a dispersing element (an objective prism, grating, or a grism – grating ruled on a prism) is placed in the front of the telescope optics, thus providing wavelength-dispersed images for all objects. This approach tends to work only for the bright sources in ground-based observations, since any given pixel gets the signal from the object only at the corresponding wavelength, but the sky background that dominates the noise has contributions from all wavelengths. For space-based observations, this is much less of a problem. Crowding can also be a problem due to the overlap of adjacent spectra. Traditionally, slitless spectroscopy surveys have been used to search for emission-line objects (e.g., star-forming or active galaxies and AGN) or objects with a particular continuum signature (e.g., low metallicity stars).

One good example is the set of surveys done at the Hamburg Observatory, covering both the Northern sky from the Calar Alto observatory in Spain and the Southern sky from ESO (http://www.hs.uni-hamburg.de/EN/For/Exg/Sur) that used a wide-angle objective prism survey to look for AGN candidates, hot stars, and optical counterparts to ROSAT X-ray sources. For each $5.5 \times 5.5\ \mathrm{deg}^2$ field in the survey area, two objective prism plates were taken as well as an unfiltered direct plate to determine accurate positions and recognize overlaps. All plates were subsequently digitized and, under good seeing conditions, a spectral resolution of 45A at Hγ was be achieved. Candidates selected from these slitless spectra were followed by targeted spectroscopy, producing ~500 AGN and over 2,000 other emission-line sources, and a large number of extremely metal-poor stars.

4.6 Figures of Merit

Since sky surveys cover so many dimensions of the OPS, but only a subset of them, comparisons of surveys can be difficult if not downright meaningless. What really matters is a scientific discovery potential, which very much depends on what are the primary science goals; for example, a survey optimized to find supernovae may not be very efficient for studies of galaxy clusters, and vice versa. However, for just about any scientific application, area coverage and depth, and, in the case of synoptic sky surveys, also the number of epochs per field are the most basic characteristics and can be compared fairly, at least in any given wavelength regime. Here we attempt to define some general indicators of the scientific discovery potential for sky surveys based on these general parameters.

Often, but misleadingly, surveys are compared using the *etendue*, the product of the telescope area, A, and the solid angle subtended by the individual exposures, Ω. However, $A\Omega$ simply reflects the properties of a telescope and the instrument optics. It implies nothing whatsoever about a survey that may be done using that telescope and instrument combination, as it does not say anything about the depth of the individual exposures, the total area coverage, and the number of exposures per field. The $A\Omega$ is the same for a single, short exposure, and for a survey that has thousands of fields, several bandpasses, deep exposures, and hundreds of epochs for each field. Clearly, a more appropriate figure of merit (FoM), or a set thereof, is needed.

We propose that a fair measure of a general scientific potential of a survey would be a product of a measure of its average depth and its coverage of the relevant dimensions of the OPS or how deep, how wide, how often, and for how long.

As a quantitative measure of depth in an average single observation, we can define a quantity that is roughly proportional to the S/N for background-limited observations for an unresolved source, namely:

$$D = [A \times t_{exp} \times \varepsilon]^{1/2} / FWHM$$

where A is the effective collecting area of the telescope in m^2, t_{exp} is the typical exposure length in sec, ε is the overall throughput efficiency of the telescope+instrument, and $FWHM$ is the typical PSF or beam size full-width at half-maximum (i.e., the seeing, for the ground-based visible and NIR observations) in arcsec.

The coverage of the OPS depends on the number of bandpasses (or, in the case of radio or high energy surveys, octaves or, in the case of spectroscopic surveys, the number of independent resolution elements in a typical survey spectrum), N_b, and the total survey area covered, $\Omega_{\rm tot}$, expressed as a fraction of the entire sky. In the case of synoptic sky surveys, additional relevant parameters include the area coverage rate regardless of the bandpass (in $\deg^2$/night, or as a fraction of the entire sky per night), $R = d\Omega/dt$, the number of exposures per field per night regardless of the bandpass, N_e, or, for the survey as a whole, the average total number of visits per field in a given bandpass, N_{avg}. For a single-epoch imaging survey, $N_{avg} = 1$. The total number of all exposures for a given field, regardless of the bandpass is $N_{tot} = N_b \times N_{avg}$. The coverage along the time axis of the OPS is roughly proportional to N_{tot}.

Thus, we define the Scientific Discovery Potential (*SDP*) FoM as:

$$SDP = D \times \Omega_{\rm tot} \times N_b \times N_{avg}$$

It probably makes little sense to compare single-pass and synoptic surveys using this metric: single-pass surveys are meaningfully compared mainly by the depth, the area coverage, and the number of bandpasses, whereas for the synoptic sky surveys the number of passes per field matters a lot since their focus is on the exploration of the time domain.

These FoMs pertain to the survey as a whole and are certainly applicable for an archival research. If what matters is a discovery *rate* of transient events *(TDR)*, we define another FoM:

$$TDR = D \times R \times N_e$$

That is, the area covered in a given night is observed N_e times, in any bandpass. Note: *TDR* is not the actual number of transients per night, as that depends on many other factors, but it should be roughly proportional to it. Obviously, the longer one runs the survey, the more transient events are found.

Both of these indicators are meaningful for the imaging surveys. For a spectroscopic (targeted) survey, a useful FoM may be a product of the depth parameter, D, and the number of targets covered.

These FoMs are in some sense the bare minimum of information needed to characterize or compare surveys. They do not account for things like the sky background and transparency, the total numbers of sources detected (which clearly depends strongly on the Galactic latitude), the width of the bandpasses (wavelength resolution), the dynamical range of the data, the quality of the calibration, instrumental noise and artifacts, the angular resolution, the uniformity of coverage as a function of both position on the sky and time, and, in the case of synoptic sky surveys, the cadences. These FoMs also do not account for the operational parameters such as the data availability and access, the time delay between the observations and the event publishing, etc., all of which affect the science produced by the surveys.

They also implicitly assume that surveys in the same wavelength regime and with the same type of data, for example, direct imaging, are being compared. A comparison of surveys at different wavelengths makes sense only in some specific scientific context. For example, large collecting areas of radio telescopes do not make them that much more scientifically effective than, say, X-ray telescopes, with their small effective collecting areas. A fair comparison of surveys with different types of data, for example, images and spectra, would be even more difficult and context dependent.

Another approach is to compare the hypervolumes covered by the surveys in some subset of the OPS. For example, we can define the 4-dimensional survey hypervolume (*SHV*) as:

$$SHV = \Omega_{\text{tot}} \times \log(f_{max}/f_{min}) \times (\lambda_{max}/\lambda_{min}) \times N_{tot}$$

where Ω_{tot} is the total sky area covered, now expressed as a fraction of the entire sky; (f_{max}/f_{min}) is the dynamical range of flux measurements, with f_{min} being the limiting flux for significant detections and f_{max} the saturation level; $(\lambda_{max}/\lambda_{min})$ is the dynamical range of wavelengths covered; and N_{tot} is the mean number of exposures per field; $N_{tot} = N_b$ for the single-pass surveys. Thus, we are carving out a hypervolume along the spatial, flux, wavelength, and possibly also temporal axes. A further refinement would be to divide each of the terms in this product by the resolution of measurements on the corresponding axis. That would give a number of the independent *SHV* elements in the survey.

The *SHV* as defined above provides an FoM that *can* be used, at least in principle, to compare surveys on different wavelengths and even with different types of data (images, spectra, time series, etc.). However, it does not provide much of the other potentially relevant information described above, and it does not reflect the *depth* of the survey. Thus the SHV favors smaller telescopes.

All of these FoMs provide a generality at the cost of a scientific specificity. They are more appropriate for surveys with a broad range of scientific goals than for the narrowly focused ones. Like any tool, they should be used with a caution.

➋ *Table 5-2* gives our best estimates of these FoMs for a number of selected surveys. It is meant to be illustrative rather than definitive. The survey parameters used are based on the information available in the literature and/or the web, and sometimes our informed estimates.

An alternative, probably far too simple, way of comparing surveys is by their quantity of data collected, for example, in TB. We do not advocate such simplemindedness, since not all bits are of an equal value, but many important properties of surveys, including their scientific potential, *do* correlate at least roughly with the size of the data sets.

If the main scientific goals are statistical studies that require large samples of objects, then a reasonable figure of merit may be the product of the number of detected sources, N_{src}, and the independent measured parameters (attributes) per source, N_{param}. The later is also a measure of the complexity of the data set. For a spectroscopic survey, simply the number of spectra obtained is a reasonable indicator of the survey's scope.

5 From the Raw Data to Science-Ready Archives

Surveys, being highly data-intensive ventures where uniformity of data products is very important, pose a number of data processing and analysis challenges (Djorgovski and Brunner 2001). Broadly speaking, the steps along the way include: obtaining the data (telescope, observatory

■ Table 5-2

Parameters and figures of merit for some of the wide-field surveys. The parameters listed are our best estimates on the basis of the published information. See the text for more details and websites. For PQ/NEAT, some data were taken in 4 bandpasses, and some in 1; we use 2 as an effective average. The PTF takes ~70% of the data in the *R* band, and ~30% in the *g* band; we averaged the numbers. For CRTS, the three sub-surveys were combined, and we use the effective duration of 4 years, as already accomplished as of this writing. The PTF and the PS1 are currently in progress; the numbers for the SkyMapper and the LSST are the projections for the future. We assumed the quoted durations of 3 years for the PS1 survey, the PTF, and the SkyMapper survey, and 10 years for the LSST. These durations scale directly with the assumed N_{tot}, and thus affect the SDP and the SHV figures of merit. For a more fair comparison, the ongoing and the proposed synoptic surveys should probably be compared on a per-year basis, an exercise we leave to the reader

Survey	A [m^2]	t_{exp} [s]	ε	FWHM arcsec	D	N_b	N_{avg}	Ω_{tot} [deg^2] (frac)	SDP	N_e	R deg^2/nt (frac)	TDR	log f_{max}/f_{min}	λ_{max}/λ_{min}	SHV
SDSS (imaging)	4.0	54	0.4	1.5	6.2	5	1	14,500 (0.35)	10.85	1	…	…	4.2	3	22
2MASS	1.3	45	0.35	2.5	1.8	3	1	41,250 (1.0)	5.4	1	…	…	4.5	2	27
UKIDSS	11.3	40	0.35	0.8	15.7	4	1	7,200 (0.175)	11.0	1	…	…	4.5	2.5	7.9
NVSS	12, 200	30	0.7	45	11.2	1	1	34,000 (0.82)	9.2	1	…	…	5	1.1	4.5
FIRST	12, 200	165	0.7	5	237.4	1	1	9,900 (0.24)	57	1	…	…	5	1.1	1.3
SDSS stripe 82	4.0	54	0.4	1.5	6.2	5	300	300 (0.007)	67	1	…	…	4.2	3	137

PanSTARRS PS1, 3 years	2.5	30	0.5	1.0	6.1	6	12	30,000 (0.73)	320	1	6,000 (0.145)	0.887	4.2	3	660
PQ/NEAT	1.0	150	0.4	2	3.9	4,1 ⟨2⟩	50	20,000 (0.48)	190	2	500 (0.012)	0.095	4.0	2.5	480
CRTS 4 year	2.32	30	0.4	3	1.8	1	400	34,000 (0.82)	590	4	2,000 (0.048)	0.345	4.0	2.5	3, 280
PTF 3 year	1.0	60	0.4	1.8	2.7	2	50	22,000 (0.53)	145	2	1,000 (0.024)	0.131	4.5	1.5	360
SkyMapper 3 year (est.)	1.1	110	0.5	2	3.3	6	36	22,000 (0.53)	380	2	800 (0.019)	0.128	4.2	3	1, 440
LSST 10 year (est.)	35.3	15	0.5	0.8	20.3	5	2,000	25,000 (0.60)	122, 000	2	4,000 (0.097)	3.94	4.2	2.5	63, 000

and instrument control software), on-site data processing, if any, detection of discrete sources and measurements of their parameters in imaging surveys, or extraction of 1-dimensional spectra and measurements of the pertinent features in the spectroscopic ones, data calibrations, archiving and dissemination of the results, and finally the scientific analysis and exploration. Typically, there is a hierarchy of ever more distilled and value-added data products, starting from the raw instrument output and ending with ever more sophisticated descriptions of detected objects.

The great diversity of astronomical instruments and types of data with their specific processing requirements are addressed elsewhere in these volumes. Likewise, the data archives, Virtual Observatory and Astroinformatics issues, data mining, and the problem-specific scientific analysis are beyond the scope of this review. Here we address the intermediate steps that are particularly relevant for the processing and dissemination of survey data.

Many relevant papers for this subject can be found in the Astronomical Data Analysis and Software Systems (ADASS) and Astronomical Data Analysis (ADA) conference series and in the SPIE volumes that cover astronomical instruments and software. Another useful reference is the volume edited by Graham et al. (2008).

5.1 Data Processing Pipelines

The actual gathering and processing of the raw survey data encompasses many steps, which can be often performed using a dedicated software pipeline that is usually optimized for the particular instrument, and for the desired data output, that by itself may introduce some built-in biases, but if the original raw data are kept, they can always be reprocessed with improved or alternative pipelines.

Increasingly, we see surveys testing their pipelines extensively with simulated data well before the actual hardware is built. This may reflect a cultural influence of the high-energy physics, as they are increasingly participating in the major survey projects, and data simulations are essential in their field. However, one cannot simulate the problems that are discovered only when the real data are flowing.

The first step involves hardware-specific data acquisition software used to operate the telescopes and the instruments themselves. In principle this is not very different from the general astronomical software used for such purposes, except that the sky surveying generally requires a larger data throughput, a very stable and reliable operation over long stretches of time, and considerably greater data flows than is the case for most astronomical observing. In most cases, additional flux calibration data are taken, possibly with separate instruments or at different times. Due to the long amounts of time required to complete a survey (often several years), a great deal of care must be exercised to monitor the overall performance of the survey in order to ensure a uniform data quality.

Once the raw images, spectra, data cubes, or time series are converted in a form that has the instrumental signatures removed and the data are represented as a linear intensity as a function of the spatial coordinates, wavelength, time, etc., the process of source detection and characterization starts. This requires a good understanding of the instrumental noise properties, which determines some kind of a detection significance threshold: one wants to go as deep as possible, but not count the noise peaks. In other words, we always try to maximize the completeness (the fraction of the real sources detected) while minimizing the contamination (the fraction of the noise peaks mistaken for real sources). In the linear regime of a given detector, the former should be as close to unity, and the latter as close to zero as possible. Both deteriorate at the fainter flux levels as the S/N drops. Typically, a detection limit is taken as a flux level where the completeness falls below 90% or so and contamination increases above 10% or so. However, *significant* detections actually occur at some higher flux level.

Most source detection algorithms require a certain minimum number of adjacent or connected pixels above some signal-to-noise thresholds for detection. The optimal choice of these thresholds depends on the power spectrum of the noise. In many cases, the detection process involves some type of smoothing or optimal filtering, for example, with a Gaussian whose width approximates that of an unresolved point source. Unfortunately, this also builds in a preferred scale for source detection, usually optimized for the unresolved sources (e.g., stars) or the barely resolved ones (e.g., faint galaxies), which are the majority. This is a practical solution, but with the obvious selection biases, with the detection of sources depending not only on their flux but also on their shape or contrast: there is almost always a *limiting surface brightness* (averaged over some specific angular scale) in addition to the *limiting flux*. A surface brightness bias is always present at some level whether it is actually important or not for a given scientific goal. Novel approaches to source or, more accurately, structure detection involve so-called multi-scale techniques (e.g., Aragon-Calvo et al. 2007).

Once individual sources are detected, a number of photometric and structural parameters are measured for them, including fluxes in a range of apertures, various diameters, radial moments of the light distribution, etc., from which a suitably defined, intensity-weighted centroid is computed. In most cases, the sky background intensity level is determined locally, for example, in a large aperture surrounding each source; crowding and contamination by other nearby sources can present problems and create detection and measurement biases. Another difficult problem is deblending or splitting of adjacent sources, typically defined as a number of distinct, adjacent intensity peaks connected above the detection surface brightness threshold. A proper approach keeps track of the hierarchy of split objects, usually called the parent object (the blended composite), the children objects (the first level splits), and so on. Dividing the total flux between them and assigning other structural parameters to them are nontrivial issues and depend on the nature of the data and the intended scientific applications.

Object detection and parameter measurement modules in survey processing systems often use (or are based on) some standard astronomical program intended for such applications, for example, *FOCAS* (Jarvis and Tyson 1981), *SExtractor* (Bertin and Arnouts 1996) or *DAOPHOT* (Stetson 1987), to mention just a few popular ones. Such programs are well documented in the literature. Many surveys have adopted modified versions of these programs, optimized for their own data and scientific goals.

Even if custom software is developed for these tasks, the technical issues are very similar. It is generally true that all such systems are built with certain assumptions about the properties of sources to be detected and measured, and optimized for a particular purpose, for example, detection of faint galaxies or accurate stellar photometry. Such data may serve most users well, but there is always a possibility that a custom reprocessing for a given scientific purpose may be needed.

At this point (or further down the line), astrometric and flux calibrations are applied to the data, using the measured source positions and instrumental fluxes. Most surveys are designed so that improved calibrations can be reapplied at any stage. In some cases, it is better to apply such calibration after the object classification (see below), as the transformations may be different for the unresolved and the resolved sources. Once the astrometric solutions are applied, catalogs from adjacent or overlapping survey images can be stitched together.

In the mid-1990s, the rise of the TB-scale surveys brought the necessity of dedicated, optimized, and highly automated pipelines and databases to store, organize, and access the data. One example is DPOSS, initially processed using *SKICAT* (Weir et al. 1995c), a system that incorporated databases and machine learning, which was still a novelty at that time. SDSS developed a set of pipelines for the processing and cataloguing of images, their astrometric and photometric calibration, and for the processing of spectra; additional, specialized pipelines were added later, to respond to particular scientific needs. For more details, see, for example, York et al. (2000), Lupton et al. (2001), Stoughton et al. (2002), and the documentation available at the SDSS website.

A major innovation of SDSS (at least for the ground-based data; NASA missions and data centers were also pioneering such practices in astronomy) was the effective use of databases for data archiving and the web-based interfaces for the data access, and in particular the *SkyServer* (Szalay et al. 2001, 2002). Multiple public data releases were made using this approach, with the last one (DR8) in 2011, covering the extensions (SDSS-II and SDSS-III) of the original survey. By the early 2000s, similar practices were established as standard for most other surveys; for example, the UKIDSS data processing is described by Dye et al. (2006) and Hodgkin et al. (2009).

Synoptic sky survey added the requirement of data processing in real time, or as close to it as possible, so that transient events can be identified and followed in a timely fashion. For example, the PQ survey (2003–2008) had 3 independent pipelines, a traditional one at Yale University (Andrews et al. 2008), an image subtraction pipeline optimized for SN discovery at the LBNL *Nearby Supernova Factory* (Aldering et al. 2002), and the *Palomar-Quest Event Factory* (Djorgovski et al. 2008) pipeline at Caltech (2005–2008), optimized for a real-time discovery of transient events. The latter served as a basis for the CRTS survey pipeline (Drake et al. 2009). Following in the footsteps of the PQ survey, the PTF survey operates in a very similar manner, with the updated version of the NSNF near-real-time image subtraction pipeline for discovery of transients and a non-time-critical pipeline for additional processing at IPAC.

An additional requirement for the synoptic sky surveys is a timely and efficient dissemination of transient events, now accomplished through a variety of electronic publishing mechanisms. Perhaps the first modern example was the *Gamma-Ray Coordinates Network* (GCN; Barthelmy et al. 2000; http://gcn.gsfc.nasa.gov) that played a key role in cracking the puzzle of the GRBs. For the ground-based surveys, the key effort was the *VOEventNet* (VOEN; Williams and Seaman 2006; http://voeventnet.caltech.edu) that developed the *VOEvent*, now an adopted standard protocol for astronomical event electronic publishing and communication and deployed it in an experimental robotic telescope network with a feedback, using the PQEF as a primary testbed. This effort currently continues through the *SkyAlert* facility (http://skyalert.org; Williams et al. 2009) that uses the CRTS survey as its primary test bed. A variety of specific event dissemination mechanism have been deployed, using the standard webpages, RSS feeds, and even the mobile computing and social media.

5.2 Source and Event Classification

Object classification, for example, as stars or galaxies in the visible and near-IR surveys, but more generally as resolved and unresolved sources, is one of the key issues. Classification of objects is an important aspect of characterizing the astrophysical content of a given sky survey, and for many scientific applications one wants either stars (i.e., unresolved objects) or galaxies; consider, for example, studies of the Galactic structure and studies of the large-scale structure in the universe. More detailed morphological classification, for example, Hubble types of detected galaxies, may be also performed if the data contain sufficient discriminating information to enable it. Given the large data volumes involved in digital sky surveys, object classification must be automated, and in order to make it really useful, it has to be as reliable and objective as possible and homogeneous over the entire survey. Often, the *classification limit* is more relevant than the detection limit for definition of statistical samples of sources (e.g., stars, galaxies, quasars).

In most cases, object classification is based on some quantitative measurements of the image morphology for the detected sources. For example, star-galaxy separation in optical and near-IR surveys uses the fact that all stars (and also quasars) would be unresolved point sources and that the observed shape of the light distribution would be given by the point-spread function, whereas galaxies would be more extended. This may be quantified through various measures of the object radial shape or concentration, for example, moments of the light distribution in various combinations. The problem of star-galaxy separation thus becomes a problem of defining a boundary in some parameter space of observed object properties, which would divide the two classes. In simplest approaches, such a dividing line or surface is set empirically, but more sophisticated techniques use artificial intelligence methods, such as the Artificial Neural Nets or Decision Trees (e.g., Weir et al. 1995a; Odewahn et al. 2004; Ball et al. 2006; Donalek et al. 2008). They require a training data set of objects for which the classification is known accurately from some independent observations. Because of this additional information input, such techniques can outperform the methods where survey data alone are used to decide on the correct object classifications.

There are several practical problems in this task. First, fainter galaxies are smaller in angular extent, thus approaching stars in their appearance. At the fainter flux levels, the measurements are noisier, and thus the two types of objects become indistinguishable. This sets a classification limit to most optical and near-IR surveys, which is typically at a flux level a few times higher than the detection limit. Second, the shape of the point-spread function may vary over the span of the survey, for example, due to the inevitable seeing variations. This may be partly overcome by defining the point-spread function locally and normalizing the structural parameters of objects so that the unresolved sources are the same over the entire survey. In other words, one must define the unresolved source template that would be true locally, but may (and usually does) vary globally. Furthermore, this has to be done automatically and reliably over the entire survey data domain, which may be very heterogeneous in depth and intrinsic resolution. Additional problems include object blending, saturation of signal at bright flux levels, detector nonlinearities, etc., all of which modify the source morphology and thus affect the classification.

The net result is that the automated object classification process is always stochastic in nature. Classification accuracies better than 90% are usually required, but accuracies higher than about 95% are generally hard to achieve, especially at faint flux levels.

In other situations, for example, where the angular resolution of the data is poor, or where nonthermal processes are dominant generators of the observed flux, morphology of the objects may have little meaning, and other approaches are necessary. Flux ratios in different bandpasses, i.e., the spectrum shape, may be useful in separating different physical classes of objects.

A much more challenging task is the automated classification of transient events discovered in synoptic sky surveys (Mahabal et al. 2005, 2008a, b; Djorgovski et al. 2006, 2011b; Bloom et al. 2012). Physical classification of the transient sources is the key to their interpretation and scientific uses, and in many cases scientific returns come from the follow-up observations that depend on scarce or costly resources (e.g., observing time at larger telescopes). Since the transients change rapidly, a rapid (as close to the real time as possible) classification, prioritization, and follow-up are essential, the time scale depending on the nature of the source, which is initially unknown. In some cases the initial classification may remove the rapid-response requirement, but even an archival (i.e., not time-critical) classification of transients poses some interesting challenges.

This entails some special challenges beyond traditional automated classification approaches, which are usually done in some feature vector space, with an abundance of self-contained data derived from homogeneous measurements. Here, the input information is generally sparse and heterogeneous: there are only a few initial measurements, and the types differ from case to case, and the values have differing variances; the contextual information is often essential and yet difficult to capture and incorporate in the classification process; many sources of noise, instrumental glitches, etc., can masquerade as transient events in the data stream; new, heterogeneous data arrive, and the classification must be iterated dynamically. A high completeness, a low contamination, and the need to complete the classification process and make an optimal decision about expending valuable follow-up resources (e.g., obtain additional measurements using a more powerful instrument at a certain cost) in real time are challenges that require some novel approaches.

The first challenge is to associate classification probabilities that any given event belongs to a variety of known classes of variable astrophysical objects and to update such classifications as more data come in, until a scientifically justified convergence is reached. Perhaps an even more interesting possibility is that a given transient represents a previously unknown class of objects or phenomena that may register as having a low probability of belonging to any of the known data models. The process has to be *as automated as possible, robust, and reliable*; it has to operate from *sparse and heterogeneous data*; it has to maintain a *high completeness* (not miss any interesting events) yet a *low false alarm rate*; and it has to *learn* from the past experience for an ever improving, evolving performance.

Much of the initial information that may be used for the event classification is archival, implying a need for a good VO-style infrastructure. Much of the relevant information is also contextual: for example, the light curve and observed properties of a transient might be consistent with both it being a cataclysmic variable star, a blazar, or a supernova. If it is subsequently known that there is a galaxy in close proximity, the supernova interpretation becomes much more plausible. Such information, however, can be characterized by high uncertainty and absence, and by a rich structure – if there were two candidate host galaxies, their morphologies, distances, and luminosities become important, for example, is this type of supernova more consistent with being in the extended halo of a large spiral galaxy or in close proximity to a faint dwarf galaxy? The ability to incorporate such contextual information in a quantifiable fashion is essential. There is a need to find a means of harvesting the human pattern recognition skills, especially in the context of capturing the relevant contextual information and turning them into machine-processible algorithms.

These challenges are still very much a subject of an ongoing research. Some of the relevant papers and reviews include Mahabal et al. (2010a, b, c), Djorgovski et al. (2011b), Richards et al. (2011), and Bloom and Richards (2012), among others.

5.3 Data Archives, Analysis, and Exploration

In general, the data processing flow is from the pixel (image) domain to the catalog domain (detected sources with measured parameters). This usually results in a reduction of the data volume by about an order of magnitude (this factor varies considerably, depending on the survey or the data set), since most pixels do not contain statistically significant signal from resolved sources. However, the ability to store large amounts of digital image information online opens up interesting new possibilities, whereby one may want to go back to the pixels and remeasure fluxes or other parameters on the basis of the catalog information. For example, if a source was detected (i.e., cataloged) in one bandpass, but not in another, it is worth checking if a marginal detection is present even if it did not make it past the statistical significance cut the first time; even the absence of flux is sometimes useful information.

Once all of the data has been extracted from the image pixels by the survey pipeline software, it must be stored in some accessible way in order to facilitate scientific exploration. Simple user file systems and directories are not suitable for really large data volumes produced by sky surveys. The transition to Terascale data sets in the 1990s necessitated use of dedicated database software. Using a database system provides significant advantages (e.g., powerful and complex query expressions) combined with a rapid data access. Fortunately, commercially available database systems can be adopted for astronomical uses. Relational databases accessed using the *Structured Query Language* (SQL) tend to dominate at this time, but different architectures may be better scalable for the much larger data volumes in the future.

A good example of a survey archive is the *SkyServer* (Szalay et al. 2001, 2002; http://skyserver.org/) that provides access to data (photometry and spectra) for objects detected in the different SDSS data sets. This supports more than just positional searching – it offers the ability to pose arbitrary queries (expressed in SQL) against the data so that, for example, one can find all merging galaxy pairs or quasars with a broad absorption line and a nearby galaxy within 10 arcsec. Users can get their own work areas so that query results can be saved and files uploaded to use in queries (as user-supplied tables) against the SDSS data.

Currently, most significant surveys are stored in archives that are accessible through the Internet using a variety of web service interfaces. Their interoperability is established through the Virtual Observatory framework. Enabling access to such survey archives via web services and not just web pages means that programs can be written to automatically analyze and explore vast amounts of data. Whole pipelines can be launched to coordinate and federate multiple queries against different archives, potentially taking hundreds of hours to automatically find the rarest species of objects. Of course, the utility of any such archive is only as good as the metadata provided and the hardest task is often figuring out exactly how the same concept is represented in different archives, for example, one archive might report flux in a particular passband and another magnitude and manually reconciling these.

The *Semantic Web* is an emerging technology that can help solve these challenges (Antoniou and Van Harmelen 2004). It is based on machine-processible descriptions of concepts, and it goes beyond simple term matching with expressions of concept hierarchies, properties, and relationships allowing knowledge discovery. It is a way of encoding a domain expertise (e.g., in astronomy) in a way that may be used by a machine. Ultimately, it may lead to data inferencing by artificial intelligence (AI) engines. For example, discovering that a transient detection has no previous outburst history, is near a galaxy and has a spectrum with silicon absorption but no hydrogen, a system could reason that it is likely to be a type Ia supernova and therefore its progenitor was a white dwarf and so perform an appropriate archival search to find it.

Cloud computing is an emerging paradigm that may well change the ways we approach data persistence, access, and exploration. Commodity computing brings economies of scale, and it effectively outsources a number of tedious tasks that characterize data-intensive science. It is possible that in the future most of our data, survey archives included, and data mining and exploration services for knowledge discovery will reside in the Cloud.

Most of the modern survey data sets are so information rich, that a wide variety of different scientific studies can be done with the same data. Therein lies their scientific potential (Djorgovski et al. 1997b, 2001a, b, c, 2002; Babu and Djorgovski 2004, and many others). However, this requires some powerful, general tools for the exploration, visualization, and analysis of large survey data sets. Reviewing them is beyond the scope of this chapter, but one recent example is the *Data Mining and Exploration* system (DAME; Brescia et al. 2010, 2011; http://dame.dsf.unina.it); see also the review by Ball and Brunner (2010). The newly emerging discipline of Astroinformatics may provide a research framework and environment that would foster development of such tools.

6 Concluding Comments

In this chapter we have attempted to summarize, albeit briefly, the history and the state of the art of sky surveys. However, this is a very rapidly evolving field, and the reader is advised to examine the subsequent literature for the updates and descriptions of new surveys and their science.

In some fields, surveys completely dominate the observational approaches; for example, cosmology, either as a quest to describe the global properties of the universe, the nature of the dark energy, etc., or the history of structure and galaxy formation and evolution, is now tackled largely through large surveys, both from ground and space. Surveys discover cosmic explosions, extrasolar planets, and even new or predicted phenomena.

Sky surveys have transformed the ways in which astronomy is done and pushed it from the relative data poverty to a regime of an immense data overabundance. They are the by far the largest generators of data in astronomy, and they have already enabled a lot of important science and will undoubtedly continue to do so. They have also fostered the emergence of the Virtual Observatory framework and Astroinformatics as means of addressing both the challenges and the opportunities brought by the exponential data growth. They also represent a superb starting point for education and public outreach, for example, with the *Google Sky* and the *WorldWide Telescope* (WWT; http://www.worldwidetelescope.org) sky browsers.

Surveys have also revitalized the role of small telescopes in the era of giant ones, both for the surveying itself and for the immediate imaging and photometric follow-up (Djorgovski 2002). Small telescopes do not imply a small science. Survey-based astronomy is inherently systemic, requiring a full hierarchy of observational facilities, since much of the survey-based science is in the follow-up studies of selected sources. Mutual leveraging of survey and follow-up telescopes, online archives, and cyber-infrastructure creates an added value for all of them.

There is, however, one significant bottleneck that we can already anticipate in the survey-driven science: the follow-up spectroscopy of interesting sources selected from imaging surveys. While there seems to be a vigorous ongoing and planned activity to map and monitor the sky in many ways and many wavelengths, spectroscopic surveys will be necessary in order to interpret and understand the likely overabundance of potentially interesting objects. This looming crisis may seriously limit the scientific returns from the ongoing and future surveys.

Another important lesson is that the cost of these data-intensive projects is increasingly dominated by the cost of software development, implementation, and maintenance. Nobody has ever *under*estimated the cost of software. Our community has to develop more effective ways of sharing and leveraging software efforts. This remains as one of the key motivations behind the VO and Astroinformatics.

In addition to their roles as scientific and technological catalysts, surveys have also started to change the sociology and culture of astronomy by opening new modes of research; new kinds of problems to be addressed, requiring new skills; and new modes of scientific publishing and knowledge preservation. This cultural shift is both inevitable and profoundly transformational. Other sciences have undergone comparable or greater changes, driven by the ways in which problems are defined and data are obtained and analyzed; biology is a good example.

Some sociological changes may be a mixed blessing. By their nature, surveys tend to require large teams, since that can help secure the necessary resources (funding, observing time) and the manpower. Many astronomers are uneasy about this trend toward the high-energy physics mode of research. Generating large data sets requires large-scale efforts. However, important discoveries are still being made at all scales, from individuals and small groups to large collaborations. Proposed survey science tends to be a committee-designed science and thus often safe, but unimaginative; actual survey science tends to be dominated by the unexpected uses of the data and surprises.

One important way in which surveys have changed astronomy is their role as an intermediary step between the sky and the scientist. The large information content of modern sky surveys enables numerous studies that go well beyond the original purposes. The traditional approach where we observe selected targets and make new discoveries using such primary observational data is still with us and will remain. However, there is now a new way of observing the sky, through its representation in the survey archives, using software instruments. It is now possible to make significant observational advances and discoveries without ever going to a telescope. Thus we see a rise in prominence of archival research, which can be as cutting-edge as any observations with the world's largest telescopes and space observatories.

This type of research requires new computational science skills, from data farming (databases, their interoperability, web services, etc.) to data mining and knowledge discovery. The methods and the tools that work efficiently in the Megabyte to Gigabyte regime usually do not scale to Terabytes and Petabytes of data, let alone the greatly increased complexity of the modern data sets. Effective visualization tools and techniques for high-dimensionality parameter sets are another critical issue. We need new kinds of expertise for the new, data-rich and data-intensive astronomy in the twenty-first century. As the science evolves, so does its methodology: we need both new kinds of tools, and the people who know how to use them.

Unfortunately, we are currently neither training properly the new generations of researchers in these requisite skills nor rewarding the career paths that bridge astronomy and ICT. The culture of academia changes slowly, and these educational and professional recognition issues may be among the key obstacles in our path toward the full scientific utilization of the great and growing data abundance brought by the modern sky surveys.

Astronomy is not alone among the sciences facing these challenges. Interdisciplinary exchanges in the context of e-Science, cyber-infrastructure, and science informatics can help us tackle these important issues more efficiently. All of them signal a growing virtualization of science, as most of our work moves into the cyberspace.

To end on a positive note, we are likely entering a new golden age of discovery in astronomy, enabled by the exponential growth of the ICT and the resulting exponential growth of data rates, volumes, and complexity. Any science must rest on the data as its empirical basis, and sky surveys are increasingly playing a fundamental role in this regard in astronomy. We have really just started to exploit them, and the future will bring many new challenges and opportunities for discovery.

Acknowledgments

We are indebted to many colleagues and collaborators over the years, especially the key members of the survey teams: Nick Weir, Usama Fayyad, Joe Roden, Reinaldo de Carvalho, Steve Odewahn, Roy Gal, Robert Brunner, and Julia Kennefick in the case of DPOSS; Eilat Glikman, Roy Williams, Charlie Baltay, David Rabinowitz, and the rest of the Yale team in the case of PQ; and Steve Larson, Ed Beshore, and the rest of the Arizona and Australia team in the case of CRTS. Likewise, we acknowledge numerous additional colleagues and collaborators in the Virtual Observatory and the Astroinformatics community, especially Alex Szalay, Jim Gray, Giuseppe Longo, Yan Xu, Tom Prince, Mark Stalzer, and many others. Several tens of excellent undergraduate research students at Caltech contributed to our work through the years, many of them supported by the Caltech's SURF program. And last, but not least, the staff of Palomar, Keck, and other observatories who helped the data flow. Our work on sky surveys and their exploration has been supported in part by the NSF grants AST-0122449, AST-0326524, AST-0407448, CNS-0540369, AST-0834235, AST-0909182, and IIS-1118041; the NASA grant 08-AISR08-0085; and by the Ajax and Fishbein Family Foundations. Some of the figures in this chapter have been produced using an immersive VR visualization software, supported in part by the NSF grant HCC-0917814. We thank numerous colleagues, and in particular H. Bond, G. Longo, M. Strauss, and M. Kurtz, whose critical reading improved the text. Finally, we thank The Editors for their saintly patience while waiting for the completion of this chapter.

Cross-References

➋ TBD

Appendix: A Partial List of Sky Surveys, Facilities, and Archives as of 2011

We provide this listing as a handy starting point for a further exploration. The listing is probably biased and subjective, and is certainly incomplete, and we apologize for any omissions. The websites given are the best available as of 2011, and a judicious use of search engines can be used to provide updates and corrections. Likewise, VO-related services can be an effective way to discover available data resources for a particular problem.

An exhaustive listing of surveys is maintained by the *NASA/IPAC Extragalactic Database* (NED), currently available at http://ned.ipac.caltech.edu/samples/NEDmdb.html

2MASS Two Micron All Sky Survey, http://www.ipac.caltech.edu/2mass

ADS Astrophysics Data System, http://adswww.harvard.edu/

AGAPE Andromeda Galaxy Amplified Pixels Experiment, http://www.ing.iac.es/PR/SH/SH2006/agape.html

AGILE Astro-rivelatore Gamma a Immagini LEggero, http://agile.rm.iasf.cnr.it/publ02.html

Akari http://www.ir.isas.jaxa.jp/ASTRO-F/Outreach/index_e.html

ALFALFA Arecibo Legacy Fast ALFA Survey, http://egg.astro.cornell.edu

ALMA Atacama Large Millimeter/sub-millimeter Array, http://www.eso.org/public/teles-instr/alma.html

APASS The AAVSO Photometric All-Sky Survey, http://www.aavso.org/apass
ASAS-1,2,3 All Sky Automated Survey, http://www.astrouw.edu.pl/asas/
ASKAP Australian Square Kilometer Array Pathfinder, http://www.atnf.csiro.au/projects/askap
ATA Allen Telescope Array, http://www.seti.org/ata
CANDELS Cosmic Assembly Near-infrared Deep Extragalactic Legacy Survey, http://candels.ucolick.org
CARMA Combined Array for Research in Millimeter-wave Astronomy, http://www.mmarray.org/
CDFS Chandra Deep Field South, http://www.eso.org/~vmainier/cdfs_pub/
CDS Centre de Données astronomiques de Strasbourg, http://cdsweb.u-strasbg.fr/
COBE Cosmic Background Explorer, http://lambda.gsfc.nasa.gov/product/cobe
CoRoT COnvection ROtation and planetary Transits, http://smsc.cnes.fr/COROT/
COSMOS Cosmic Evolution Survey, http://cosmos.astro.caltech.edu
CRTS Catalina Real-Time Transients Survey, http://crts.caltech.edu
CSS Catalina Sky Survey, http://www.lpl.arizona.edu/css
CXC Chandra X-Ray Center, http://cxc.harvard.edu/
DASCH Digital Access to a Sky Century at Harvard, http://hea-www.harvard.edu/DASCH
DENIS Deep Near Infrared Survey of the Southern Sky, http://cdsweb.u-strasbg.fr/online/denis.html
DES Dark Energy Survey, http://www.darkenergysurvey.org
DLS Deep Lens Survey, http://dls.physics.ucdavis.edu
DPOSS Palomar Digital Sky Survey, http://www.astro.caltech.edu/~george/dposs
DSS Digitized Sky Survey, http://archive.stsci.edu/cgi-bin/dss_form
EROS-1,2 Expérience pour la Recherche d'Objets Sombres, http://eros.in2p3.fr/
FGST Fermi Gamma-ray Space Telescope, http://fermi.gsfc.nasa.gov
FIRST Faint Images of the Radio Sky at Twenty-Centimeters, http://sundog.stsci.edu
Gaia http://gaia.esa.int
GALEX Galaxy Evolution Explorer, http://www.galex.caltech.edu
GAMA Galaxy And Mass Assembly survey, http://gama-survey.org/
GLIMPSE Spitzer Galactic Legacy Infrared Mid-Plane Survey Extraordinaire, http://www.astro.wisc.edu/sirtf/
GMAN Global Microlensing Alert Network, http://darkstar.astro.washington.edu/
GMRT Giant Metrewave Radio Telescope, http://gmrt.ncra.tifr.res.in/
GOODS The Great Observatories Origins Deep Survey, http://www.stsci.edu/science/goods
GSC Guide Star Catalog, http://gsss.stsci.edu/Catalogs/Catalogs.htm
HDF Hubble Deep Field, http://www.stsci.edu/ftp/science/hdf/hdf.html
HEASARC High Energy Astrophysics Science Archive Research Center, http://heasarc.gsfc.nasa.gov/
HIPASS HI Parkes All Sky Survey, http://aus-vo.org/projects/hipass
HST Hubble Space Telescope, http://www.stsci.edu/hst
HUDF Hubble Ultra-Deep Field, http://www.stsci.edu/hst/udf
IceCube http://icecube.wisc.edu/, a neutrino experiment in Antarctica.
IRAS Infrared Astronomical Satellite, http://iras.ipac.caltech.edu/IRASdocs/iras.html
IRSA http://irsa.ipac.caltech.edu/
Kepler http://kepler.nasa.gov/
LAMOST Large Sky Area Multi-Object Spectroscopic Telescope, http://www.lamost.org

LINEAR Lincoln Near Earth Asteroid Research, http://www.ll.mit.edu/mission/space/linear/
LOFAR LOw Frequency Array, http://www.lofar.org/
LSST Large Synoptic Survey Telescope, http://www.lsst.org
MACHO Massive Astrophysical Compact Halo Object, http://wwwmacho.anu.edu.au/
MAST Multimission Archive at STScI, http://archive.stsci.edu/index.html
MAXI Monitor of All-sky X-ray Image, http://kibo.jaxa.jp/en/experiment/ef/maxi/
MeerKAT http://www.ska.ac.za/meerkat/index.php
MicroFun Microlensing Follow-Up Network, http://www.astronomy.ohio-state.edu/~microfun/
MILAGRO http://scipp.ucsc.edu/personnel/milagro.html
MOA Microlensing Observations in Astrophysics, http://www.phys.canterbury.ac.nz/moa/
MOST Microvariability and Oscillations of Stars, http://www.astro.ubc.ca/MOST/
NDWFS The NOAO Deep Wide-Field Survey, http://www.noao.edu/noao/noaodeep
NEAT Near-Earth Asteroid Tracking team, http://neat.jpl.nasa.gov
NED NASA/IPAC Extragalctic Database, http://ned.ipac.caltech.edu/
NOMAD Naval Observatory Merged Astrometric Dataset http://www.usno.navy.mil/USNO/astrometry/optical-IR-prod/nomad).
NVSS NRAO VLA Sky Survey, http://www.cv.nrao.edu/nvss
OGLE-I,II,II,IV Optical Gravitation Lensing Experiment, http://ogle.astrouw.edu.pl/
PANDAS Pan-Andromeda Archaeological Survey, http://www.nrc-cnrc.gc.ca/eng/projects/hia/pandas.html
PanSTARRS Panoramic Survey Telescope and Rapid Response System, http://pan-starrs.ifa.hawaii.edu
Pi of the Sky http://grb.fuw.edu.pl/
PKS Parkes Radio Sources, http://archive.eso.org/starcat/astrocat/pks.html
PLANET Probing Lensing Anomalies NETwork, http://www.planet-legacy.org/
PMN Parkes-MIT-NRAO survey, http://www.parkes.atnf.csiro.au/observing/databases/pmn/pmn.html
POSS Palomar Observatory Sky Survey, http://www.astro.caltech.edu/~wws/poss2.html
PQ The Palomar-Quest Sky Survey, http://palquest.org
PTF Palomar Transient Factory, http://www.astro.caltech.edu/ptf
RAPTOR Rapid Telescopes for Optical Response, http://www.lanl.gov/quarterly/q_fall03/observatories.shtml
RASS ROSAT All Sky Survey, http://heasarc.nasa.gov/docs/rosat/rass.html
RoboNet-I,II http://www.astro.ljmu.ac.uk/RoboNet/, http://robonet.lcogt.net/
ROTSE Robotic Optical Transient Search Experiment, http://www.rotse.net
SASSy SCUBA-2 All-Sky Survey, http://www.jach.hawaii.edu/JCMT/surveys/sassy/
SDSS Sloan Digital Sky Survey, http://www.sdss.org, http://www.sdss3.org
SKA Square Kilometer Array, http://www.skatelescope.org
SNFactory The Nearby Supernova Factory, http://snfactory.lbl.gov/
SpaceWatch http://spacewatch.lpl.arizona.edu/
SuperWASP Super Wide Angle Search for Planets, http://www.superwasp.org/
SWIRE Spitzer Wide-area InfraRed Extragalactic survey, http://swire.ipac.caltech.edu
UKIDSS UKIRT Infrared Deep Sky Survey, http://www.ukidss.org
USNO United States Naval Observatory, http://www.usno.nasa.mil/USNO
VISTA Visible and Infrared Survey Telescope for Astronomy, http://www.vista.ac.uk/
VLA NRAO Very Large Array, http://www.vla.nrao.edu
VLT Very Large Telescope, http://www.eso.org/public/teles-instr/vlt.html

VO Virtual Observatory, http://www.ivoa.net
VST VLT Survey Telescope, http://www.eso.org/public/teles-instr/surveytelescopes/vst.html
WISE Wide-Field Infrared Survey Explorer, http://wise.ssl.berkeley.edu
WMAP Wilkinson Microwave Anisotropy Probe, http://map.gsfc.nasa.gov

References

Abdo, A., et al. 2009, ApJS, 183, 4
Abdo, A., et al. 2010, ApJS, 188, 405
Abell, G. 1958, ApJSS, 3, 211
Akerlof, C., et al. 2003, PASP, 115, 132
Alcock, C., et al. 1993, Nature, 365, 621
Alcock, C., et al. ApJ 2000, 542, 281
Aldering, G., et al. (the NSNF team) 2002, Overview of the nearby supernova factory, in Proc. SPIE 4836, Survey and Other Telescope Technologies and Discoveries, eds. J. A. Tyson, & S. Wolff (Bellingham/Wash: SPIE), 61
Andrews, P., et al. (the PQ team) 2008, PASP, 120, 703
Ansari, R., et al. 1997, A&A, 324, 843
Antoniou, G., & Van Harmelen, F. 2004, A Semantic Web Primer (Cambridge, MA: MIT Press)
Aragon-Calvo, M., Jones, B., van de Weygaert, R., & van der Hulst, J. 2007, A&A, 474, 315
Astier, P., et al. 2006, A&A, 447, 31
Atkins, D., et al. 2003, Revolutionizing Science and Engineering Through Cyberinfrastructure (Washington, DC: The National Science Foundation)
Aubourg, E., et al. 1993, Nature, 365, 623
Baade, W., & Zwicky, F. 1927, CoMtW, 3, 73
Babu, G. J., & Djorgovski, S. G. 2004, Stat. Sci., 19, 322
Ball, N., & Brunner, R. J. 2010, Int. J. Mod. Phys. D, 19, 1049
Ball, N., et al. 2006, ApJ, 650, 497
Baltay, C., et al. 2007, PASP, 119, 1278
Barthelmy, S., et al. 2000, in APS Conf. Ser. 526, Proceedings of the 5th Huntsville GRB Workshop, New York, ed. R. M. Kippen et al., 731
Becker, R., White, R., & Helfand, D. 1995, ApJ, 450, 559
Becker A., et al. 2004, ApJ, 611, 418
Beckwith, S., et al. (the HUDF team) 2006, AJ, 132, 1729
Bell, G., Hey, T., & Szalay, A. 2009, Science, 323, 1297
Bertin, E., & Arnouts, S. 1996, A&AS, 117, 393
Bhatti, W., et al. 2010, ApJS, 186, 233
Bloom, J. S., & Richards, J. W. 2012, in Advances in Machine Learning and Data Mining for Astronomy, eds. M. Way et al. (Boca Raton : Chapman & Hall)
Bloom, J. S., et al. (the PTF team) 2012, PASP (submitted)
Bond, I., et al. 2004, ApJL, 606, L155
Bower, G., et al. 2010, ApJ, 725, 1792
Bower, G., et al. 2011, ApJ, 739, 76
Bowyer, S., Drake, J., & Vennes, S. 2000, ARAA, 38, 231
Brandt, N., & Hassinger, G. 2005, ARAA, 43, 827
Brescia, M., et al. 2010, in INGRID 2010 Workshop on Instrumenting the GRID, eds. F. Davoli et al. (Berlin: Springer)
Brescia, M., Cavuoti, S., Djorgovski, S. G., Donalek, C., Longo, G., & Paolillo, M. 2012, in Springer Series on Astrostatistics, Astrostatistics and Data Mining in Large Astronomical Databases, eds. L. M. Barrosaro et al. (in press)
Brunner, R., Djorgovski, S. G., & Szalay, A. (eds.) 2001a, Virtual Observatories of the Future, ASPCS, Vol. 225 (San Francisco: ASP)
Brunner, R., Djorgovski, S. G., Gal, R., Mahabal, A., & Odewahn, S. 2001b, ASPCS, 225, 64
Brunner, R., Djorgovski, S. G., Prince, T., & Szalay, A. 2001c, in Handbook of Massive Data Sets, eds. J. Abello et al. (Boston: Kluwer), 931
Cabanela, J., et al. 2003, PASP, 115, 837
Centrella, J. 2010, in AIP Conf. Proc. 1381, Proceedings of the 25th Texas Symposium on Relativistic Astrophysics (Melville: AIP), 98
Colless, M. 1999, Phil. Trans. Roy. Soc. Lond. A, 357, 105
Colless, M., et al. (the 2dF survey team) 2001, MNRAS, 328, 1039
Condon, J., et al. (the NVSS survey team) 1998, AJ, 115, 1693
Cristiani, S., Renzini, A., & Williams, R. (eds.) 2002, ESO Astrophysics Symposia, Deep Fields: Proceedings of the ESO Workshop (Berlin: Springer)
Croft, S., et al. 2010, ApJ, 719, 45
Croft, S., et al. 2011, ApJ, 731, 34
Croom, S., et al. 2004, MNRAS, 349, 1397
Davis, M., Huchra, J., Latham, D., & Tonry, J. 1982, ApJ, 253, 423
Davis, M., et al. 2003, SPIE, 4834, 161
Davis, M., et al. (the AEGIS team) 2007, ApJ, 660, L1
de Carvalho, R., Djorgovski, S. G., Weir, N., Fayyad, U., Cherkauer, K., Roden, J., & Gray, A. 1995,

in ASPCS 77, Proceedings of the ADASS IV, eds. R. Shaw et al. (San Francisco: ASP), 272
Dickinson, M., et al. (the GOODS team) 2003, in ESO Astrophysics Symposia, The Mass of Galaxies at Low and High Redshift, eds. R. Bender, & A. Renzini (Berlin: Springer), 324
Diercks, A. 2001, ASPCS, 225, 46
Djorgovski, S. G. 1992a, in Morphological and Physical Classification of Galaxies, eds. G. Longo et al. (Dordrecht: Kluwer), 337
Djorgovski, S. G. 1992b, in ASPCS 24, Cosmology and Large-Scale Structure in the Universe, ed. R. de Carvalho (San Francisco: ASP), 19
Djorgovski, S. G. 2000, in ASPCS 213, Bioastronomy 1999, eds. K. Meech, & G. Lemarchand (San Francisco: ASP), 519
Djorgovski, S. G. 2002, in Small Telescopes in the New Millenium. I. Perceptions, Productivity, and Priorities, ed. T. Oswalt (Dordrecht: Kluwer), 85
Djorgovski, S. G. 2005, in IEEE Proceeidngs of CAMP05: Computer Architectures for Machine Perception, eds. V. Di Gesu, & D. Tegolo (Los Alamitos: IEEE), 125
Djorgovski, S. G., & Brunner, R. 2001, Software: digital sky surveys, in Encyclopedia of Astronomy and Astrophysics, ed. S. Maran (London: Institute of Physics Publications)
Djorgovski, S. G., & Davis, M. 1987, ApJ, 313, 59
Djorgovski, S. G., & Williams, R. 2005, ASPCS, 345, 517
Djorgovski, S., Weir, N., & Fayyad, U. 1994, in ASPCS 61, Proceedinges of the ADASS III, eds. D. Crabtree et al. (San Francisco: ASP), 195
Djorgovski, S. G., de Carvalho, R., Gal, R., Pahre, M., Scaramella, R., & Longo, G. 1997a, in Proc. IAU Symp. 179, New Horizons From Multi-Wavelength Sky Surveys, eds. B. McLean et al. (Dordrecht: Kluwer), 424
Djorgovski, S. G., de Carvalho, R., Odewahn, S., Gal, R., Roden, J., Stolorz, P., & Gray, A. 1997b, in Proc. SPIE 3164, Applications of Digital Image Processing XX, ed. A. Tescher (Bellingham: SPIE), 98
Djorgovski, S. G., Gal, R., Odewahn, S., de Carvalho, R., Brunner, R., Longo, G., & Scaramella, R. 1999, in Wide Field Surveys in Cosmology, eds. S. Colombiet et al. (Gif sur Yvette: Editions Frontieres), 89
Djorgovski, S. G., et al. 2001a, in ASPCS 225, Virtual Observatories of the Future, eds. R. Brunner et al. (San Francisco: ASP), 52
Djorgovski, S. G., et al. 2001b, in ESO Astroph. Symp., Mining the Sky, eds. A. Banday et al. (Berlin: Springer), 305
Djorgovski, S. G., et al. 2001c in Proc. SPIE 4477, Astronomical Data Analysis, eds. J.-L. Starck, & F. Murtagh (Bellingham: SPIE), 43
Djorgovski, S. G., Brunner, R., Mahabal, A., Williams, R., Granat, R., & Stolorz, P. 2002, in Statistical Challenges in Astronomy III, eds. E. Feigelson, & J. Babu (New York: Springer), 125
Djorgovski, S. G., et al. 2006, in Proceedings of the 18th International Conference on Pattern Recognition, Hong Kong, Vol. 1, eds. Y.Y. Tang, et al. (IEEE Press), 856
Djorgovski, S. G., et al. (the PQ team) 2008, AN, 329, 263
Djorgovski, S. G., et al. 2011a, in The First Year of MAXI: Monitoring Variable X-ray Sources, eds. T. Mihara, & N. Kawai, (Tokyo: JAXA Special Publ). In press
Djorgovski, S. G., Donalek, C., Mahabal, A., Moghaddam, B., Turmon, M., Graham, M., Drake, A., Sharma, N., & Chen, Y. 2011b, in Proceedings of the CIDU 2011 Conference, eds. A. Srivasatva, et al. (Mountain View, CA), 174
Djorgovski, S.G., Mahabal, A., Drake, A., Graham, M., Donalek, C., & Williams, R. 2012, in Proc. IAU Symp. 285, New Horizons in Time Domain Astronomy, eds. E. Griffin et al. (Cambridge: Cambridge University Press). In press
Donalek, C., et al. 2008, AIPC, 1082, 252
Drake, A. J., et al. 2009, ApJ, 696, 870
Drake, A. J., et al. 2012, in Proc. IAU Symp. 285, New Horizons in Time Domain Astronomy, eds. E. Griffin et al. (Cambridge: Cambridge University Press). In press
Dreyer, J. L. E., 1888, Mem. R. Astron. Soc., 49, 1
Dreyer, J. L. E., 1895, Mem. R. Astron. Soc., 51, 185
Dye, S., et al. (the UKIDSS team) 2006, MNRAS, 372, 1227
Falco, E., Kurtz, M., Geller, M., Huchra, J., Peters, J., Berlind, P., Mink, D., Tokarz, S., & Elwell, B. 1999, PASP, 111, 438
Ferguson, H., Dickinson, M., & Williams, R. 2000, ARAA, 38, 667
Filippenko, A., Riess, A., 1998, PhR, 307, 31
Filippenko, A., Li, W., Treffers, R., & Modjaz, M. 2001, ASPCS, 246, 121
Fisher, K., Huchra, J., Strauss, M., Davis, M., Yahil, A., & Schlegel, D. 1995, ApJS, 100, 69
Fukugita, M., et al. (the SDSS team) 1996, AJ, 111, 1748
Gal, R., de Carvalho, R., Odewahn, S., Djorgovski, S.G., Mahabal, A., Brunner, R., & Lopes, P. 2005, AJ, 128, 3082
Gal, R., Lopes, P., de Carvalho, R., Kohl-Moreira, J., Capelato, H., & Djorgovski, S. G. 2009, AJ, 137, 2981
Geller, M., & Huchra, J. 1989, Science, 246, 897

Geller, M., Diaferio, A., & Kurtz, M. 2011, ApJ, 142, 133
Giacconi, R., et al. (the CDF-S team) 2001, ApJ, 551, 624
Giavalisco, M., et al. (the GOODS team) 2004, ApJ, 600, L93
Giovanelli, R., & Haynes, M. 1991, ARAA, 29, 499
Giovanelli, R., et al. (the ALFALFA team) 2005, AJ, 130, 2598
Graham, M., Fitzpatrick, M., & McGlynn, T. (eds.) 2008, in ASPCS 382, The National Virtual Observatory: Tools and Techniques for Astronomical Research (San Francisco: ASP)
Gray, J., & Szalay, A. 2006, Nature, 440, 23
Green, R. 1976, PASP, 88, 665
Griffin, E., Hanisch, R., & Seaman, R. (eds.) 2012, Proc. IAU Symp. 285, New Horizons in Time Domain Astronomy (Cambridge: Cambridge University Press). In press
Grindlay, J., et al. 2009, in ASPCS 410, Preserving Astronomy's Photographic Legacy: Current State and the Future of North American Astronomical Plates, eds. W Osborn, & L. Robbins (San Francisco: ASP), 101
Gunn, J., et al. (the SDSS team) 1998, AJ, 116, 3040
Gunn, J., et al. (the SDSS team) 2006, AJ, 131, 2332
Halzen, F., & Klein, S. 2010, Rev. Sci. Instrum., 81, 081101
Hambly, N., et al. (the SuperCOSMOS group) 2001, MNRAS, 326, 1279
Hamuy, M., et al. (the Calan/Tololo Supernova Search team) 1993, AJ, 106, 2392
Hanisch, R. 2001, AIP Conf. Proc., 575, 45
Hanisch, R. 2010, in ASPCS 434, Proceedings of ADASS XIX, eds. Y. Mizumoto et al. (San Francisco: ASP), 65
Harwit, M. 1975, QJRAS, 16, 378
Harwit, M. 2003, Phys. Today, 56, 38
Harwit, M., & Hildebrand, R. 1986, Nature, 320, 724
Hey, T., & Trefethen, A. 2003, Phil. Trans. R. Soc. London, 361, 1809
Hey, T., & Trefethen, A. 2005, Science, 308, 817
Hey, T., Tansley, S., & Tolle, K. (eds.) 2009, The Fourth Paradigm: Data-Intensive Scientific Discovery (Redmond, WA: Microsoft Research)
Hodgkin, S. T., Irwin, M. J., Hewett, P. C., & Warren, S. J. 2009, MNRAS, 394, 675
Huchra, J., Davis, M., Latham, D., & Tonry, J. 1983, ApJSS, 52, 89
Humason, M., Mayall, N., & Sandage, A. 1956, AJ, 61, 97
Impey, C., & Bothun, G., 1997, ARAA, 35, 267
Ivezic, Z., et al. (the SDSS team) 2008, AJ, 134, 973
Ivezic, Z., et al. (the LSST team) 2009, The LSST Science Book, v2.0. arXiv:0912.0201, http://www.lsst.org/lsst/scibook
Jannuzi, B., & Dey, A. 1999, ASPCS, 191, 111
Jarvis, J.F., & Tyson, J. A. 1981, AJ, 86, 476
Kaiser, N., et al. 2002, SPIE, 4836, 154
Kashikawa, N., et al. (the Subaru Deep Field team) 2004, PASJ, 56, 1011
Koekemoer, A., et al. (the CANDELS team) 2011, ApJS, 197, 36
Kohonen, T. 1989, Self-Organisation and Associative Memory (Berlin: Springer)
Kormendy, J. 1985, ApJ, 295, 73
Kotera, K., & Olinto, A. 2011, ARAA, 49, 119
Kurtz, M. 1995, PASP, 107, 776
La Barbera, F., Busarello, G., Merluzzi, P., de la Rosa, I., Coppola, G., & Haines, C. 2009, ApJ, 689, 913
Lahav, O., & Suto, Y. 2004, Measuring our universe from galaxy redshift surveys. Living Rev. Relat., 7, 8
Lasker, B., et al. (the HST GSC team) 2008, AJ, 136, 735
Lauberts, A. 1982, The ESO/Uppsala Survey of the ESO(B) Atlas (Munich: European Southern Observatory)
Law, N., et al. (the PTF team) 2009, PASP, 123, 1395
Lawrence, A., et al. (the UKIDSS team) 2007, MNRAS, 379, 1599
Leaman, J., et al. 2011, MNRAS, 412, 1419
Leavitt, H. S., & Pickering, E. C. 1912, Harv. Coll. Obs. Circ., 173, 1
Le Fevre, O., et al. 2005, A&A, 439, 845
Li, W. D., et al. 1999, AIP Conf. Proc. 522, 103
Lilly, S. J., et al. 2007, ApJS, 172, 70
Lonsdale, C., et al. (the SWIRE team) 2003, PASP, 115, 897
Lupton, R., et al. 2001, in ASPCS 238, Proceedings of the ADASS X (San Francisco: ASP), 269
MacGillivray, H., et al. (eds.) 1994, Proc. IAU Symp. 161, Astronomy From Wide-Field Imaging (Dordrecht: Kluwer)
Maddox, S., Efstathiou, G., Sutherland, W., & Loveday, J. 1990a, MNRAS, 242, 43
Maddox, S., Efstathiou, G., Sutherland, W., & Loveday, J. 1990b, MNRAS, 243, 692
Mahabal, A., et al. (the PQ team) 2005, in ASPCS 347, Proceedings of the ADASS XIV, eds. P. Shopbell et al. (San Francisco: ASP), 604
Mahabal, A., et al. (the PQ Team) 2008a, AN, 329, 288
Mahabal, A., et al. 2008b, in AIP Conf. Ser. 1082, Proceedings of the International Conference on Classification and Discovery in Large Astronomical Surveys (Melville: American Institute of Physics), 287
Mahabal, A., Djorgovski, S.G., Donalek, C., Drake, A., Graham, M., Moghaddam, B., Turmon, M., & Williams, R. 2010a, in ASPCS 434, Proceedings of the ADASS XIX, eds. Y. Mizumoto et al. (San Francisco: ASP), 115

Mahabal, A., Djorgovski, S.G., Donalek, C., Drake, A., Graham, M., Williams, R., Moghaddam, B., & Turmon, M. 2010b, in EAS Publ. Ser. 45, Gaia: At the Frontiers of Astrometry, eds. C. Turon et al. (Paris: EDP Sciences), 173
Mahabal, A., Djorgovski, S.G., & Donalek, C. 2010c, in Hotwiring the Transient Universe, eds. S. Emery Bunn et al., 31. Internet: Lulu Enterprises http://www.lulu.com
Mahabal, A. A., et al. 2011, BASI, 39, 38
Mao, S., & Paczynski, B. 1991, ApJ, 374, 37
Martin, C., et al. (the GALEX team) 2005, ApJ, 619, L1
McGlynn, T., Scollick, K., & White, N. 1997, in Proc. IAU Symp. 179, New Horizons From Multi-Wavelength Sky Surveys, eds. B. McLean et al. (Dordrecht: Kluwer), 465
McLean, B., Golombek, D., Haynes, J., & Payne, H. (editors) 1998, Proc. IAU Symp. 179, New Horizons From Multi-Wavelength Sky Surveys, (Dordrecht: Kluwer)
Minkowski, R., & Abel, G. 1963, The National Geographic Society-Palomar Observatory Sky Survey, in Basic Astronomical Data: Stars and Stellar Systems, ed. K.A. Strand (Chicago: University of Chicago Press), 481
Monet, D.G., et al. 2003, AJ, 125, 984
Mottram, C.J., & Fraser, S. N. 2008, AN, 329, 317
Neugebauer, G., & Leighton, R. 1969, Two-Micron Sky Survey. A Preliminary Catalogue (Washington, DC: NASA)
Nilson, P. 1973, The Uppsala General Catalogue of Galaxies, (Uppsala: Royal Society of Science of Uppsala)
Odewahn, S. C., et al. (the DPOSS team) 2004, AJ, 128, 3092
Ofek, E. O., et al. 2011, ApJ, 740, 65
Paczynski, B. 1986, ApJ, 304, 1
Paczynski, B. 2000, PASP, 112, 1281
Perlmutter, S., et al. 1997, ApJ, 483, 565.
Perlmutter, S., et al. 1998, Nature, 391, 51
Perlmutter, S., Turner, M. S., & White, M. 1999, PhRvL, 83, 670
Pojmanski, G., 1997, Acta Astron., 47, 467
Rau, A., et al. (the PTF team) 2009, PASP, 121, 1334
Reid, I. N., et al. 1991, PASP, 103, 661
Richards, G., et al. (the SDSS team) 2009, AJ, 137, 3884
Richards, J. W., et al. 2011, ApJ, 733, 10
Riess, A., et al. 1998, AJ, 116, 1009
Rykoff, E. S., et al. 2005, ApJ, 631, 1032
Sako, M., et al. (the SDSS team) 2008, AJ, 135, 348
Salzer, J., & Haynes, M. 1996, ASPCS, 106, 357
Saunders, W., et al. (the PSCz team) 2000, MNRAS, 317, 55
Scoville, N., et al. 2007, ApJS, 172, 1
Sesar, B., et al. 2007, AJ, 134, 2236
Shectman, S. A., et al. 1996, ApJ, 470, 172
Siemion A. P. V., et al. 2012, ApJ, 744, 109
Skrutskie, M., et al. (the 2MASS team) 2006, AJ, 131, 1163
Stetson, P. B. 1987, PASP, 99, 191
Stoughton, C., et al. (the SDSS team) 2002, AJ, 123, 485
Strauss, M., Huchra, J., Davis, M., Yahil, A., Fisher, K., & Tonry. J. 1992, ApJS, 83, 29
Sullivan, W. 2009, Cosmic Noise: A History of Early Radio Astronomy (Cambridge: Cambridge University Press)
Szalay, A., & Gray, J. 2001, Science, 293, 2037
Szalay, A., Gray, J., et al. 2001, The SDSS SkyServer, Public Access to the Sloan Digital Sky Server. Microsoft Research Techinical Report 2001-104
Szalay, A., Gray, J., et al. 2002, in: Proceedings of the 2002 ACM SIGMOD International Conferences on Management of Data, eds. M. Franklin et al., (New York: ACM Publ)
Tarter, J. 2001, ARAA, 39, 511
Tonry, J. 2011, PASP, 123, 58
Tonry, J., Burke, B., & Schechter, P. 1997, PASP, 109, 1154
Tsapras, Y., et al. 2009, AN, 330, 4
Tyson, J.A., at al (the LSST team) 2002, Proc. SPIE 4836, 10
Udalski, A., et al. 1993, AcA, 43, 289
Udalski, A., et al. 2002, AcA, 52, 115
Viggh, H. E. M., Stokes, G. H., Shelly, F. C., Blythe, M. S., Stuart, J. S. 1997, PASP, 29, 959
Voges, W., et al. (the RASS team) 1999, A&A, 349, 389
Vogt, N., et al. 2005, ApJS, 159, 41
Vorontsov-Velyaminov, B., & Arkhipova, V. 1974, Morphological Catalog of Galaxies (Moscow: Moscow State univ)
Weir, N., Fayyad, U., & Djorgovski, S.G. 1995a, AJ, 109, 2401
Weir, N., Djorgovski, S.G., & Fayyad, U. 1995b, AJ, 110, 1
Weir, N., Fayyad, U., Djorgovski, S. G., & Roden, J. 1995c, PASP, 107, 1243
Williams, R., et al. (the HDF team) 1996, AJ, 112, 1335
Williams, R., & Seaman, R. 2006, in ASPCS 351, Proceedings of the ADASS XV (San Francisco: ASP), 637
Williams, R., Djorgovski, S. G., Drake, A., Graham, M., & Mahabal, A. 2009, in ASPCS 411, Proceedings of the ADASS XVII (San Francisco: ASP), 115
Wittman, D., et al. (the DLS team) 2002, Proc. SPIE, 4836, 73
Wright, E., et al. (the WISE team) 2010, AJ, 140, 1868

Yock, P., et al. 2000, PASA, 104, 35
Yoo, J., Gray, A., Roden, J., Fayyad, U., de Carvalho, R., & Djorgovski, S. G. 1996, in ASPCS 101, Proceedings of the ADASS V, eds. G. Jacoby, & J. Barnes (San Francisco: ASP), 41
York, D. G., et al. (the SDSS team) 2000, AJ, 120, 1579
Yost, S. A., et al. 2007, ApJ, 669, 1107
Zwicky, F. 1957, Morphological astronomy (Berlin: Springer)

6 Techniques of Radio Astronomy

T. L. Wilson
Naval Research Laboratory, Washington, DC, USA

T.D. Oswalt, H.E. Bond (eds.), *Planets, Stars and Stellar Systems. Volume 2: Astronomical Techniques, Software, and Data*, DOI 10.1007/978-94-007-5618-2_6, © Springer Science+Business Media Dordrecht 2013

Abstract: This chapter provides an overview of the techniques of radio astronomy. This study began in 1931 with Jansky's discovery of emission from the cosmos, but the period of rapid progress began 15 years later. From then to the present, the wavelength range has expanded from a few meters to the sub-millimeters, the angular resolution increased from degrees to finer than milli arc seconds, and the receiver sensitivities have improved by large factors. Today, the technique of aperture synthesis produces images comparable to or exceeding those obtained with the best optical facilities. In addition to technical advances, the scientific discoveries made in the radio range have contributed much to opening new visions of our universe. There are numerous national radio facilities spread over the world. In the near future, a new era of truly global radio observatories will begin. This chapter contains a short history of the development of the field, details of calibration procedures, coherent/heterodyne and incoherent/bolometer receiver systems, observing methods for single apertures and interferometers, and an overview of aperture synthesis.

Keywords: Aperture Synthesis, Bolometers, Coherent Receivers, Heterodyne Receivers, High Angular Resolution, Imaging, Incoherent Receivers, Polarimeters, Radio Astronomy, Spectrometers

1 Introduction

Following a short introduction, the basics of simple radiative transfer, propagation through the interstellar medium, polarization, receivers, antennas, interferometry, and aperture synthesis are presented. References are given mostly to more recent publications, where citations to earlier work can be found; no internal reports or web sites are cited. The units follow the usage in the astronomy literature. For more details, see Thompson et al. (2001), Gurvits et al. (2005); Wilson et al. (2008), and Burke and Graham-Smith (2009).

The origins of optical astronomy are lost in prehistory. In contrast radio astronomy began recently, in 1931, when K. Jansky showed that the source of excess radiation at $\nu = 20.5$ MHz ($\lambda = 14.6$ m) arose from outside the solar system. G. Reber followed up and extended Jansky's work, but the most rapid progress occurred after 1945, when the field developed quickly. The studies included broadband radio emission from the Sun, as well as emission from extended regions in our galaxy, and later other galaxies. In wavelength, the studies began at a few meters where the emission was rather intense and more easily measured (see Sullivan 2005, 2009). Later, this was expanded to include centimeter, millimeter, and then submillimeter wavelengths. In ➲ *Fig. 6-1*, a plot of transmission through the atmosphere as a function of frequency ν and wavelength, λ is presented. The extreme limits of the earth-bound radio window extend roughly from a lower frequency of $\nu \cong 10$ MHz ($\lambda \cong 30$ m) where the ionosphere sets a limit, to a highest frequency of $\nu \cong 1.5$ THz ($\lambda \cong 0.2$ mm), where molecular transitions of atmospheric H_2O and N_2 absorb astronomical signals. There is also a prominent atmospheric feature at ~55 GHz, or 6 mm, from O_2. The limits shown in ➲ *Fig. 6-1* are not sharp since there are variations with altitude, geographic position, and time. Reliable measurements at the shortest wavelengths require remarkable sites on earth. Measurements at wavelengths shorter than $\lambda = 0.2$ mm require the use of high flying aircraft, balloons, or satellites. The curve in ➲ *Fig. 6-1* allows an estimate of the height above sea level needed to carry out astronomical measurements.

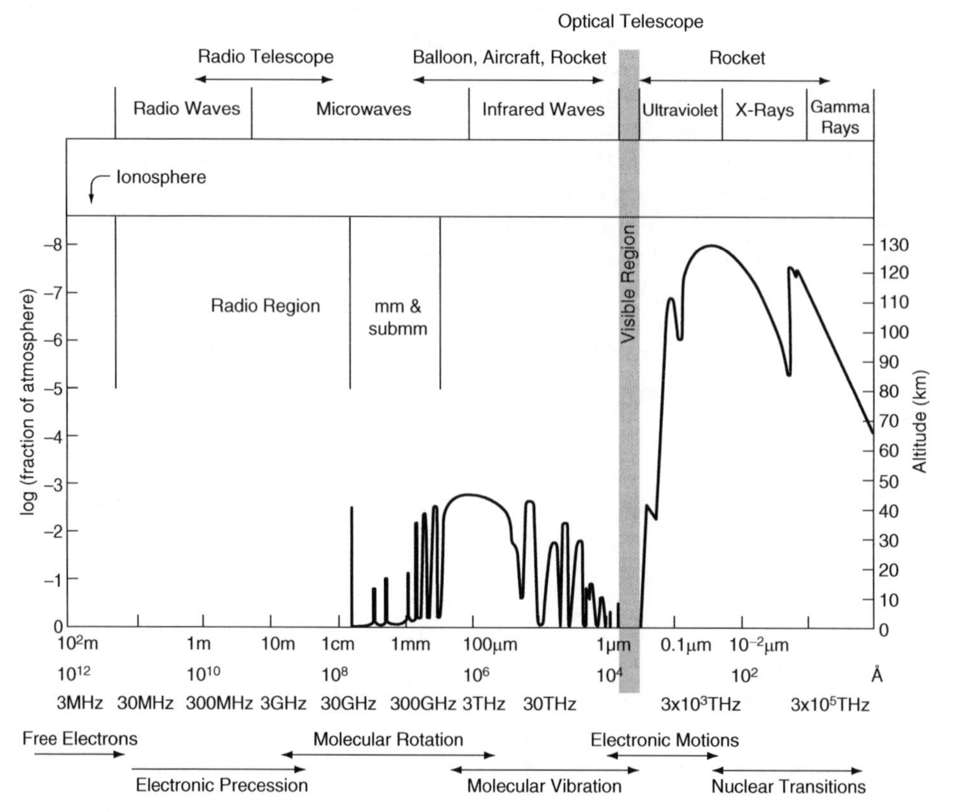

Fig. 6-1

A plot of transmission through the atmosphere versus wavelength, λ in metric units and frequency, ν, in hertz. The thick curve gives the fraction of the atmosphere (*left vertical axis*) and the altitude (*right axis*) needed to reach a transmission of 0.5. The fine scale variations in the thick curve are caused by molecular transitions (see Townes and Schawlow 1975). The thin vertical line on the left (~10 MHz) marks the boundary where ionospheric effects impede astronomical measurements. The labels above indicate the types of facilities needed to measure at the frequencies and wavelengths shown. For example, from the thick curve, at $\lambda = 100$ μm, one half of the astronomical signal penetrates to an altitude of 45 km. In contrast, at $\lambda = 10$ cm, all of this signal is present at the earth's surface. The arrows at the bottom of the figure indicate the type of atomic or nuclear process that gives rise to the radiation at the frequencies and wavelengths shown above (From Wilson et al. 2008)

The broadband emission mechanism that dominates at meter wavelengths has been associated with the synchrotron process. Thus, although the photons have energies in the micro electron volt range, this emission is caused by highly relativistic electrons (with γ factors of more than 10^3) moving in microgauss fields. In the centimeter and millimeter wavelength ranges, some broadband emission is produced by the synchrotron process, but additional emission arises from free–free Bremsstrahlung from ionized gas near high mass stars and quasi-thermal broadband emission from dust grains. In the millimeter/submillimeter range, emission from dust grains dominates, although free–free and synchrotron emission may also contribute. Spectral lines of molecules become more prominent at millimeter/submillimeter wavelengths (see Rybicki and Lightman 1979; Lequeux 2004; Tielens 2005).

Radio astronomy measurements are carried out at wavelengths vastly longer than those used in the optical range (see *Fig. 6-1*), so extinction of radio waves by dust is not an important effect. However, the longer wavelengths lead to lower angular resolution, θ, since this is proportional to λ/D where D is the size of the aperture (see Jenkins and White 2001). In the 1940s, the angular resolutions of radio telescopes were on scales of many arc minutes, at best. In time, interferometric techniques were applied to radio astronomy, following the method first used by Michelson. This was further developed, resulting in Aperture Synthesis, mainly by M. Ryle and associates at Cambridge University (for a history, see Kellermann and Moran 2001). Aperture synthesis has allowed imaging with angular resolutions finer than milli arc seconds with facilities such as the Very Long Baseline Array (VLBA).

Ground-based measurements in the submillimeter wavelength range have been made possible by the erection of facilities on extreme sites such as Mauna Kea, the South Pole and the 5 km high site of the Atacama Large Millimeter/submillimeter Array (ALMA). Recently there has been renewed interest in high-resolution imaging at meter wavelengths. This is due to the use of corrections for smearing by fluctuations in the electron content of the ionosphere and advances that facilitate imaging over wide angles (see, e.g., Venkata 2010). With time, the general trend has been toward higher sensitivity, shorter wavelength, and higher angular resolution.

Improvements in angular resolution have been accompanied by improvements in receiver sensitivity. Jansky used the highest quality receiver system then available. Reber had access to excellent systems. At the longest wavelengths, emission from astronomical sources dominates. At millimeter/submillimeter wavelengths, the transparency of the earth's atmosphere is an important factor, adding both noise and attenuating the astronomical signal, so both lowering receiver noise and measuring from high, dry sites are important. At meter and centimeter wavelengths, the sky is more transparent and radio sources are weaker.

The history of radio astronomy is replete with major discoveries. The first was implicit in the data taken by Jansky. In this, the intensity of the extended radiation from the Milky Way exceeded that of the quiet Sun. This remarkable fact shows that radio and optical measurements sample fundamentally different phenomena. The radiation measured by Jansky was caused by the synchrotron mechanism; this interpretation was made more than 15 years later (see Rybicki and Lightman 1979). The next discovery, in the 1940s, showed that the active Sun caused disturbances seen in radar receivers. In Australia, a unique instrument was used to associate this variable emission with sunspots (see Dulk 1985; Gary and Keller 2004). Among later discoveries have been: (1) discrete cosmic radio sources, at first, supernova remnants and radio galaxies (in 1948, see Kirshner 2004); (2) the 21 cm line of atomic hydrogen (in 1951, see Sparke and Gallagher 2007; Kalberla et al. 2005); (3) Quasi Stellar Objects (in 1963, see Begelman and Rees 2009); (4) the Cosmic Microwave Background (in 1965, see Silk 2008); (5) interstellar molecules (see Herbst and Dishoeck 2009) and the connection with Star Formation, later including circumstellar and protoplanetary disks (in 1968, see Stahler and Palla 2005; Reipurth et al. 2007); (6) pulsars (in 1968, see Lyne and Graham-Smith 2006); (7) distance determinations using source proper motions determined from Very Long Baseline Interferometry (see Reid 1993); and (8) molecules in high redshift sources (see Solomon and Vanden Bout 2005). These areas of research have led to investigations such as the dynamics of galaxies, dark matter, tests of general relativity, Black Holes, the early universe, and gravitational radiation (for overviews see Longair 2006; Harwit 2006). Radio astronomy has been recognized by the physics community in that four Nobel Prizes (1974, 1978, 1993, and 2006) were awarded for work in this field. In chemistry, the community has been made aware of the importance of a more general chemistry involving ions and molecules (see Herbst 2001). Two Nobel Prizes for chemistry were awarded to persons actively engaged in molecular line astronomy.

Over time, the trend has been away from small groups of researchers constructing special purpose instruments toward the establishment of large facilities where users propose projects carried out by specialized staffs. These large facilities are in the process of becoming global. Similarly, the evolution of data reduction has been toward standardized packages developed by large teams. In addition, the demands of the interpretation of astronomical phenomena have led to multiwavelength analyses interpreted with the use of detailed models.

Outside the norm are projects designed to measure a particular phenomenon. A prime example is the study of the cosmic microwave background (CMB) emission from the early universe. CMB data were taken with the COBE and WMAP satellites. These results showed that the CMB is a Black Body (see ➋ Eq. 6.6) with a temperature of 2.73 K. Aside from a dipole moment caused by our motion, there is angular structure in the CMB at a very low level; this is being studied with the PLANCK satellite. Much effort continues to be devoted to measurements of the polarization of the CMB with ground-based experiments such as BICEP, CBI, DASI, and QUIET. For details and references to other CMB experiments, see their web sites. In spectroscopy, there have been extensive surveys of the 21 cm line of atomic hydrogen, H I (see Kalberla et al. 2005) and the rotational $J = 1 - 0$ line from the ground state of carbon monoxide (see Dame et al. 1987). These surveys have been extended to external galaxies (see Giovanelli and Haynes 1991). During the Era of Reionization (redshift $z \sim 10$ to 15), the H I line is shifted to meter wavelengths. The detection of such a feature is the goal of a number of individual groups, under the name HERA (Hydrogen Epoch of Reionization Arrays).

1.1 A Selected List of Radio Astronomy Facilities

There are a large number of existing facilities; a selection is listed here. General purpose instruments include the largest single dishes: the Parkes 64-m, the Robert C. Byrd Green Bank Telescope, hereafter GBT, the Effelsberg 100 m, the 15-m James Clerk Maxwell Telescope (JCMT), the IRAM 30-m millimeter telescope, and the 305-m Arecibo instrument. All of these have been in operation for a number of years. Interferometers form another category of instruments. The Expanded Very Large Array, the EVLA, is now in the test phase with "shared risk" observing. Other large interferometer systems are the VLBA, the Westerbork Synthesis Radio Telescope in the Netherlands, the Australia Telescope, the Giant Meter Wave Telescope in India, the MERLIN array a number of arrays at Cambridge University in the UK and the MOST facility in Australia. In the millimeter range, CARMA in California and Plateau de Bure in France are in full operation, as is the Sub-Millimeter Array of the Harvard-Smithsonian CfA and ASIAA on Mauna Kea, Hawaii. At longer wavelengths, the Low-Frequency Array, LOFAR, has started the first measurements and will expand by adding stations throughout Europe. The Square Kilometer Array, the SKA, is in the planning phase as is the FASR solar facility, while the Australian SKA Precursor (ASKAP), the South African SKA precursor, (MeerKAT), the Murchison Widefield Array in Western Australia and Long Wavelength Array in New Mexico are under construction. The first station of the Long Wavelength Array is now in operation. A portion of the Allen Telescope Array, ATA, is in operation. At submillimeter wavelengths, the Herschel Satellite Observatory has been delivering data. The Five Hundred Meter Aperture Spherical

Telescope, FAST, a design based on the Arecibo instrument, is being planned in China. This will be the world's largest single aperture. The Large Millimeter Telescope, LMT, a joint Mexican-US project, will soon begin science operations as will the Stratospheric Far-Infrared Observatory (SOFIA) operated by NASA and the German DLR organization. Descriptions of these instruments are to be found in the Internet. Finally, the most ambitious ground-based astronomy project to date is ALMA which started science operations in late 2011 (for an account of the variety of ALMA science goals, see Bachiller and Cernicharo 2008).

2 Radiative Transfer and Black Body Radiation

The total flux of a source is obtained by integrating Intensity (in Watts m^{-2} Hz^{-1} $steradian^{-1}$) over the total solid angle Ω_s subtended by the source

$$S_\nu = \int_{\Omega_s} I_\nu(\theta, \varphi) \cos\theta \, d\Omega. \tag{6.1}$$

The flux density of astronomical sources is given in units of the Jansky (hereafter Jy), that is, $1\,Jy = 10^{-26}\,W\,m^{-2}Hz^{-1}$.

The *equation of transfer* is useful in interpreting the behavior of astronomical sources, receiver systems, the effect of the earth's atmosphere on measurements. Much of this analysis is based on a one-dimensional version of the general expression as (see Lequeux 2004 or Tielens 2005):

$$\frac{dI_\nu}{ds} = -\kappa_\nu I_\nu + \varepsilon_\nu \quad . \tag{6.2}$$

The linear absorption coefficient κ_ν and the emissivity ε_ν are independent of the intensity I_ν. From the *optical depth* definition $d\tau_\nu = -\kappa_\nu\,ds$, the Kirchhoff relation $\varepsilon_\nu/\kappa_\nu = B_\nu$ (see Eq. 6.6), and the assumption of an isothermal medium, the result is:

$$I_\nu(s) = I_\nu(0)\,e^{-\tau_\nu(s)} + B_\nu(T)\,(1 - e^{-\tau_\nu(s)}) \quad . \tag{6.3}$$

For a large optical depth, that is, for $\tau_\nu(0) \to \infty$, (Eq. 6.3) approaches the limit

$$I_\nu = B_\nu(T). \tag{6.4}$$

This is case for planets and the 2.73 K CMB. From (Eq. 6.3), the difference between $I_\nu(s)$ and $I_\nu(0)$ gives

$$\Delta I_\nu(s) = I_\nu(s) - I_\nu(0) = (B_\nu(T) - I_\nu(0))(1 - e^{-\tau}). \tag{6.5}$$

this represents the result of an on-source minus off-source measurement, which is relevant for discrete sources.

The spectral distribution of the radiation of a black body in thermodynamic equilibrium is given by the Planck law

$$B_\nu(T) = \frac{2h\nu^3}{c^2}\frac{1}{e^{h\nu/kT} - 1} \quad . \tag{6.6}$$

If $h\nu \ll kT$, the *Rayleigh-Jeans Law* is obtained:

$$B_{RJ}(\nu, T) = \frac{2\nu^2}{c^2} kT \quad . \tag{6.7}$$

In the Rayleigh-Jeans relation, the brightness and the thermodynamic temperatures of Black Body emitters are strictly proportional (Eq. 6.7). This feature is useful, so the normal expression of brightness of an extended source is *brightness temperature* T_B:

$$T_B = \frac{c^2}{2k}\frac{1}{\nu^2}I_\nu = \frac{\lambda^2}{2k}I_\nu \,. \tag{6.8}$$

If I_ν is emitted by a black body and $h\nu \ll kT$ then (Eq. 6.8) gives the thermodynamic temperature of the source, a value that is independent of ν. If other processes are responsible for the emission of the radiation (e.g., synchrotron, free–free, or broadband dust emission), T_B will depend on the frequency; however; (Eq. 6.8) is still used. If the condition $\nu(\text{GHz}) \ll 20.84\,(\text{T(K)})$ is not valid, (Eq. 6.8) can still be applied, but T_B will differ from the thermodynamic temperature of a black body. However, corrections are simple to obtain.

If (Eq. 6.8) is combined with (Eq. 6.5), the result is an expression for brightness temperature:

$$J(T) = \frac{c^2}{2k\nu^2}(B_\nu(T) - I_\nu(0))(1 - e^{-\tau_\nu(s)}) \,.$$

The expression $J(T)$ can be expressed as a temperature in most cases. This quantity is referred to as T_R^*, the *radiation temperature* in the millimeter/submillimeter range, or the *brightness temperature*, T_B for longer wavelengths. In the Rayleigh-Jeans approximation the equation of transfer is:

$$\frac{dT_B(s)}{d\tau_\nu} = T_{bk}(0) - T(s) \quad , \tag{6.9}$$

where T_B is the measured quantity, $T_{bk}(s)$ is the background source temperature and $T(s)$ is the temperature of the intervening medium. If the medium is isothermal, the general (one-dimensional) solution becomes

$$T_B = T_{bk}(0)\, e^{-\tau_\nu(s)} + T\,(1 - e^{-\tau_\nu(s)}) \quad . \tag{6.10}$$

2.1 The Nyquist Theorem and Noise Temperature

This theorem relates the thermodynamic quantity temperature to the electrical quantities voltage and power. This is essential for the analysis of noise in receiver systems. The average power per unit bandwidth, P_ν (also referred to as Power Spectral Density, PSD), produced by a resistor R is

$$P_\nu = \langle i\nu \rangle = \frac{\langle \nu^2 \rangle}{2R} = \frac{1}{4R}\langle \nu_N^2 \rangle \,, \tag{6.11}$$

where $\nu(t)$ is the voltage that is produced by i across R, and $\langle \cdots \rangle$ indicates a time average. The first factor $\frac{1}{2}$ arises from the condition for the transfer of maximum power from R over a broad range of frequencies. The second factor $\frac{1}{2}$ arises from the time average of ν^2. Then

$$\langle \nu_N^2 \rangle = 4R\,k\,T \,. \tag{6.12}$$

When inserted into (Eq. 6.11), the result is

$$P_\nu = k\,T \,. \tag{6.13}$$

(➤ Eq. 6.13) can also be obtained by a reformulation of the Planck law for one dimension in the Rayleigh-Jeans limit. Thus, the available noise power of a resistor is proportional to its temperature, the *noise temperature* T_N, independent of the value of R and of frequency.

Not all circuit elements can be characterized by thermal noise. For example, a microwave oscillator can deliver 1 μW, the equivalent of more than 10^{16} K, although the physical temperature is ~300 K. This is an example of a very *nonthermal* process, so temperature is not a useful concept in this case.

2.2 Overview of Intensity, Flux Density, and Main Beam Brightness Temperature

Temperatures in radio astronomy have given rise to some confusion. A short summary with references to later sections is given here. Power is measured by an instrument consisting of an antenna and receiver. The power input can be calibrated and expressed as Flux Density or Intensity. For very extended sources, Intensity (see ➤ Eq. 6.8) can be expressed as a temperature, the *main beam brightness temperature*, T_{MB}. To obtain T_{MB}, the measurements must be calibrated (➤ Sect. 5.3) and corrected using the appropriate efficiencies (see ➤ Eq. 6.37 and following). For discrete sources, the combination of (➤ Eq. 6.1) with (➤ 6.8) gives:

$$S_\nu = \frac{2\,k\,\nu^2}{c^2}\,T_B\,\Delta\Omega \quad . \tag{6.14}$$

For a source with a Gaussian spatial distribution, this relation is

$$\left[\frac{S_\nu}{\text{Jy}}\right] = 0.0736\,T_B\left[\frac{\theta}{\text{arc seconds}}\right]^2\left[\frac{\lambda}{\text{mm}}\right]^{-2} \tag{6.15}$$

if the flux density S_ν and the actual (or "true") source size are known, then the *true brightness temperature*, T_B, of the source can be determined. For Local Thermodynamic Equilibrium (LTE), T_B represents the physical temperature of the source. If the *apparent* source size, that is, the source angular size as measured with an antenna is known, (➤ Eq. 6.15) allows a calculation of T_{MB}. For discrete sources, T_{MB} depends on the angular resolution. If the antenna beam size (see ➤ *Fig. 6-3* and discussion) has a Gaussian shape θ_b, the relation of actual θ_s and apparent size θ_o is:

$$\theta_o^2 = \theta_s^2 + \theta_b^2 . \tag{6.16}$$

then from (➤ Eq. 6.14), the relation of T_{MB} and T_B is:

$$T_{MB}\left(\theta_s^2 + \theta_b^2\right) = T_B\,\theta_s^2 \tag{6.17}$$

Finally, the PSD entering the receiver (➤ Eq. 6.13) is antenna temperature, T_A; this is relevant for estimating signal-to-noise ratios (see (➤ Eq. 6.39) and (➤ Eq. 6.42)). To establish temperature scales and relate received power to source parameters for filled apertures, see ➤ Sect. 5.3. For interferometry and Aperture Synthesis, see ➤ Sect. 6.

2.3 Interstellar Dispersion and Polarization

Pulsars emit radiation in a short time interval (see Lorimer and Kramer 2004; Lyne and Graham-Smith 2006). If all frequencies are emitted at the same instant, the arrival time delay of different frequencies is caused by the ionized Interstellar Medium (ISM). This is characterized by the quantity $\int_0^L N(l)\,\mathrm{d}l$, which is the column density of the electrons to a distance L. Since distances in astronomy are measured in parsecs it has become customary to express the *dispersion measure* as:

$$\mathrm{DM} = \int_0^L \left(\frac{N}{\mathrm{cm}^{-3}}\right) \mathrm{d}\left(\frac{l}{\mathrm{pc}}\right) \tag{6.18}$$

The lower frequencies are delayed more in the ISM, so the relative time delay is:

$$\frac{\Delta\tau_\mathrm{D}}{\mu\mathrm{s}} = 1.34 \times 10^{-9} \left[\frac{\mathrm{DM}}{\mathrm{cm}^{-2}}\right] \left[\frac{1}{\left(\frac{\nu_1}{\mathrm{MHz}}\right)^2} - \frac{1}{\left(\frac{\nu_2}{\mathrm{MHz}}\right)^2}\right] \tag{6.19}$$

Since both time delay $\Delta\tau_\mathrm{D}$ and observing frequencies $\nu_1 < \nu_2$ can be measured with high precision, a very accurate value of DM for a given pulsar can be determined. Provided the distance to the pulsar, L, is known, a good estimate of the average electron density between observer and pulsar can be found. However since L is usually known with moderate accuracy, only approximate values for N can be obtained. Often the opposite procedure is used: From reasonable values for N, a measured DM provides information on the unknown distance L to the pulsar.

Broadband linear polarization is caused by nonthermal processes (see Rybicki and Lightman 1979) including Pulsar radiation, quasi-thermal emission from aligned, nonspherical dust grains (see Hildebrand 1983), and scattering from free electrons. Faraday rotation will change the position angle of linear polarization as the radiation passes through an ionized medium; this varies as λ^2, so this effect is larger for longer wavelengths. It is usual to characterize polarization by the four Stokes Parameters, which are the sum or difference of measured quantities. The total intensity of a wave is given by the parameter I. The amount and angle of linear polarization are given by the parameters Q and U, while the amount and sense of circular polarization is given by the parameter V. Hertz dipoles are sensitive to a single linear polarization. By rotating the dipole over an angle perpendicular to the direction of the radiation, it is possible to determine the amount and angle of linearly polarized radiation. Helical antennas or arrangements of two Hertz dipoles are sensitive to circular polarization. Generally, polarized radiation is a combination of linear and circular, and is usually less than 100% polarized, so four Stokes parameters must be specified. The definition of the sense of circular polarization in radio astronomy is the same as in Electrical Engineering but *opposite* to that used in the optical range; see Born and Wolf (1965) for a complete analysis of polarization, using the *optical* definition of circular polarization. Poincaré introduced a representation that permits an easy visualization of all the different states of polarization of a vector wave. See Thompson et al. (2001), Crutcher (2008), Thum et al. (2008), or Wilson et al. (2008) for more details.

3 Receiver Systems

3.1 Coherent and Incoherent Receivers

Receivers are assumed to be linear power measuring devices, that is, any nonlinearity is a small quantity. There are two types of receivers: coherent and incoherent. Coherent receivers are preserve the phase of the input radiation while incoherent do not. Heterodyne (technically "superheterodyne") receivers are those which shift the frequency of the input but preserve phase. The most commonly used coherent receivers employ heterodyning, that is, frequency shifting (see ➋ Sect. 4.2.1). The most commonly used incoherent receivers are bolometers (➋ Sect. 4.1); these are direct detection receivers, that is, operate at sky frequency. Both coherent and incoherent receivers add noise to the astronomical input signal; it is assumed that the noise of both the input signal and the receiver follow Gaussian distributions. The noise contribution of coherent receivers is expressed in Kelvins. Bolometer noise is characterized by the *Noise Equivalent Power*, or NEP, in units of Watts $Hz^{-1/2}$ (see ➋ Sects. 3.1.1 and ➋ 5.3.3). NEP is the input power level which doubles the output power. More extensive discussions of receiver properties are given in Rieke (2002) or Wilson et al. (2008).

To analyze the performance of a receiver, the commonly accepted model is an ideal receiver with no internal noise, but connected to two noise sources, one for the external noise (including the astronomical signal) and a second for the receiver noise. To be useful, receiver systems must increase the input power level. The noise contribution is characterized by the *Noise Factor*, *F*. If the signal-to-noise ratio at the input is expressed as (S_1/N_1) and at the output as (S_2/N_2), the noise factor is:

$$F = \frac{S_1/N_1}{S_2/N_2}. \tag{6.20}$$

A further step is to assume that the signal is amplified by a gain factor G but otherwise unchanged. Then $S_2 = G\,S_1$ and:

$$F = \frac{N_2}{G\,N_1}. \tag{6.21}$$

For a direct detection system such as a bolometer, $G = 1$. For coherent receivers, there must be a minimum noise contribution (see ➋ Sect. 4.2.4), so $F > 1$. For coherent receivers, F is expressed in temperature units as T_R using the relation

$$T_R = (F - 1) \cdot 290\mathrm{K}. \tag{6.22}$$

3.1.1 Receiver Calibration

Heterodyne receiver noise performance is usually expressed in degrees Kelvin. In the calibration process, a power scale (the PSD) is established at the receiver input. This is measured in terms of the noise temperature. To calibrate a receiver, the noise temperature increment ΔT at the receiver input must be related to a given measured receiver output increment Δz (this applies to coherent receivers which have a wide dynamic range and a total power or "DC" response). Usually resistive loads at two known (thermodynamic) temperatures T_{L} and T_{H} are used. The receiver outputs are z_{L} and z_{H}, while T_{L} and T_{H} are the resistive loads at two temperatures. The relations are:

$$z_{\mathrm{L}} = (T_{\mathrm{L}} + T_{\mathrm{R}})\,G,$$
$$z_{\mathrm{H}} = (T_{\mathrm{H}} + T_{\mathrm{R}})\,G,$$

taking

$$y = z_{\mathrm{H}}/z_{\mathrm{L}}. \tag{6.23}$$

The result is:

$$T_{rx} = \frac{T_H - T_L\, y}{y - 1} \quad . \tag{6.24}$$

This is known as the "y-factor"; the procedure is a "hot-cold" measurement. The determination of the y-factor is calculated in the Rayleigh-Jeans limit. Absorbers at temperatures of T_H and T_L are used to produce the inputs. Often these are chosen to be at the ambient temperature ($T_H \cong 293$ K or 20°C) and at the temperature of liquid nitrogen ($T_L \cong 78$ K or −195°C). When receivers are installed on antennas, such "hot-cold" calibrations are done only infrequently. As will be discussed in ❯ Sect. 5.3.2, in the centimeter and meter range, calibration signals are provided by noise diodes; from measurements of sources with known flux densities intensity scales are established. Any atmospheric corrections are assumed to be small at these wavelengths. As will be discussed in ❯ Sect. 5.3.3, in the millimeter/submillimeter wavelength range, from measurements of an ambient load (or two loads at different temperatures), combined with measurements of emission from the atmosphere and models of the atmosphere, estimates of atmospheric transmission are made.

Bolometer performance is characterized by the *Noise Equivalent Power*, or NEP, given in units of Watts $\mathrm{Hz}^{-1/2}$. The expression for NEP can be related to a receiver noise temperature. For ground-based bolometer systems, background noise dominates. For these, the background noise is given as T_{BG}:

$$\mathrm{NEP} = 2\varepsilon\, k\, T_{BG} \sqrt{\Delta\nu} \quad . \tag{6.25}$$

here ε is the emissivity of the background and $\Delta\nu$ is the bandwidth. Typical values for ground-based millimeter/submillimeter bolometers are $\varepsilon = 0.5$, $T_{BG} = 300$ K and $\Delta\nu = 100$ GHz. For these values, NEP $= 1.3 \times 10^{-15}$ Watts $\mathrm{Hz}^{-1/2}$. With the collecting area of the IRAM 30 m or the JCMT telescopes, sources in the milli-Jansky (mJy) range can be measured.

Usually bolometers are "A. C." coupled, that is, the output responds to *differences* in the input power, so hot-cold measurements are not useful for characterizing bolometers. The response of bolometers is usually determined by measurements of sources with known flux densities, followed by measurements at, for example, elevations of 20°, 30°, 60°, and 90° to determine the atmospheric transmission (see ❯ Sect. 5.3.4).

3.1.2 Noise Uncertainties Due to Random Processes

The noise contributions from source, atmosphere, ground, telescope surface and receiver are always additive:

$$T_{sys} = \sum T_i \tag{6.26}$$

From Gaussian statistics, the Root Mean Square, RMS, noise is given by the mean value divided by the square root of the number of samples. From the estimate that the number of samples is given by the product of receiver bandwidth multiplied by the integration time, the result is:

$$\Delta T_{RMS} = \frac{T_{sys}}{\sqrt{\Delta\nu\, \tau}} \quad . \tag{6.27}$$

A much more elaborate derivation is to be found in Chapter 4 of Rohlfs and Wilson (2004), while a somewhat simpler account is in Wilson et al. (2008). The calibration process in ❯ Sect. 3.1.1 allows the receiver noise to be expressed in degrees Kelvin. The relation of T_{sys}

to T_{rx} is $T_{sys} = T_A + T_{rx}$, where T_A represents the power entering the receiver; at some wavelengths T_A will dominate T_{rx}. In the millimeter/submillimeter range, use is made of T^*_{sys}, the system noise outside the atmosphere, since the attenuation of astronomical radiation is large. This will be presented in ➋ Sect. 5.3.1 and following.

3.1.3 Receiver Stability

Sensitive receivers are designed to achieve a low value for T_{rx}. Since the signals received are of exceedingly low power, receivers must also provide large receiver gains, G (of order 10^{12}), for sufficient output power. Thus even very small gain instabilities can dominate the thermal receiver noise. Since receiver stability considerations are of prime importance, comparison switching was necessary for early receivers (Dicke 1946). Great advances have been made in improving receiver stability since the 1960s so the need for rapid switching is lessened. In the meter and centimeter wavelength range, the time between reference measurements has increased. However in the millimeter/submillimeter range, instabilities of the atmosphere play an important role; to insure that noise decreases following (➋ Eq. 6.27), the effects of atmospheric and/or receiver instabilities must be eliminated. For single dish measurements, atmospheric changes can be compensated for by rapidly differencing a measurement of the target source and a reference. Such *comparison or "Dicke" switched* measurements are necessary for ground-based observations. If a typical procedure consists of using a total power receiver to measure on-source for 1/2 of the total time, then an off-source comparison for 1/2 of the time and taking the difference of on-source minus off-source measurements, the ΔT_{RMS} will be a factor of 2 larger than the value given by (➋ Eq. 6.27).

4 Practical Aspects of Receivers

This section concentrates on receivers that are currently in use. For more details see Goldsmith (1988), Rieke (2002), or Wilson et al. (2008).

4.1 Bolometer Radiometers

Bolometers operate by use of the effect that the resistance, R, of a material varies with the temperature. In the 1970s, the most sensitive bolometers were semiconductor devices pioneered by F. Low. This is achieved when the bolometer element is cooled to very low temperatures. When radiation is incident, the characteristics change, so this is a measure of the intensity of the incident radiation. Because this is a thermal effect, it is independent of the frequency and polarization of the radiation absorbed. Thus bolometers are intrinsically broadband devices. It is possible to mount a polarization-sensitive device before the bolometer and thereby measure the direction and degree of linear polarization. Also, it is possible to carry out spectroscopy, if frequency sensitive elements, either filters, Michelson or Fabry-Perot interferometers, are placed before the bolometer element. Since these spectrometers operate at the sky frequency, the fractional resolution $(\Delta\nu/\nu)$ is at best ~10^{-4}. The data from each bolometer detector element (pixel) must be read out and then amplified.

For single dish (i.e., filled apertures) broadband continuum measurements at $\lambda < 2$ mm, multi-beam bolometers are the most common systems, and such systems can have a large number of beams. A promising new development in bolometer receivers is *Transition Edge Sensors* referred to as TES bolometers. These superconducting devices may allow more than an order of magnitude increase in sensitivity, if the bolometer is not background limited. For bolometers used on earth-bound telescopes, the improvement with TES systems may be only about two to three times more sensitive than the semiconductor bolometers, but TESs will allow readouts from a much larger number of pixels.

A number of large bolometer arrays have produced numerous publications: (1) MAMBO2 (MAx-Planck-Millimeter Bolometer) used on the IRAM 30-m telescope at 1.3 mm, (2) SCUBA (Submillimeter Common User Bolometer Array; Holland et al. 1999) on the JCMT, (3) the LABOCA (LArge Bolometer CAmera) array on the APEX 12 m (4) telescope, SHARC (Submillimeter High Angular Resolution Camera) on the Caltech Submillimeter Observatory 10-m telescope and (5) MUSTANG (MUtiplexed Squid TES Array) on the GBT. SCUBA will be replaced with SCUBA-2 now being constructed at the UK Astronomy Technology Center, and there are plans to replace the MUSTANG array by MUSTANG-2, which is a larger TES system.

4.2 Coherent Receivers

Usually, coherent receivers make use of heterodyning to shift the signal input frequency without changing other properties of the input signal; in practice, this is carried out by the use of mixers (➋ Sect. 4.2.2). The heterodyne process is used in all branches of communications technology; use of heterodyning allows measurements with unlimited spectral resolution. Although heterodyne receivers have a number of components, these systems have more flexibility than bolometers.

4.2.1 Noise Contributions in Coherent Receivers

The noise generated in the first element dominates the system noise. The mathematical expression is given by the *Friis* relation which accounts for the effect of cascaded amplifiers on the noise performance of a receiver:

$$T_S = T_{S1} + \frac{1}{G_1} T_{S2} + \frac{1}{G_1 G_2} T_{S3} + \cdots + \frac{1}{G_1 G_2 \ldots G_{n-1}} T_{Sn} \tag{6.28}$$

where G_1 is the gain of the first element and T_{S1} is the noise temperature of this element. For $\lambda > 3$ mm ($\nu < 115$ GHz), the best cooled first elements, High Electron Mobility Transistors (HEMTs), typically have $G_1 = 10^3$ and $T_{S1} = 50$ K; for $\lambda < 0.8$ mm, the best cooled first elements, superconducting mixers, typically have $G_1 \leq 1$, that is, a small loss, and $T_{S1} \leq 500$ K. The stage following the mixer should have the lowest noise temperature and high gain.

4.2.2 Mixers

Mixers have been used in heterodyne receivers since Jansky's time. At first these were metal-oxide-semiconductor or *Schottky* mixers. Mixers allow the signal frequency to be changed without altering the characteristics of the signal. In the mixing process, the input signal is multiplied by an intense monochromatic signal from a *local oscillator*, LO. The frequency stability of the LO signal is maintained by a stabilization device in which the LO signal is compared with a stable input, in recent times, an atomic standard. These phaselock loop systems produce a pure, highly stable, monochromatic signal. The mixer can be operated in the Double Sideband (DSB) mode, in which two sky frequencies, "signal" and "image" at equal separations from the LO frequency (equal to the IF frequency) are shifted into intermediate (IF) frequency band. For spectral line measurements, usually one sideband is wanted, but the other not. DSB operation adds both noise and (usually) unwanted spectral lines; for spectral line measurements, single sideband (SSB) operation is preferred. In SSB operation, the unwanted sideband is suppressed, at the cost of more complexity. In the submillimeter wavelength ranges, DSB mixers are still commonly used as the first stage of a receiver; in the millimeter and centimeter ranges, SSB operation is now the rule.

A significant improvement can be obtained if the mixer junction is operated in the superconducting mode. The noise temperatures and LO power requirements of superconducting mixers are much lower than Schottky mixers. Finally, the physical layout of such devices is simpler since the mixer is a planar device, deposited on a substrate by lithographic techniques. SIS mixers consist of a superconducting layer, a thin insulating layer and another superconducting layer (see Phillips and Woody 1982).

Superconducting Hot Electron Bolometer mixers (HEB) are heterodyne devices, in spite of the name. These mixers make use of superconducting thin films which have submicron sizes (see Kawamura et al. 2002).

A number of multi-beam heterodyne cameras are in operation in the centimeter range, but only a few in the millimeter/submillimeter range. The first millimeter multi-beam system was the SEQUOIA array receiver pioneered by S. Weinreb; such devices are becoming more common. In contrast, multi-beam systems that use SIS front ends are rare. Examples are a nine beam Heterodyne Receiver Array of SIS mixers at 1.3 mm, HERA, on the IRAM 30-m telescope, HARP-B, a 16 beam SIS system in operation at the JCMT for 0.8 mm, and the CHAMP+ receiver at the Max-Planck-Inst. für Radioastronomy on the APEX 12-m telescope.

4.2.3 Square-Law Detectors

For heterodyne receivers the input is normally amplified (for $\nu < 115$ GHz), translated in frequency, and then detected in a device that produces an output signal $y(t)$ which is proportional to the square of $v(t)$:

$$y(t) = a\,v^2(t) \tag{6.29}$$

Once detected, phase information is lost. For interferometers, the output of each antenna is a voltage, shifted in frequency and then digitized. This output is brought to a central location for correlation.

4.2.4 The Minimum Noise in a Coherent System

The ultimate limit for coherent receivers or amplifiers is obtained by an application of the *Heisenberg uncertainty principle* involving phase and number of photons. From this, the

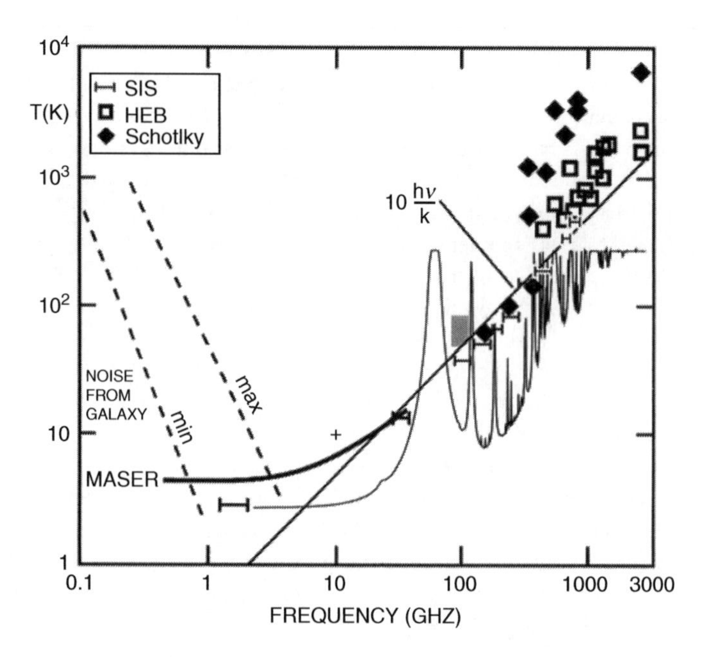

Fig. 6-2
Receiver noise temperatures for coherent amplifier systems compared to the temperatures from the Milky Way galaxy (at long wavelengths, on *left part* of figure) and the atmosphere (at millimeter/submillimeter wavelengths on the *right side*). The atmospheric emission is based on a model of zenith emission for 0.4 mm of water vapor, that is, excellent weather (plot from B. Nicolic (Cambridge Univ.) using the "AM" program of S. Paine (Harvard-Smithsonian Center for Astrophysics)). This does not take into account the absorption of the astronomical signal. In the 1–26 GHz range, the two horizontal lines represent the noise temperatures of the best HEMT amplifiers, while the solid line represents the noise temperatures of maser receivers. The shaded region between 85 and 115.6 GHz is the receiver noise for the SEQUOIA array (Five College Radio Astronomy Observatory) which consists of monolithic millimeter integrated circuits (MMIC). The meaning of the other symbols is given in the upper left of the diagram (SISs are Superconductor-Insulator-Superconductor mixers, HEBs are Hot Electron Bolometer mixers). The double sideband (DSB) mixer noise temperatures were converted to single sideband (SSB) noise temperatures by doubling the receiver noise. The ALMA mixer noise temperatures are SSB, as are the HEMT values. The line marked "10 hν/kT" refers to the limit described in (Eq. 6.30). Some data used in this diagram are taken from Rieke (2002). (The figure is from Wilson et al. 2008)

minimum noise of a coherent amplifier results in a receiver noise temperature of

$$T_{\rm rx}(\text{minimum}) = \frac{h\nu}{k} \quad . \tag{6.30}$$

For *incoherent* detectors, such as bolometers, phase is not preserved, so this limit does *not* exist. In the millimeter wavelength region, this noise temperature limit is quite small; at $\lambda = 2.6$ mm ($\nu = 115$ GHz), this limit is 5.5 K. The value for the ALMA receiver in this range is about five to six times the minimum. A significant difference between radio and optical regimes is that the minimum noise in the radio range is small, so that the power from a single receiver can be amplified and then divided. For example, for the EVLA, the voltage output of all 351 antenna pairs are combined with little or no loss in the signal-to-noise ratio. Another example is given in ➋ Sect. 4.3.1, where a radio polarimeter can produce all four Stokes parameters from two inputs without a loss of the signal-to-noise ratio.

4.3 Back Ends: Polarimeters and Spectrometers

The term "Back End" is used to specify the devices following the IF amplifiers. Many different back ends have been designed for specialized purposes such as continuum, spectral, or polarization measurements.

For a single dish continuum correlation receiver, the (identical) receiver input is divided, amplified in two identical systems, and then the outputs are multiplied. The gain fluctuations are not correlated but the signals are, so the effect on the output is the same as with a Dicke switched system, but with *no* time spent on a reference.

4.3.1 Polarimeters

A typical heterodyne dual polarization receiver consists of two identical systems, each sensitive to one of the two orthogonal polarizations, linear or circular. Both systems must be connected to the same local oscillator to insure that the phases have a definite relation. Given this arrangement, a polarimeter can provide values of all four Stokes parameters simultaneously. All Stokes parameters can also be measured using a single receiver whose input is switched from one sense of polarization to the other, but then the integration time for each polarization will be halved.

4.3.2 Spectrometers

Spectrometers analyze the spectral information contained in the radiation field. To accomplish this, the spectrometer must be SSB and the frequency resolution $\Delta\nu$ is usually very good, sometimes in the kHz range. In addition, the time stability must be high. If a resolution of $\Delta\nu$ is to be achieved for the spectrometer, all those parts of the system that enter critically into the frequency response have to be maintained to better than $0.1\,\Delta\nu$. For an overview of the current state of spectrometers, see Baker et al. (2007).

Conceptually, the simplest spectrometer is composed of a set of n adjacent filters, each with a bandwidth $\Delta\nu$. Following each filter is a square-law detector and integrator. For a finer resolution, another set of n filters must be constructed.

Another approach to spectral analysis is to Fourier transform (FT) the input, $v(t)$ to obtain $v(\nu)$ and then square $v(\nu)$ to obtain the Power Spectral Density. The maximum bandwidth is limited by the sampling rate. From (another!) Nyquist theorem, it is necessary to sample at a rate equal to twice the bandwidth. In the simplest scheme, for a bandwidth of 1 GHz, the sampling must occur at a rate of 2 GHz. After sampling and Fourier transforming, the output is squared to produce power in an "FX" autocorrelator. For 10^3 samples, each channel will have a 1 MHz resolution.

For "XF" systems, the input $v(t)$ is multiplied (the "X") with a delayed signal $v(t-\tau)$ to obtain the autocorrelation function $R(\tau)$. This is then Fourier transformed to obtain the spectrum. For 10^3 samples, there will be 10^3 frequency channels. For an XF system the time delays are performed in a set of serial digital shift registers with a sample delayed by a time τ. Autocorrelation can also be carried out with the help of analog devices using a series of cable delay lines; these can provide very large bandwidths. The first XF system for astronomy was a digital autocorrelator built by S. Weinreb in 1963.

The two significant advantages of digital spectrometers are: (1) flexibility and (2) a noise behavior that follows $1/\sqrt{t}$ after many hours of integration. The flexibility allows the choice of many different frequency resolutions and bandwidths or even to employ a number of different spectrometers, each with different bandwidths, simultaneously.

A serious drawback of digital auto and cross-correlation spectrometers had been limited bandwidths. However, advances in digital technology in recent years have allowed the construction of autocorrelation spectrometers with several 10^3 channels covering instantaneous bandwidths of several GHz.

Autocorrelation systems are used in single antennas. The calculation of spectra makes use of the symmetric nature of the autocorrelation function, ACF, so the number of delays gives the number of spectral channels.

Cross-correlators are used in interferometers and in some single dish applications. When used in an interferometer, the cross-correlation is between different inputs so will not necessarily be symmetric. Thus, the zero delay of the cross-correlator is placed in channel $N/2$. The number of delays, N, allows the determination of $N/2$ spectral intensities, and $N/2$ phases. The cross-correlation hardware can employ either an XF or an FX correlator. For more details about the use of cross-correlation, see Sect. 6.

Until recently, spectrometers with bandwidths of several gigahertz often made use of Acoustic Optical analog techniques. The Acoustic Optical Spectrometer (AOS) makes use of the diffraction of light by ultrasonic waves: these cause periodic density variations in the crystal through which it passes. These density variations in turn cause variations in the bulk constants of the crystal, so that a plane light wave passing through this medium will be modulated by the interaction with the crystal. The modulated light is detected in a charge coupled device. Typical AOS's have an instantaneous bandwidth of 2 GHz and 2,000 spectral channels.

In all cases, the spectra of the individual channels of a spectrometer are expressed in terms of temperature with the relation:

$$T_i = [(S_i - R_i)/R_i] \cdot T_{\text{sys}} \tag{6.31}$$

where S_i is the normalized spectrum of channel i for the on-source measurement and R_i is the corresponding reference for this channel. For millimeter/submillimeter spectra, T_{sys} is replaced by T^*_{sys} (corrected for atmospheric losses; see Sect. 5.3.3). For cross-correlators, as used in interferometers, the signals from two antennas are multiplied. In this case, the value of T_{sys} is the square root of the product of the system noise temperatures of the two systems.

5 Antennas

The antenna serves to focus power into the feed, a device that efficiently transfers power in the electromagnetic wave to the receiver. According to the principle of *reciprocity*, the properties of antennas such as beam sizes, efficiencies, etc., are the same whether these are used for receiving or transmitting. Reciprocity holds in astronomy, so it is usual to interchangeably use expressions that involve either transmission or reception when discussing antenna properties. All of the following applies to the far-field radiation.

5.1 The Hertz Dipole

The total power radiated from a Hertz dipole carrying an oscillating current I at a wavelength λ is

$$P = \frac{2c}{3}\left(\frac{I\Delta l}{2\lambda}\right)^2 \quad . \tag{6.32}$$

For the Hertz dipole, the radiation is linearly polarized with the electric field along the direction of the dipole. The radiation pattern has a donut shape, with the cylindrically symmetric maximum perpendicular to the axis of the dipole. Along the direction of the dipole, the radiation field is zero. To improve directivity, reflecting screens have been placed behind a dipole, and in addition, collections of dipoles, driven in phase, are used. Hertz dipole radiators have the best efficiency when the size of the dipole is $1/2\,\lambda$.

5.2 Filled Apertures

This Section is a simplified description of antenna properties needed for the interpretation of astronomical measurements (for more detail, see Baars 2007). At centimeter and shorter wavelengths, flared waveguides ("feed horns") or dipoles are used to convey power focused by the antenna (i.e., electromagnetic waves in free space) to the receiver (voltage). At the longest wavelengths, dipoles are used as the antennas. Details are to be found in Love (1976) and Goldsmith (1988, 1994).

5.2.1 Angular Resolution and Efficiencies

From diffraction theory (see Jenkins and White 2001), the angular resolution of a reflector of diameter D at a wavelength λ is

$$\theta = k\frac{\lambda}{D} \quad . \tag{6.33}$$

where k is of order unity. This universal result gives a value for θ (here in radians when D and λ have the same units). Diffraction theory also predicts the unavoidable presence of sidelobes, that is, secondary maxima. The sidelobes can be reduced by *tapering* the antenna illumination. Tapering lowers the response to very compact sources and increases the value of θ, that is, widens the beam.

The reciprocity concept provides a method to measure the power pattern (response pattern or Point Spread Function, PSF) using transmitters. However, the distance from a large antenna A (diameter $D \gg \lambda$) to a transmitter B (small in size) must be so large that B produces plane waves across the aperture D of antenna A, that is, B is in the far field of A. This is the *Rayleigh* distance; it requires that the curvature of a wavefront emitted by B is much less than $\lambda/16$ across the geometric dimensions of antenna A. By definition, at the Rayleigh distance $\mathcal{D}$, the curvature must be $\gg D^2/8\lambda$ for an antenna of diameter D.

Often, the *normalized power pattern* is measured:

$$P_{\rm n}(\vartheta,\varphi) = \frac{1}{P_{\rm max}} P(\vartheta,\varphi) \quad . \tag{6.34}$$

For larger apertures, the transmitter is usually replaced by a small diameter radio source of known flux density (see Baars et al. 1977; Ott et al. 1994). The flux densities of a few primary calibration sources are determined by measurements using horn antennas at centimeter and millimeter wavelengths. At millimeter/submillimeter wavelengths, it is usual to employ planets, or moons of planets, whose surface temperatures are known (see Altenhoff 1985; Sandell 1994).

The *beam solid angle* $\Omega_{\rm A}$ of an antenna is given by

$$\Omega_{\rm A} = \iint_{4\pi} P_{\rm n}(\vartheta,\varphi)\,{\rm d}\Omega = \int_0^{2\pi}\int_0^{\pi} P_{\rm n}(\vartheta,\varphi)\sin\vartheta\,{\rm d}\vartheta\,{\rm d}\varphi \tag{6.35}$$

this is measured in steradians (sr). The integration is extended over all angles, so $\Omega_{\rm A}$ is the solid angle of an ideal antenna having $P_{\rm n} = 1$ for $\Omega_{\rm A}$ and $P_{\rm n} = 0$ everywhere else. For most antennas the (normalized) power pattern has much larger values for a limited range of both ϑ and φ than for the remainder; the range where $\Omega_{\rm A}$ is large is the main beam of the antenna; the remainder are the sidelobes or backlobes (➔ *Fig. 6-3*).

In analogy to (➔ Eq. 6.35) the *main beam solid angle* $\Omega_{\rm MB}$ is defined as

$$\Omega_{\rm MB} = \iint_{\substack{\rm main\\ \rm lobe}} P_{\rm n}(\vartheta,\varphi)\,{\rm d}\Omega \quad . \tag{6.36}$$

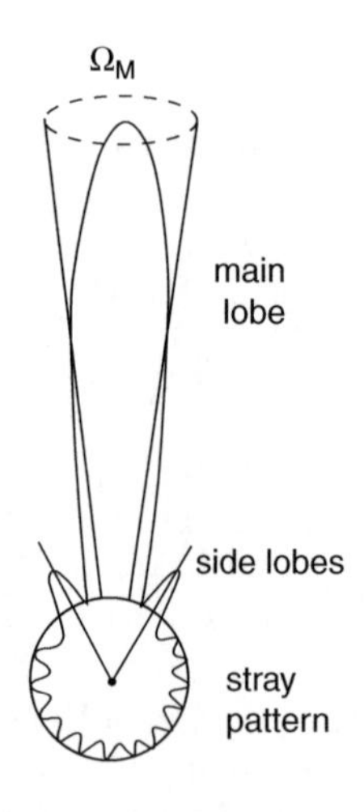

Fig. 6-3

A polar power pattern showing the main beam, and near and far sidelobes. The weaker far sidelobes have been combined to form the stray pattern

The quality of a single antenna depends on how well the power pattern is concentrated in the main beam. The definition of *main beam efficiency* or *beam efficiency*, η_B, is:

$$\eta_B = \frac{\Omega_{MB}}{\Omega_A} \quad . \tag{6.37}$$

η_B is the fraction of the power that is concentrated in the main beam. The main beam efficiency can be modified (within limits) for parabolic antennas by changing the illumination of the main reflector. An underilluminated antenna has a wider main beam but lower sidelobes. The angular extent of the main beam is usually described by the *full width to half power* (FWHP), the angle between points of the main beam where the normalized power pattern falls to 1/2 of the maximum. For elliptically shaped main beams, values for widths in orthogonal directions are needed. The beamwidth, θ, is given by (➋ Eq. 6.33). If the FWHP beamwidth is well defined, the location of an isolated source is determined to the accuracy given by the FWHP divided by the S/N ratio. Thus, it is possible to determine positions to small fractions of the FWHP beamwidth, if the signal-to-noise ratio is high and noise is the only limit.

If a plane wave with the power density $|\langle\vec{S}\rangle|$ in Watts m^{-2} is intercepted by an antenna, a certain amount of power is extracted from this wave. This power is P_e and the fraction is:

$$A_e = P_e / |\langle\vec{S}\rangle| \tag{6.38}$$

the *effective aperture* of the antenna. A_e has the dimension of m^2. Compared to the *geometric aperture* A_g an aperture efficiency η_A can be defined by:

$$A_e = \eta_A A_g \quad . \tag{6.39}$$

If an antenna with a normalized power pattern $P_n(\vartheta,\varphi)$ is used to receive radiation from a brightness distribution $B_\nu(\vartheta,\varphi)$ in the sky, at the output terminals of the antenna the power per unit bandwidth (PSD), in Watts Hz^{-1}, P_ν is:

$$P_\nu = \tfrac{1}{2} A_e \int\!\!\int B_\nu(\vartheta,\varphi)\, P_n(\vartheta,\varphi)\, d\Omega \,. \tag{6.40}$$

By definition, this operates in the Rayleigh-Jeans limit, so the equivalent distribution of brightness temperature can be replaced by an equivalent *antenna temperature* T_A (➋ 6.13):

$$P_\nu = k\, T_A \,. \tag{6.41}$$

This definition of *antenna temperature* relates the output of the antenna to the power from a matched resistor. When these two power levels are equal, then the antenna temperature is given by the temperature of the resistor. The effective aperture A_e can be replaced by the beam solid angle $\Omega_A \cdot \lambda^2$. Then (➋ 6.40) becomes

$$T_A(\vartheta_0,\varphi_0) = \frac{\int T_B(\vartheta,\varphi) P_n(\vartheta-\vartheta_0,\varphi-\varphi_0)\sin\vartheta\, d\vartheta\, d\varphi}{\int P_n(\vartheta,\varphi)\, d\Omega} \tag{6.42}$$

From (➋ Eq. 6.42), $T_A < T_B$ in all cases. The numerator is the *convolution* of the brightness temperature with the beam pattern of the telescope (Fourier methods are of great value in this analysis; see Bracewell 1986). The brightness temperature $T_b(\vartheta,\varphi)$ corresponds to the thermodynamic temperature of the radiating material *only* for thermal radiation in the Rayleigh-Jeans limit from an optically thick source; in all other cases T_B is a convenient quantity that represents source intensity at a given frequency. The quantity T_A in (➋ Eq. 6.42) was obtained for an antenna in which ohmic losses and absorption in the earth's atmosphere were neglected. These losses can be corrected in the calibration process. Since T_A is the quantity measured while T_B is desired, (Eq. 6.42) must be inverted. (Eq. 6.42) can be solved only if $T_A(\vartheta, \varphi)$ and $P_n(\vartheta, \varphi)$ are known exactly over the full range of angles. In practice this inversion is possible only approximately, since both $T_A(\vartheta, \varphi)$ and $P_n(\vartheta, \varphi)$ are known only for a limited range of ϑ and φ values, and the measured data are affected by noise. Therefore only an approximate deconvolution can be performed. If the source distribution $T_B(\vartheta, \varphi)$ has a small extent compared to the telescope beam, the best estimate for the upper limit to the actual FWHP source size is 1/2 of the FWHP of the telescope beam.

5.2.2 Efficiencies for Compact Sources

For a source small compared to the beam, (Eq. 6.40) and (Eq. 6.41) give:

$$P_\nu = \tfrac{1}{2} A_e \, S_\nu = k \, T_A \tag{6.43}$$

T_A is the antenna temperature at the receiver, while T'_A is this quantity corrected for effect of the earth's atmosphere. In the meter and centimeter range $T_A = T'_A$, so in the following, T'_A will be used:

$$T'_A = \Gamma S_\nu \tag{6.44}$$

where Γ is the *sensitivity* of the telescope measured in K Jy^{-1}. Introducing the aperture efficiency η_A according to (Eq. 6.39) we find

$$\Gamma = \eta_A \frac{\pi D^2}{8k} \quad . \tag{6.45}$$

Thus Γ or η_A can be measured with the help of a calibrating source provided that the diameter D and the noise power scale in the receiving system are known. When (Eq. 6.44) is solved for S_ν, the result is:

$$S_\nu = 3,520 \, \frac{T'_A[\mathrm{K}]}{\eta_A [D/\mathrm{m}]^2} \, . \tag{6.46}$$

The *brightness temperature* is defined as the Rayleigh-Jeans temperature of an equivalent black body which will give the same power per unit area per unit frequency interval per unit solid angle as the celestial source. Both T'_A and T_{MB} are defined in the Rayleigh-Jeans limit, but the brightness temperature scale has to be corrected for antenna efficiency. The conversion from source flux density to source brightness temperature for sources with sizes small compared to the telescope beam is given by (Eq. 6.15).

For sources small compared to the beam, the antenna and main beam brightness temperatures are related by the main beam efficiency, η_B:

$$\eta_B = \frac{T'_A}{T_{MB}} \, . \tag{6.47}$$

This is valid for sources where sidelobe structure is not important (see the discussion after (Eq. 6.42)). Although a source may not have a Gaussian shape, fits of multiple Gaussians can be used to obtain an accurate representation.

What remains is a calibration of the temperature scales and a correction for absorption in the earth's atmosphere. This is dealt with in Sect. 5.3.

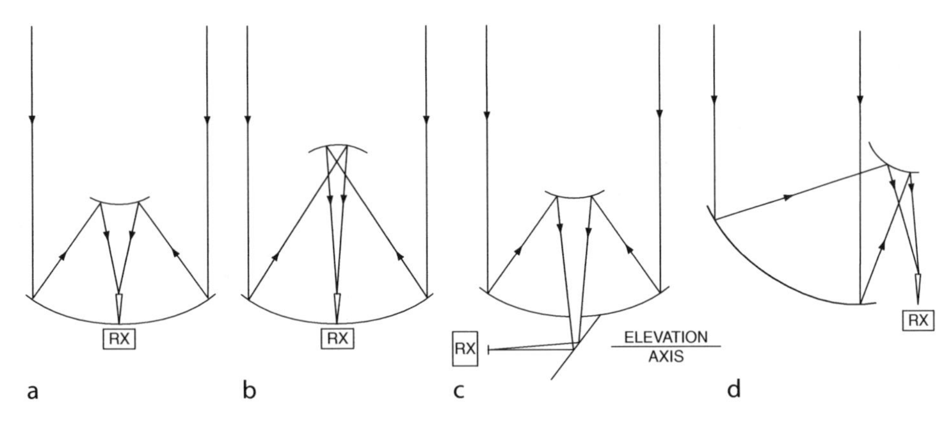

Fig. 6-4
The geometry of parabolic apertures: (a) Cassegrain, (b) Gregorian, (c) Nasmyth and, (d) offset Cassegrain systems (From Wilson et al. 2008)

5.2.3 Foci, Blockage, and Surface Accuracy

If the size of a radio telescope is more than a few hundred wavelengths, designs are similar to those of optical telescopes. Cassegrain, Gregorian, and Nasmyth systems have been used. See *Fig. 6-4* for a sketch of these focal systems. In a Cassegrain system, a convex hyperbolic reflector is introduced into the converging beam immediately in front of the prime focus. This reflector transfers the converging rays to a secondary focus which, in most practical systems is situated close to the apex of the main dish. A Gregorian system makes use of a concave reflector with an elliptical profile. This must be positioned behind the prime focus in the diverging beam. In the Nasmyth system, this secondary focus is situated in the elevation axis of the telescope by introducing another, usually flat, mirror. The advantage of a Nasmyth system is that the receiver front ends remain horizontal while the telescope is pointed toward different elevations. This is an advantage for receivers cooled with liquid helium, which may become unstable when tipped. Cassegrain and Nasmyth foci are commonly used in the millimeter/submillimeter wavelength ranges.

In a secondary reflector system, feed illumination beyond the edge receives radiation from the sky, which has a temperature of only a few K. For low-noise systems, this results in only a small overall system noise temperature. This is significantly less than for prime focus systems. This is quantified in the so-called G/T value, that is, the ratio of antenna gain to system noise. Any telescope design must aim to minimize the excess noise at the receiver input while maximizing gain. For a specific antenna, this maximization involves the design of feeds and the choice of foci.

The secondary reflector and its supports block the central parts in the main dish from reflecting the incoming radiation, causing some significant differences between the actual beam pattern and that of an unobstructed antenna. Modern designs seek to minimize blockage due to the support legs and sub-reflector.

The beam pattern differs from a uniformly illuminated unblocked aperture for three reasons:

1. The illumination of the reflector will not be uniform but has a taper by 10 dB, that is, a factor of 10 or more at the edge of the reflector. This is in contrast to optical telescopes which have no taper.
2. The sidelobe level is strongly influenced by this taper: a larger taper lowers the sidelobe level.
3. The secondary reflector must be supported by three or four support legs, which will produce aperture blocking and thus affect the shape of the beam pattern.

Feed leg blockage will cause deviations from circular symmetry. For altitude-azimuth telescopes, these sidelobes will change position on the sky with hour angle (see Reich et al. 1978). This may be a serious defect, since these effects will be significant for maps of low-intensity regions near an intense source. The sidelobe response may depend on the polarization of the incoming radiation (see Sect. 5.3.6).

A disadvantage of on-axis systems, regardless of focus, is that they are often more susceptible to instrumental frequency baselines, so-called *baseline ripples* across the receiver band than primary focus systems (see Morris 1978). Part of this ripple is caused by reflections of noise from source or receiver in the antenna structure. Ripples from the receiver can be removed if the amplitude and phase are constant in time. Baseline ripples caused by the source, sky, or ground radiation are more difficult to eliminate since these will change over short times. It is known that large amounts of blockage and larger feed sizes lead to large baseline ripples. The influence of baseline ripples on measurements can be reduced to a limited extent by appropriate observing procedures. A possible solution is an off-axis system such as the GBT of the National Radio Astronomy Observatory. In contrast to the GBT, the Effelsberg 100-m has a large amount of blocking from massive feed support legs and, as a result, show large instrumental frequency baseline ripples. These ripples might be mitigated by the use of scattering cones in the reflector.

The gain of a filled aperture antenna with small-scale surface irregularities ε cannot increase indefinitely with increasing frequency but reaches a maximum at $\lambda_m = 4\pi\varepsilon$, and this gain is a factor of 2.7 below that of an error-free antenna of identical dimensions. The usual rule-of-thumb is that the irregularities should be 1/16 of the shortest wavelength used. Larger filled aperture radio telescopes are made up of panels. For these, the irregularities are of two types: (1) roughness of the individual panels and (2) misadjustment of panels. The second irregularity gives rise to an *error beam*. The FWHP of the error beam is given approximately by the ratio of wavelength to panel size. In addition, if the surface material is not a perfect conductor, there will be some loss and consequently additional noise.

5.3 Single Dish Observational Techniques

5.3.1 The Earth's Atmosphere

For ground-based facilities, the amplitudes of astronomical signals have been attenuated and the phases have been altered by the earth's atmosphere. In addition to attenuation, the receiver noise is increased by atmospheric emission, the signal is refracted and there are changes in the path length. These effects may change slowly with time, but there can also be rapid changes such as scintillation and anomalous refraction. Thus propagation properties must be taken into account if the astronomical measurements are to be correctly interpreted. At meter wavelengths, these effects are caused by the ionosphere. In the millimeter/submillimeter range, tropospheric effects are especially important. The various constituents of the atmosphere absorb by different amounts. Because the atmosphere can be considered to be in LTE, these constituents also emit radiation.

The total amount of precipitable water (usually measured in mm) is an integral along the line-of-sight to a source. Frequently, the amount of H_2O is determined by measurements of the continuum emission of the atmosphere with a small dish at 225 GHz. For a set of measurements at elevations of 20°, 30°, 60°, and 90°, combined with models, rather accurate values of the atmospheric τ can be obtained. For extremely dry millimeter/submillimeter sites, measurements of the 183 GHz spectral line of water vapor can be used to estimate the total amount of H_2O in the atmosphere. For sea level sites, the 22.235 GHz line of water vapor has been used for this purpose. The scale height $H_{\mathrm{H_2O}} \approx 2$ km, is considerably less than $H_{\mathrm{air}} \approx 8$ km of dry air. For this reason, sites for submillimeter radio telescopes are usually mountain sites with elevations above ≈3,000 m. For ionospheric effects, even the highest sites on earth provide no improvement.

The effect on the intensity of a radio source due to propagation through the atmosphere is given by the standard relation for radiative transfer (from (➋ Eq. 6.10)):

$$T_{\mathrm{B}}(s) = T_{\mathrm{B}}(0)\,\mathrm{e}^{-\tau_\nu(s)} + T_{\mathrm{atm}}\left(1 - \mathrm{e}^{-\tau_\nu(s)}\right) \quad . \tag{6.48}$$

Here s is the (geometric) path length along the line-of-sight with $s = 0$ at the upper edge of the atmosphere and $s = s_0$ at the antenna, $\tau_\nu(s)$ is the optical depth, T_{atm} is the temperature of the atmosphere and $T_{\mathrm{B}}(0)$ is the temperature of the astronomical source above the atmosphere. Both the (volume) absorption coefficient κ and the gas temperature T_{atm} will vary with s. Introducing the mass absorption coefficient k_ν by

$$\kappa_\nu = k_\nu \cdot \varrho\,, \tag{6.49}$$

where ϱ is the gas density; this variation of κ can mainly be related to that of ϱ as long as the gas mixture remains constant along the line of sight. This is a simplified relation. For a more detailed calculation, a multilayer model is needed.

Models can provide corrections for average effects; fluctuations and detailed corrections needed for astronomy must be determined from real-time measurements.

5.3.2 Meter and Centimeter Calibration Procedures

This involves a three step procedure: (1) the measurements must be corrected for atmospheric effects, (2) relative calibrations are made using secondary standards, and (3) if needed, gain versus elevation curves for the antenna must be established.

In the centimeter wavelength range, atmospheric effects are usually small. For steps (2) and (3) the calibration is carried out with the use of a pulsed signal injected before the receiver. This pulsed signal is added to the receiver input. The calibration signal must be stable, broadband, and of reasonable size. Often noise diodes are used as pulsed broadband calibration sources. These are secondary standards that provide broadband radiation with effective temperatures $>10^5$ K. With a pulsed calibration, the receiver outputs are recorded separately as (1) receiver only, (2) receiver plus calibration, and (3) repeat of this cycle. If the calibration signal has a known value and the zero point of the receiver system is measured, the receiver noise is determined (see ➋ Eq. 6.24). Most often the calibration value in either Jy/beam or T_{MB} units is determined by a continuum scan through a non-time variable compact discrete source of known flux density. Lists of calibration sources are to be found in Baars et al. (1977), Altenhoff (1985), Ott et al. (1994), and Sandell (1994).

5.3.3 Millimeter and Submillimeter Calibration Procedures

In the millimeter/submillimeter wavelength range, the atmosphere has a larger influence and can change on timescales of seconds, so more complex corrections are needed. Also, large telescopes may operate close to the limits caused by their surface accuracy, so that the power received in the error beam may be comparable to that received in the main beam. In addition, many sources such as molecular clouds are rather extended. Thus, relevant values of telescope efficiencies must be used (see Downes 1989). The calibration procedure used in the millimeter/submillimeter range is referred to as the *chopper wheel* method (Penzias and Burrus 1973). This consists of two steps:

1. The measurement of the receiver noise (the method is very similar to that in Sect. 3.1.1).
2. The measurement of the receiver response when directed toward cold sky at a certain elevation.

In the following it is assumed that the receiver is operated in the SSB mode. For (1), the output of the receiver while measuring an ambient load, T_{amb}, is denoted by V_{amb}:

$$V_{amb} = G\,(T_{amb} + T_{rx})\,. \tag{6.50}$$

where G is the system gain. This is sometimes repeated with a second load at a different temperature. The result is a determination of the receiver noise as in Sect. 3.1.1. For step (2), the load is removed; then the output refers to noise from a source-free sky (T_{sky}), ground ($T_{gr} = T_{amb}$), and receiver:

$$V_{sky} = G\left[F_{eff}\,T_{sky} + (1 - F_{eff})\,T_{gr} + T_{rx}\right]. \tag{6.51}$$

where F_{eff} is the *forward efficiency*. This is the fraction of power in the forward beam of the feed. This can be interpreted as the response to a source with the angular size of the Moon (it is assumed that F_{eff} is appropriate for an extended molecular cloud). Taking the difference between V_{amb} and V_{sky}:

$$\Delta V_{cal} = V_{amb} - V_{sky} = G\,F_{eff}\,T_{amb}\,e^{-\tau_\nu}\,, \tag{6.52}$$

where τ_ν is the atmospheric absorption at the frequency of interest. If it is assumed that $T_{sky}(s) = T_{atm}\,(1 - e^{-\tau_\nu})$ describes the emission of the atmosphere, and, as in (6.48), τ_ν in is the same for emission and absorption, emission measurements can provide the value of τ_ν. If $T_{atm} = T_{amb}$, the correction is simplified. For more complex situations, models of the atmosphere are needed (see, e.g., Pardo et al. 2009). Once τ_ν is known, the signal from the radio source after passing through the earth's atmosphere, is

$$\Delta V_{sig} = G\,T'_A\,e^{-\tau_\nu}$$

or

$$T'_A = \frac{\Delta V_{sig}}{\Delta V_{cal}}\,F_{eff}\,T_{amb}$$

where T'_A is the antenna temperature of the source outside the earth's atmosphere. We define

$$T_A^* = \frac{T'_A}{F_{eff}} = \frac{\Delta V_{sig}}{\Delta V_{cal}}\,T_{amb} \tag{6.53}$$

The right side involves only measured quantities. T_A^* is commonly referred to as the *corrected antenna temperature*, but it is really a *forward beam brightness temperature*. An analogous temperature is T_{sys}^*, the system noise correcting for all atmospheric effects:

$$T_{sys}^* = \left(\frac{T_{rx} + T_{sky}}{F_{eff}} \right) e^{\tau} \tag{6.54}$$

This result is used to determine continuum or line temperature scales (➋ Eq. 6.31). A typical set of values for $\lambda = 3$ mm are: $T_{rx} = 40$ K, $T_{sky} = 50$ K, $\tau = 0.3$. Using these, the $T_{sys}^* = 135$ K.

For sources $\ll 30'$, there is an additional correction for the telescope beam efficiency, which is commonly referred to as B_{eff}. Then

$$T_{MB} = \frac{F_{eff}}{B_{eff}} T_A^*$$

Typical values of F_{eff} are $\cong 0.9$, and at the shortest wavelengths used for a telescope, $B_{eff} \cong 0.6$. In general, for extended sources, the brightness temperature corrected for absorption by the earth's atmosphere, T_A^*, should be used.

5.3.4 Bolometer Calibrations

Since most bolometers are AC coupled (i.e., respond to differences), the DC response (i.e., respond to total power) used in "hot–cold" or "chopper wheel" calibration methods cannot be used. Instead, astronomical data are calibrated in two steps:

1. Measurements of atmospheric emission at a number of elevations to determine the opacities at the azimuth of the target source
2. The measurement of the response of a nearby source with a known flux density; immediately after this, a measurement of the target source is carried out

5.3.5 Continuum Observing Strategies

1. *Position switching and wobbler switching.* Switching against a load or absorber is used only in exceptional circumstances, such as studies of the 2.73 K cosmic microwave background. For the CMB, Penzias and Wilson (1965) used a helium cooled load with a precisely known temperature. For compact regions, compensation of transmission variations of the atmosphere is possible if double beam systems can be used. At higher frequencies, in the millimeter/submillimeter range, rapid movement of the telescope beam (by small movements of the sub-reflector or a mirror in the path from antenna to receiver) over small angles is referred to as "beam switching," "wobbling," or "wobbler switching." This is used to produce two beams on the sky for a single pixel receiver. The individual telescope beams should be spaced by a distance of 3 FWHP beamwidths.
2. *Mapping of extended regions and on-the-fly mapping.* Multi-beam bolometer systems are preferred for continuum measurements at $\nu > 100$ GHz. Usually, a wobbler system is needed for such arrays. With these, it is possible to measure a fairly large region and to better cancel sky noise due to weather. Some details of more recent data taking and reduction methods are given in, for example, Johnstone et al. (2000) or Motte et al. (2006).

If extended areas are to be mapped, scans are made along one direction (e.g., Azimuth or Right Ascension). Then the antenna is offset in the orthogonal direction by 1/2–1/3 of a beamwidth, and the scanning is repeated until the region is completely mapped. This is referred to as a "raster scan." There should be reference positions free of sources at the beginning and at the end of each scan, to allow the determination of zero levels, and calibrations should be made before the scans are begun. For more secure results, the map is then repeated by scanning in the orthogonal direction (e.g., Elevation or Declination). Then both sets of results are placed on a common grid, and averaged; this is referred to as "basket weaving."

Extended emission regions can also be mapped using a double beam system, with the receiver input periodically switched between the first and second beam. In this procedure, there is some suppression of very extended emission. A summation of the beam switched data along the scan direction has been used to reconstruct infrared images. More sophisticated schemes can recover most, but not all, of the information (Emerson et al. 1979; "EKH"). Most millimeter/submillimeter antennas employ wobbler switching in azimuth to cancel ground radiation. By measuring a source using scans in azimuth at different hour angles and then transforming the positions to an astronomical coordinate frame and combining the maps it is possible to reduce the effect of sidelobes caused by feed legs and suppress sky noise (Johnstone et al. 2000).

5.3.6 Additional Requirements for Spectral Line Observations

In addition to the requirements placed on continuum receivers, there are three additional requirements for spectral line receiver systems.

If the observed frequency of a line is compared to the known rest frequency, the relative radial velocity of the source and the receiving system can be determined. But this velocity contains the motion of the source as well as that of the receiving system, so the velocity measurements are referred to some standard of rest. This velocity can be separated into several independent components: (1) Earth rotation with a maximum velocity $v = 0.46\,\mathrm{km\,s^{-1}}$ and (2) the motion of the center of the Earth relative to the barycenter of the Solar System is said to be reduced to the *heliocentric* system. Correction algorithms are available for observations of the earth relative to center of mass of the solar system. The *standard solar motion* is the motion relative to the mode of the velocity of the stars in the solar neighborhood. Data where the standard solar motion has been taken into account are said to refer to the *local standard of rest* (LSR). Most extragalactic spectral line data do *not* include the LSR correction but are referred to the heliocentric velocity. For high redshift sources, special relativity corrections must be included.

For larger bandwidths, there is an instrumental spectrum and a "baseline" must be subtracted from the (on-off)/off spectrum. Often a linear fit to spectrum is sufficient, but if curvature is present, polynomials of second or higher order must be subtracted. At high galactic latitudes, more intense 21 cm line radiation from the galactic plane can give rise to artifacts in spectra from scattering of radiation within the antenna (see Kalberla et al. 2010). This is apparently less of a problem in surveys of galactic carbon monoxide (see Dame et al. 1987).

5.3.7 Spectral Line Observing Strategies

Astronomical radiation is often only a small fraction of the total power received. To avoid stability problems, the signal of interest must be compared with another that contains approximately the same total power and differs only in the fact that it contains no source. The receiver must be stable so that any gain or bandpass changes occur over timescales long compared to the time needed for position change. To detect an astronomical source, three observing modes are used to produce a suitable comparison.

1. *Position switching and wobbler switching.* The signal "on source" is compared with a measurement obtained at a nearby position in the sky. For spectral lines, there must be little line radiation at the comparison region. This is referred to as the "total power" observing mode. A variant of this method is wobbler switching. This is very useful for compact sources, especially in the millimeter/submillimeter range.
2. *On-the-fly mapping.* This very important observing method is an extension of method (1). In this procedure, spectral line data is taken at a rate of perhaps one spectrum or more per second.
3. *Frequency switching.* For many sources, spectral line radiation at ν_0 is restricted to a narrow band, that is, present only over a small frequency interval, $\Delta\nu$, for example, $\Delta\nu/\nu_0 \approx 10^{-5}$. If all other effects vary very little over $\Delta\nu$, changing the frequency of a receiver on a short time by up to 10 $\Delta\nu$ produces a comparison signal with the line well shifted. The line is measured all of the time, so this is an efficient observing mode.

6 Interferometers and Aperture Synthesis

From diffraction theory, the angular resolution is given by (➋ Eq. 6.33). However, as shown by Michelson (see Jenkins and White 2001), a much higher resolving power can be obtained by coherently combining the output of two reflectors of diameter $d \ll B$ separated by a distance B yielding $\theta \approx \lambda/\mathrm{B}$. In the radio/millimeter/submillimeter range, from (➋ Eq. 6.30), the outputs can be amplified without seriously degrading the signal-to-noise ratio. This amplified signal can be divided and used to produce a large number of cross-correlations.

Aperture synthesis is a further development. This is the procedure to produce high-quality images of sources by combining a number of measurements for different antenna spacings up to the maximum B. The longest spacing gives the angular resolution of an equivalent large aperture. This has become the method to obtain high-quality, high-angular resolution images. The first practical demonstration of aperture synthesis in radio astronomy was made by M. Ryle and his associates (see Sect. 3 in Kellermann and Moran 2001). Aperture synthesis allows the reproduction of the imaging properties of a large aperture by sampling the radiation field at individual positions within the aperture. Using this approach, a remarkable improvement of the radio astronomical imaging was made possible. More detailed accounts are to be found in Taylor et al. (1999), Thompson et al. (2001), or Dutrey (2001).

The simplest case is a two-element system in which electromagnetic waves are received by two antennas. These induce the voltage V_1 at A_1:

$$V_1 \propto E\,\mathrm{e}^{\,\mathrm{i}\,\omega t}\,, \tag{6.55}$$

while at A_2:

$$V_2 \propto E\,\mathrm{e}^{\,\mathrm{i}\,\omega\,(t-\tau)}\,, \tag{6.56}$$

where E is the amplitude of the incoming electromagnetic plane wave and τ is the geometric delay caused by the relative orientation of the interferometer baseline $\vec{B}$ and the direction of the wave propagation. For simplicity, receiver noise and instrumental phase were neglected in (➋ Eq. 6.55) and (➋ Eq. 6.56). The outputs will be correlated. Today all radio interferometers use direct correlation followed by an integrator. The output is proportional to:

$$R(\tau) \propto \frac{E^2}{T} \int_0^T e^{\,i\,\omega t}\, e^{-i\,\omega(t-\tau)}\; dt\,.$$

If T is a time much longer than the time of a single full oscillation, that is, $T \gg 2\pi/\omega$ then the average over time T will not differ much from the average over a single full period, resulting in

$$R(\tau) \propto \tfrac{1}{2}E^2\, e^{\,i\,\omega\tau} \quad . \tag{6.57}$$

The output of the correlator + integrator varies periodically with τ, the delay. Since $\vec{s}$ is slowly changing due to the rotation of the earth, τ will vary, producing *interference fringes* as a function of time.

The basic components of a two-element system are shown in ➋ *Fig. 6-5*. If the radio brightness distribution is given by $I_\nu(\vec{s})$, the power received per bandwidth $d\nu$ from the source element $d\Omega$ is $A(\vec{s})I_\nu(\vec{s})\, d\Omega\, d\nu$, where $A(\vec{s})$ is the effective collecting area in the direction $\vec{s}$;

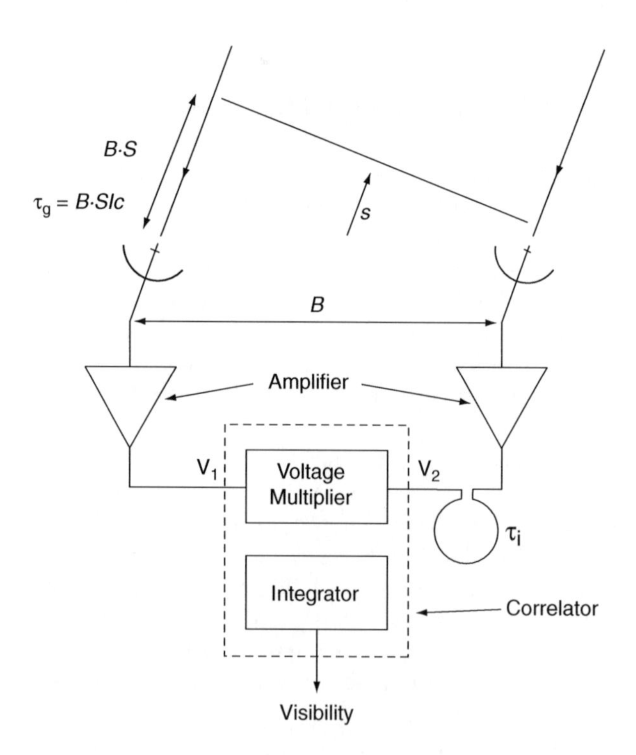

Fig. 6-5

A schematic diagram of a two-element correlation interferometer. The antenna output voltages are V_1 and V_2, the instrumental delay is τ_i, and the geometric delay is τ_g. s is the direction to the source. Perpendicular to s is the projection of the baseline B. The signal is digitized after conversion to an intermediate frequency. Time delays are introduced using digital shift registers (From Wilson et al. 2008)

the same $A(\vec{s})$ is assumed for each of the antennas. The amplifiers are assumed to have constant gain and phase factors (neglected here for simplicity).

The output of the correlator for radiation from the direction $\vec{s}$ (Fig. 6-5) is

$$r_{12} = A(\vec{s})\, I_\nu(\vec{s})\, \mathrm{e}^{\mathrm{i}\,\omega\tau}\, \mathrm{d}\Omega\, \mathrm{d}\nu \tag{6.58}$$

where τ is the difference between the geometrical and instrumental delays τ_g and τ_i. If $\vec{B}$ is the baseline vector between the two antennas

$$\tau = \tau_g - \tau_i = \frac{1}{c}\,\vec{B}\cdot\vec{s} - \tau_i \tag{6.59}$$

the total response is obtained by integrating over the source S

$$R(\vec{B}) = \iint_{\Omega} A(\vec{s}) I_\nu(\vec{s})\, \mathrm{e}^{2\pi\,\mathrm{i}\,\nu(\frac{1}{c}\vec{B}\cdot\vec{s}-\tau_i)}\, \mathrm{d}\Omega\, \mathrm{d}\nu \tag{6.60}$$

The function $R(\vec{B})$, the *Visibility Function*, is closely related to the mutual coherence function (see Born and Wolf 1965; Thompson et al. 2001; Wilson et al. 2008) of the source. For parabolic antennas, it is usually assumed that $A(\vec{s}) = 0$ outside the main beam area so that (Eq. 6.60) is integrated only over this region. A one-dimensional version of (Eq. 6.60), for a baseline B, frequency $\nu = \nu_0$ and instrumental time delay $\tau_i = 0$, is

$$R(B) = \int A(\theta)\, I_\nu(\theta)\, \mathrm{e}^{2\pi\,\mathrm{i}\,\nu_0(\frac{1}{c}B\,\theta)}\, \mathrm{d}\theta \tag{6.61}$$

With $\theta = x$ and $B_x/\lambda = u$, this is

$$R(B) = \int A(\theta)\, I_\nu(\theta)\, \mathrm{e}^{2\pi\,\mathrm{i}\,u\,x}\, \mathrm{d}\theta \tag{6.62}$$

This form of (Eq. 6.60) illustrates more clearly the Fourier Transform relation of u and x. This simplified version will be used to provide illustrations of interferometer responses (see Sect. 6.2). In two dimensions, (Eq. 6.60) takes on a similar form with the additional variables y and $B_y/\lambda = v$. The image can be obtained from the inverse Fourier transform of Visibilities; see (Eq. 6.65).

6.1 Calibration

Amplitude and phase must be calibrated for all interferometer measurements. In addition, the instrumental passband must be calibrated for spectral line measurements. The amplitude scale is calibrated by a determination of the system noise at each antenna using methods presented for single dish measurements (see Sect. 5.3.2 and following). In the centimeter range, the atmosphere plays a small role while in the millimeter and submillimeter wavelength ranges, the atmospheric effects must be accounted for. For phase measurements, a suitable point-like source with an accurately known position is required to determine $2\pi\nu\tau_i$ in (Eq. 6.60). For interferometers, the best calibration sources are usually unresolved or point-like sources. Most often these are extragalactic time variable sources. To calibrate the response in units of flux density or brightness temperatures, these amplitude measurements must be referenced to primary calibrators (see a list of non-variable sources of known flux densities in Ott et al. 1994 or Sandell 1994).

The calibration of the instrumental passband is carried out by a longer integration on an intense source to determine the channel-to-channel gains and offsets. The amplitude, phase, and passband calibrations are carried out before the source measurements. The passband calibration is usually carried out every few hours or once per observing session. The amplitude and phase calibrations are made more often; the time between such calibrations depends on the stability of the electronics and weather. If weather conditions require frequent measurements of calibrators (perhaps less than once per minute for "fast switching"), integration time is reduced. In case of even more rapid weather changes, the ALMA project will make use of water vapor radiometers mounted on each antenna (see Sect. 5.3.1). These will be used to determine the total amount of H_2O vapor above each antenna, and use this to make corrections to phase.

6.2 Responses of Interferometers

6.2.1 Time Delays and Bandwidth

The instrumental response is reduced if the bandwidth at the correlator is large compared to the delay caused by the separation of the antennas. For large bandwidths, the loss of correlation can be minimized by adjusting the phase delay so that the difference of arrival time between antennas is negligible. In practice, this is done by inserting a delay between the antennas so that $\frac{1}{c}\,\vec{B}\cdot\vec{s}$ equals $\tau_{\rm i}$. This is equivalent to centering the response on the central, or *white light fringe*.

Similarly, the reduction of the response caused by finite bandwidth can be estimated by an integration of (Eq. 6.60) over frequency, taking $A(\vec{s})$ and $I_\nu(\vec{s})$ as constants. The result is a factor, $\sin(\Delta\nu\tau)/\Delta\nu\tau$ in (Eq. 6.60). This will reduce the interferometer response if $\Delta\nu\tau \sim 1$. For typical bandwidths of 100 MHz, the offset from the zero delay must be $\ll 10^{-8}$ s. This adjustment of delays is referred to as *fringe stopping*. The exponent in (Eq. 6.60) has both sine and cosine components, but digital cross-correlators record both components, so that the entire response can be recovered.

6.2.2 Beam Narrowing

The *white light fringe* the delay compensation must be set with a high accuracy to prevent a reduction in the interferometer response. For a finite primary antenna beamwidth, θ_b, this cannot be the case over the entire beam. For a bandwidth $\Delta\nu$ there will be a phase difference. Converting the wavelengths to frequencies and using $\sin\theta \cong \theta$ the result is

$$\Delta\phi = 2\pi\,\frac{\theta_{\rm offset}}{\theta_b}\,\frac{\Delta\nu}{\nu} \tag{6.63}$$

This effect can be important for continuum measurements made with large bandwidths, but can be reduced if the cross correlation is carried out using a series of narrow contiguous IF sections. For each of these IF sections, an extra delay is introduced to center the response at the value which is appropriate for that wavelength before correlation.

6.2.3 Source Size

From an idealized source, of shape $I(\nu_0) = I_0$ for $\theta < \theta_0$ and $I(\nu_0) = 0$ for $\theta > \theta_0$, we take the primary beam size of each antenna to be much larger, and define the fringe width for a baseline

$B\,\theta_b$ to be $\frac{\lambda}{B}$. The result is

$$R(B) = A\,I_0 \cdot \theta_0\;\mathrm{e}^{\mathrm{i}\,\pi\frac{\theta_0}{\theta_b}}\left[\frac{\sin\left(\pi\theta_0/\theta_b\right)}{\left(\pi\theta_0/\theta_b\right)}\right] \tag{6.64}$$

The first terms are normalization and phase factors. The important term is in brackets. If $\theta_0 >> \theta_b$, the interferometer response is reduced. This is sometimes referred to as the problem of "missing short spacings." To correct for the loss of source flux density, the interferometer data must be supplemented by single dish measurements. The diameter of the single dish antenna should be larger than the shortest interferometer spacing. This single dish image must extend to the FWHP of the smallest of the interferometer antennas. When Fourier transformed and appropriately combined with the interferometer response, the resulting data set has *no* missing flux density.

6.3 Aperture Synthesis

To produce an image, the integral equation (➲ Eq. 6.60) must be inverted. A number of approximations may have to be applied to produce high-quality images. In addition, the data are affected by noise. The most important steps of this development will be presented.

For imaging over a limited region of the sky, rectangular coordinates are adequate, so relation (➲ Eq. 6.60) can be rewritten with coordinates (x, y) in the image plane and coordinates (u, v) in the Fourier plane. The coordinate w, corresponding to the difference in height, is set to zero. Then the relevant relation is:

$$I'(x, y) = A(x, y)\,I(x, y) = \int_{-\infty}^{\infty} V(u, v, 0)\,\mathrm{e}^{-2\pi\,\mathrm{i}\,(ux+vy)}\,\mathrm{d}u\,\mathrm{d}v \tag{6.65}$$

where $I'(x, y)$ is the intensity $I(x, y)$ modified by the primary beam shape $A(x, y)$. It is easy to correct $I'(x, y)$ by dividing by $A(x, y)$. Usually data present beyond the half power point is excluded.

The most important definitions are:

1. *Dynamic range*: The ratio of the maximum to the minimum intensity in an image. In images made with an interferometer array, it is assumed that corrections for primary beam taper have been applied. If the minimum intensity is determined by the random noise in an image, the dynamic range is defined by the signal-to-noise ratio of the brightest feature in the image. The dynamic range is an indication of the ability to recognize low-intensity features in the presence of intense features. If the minimum noise is determined by artifacts, that is, noise in excess of the theoretical value, "image improvement techniques" should be applied.
2. *Image fidelity*: This is defined by the agreement between the measured results and the actual ("true") source structure. A quantitative assessment of fidelity is:

   $$F = |(S - R)|/S$$

   where F is the fidelity, R is the resulting image obtained from the measurement, and S is the actual source structure. The highest fidelity is $F = 0$. Usually errors can only be estimated using a priori knowledge of the correct source structure. In many cases, S is a source model, while R is obtained by processing S with a model of the instrumental response. This relation can only be applied when the value of R is more than five times the RMS noise.

6.3.1 Interferometric Observations

Usually measurements are carried out in one of four ways.

1. *Measurements of a single target source*. This is similar to the case of single telescope position switching. Two significant differences with single dish measurements are that the interferometer measurement may have to extend over a wide range of hour angles to provide a better coverage of the (u, v) or Fourier plane, and that instrumental phase must be determined also. After the measurement of a calibration source or reference source, which has a known position and size, the effect of instrumental phases in the instrument and atmosphere is removed and a calibration of the amplitudes of the source is made. Target sources and calibrators are usually observed alternately; the calibrator should be close to the target source. The time variations caused by instrumental and weather effects must be slower than the time between measurements of source and calibrator. If, as is the case for millimeter/submillimeter wavelength measurements, weather is an important influence, target and calibration source must be measured often. For ALMA (see ➋ *Fig. 6-6*), observing will

Fig. 6-6
A photo of ALMA antennas on the 5km high site. To date, this is the most ambitious construction project in ground-based astronomy. ALMA is now being built in north Chile on a 5 km high site. It will consist of fifty-four 12-m and twelve 7-m antennas, operating in 10 bands between wavelength 1 cm and 0.3 mm. In Early Science, four receiver bands at 3, 1.3, 0.8, and 0.6 mm will be available. The high ALMA sensitivity is due to the extremely low noise receivers, the highly accurate antennas, and the high altitude site. At the largest antenna spacing, and shortest wavelength, the angular resolution will be ~5 milliarcseconds (courtesy ESO/NRAO/NAOJ from the NRAO newsletter)

follow a two-part scheme. For *fast switching* there will be integrations of perhaps 10 s on a nearby calibrator, then a few minutes on-source. This method will reduce the amount of phase fluctuations, at the cost of on-source observing time. For more rapid changes in the earth's atmosphere, phases will be corrected using measurements of atmospheric water vapor from measurements of the 183 GHz line.

2. *Snapshot mode.* A series of short observations (at different hour angles) of one source after another, and then the measurements are repeated. For sensitivity reasons, snapshots are usually made in the radio continuum or more intense spectral lines. As in observing method (1), measurements of source and calibrator are interspersed to remove the effects of instrumental phase drifts and to calibrate the amplitudes of the sources in question. The images will be affected by the shape of the synthesized beam since there is sparse coverage in the (u,v) plane. If the size of the source to be imaged is comparable to the primary beam of the individual antennas there should be a correction for the power pattern.
3. *Multi-configuration imaging.* Here the goal is the image of a source either with high dynamic range or high sensitivity. Measurements with a number of different interferometer configurations better fill the uv plane. These measurements are taken at different epochs and after calibration, the visibilities are entered into a common data set.
4. *Mosaicing.* An extension of procedure (1) can be used for sources with an extent much larger than the primary antenna beam. These images require measurements at adjacent pointings. This is spoken of as *mosaicing*. In a mosaic, the antennas are pointed at nearby positions. These positions should overlap at the half power point. The images can be formed separately and then combined to produce an image of the larger region. Another method is to combine the data in the (u,v) plane and then form the image.

6.4 Interferometer Sensitivity

The random noise limit to an interferometer system can be calculated following the method used for a single telescope (Eq. 6.27). The use of (Eq. 6.43) provides a conversion from ΔT_{RMS} to ΔS_ν. For an array of n identical antennas, there are $N = n(n-1)/2$ simultaneous pairwise correlations, so the RMS variation in flux density is:

$$\Delta S_\nu = \frac{2\,M\,k\,T^*_{\mathrm{sys}}}{A_{\mathrm{e}}\sqrt{2\,N\,t\,\Delta\nu}}\,. \tag{6.66}$$

with $M \cong 1$, A_{e} the effective area of each antenna and T^*_{sys} given by (Eq. 6.54). This relation can be recast in the form of brightness temperature fluctuations using the Rayleigh-Jeans relation; then the RMS noise in brightness temperature units is:

$$\Delta T_{\mathrm{B}} = \frac{2\,M\,k\,\lambda^2\,T^*_{\mathrm{sys}}}{A_{\mathrm{e}}\Omega_{\mathrm{b}}\sqrt{2\,N\,t\,\Delta\nu}}\,. \tag{6.67}$$

For a Gaussian beam, $\Omega_{\mathrm{mb}} = 1.133\,\theta^2$, so the RMS temperature fluctuations can be related to observed properties of a synthesis image.

Aperture synthesis is based on discrete samples of the visibility function $V(u,v)$, with the goal of the densest possible coverage of the (u,v) or Fourier plane. It has been observed that the RMS noise in a synthesis image obtained by Fourier transforming the (u,v) data is often higher than given by (Eq. 6.66) or (Eq. 6.67). Possible causes are: (1) phase fluctuations caused by atmospheric or instrumental instabilities; (2) incomplete sampling of the (u, v) plane, which gives rise to artifacts such as stripe-like features in the images; or (3) grating rings around more intense sources; these are analogous to *high sidelobes* in single dish diffraction patterns.

6.5 Corrections of Visibility Functions

6.5.1 Amplitude and Phase Closure

The relation between the measured visibility $\widetilde{V_{ik}}$ and *actual* visibility V_{ik} is considered linear:

$$\widetilde{V_{ik}}(t) = g_i(t)\, g_k^*(t)\, V_{ik} + \varepsilon_{ik}(t)\,. \tag{6.68}$$

Values for the complex antenna gain factors g_k and the noise term $\varepsilon_{ik}(t)$ are determined by measuring calibration sources as frequently as possible. Actual values for g_k are then computed by linear interpolation. The (complex) gain of the array is obtained by the multiplication of the gains of the individual antennas. If the array consists of n such antennas, $n(n-1)/2$ visibilities can be measured simultaneously, but only $(n-1)$ independent gains g_k are needed since one antenna in the array can be taken as a reference. So in an array with many antennas, the number of antenna pairs greatly exceeds the number of antennas. For phase, one must determine n phases. Often these conditions can be introduced into the solution in the form of *closure errors*. Defining the phases φ, θ, and ψ by

$$\begin{aligned} \widetilde{V_{ik}} &= |\widetilde{V_{ik}}|\, \mathrm{e}^{\mathrm{i}\,\varphi_{ik}}\,, \\ G_{ik} &= |g_i|\,|g_k|\, \mathrm{e}^{\mathrm{i}\,\theta_i}\, \mathrm{e}^{-\mathrm{i}\,\theta_k}\,, \\ V_{ik} &= |V_{ik}|\, \mathrm{e}^{\mathrm{i}\,\psi_{ik}}\,. \end{aligned} \tag{6.69}$$

from (Eq. 6.68) the visibility phase ψ_{ik} on the baseline ik will be related to the observed phase φ_{ik} by

$$\varphi_{ik} = \psi_{ik} + \theta_i - \theta_k + \varepsilon_{ik}\,, \tag{6.70}$$

where ε_{ik} is the phase noise. Then the *closure phase* Ψ_{ikl} around a closed triangle of baseline ik, kl, li,

$$\Psi_{ikl} = \varphi_{ik} + \varphi_{kl} + \varphi_{li} = \psi_{ik} + \psi_{kl} + \psi_{li} + \varepsilon_{ik} + \varepsilon_{kl} + \varepsilon_{li}\,, \tag{6.71}$$

will be independent of the phase shifts θ introduced by the individual antennas and the time variations. With this procedure, phase errors can be minimized.

If four or more antennas are used simultaneously, then the *closure amplitudes* can be formed. These are independent of the antenna gain factors:

$$A_{klmn} = \frac{|V_{kl}||V_{mn}|}{|V_{km}||V_{ln}|}\,. \tag{6.72}$$

Both phase and closure amplitudes can be used to improve the quality of the complex visibility function.

At each antenna there is an unknown complex gain factor g with amplitude and phase, the total number of unknowns can be reduced significantly by measuring closure phases and amplitudes. If four antennas are available, 50% of the phase information and 33% of the amplitude information can thus be recovered; in a ten-antenna configuration, these ratios are 80% and 78% respectively.

6.5.2 Calibrations, Gridding, FFTs, Weighting, and Self-Calibration

For two-antenna interferometers, phase calibration can only be made pair-wise. This is referred to as "baseline-based" solutions for the calibration. For a multi-antenna system, "antenna-based" solutions are preferred. These are determined by applying phase and amplitude closure for subsets of antennas and then solving for the best fit for each.

Normally the Cooley-Tukey fast Fourier transform algorithm is used to invert (Eq. 6.65). To apply the simplest version of the FFT, the visibilities must be placed on a regular grid with sizes that are powers of two of the sampling interval. Since the data seldom lie on such regular grids, an interpolation scheme must be used. From the gridded (u, v) data, an image with a resolution corresponding to λ/D, where D is the array size, is obtained. However, this may still contain artifacts caused by the observing procedure, especially the limited coverage of the (u, v) plane. Therefore the dynamic range of such so-called *dirty* maps is rather small. This can be improved by further analysis.

If the calibrated visibility function $V(u, v)$ is known for the full (u, v) plane both in amplitude and in phase, this can be used to determine the modified (i.e., structure on angular scales finer than λ/D are lost) intensity distribution $I'(x, y)$ by performing the Fourier transformation (Eq. 6.65). However, in a realistic situation $V(u, v)$ is only sampled at discrete points and in some regions of the (u, v) plane, $V(u, v)$ is not measured at all. The visibilities can be weighted by a grading function, g. For a discrete number of visibilities, a version of (Eq. 6.65) involving a summation, not an integral, is used to obtain an image with the use of a discrete Fourier transform (DFT):

$$I_D(x, y) = \sum_k g(u_k, v_k) V(u_k, v_k) \, e^{-2\pi i (u_k x + v_k y)} , \qquad (6.73)$$

where $g(u, v)$ is a weighting function referred to as the grading or apodization. $g(u, v)$ can be used to change the effective beam shape and sidelobe level. There are two widely used weighting functions: uniform and natural. Uniform weighting uses $g(u_k, v_k) = 1$, while natural weighting uses $g(u_k, g_k) = 1/N_s(k)$, where $N_s(k)$ is the number of data points within a symmetric region of the (u, v) plane. Data which are naturally weighted result in lower angular resolution but give a better signal-to-noise ratio than uniform weighting. But these are only extreme cases. Intermediate weighting schemes are referred to as *robust* weighting.

Often the reconstructed image I_D may not be a particularly good representation of I', but these are related by:

$$I_D(x, y) = P_D(x, y) \otimes I'(x, y) , \qquad (6.74)$$

where $I'(x, y)$ is the best representation of the source intensity modified by the primary beam shape; it contains only those spatial frequencies (u_k, v_k) where the visibility function has been measured (see (Eq. 6.65)). The expression for P_D is:

$$P_D = \sum_k g(u_k, v_k) \, e^{-2\pi i (u_k x + v_k y)} \qquad (6.75)$$

which is the response to a point source, or the *point spread function* PSF for the dirty beam. Thus P_D is a transfer function that distorts the image; P_D is produced assuming an amplitude of unity and phase zero at each point sampled. This is the response of the interferometer system to a point source. The sum in (Eq. 6.75) extends over the same positions (u_k, v_k) as in (Eq. 6.73); the sidelobe structure of the beam depends on the distribution of these points.

Amplitude and phase errors scatter power across the image, giving the appearance of enhanced noise. This problem can be alleviated to an impressive extent by the method of *self-calibration*. This process can be applied if there is a sufficiently intense compact feature in the field contained within the primary beam of the interferometer system. If self-calibration can be applied, the positional information is usually lost. Self-calibration can be restricted to an improvement of phase alone or to both phase and amplitude. Normally, self-calibration is carried in the (u, v) plane. If this method is used on objects with low signal-to-noise ratios, this may lead to a concentration of random noise into one part of the interferometer image (see Cornwell and Fomalont 1989). For measurements of weak spectral lines, self-calibration is carried out using a continuum source in the field. The corrections are then applied to the spectral line data. In the case of intense lines, one of the frequency channels containing the emission is used.

6.5.3 More Elaborate Improvements of Visibility Functions: The CLEANing Procedure

CLEANing is the most commonly used technique to improve single radio interferometer images (Högbom 1974). In addition to its inherent low dynamic range, the dirty map often contains features such as negative intensity artifacts that cannot be real. Another unsatisfactory aspect is that the solution is quite often rather unstable in that it can change drastically when more visibility data are added.

The CLEAN method approximates the intensity distribution that represents the best image of the source (subject to angular resolution, noise, etc.), $I(x, y)$, by the superposition of a finite number of point sources with positive intensity A_i placed at positions (x_i, y_i). The goal of CLEAN is to determine the $A_i(x_i, y_i)$, such that

$$I''(x, y) = \sum_i A_i \, P_\mathrm{D}(x - x_i, y - y_i) + I_\varepsilon(x, y) \tag{6.76}$$

where I'' is the dirty map obtained from the inversion of the visibility function and P_D is the dirty beam (Eq. 6.75). $I_\varepsilon(x, y)$ is the residual brightness distribution after decomposition. Approximation (Eq. 6.76) is considered successful if I_ε is of the order of the noise in the measured intensities. This decomposition must be carried out iteratively.

The CLEAN algorithm is most commonly applied in the image plane. This is an iterative method which functions in the following fashion: (1) find the peak intensity of the dirty image, then subtract a fraction γ (the so-called "loop gain") having the shape of the dirty beam from the image, and (2) repeat this n times.

This *loop gain* has values $0 < \gamma < 1$ while n is often taken to be 10^4. The goal is that the intensities of the residuals are comparable to the noise limit. Finally, the resulting model is convolved with a *clean beam* of Gaussian shape with an FWHP given by the angular resolution expected from λ/D where D is the maximum baseline length. Whether this algorithm produces a realistic image depends on the quality of the data and other variables.

6.5.4 More Elaborate Improvements of Visibility Functions: The Maximum Entropy Procedure

The Maximum Entropy Deconvolution Method (MEM) is commonly used to produce a single optimal image from a set of separate but contiguous images (Gull and Daniell 1978). The problem of how to select the "best" image from many possible images which all agree with the measured visibilities is solved by MEM. Using MEM, those values of the interpolated visibilities are selected, so that the resulting image is consistent with all previous relevant data. In addition, the MEM image has maximum smoothness. This is obtained by maximizing the *entropy* of the image. One definition of entropy is given by

$$\mathcal{H} = -\sum_i I_i \left[\ln\left(\frac{I_i}{M_i}\right) - 1 \right], \tag{6.77}$$

where I_i is the deconvolved intensity and M_i is a reference image incorporating all "a priori" knowledge. In the simplest case M_i is the empty field, $M_i = \text{const} > 0$, or perhaps a lower angular resolution image.

Additional constraints might require that all measured visibilities should be reproduced exactly, but in the presence of noise such constraints are often incompatible with $I_i > 0$ everywhere. Therefore the MEM image is usually constrained to fit the data such that

$$\chi^2 = \sum \frac{|V_i - V_i'|^2}{\sigma_i^2} \tag{6.78}$$

has the expected value, where V_i is the measured visibility, V_i' is a visibility corresponding to the MEM image and σ_i is the error of the measurement.

Acknowledgments

K. Weiler made a thorough review of the text and H. Bond suggested a number of improvements.

References

Altenhoff, W. J. 1985, in The Solar System: (Sub)mm Continuum Observations in ESO Conf & Workshop Proc No 22, eds. P. Shaver & K. Kjar (Garching: European Southern Observatory), 591

Baars, J. W. M. 2007, in The Parabolic Reflector Antenna in Radio Astronomy and Communication Astrophysics Space Science Library (Heidelberg: Springer)

Baars, J. W. M., Genzel, R., Pauliny-Toth, I. I. K., & Witzel, A. 1977, A&A, 61, 99

Bachiller, R., & Cernicharo, J., ed. 2008, in Science with the Atacama Large Millimeter Array: A New Era for Astrophysics (Heidelberg: Springer)

Baker, A. J., Glenn, J., Harris, A. I., Mangum, J. G., & Yun, M. S., eds. 2007, in From Z Machines to ALMA: (Sub)millimeter Spectroscopy of Galaxies Conf. Ser., Vol. 75 (San Francisco: Publications of the Astronomical Society of the Pacific)

Begelman, M., & Rees, M. 2009, in Gravity's Fatal Attraction: Black Holes in the Universe (2nd ed.; Cambridge: Cambridge University Press)

Born, M., & Wolf, E. 1965, in Principles of Optics (Oxford: Pergamon)

Bracewell, R. N. 1986, in The Fourier Transform and its Application (2nd ed.; New York: McGraw Hill)

Burke, B. F., & Graham-Smith, F. 2009, in An Introduction to Radio Astronomy (3rd ed.; Cambridge: Cambridge University Press)

Cornwell, T., & Fomalont, E. B. 1989, in Self Calibration in Synthesis Imaging in Radio Astronomy, Conf Series, Vol. 6, eds. R. Perley et al. (San Francisco: Publications of the Astronomical Society of the Pacific) 185

Crutcher, R. M. 2008, Astrophysics and Space Science, 313, 141

Dame, T. M., Ungerechts, H., Cohen, R. S., de Geus, E., Grenier, I. A., May, J., Murphy, D. C., Nyman, L.-A., & Thaddeus, P. 1987, ApJ, 322, 706

Dicke, R. H. 1946, Rev Sci Instrum, 17, 268

Downes, D. 1989, in Radio Telescopes: Basic Concepts in Diffraction-Limited Imaging with Very Large Telescopes, NATO ASI Series, Vol. 274, eds. D. M. Alloin & J. M. Mariotti (Dordrecht: Kluwer), 53

Dulk, G. 1985, Ann Rev A&A, 23, 169

Dutrey, A., ed. 2001, in IRAM Millimeter Interferometry Summer School 2 (Grenoble: IRAM)

Emerson, D., Klein, U., & Haslam, C. G. T. 1979, A&A, 76, 92

Gary, D. E., & Keller, C. U., eds. 2004, in Solar & Space Weather Astrophysics, Astrophysics & Space Science Lib., Vol. 134 (Dordrecht: Kluwer)

Giovanelli, R., & Haynes, M. P. 1991, Ann Rev A&A, 29, 499

Goldsmith, P. F., ed. 1988, in Instrumentation and Techniques for Radio Astronomy (New York: IEEE)

Goldsmith, P. F. 1994, in Quasioptical Systems: Gaussian Beam Quasioptical Propagation and Applications (New York: Wiley/IEEE)

Gull, S. F., & Daniell, G. J. 1978, Nature, 272, 68

Gurvits, L., Frey, S., & Rawlings, S., eds. 2005, in Radio Astronomy from Karl Jansky to Microjansky (Paris: EDP Sciences)

Harwit, M. 2006, in Astrophysical Concepts (4th ed.; Heidelberg: Springer)

Herbst, E. 2001, Chem Soc Rev, 30, 168

Herbst, E., & Dishoeck, E. van 2009, Ann Rev A&A, 47, 247

Hildebrand, R. 1983, Q J R Astron Soc, 24, 267

Högbom, J. 1974, A&A Suppl, 15, 417

Holland, W. S. et al. 1999, MNRAS, 303, 659

Jenkins, F. A., & White, H. E. 2001, in Fundamentals of Optics (4th ed.; New York: McGraw-Hill)

Johnstone, D. et al. 2000, ApJ Suppl, 131, 505

Kalberla, P. M. W., Burton, W. B., Hartmann, D., Arnal, E. M., Bajaja, E., Morras, R., & Pöppel, W .G. L. 2005, A&A, 440, 775

Kalberla, P. M. W. et al. 2010, A&A, 521, 17

Kawamura, J. et al. 2002, A&A, 394, 271

Kellermann, K. I., & Moran, J. M. 2001, Ann Rev A&A, 39, 457

Kirshner, R. P. 2004, in The Extravagant Universe (Princeton: Princeton University Press)

Lequeux, J. 2004, in The Interstellar Medium (Heidelberg: Springer)

Longair, M. 2006, in The Cosmic Century (Cambridge: Cambridge University Press)

Lorimer, D., & Kramer, M. 2004, in Handbook of Pulsar Astronomy (Cambridge: Cambridge University Press)

Love, A. W., ed. 1976, in Electromagnetic Horn Antennas (New York: IEEE)

Lyne, A. G., & Graham-Smith, F. 2006, in Pulsar Astronomy (3rd ed.; Cambridge: Cambridge University Press)

Morris, D. 1978, A&A, 67, 221

Motte, F., Bontemps, S., Schneider, N., Schilke, P., Menten, K. M., & Broguierè, D. 2006, A&A, 476, 1243

Ott, M. et al. 1994, A&A, 284, 331

Pardo, J. R., Cernicharo, J., Serabyn, E., & Wiedner, M. C. 2009, in ASP Conf. Ser. 417, Submillimeter Astrophysics & Technology, eds. D. C. Lis et al. (San Francisco: Publications of the Astronomical Society of the Pacific) 125

Penzias, A. A., & Burrus, C. A. 1973, Ann Rev A&A, 11, 51

Penzias, A. A., & Wilson, R. W. 1965, ApJ, 142, 419

Phillips, T. G., & Woody, D. P. 1982, Ann Rev A&A, 20, 285

Reipurth, B., Jewett, D., Keil, K., eds. 2007, in Protostars and Planets V (Tucson: University of Arizona Press)

Reich, W., Kalberla, P., Reif, K., & Neidhöfer, J. 1978, A&A, 76, 92

Reid, M. J. 1993, Ann Rev A&A, 31, 345

Rieke, G. H. 2002, in Detection of Light: From Ultraviolet to the Submillimeter (2nd ed.; Cambridge: Cambridge University Press)

Rohlfs, K., & Wilson, T. L. 2004, in Tools of Radio Astronomy (4th ed.; Heidelberg: Springer)

Rybicki, G. B., & Lightman, A. P. 1979, in Radiative Processes in Astrophysics (New York: Wiley)

Sandell, G. 1994, MNRAS, 271, 75

Silk, J. 2008, in The Infinite Cosmos (Oxford: Oxford University Press)

Solomon, P. M., & Vanden Bout, P. A. 2005, Ann Rev A&A, 43, 677

Sparke, L., & Gallagher, J. S. III 2007, in Galaxies in the Universe: An Introduction (2nd ed.; Cambridge: Cambridge University Press)

Stahler, S. W., & Palla, F. 2005, in The Formation of Stars (New York: Wiley-VCH)

Sullivan, W. T. III 2005, in The Early Years of Radio Astronomy: Reflections 50 Years after Jansky's

Discovery (Cambridge: Cambridge University Press)
Sullivan, W. T. III 2009, in Cosmic Noise, A History of Early Radio Astronomy (Cambridge: Cambridge University Press)
Taylor, G. B., Carilli, C. L., & Perley, R. A., eds. 1999, in ASP Conf. Ser. 180, Synthesis Imaging in Radio Astronomy II (San Francisco: Publications of the Astronomical Society of the Pacific)
Thum, C., Wiesemeyer, H., Paubert, G., Navarro, S., & Morris, D. 2008, Publ Astron Soc Pac, 120, 777
Tielens, A. G. G. M. 2005, in The Physics and Chemistry of the Interstellar Medium (Cambridge: Cambridge University Press)
Thompson, A. R., Moran, J. M., & Swenson, G. W. 2001, in Interferometry and Synthesis in Radio Astronomy (2nd ed.; New York: Wiley-VCH)
Townes, C. H., & Schawlow, A. H. 1975, in Microwave Spectroscopy (New York: Dover)
Venkata, U. R. 2010, in PhD thesis: Parameterized Deconvolution for Wide-Band Radio Synthesis Imaging (Socorro: New Mexico Institute of Mining & Technology)
Wilson, T. L., Rohlfs, K., & Hüttemeister, S. 2008, in Tools of Radio Astronomy (5th ed.; Heidelberg: Springer)

7 Radio and Optical Interferometry: Basic Observing Techniques and Data Analysis

John D. Monnier[1] · *Ronald J. Allen*[2]
[1]Astronomy Department, Experimental Astrophysics, University of Michigan, Ann Arbor, MI, USA
[2]Science Mission Office, Space Telescope Science Institute, Baltimore, MD, USA

T.D. Oswalt, H.E. Bond (eds.), *Planets, Stars and Stellar Systems. Volume 2: Astronomical Techniques, Software, and Data*, DOI 10.1007/978-94-007-5618-2_7, © Springer Science+Business Media Dordrecht 2013

Abstract: Astronomers usually need the highest angular resolution possible when observing celestial objects, but the blurring effect of diffraction imposes a fundamental limit on the image quality from any single telescope. Interferometry allows light collected at widely separated telescopes to be combined in order to synthesize an aperture much larger than an individual telescope, thereby improving angular resolution by orders of magnitude. Because diffraction has the largest effect for long wavelengths, radio and millimeter wave astronomers depend on interferometry to achieve image quality on par with conventional large-aperture visible and infrared telescopes. Interferometers at visible and infrared wavelengths extend angular resolution below the milliarcsecond level to open up unique research areas in imaging stellar surfaces and circumstellar environments.

In this chapter, the basic principles of interferometry are reviewed with an emphasis on the common features for radio and optical observing. While many techniques are common to interferometers of all wavelengths, crucial differences are identified that will help new practitioners to avoid unnecessary confusion and common pitfalls. The concepts essential for writing observing proposals and for planning observations are described, depending on the science wavelength, the angular resolution, and the field of view required. Atmospheric and ionospheric turbulence degrades the longest-baseline observations by significantly reducing the stability of interference fringes. Such instabilities represent a persistent challenge, and the basic techniques of phase referencing and phase closure have been developed to deal with them. Synthesis imaging with large observing datasets has become a routine and straightforward process at radio observatories, but remains challenging for optical facilities. In this context, the commonly used image reconstruction algorithms CLEAN and MEM are presented. Lastly, a concise overview of current facilities is included as an appendix.

1 Interferometry in Astronomy

1.1 Introduction

The technique of interferometry is an indispensable tool for modern astronomy. Typically the telescope diameter D limits the angular resolution for an imaging system to $\Theta \approx \frac{\lambda}{D}$ owing to diffraction, but interferometry allows the achievement of angular resolutions $\Theta \approx \frac{\lambda}{B}$ where the baseline B is set by the distance between telescopes. Interferometry has permitted the angular resolution at radio wavelengths to initially reach, and now to significantly surpass, the resolution available with both ground- and space-based optical telescopes. Indeed, radio astronomers routinely create high-quality images with high sensitivity, high angular resolution, and a large field-of-view using arrays of telescopes such as the very large array (VLA), the combined array for research in millimeter-wave astronomy (CARMA), and now the atacama large millimeter array (ALMA). Interferometer arrays are now the instruments of choice for imaging the wide range of spatial structures found both for galactic and for extragalactic targets at radio wavelengths.

At optical wavelengths, interferometry can improve the angular resolution down to the milliarcsecond level, an order-of-magnitude better than even the hubble space telescope. While atmospheric turbulence limits the sensitivity much more dramatically than for the radio, optical interferometers can nevertheless measure the angular sizes of tens of thousands of nearby galactic objects and even a growing sample of distant active galactic nuclei (AGN). Recently, optical synthesis imaging of complex objects has been demonstrated with modern arrays of four to six telescopes, producing exciting results and opening new avenues for research.

Both radio and optical interferometers also excel at precision astrometry, with the potential for *microarcsecond*-level precision for some applications. Currently, it is ground-based radio interferometry (e.g., the VLBA) that provides the highest astrometric performance, although ground-based near-IR interferometers are improving and measure different astronomical phenomena.

This chapter will provide an overview of interferometry theory and present some practical guidelines for planning observations and for carrying out data analysis at the premier ground-based radio and optical interferometer facilities currently available for research in astronomy. In this chapter, the term "radio" will be used as a shorthand for the whole class of systems from sub-mm to decametric wavelengths which usually employ coherent high-frequency signal amplification, superheterodyne signal conversion, and digital signal processing, although detailed instrumentation can vary substantially. Likewise, the term "optical" generally describes systems employing direct detection, that is, the direct combination of the signals from each collector without amplification or mixing with locally generated signals.[1]

Historically, radio and optical interferometry have usually been discussed and reviewed independently from each other, leaving the student with the impression that there is something fundamentally different between the two regimes of wavelength. Here a different approach is taken, presenting a unified and more wavelength-independent view of interferometry, nonetheless noting important practical differences along the way. This perspective will demystify some of the differing terminology and techniques in a more natural way, and hopefully will be more approachable for a broad readership seeking general knowledge. For a more detailed treatment of radio interferometry specifically, refer to the classic text by Thompson et al. (2001) and the series of lectures in the NRAO Summer School on Synthesis Imaging (e.g., Taylor et al. 1999). Optical interferometry basics have been covered in individual reviews by Quirrenbach (2001) and by Monnier (2003), and a useful collection of course notes can be found in the NASA-Michelson Course Notes (Lawson 2000) and ESO-VLTI summer school proceedings (Malbet and Perrin 2007). Recently, a few textbooks have been published on the topic of optical interferometry specifically, including Labeyrie et al. (2006), Glindemann (2010), and Saha (2011). Further technical details can also be found in ➋ Chaps. 2 and ➋ 10 of the first volume in this series.

This chapter begins with a brief history of interferometry and its scientific impact on astronomy, a basic scientific context for newcomers that illustrates why the need for better angular resolution has been and continues to be one of the most important drivers for technical innovation in astronomy.

1.2 Scientific Impact

Using interferometers in a synthesis imaging array allows designers to decouple the diffraction-limited angular resolution of a telescope (which improves linearly with the telescope size) from

[1] It should be emphasized, however, that this distinction is somewhat artificial; the first meter-wave radio interferometers ≈ 65 years ago were simple Michelson adding interferometers employing direct detection without coherent high-frequency signal amplification. At the other extreme, superheterodyne systems are currently routinely used at wavelengths as short as 10 μm, such as the UC Berkeley ISI facility.

its collecting area (which, for a filled aperture, grows quadratically with the size). In the middle of the twentieth century, radio astronomers faced a challenge in their new science; the newly discovered "radio stars" were bright enough to be observed with radio telescopes of modest collecting area, but the resolution of conventional "filled aperture" reflecting telescopes was woefully inadequate (by one or more orders of magnitude) to measure the positions and angular sizes of these enigmatic new cosmological objects with a precision sufficient to permit an identification with an optical object. Thus, separated-element interferometry, although first applied in astronomy at optical wavelengths (Michelson and Pease 1921), began to be applied in the radio with revolutionary results.

At radio wavelengths, the epoch of rapid technological development began more than 50 years ago, and now interferometry is the "workhorse" technique of choice for most radio astronomers in the world. A steady stream of exciting new results has flowed from these instruments even until today, and a complete census of the major discoveries to date would be very lengthy indeed. Here, our attention is focused on the earliest historical discoveries that *required* radio interferometers, and we have listed our nominations in ➋ *Table 7-1*. In ➋ *Fig. 7-1a*, the spectacular image of radio jets in quasar 3C175 by the VLA is shown to illustrate the high-fidelity imaging that is possible using today's radio facilities.

Modern long-baseline optical interferometry started approximately 30 years after radio interferometry, following the pioneering experiments and important scientific results with the Narrabri intensity interferometry (e.g., Hanbury Brown et al. 1974) and the heterodyne work of the Townes' group at Berkeley (Johnson et al. 1974). The first successful direct interference of stellar light beams from separated telescopes was achieved in 1974 (Labeyrie 1975) and this was followed by about 20 years of two-telescope (i.e., single baseline) experiments which measured the angular diameters of a variety of objects for the first time. The first imaging arrays with more than two telescopes were constructed in the 1990s, and the COAST interferometer was first to make a true optical synthesis image using techniques familiar to radio astronomers (Baldwin et al. 1996). Keck and VLT interferometers both include 8-m class telescopes, making them the most sensitive facilities in the world. Recently, the CHARA array has produced a large number of new images in the infrared using combinations of four telescopes simultaneously. ➋ *Table 7-1* lists a few major scientific accomplishments in the history of optical interferometry showing the diversity of contributions in many areas of stellar astronomy and even recent extragalactic observations of active galactic nuclei. With technical and algorithm advances, model-independent imaging has become more powerful and a state-of-the-art image from the CHARA array is presented in ➋ *Fig. 7-1b*, showing the surface of the rapidly rotating star Alderamin.

2 Interferometry in Theory and Practice

2.1 Introduction

The most basic interferometer used in observational astronomy consists of two telescopes configured to observe the same object and connected together as a Michelson interferometer. Photons collected at each telescope are brought together to a central location and combined coherently at the "correlator" (radio term) or the "combiner" (optical term). For wavelengths

Table 7-1

Some historically important astronomical results made possible by interferometry

Astronomical result	Date	Facility	References[a]
Radio interferometry[b]			
Solar radio emission from sunspots	1945–1946	Australia, Sea cliff interferometer	R1
First Radio Galaxies NGC 4486 and NGC 5128	1948	New Zealand, Sea cliff interferometer	R2
Identification of Cygnus A	1951–1953	Cambridge, Würzburg antennas	R3
Cygnus A double structure	1953	Jodrell Bank, Intensity interferometer	R4
AGN superluminal motions	1971	Haystack-Goldstone VLBI	R5
Dark matter in spiral galaxies	1972–1978	Caltech interferometer, Westerbork SRT	R6
Spiral arm structure and kinematics	1973–1980	Westerbork SRT	R7
Compact source in Galactic center	1974	NRAO Interferometer	R8
Gravitational lenses	1979	Jodrell Bank Mk1 + Mk2 VLBI	R9
NGC 4258 black hole	1995	NRAO VLBA	R10
Optical interferometry			
Physical diameters of hot stars	1974	Narrabri intensity interferometer	O1
Empirical effective temperature scale for giants	1987	I2T/CERGA	O2
Survey of IR Dust Shells	1994	ISI	O3
Geometry of Be star disks	1997	Mark III	O4
Near-IR Sizes of YSO disks	2001	IOTA	O5
Pulsating Cepheid ζ Gem	2001	PTI	O6
Crystalline silicates in inner YSO disks	2004	VLTI	O7
Vega is a rapid rotator	2006	NPOI	O8
Imaging gravity-darkening on Altair	2007	CHARA	O9
Near-IR sizes of AGN	2009	Keck-I	O10

[a]References: R1: Pawsey et al. (1946) and McCready et al. (1947). R2: Bolton et al. (1949). R3: Smith (1951) and Baade and Minkowski (1954). R4: Jennison and Das Gupta (1953). R5: Whitney et al. (1971) and Cohen et al. (1971). R6: Rogstad and Shostak (1972), Bosma (1981a), and Bosma (1981b). R7: Allen et al. (1973), Rots and Shane (1975), Rots (1975), Visser (1980b), and Visser (1980a). R8: Goss et al. (2003). R9: Porcas et al. (1979) and Walsh et al. (1979). R10: Miyoshi et al. (1995). O1: Hanbury Brown et al. (1974). O2: di Benedetto and Rabbia (1987). O3: Danchi et al. (1994). O4: Quirrenbach et al. (1997). O5: Millan-Gabet et al. (2001). O6: Lane et al. (2000). O7: van Boekel et al. (2004). O8: Peterson et al. (2006). O9: Monnier et al. (2007). O10: Kishimoto et al. (2009)

[b]Radio list in part from Wilkinson et al. (2004) and Ekers (2010, private communication), with additions by one of the authors (RJA). Historical material prior to 1954 is also from Goss (2011, private communication) and Sullivan (2009)

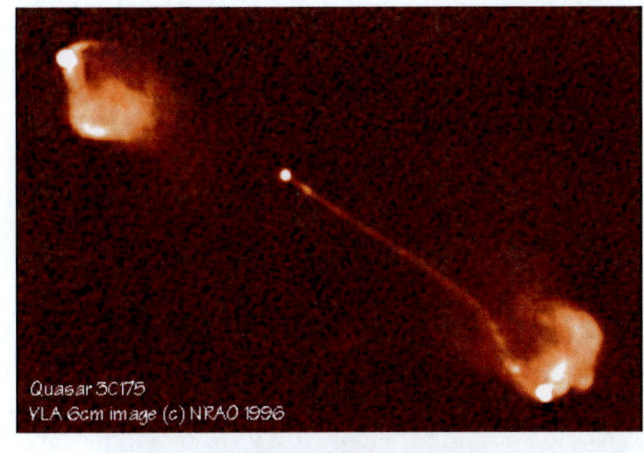

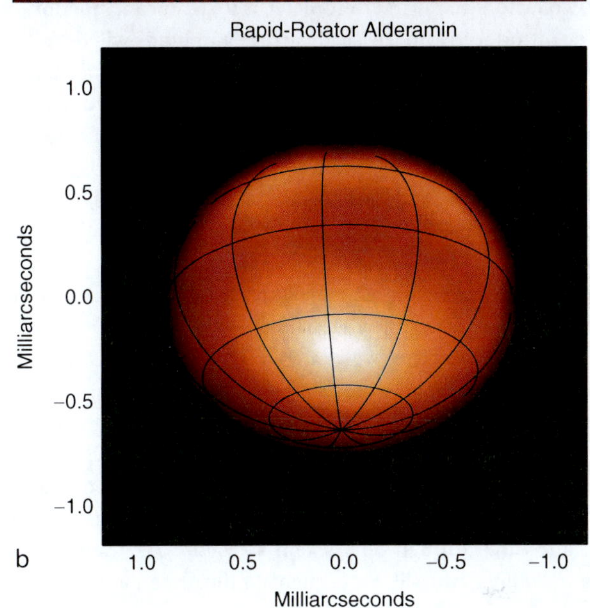

Fig. 7-1

Examples of imaging with interferometric arrays. (a): Radio image of jets in quasar 3C175 produced by the very large array with field of view ~1 arcmin ~200 kpc and angular resolution of 0.35 arcsec (Reprinted with permission Bridle et al. 1994). (b): Near-infrared image of the rapidly rotating star Alderamin produced by the CHARA Array with field of view of 7 $R_\odot$ ~2.5 milliarcsec and angular resolution of 0.6 milliarcsec (Reprinted with permission Zhao et al. 2009). The hot polar region and the cool equator is caused by the effect of "gravity darkening"

longer than ~0.2 mm, the free-space electric field is usually converted into cabled electrical signals and coherently amplified at the focus of each telescope. The celestial signal is then mixed with a local oscillator signal sent to both telescopes from a central location, and the difference frequency transmitted in cables back to the centrally located correlator. For shorter wavelengths, cable losses increase, and signal transmission moves eventually into free space in a more "optical" mode, using mirrors and long-distance transmission of light beams in (sometimes evacuated) pipes.

Depending on the geometry, the light from an astronomical object will in general be received at one telescope before it arrives at the other. If the fractional signal bandwidth $\Delta\nu$ is very narrow (either because of the intrinsic emission properties of the source, e.g., a spectral line, or because of imposed bandwidth limitations in amplifiers and/or filters), then the signal has a high degree of "temporal coherence," which is to say that the wave packet describing all the photons in the signal is extended in time by $\tau \approx 1/\Delta\nu$ s. Expressing the bandwidth in terms of wavelength, $\Delta\nu = (c/\lambda_0^2) \cdot \Delta\lambda$ where λ_0 is the band center and c is the speed of light. The coherence time is then $\tau = (1/c) \cdot (\lambda_0^2/\Delta\lambda)$, and $c \cdot \tau$ is a scale size of the wave packet called the *coherence length*, $L_c = c \cdot \tau = \lambda_0^2/\Delta\lambda$ (e.g., Hecht 2002, Chapter 7). If the path difference between the two collectors in an interferometer is a significant fraction of L_c, an additional time delay must be introduced; otherwise, the fringe amplitude will decrease or even disappear. For ground-based systems, the geometry is continually changing for all directions in the sky (except in the directions to the equatorial poles), requiring a continually changing additional delay to maintain the temporal coherence. The special location on the sky where the adjusted time delay is matched perfectly is often called the "phase center" or point of zero optical path delay (OPD), although such a condition actually defines the locus of a plane passing through the midpoint between the collectors and perpendicular to the baseline, and cutting the celestial sphere in a great circle. Since the telescope optics usually limits the field of view to only a tiny portion of this great circle, adjusting the phase center is the equivalent of "pointing" the interferometer at a given object within that field of view.

The final step is to interfere the two beams to measure the *spatial coherence* (often called the mutual coherence) of the electric field as sampled by the two telescopes. If the object observed is much smaller than the angular resolution of the interferometer, then interference is complete and one observes 100% coherence at the correlator/combiner. However, objects that are *resolved* (i.e., much larger than the angular resolution of the interferometer) will show less coherence due to the fact that different patches of emission on the object do not interfere at the same time through our system. ➲ *Figure 7-2* shows two simple cases of an interferometer as a Young's two-slit experiment to illustrate basic principles. At the left, the interferometer is made up of two slits and the response for a monochromatic point source (i.e., incoming plane waves) is shown. The result should be familiar: an interference fringe modulating the intensity from 100% to 0% with a periodicity that corresponds to a fringe spacing of $\frac{\lambda}{B}$ on the sky. Next to this panel is shown an example of two equal-brightness point sources separated by $\frac{1}{2}\frac{\lambda}{B}$, half the fringe spacing. The location of constructive interference for one point coincides with the location of destructive interference for the other source. Since the two sources are mutually incoherent, the superposition of the two fringe results in an even light distribution, that is, no fringe at all! In optical interferometry language, the first example fringe has a fringe contrast (or visibility) of 1 while the second example fringe has a visibility of 0.

➲ *Figure 7-3* contains a schematic of a basic interferometer as typically realized for both radio and optical configurations. While instrumental details vary immensely in how one transmits and interferes the signals for radio, millimeter, infrared, and visible-light interferometers, the basic principles are the same. The foundational theory common to all interferometers will be introduced next.

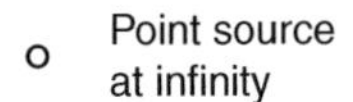

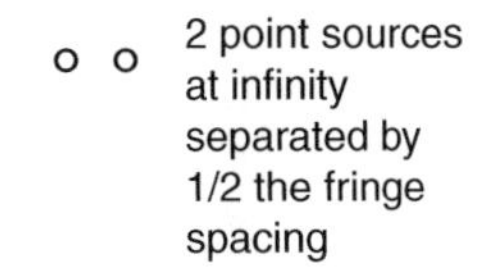

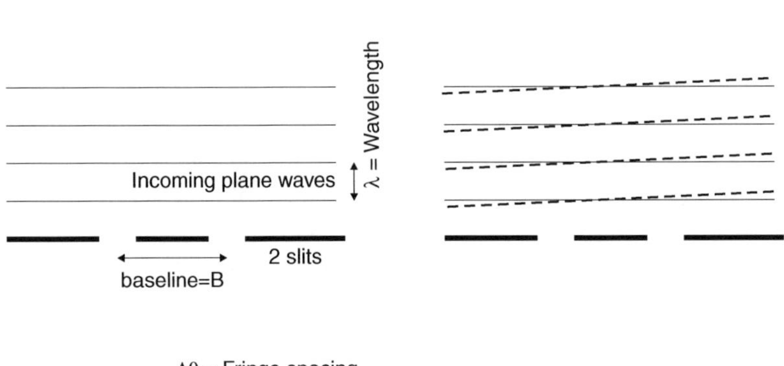

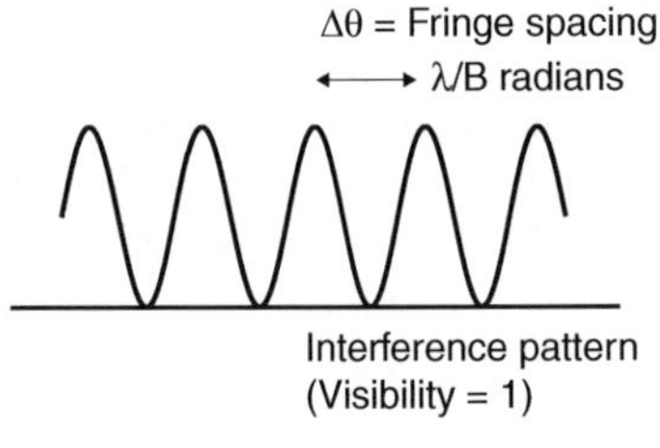

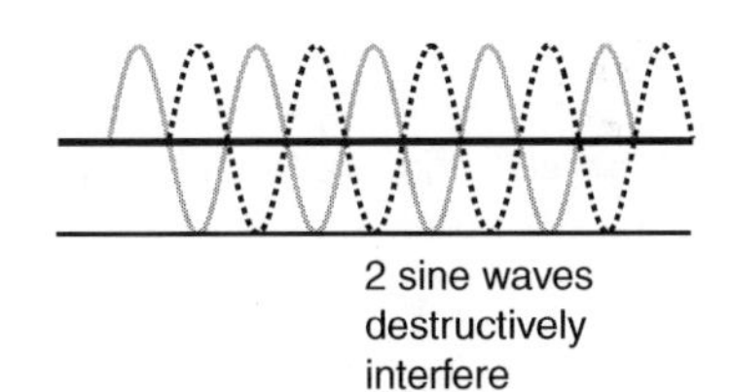

Fig. 7-2
Basic operating principle behind interferometry as illustrated by Young's two-slit experiment (Adapted from Monnier 2003)

2.2 Interferometry in Theory

The fundamental equation of interferometry is typically derived by introducing the van-Cittert Zernike Theorem and a complete treatment can be found in Chapter 3 of the book by Thompson et al. (2001). Here the main result will be presented without proof, beginning by defining an interferometric observable called the *complex visibility*, $\tilde{\mathcal{V}}$. The visibility can be derived from the intensity distribution on the sky $I(\vec{\sigma})$ using a given interferometer baseline $\vec{B}$ (which is the separation vector between two telescopes) and the observing wavelength λ:

$$\tilde{\mathcal{V}} = |\mathcal{V}|e^{i\phi_{\mathcal{V}}} = \int_{\text{sky}} A_N(\vec{\sigma}) I(\vec{\sigma}) e^{-\frac{2\pi i}{\lambda}\vec{B}\cdot\vec{\sigma}} d\Omega \tag{7.1}$$

Here, the $\vec{\sigma}$ represents the vector pointing from the center of the field-of-view (called the "phase center") to a given location on the celestial sphere using local (East, North) equatorial coordinates and the telescope separation vector $\vec{B}$ also using east and north coordinates. The modulus

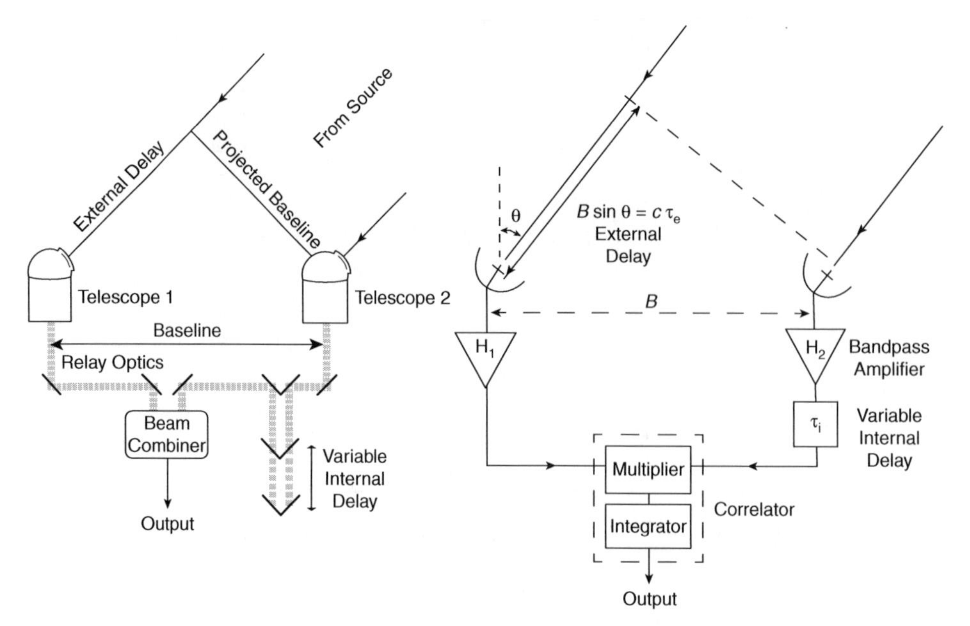

Fig. 7-3
Here are more realistic examples of long-baseline interferometers, both optical and radio, including the light collectors and delay line. *Left panel*: Optical interferometer (Adapted from Monnier 2003). *Right panel*: Radio interferometer (Adapted from *Fig. 2.3* in Thompson et al. 2001). See text for further discussion

of the complex visibility $|\mathcal{V}|$ is referred to as the *fringe amplitude or visibility* while the argument $\phi_\mathcal{V}$ is the *fringe phase.* $A_N(\vec{\sigma})$ represents the normalized pattern that quantifies how off-axis signals are attenuated as they are received by a given antenna or telescope. In this treatment, the astronomical object is assumed to be small in angular size in order to ignore the curvature of the celestial sphere.

The physical baseline $\vec{B}$ can be decomposed into components $\vec{u} = (u, v)$ in units of observing wavelength along the east and north directions (respectively) as projected in the direction of our target. The vector $\vec{\sigma} = (l, m)$ also can be represented in rectilinear coordinates on the celestial sphere, where l points along local east and m points north.[2] Here, l and m both have units of radians. Equation 7.1 now becomes:

$$\tilde{\mathcal{V}}(u,v) = |\mathcal{V}|e^{i\phi_\mathcal{V}} = \int_{l,m} A_N(l,m)I(l,m)e^{-2\pi i(ul+vm)}\,dl\,dm \tag{7.2}$$

The fundamental insight from (7.2) is that an interferometer is a Fourier Transform machine – it converts an intensity distribution $I(l, m)$ into measurements of Fourier components $\tilde{\mathcal{V}}(u, v)$ for all the baselines in the array represented by the (u, v) coverage. Since an intensity distribution can be described fully in either image space or Fourier space, the collection of sufficient Fourier components using interferometry allows for an image reconstruction through

[2]There are several different coordinate systems in use to describe the geometry of ground-based interferometers used in observing the celestial sphere (see, e.g., Thompson et al. 2001, Chapter 4 and Appendix 4.1).

an inverse Fourier Transform process, although practical limitations lead to compromises in the quality of such images.

2.3 Interferometry in Practice

In this section, the similarities and differences between radio and optical interferometers are summarized along with the reasons for the main differences. Interested readers can find more detailed on specific hardware implementations in Volume I of this series.

Modern radio and optical interferometers typically use conventional steerable telescopes to collect photons from the target. In the radio, a telescope is often called an antenna; it is typically a parabolic reflector with a very short focal length (f/D ≈ 0.35 is common), with signal collection and initial amplification electronics located at the prime focus. Owing to the large value of $\Delta\Theta \sim \frac{\lambda}{\text{Diameter}}$, the diffraction pattern of the antenna aperture is physically a relatively large region at the prime focus. This fact, coupled with the cost and complexity of duplicating and operating many low-noise receivers in close proximity to each other, has meant that antennas used in radio astronomy typically have only a "single-pixel" signal collection system (a dipole or a "feed horn") at the prime focus.[3] Light arriving from various directions on the sky is attenuated depending on the shape of the diffraction pattern, written as A_N in (7.2) and often called the "antenna pattern" or the "primary beam." The signal collection system may be further limited to a single polarization mode, although systems are common that simultaneously accept both linear (or both circular) polarization states. After initial amplification, the signal is usually mixed with a local oscillator to "down-convert" the high frequencies to lower frequencies that can more easily be amplified and processed further. These lower-frequency signals from the separate telescopes can also be more easily transported over large distances to a common location using, for example, coaxial cable, or by modulating an optical laser and using fiber optics. This common location houses the "backend" of the receiver, where the final steps in signal analysis are carried out including band definition, correlation, digitizing, and spectral analysis. In some cases, the telescope signals are recorded onto magnetic media and correlated at a later time and in a distant location (e.g., the "Very Long Baseline Array" or global VLBI).

In the optical, the light from the object is generally focused by the telescope, re-collimated into a compressed beam for free-space transport, and then sent to a central location in a pipe which is typically evacuated to avoid introducing extra air dispersion and ground-level turbulence. In rare cases, the light at the telescope is focused directly into a single-mode fiber, which is the dielectric equivalent to the metallic waveguides used in radio and millimeter receivers. Note that atmospheric seeing is very problematic for even small visible and infrared telescopes while it is usually negligible compared to the diffraction limit for even the largest radio and mm-wave telescopes.

Both radio and optical interferometers must delay the signals from some telescopes to match the optical paths. After mixing to a lower frequency, radio interferometers can use switchable lengths of coaxial cable in order to introduce delays. More recently, the electric fields can be directly digitized with bandwidths of >5 GHz, and these "bits" can be saved in physical memory and then recalled at a later time. For visible and infrared systems, switchable fiber optics are not practical due to losses and glass dispersion; the only solution is to use an optical "free-space"

[3] Very recently, several radio observatories have begun to equip their antennas (and at least one entire synthesis telescope) with arrays of such feeds.

delay line consisting of a retroreflector moving on a long track and stabilized through laser metrology to compensate for air path disturbances and vibrations in the building.

In a radio interferometer, once all the appropriate delays have been introduced, the signals from each telescope can be combined. Early radio signal correlators operated in an "optical" mode as simple adding interferometers, running the sum of the signals from the two arms through a square-law detector. The output of such a detector contains the product of the two signals. Unfortunately, the desired correlation product also comes with a large total power component caused by temporally uncorrelated noise photons contributed (primarily) by the front-end amplifiers in each arm of the interferometer plus atmosphere and ground noise. This large signal demanded excellent DC stability in the subsequent electronics, and it was not long in the history of radio interferometry before engineers found clever switching and signal-combination techniques to suppress the DC component. These days signal combiners deliver only the product of the signals from each arm, and are usually called "correlators."[4] Most modern radio/millimeter arrays use digital correlators that introduce time lags between all pairs of telescopes in order to do a full temporal cross-correlation. This allows a detailed wavelength-dependent visibility to be measured, that is, an interferometric spectrum with $R = \frac{\lambda}{\Delta\lambda} > 100,000$ if necessary. By most metrics, radio correlators have reached their fundamental limit in terms of extracting full spectral and spatial information and can be fairly sophisticated and complex to configure when correlating multiple bandpasses simultaneously with high spectral resolution.[5]

In the visible and infrared, the electric fields cannot be further amplified without strongly degrading the signal-to-noise ratio, and so parsimonious beam combining strategies are common that split the signal using, for example, partly reflecting mirrors into a small number of pairs or triplets. Furthermore, most optical systems have only modest spectral resolutions of typically $R \sim 40$ in order to maintain high signal-to-noise ratio, although a few specialized instruments exist that reach $R > 1,000$ or even $R > 30,000$. Signal combination finally takes place simply by mixing the light beams together and modulating the relative optical path difference, either using spatial or temporal encoding. The total power measurement in a visible-light or infrared detector will reveal the interference fringe, and a Fourier analysis can be used to extract the complex visibility $\tilde{\mathcal{V}}$.

Because the ways of measuring visibilities are quite different, radio and optical interferometrists typically report results in different units. Radio/mm interferometers measure correlated flux density in units of Jansky (10^{-26} W m^{-2} Hz^{-1}), just as suggested by (➋ 7.2).[6] In the optical however, interferometers tend to always measure a normalized visibility that varies from 0 to 1 – this is simply the correlated signal normalized by the total power. One can convert the latter to correlated flux density by simply multiplying by the known total flux density of the target at the observed wavelengths, or otherwise by carrying out a calibration of the system by a target of known flux density.

[4]This is not all advantageous; if the data is intended to be used in an imaging synthesis, the absence of the total power component means that the value of the map made from the data will integrate to zero. In other words, without further processing, the image will be sitting on a slightly negative "floor." If more interferometer spacings around zero are also missing, the floor becomes a "bowl." All this is colloquially called "the short-spacing problem," and it adversely affects the photometric accuracy of the image. A significant part of the computer processing "bag of tricks" used to "restore" such images is intended to address this problem, although the only proper way to do that is to obtain the missing data and incorporate it into the synthesis.

[5]At millimeter and submillimeter wavelengths, correlators still do not attain the maximum useful bandwidths for continuum observations.

[6]Recall that an integration of specific intensity over solid angle results in a flux density, often expressed in Jansky.

2.3.1 Quantum Limits of Amplifiers

The primary reason why radio and optical interferometers differ so much in their detection scheme is because coherent amplifiers would introduce too much extraneous noise at the high frequencies encountered in the optical and infrared. This difference is fundamental and is explored in more detail in this section.

At radio frequencies, there are huge numbers of photons in a typical sample of the electromagnetic field, so the net phase of a packet of radio photons (either from the source or from a noisy receiver) is well defined and amplifiers can operate coherently. The ultimate limits which apply to such amplifiers are dictated by the uncertainty principle as stated by Heisenberg. Beginning with the basic "position – uncertainty" relation $\Delta x\ \Delta p_x \geq h/4\pi$, it is easy to derive the "energy – time" relation $\Delta E\ \Delta t \geq h/4\pi$. Since the uncertainty in the energy of the n photons in a wave packet can be written as $\Delta E = h\nu\ \Delta n$ and the uncertainty in the phase of the aggregate as $\Delta\phi = 2\pi\nu\ \Delta t$, this leads to the equivalent uncertainty relation $\Delta\phi\ \Delta n \geq 1/2$.

An ideal amplifier which adds no noise to the input photon stream leads to a contradiction of the uncertainty principle. The following argument shows how this happens (adapted from Heffner 1962): Consider an ideal coherent amplifier of gain G which creates new photons in phase coherence with the input photons, and assume it adds no incoherent photons of its own to the output photon stream. With n_1 photons going into such an amplifier, there will be $n_2 = Gn_1$ photons at the output, all with the same phase uncertainty $\Delta\phi_2 = \Delta\phi_1$ with which they went in. In addition, in this model, it is expected that $\Delta n_2 = G\Delta n_1$ (no additional "noise" photons unrelated to the signal). But, according to the same uncertainty relation, the photon stream coming out of the amplifier must also satisfy $\Delta\phi_2\ \Delta n_2 \geq 1/2$. This would imply that $\Delta\phi_1\ \Delta n_1 \geq \frac{1}{2G}$, which for large G says that the input photon number and wave packet phase could be measured with essentially no noise. But this contradicts the same uncertainty relation for the input photon stream, which requires that $\Delta\phi_1\ \Delta n_1 \geq 1/2$. This contradiction shows that one or more of our assumptions must be wrong. The argument can be saved if the amplifier itself is required to add noise of its own to the photon stream; the following heuristic construction shows how. Using the identity $\Delta n_2 = (G - N)\cdot\Delta n_1 + N\Delta n_1$ at the output (where N is an integer $N \geq 1$), and referring this noise power back to the input by dividing it with the amplifier gain G, this leads to $(1 - N/G)\cdot\Delta n_1 + (N/G)\cdot\Delta n_1$ at the input to the amplifier, which for large G is Δn_1. The smallest possible value of N is 1. This preserves the uncertainty relation at the expense of an added minimum noise power of $h\nu$ at the input. Oliver (1965) has elaborated and generalized this argument to include all wavelength regimes, and has shown that the minimum total noise power spectral density ψ_ν of an ideal amplifier (relative to the input) is:

$$\psi_\nu = \frac{h\nu}{e^{(h\nu/kT)} - 1} + h\nu\ \mathrm{W/Hz}\,, \tag{7.3}$$

where T is the kinetic temperature that the amplifier input faces in the propagation mode to which the amplifier is sensitive. For $h\nu < kT$ this reduces to $\psi_\nu \approx kT$ W/Hz, which can be called the "thermal" regime of radio astronomy. For $h\nu > kT$ this becomes $\psi_\nu \approx h\nu$ W/Hz in the "quantum" regime of optical astronomy. The crossover point where the two contributions are equal is where $h\nu/kT = \ln 2$, or at $\lambda_c \cdot T_c = 20.75$ (mm K). As an illustration of the use of this equation, consider this example: The sensitivity of high-gain radio-frequency amplifiers can usually be improved by reducing their thermodynamic temperatures. However, for instance at a wavelength of 1 mm, it might be unnecessary (depending on details of the signal chain) to aim for a high-gain amplifier design to lower the thermodynamic temperature below about 20 K, since at that point, the sensitivity is in any case limited by quantum noise. At even shorter wavelengths, the rationale for cooled amplifiers disappears, and at optical wavelengths, amplifiers are clearly not useful since the noise is totally dominated by spontaneous emission[7] and is equivalent to thermal emission temperatures of thousands of degrees. The extremely faint signals common in modern optical observational astronomy translate into very low photon rates, and the addition of such irrelevant photons into the data stream by an amplifier would not be helpful.

2.4 Atmospheric Turbulence

So far, the analysis of interferometer performance has assumed a perfect atmosphere. However, the electromagnetic signals from cosmic sources are distorted as they pass through the intervening media on the way to the telescopes. These distortions occur first in the interstellar medium, followed by the interplanetary medium in the solar system, then the Earth's ionosphere, and finally the Earth's lower atmosphere (the troposphere) extending from an altitude of $\approx$11 km down to ground level. The media involved in the first three sources of distortion contain ionized gas and magnetic fields, and their effects on signal propagation depend strongly on wavelength (generally as $\propto \lambda^2$) and polarization. At wavelengths shorter than about 10 cm, the troposphere begins to dominate. Molecules in the troposphere (especially water vapor) become increasingly troublesome at frequencies above 30 GHz (1 cm wavelength), and the atmosphere is essentially opaque beyond 300 GHz except for two rather narrow (and not very clear) "windows" from 650–700 and 800–900 GHz which are usable only at the highest-altitude sites. The next atmospheric windows appear in the IR at wavelengths less than about 15 μm. The optical window opens around 1 μm, and closes again for wavelengths shorter than about 350 nm.

The behavior of the troposphere is thus of prime importance to ground-based astronomy at wavelengths from the decimeter-radio to the optical. Interferometers are used in the study of structure in the troposphere, and a summary of approaches and results with many additional references is given in Thompson et al. (2001), Carilli and Holdaway (1999), and Sutton and Hueckstaedt (1996, Chapter 13). A discussion oriented toward optical wavelengths can be found in Quirrenbach (2000). Since the main focus here is on using interferometers to measure the properties of the cosmic sources themselves, our discussion is limited to some "rules of thumb" for choosing the interferometer baseline length and the time interval between measurements of the source and of a calibrator in order to minimize the deleterious effects of propagation on the fringe amplitudes and (especially) fringe phases.

2.4.1 Phase Fluctuations: Length Scale

Owing to random changes in the refractive index of the atmosphere and the size distribution of these inhomogeneities, the path length for photons will be different along different parallel lines of sight. This fluctuating path length difference grows almost linearly with the separation d of the two lines of sight for separations up to some maximum, called the outer scale length (typically tens to hundreds of meters, with some weak wavelength dependence), and is

[7] Although amplifiers are currently used in the long-distance transmission of near-IR (digital) communication signals in optical fibers, the signal levels are relatively large and low noise is not an important requirement.

roughly constant beyond that. Surprisingly, in spite of the differences in the underlying physical processes causing refraction, variations in the index of refraction are quite smooth across the visible and all the way through to the radio. At short radio wavelengths, the fluctuations are dominated by turbulence in the water vapor content; at optical/IR wavelengths, it is temperature and density fluctuations in dry air that dominate.

Using a model of fully developed isotropic Kolmogorov turbulence for the Earth's atmosphere, the rms path length difference grows according to $\sigma_d \propto d^{5/6}$ for a path separation d (see Thompson et al. 2001, Chapter 13, for references). High altitude sites show smaller path length differences as the remaining vertical thickness of the water vapor layer decreases. Relatively large seasonal and diurnal variations also exist at high mountain sites as the atmospheric temperature inversion layer generally rises during the summer and further peaks during midday. Variations in σ_d by factors of ~10 are not unusual (see Thompson et al. 2001, *Fig. 13.13*), but a rough average value for a good observing site is $\sigma_d \approx 1$ mm for baselines $d \approx 1$ km at millimeter wavelengths, and $\sigma_d \approx 1\,\mu$m for baselines $d \approx 50$ cm at infrared wavelengths.

The length scale fluctuations translate into fringe phase fluctuations of $\sigma_\phi = 2\pi\sigma_d/\lambda$ in radians. The *maximum coherent baseline* d_0 is defined as that baseline length for which the rms phase fluctuations reach 1 rad. Using the expressions in the previous paragraph and coefficients suitable for the radio and optical ranges at the better observing sites, two useful approximations are $d_0 \approx 140 \cdot \lambda^{6/5}$ m for λ in millimeters (useful at millimeter radio wavelengths), and $d_0 \approx 10 \cdot \lambda^{6/5}$ cm for λ in microns (useful at IR wavelengths). These two expressions are in fact quite similar; using the "millimeter expression" to calculate d_0 in the IR underestimates the value obtained from the "IR expression" by a factor of 2.8, which is at the level of precision to be expected.

At shorter wavelengths (visible and near-infrared), atmospheric turbulence limits even the image quality of small telescopes. This has led to a slightly different perspective for the length scale that characterizes atmospheric turbulence, although it is closely related to the previous description. The Fried length r_0 (Fried 1965) is the equivalent-sized telescope diameter whose diffraction limit matches the image quality through the atmosphere due to *seeing*. It turns out that this quantity is proportional to the length scale where the rms phase error over the telescope aperture is ≈ 1 rad. In other words, apertures with diameters small compared to r_0 are approximately diffraction limited, while larger apertures have resolution limited by turbulence to $\approx \lambda/r_0$. It can be shown that, for an atmosphere with fully developed Kolmogorov turbulence, $r_0 \approx 3.2 d_0$ (Thompson et al. 2001, Chapter 13).

2.4.2 Phase Fluctuations: Time Scale

Although fluctuations of order 1 rad may be no more than a nuisance at centimeter wavelengths, requiring occasional phase calibration (see Sect. 3.1.3), they will be devastating at IR and visible wavelengths owing to their rapid variations in time. In order to relate the temporal behavior of the turbulence to its spatial structure, a model of the latter is required along with some assumption for how that structure moves over the surface of the Earth. One specific set of assumptions is described in Thompson et al. (2001, Chapter 13); however, for the purposes here it is sufficient to use Taylor's "frozen atmosphere" model with a nominally static phase screen that moves across the Earth's surface with the wind at speed v_s. This phase screen traverses the interferometer baseline d in a time $\tau_d = d/v_s$, at the conclusion of which the total path length variation is σ_d. Taking the critical time scale τ_c to be when the rms phase error

Table 7-2

Approximate baseline length, Fried length, and time scales for a 1-rad rms phase fluctuation in the Earth's troposphere and a wind speed of 10 m/s[a]

Wavelength	Max. coherent baseline d_0	Fried length r_0	Time scale τ_c	Isoplanatic angle at zenith Θ_{iso}
0.5 μm (visible)	4.4 cm	14 cm	4.4 ms	5.5″
2.2 μm (near-IR)	26 cm	83 cm	26 ms	33″
1 mm (millimeter)	140 m	450 m	14 s	3.5°
10 cm (radio)	35 km	112 km	58 min	large[b]

[a]From parameters for Kolmogorov turbulence given in Thompson et al. (2001, Chapter 13), and in Woolf (1982, Table 2). The inner and outer scale lengths are presumed to remain constant in these rough approximations. Values are appropriate for a good observing site and improve at higher altitudes. See Sect. 2.4 for more discussion
[b]Limited in practice by observing constraints such as telescope slew rates and elevation limits, and source availability

reaches 1 rad, then $\tau_c \approx d_0/v_s$ with d_0 given in the previous paragraph. As an example consider a wind speed of 10 m/s; this leads to $\tau_c \approx 14$ s at $\lambda = 1$ mm, and ≈ 10 ms at $\lambda = 1\,\mu$m. Clearly the techniques required to manage these variations will be very different at the two different wavelength regimes, even though the magnitude of the path length fluctuations (in radians of phase) is similar. Representative values of these quantities are collected in *Table 7-2*.

2.4.3 Calibration: Isoplanatic Angle

The routine calibration of interferometer phase and amplitude is usually done by observing a source with known position and intensity interleaved in time with the target of interest. At centimeter wavelengths and longer, the discussion in the previous section indicates that such measurements can be done on time scales of minutes to hours, providing ample time to reposition telescopes elsewhere on the sky in order to observe a calibrator. But how close to the target of interest does such a calibrator have to be? Ideally, the calibrator ought to be sufficiently nearby on the celestial sphere that the line of sight traverses a part of the atmosphere with substantially the same phase delay as the line of sight to the target. This angle is called the *isoplanatic angle* Θ_{iso}; it characterizes the angular scale size over which different parts of the incoming wavefront from the target encounter closely similar phase shifts, thereby minimizing the image distortion. The isoplanatic angle can be roughly estimated by calculating the angle subtended by an r_0-sized patch at a height h that is characteristic for the main source of turbulence; hence, roughly $\Theta_{iso} \approx \frac{r_0}{h}$. Within a patch on the sky with this angle, the telescope/interferometer PSF remains substantially constant, retaining the convolution relation between the source brightness distribution and the image. Some approximate values are given in *Table 7-2* as a guide.

At visible and near-IR wavelengths, *Table 7-2* shows that the isoplanatic angle is very small, smaller than an arcminute. Unfortunately, the chance of having a suitably bright and point-like object within this small patch of the sky is very low. Even if an object did exist, it would be nearly impossible to repetitively reposition the telescope and delay line at the millisecond level timescale needed to "freeze" the turbulence between target and calibrator measurements. Special techniques to deal with this problem will be discussed further in Sect. 3.1.3.

3 Planning Interferometer Observations

The issues to consider when writing an interferometer observing proposal or planning the observations themselves include: the desired sensitivity (i.e., the unit telescope collecting area, the number of telescopes to combine at once, the amount of observing time), the required field-of-view and angular resolution (i.e.,the shortest and longest baselines), calibration strategy and expected systematic errors (i.e., choosing phase and amplitude calibrators), the expected complexity in the image (i.e., the completeness of u,v coverage, do science goals demand model-fitting or model-independent imaging), and the spectral resolution (i.e., correlator settings, choice of combiner instrument). Many of these issues are intertwined, and the burden on the aspiring observer to reach a compatible set of parameters can be considerable. Prospective observers planning to use the VLA are fortunate to have a wide variety of software planning tools and user's guides already at their disposal, but those hoping to use more experimental facilities or equipment which is still in the early phases of commissioning will find their task more challenging.

Here, the most common issues encountered during interferometer observations will be introduced. In many ways, this is more of a list of things to worry about rather than a compendium of solutions. The basic equations and considerations have been collected in ➲ *Table 7-3*. In order to obtain the latest advice on optimizing a request for observing time, or to plan an observing run, observers ought to consult the Web sites, software tools, and human assistants available for them at each installation (see Appendix for a list of current facilities).

3.1 Sensitivity

Fortunately, modern astronomers can find detailed documentation on the expected sensitivities for most radio and optical interferometers currently available. Indeed, the flexibility of modern instrumentation sometimes defies a back-of-the-envelope estimation for the true signal-to-noise ratio (SNR) expected for a given observation. In order to better understand what limits sensitivity for real systems, the dominant noise sources and the key parameters affecting signal strength are introduced. Most of the focus will be for observations of point sources since

Table 7-3
Planning interferometer observations

Consideration	Equation
Angular resolution	$\Theta = \frac{1}{2}\frac{\lambda}{B_{max}}$
Spectral resolution	$R = \frac{\lambda}{\Delta\lambda} = \frac{c}{\Delta v}$
Field-of-view	
Primary beam	$\Delta\Theta \sim \frac{\lambda}{D_{Telescope}}$
Bandwidth-smearing	$\Delta\Theta \sim R \cdot \frac{\lambda}{B_{max}}$
Time-smearing	$\Delta\Theta \sim \frac{230}{\Delta t_{minutes}} \frac{\lambda}{B_{max}}$
Phase referencing	
Coherence time	see ➲ *Table 7-2*
Isoplanatic angle	see ➲ *Table 7-2*

resolved sources do not contribute signal to all baselines in an array and this case must be treated with some care.

Here, the discussions of the radio and optical cases are separate because of the large differences in the nature of the noise processes (e.g., see Sect. 2.3.1) and the associated nomenclature. Radio and optical observations lie at the two limits of Bose-Einstein quantum statistics that govern photon arrival rates (e.g., Pathria 1972, see Sect. 6.3). At long wavelengths, the occupation numbers are so high that the statistics evolve into the Gaussian limit and where the root-mean-square (rms) fluctuation in the detected power ΔP is proportional to the total power P itself (e.g., ΔPower $\propto$ Power). On the other hand, in the optical limit, the sparse occupation of photon states results in the familiar Poisson statistics where the level of photon fluctuations ΔN is proportional to $\sqrt{N}$. Most of the SNR considerations for interferometers are in common with single-dish radio and standard optical photometry, and so interested readers are referred to the relevant chapters in Volumes 1 and 2 of this series.

3.1.1 Radio Sensitivity

The signal power spectral density P_ν received by a radio telescope of effective area A_e (m^2) from a celestial point source of flux density S_ν (Jansky = W/m^2/Hz) is $P_\nu = A_e \cdot S_\nu$ (W/Hz). It is common to express this as the power which would be delivered to a radio circuit (wire, coaxial cable, or waveguide) by a matched termination at a physical temperature T_A, called the "antenna temperature," so that $T_A = A_e S_\nu / 2k$ (Kelvin) where k = Boltzmann's constant and the factor 1/2 accounts for the fact that, although the telescope's reflecting surface concentrates both states of polarization at a focus, the "feed" collects the polarization states separately. As described in Sect. 2.3.1, the amplifier which follows must add noise; this additional noise power (along with small contributions from other extraneous sources in the telescope field of view) P_ν^s can likewise be expressed as $P_\nu^s = kT_s/2$, where T_s is the "system temperature." The rms fluctuations in this noise power will limit the faintest signals that can be distinguished. As mentioned in the previous paragraph, these fluctuations are directly proportional to the receiver noise power itself, so $\Delta T_s \propto T_s$. They will also be inversely proportional to the square root of the number of samples of this noise present in the receiver passband. The coherence time of a signal in a bandwidth $\Delta\nu$ is proportional to $1/\Delta\nu$, so in an integration time τ there are of order $\tau\Delta\nu$ independent samples of the noise, and the statistical uncertainty will improve as $1/\sqrt{\tau\Delta\nu}$. The ratio of the rms receiver noise power fluctuations to the signal power is therefore:

$$\Delta T_s / T_A \propto \frac{2kT_s}{A_e S \sqrt{\tau \Delta \nu}} . \tag{7.4}$$

The minimum detectable signal ΔS is defined as the value of S for which this ratio is unity. For this "minimum" value of S, the equation becomes:

$$\Delta S = \frac{f_c \cdot kT_s}{A_e \sqrt{\tau \Delta \nu}} , \tag{7.5}$$

The coefficient of proportionality f_c for this equation is of order unity, but the precise value depends on a number of details of how the receiver operates. These details include whether the receiver output contains both polarization states, whether both the in-phase and the quadrature channels of the complex fringe visibility are included, whether the receiver operates in single- or double-sideband mode, and how precisely the noise is quantized if a digital correlator is used.

Further discussion of the various possibilities is given in Thompson et al. (2001, Chapter 6). For the present purpose, it suffices to notice that the sensitivity for a specific radio interferometer system improves only slowly with integration time and with further smoothing of the frequency (radial velocity) resolution. The most effective improvements are made by lowering the system temperature and by increasing the collecting area.

The point-source sensitivity continues to improve as telescopes are added to an array. An array of n identical telescopes contains $N_b = n(n-1)/2$ distinct baselines. If the signals from each telescope are split into multiple copies, N_b interference pairs can be made. The rms noise in the flux density on a point source including all the data is then:

$$\Delta S = \frac{f_c \cdot kT_s}{A_e\sqrt{N_b \tau \Delta\nu}} . \tag{7.6}$$

So far the discussion has been made for isolated point sources. Extended sources are physically characterized by their surface brightness power spectral density $B_{\text{surf}}(\alpha, \delta, \nu)$ (Jansky/steradian) and by the angular resolution of the observation as expressed by the solid angle Ω_b of the synthesized beam in steradians (see ➋ Sect. 5). By analogy with the discussion of rms noise power from thermal sources given earlier, it is usual to express the surface brightness power spectral density for an extended sources in terms of temperature. This conversion of units to Kelvins is done using the Rayleigh-Jeans approximation to the Planck black-body radiation law, although the radiation observed in the image is only rarely thermally generated. The conversion from $B_{\text{surf}}(\alpha, \delta, \nu)$ (Jansky/steradian) to T_b in Kelvins is:

$$T_b = \frac{\lambda^2 B_{\text{surf}}}{2k\Omega_b} , \tag{7.7}$$

which requires $(h\nu/kT \ll 1)$ if the radiation is thermal; otherwise, this conversion can be viewed merely as a convenient change of units. The rms brightness temperature sensitivity in a radio synthesis image from receiver noise alone is then:

$$\Delta T_b = \frac{f_c \lambda^2 T_s}{2A_e \Omega_b \sqrt{N_b \tau \Delta\nu}} . \tag{7.8}$$

The final equations above for the sensitivity on synthesis imaging maps show that the more elements one has, the better the flux density sensitivity will be. For example, if one compares an array of $N_b = 20$ baselines with an array containing $N_b = 10$ baselines, the flux density SNR is improved by a factor $\sqrt{2}$ no matter where the additional ten baselines are located in the u, v plane. However, the brightness temperature sensitivity does depend critically on the actual distribution of baselines used in the synthesis. For instance, if the same number of telescopes is "stretched out" to double the maximum extent on the ground, the equations above show that the flux density sensitivity ΔS remains the same, but the brightness temperature sensitivity ΔT_b is worse by a factor of 4 since the synthesized beam is now four times smaller in solid angle. This is a serious limitation for spectral line observations where the source of interest is (at least partially) resolved and where the maximum surface brightness is modest. For instance, clouds of atomic hydrogen in the galactic ISM never seem to exceed surface brightness temperatures of ≈ 80 K, so the maximum achievable angular resolution (and hence the maximum useable baseline in the array) is limited by the receiver sensitivity. This can only be improved by lowering the system temperature on each telescope or by increasing the number of interferometer measurements with more telescopes and/or more observing time.

A cautionary note is appropriate here. In the case of an optical image of an extended object taken, for example, with charge-coupled device (CCD) camera on a filled aperture telescope, a simple way of improving the SNR is to average neighboring pixels together, thereby creating a smoothed image of higher brightness sensitivity. At first sight, the equation for ΔT_b above suggests that this should also happen with synthesis images, but here the improvement is not as dramatic as it may seem at first sight. The reason is that the action of smoothing is equivalent to discarding the longer baselines in the u, v plane; for instance, reducing the longest baseline used in the synthesis by a factor of 2 would indeed lead to an image with brightness temperature sensitivity which is better by a factor of 2^2, but the effective reduction of the number of interferometers from N to $N/2$ means that the net improvement is only $2^{1.5}$. A better plan would have been to retain all the interferometers but to shrink the array diameter with the factor 2 by moving the telescopes into a more compact configuration. This is one reason why interferometer arrays are usually constructed to be reconfigurable.

3.1.2 Visible and Infrared Sensitivity

As mentioned earlier, the visible and infrared cases deviate substantially from the radio case. While the sensitivity is still dependent on the collecting area of the telescopes (A_e), the dominant noise processes behave quite differently. In the visible and infrared (V/IR), noise is generated by the random arrival times of the photons governed by Poisson statistics $\Delta N = \sqrt{N}$, where N is the mean number of photons expected in a time interval τ and ΔN is the rms variation in the actual measured number of photons. Depending on the observing setup (e.g., the observing wavelength, spectral resolution, high visibility case or low visibility case), the dominant noise term can be Poisson noise from the source itself, Poisson noise from possible background radiation, or even detector noise. Because of the centrality of Poisson statistic, it is common to work in units of total detected photoelectrons N within a time interval τ, rather than power spectral density P_ν or system temperature T_S. This conversion is straightforward:

$$N = \eta \frac{P_\nu \Delta\nu}{h\nu} \tau \tag{7.9}$$

$$= \eta \frac{S_\nu A_e \Delta\nu}{h\nu} \tau \tag{7.10}$$

where η represents the total system detection efficiency which is the combination of optical transmission of system and the quantum efficiency of the detector and the other variables are the same as for the radio case introduced in the last section.

For the optical interferometer, atmospheric turbulence limits the size of the aperture that can be used without adaptive optics. (The atmosphere does not limit the useful size of the current generation of single-dish mm-wave and radio telescopes.) The Fried parameter r_0 sets the coherence length and thus the max$(A_e) \sim r_0^2$. Likewise without corrective measures, the longest useful integration time is limited to the atmospheric coherence time $\tau \sim \tau_c$. There exists a *coherent volume* of photons that can be used for interferometry, scaling like $r_0 \cdot r_0 \cdot c\tau_c$. As an example, consider the coherent volume of photons for decent seeing conditions in the visible ($r_0 \sim 10$ cm, $\tau_c \sim 3.3$ ms). From this, the limiting magnitude can be estimated by requiring at least 10 photons to be in this coherent volume. Assuming a bandwidth of 100 nm, 10 photons ($\lambda \sim 550$ nm) in the above coherent volume corresponds to a V magnitude of 11.3, which is the best limit one could hope to achieve.[8] This is more than 14 magnitudes worse than faint sources observed by today's 8-m class telescopes that can benefit from integration times measured in hours instead of milliseconds. Because the atmospheric coherence lengths and timescales behave approximately like $\lambda^{\frac{6}{5}}$ for Kolmogorov turbulence, the coherent volume $\propto \lambda^{\frac{18}{5}}$. Until the deleterious atmospheric effects can be neutralized, ground-based optical interferometers will never compete with even small single-dish telescopes in raw point-source sensitivity.

Under the best case, the only source of noise is Poisson noise from the object itself. Indeed, this limit is nearly achieved with the best visible-light detectors today that have read-noise of only a few electrons. More commonly, especially in the infrared, detectors introduce the noise that limits sensitivity, typically 10–15 electrons of read-noise in the near-IR for the short exposures required to effectively freeze the atmospheric turbulence. For wavelengths longer than about 2.0 μm (i.e., K, L, M, N bands), Poisson noise from the thermal background begins to dominate over other sources of noise. Highly sensitive infrared interferometry will require a space platform that will allow long coherence times and low thermal background. Please consult the observer manual for each specific interferometer instrumentation to determine point-source sensitivity.

Another important issue to consider is that a low visibility fringe ($\mathcal{V} \ll 1$) is harder to detect than a strong one. Usually fringe detection sets the limiting magnitude of an interferometer/instrument, and this limit often scales like $N\mathcal{V}$, the number of "coherent" photons. For readnoise or background noise dominant situations (common in NIR), this means that if the point-source ($\mathcal{V} = 1$) limiting magnitude is 7.5, then a source with $\mathcal{V} = 0.1$ would need to be as bright as magnitude 5.0 to be detected. The magnitude limit worsens even more quickly for low visibility fringes when noise from the source itself dominates, since brighter targets bring along greater noise. Another common expression found in the literature is that the SNR for a visible-light interferometer scales like $N\mathcal{V}^2$. This latter result can be derived by assuming that the "signal" is the average power spectrum $(N\mathcal{V})^2$ and the dominant noise process is photon noise which has a power spectrum that scales like N here.

3.1.3 Overcoming the Effects of the Atmosphere: Phase Referencing, Adaptive Optics, and Fringe Tracking

As discussed above, the limiting magnitude will strongly depend on the maximum coherent integration time that is set by the atmosphere. Indeed, this limitation is very dramatic, restricting visible-light integrations to mere milliseconds and millimeter radio observations to a few dozen minutes. For mm-wave and radio observations, the large isoplanatic angle and long atmospheric coherence times allow for real-time correction of atmospheric turbulence by using *phase referencing*.

In a phase-referencing observing sequence, the telescopes in the array will alternate between the (faint) science target and a (bright) phase calibrator nearby in the sky. If close enough in angle (within the isoplanatic patch), then the turbulence will be the same between the target and bright calibrator; thus, the high SNR measurement of fringe phase on the calibrator can be used

[8]Real interferometers will have a realistic limit about 1–2 orders of magnitude below the theoretical limit due to throughput losses and nonideal effects such as loss of system visibility.

to account for the atmospheric phase changes. Another key aspect is that the switching has to be fast enough so that the atmospheric turbulence does not change between the two pointings. With today's highly sensitive radio and mm-wave receivers, enough bright targets exist to allow nearly full sky coverage so that most faint radio source will have a suitable phase calibrator nearby.[9] In essence, phase referencing means that a fringe does not need to be detected within a single coherence time τ_c but rather one can coherently integrate for as long as necessary with sensitivity improving as $1/\sqrt{t}$. In ➋ Sect. 4 a simple example is presented that demonstrates how phase-referencing works with simulated data.

In the visible and infrared, phase referencing by alternate target/calibrator sequences is practically impossible since $\tau_c \ll 1$ s and $\Theta_{\text{iso}} \ll 1$ arcmin isoplanatic patch size. In V/IR interferometry, observations still alternate between a target and calibrator in order to calibrate the statistics of the atmospheric turbulence but not for phase referencing. A special case exists for dual-star narrow-angle astrometry (Shao and Colavita 1992) where a "Dual Star" module located at each telescope can send light from two nearby stars down two different beam trains to be interfered simultaneously. At K band, the stars can be as far as $\sim 30''$ apart for true phase referencing. This approach is being attempted at the VLT (PRIMA, Delplancke et al. 2006) and Keck Interferometers (ASTRA, Eisner et al. 2010). This technique can be applied to only a small fraction of objects owing to the low sky density of bright phase reference calibrators.

Adaptive optics (AO) can be used on large visible and infrared telescopes to effectively increase the collecting area A_e term in our signal equation, allowing the full telescope aperture to be used for interferometry. AO on a 10-m class telescope potentially boosts infrared sensitivity by $\times 100$ over the seeing limit; however, this method still requires a bright enough AO natural or laser guide star to operate. Currently, only the VLT and Keck Interferometers have adaptive optics implemented for regular use with interferometry. A related technique of *fringe tracking* is in more widespread use, whereby the interferometer light is split into two channels so that light from one channel is used exclusively for measuring the changing atmospheric turbulence and driving active real-time path length compensation. In the meantime, the other channel is used for longer science integrations (at VLTI, Keck, CHARA). This method improves the limiting magnitude of the system at some wavelengths if the object is substantially brighter at the fringe tracking wavelength, such as for dusty reddened stars. Fringe tracking sometimes can be used for very high spectral observations of stars ordinarily too faint to observe at high dispersion.

It is important to mention these other optical interferometer subsystems (e.g., AO, fringe tracker) here because they are crucial for improving sensitivity, but the additional complexities do pose a challenge for observers. Each subsystem has its own sensitivity limit and now multiple wavelengths bands are needed to drive the crucial subsystems. As an extreme example, consider the Keck Interferometer Nuller (Colavita et al. 2009). The R-band light is used for tip-tilt and adaptive optics, the H band is used to correct for turbulence in air-filled coude path, the K band is used to fringe track, and finally, the 10 μm light is used for the nulling work. If the object of interest fails to meet the sensitivity limit of any of these subsystems, then observations are not possible – most strongly affecting highly reddened sources like young stellar objects and evolved stars.

[9] At the shortest sub-mm wavelengths, phase referencing is quite difficult due to strong water vapor turbulence, but can be partially corrected using "water-vapor monitoring" techniques (e.g., Wiedner et al. 2001).

3.2 (u,v) Coverage

One central difference between interferometer and conventional single-telescope observations is the concept of (u,v) coverage. Instead of making a direct "image" of the sky at the focal plane of a camera, the individual fringe visibilities for each pair of telescopes are obtained. As discussed in ➋Sect. 2.2, each measured complex visibility is a single Fourier component of a portion of the sky. The goal of this subsection is to understand how to estimate (u,v) coverage from the array geometry and which characteristics of Fourier coverage affect the final reconstructed image.

For a given layout of telescopes in an interferometer array, the Fourier coefficients that can be measured are determined by drawing baselines between each telescope pair. To do this, an (x,y) coordinate system is first constructed to describe the positions of each element of the array; for ground-based arrays in the northern hemisphere, the convention is to orient the +x axis toward the east and the +y axis toward north. The process of determining the complete ensemble of (u,v) points provided by any given array can be laborious for arrays with a large number of elements. A simple method of automating the procedure is as follows. First, construct a distribution in the (x,y) plane of delta functions of unit strength at the positions of all elements. The (u,v) plane coverage can be obtained from the two-dimensional autocorrelation of this distribution, as illustrated in ➋ *Fig. 7-4* for four simple layouts of array elements. The delta functions for each array element are shown as dots in the upper row of sketches in this figure, and the corresponding dots in the u,v distributions are shown in the lower row of autocorrelations. Note that each point in the (u,v) plane is repeated on the other side of the origin owing to symmetry; of course the values of amplitude and phase measured on a source at one baseline will be the same whether one thinks of the baseline as extending from telescope 1 to

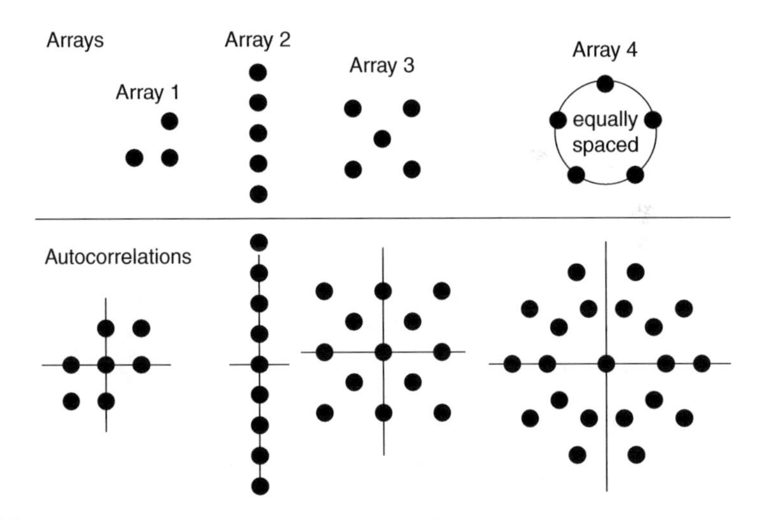

Fig. 7-4
Here the "snapshot" coverages of a few simple interferometer layouts are shown for an object located at zenith. The top portion shows the physical layout of four examples arrays while the bottom portion shows the corresponding autocorrelation function, which is the same as the (u,v) coverage

telescope 2, or the converse. For an array of N telescopes, one can measure $\binom{N}{2} = \frac{(N)(N-1)}{2}$ independent Fourier components.

Sometimes the array geometry may result in the (near-)duplication of baselines in the (u,v) plane. This is the case for array #2 in the ➲ *Fig. 7-4*, where the shortest spacing is duplicated four times, the next spacing is duplicated three times, the following spacing is duplicated twice, and only the longest spacing of this array is unique. While each of these interferometers does contribute statistically independent data as far as the noise is concerned, it is an inefficient use of hardware since the astrophysical information obtained from such redundant baselines is essentially the same. In order to optimize the Fourier coverage for a limited number of telescopes, a layout geometry should be *nonredundant*, with no baseline appearing more than once, so that the maximum number of Fourier components can be measured for a given array of telescopes. A number of papers have been written on how to optimize the range and uniformity of (u,v) coverage under different assumptions (Golay 1971; Holdaway and Helfer 1999; Keto 1997). Note that in the sketches of ➲ *Fig. 7-4*, array #4 provides superior coverage in the u,v plane compared to arrays #3 and #2 with the same number of array elements.

Finally note that the actual (u,v) coverage depends not on the *physical baseline separations* of the telescopes but on the *projected baseline separations* in the direction of the target. For ground-based observing, a celestial object moves across the sky along a line of constant declination, so the (u,v)-coverage is actually constantly changing with time. This is largely a benefit since earth rotation dramatically increases the (u,v)-coverage without requiring additional telescopes. This type of synthesis imaging is often called *Earth rotation aperture synthesis*. The details depend on the observatory latitude and the target declination, and a few simple cases are presented in ➲ *Fig. 7-5*. In general, sources with declinations very different from the local

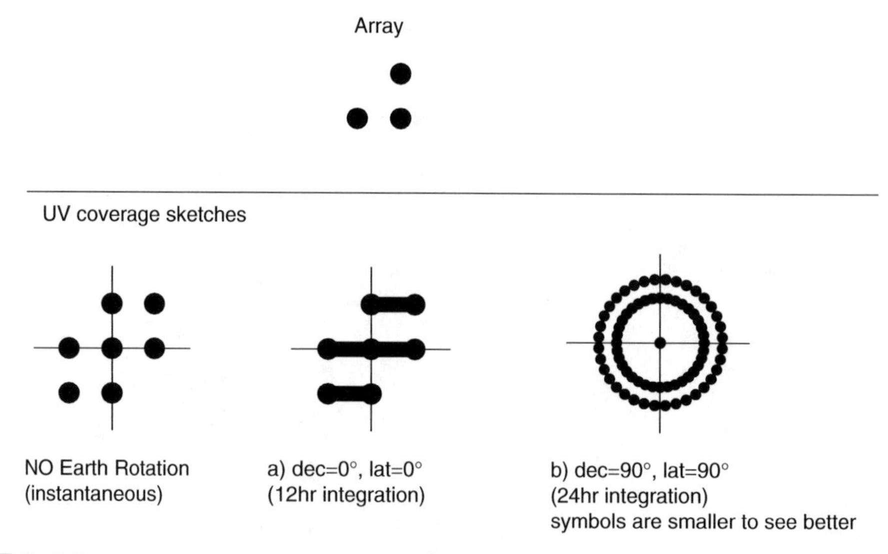

Fig. 7-5
The rotation of the Earth introduces changing baseline projections as an object traverses the sky. This figure take a simple three-telescope array and shows the "(u,v) tracks" for a few different examples of differing observatory latitudes and target declinations.

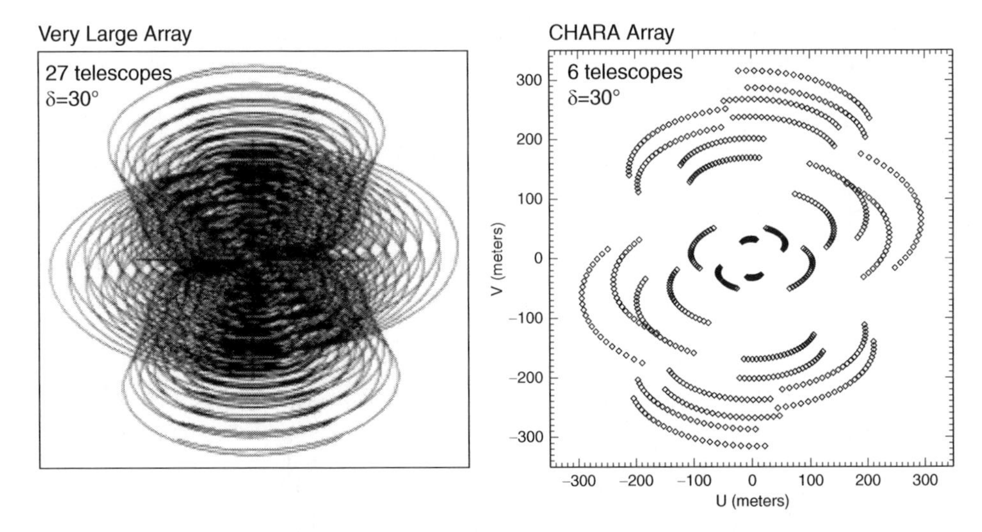

Fig. 7-6
This figure shows the dramatic advantage in Fourier coverage that the Very Large Array (*VLA*) possesses (*left panel*) compared to the CHARA array (*right panel*). The radio interferometer VLA, located in Socorro, NM, has 27 (movable) elements and the coverage shown here corresponds to observations ± 4.5 h around source transit (Adapted from Thompson et al. 1980). The optical interferometer CHARA, located on Mt. Wilson, CA, has six (fixed) elements and the coverage shown here corresponds to ±3 h from transit (Adapted from ten Brummelaar et al. 2005). The large gaps in Fourier coverage for CHARA limit its ability to reconstruct highly complex images

latitude will never reach a high elevation in the sky, such that the north-south (u,v) coverage will be foreshortened and the angular resolution in that direction correspondingly reduced.

Figure 7-6 shows the actual Fourier coverage for the 27-telescope very large array (VLA) and for the 6-telescope CHARA Array. For $N = 27$, the VLA can measure 351 Fourier components while CHARA ($N = 6$) can measure only 15 simultaneously. Notice also in this figure that the ratio between the maximum baseline and the minimum baseline is much larger for the VLA (factor of 50, A array) compared to CHARA (factor of 10).

The properties of the (u,v)-coverage can be translated into some rough characteristics of the final reconstructed image. The final image will have an angular resolution of $\sim \frac{\lambda}{B_{\max}}$, and note that the angular resolution may not be the same in all directions. It is crucial to match the desired angular resolution with the maximum baseline of the array because longer baselines will over-resolve your target and have very poor (or non existent) signal-to-noise ratio (see discussion Sect. 3.1.1). This functionally reduces the array to a much smaller number of telescopes which dramatically lowers both overall signal-to-noise ratio and the ability to image complex structures. For optical arrays that combine only three or four telescopes, relatively few (u,v) components are measured concurrently and this limits how much complicated structure can be reconstructed.[10] From basic information theory under best case conditions, one needs at least as many independent visibility measurements as the number of independent pixels in

[10]Fortunately, targets of optical interferometers are generally spatially compact and so sparser (u,v) coverage can often be acceptable.

the final image. For instance, it will take hundreds of components to image a star covered with dozens of spots of various sizes, while only a few data points can be used to measure a binary system with unresolved components.

3.3 Field-of-View

While the (u,v) coverage determines the angular resolution and quality of image fidelity, the overall imaging field-of-view is constrained by a number of other factors.

A common limitation for field-of-view is the primary beam pattern of each individual telescope in the array and this was already discussed in Sect. 2.3: $\Delta\Theta \sim \frac{\lambda}{\text{Diameter}}$. This limit can be addressed by *mosaicing*, which entails repeated observations over a wide sky area by coordinating multiple telescope pointings within the array and stitching the overlapping regions together into a single wide-field image. This practice is most common in the mm-wave where the shorter wavelengths result in a relatively small primary beam. A useful rule of thumb is that your field-of-view (in units of the fringe spacing) is limited to the ratio of the baseline to the telescope diameter. Most radio and mm-wave imaging is limited by their primary beam; however, there is a major push to begin using "array feeds" to allow imaging in multiple primary beams simultaneously.

Another limitation to field-of-view is the spectral resolution of the correlator/combiner. The spectral resolution of each channel can be defined as $R = \frac{\lambda}{\Delta\lambda}$. A combiner or correlator cannot detect a fringe that is outside the system coherence envelope, which is simply related to the spectral resolution R. The maximum observable field of view is R times the finest fringe spacing, or $\Delta\Theta \sim R \cdot \frac{\lambda}{B_{\text{max}}}$, often referred to as the *bandwidth-smearing* limit. Most optical interferometers and also very long baseline interferometry (VLBI) are limited by bandwidth smearing.

A last limitation to field-of-view arises from temporal smearing of data by integrating for too long during an observation. Because the (u,v) coverage is constantly changing due to Earth, rotation, time averaging removes information in the (u,v)-plane resulting in reduced field-of-view. A crude field-of-view limit based on this effect is $\Delta\Theta \sim \frac{230}{\Delta t_{\text{minutes}}} \frac{\lambda}{B_{\text{max}}}$. Both radio and V/IR interferometric data can be limited by temporal smearing if care is not taken in setting up the data collection, although this limitation is generally avoidable.

3.4 Spectroscopic Capabilities

As for regular radio and optical astronomy, one tries to observe at the crudest spectral resolution that is suitable for the science goal in order to achieve maximum signal-to-noise ratio. However as just discussed, spectral resolution does impact the imaging field-of-view, bringing in another dimension to preparations. While each instrument has unique capabilities that cannot be easily generalized, most techniques will require dedicated spectral calibrations as part of observing procedures.

"Spectro-interferometry" is an exciting tool in radio and (increasingly) optical interferometry. In this application, the complex visibilities are measured in many spectral channels simultaneously, often across a spectrally resolved spectral line. This allows the different velocity components to be imaged or modeled independently. For example, this technique can be used for observing emitting molecules in a young stellar object to probe and quantify Keplerian motion around the central mass or for mapping differential rotation on the surface of a rotating star using photospheric absorption lines (e.g., Kraus et al. 2008). Spectro-interferometry is analogous to "integral field spectroscopy" on single aperture telescopes, where each pixel in the image has a corresponding measured spectrum. Another clever example of spectro-interferometry pertains to maser sources in the radio: A single strong maser in one spectral channel can be used as a phase calibrator for the rest of the spectral channels (e.g., Greenhill et al. 1998.

4 Data Analysis Methods

After observations have been completed, the data must be analyzed. Every instrument will have a customized software pipeline to take the recorded electrical signals and transform into useful astronomical quantities. That said, the data reduction process is similar for most systems and here the basic steps are outlined.

4.1 Data Reduction and Calibration Overview

The goal of the data reduction is to produce calibrated complex visibilities and related observables such as closure phases (see ➋ Sect. 5.2.1). As discussed in ➋ Sect. 3, the basic paradigm for interferometric observing is to switch between data of well-known system calibrators and the target of interest. This allows for calibration of changing atmospheric conditions by monitoring the actual phase delay through the atmosphere (in radio) or by statistically correcting for decoherence from turbulence (in optical).

One begins by plotting the observed fringe amplitude versus time. ➋ *Figure 7-7* shows a schematic example of how data reduction might proceed for the case of high-quality radio interferometry observations, such as taken with the EVLA. Here the observed fringe amplitude and phase for a calibrator-target-calibrator sequence is presented. Notice that in this example, the fringe amplitude of the calibrator is drifting up with time, as is the observed phase. As long as the switching time between target and calibrator is faster than instrumental gain drifts and atmospheric piston shifts, a simple function can be fitted to the raw calibrator measurements and then interpolated to produce the calibration curves for the target. Here a second order polynomial has been used to approximate the changing amplitude and phase response. This figure contains an example for only a single baseline, polarization, and spectral channel; there will be hundreds or thousands of panels like this in a dataset taken with an instrument as the EVLA or ALMA.

The justification for this fitting procedure can be expressed mathematically. As the wave traverses the atmosphere, telescope, and interferometer beamtrain/waveguides, the electric field can have its phase and amplitude modified.[11] These effects can be grouped together into a net complex *gain* of the system for each beam, $\tilde{\mathcal{G}}_i$ – the amplitude of $\tilde{\mathcal{G}}_i$ encodes the net amplification or attenuation of the field strength and the phase term corresponds to a combination of time

[11] In general, also the polarization states and wavefront coherence can also be modified.

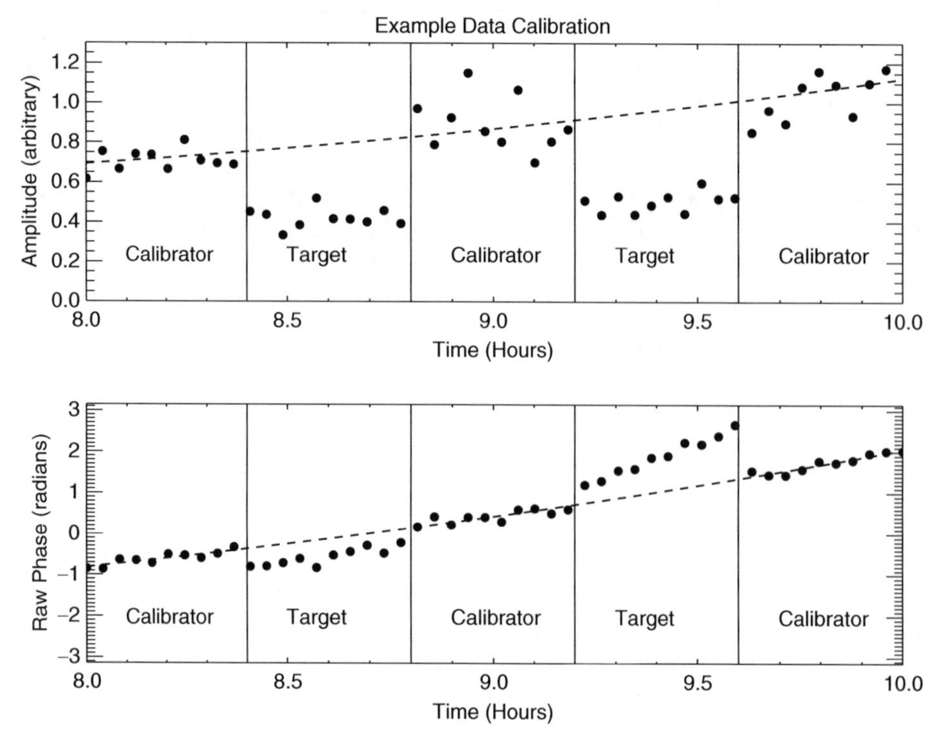

Fig. 7-7
First step of data reduction

delays in the system and effects from amplifiers in the signal chain. Thus, the measured electric field $\tilde{E}'$ can be written as a product of the original field $\tilde{E}$ times this complex gain:

$$\tilde{E}' = \tilde{\mathcal{G}}\tilde{E} \tag{7.11}$$

Since the observed complex visibility $\tilde{\mathcal{V}}_{12}$ for a baseline between telescope 1 to telescope 2 is related to the product $\tilde{E}_1\tilde{E}_2^*$, then:

$$\tilde{\mathcal{V}}'_{12} \propto \tilde{E}'_1\tilde{E}'^*_2 \tag{7.12}$$

$$\propto \tilde{\mathcal{G}}_1\tilde{E}_1 \cdot \tilde{\mathcal{G}}_2^*\tilde{E}_2^* \tag{7.13}$$

$$\propto \tilde{\mathcal{G}}_1\tilde{\mathcal{G}}_2^*\tilde{\mathcal{V}}_{12} \tag{7.14}$$

Thus, the measured complex visibility $\tilde{\mathcal{V}}'_{12}$ is closely related to the true $\tilde{\mathcal{V}}_{12}$, differing only by complex factor $\tilde{\mathcal{G}}_1\tilde{\mathcal{G}}_2^*$. By observing a calibrator with known structure, this gain factor can be measured, even if the calibrator is not a point source for the interferometer. For a radio array, the gain factors are mainly associated with the individual telescope collectors and not the baseline, and so the same gain factors appear in many baselines. This redundancy has led to the development of additional off-line procedures to "self-calibrate" radio imaging data using "closure amplitude" techniques (see Sect. 5.2.1).

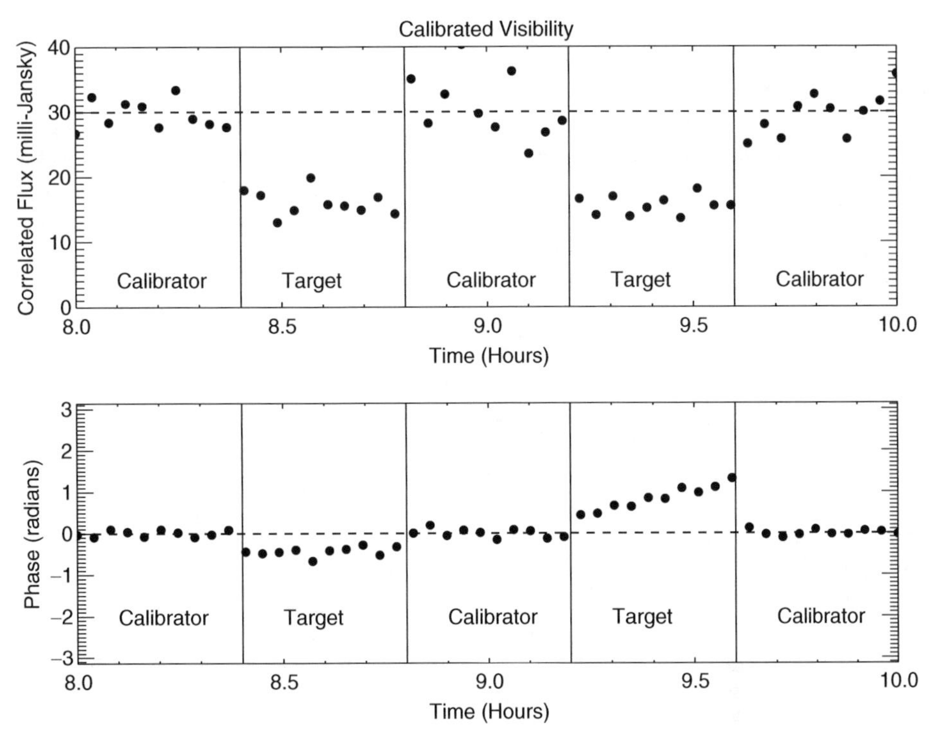

Fig. 7-8
Calibrator data following estimate of time-varying system visibilites

Once the system drifts have been estimated by measurements of the calibrator, this correction can be applied to the whole dataset. *Figure 7-8* shows the calibrated result, where the calibrator flux was assumed to be 30 Jy. In practice, radio phase calibrators are time-variable in flux and so each dataset typically includes an "amplitude calibrator," a well-studied object with known flux as a reference. These calibrated data can now be averaged and used for further model fitting or synthesis imaging. In the example shown here, both the target and calibrator have reasonable signal-to-noise-ratio. In a more realistic case, the signal-to-noise of the target will be orders of magnitude worse – indeed, in one observing block, there may be no discernable signal at all! The calibrator measurements are used to phase up the array and allow for very long phase-coherent integrations (averaging in the complex (u,v) plane). Unfortunately, this "blind" phase referencing cannot generally be used in optical interferometry (see Sect. 3.1.3) where the short atmospheric coherence time and the worse turbulence requires active fringe tracking for both target and calibrator at all times.

Note that actual data will not look quite like this simplified schematic. First, raw data might have random data glitches or bad values that need to be flagged. Also, one tends to only observe the calibrator for a short time, just enough to measure the phase. In fact, the time to slew between targets can be similar to the length of time spent integrating on each calibrator.

The time spent on the target during each visit is generally as long as possible given the atmospheric coherence time which can vary greatly with baseline length, observing conditions, and wavelength (see ➋ Sect. 2.4).

A common complication is that the calibrator may not be an unresolved object nor constant in flux. NRAO maintains a calibrator database that is used to determine the suitability of each calibrator for different situations. As long as the calibrator morphology is known, the observer can apply a visibility amplitude and phase correction to account for the known structure. After this correction, the calibration procedure is the same.

For visible and infrared interferometry, the procedure is very similar. In general, optical interferometers measure a time-averaged squared visibility and not visibility amplitude since the $\mathcal{V}^2$ can be bias corrected more easily for low signal-to-noise ratio data when observing with no phase referencing (Colavita 1999). As discussed earlier, optical interferometers cannot employ phase referencing between two targets [12] due to the tiny isoplanatic patch and short temporal coherence times. Instead of averaging fringe phases, closure phases (see ➋ Sect. 5.2.1) are formed and averaged over longer time frames following a similar interpolation of calibration data. Lastly, calibrators tend to be stars with known or well-estimated diameters. For a given baseline, the observed raw calibrator $\mathcal{V}^2$ are boosted to account for partially resolving them during the observation before the system visibility is estimated.

When carrying out spectral line and/or polarization measurements, additional calibrations are required. As for single telescope observations, one must observe a source with known spectrum and/or polarization signature in order to correct for system gains. These procedures add steps to the data reduction but are straightforward.

A diversity of packages and data analysis environments are in use for data reduction of interferometer observations. In the radio and mm-wave regime, the most popular packages are AIPS and Miriad. In addition, the CASA package will be used for ALMA and supports many EVLA operations now too. In the visible and infrared, the data reduction packages are usually closely linked to the instrument and are provided by the instrument builders. In most cases, there are data analysis "cookbooks" that provide step-by-step examples of how to carry out all steps in the data reduction. Few instruments have complete pipelines that require no user input, although improved scripting is a high priority for future development. A number of summer schools are offered that train new users of interferometer facilities in the details of observation planning and data reduction.

The data products from this stage are calibrated complex visibilities and/or closure phases. No astronomical interpretation has occurred yet. The de facto standard data format for radio data is UVFITS. Unfortunately, this format is not strictly defined but rather represents the data supported by the NRAO AIPS package and importable into CASA (which uses a new format called the *Measurement Set*). Recently the VLBA community fully documented and registered the FITS Interferometry Data Interchange (FITS-IDI) format (see *"FITS-IDI definition document" AIPS Memo 114*). The optical interferometry community saw the problems of radio in having a poorly defined standard and, through IAU-sanctioned activities, crafted a common FITS-based data standard called OIFITS whose specifications were published by Pauls et al. (2005). This standard is in wide use by most optical interferometers in the world today.

[12] Phase referencing is possible using "Dual Star Feeds" which allows truly simultaneous observing of a pair of objects in the same narrow isoplanatic patch on the sky. This capability has been demonstrated on PTI, Keck, and VLTI.

4.2 Model Fitting for Poor (u,v) Coverage

The ultimate goal of most interferometric observations is to have sufficient data quality and (u,v) coverage to make a synthesis image. With high image fidelity, an astronomer can interact with the image just as one would had it come from a standard telescope imager.

Still, cases are plentiful where this ideal situation is not achievable and one will fit a model to the visibility data directly. There are two classes of models: *geometric* models and *physical* models. *Geometric models* are simple shapes that describe the emission but without any physics involved. Common examples include Gaussians, uniform disks, binary system of two uniform disks, etc. *Physical models* start with a physical picture including densities, opacities, and sources of energy. Typically a radiative transfer calculation is used to create a synthetic image which can be Fourier transformed (following (➋ 7.2)) to allow fitting to complex visibilities. Geometric models are useful for very simple cases when an object is marginally resolved or when physical models are not available (or not believable!). Physical models are required to connect observations with "real" quantities like densities and temperatures, although size scales can be extracted with either kind of model.

In radio and mm-wave, model fitting is now relatively rare[13] since high-fidelity imaging is often achievable. However in the optical, model fitting is still the most common tool for interpreting interferometry data. In many cases, a simple uniform disk or Gaussian is adequate to express the characteristic size scale of an object. By directly fitting to visibility amplitudes, the data can be optimally used and proper error analysis can be performed. The fitting formulae for the two most common functions can be expressed in closed form as a function of baseline B and wavelength λ:

$$|\mathcal{V}| = 2\frac{J_1(\pi B\Theta_{\rm diameter}/\lambda)}{\pi B\Theta_{\rm diameter}/\lambda} \qquad \text{case : Uniform Disk} \tag{7.15}$$

$$|\mathcal{V}| = e^{-\frac{\pi^2}{4\ln 2}(\Theta_{\rm FWHM}B/\lambda)^2} \qquad \text{case : Gaussian} \tag{7.16}$$

➋ *Figure 7-9a* illustrates both the model-fitting process and the importance of choosing the most physically plausible function. Here some simulated visibility data spanning baselines from 0 to 60 m are plotted along with curves for five common brightness distributions – a uniform disk, a Gaussian disk, two binary models, and a ring model. For the case of marginally resolved objects, only the *characteristic scale* of a given model can be constrained and there is no way to distinguish *between models* without longer baseline information (for more elaborate discussion, see Lachaume 2003). Note how all the curves fit the data equally well at short baselines and high visibilities, but that the *interpretation* of each curve is quite different. Without longer baselines that can clearly distinguish between these models, the observer must rely on theoretical expectations to guide model choice. For example, a normal G star should closely resemble a uniform disk in the near-infrared while disk emission from a young stellar object in the sub-mm might be more Gaussian. Despite the uncertainties when fitting to marginally resolved targets, fitting interferometric data allows very precise determinations of model parameters as can be seen in ➋ *Fig. 7-9b* where the CHARA Array was used to monitor the variations in diameter of the Cepheid variable δ Cep.

[13]Historically, model fitting was the way data was handled in order to discern source structure in the early days of radio interferometry. A classic example is the model of Cygnus A, which Jennison and Das Gupta fitted to long-baseline intensity-interferometry data at 2.4-m wavelength in 1953 (Jennison and Das Gupta 1953). Sullivan (2009, p. 353 et. seq) gives more details of this fascinating story.

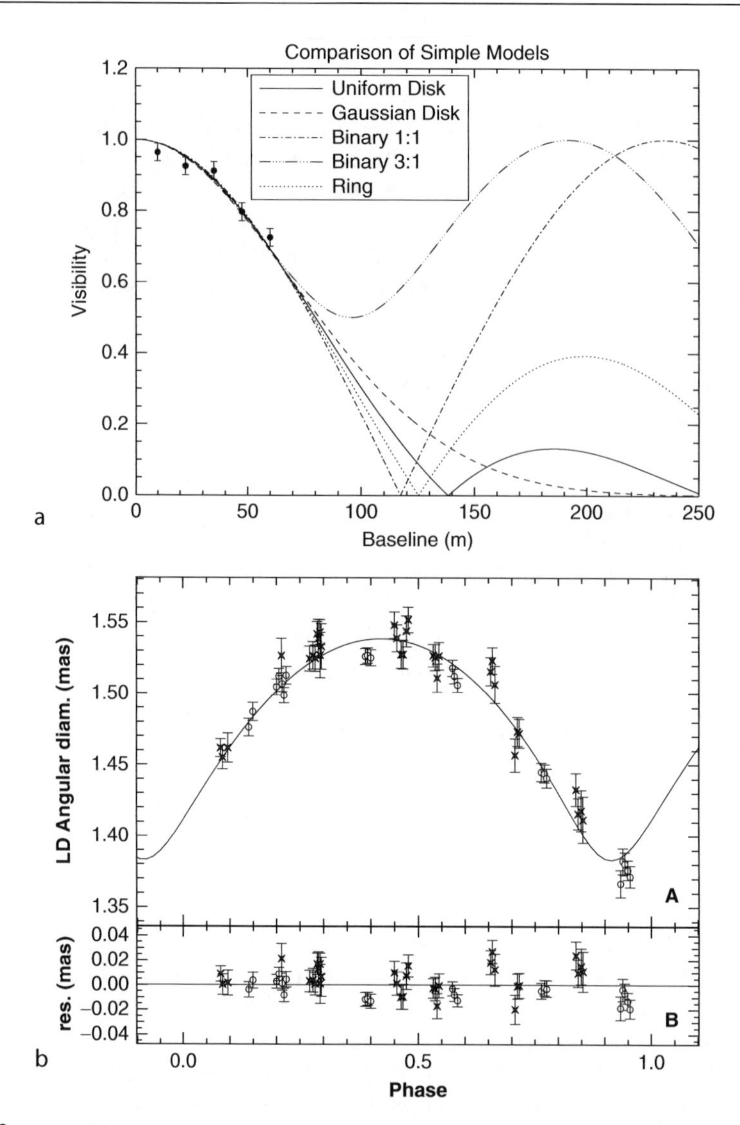

Fig. 7-9

(a) This figure shows visibility data and model fits for various brightness distributions, including uniform disk, Gaussian disk, binary system, and ring. Notice that all the models fit the data nearly identically for visibility about 50%, corresponding here to baselines shorter than 50 m. One *requires* long baselines (or exceedingly high signal-to-noise ratios) to distinguish between any of these various models, illustrating the importance of a priori knowledge of the nature of the target before deciding on appropriate geometrical model description. This figure essentially illustrates in Fourier Space the common sense idea that blurry objects all look the same unless the source structure is larger than the blur – here, longer baselines essentially reduce the blur of diffraction. (b) The *right panel* shows precision limb-darkened (LD) diameter measurements of the classical Cepheid δ Cep at K band through a full pulsational cycle. These tiny (±4%) changes were easily tracked by the FLUOR combiner at CHARA using ~250–300 m baselines (Reprinted with permission, Mérand et al. 2005)

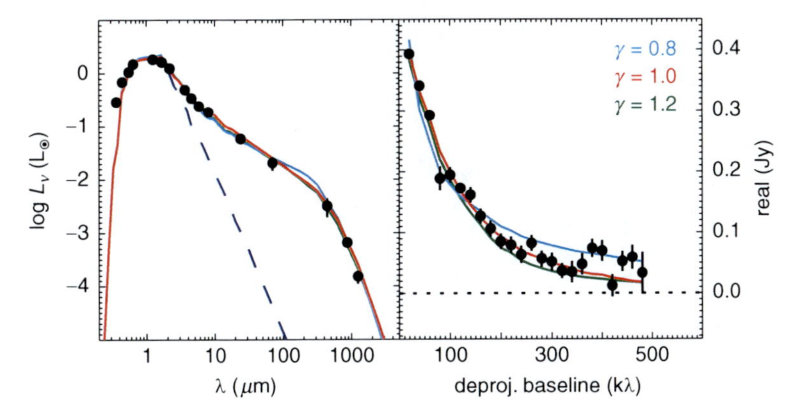

Fig. 7-10

This figure illustrates how one can fit spectral energy distributions and visibility simultaneously using simple physical models, as opposed to using purely descriptive or geometric models. Models for protoplanetary disk emission with various radial mass density profiles (*left panel*) are represented by *different color lines*. The visibility data (*right panel*) were obtained using the Submillimeter Array at 0.87 mm (Reprinted with permission, Andrews et al. 2009)

A recently published example of model fitting at longer wavelengths is shown in *Fig. 7-10*. Here, a semi-analytic physical model (in this case of a circumstellar disk) was used to simultaneously fit the spectral energy distribution along with the visibility data. When realistic physical models are available, multi-wavelength constraints can make a dramatic improvement to the power of high angular resolution data and should be included whenever possible.

5 Synthesis Imaging

Interferometer data in their raw form are not easy to visualize. Fortunately, as discussed in Sect. 2.2, the measured complex visibilities can be transformed into an equivalent brightness distribution on the sky – *an image*. This procedure is called "aperture synthesis imaging" or more generally "synthesis imaging." In this section, the critical data analysis steps are described for creating an image in both the ideal case as well as more challenging scenarios when faced with poor (u,v) coverage and phase instability.

5.1 Ideal Case

Under ideal conditions, the astronomer would have collected interferometer data with a large number of telescopes including some Earth rotation to fill in gaps in (u,v) coverage (see Sect. 3.2). In addition, each datum will consist of a fully calibrated complex visibility – both amplitude and phase information. Modern radio arrays such as the VLA and ALMA produce data of this quality when proper phase-referencing procedures (see Sect. 3.1.3) are employed.

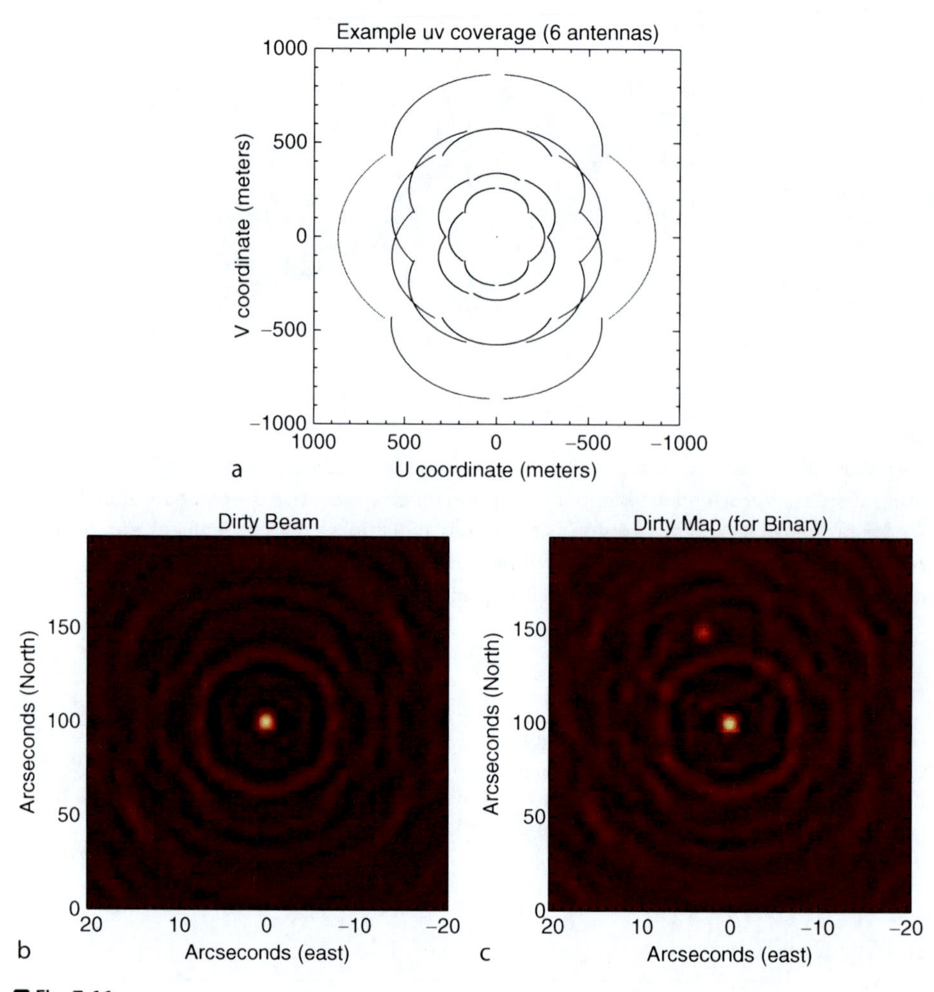

Fig. 7-11

The basic elements of the CLEAN algorithm consists of (a) the (u,v) coverage of the observations, (b) the Dirty beam made from the Fourier Transform of the gridded (u,v) support, and (c) the Dirty Map made from Fourier Transform of the observed (u,v) plane

Figure 7-11 depicts the Fourier coverage for a six telescope interferometer along with the resulting image from a direct Fourier transform of this coverage for a perfect point source (by construction, the image will be purely real). In this procedure, the values in the (u,v) grid are set to unity where data exists and to zero where no data exists. This resulting image shows artifacts because of the missing (u,v) data that were zeroed out. These artifacts are often called "sidelobes" and show both positive and negative excursions – note that negative flux density is usually a strong sign of a sidelobe since negative values in an image are typically unphysical (except for special applications like absorption line studies or polarization stokes mapping). Note that the central core of the image represents the diffraction-limited angular resolution of this observation. Often in a practical situation, the central core will be elongated because (u,v) coverage is not perfectly symmetric, with longer baselines in some directions than others. The last panel in this figure shows the resulting image for a binary star with 2:1 flux ratio. Notice that the image contains two sources; however, both sources show the same pattern of artifacts as the simple point source. Indeed, this supports the previous assertion that missing (u,v) coverage is the main origin of this pattern and suggests that the image quality can be improved by correcting for the sidelobe effect.

To proceed, the well-known convolution theorem must be introduced: *Multiplication in the "(u,v) space" is equivalent to a convolution in "image space."* Mathematically, this as can be expressed:

$$\mathrm{FT}(\tilde{\mathcal{V}}(u,v)\cdot M(u,v)) = (\mathrm{FT}\,\tilde{\mathcal{V}})\otimes(\mathrm{FT}\,M) \tag{7.17}$$

$$\tilde{\mathcal{V}}(u,v)\cdot M(u,v) \Leftrightarrow I(x,y)\otimes B(x,y) \tag{7.18}$$

where $\tilde{\mathcal{V}}$ is the full underlying complex visibility that could in principle be measured by the interferometer, M is the (u,v) mask that encodes whether data exists (1) or is missing (0),[14] FT and $\Leftrightarrow$ denote a Fourier Transform, $I = \mathrm{FT}\,\tilde{\mathcal{V}}$ is the true image distribution , and $B = \mathrm{FT}\,M$ is called the "convolving beam."

Application of the convolution theorem permits an elegant reformulation of the imaging problem into a "deconvolution" problem, where the convolving beam is a complicated function but derived directly from M, the observed (u,v) coverage. Since M is exactly known, the convolving beam B is known as well. Note that this deconvolution problem contrasts sharply with the deconvolution problem in adaptive optics imaging where the point-source function (PSF) varies in time and is never precisely known.

5.1.1 CLEAN Algorithm

One of the earliest methods developed for deconvolution was the CLEAN algorithm (Högbom 1974), which is still widely used in radio interferometry. In CLEAN, the Fourier transform of the gridded complex visibility data is called the "dirty image" (or sometimes "dirty map") and the Fourier transform of the (u,v) plane mask is called the "dirty beam." One first needs to deconvolve the dirty image with the dirty beam. To do this, the true image I is iteratively constructed by locating the peak in the dirty image and subtracting from this a scaled version of the dirty beam centered at this location. Here, the scaling of the dirty beam is typically tuned to remove a certain fraction of intensity from the peak, often 5%. One keeps track of how much one removes by collecting the "CLEAN" components in a list.

Consider this example. The dirty image has peak of 1.0 Jy at pixel location (3,10). One creates a scaled dirty beam with a peak of 0.05 Jy, shifts the peak to the position (3,10), and subtracts this scaled dirty beam from the dirty image. This CLEAN component is collected and labeled by location (3,10) and flux contribution (0.05 Jy). Continue this procedure, flux is removed from dirty image and CLEAN components are collected. This procedure is halted when the residual dirty image contains only noise. Since the intensity in the true image is expected to be positive definite, one common criterion for halting the CLEAN cycles is when the largest negative value in the image is comparable to the largest positive value in the image, thereby avoiding any CLEAN components having negative flux.

[14] In general, one can consider a full "Spatial Transfer Function" which can have weights between 0 and 1. Here, consider just a simple binary mask in the (u,v) plane for simplicity.

In principle, this collection of delta function point sources (all the CLEAN components) *is* the best estimate of the image distribution. However, an image of point-sources is not visually appealing and a common procedure is to convolve the point source distribution with a "CLEAN" beam, a perfect 2-D Gaussian with a core that matches the FWHM of the dirty beam. Commonly, a filled ellipse will be included in the corner of a CLEANed image showing the 2D FWHM of the restoring beam. Lastly, one adds back the residual image from the dirty map (which should contain only noise) so that the noise level is apparent in the final image and remnant uncorrected sidelobe artifacts, if present, can be readily identified. These steps are illustrated in *Fig. 7-12* where the CLEAN procedure has been applied to the examples shown in *Fig. 7-11*.

The steps of the CLEAN algorithm are summarized below.

1. Create dirty map and beam.
2. Find peak of dirty map.
3. Subtract scaled version of dirty beam from dirty map, removing a small percentage (e.g., ~5%) of peak intensity. Collect CLEAN components.
4. Repeat last step until negative residuals are comparable to positive residuals.
5. Convolve CLEAN components with CLEAN beam.
6. Add back the residuals.

While it is beyond the scope to address the weaknesses of the CLEAN algorithm in detail, a few issues are worth mentioning in passing. Typical CLEAN algorithms do not naturally deal with errors unless all the visibility data are of similar quality. The case when each visibility point is weighted equally is called *natural weighting* (which gives best SNR for detecting faint objects) and the case when each portion of the (u,v) plane is equally weighted, so called *uniform weighting* (which gives somewhat higher angular resolution but at some loss of sensitivity). Briggs (1995) introduced a "ROBUST" parameter than can naturally span these two extremes. Other problems with CLEAN include difficulty reconstructing low surface brightness regions, large scale emission and the fact that the final reconstructed image is of a degraded resolution because the convolving PSF actually suppresses the longest baseline visibilities from the final image. Finally, if the imaging step makes use of the Fast Fourier Transform, several additional artifacts can appear in the image as a consequence of the necessity to grid the input u, v data.

5.1.2 Maximum Entropy Method (MEM)

Another common method for reconstructing images from interferometric data is called the *maximum entropy method* (Gull and Skilling 1983; Skilling and Bryan 1984). This approach asks the question: "How to choose which reconstructed image is best, considering that there are an infinite number of images that can fit the interferometer dataset within the statistical uncertainties?" The simple answer here is that the "best" image is the one that both fits the data acceptably and also maximizes the *Entropy S* defined as:

$$S = -\sum_i f_i \ln \frac{f_i}{m_i} \tag{7.19}$$

where f_i is the (positive definite) fraction of flux in pixel i, and m_i is called the *image prior* which can encapsulate prior knowledge of the flux distribution (e.g., known from physical considerations or lower resolution observations). Narayan and Nityananda (1986) describe general

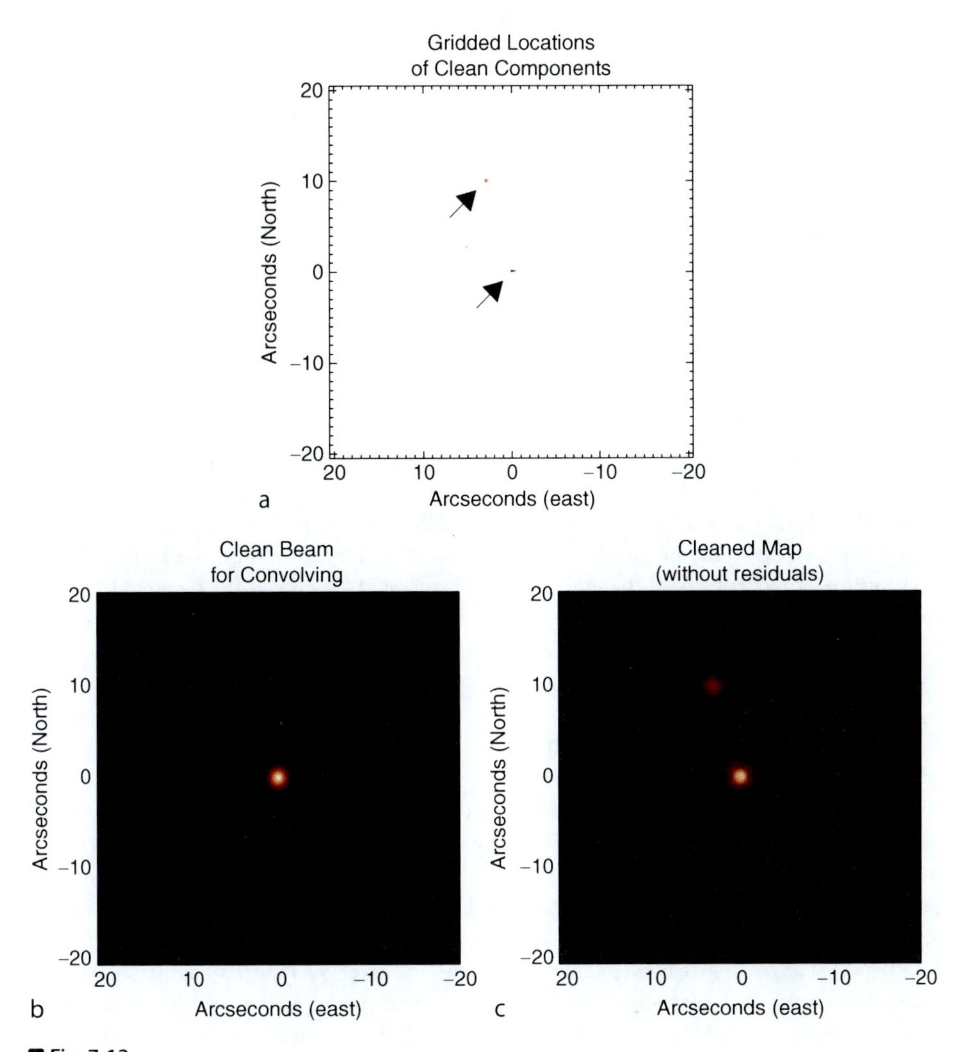

Fig. 7-12

The basic restoration steps of the CLEAN algorithm consist of (a) collecting the locations of the CLEAN components, (b) creating a CLEAN beam which is a Gaussian beam with same angular resolution of the central portion of the Dirty Beam, and (c) convolving the CLEAN components with the CLEAN beam. Here, the final step of adding back the residuals of the dirty map has been left off. The *arrows* in panel (a) mark the location of the two clusters of clean components near the locations of the two components of the binary

properties of this algorithm in a lucid review article and motivate the above methodology using *Bayes' Theorem*, the cornerstone of so-called *Bayesian Statistics*.

Entropy is an interesting statistic since it quantifies the amount of complexity in a distribution. It is often stated that MEM tries to find the "smoothest image" consistent with the data. This indeed is a highly desirable feature since any structures in the reconstructed image should be based on the data itself and not artifacts from the reconstruction process. However, MEM does not actually select the smoothest image but rather one with "most equal" uniform set of values – MEM does not explicitly take into account spatial structure but only depends on the distribution of pixel values. Indeed with more study, the maximum entropy functional has been found not to be that special, except for its privileged appearance in physics. From a broader perspective, maximum entropy can be considered one member of a class of regularizers that allow the inverse problem to be well defined, and MEM is not necessarily the best nor even suitable for some imaging problems (e.g., other regularizers include total variation or maximum likelihood).

MEM performs better for reconstructing smooth large scale emission than CLEAN, although MEM is much more computationally demanding. MEM can naturally deal with heterogeneous data with varying errors since the data is essentially fitted using a χ^2-like statistic. This involves only a "forward" transform from image space to (u,v) space, thus avoiding all the issues of zeros in the (u,v) grid and the need for deconvolution. In addition, MEM images possess *super-resolution* beyond the traditional $\sim \frac{\lambda}{D}$ diffraction limit since smoothness is introduced in the process only indirectly through the entropy statistic. For instance, if the object is a point-like object, then the FWHM of the reconstructed MEM image depends on the signal-to-noise of the data, not just the length of the longest baseline. Super-resolution is viewed with some skepticism by practitioners of CLEAN because structures beyond the formal diffraction limit may be artifacts of the entropy functional. MEM has been implemented in many radio interferometry data processing environments, such as AIPS for VLA/VLBA, Caltech VLBI package (Sivia 1987), and for optical interferometry (e.g., BSMEM, Buscher 1994).

5.2 Nonideal Cases for Imaging

Most modern interferometric arrays in the radio (VLA) and millimeter (ALMA, CARMA) have a sufficient number of elements for good (u,v) coverage and also employ rapid phase referencing for absolute phase calibration. This allows either CLEAN or MEM methods to make imaging possible, although mosaicing large fields can still pose a computational challenge.

In optical interferometry and perhaps also at some sub-mm wavelengths, atmospheric turbulence changes the fringes phases so quickly and over such small angular scales that phase referencing is not practical. As discussed earlier, modeling can still be done for the visibility amplitudes but the turbulence scrambles the phase information beyond utility. Fortunately, there is a clever method to recover some of the lost phase information and this is discussed next.

5.2.1 Closure Phase, the Bispectrum, and Closure Amplitudes

As discussed earlier, phase referencing is used to correct for drifting phases. So what can be done when phase referencing is not possible? Without valid phase information accompanying the visibility amplitude measurements, one cannot carry out the inverse Fourier transform that lies at the core of synthesis imaging and the CLEAN algorithm specifically. While in some cases, the fringe phases do not carry much information (e.g., for symmetrical objects), in general, phases carry most of the information for complex scenes.

Early in the history of radio interferometry a clever idea, now referred to as "closure phase," was discovered to recover some level of phase information when observing with three telescopes (Jennison 1958). The method was introduced to partly circumvent the combination of poor receiver stability and variable multi-path propagation in early radio-linked long-baseline ($\gtrsim$2 km) interferometer systems at Jodrell Bank. The term "closure phase" itself appeared later on, in the paper by Rogers et al. (1974) describing an application at centimeter radio wavelengths using very stable and accurate, but independent, reference oscillators at the three stations in a so-called very long baseline interferometer (VLBI) array. Closure phase was critical for VLBI work in the 1980s although it became less necessary as phase referencing became feasible. Application at optical wavelengths was first mentioned by Rogstad (1968), but carried out only much later in the optical range through aperture masking experiments (e.g., Baldwin et al. 1986). By 2006, nearly all separated-element optical arrays with three or more elements have succeeded in obtaining closure phase measurements (COAST, NPOI, IOTA, ISI, VLTI, CHARA). An optical observer now can expect closure phases to be a crucial observable for most current instrumentation.

The principle behind the power of closure phases is briefly described and the interested reader is referred to Monnier et al. (2007) for more detailed information on taking advantage of such phase information in optical interferometry.

Consider ❯ *Fig. 7-13a* in which a time delay is introduced above one slit in a Young's interferometer. This time delay introduces a phase shift for the detected fringe and the magnitude of the phase shift is independent of the baseline length. For the case of three telescopes (see ❯ *Fig. 7-13b*), a delay above one telescope will introduce phase shifts in *two fringes*. For instance, a delay above telescope 2 will show up as an equal phase shift for baseline 1–2 and baseline 2–3, but with *opposite* signs. Hence, the sum of three fringe phases, between 1–2, 2–3, and 3–1, will be insensitive to the phase delay above telescope 2. This argument holds for arbitrary phase delays above any of the three telescopes. In general, the sum of three phases around

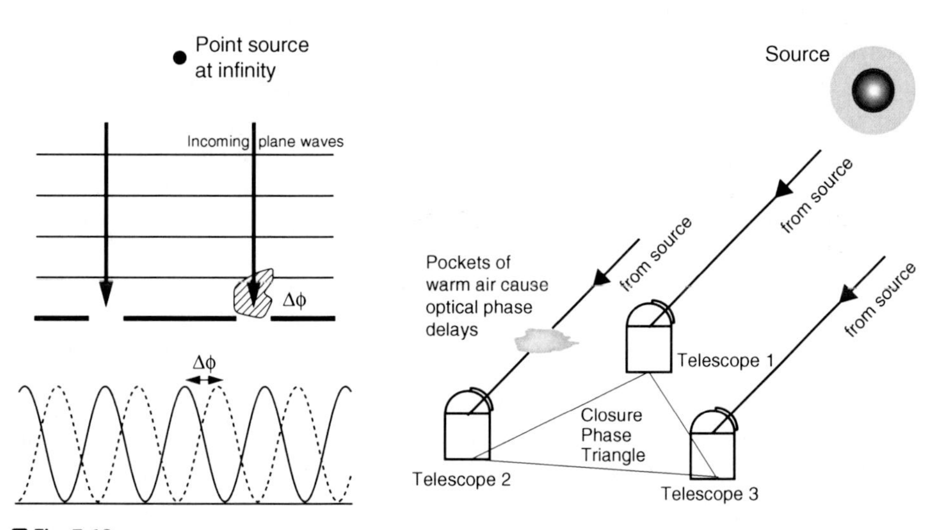

Fig. 7-13
***Left*: Atmospheric turbulence introduces extra path length fluctuations that induce fringe phase shifts. At optical wavelengths, these phase shifts vary by many radians over short time scales ($\ll$1 s) effectively scrambling the Fourier phase information. *Right* Phase errors introduced at any telescope cause equal but opposite phase shifts in adjoining baselines, canceling out in the *closure phase* (See also Monnier et al. 2006; Readhead et al. 1988) (Figures reprinted from Monnier 2003)**

a closed triangle of baselines, the *closure phase*, is a good interferometric observable; that is, it is independent of telescope-specific phase shifts induced by the atmosphere or optics.

The closure phase Φ_{ijk} can thus be written in terms of the three telescopes i,j,k in the triangle:

$$\Phi_{ijk} = \phi_{ij} + \phi_{jk} + \phi_{ki} \tag{7.20}$$

where ϕ_{ij} represents the measured Fourier phase for the baseline connecting telescopes i,j. Alternatively, the closure phase can be written in terms of the (u_0,v_0,u_1,v_1) in the Fourier (hyper-)plane where (u_0,v_0) represents the (u,v) coverage for baseline i, j in the triangle, (u_1,v_1) represents the (u,v) coverage for baseline j, k in the triangle, and the last leg of the triangle can be calculated from others since the sum of the three baselines must equal zero to be a "closure triangle." See definition and explanation put forward in documentation of the OI-FITS data format (Pauls et al. 2005).

Another method to derive the invariance of the closure phase to telescope-specific phase shifts is through the *bispectrum*. The bispectrum $\tilde{B}_{ijk} = \tilde{\mathcal{V}}_{ij}\tilde{\mathcal{V}}_{jk}\tilde{\mathcal{V}}_{ki}$ is formed through triple products of the complex visibilities around a closed triangle, where ijk specifies the three telescopes. Using (➋ 7.14) and using the concept of telescope-specific complex gains G_i, it can be seen how the telescope-specific errors affect the measured bispectrum:

$$\tilde{B}_{ijk} = \tilde{\mathcal{V}}_{ij}^{\text{measured}}\tilde{\mathcal{V}}_{jk}^{\text{measured}}\tilde{\mathcal{V}}_{ki}^{\text{measured}} \tag{7.21}$$

$$= |G_i||G_j|e^{i(\Phi_i^G-\Phi_j^G)}\tilde{\mathcal{V}}_{ij}^{\text{true}} \cdot |G_j||G_k|e^{i(\Phi_j^G-\Phi_k^G)}\tilde{\mathcal{V}}_{jk}^{\text{true}} \cdot |G_k||G_i|e^{i(\Phi_k^G-\Phi_i^G)}\tilde{\mathcal{V}}_{ki}^{\text{true}} \tag{7.22}$$

$$= |G_i|^2|G_j|^2|G_k|^2\tilde{\mathcal{V}}_{ij}^{\text{true}} \cdot \tilde{\mathcal{V}}_{jk}^{\text{true}} \cdot \tilde{\mathcal{V}}_{ki}^{\text{true}} \tag{7.23}$$

From the above derivation, the bispectrum is a complex quantity whose phase is identical to the closure phase, while the individual telescope gains affect only the bispectrum amplitude. The use of the bispectrum for reconstructing diffraction-limited images from speckle data was developed independently (Weigelt 1977) of the closure phase techniques, and the connection between the approaches elucidated only later (Cornwell 1987; Roddier 1986).

A three-telescope array with its one triangle can provide a single closure phase measurement, a paltry substitute for the three Fourier phases available using phase referencing. However, as one increases the number of elements in the array from three telescopes to seven telescopes, the number of independent closure phases increases dramatically to 15, about 70% of the total 21 Fourier phases available. An array the size of the VLA with 27 antennae is capturing 93% of the phase information. Indeed, imaging of bright objects does not require phase referencing and the VLA can make high-quality imaging through closure phases alone. Note that imaging using closure phases alone retains no absolute astrometry information; astrometry requires phase referencing.

A related quantity useful in radio is the *closure amplitude* (which requires sets of four telescopes) and this can be used to compensate for unstable amplifier gains and varying antenna efficiencies (e.g., Readhead et al. 1980). Closure amplitudes are not practical for current optical interferometers partially because most fringe amplitude variations are not caused by telescope-specific gain changes but rather by changing coherence (e.g., due to changing atmosphere).

Closure phases (and closure amplitudes) can be introduced into the imaging process in a variety of ways. For the CLEAN algorithm, the closure phases cannot be directly used because CLEAN requires estimates of the actual Fourier phases in order to carry out a Fourier transform. A clever iterative scheme known as *self-calibration* was described by Readhead and Wilkinson

(1978) and Cornwell and Wilkinson (1981) which alternates between a CLEANing stage and a self-calibration stage that estimates Fourier phases from the closure phases and most recent CLEANed image. As for standard CLEAN itself, self-calibration cannot naturally deal with errors in the closure phases and closure amplitudes, and thus is recognized as not optimal. That said, self-calibration is still widely used along with CLEAN even in the case of phase referencing to dramatically improve the imaging dynamic range for imaging of and around bright objects.

Closure phases, the bispectrum, and closure amplitudes can be quite naturally incorporated into "forward-transform" image reconstruction schemes such as the maximum entropy method. Recall that MEM basically performs a minimization of a regularizer constrained by some goodness-of-fit to the observed data. Thus, the bispectral quantities can be fitted just like all other observables. That said, the mathematics can be difficult and the program BSMEM (Buscher 1994) was one of the first useful software suitable for optical interferometers that successfully solved this problem in practice. Currently, the optical interferometer community has produced several algorithms to solve this problem, including the Building Block Method (Hofmann and Weigelt 1993), MACIM (Ireland et al. 2006), MIRA (Thiébaut 2008), WISARD (Meimon et al. 2008), and SQUEEZE (Baron et al. 2010). See Malbet et al. (2010) for a description of a recent blind imaging competition between some of these algorithms.

5.3 Astrometry

Astrometry is still a specialized technique within interferometry and a detailed description is beyond the scope of this chapter. Typically, the precise separation between two objects on the sky, the *relative astrometry*, is needed to be known for some purpose, such as a parallax measurement. If both objects are known point-like objects with no asymmetric structure, then the precise knowledge of the baseline geometry can be used, along with detailed measurements of interferometric fringe phase, to estimate their angular separation. In general, the astrometric precision $\Delta\Phi$ on the sky is related to the measured fringe phase SNR as follows:

$$\Delta\Phi \sim \frac{\lambda}{\text{Baseline}} \cdot \frac{1}{\text{SNR}} \tag{7.24}$$

Hence, if one measures a fringe with a signal-to-noise of 100 (which is quite feasible) then one can determine the relative separation of two sources with a precision 100× smaller than the fringe spacing. Since VLBI and optical interferometers have fringe spacings of ~1 milliarcsec, this allows for astrometric precision at the 10 *microarcsec* level allowing parallax measurements at many kiloparsecs and also allows us to take a close look around the black hole at the center of the galaxy. Unfortunately, at this precision level, many systematic effects become critical and knowledge of the absolute baseline vector between telescopes is crucial and demanding. The interested reader should consult specific instruments and recent results for more information (Muterspaugh et al. 2010; Reid et al. 2009, for radio and optical respectively).

6 Concluding Remarks

In barely more than 50 years, separated-element interferometry has come to dominate radio telescope design as the science has demanded ever-increasing angular resolution. The addition of more elements to an array has provided improvements both in mapping speed and in sensitivity. The early problems with instability in the electronics were solved years ago by improvements in radio and digital electronics, and it is presently not unusual to routinely achieve interferometer phase stability of order $\lambda/1,000$ over time spans of many tens of minutes. Sensitivity improvements by further reductions in receiver temperature are reaching a point of diminishing returns as the remaining contributions from local spillover, the Galactic background, and atmospheric losses come to dominate the equation. Advances in the speed and density of digital solid-state devices are presently driving the development of increased sensitivity, with flexible digital signal processors providing wider bands for continuum observations and high-frequency resolution for precision spectroscopy. With the commissioning of the billion-dollar international facility ALMA underway and long-term plans for ambitious arrays at longer wavelengths (SKA, LOFAR), the future of "radio" interferometry is bright and astronomers are eager to take advantage of the new capabilities, especially vast improvements to sensitivity.

The younger field of optical interferometry is still rapidly developing with innovative beam combination, fringe tracking methods, and extensions of adaptive optics promising significant improvements in sensitivity for years to come. In the long run, however, the atmosphere poses a fundamental limit to the ultimate astrometric precision and absolute sensitivity for visible and infrared interferometry that can only be properly overcome by placing the telescopes into space or through exploring exotic sites such as Dome A, Antarctica. Currently however, the emphasis of the majority of the scientific community favors ever-larger collecting area in the near- and mid-IR (e.g., JWST) as astronomers reach to ever greater distances and earlier times, rather than higher angular resolution.

Eventually, the limits of diffraction are likely to limit the science return of space telescopes, and the traditional response of building a larger aperture will no longer be affordable. Filled apertures increase in weight and in cost as a power >1 of the diameter (Bely 2003). For ground-based telescopes, cost $\sim D^{2.6}$. In space, the growth is slower, $\sim D^{1.6}$, although the coefficient of proportionality is much larger.[15] Over the last two decades, the prospects have become increasingly bleak for building ever larger filled-aperture telescopes within the anticipated space science budgets. Interferometry permits the connection of individual collectors of modest size and cost into truly gigantic space-based constellations with virtually unlimited angular resolution (Allen 2007). Ultimately, the pressure of discovery is likely to make interferometry in space at optical, UV, and IR wavelengths a necessity, just as it did at radio wavelengths more than half a century earlier.

Appendix A: Current Facilities

Table 7-4 provides a comprehensive list of existing and planned interferometer facilities. These facilities span the range from "private" instruments, currently available only to their developers, to general purpose instruments. An example of the latter is the Square Kilometer Array; this is a major next-generation facility for radio astronomy being planned by an

[15] It's interesting to note that at some large-enough diameter, ground-based telescopes are likely to become similar in cost to those in space, especially if one considers life-cycle costs.

■ Table 7-4
Alphabetical list of current and planned interferometer arrays

Acronym	Full name	Lead institution(s)	Location
ALMA	Atacama Large Millimeter Array	International/NRAO	Chajnantor, Chile
ATA	Allen Telescope Array	SETI Institute	Hat Creek Radio Observatory, CA, USA
ATCA	Australia Telescope Compact Array	CSIRO/ATNF	Narrabri, Australia
CARMA	Combined Array for Research in Millimeter Astronomy	Caltech, UCB, UChicago, UIUC, UMD	Big Pine, California USA
CHARA	Center for High Angular Resolution Astronomy	Georgia State University	Mt. Wilson, CA, USA
DRAO	Dominion Radio Astrophysical Observatory	Herzberg Institute of Astrophysics	Penticton BC, Canada
EVLA	Expanded Very Large Array	NRAO	New Mexico USA
EVN	The European VLBI Network	International	Europe, UK, US, S. Africa, China, Russia
FASR	Frequency Agile Solar Radiotelescope	National Radio Astronomy Observatory + six others	Owens Valley, CA, USA
ISI	Infrared Spatial Interferometer	Univ. California at Berkeley	Mt. Wilson, CA, USA
Keck-I	Keck Interferometer (Keck-I to Keck-II)	NASA-JPL	Mauna Kea, HI, USA
LBTI	Large Binocular Telescope Interferometer	LBT Consortium	Mt. Graham, AZ, USA
LOFAR	Low Frequency ARray	Netherlands Institute for Radio Astronomy – ASTRON	Europe
LWA	Long Wavelength Array	US Naval Research Laboratory	VLA Site, NM, USA
MOST	Molonglo Observatory Synthesis Telescope	School of Physics, Univ. Sydney	Canberra, Australia
MRO	Magdalena Ridge Observatory	Consortium of New Mexico Institutions,	Magdalena Ridge, NM, USA
MWA	Murchison Widefield Array	International/MIT-Haystack	Western Australia

Table 7-4
(Continued)

Acronym	Full name	Lead institution(s)	Location
NPOI	Navy Prototype Optical Interferometer	Naval Research Laboratory/US Naval Observatory	Flagstaff, AZ, USA
OHANA	Optical Hawaiian Array for Nanoradian Astronomy	Consortium (mostly French Institutions), Mauna Kea Observatories, others	Mauna Kea, HI, USA
PdB	Plateau de Bure interferometer	International/IRAM	Plateau de Bure, FR
SKA	Square Kilometer Array	see Appendix 6	under study
SMA	Submillimeter Array	Smithsonian Astrophysical Observatory	Mauna Kea, HI, USA
SUSI	Sydney University Stellar Interferometer	Sydney University	Narrabri, Australia
VLBA	Very Long Baseline Array	NRAO	Hawaii to St. Croix, US Virgin Islands
VLTI-UT	VLT Interferometer (Unit Telescopes)	European Southern Observatory	Paranal, Chile
VLTI-AT	VLT Interferometer (Auxiliary Telescopes)	European Southern Observatory	Paranal, Chile
VERA	VLBI Exploration of Radio Astrometry	Nat. Astron. Obs. Japan (NAOJ)	Japan
VSOP-2	VLBI Space Observatory Programme-2	Japan Aerospace Exploration Agency	Space-Ground VLBI
WSRT	Westerbork Synthesis Radio Telescope	Netherlands Institute for Radio Astronomy – ASTRON	Westerbork, NL

Table 7-5

Subset of those interferometer arrays from *Table 7-4* which are presently open (or will soon be open) for use by qualified researchers from the general astronomy community

Acronym	Telescope		Maximum baseline	Wavelength coverage	Observer resources web page link
	Number	Size (m)			
SUSI	2	0.14	64 (640[a]) m	0.5–1.0 μm	http://www.physics.usyd.edu.au/sifa/Main/SUSI
NPOI	6	0.12	64 (> 250[a]) m	0.57–0.85 μm	http://www.lowell.edu/npoi/
CHARA	6	1.0	330 m	0.6–2.4 μm	http://www.chara.gsu.edu/CHARA/
MRO[a]	~10	~1.5	~350 m	0.6–2.5 μm	http://www.mro.nmt.edu/Home/index.htm
VLTI	4	1.8–8.0	130 (202[a]) m	1–2.5, 8–13 μm	http://www.eso.org/sci/facilities/paranal/telescopes/vlti/
Keck-I	2	10.0	85 m	1.5–4, 8–13 μm	Http://planetquest.jpl.nasa.gov/Keck/
LBTI[a]	2	8.4	23[a] m	1–20 μm	http://lbti.as.arizona.edu/
ISI	3	1.65	85 (> 100[a]) m	8–12 μm	http://isi.ssl.berkeley.edu/
ALMA[a]	66	7–12	15 km	0.3–3.6 mm	http://science.nrao.edu/alma/index.shtml
SMA	8	6	500 m	0.4–1.7 mm	http://www.cfa.harvard.edu/sma/
CARMA	23	3.5–10.4	2,000 m	1.3, 3, 7 mm	http://www.mmarray.org/
PdB	6	15	760 m	1.3, 2, 3 mm	http://www.iram-institute.org/EN/
VLBA	10	25	8,000 km	3 mm–28 cm	http://science.nrao.edu/vlba/index.shtml
eMERLIN	7	25–76	217 km	13 mm–2 m	http://www.e-merlin.ac.uk/
EVN	27	14–305	> 10,000 km	7 mm–90 cm	http://www.evlbi.org/
VERA	4	20	2,270 km	7 mm–1.4 cm	http://veraserver.mtk.nao.ac.jp/index.html
ATCA	6	22	6 km	3 mm–16 cm	http://www.narrabri.atnf.csiro.au/
EVLA	27	25	27 km	6 mm–30 cm	http://science.nrao.edu/evla/index.shtml
WSRT	14	25	2,700 m	3.5 cm–2.6 m	http://www.astron.nl/radio-observatory/astronomers/wsrt-astronomers
DRAO	7	9	600 m	21, 74 cm	e-mail: Tom.Landecker@nrc-cnrc.gc.ca
GMRT	30	45	25 km	21 cm–7.9 m	http://www.gmrt.ncra.tifr.res.in
MOST	64	~ 12	1,600 m	36 cm	http://www.physics.usyd.edu.au/sifa/Main/MOST
LOFAR	many	simple	1,500 km	1.2–30 m	http://www.astron.nl/radio-observatory/astronomers/lofar-astronomers

[a]Indicates capabilities under development

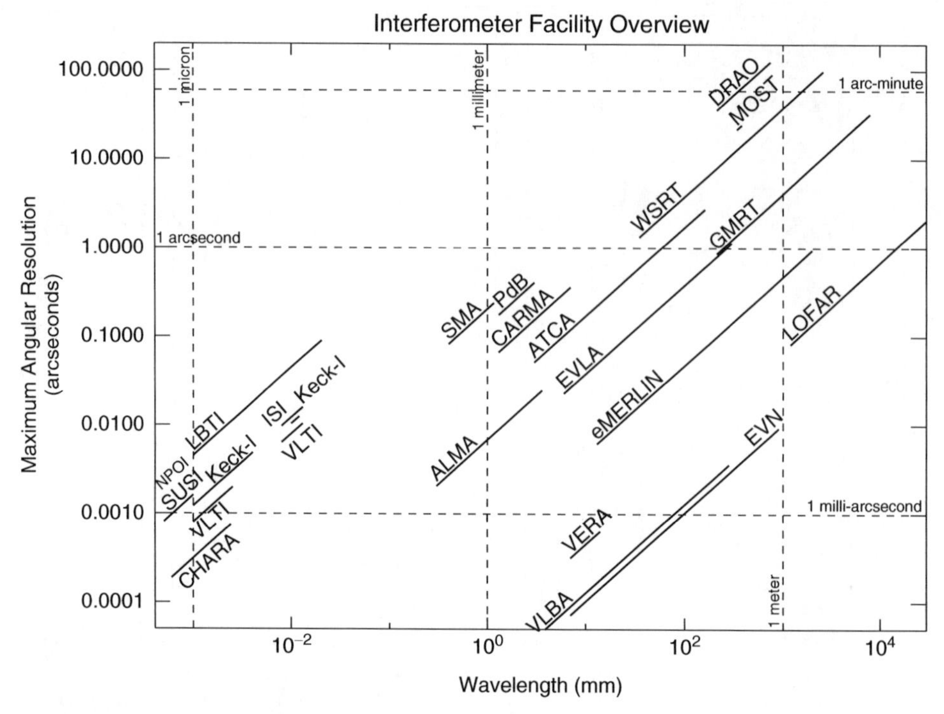

Fig. 7-14

Graphical representation of the wavelength coverage and maximum angular resolution available using the radio and optical interferometers of the world. See *Table 7-5* **for more information**

international consortium, with a Project Office at the Jodrell Bank Observatory in the UK (http://www.skatelescope.org/). The location for the SKA has not yet been finalized, but (as of 2011) sites in Australia and South Africa are currently being discussed. Plans for the SKA have spawned several "proof-of-concept" or "pathfinder" instruments which are being designed and built including LOFAR (Europe), MeerKAT (South Africa), ASKAP, MWA, and SKAMP (Australia), and LWA (USA).

Table 7-5 summarizes the wavelength and angular resolution of currently operating arrays open to the general astronomer. A few of the facilities in this list are still under construction. *Figure 7-14* summarizes the vast wavelength range and angular resolution available using the radio and optical interferometers of the world.

Acknowledgments

We are grateful to the following colleagues for their comments and contributions: W.M. Goss, R.D. Ekers, T.L. Wilson, S. Kraus, M. Zhao, T. ten Brummelaar, A. King, and P. Teuben.

References

Allen, R. J. 2007, PASP, 119, 914
Allen, R. J., Goss, W. M., & van Woerden, H. 1973, A&A, 29, 447
Andrews, S. M., Wilner, D. J., Hughes, A. M., Qi, C., & Dullemond, C. P. 2009, ApJ, 700, 1502
Baade, W., & Minkowski, R. 1954, ApJ, 119, 206
Baldwin, J. E., Haniff, C. A., Mackay, C. D., & Warner, P. J. 1986, Nature, 320, 595
Baldwin, J. E., Beckett, M. G., Boysen, R. C., Burns, D., Buscher, D. F., Cox, G. C., Haniff, C. A., Mackay, C. D., Nightingale, N. S., Rogers, J., Scheuer, P. A. G., Scott, T. R., Tuthill, P. G., Warner, P. J., Wilson, D. M. A., & Wilson, R. W. 1996, A&A, 306, L13+
Baron, F., Monnier, J. D., & Kloppenborg, B. 2010, A novel image reconstruction software for optical/infrared interferometry. In Society of Photo-Optical Instrumentation Engineers (SPIE) Conference Series, Vol. 7734
Bely, P. Y. 2003, The Design and Construction of Large Optical Telescopes (New York, NY: Springer)
Bolton, J. G., Stanley, G. J., & Slee, O. B. 1949, Nature, 164, 101
Bosma, A. 1981a, AJ, 86, 1791
Bosma, A. 1981b, AJ, 86, 1825
Bridle, A. H., Hough, D. H., Lonsdale, C. J., Burns, J. O., & Laing, R. A. 1994, AJ, 108, 766
Briggs, D. S. 1995, High Fidelity Interferometric Imaging: Robust Weighting and NNLS Deconvolution. In American Astronomical Society Meeting Abstracts, Vol. 27, pp. 112.02
Buscher, D. F. 1994, in IAU Symposium, Vol. 158, Very High Angular Resolution Imaging, ed. J. G. Robertson, & W. J. Tango (Dordrecht: Kluwer Academic), 91–+
Carilli, C. L., & Holdaway, M. A. 1999, Radio Sci, 34, 817
Cohen, M. H., Cannon, W., Purcell, G. H., Shaffer, D. B., Broderick, J. J., Kellermann, K. I., & Jauncey, D. L. 1971, ApJ, 170, 207
Colavita, M. M. 1999, PASP, 111, 111
Colavita, M. M., Serabyn, E., Millan-Gabet, R., Koresko, C. D., Akeson, R. L., Booth, A. J., Mennesson, B. P., Ragland, S. D., Appleby, E. C., Berkey, B. C., Cooper, A., Crawford, S. L., Creech-Eakman, M. J., Dahl, W., Felizardo, C., Garcia-Gathright, J. I., Gathright, J. T., Herstein, J. S., Hovland, E. E., Hrynevych, M. A., Ligon, E. R., Medeiros, D. W., Moore, J. D., Morrison, D., Paine, C. G., Palmer, D. L., Panteleeva, T., Smith, B., Swain, M. R., Smythe, R. F., Summers, K. R., Tsubota, K., Tyau, C., Vasisht, G., Wetherell, E., Wizinowich, P. L., & Woillez, J. M. 2009, PASP, 121, 1120
Cornwell, T. J. 1987, A&A, 180, 269
Cornwell, T. J., & Wilkinson, P. N. 1981, MNRAS, 196, 1067
Danchi, W. C., Bester, M., Degiacomi, C. G., Greenhill, L. J., & Townes, C. H. 1994, AJ, 107, 1469
Delplancke, F., Derie, F., Lév"que, S., Ménardi, S., Abuter, R., Andolfato, L., Ballester, P., de Jong, J., Di Lieto, N., Duhoux, P., Frahm, R., Gitton, P., Glindemann, A., Palsa, R., Puech, F., Sahlmann, J., Schuhler, N., Duc, T. P., Valat, B., & Wallander, A. 2006, in Presented at the SPIE Conf. , Vol. 6268, SPIE Conf. Ser. (Bellingham WA: SPIE)
di Benedetto, G. P., & Rabbia, Y. 1987, A&A, 188, 114
Eisner, J. A., Akeson, R., Colavita, M., Ghez, A., Graham, J., Hillenbrand, L., Millan-Gabet, R., Monnier, J. D., Pott, J. U., Ragland, S., Wizinowich, P., & Woillez, J. 2010, Science with the Keck Interferometer ASTRA program. In Society of Photo-Optical Instrumentation Engineers (SPIE) Conference Series, Vol. 7734
Fried, D. L. 1965, J Opt Soc Am, 55, 1427
Glindemann, A. 2010, Principles of Stellar Interferometry (Berlin: Springer)
Golay, M. 1971, J Opt Soc Am, 61, 272+
Goss, W. M., Brown, R. L., & Lo, K. Y. 2003, Astron Nachr Suppl, 324, 497
Greenhill, L. J., Gwinn, C. R., Schwartz, C., Moran, J. M., & Diamond, P. J. 1998, Nature, 396, 650
Gull, S. F. & Skilling, J. 1983, in Proceedings of an International Symposium held in Sydney, Australia, Indirect Imaging. Measurement and Processing for Indirect Imaging, ed. J. A. Roberts (Cambridge, UK: Cambridge University Press), 267+
Hanbury Brown, R., Davis, J., & Allen, L. R. 1974, MNRAS, 167, 121
Hecht, E. 2002, Optics (Boston, MA: Addison-Wesley)
Heffner, H. 1962, Proc IRE, 50, 1604
Hofmann, K., & Weigelt, G. 1993, A&A, 278, 328
Högbom, J. A. 1974, A&AS, 15, 417
Holdaway, M. A., & Helfer, T. T. 1999, in ASP Conf. Ser., Vol. 180, Synthesis Imaging in Radio Astronomy II, ed. G. B. Taylor, C. L. Carilli, & R. A. Perley (San Francisco, CA: ASP), 537–+
Ireland, M. J., Monnier, J. D., & Thureau, N. 2006, Monte-Carlo imaging for optical interferometry. In Society of Photo-Optical Instrumentation Engineers (SPIE) Conference Series, Vol. 6268
Jennison, R. C. 1958, MNRAS, 118, 276+

Jennison, R. C., & Das Gupta, M. K. 1953, Nature, 172, 996
Johnson, M. A., Betz, A. L., & Townes, C. H. 1974, Phys Rev Lett, 33, 1617
Keto, E. 1997, ApJ, 475, 843
Kishimoto, M., Hönig, S. F., Antonucci, R., Kotani, T., Barvainis, R., Tristram, K. R. W., & Weigelt, G. 2009, A&A, 507, L57
Kraus, S., Hofmann, K., Benisty, M., Berger, J., Chesneau, O., Isella, A., Malbet, F., Meilland, A., Nardetto, N., Natta, A., Preibisch, T., Schertl, D., Smith, M., Stee, P., Tatulli, E., Testi, L., & Weigelt, G. 2008, A&A, 489, 1157
Labeyrie, A. 1975, ApJ Lett, 196, L71
Labeyrie, A., Lipson, S. G., & Nisenson, P. 2006, An Introduction to Optical Stellar Interferometry (Cambridge, UK: Cambridge University Press)
Lachaume, R. 2003, A&A, 400, 795
Lane, B. F., Kuchner, M. J., Boden, A. F., Creech-Eakman, M., & Kulkarni, S. R. 2000, Nature, 407, 485
Lawson, P. R., ed. 2000, Principles of Long Baseline Stellar Interferometry (Pasadena, CA: NASA/JPL)
Malbet, F., & Perrin, G., eds. 2007, Foreword: Proceedings of the Euro Summer School "Observation and Data Reduction with with the VLT Interferometer". http://vltischool.obs.ujf-grenoble.fr. New A Rev., 51:563–564
Malbet, F., Cotton, W., Duvert, G., Lawson, P., Chiavassa, A., Young, J., Baron, F., Buscher, D., Rengaswamy, S., Kloppenborg, B., Vannier, M., & Mugnier, L. 2010, The 2010 interferometric imaging beauty contest. In Society of Photo-Optical Instrumentation Engineers (SPIE) Conference Series, 7734
McCready, L. L., Pawsey, J. L., & Payne-Scott, R. 1947, R Soc Lond Proc Ser A, 190, 357
Meimon, S., Mugnier, L. M., & Le Besnerais, G. 2008, J Opt Soc Am A, 26, 108
Mérand, A., Kervella, P., Coudé du Foresto, V., Ridgway, S. T., Aufdenberg, J. P., ten Brummelaar, T. A., Berger, D. H., Sturmann, J., Sturmann, L., Turner, N. H., & McAlister, H. A. 2005, A&A, 438, L9
Michelson, A. A., & Pease, F. G. 1921, ApJ, 53, 249
Millan-Gabet, R., Schloerb, F. P., & Traub, W. A. 2001, ApJ, 546, 358
Miyoshi, M., Moran, J., Herrnstein, J., Greenhill, L., Nakai, N., Diamond, P., & Inoue, M. 1995, Nature, 373, 127
Monnier, J. D. 2003, Rep Prog Phys, 66, 789
Monnier, J. D., Berger, J.-P., Millan-Gabet, R., Traub, W. A., Schloerb, F. P., Pedretti, E., Benisty, M., Carleton, N. P., Haguenauer, P., Kern, P., Labeye, P., Lacasse, M. G., Malbet, F., Perraut, K., Pearlman, M., & Zhao, M. 2006, ApJ, 647, 444
Monnier, J. D., Zhao, M., Pedretti, E., Thureau, N., Ireland, M., Muirhead, P., Berger, J., Millan-Gabet, R., Van Belle, G., ten Brummelaar, T., McAlister, H., Ridgway, S., Turner, N., Sturmann, L., Sturmann, J., & Berger, D. 2007, Science, 317, 342
Muterspaugh, M. W., Lane, B. F., Kulkarni, S. R., Konacki, M., Burke, B. F., Colavita, M. M., Shao, M., Wiktorowicz, S. J., & O'Connell, J. 2010, AJ, 140, 1579
Narayan, R., & Nityananda, R. 1986, ARAA, 24, 127
Oliver, B. M. 1965, Proc IEEE, 53, 436
Pathria, R. K. 1972, Statistical Mechanics (Oxford, NY: Pergamon Press)
Pauls, T. A., Young, J. S., Cotton, W. D., & Monnier, J. D. 2005, PASP, 117, 1255
Pawsey, J. L., Payne-Scott, R., & McCready, L. L. 1946, Nature, 157, 158
Peterson, D. M., Hummel, C. A., Pauls, T. A., Armstrong, J. T., Benson, J. A., Gilbreath, G. C., Hindsley, R. B., Hutter, D. J., Johnston, K. J., Mozurkewich, D., & Schmitt, H. R. 2006, Nature, 440, 896
Porcas, R. W., Booth, R. S., Browne, I. W. A., Walsh, D., & Wilkinson, P. N. 1979, Nature, 282, 385
Quirrenbach, A. 2000, in Principles of Long Baseline Stellar Interferometry (Pasadena, CA: NASA/JPL), 71–+
Quirrenbach, A. 2001, ARA&A, 39, 353
Quirrenbach, A., Bjorkman, K. S., Bjorkman, J. E., Hummel, C. A., Buscher, D. F., Armstrong, J. T., Mozurkewich, D., Elias, II, N. M., & Babler, B. L. 1997, ApJ, 479, 477
Readhead, A. C. S., & Wilkinson, P. N. 1978, ApJ, 223, 25
Readhead, A. C. S., Walker, R. C., Pearson, T. J., & Cohen, M. H. 1980, Nature, 285, 137
Readhead, A., Nakajima, T., Pearson, T., Neugebauer, G., Oke, J., & Sargent, W. 1988, AJ, 95, 1278
Reid, M. J., Menten, K. M., Brunthaler, A., Zheng, X. W., Moscadelli, L., & Xu, Y. 2009, ApJ, 693, 397
Roddier, F. 1986, Opt Commun, 60, 145+
Rogers, A. E. E., Hinteregger, H. F., Whitney, A. R., Counselman, C. C., Shapiro, I. I., Wittels, J. J., Klemperer, W. K., Warnock, W. W., Clark, T. A., & Hutton, L. K. 1974, ApJ, 193, 293
Rogstad, D. H. 1968, Appl Opt, 7, 585+
Rogstad, D. H., & Shostak, G. S. 1972, ApJ, 176, 315
Rots, A. H. 1975, A&A, 45, 43
Rots, A. H., & Shane, W. W. 1975, A&A, 45, 25
Saha, S. K. 2011, Aperture Synthesis, ed. Saha, S. K. (New York, NY: Springer)
Shao, M., & Colavita, M. M. 1992, A&A, 262, 353

Sivia, D. 1987, PhD thesis, Cambridge University
Skilling, J., & Bryan, R. K. 1984, MNRAS, 211, 111+
Smith, F. G. 1951, Nature, 168, 555
Sullivan, W. T. 2009, Cosmic Noise (Cambridge: Cambridge University Press)
Sutton, E. C., & Hueckstaedt, R. M. 1996, A&AS, 119, 559
Taylor, G. B., Carilli, C. L., & Perley, R. A., eds. 1999, ASP Conf. Ser., Vol. 180, Synthesis Imaging in Radio Astronomy II (San Francisco, CA: ASP)
ten Brummelaar, T. A., McAlister, H. A., Ridgway, S. T., Bagnuolo, Jr., W. G., Turner, N. H., Sturmann, L., Sturmann, J., Berger, D. H., Ogden, C. E., Cadman, R., Hartkopf, W. I., Hopper, C. H., & Shure, M. A. 2005, ApJ, 628, 453
Thiébaut, E. 2008, MIRA: an effective imaging algorithm for optical interferometry. In Society of Photo-Optical Instrumentation Engineers (SPIE) Conference Series, Vol. 7013
Thompson, A. R., Clark, B. G., Wade, C. M., & Napier, P. J. 1980, ApJ Suppl, 44, 151
Thompson, A. R., Moran, J. M., & Swenson, Jr., G. W. 2001, Interferometry and Synthesis in Radio Astronomy (2nd ed.; New York: Wiley)
van Boekel, R., Min, M., Leinert, C., Waters, L. B. F. M., Richichi, A., Chesneau, O., Dominik, C., Jaffe, W., Dutrey, A., Graser, U., Henning, T., de Jong, J., Köhler, R., de Koter, A., Lopez, B., Malbet, F., Morel, S., Paresce, F., Perrin, G., Preibisch, T., Przygodda, F., Schöller, M., & Wittkowski, M. 2004, Nature, 432, 479
Visser, H. C. D. 1980a, A&A, 88, 159
Visser, H. C. D. 1980b, A&A, 88, 149
Walsh, D., Carswell, R. F., & Weymann, R. J. 1979, Nature, 279, 381
Weigelt, G. P. 1977, Opt Commun, 21, 55
Whitney, A. R., Shapiro, I. I., Rogers, A. E. E., Robertson, D. S., Knight, C. A., Clark, T. A., Goldstein, R. M., Marandino, G. E., & Vandenberg, N. R. 1971, Science, 173, 225
Wiedner, M. C., Hills, R. E., Carlstrom, J. E., & Lay, O. P. 2001, ApJ, 553, 1036
Wilkinson, P. N., Kellermann, K. I., Ekers, R. D., Cordes, J. M., & Lazio, T. J. W. 2004, New A Rev., 48, 1551
Woolf, N. J. 1982, ARA&A, 20, 367
Zhao, M., Monnier, J. D., Pedretti, E., Thureau, N., Mérand, A., ten Brummelaar, T., McAlister, H., Ridgway, S. T., Turner, N., Sturmann, J., Sturmann, L., Goldfinger, P. J., & Farrington, C. 2009, ApJ, 701, 209

8 Absolute Calibration of Astronomical Flux Standards

Susana Deustua[1] · *Stephen Kent*[2] · *J. Allyn Smith*[3]
[1]Instruments Division, Space Telescope Science Institute, Baltimore, MD, USA
[2]Fermi National Accelerator Laboratory, Batavia, IL, USA
[3]Austin Peay State University, Clarksville, TN, USA

T.D. Oswalt, H.E. Bond (eds.), *Planets, Stars and Stellar Systems. Volume 2: Astronomical Techniques, Software, and Data*, DOI 10.1007/978-94-007-5618-2_8, © Springer Science+Business Media Dordrecht 2013

The goal of astronomical data calibration is to convert measurements recorded in some particular instrumental units into physical quantities, such as erg cm^{-2} s^{-1} Hz^{-1}, removing as much as possible all instrumental signatures. In principle, photometric calibration is a solved problem – laboratory reference standards such as blackbody furnaces achieve precisions well in excess of those needed for astrophysics. In practice, however, transferring the calibration from these laboratory standards to astronomical objects of interest is far from trivial – the transfer must reach outside the atmosphere, extend over 4π steradian of sky, cover a wide range of wavelengths, and span an enormous dynamic range in intensity.

The calibration process is never perfect, nor is a perfect calibration generally required for science – instead, imperfect knowledge of the physics and/or distances of astronomical sources are limiting factors In general, the goal of the calibration process is to achieve accuracy sufficiently high that it is not the limiting factor in a science analysis.

For many scientific problems, one only requires accurate differential measurements of a source or sources ("differential photometry") with little, if any, need for knowledge of absolute physical quantities – an example is the detection of planets when they transit the parent star. The most precise differential measurements often can be achieved only if all data are collected with the same instrument. For a broad class of problems, where the highest accuracy is not required and one can combine data collected by similar but distinct instruments and telescopes, one often relies on astronomical sources themselves to be used as calibrators, and the relative calibration of sources with respect to one another (relative photometry) is more accurate than the absolute calibration. Much of the field of classical stellar photometry falls in this category (Bessell 2005). In some cases, one does rely on converting measurements to physical quantities, even for making differential measurements – for example, tracking the long-term constancy of the solar flux. For other problems, such as measurement of the effective temperatures of stars or of hot gas in X-ray clusters, one requires spectrophotometric calibration in physical units ("relative spectrophotometry"). Finally, for a handful of science problems (primarily but not exclusively in the solar system), the absolute flux zero-point is needed as well ("absolute spectrophotometry").

In nearly all cases, one is interested in the monochromatic flux measured at a particular wavelength. In some cases (particularly at IR through UV wavelengths), calibration of the bandpass through which a measurement is made is as important as calibration of the flux itself. Often measurements retain some signature of the instrumentation used to collect the data and, sometimes (such as UBV photometry), long after the original instrumentation is no longer available. At still higher energies (X-ray and above), photons are sufficiently energetic that one can measure the energy of each one individually.

In principle, the distinction between point sources (sources that are unresolved in one's instrument) and extended sources is not of much consequence – for the latter, one calibrates the intensity of a source (flux per steradian of solid angle) rather than a simple flux itself. However, the distinction in the response of a telescope and instrument to point versus extended sources is sometimes sufficiently complex that it can complicate the calibration procedure if one's target and calibration sources are of different types. A full treatment of all calibration issues would require multiple volumes and cover topics including wavelength, astrometric, flux, and time calibrations and would cover not just electromagnetic radiation but also particles such as cosmic rays or neutrinos, other forms of radiation such as gravity waves, and calibration of the physics of astronomical sources such as the period-luminosity relation for Cepheids. These topics are generally separable, however, and can mostly be treated independently of one another.

Table 8-1
Wavelength ranges and corresponding energy range in eV (where commonly used)

Wavelength range	Energy range	Name
<1 pm	>1.2 MeV	Gamma ray
1 pm–1 nm	1.2 keV–1.2 MeV	X-ray
1 nm–0.200 μm	6.2 eV–1.6 keV	Extreme to near ultraviolet (XUV/NUV)
0.200–0.800 μm	1.6–6.2 eV	Ultraviolet to visible UVIS
0.8–2 μm		Near infrared (NIR)
2–300 μm		Mid to far infrared (MIR/FIR)
300 μm–>1mm		Millimeter to radio

This chapter will focus primarily on spectrophotometric flux calibration of electromagnetic radiation from radio to gamma rays, but emphasizing the ultraviolet to the infrared region of the spectrum. With a few exceptions, this latter region is where science drivers are the most demanding.

This article reviews the current status of absolute (spectro) photometric calibration in astronomy and astrophysics. Sect. 1 discusses the science drivers for precise SI[1]-traceable photometric calibration. Sect. 2 describes the basics of calibration. Flux standards commonly used in current calibration programs, and their limitations are presented in Sect. 3. In Sect. 4, we discuss ways of minimizing the effect of the Earth's atmosphere on ground-based observation and summarize efforts to calibrate the atmosphere. In Sect. 5, we conclude with a brief discussion on the validity (reliability) of absolute flux calibration.

To simplify the discussion, we divide, rather arbitrarily, the electromagnetic spectrum regions by wavelength as in *Table 8-1* below. We also use SI units wherever possible; though make exceptions to conform to common practice. Thus, for example, we use μm (microns) for wavelength, and flux units in Watts per meter squared (W m^{-2}).

1 Science Drivers for Calibration

This section presents a sampling of science investigations that are drivers for accurate calibration (Kent et al. 2009).

In 1998, we learned that the expansion of the universe is accelerating, implying the existence of a new component of the universe dubbed *dark energy*. Type Ia supernovae are one of four principal methods for probing the expansion history. These supernovae are considered to be *standardizable candles:* from observations of light curves and spectra, can be derived the rest-frame B-band luminosity of a supernova that is the same on average with a scatter of ≈15% for a single object. Cosmological and dark-energy parameters are determined from the shape of the Hubble brightness-redshift relationship for hundreds of supernova from $z = 0$ to $z = 1.7$. The goal is to distinguish competing dark-energy models whose predictions differ by as little

[1]SI, the International System of Units, is the system of weights and measures adopted by the international science and commerce communities. SI base units are the kilogram, meter, second, ampere, kelvin, candela, and mole.

as 2% (Albrecht et al. 2006). Since the rest-frame B-band falls in different observational band passes, depending on its redshift, the relative zero-points of all bands from 0.35 to 1.7 μm must be cross-calibrated to an accuracy of better than 1% across this wavelength range.

The fundamental parameters of stars, including mass, radius, metallicity, and age, are inferred by matching accurate models of stellar atmospheres to calibrated spectroscopic data and thus determining effective temperature, surface gravity, composition, and, if necessary, interstellar reddening. For stars with relatively simple atmospheres such as hydrogen white dwarfs, atmosphere models are thought to be quite accurate and can be used to predict photometric parameters and, in combination with stellar interior models, the radii and absolute luminosities as well. By combining these data with photometric measurements, it is possible to predict distances. A comparison of these predictions with measured trigonometric parallaxes for those stars with such measurements shows excellent agreement (Holberg et al. 2008). If calibrations can be improved to the level of 1% and with more stars (such as will be measured with Gaia, a European space mission scheduled for launch ~2013), it will be possible to make meaningful tests of 3-D spherical models, derive masses directly, and make better quantitative tests of evolutionary models.

Galaxy clusters are the largest dynamically bound systems in the universe, and information about clusters such as their masses can be determined from multiple, independent data sets, including number counts and velocities of individual galaxies, X-ray measures of the flux and temperature of a hot, intracluster medium (ICM), and weak lensing shear measurement of background galaxies. Theoretical simulations (Kravtsov et al. 2006) suggest that the parameter Y, a the product of the ICM gas mass and temperature, is a particularly sensitive measure of cluster mass and is insensitive to imperfections in clusters such as their not being in complete hydrostatic equilibrium. The scatter in the relation between Y and cluster mass is only 5%. To make full use of this method, the calibration of X-ray temperatures and fluxes should be significantly better (i.e., a few percent).

In addition to scientific problems requiring measurements in physical units, many science questions require combining data obtained with multiple instruments and/or over multiple times. Typical problems include the long-term constancy or variability of sources, such as the Sun and the isotropy of source counts over the whole sky. One may not need actual SI units to carry out one's particular science, but they are used to tie together data sets where intercalibration, such as observing the same objects at the same time by the same instrument, is not feasible. The required accuracy depends on the particular science but might credibly take advantage of calibrations that achieve subpercent accuracy.

2 Calibration Basics

In this section, we will focus on the spectrophotometric calibration of point sources. The goal is to express the monochromatic flux of such sources in units of W m^{-2} Hz^{-1} (or equivalent, e.g., ergs s^{-1}-cm^{-2} Hz^{-1}) at one or more particular frequencies or wavelengths as measured outside the Earth's atmosphere. We thus ignore any additional steps required to convert this flux to luminosity at a source, which requires knowledge of distance, redshift, intervening opacity, time variability, polarization, source beaming, radiation coherence, etc. We will discuss point versus extended sources below.

2.1 Basic Physics

Ideally, all calibrations eventually tie back to sources whose properties can be well characterized using basic physics. In the vernacular of calibration scientists, this is referred to as SI traceability – i.e., a quantity that can be traced in a direct way to the established International System of Measures.

2.1.1 Blackbodies

The intensity of light from a blackbody is given by the familiar equation:

$$I_\nu = \frac{2h\nu^3}{c^2} \frac{1}{e^{h\nu/kT} - 1} \ \mathrm{W - m^2 - Hz^{-1} - ster^{-1}}$$

The shape of spectrum is characterized uniquely by temperature T, and the area of an aperture in the blackbody cavity determines the absolute flux. If one has a perfect blackbody furnace with a small opening and observes the furnace a large distance such that the opening subtends a solid angle $\Delta\Omega$ one would measure a flux

$$F_\nu = I_\nu \Delta\Omega$$

Blackbody sources are practical calibrators between the mid-infrared to the ultraviolet region of the spectrum. At long wavelengths (typically mid-infrared or longer), the Rayleigh–Jeans law

$$I_\nu = \frac{2kT\nu^2}{c^2}$$

provides a good approximation to the spectrum. At these wavelengths, the accuracy of a calibration is directly proportional to the accuracy with which the source temperature and area of the calibrator is known. At shorter wavelengths, where the Wien cutoff is important, accurate knowledge of the temperature is much more important.

2.1.2 White Dwarf Atmospheres

Stellar atmosphere models play an important role in calibrations in the ultraviolet through infrared portion of the spectrum, since stars often serve as surrogate calibration sources and accurate models are needed to provide a calibration at arbitrary wavelengths and resolutions. Most modeling requires knowledge of element abundances and complex atomic physics, little of which is known with sufficiently high precision, as well as temperature and the surface gravity. The atmospheres of DA (pure hydrogen) white dwarfs, however, are simple enough that models of these types of stars may be accurate to 1–2% over certain wavelength ranges, the ultraviolet to infrared (Holberg and Bergeron 2006). In practice, one must also select stars that are close so that interstellar reddening is not a problem. A small set of such stars provides the fundamental calibration for the Hubble Space Telescope instruments in the UV to near-IR (Bohlin 2007).

2.1.3 Particle and-High Energy Photon Sources

At higher energies (far-UV, X-ray, and gamma-ray), blackbodies are no longer useful, but one can make use of particle beams to generate photons of a known flux and energy.

A circulating beam of monoenergetic electrons generates photons whose energy and spectrum can be calculated using standard equations for synchrotron radiation – the spectrum depends on the energy of the electrons and the curvature radius of the beam. For example, one of the beamlines at the Synchrotron Ultraviolet Radiation Facility at NIST[2] provides a calibrated light source with accuracy better than 1% over 200–400 nm (Arp et al. 2007).

A linear beam of monoenergetic electrons passing through any material substance will generate photons whose energy and spectrum can be calculated using standard equations for bremsstrahlung radiation – the spectrum depends on the energy of the electrons and the composition and thickness of the material. As an example, the LAT (Large Area Telescope) experiment on the Fermi Gamma-Ray Space Telescope was calibrated, in part, in this fashion using the CERN[3] Proton Synchrotron T9 test beam (Baldini et al. 2007). (For completeness, the energy of the electrons was not known ab initio but was determined by measuring their deflection angle in a magnetic field of known strength.)

2.1.4 Monoenergetic Photon Beams

All of the above calibration sources generate polychromatic beams of photons. Lasers, on the other hand, produce monochromatic beams – the only unknown is the absolute flux of photons. Such a flux can be calibrated by shining the laser into a temperature-controlled bath such as liquid helium; the amount of energy that needs to be extracted in order to maintain a constant temperature is a measure of the energy flux in the laser beam. The Laser Optimized Cryogenic Facility at NIST (Livigini 2003) implements this technique.

2.2 From Physics to Science Target

Conceptually, one would position a calibration source, such as a blackbody furnace, in an appropriate location so as to permit observing the source with one's telescope and detector in the same fashion as one observes a particular science target. In reality, one must rely on more complex schemes to transfer the calibration from the laboratory to the field. Space observatories usually rely on prelaunch characterization of their payloads, and, unless they carry their own calibration hardware systems, observations of celestial sources to verify the flux calibration and to monitor for changes in the instrument response. Ground-based systems have easier access to calibration sources but must contend with the transmission through the atmosphere. In practice, the transfer of calibration from a laboratory reference to a science target generally requires calibrating each component (or a combination of components) of an instrument/telescope system using multiple, redundant schemes, none of which by itself provides complete end–end calibration.

[2]NIST: The National Institute of Standards and Technology, an agency of the US Department of Commerce
[3]CERN: Conseil Eurepéen pour la Recherche Nucléaire, now the European Organization of Nuclear Research, a particle physics laboratory located in Switzerland

In spite of the range in technology used in going from radio wavelengths to gamma-ray energies, the calibration of astronomical telescope systems can often all be split into three pieces: the instrument/detector combination, the telescope, and (for ground observations) the atmosphere. One often calibrates each piece independently or (in favorable cases) the instrument and telescope in combination, with the atmosphere always being calibrated on its own.

When feasible, instrumentation should be calibrated using primary calibration facilities, for example, the Extreme UltraViolet Experiment (EUVE) satellite utilized NIST's Synchrotron Ultraviolet Radiation Facility (Sirk et al. 1997), but more often, we use secondary physics standards that are calibrated against, e.g., NIST-established standards examples include Si photodiodes, which can be calibrated to 0.2%, and NIR photodiodes, which can be calibrated to 0.5% (Larason and Houston 2008). The solar spectrum instruments, SOLSpec (on the Atlas 1 and Atlas-2 shuttle missions) and SOSP (on the Eureca satellite) used to measure the spectrum of the Sun, made use of onboard deuterium and ribbon-filament lamps as local calibrators (Thuillier et al. 2003). Thermal blackbodies require accurate thermometers to measure temperature, such as the Germanium Resistive Thermometers used by the FIRAS experiment on COBE (Fixsen 1994).

The calibration methods for radio systems are fairly mature (Findlay 1966; Baars et al. 1977). The instrument (receiver) is calibrated using accurately calibrated noise diodes, which are switched in and out of the signal chain during observing in order to track receiver gain variations. Antenna calibration is more problematic. If one is observing an extended source such as the CMB background, the effective area of the antenna depends only on wavelength and not size. For point sources, however, one additionally needs to know the antenna "gain." At long wavelengths, one typically uses simple dipole or horn antennas, for which the gain can be calculated from first principles to high accuracy. At higher wavelengths, one can make use of "artificial moons," which are blackbody disks of various temperatures placed in the far field of an antenna (Ulich et al. 1980). One cannot normally use astronomical sources as fundamental calibrators because most have nonthermal spectra with spectral shapes that are not known a priori.

In the submillimeter and infrared region, at wavelengths $\lambda > 10\ \mu\text{m}$, we can take advantage of many thermal astronomical sources that are bright enough to be used as approximate blackbody standards. For point sources such as stars, one calibrates the flux at a single wavelength and relies on the standard blackbody spectrum to extrapolate to other wavelengths. Depending on one's angular resolution, the planets and the Moon can be treated as either point sources or extended sources; for the latter case, one calibrates the temperature from observations at one or more wavelengths and again relies on the standard blackbody spectrum to extrapolate to other wavelengths. Mars is often used as a calibration source, even though its temperature varies as a function of position on the planet and time, since its distance from the Sun varies on its elliptical orbit (Wright 1976). The temperature scale is established using data from the Mariner 6 and 7 spacecraft, which in turn referenced their data to onboard thermal calibration plates whose temperatures in turn were measured using thermistors (Neugebauer et al. 1971).

In the near-infrared, optical, and ultraviolet regions, calibrations are once again often tied to blackbody references, but because blackbody furnaces operate at temperatures of order 1,000–2,600 K, the Wien cutoff of the blackbody spectral energy distribution becomes important, and temperatures of the reference sources must be determined with great care. Often lamps are used as intermediate reference sources to transfer the calibration from the blackbody furnace to the telescope. Usually, the reference source is observed directly, providing a calibration of the telescope and instrument combined. To the extent possible, atmospheric extinction is calibrated separately (see Sect. 4). Bright stars have been observed and directly calibrated against reference sources at out to approximately $\lambda = 1\,\mu m$ (Oke and Schild 1970; Hayes and Latham 1975) and indirectly, in various atmospheric windows, out to $\lambda = 10\,\mu m$ (Rieke et al. 1985). These measurements establish the absolute flux from a particular set of sources at particular wavelengths. One can then use a variety of sources with either theoretical or measured relative spectrophotometry to interpolate/extrapolate those fluxes to shorter and longer wavelengths. The use of white dwarfs for this purpose has been described in Sect. 2.1.2. The Sun is also used for this purpose. The solar spectrum has been measured numerous times and is now known, with varying degrees of accuracy, between 0.2 and 100 μm (Labs and Neckel 1968; Thuillier et al. 2003). Depending on the portion of the spectrum, observations have been made from the ground (including high-altitude observatories), aircraft, balloons, rockets, and spacecraft. The Sun is too bright to be compared directly with typical astronomical sources, so one often uses "solar analogs" as surrogates (Johnson 1965; Rieke et al. 2008). In this way, calibrations have been extended all the way to the ultraviolet (Morrissey et al. 2007).

Starting in the far ultraviolet, no sources exist to extend ground-based calibrations to shorter wavelengths. Since all observations must be made from rockets or spacecraft, on accurate prelaunch calibration of detectors and instruments entirely. As best as possible, these calibrations attempt to perform full-aperture calibrations with sources at infinite distance, but these conditions are difficult to achieve and often large corrections are required. At least in the X-ray region, multiple missions have observed the Crab Nebula repeatedly, providing some measure of cross-check (Kirsch et al. 2005).

3 Flux Standards

It is often impractical to frequently observe SI-traceable irradiance standards or recalibrate telescopes after they are deployed, so we are driven to use astronomical sources themselves as standards for calibration. In an ideal universe, we would have a spectrophotometric standards strategically located such that their emitted light goes through the optical system in exactly the same way as the target source. Further, the energy in any given passband would be known precisely and measured in SI units, e.g., watts, and, these standards would be available for all wavelengths and therefore, instruments. In reality, we rely on a set of flux standards, such as stars, whose spectral energy distribution we assume to be known and stable (i.e., NOT variable). In practice, flux standards only have a "known" spectral energy distributions in a limited wavelength range, and more often than not, turn out to be variable (on long time scales). That said, flux standards are nevertheless necessary for spectrophotometric measurements, as well as providing a means of tracking changes in the behavior of instruments.

Astrophysical flux standards are divided into two main categories: absolute or primary standards and secondary standards. In the first category are those sources which have been calibrated against SI-traceable calibrators; the second category consists of flux standards which are calibrated "in the field" against the primary standards, as per Landolt (1992) Absolute calibration can refer to both direct (e.g., comparison source is a true blackbody) and indirect (e.g., stellar atmosphere models) methods where the uncertainty in the measurement is limited solely by physics.

Historically, the primary standards are quite bright. An example is Vega with an apparent magnitude in V ~ 0. Principal reasons for this are:

1. It is easy to obtain very high signal to noise, SNR > 1,000, measurements with short exposures for bright sources.
2. The lower signal contrast between the flux from a 0 magnitude star and a furnace, which minimizes the number of optical elements needed, e.g., neutral density filters, to equalize the observed emissions.

Secondary standards are usually many orders fainter, most often to match a telescope's sensitivity and not saturate, but also to match the brightness of science targets.

3.1 Primary Flux Standards Across the Electromagnetic Spectrum

We define the primary flux standards to be sources with SI-traceable calibrations, namely, sources whose irradiance has been measured in physical units of energy per unit time per unit area per unit wavelength (or frequency).

Table 8-2 lists the most often used or referenced absolute flux standards, their identification, method of calibration, and the estimated current uncertainty in the calibration. *Table 8-3* lists the wavelength regime where the primary flux standards are used. Note that except for the gamma-ray region, more than one primary standard is available, and that overlap across wavelength regions is minimal, highlighting the fact that there is no single absolute flux standard in astronomy.

3.2 Absolute (Primary) Flux Standards in the UV, Visible, and Near Infrared

The general method for calibrating astronomical primary flux standards (a celestial source) is to make near simultaneous measurements of the target primary standard and an SI-traceable calibration source with the same instrumentation. In principle, this ensures that differences in the detected signals between the two sources are due to the differences in the emitted energy and not to changes in the behavior of the equipment. Thus, formally,

$$E(\lambda)_{\text{primary}} = \frac{S(\lambda)_{\text{primary}}}{S(\lambda)_{\text{SI-source}}} \times E(\lambda)_{\text{SI-source}}$$

where $E(\lambda)_{\text{primary}}$ and $E(\lambda)_{\text{SI-source}}$ are the emitted flux, $S(\lambda)_{\text{primary}}$ and $S(\lambda)_{\text{SI-source}}$ are the measured signals at wavelength, λ, of the celestial source and the SI standard, respectively. $E(\lambda)_{\text{SI-source}}$ is known. However, unless the celestial source and the SI irradiance standard are measured in exactly the same way, with the same total path, the above is a two (or more) step process. Namely,

$$\begin{aligned}
S(\lambda)_{\text{primary}} &= R(\lambda)_{\text{primary}} \times E(\lambda)_{\text{primary}} \\
S(\lambda)_{\text{SI-source}} &= R(\lambda)_{\text{SI-source}} \times E(\lambda)_{\text{SI-source}} \\
E(\lambda)_{\text{primary}} &= \frac{S(\lambda)_{\text{primary}}}{S(\lambda)_{\text{SI-source}}} \frac{R(\lambda)_{\text{SI-source}}}{R(\lambda)_{\text{primary}}} \times E(\lambda)_{\text{SI-source}}
\end{aligned}$$

where $R(\lambda)$ is the response of the system and can include not only the instrument response, but also, for example, the effects of observing through the Earth's atmosphere with different path lengths.

Table 8-2
Primary irradiance standards used in astronomy

Object	Calibration method		Uncertainty	References
Mars planet	Direct	Blackbody		Sinton and Strong (1960) Rieke et al. (1985)
	Direct	Transfer standard		Neugebauer et al. (1971) Weiland et al. (2011)
Sirius A1 V star	Direct	Blackbody, emissive spheres	<1%	Price et al. (2004)
	Indirect	Stellar atmosphere model		Cohen et al. (1992)
Sun G2 V star	Direct	Radiometers, pyrometers, lamps		Thuillier et al. (2003) and references therein
Vega A0 V star	Direct	Blackbodies, Lamps	~1–8%	Megessier (1995) and references therein
	Indirect	Infrared flux method		Leggett (1985)
CasA radio source	Direct	Transfer standard	1–3%	Baars et al. (1977)
Crab Nebula	Indirect	Power law model		Kirsch et al. (2005) and references therein
CygA radio source	Indirect	Transfer standard	~5%	Baars et al. (1977)
Virgo A radio source	Indirect	Transfer standard	~5%	Baars et al. (1977)
G191B2B DA WD	Indirect	Stellar Atmosphere Model	~1–2%	Bohlin et al. (1995)
Taurus A radio source	Indirect	Transfer standard	~5%	Baars et al. (1977)

Starting in the late 1940s, the wavelength range encompassed by observational astronomy expanded enormously from the visible – roughly the UV atmospheric cut off at ~0.25 μm to approximately 0.8 μm – to the X-ray, infrared and radio, and in the last decades of the twentieth century into the gamma ray. New technologies have been key, from the development of solid state detectors, e.g., photomultiplier tubes, CCDs (charge-coupled devices), and HgCdTe arrays, to the increase in computing power, as well as, of course, the investment in space observatories. Since observational astronomy has come to encompass almost the entire electromagnetic spectrum from the high-energy gamma rays to the meter-wavelength radio, the number of primary standards has necessarily increased (see *Table 8-2*). Sometime between the 1960s and the 1970s, Alpha Lyrae (Vega) became the de facto absolute flux standard in the UV-visible-infrared. More recently, Sirius A has been adopted as an absolute flux standard. With the advent

Table 8-3
Primary standards used in different wavelength regimes

Object	γ ray	X-ray	XUV	UV/VIS	NIR	MIR/FIR	mm	Radio
3C 273		X						
3C 295								X
CasA[a]								X
Crab Nebula	X	X						X
Cyg A								X
G 191 B2B[a]		X		X				
GD 153[a]			X		X			
GD 71[a]					X			
HX 43			X					
HZ 44		X		X				
Mars						X	X	
P330E[a]					X			
Sirius						X		
Taurus A								X
The sun[a]				X		X		
Uranus							X	
Vega				X	X	X		

[a]These stars are designated as the primary calibrators for an observatory/telescope. Their pedigrees trace back to an absolute standard, through one or more steps and usually rely on models; though they do not all have direct measurements against SI calibrated standards

of solar space missions, the Sun's flux has been measured against SI-traceable standards; it too can be used as a primary standard at wavelengths long ward of 1–2 μm. We now describe the current status of Vega, Sirius A, and the Sun as astronomical primary flux standards.

3.2.1 Vega

Alpha Lyrae, Vega, is one of the brightest stars in the night sky. An A0V star, it defines the zero-points of the popular Johnson-Kron-Cousins photometric system. Vega defines this photometric system such that an ideal A0V sets the zero-points of each bandpass which are, in magnitudes, B = V = R = I = J = H = K = 0. Colors, e.g., B-V, are also exactly 0. The real Vega has magnitudes in this system that differ from 0 by about ±0.1 mag, depending on the filter. Vega is the absolute flux standard for the Hubble Space Telescope instruments, for several of the Spitzer Infrared Telescope's instruments, and for the SDSS (Sloan Digital Sky Survey) photometric system. (Although the SDSS fundamental standard is BD 17 + 4708, it is calibrated against Vega.) As the quintessential primary flux standard, Vega has been an object of intense observational and theoretical studies, and for those reasons, we will discuss it in some detail.

Experiments to provide Vega with an SI-traceable calibration have been carried since the 1960s by Code (1960), Kharitonov (1963), Glushneva (1964), Willstrop (1960), Oke and Schild (1970), Hayes (1970), Hayes and Latham (1975), Tug et al. (1977), Terez and Terez (1979),

Table 8-4
Direct and indirect calibration experiments of Vega'sirradiance

Wavelength range (μm)	SI-traceable standard	References
<0.3	Indirect methods	Bohlin and Gilliland (2004)
0.3–1	Copper and platinum blackbodies Tungsten ribbon filament lamps	Code (1960), Kharitonov (1963), Glushneva (1964), Willstrop (1965), Oke and Schild (1970) Hayes (1970), Hayes et al. (1971), Hayes and Latham (1975), Tug et al. (1977) Terez and Terez (1979) Knyazeva and Kharitonov (1990)
1–5	Furnace (with transfer standard)	Selby et al. (1980, 1983) Blackwell et al. (1983)
1–5	Indirect: solar analog method	Campins et al. (1985)
>5	Indirect: infrared flux method	Leggett (1985)

Selby et al. (1980), Blackwell et al. (1983), Selby et al. (1983), Terez (1985), Mountain et al. (1985), Petford et al. (1985), Booth et al. (1989), Kharitonov et al. (1980), and, Knyazeva and Kharitonov (1990). These are listed in *Table 8-4*, and the measured values and uncertainties are given in *Table 8-5*. All of these have relied on an SI irradiance standard, whether a calibrated blackbody source, tungsten filament lamps or furnace, placed at an accurately measured distance from a telescope with which both Vega and the irradiance standard were observed.

For example, Hayes et al. (1971) measured Vega's flux between 0.68 and 1.080 μm different copper blackbodies. Kharitonov et al. (1980) and Knyazeva and Kharitonov (1990) obtained measurements between 0.320 μm and 0.750 μm calibrated against two ribbon filament tungsten lamps. Lockwood et al. (1992) used copper and platinum blackbodies to measure the absolute flux between 0.3295 and 0.9040 μm of Vega and 109Vir, while Oke and Schild (1970) calibrated Vega with three sources – a tungsten ribbon-filament lamp, a copper blackbody, and a platinum blackbody. Typical quoted errors are of order 2%.

Experiments by Selby et al. (1980), Blackwell et al. (1983), and Mountain et al. (1985) measured the absolute flux at infrared wavelengths between 1 and 5 μm against furnaces. In this range, the quoted uncertainty in the absolute flux calibration ranges between 3% and 8%, with higher uncertainty at the longer wavelengths.

For all of the ground-based irradiance experiments, the single largest source of uncertainty is atmospheric extinction. Other sources of error are due to metrology and geometry. There may be a small systematic uncertainty introduced by the initial SI-traceable calibration of the irradiance standards, as these were measured by at least four measurement standards laboratories in the United States, the United Kingdom, and the former Soviet Union.[4] Megessier (1995) carried out a meta-analysis of the absolute calibration experiments in the UV, visible and near infrared, determining the best value for Vega's monochromatic flux at 555 nm to be

[4]National Bureau of Standards in the United States, National Physics Laboratory in the United Kingdom, and All-Union Research Institute of Metrology and All-Union Research Institute of Optical and Physical Measurements in the former Soviet Union

Table 8-5
Irradiance values and the measured uncertainties for Vega as measured by the VARIOUS experiments listed in *Table 8-4*

Wavelength (μm)	Spectral irradiance ($W m^{-2} \mu m^{-1}$)	Uncertainty (%)	Reference
0.548	3.64×10^{-8}	2	Oke and Schild (1970)
0.554	3.48×10^{-8}	2	Tug et al. (1977)
0.5556	3.45×10^{-8}	2	Hayes et al. (1975)
0.5556	3.39×10^{-8}	2	Hayes and Latham (1975)
1.24	3.06×10^{-9}	3	Blackwell et al. (1983)
2.20	3.92×10^{-10}	4	Selby et al. (1983)
2.20	3.75×10^{-10}	8	Selby et al. (1980)
2.20	4.19×10^{-10}	3	Blackwell et al. (1983)
2.25	3.86×10^{-10}	5	Booth et al. (1989)
3.76	5.44×10^{-11}	3	Blackwell et al. (1983)
3.80	5.28×10^{-11}	4	Selby et al. (1983)
4.60	2.42×10^{-11}	12	Blackwell et al. (1983)
4.92	2.10×10^{-11}	7	Mountain et al. (1985)
10.6	9.4×10^{-13}	3	[a]Rieke et al. (1985)
21	6.27×10^{-14}	8	[a]Rieke et al. (1985)
12	5.84×10^{-13}	4	[b]Leggett (1985)
25	3.81×10^{-14}	6	[b]Leggett (1985)
60	5.97×10^{-15}	6	[b]Leggett (1985)
100	2.28×10^{-15}	10	[b]Leggett (1985)

[a]Solar analog method
[b]Infrared flux method

3.46×10^{-11} W m^{-2} nm^{-1}, with a 1-σ uncertainty = 0.7%. Equivalent analyses of the measurements at wavelengths beyond 1 μm show the 1-σ uncertainties are four times larger.

At wavelengths longer than 10 μm, Vega has been calibrated using indirect methods. Leggett (1985) provides flux values at 12, 25, 60, and 100 μm calculated from the infrared flux method developed by Blackwell and Shallis (1977) and Blackwell et al. (1979, 1980).

In recent years, discussion about whether Vega is a good absolute flux standard has been driven by the evidence that it is a pole-on rotator (cf. Graham et al. 2010) and has a dusty debris disk (Defrère et al. 2011; Müller et al. 2010; Selby et al. 1983). Principally, the effect of a pole-on rotator changes the interpretation of spectral features, and hence, the physics inputs into a model. The effect of a debris disk is twofold. The first is an excess of infrared flux, compared to the expected blackbody energy distribution, due to the emission characteristics of astrophysical dust.

The second is the wavelength-dependent size and structure of the star. Hanbury Brown et al. (1974) measured Vega's angular diameter to be 3.24 milliarcsec at 443 nm. Subsequently, Ciardi et al. (2000) measured the mean infrared (H and K) angular diameter of Vega to be 3.28 ± 0.06 milliarcsec; the dust disk is resolved (Su et al. 2005) at the wavelengths of MIPS

(Multiband Imaging Photometer) on the Spitzer Space Telescope. Consequently, traditional stellar atmosphere models of A stars may not be applicable to Vega at wavelengths longer than 5 μm. The implication is that extrapolation of the Vega calibration scale through models is not advisable. However, if the emitted flux of the Vega system is stable, then its status as a primary flux standard at those wavelengths that have SI-traceable measurements should not be affected. This has yet to be determined observationally.

3.2.2 Sirius

Alpha Canis Majoris, Sirius A, is an A1 V star, with V = −1.47 mag, almost four times brighter than Vega. Its binary companion, Sirius B, is a DA white dwarf, with V = 8.44 mag, a period of ~50 years, and currently has a separation of ~9.5 arcsec. In 2022, Sirius B will be at its largest angular separation from Sirius A.

Given the challenges of obtaining accurate flux densities by extrapolating the Vega calibration scale through the application of stellar atmosphere models, Sirius has been proposed as a better primary standard. The few Sirius calibration experiments are listed in ➋ *Table 8-6*. Although SI-traceable measurements of Sirius' flux are fewer, the assumption is that while similar to Vega in spectral type, it does not have infrared excess and is a stable star. Sirius does, however, have a faint white dwarf companion, Sirius B at a current separation of approximately 6 arcsec, which could affect low spatial resolution observations.

The only direct, SI-traceable experiment with published results that obtained an absolute flux of Sirius was the Midcourse Space Experiment (MSX) carried out by the US Department of Defense in 1996. Stellar flux measurements with MSX were calibrated against emissive reference spheres (Price et al. 2004), which had themselves been measured at the National Institute of Science and Technology in the United States before launch. ➋ *Table 8-7* lists the effective wavelength of MSX determined values for Sirius' irradiance as given in Price et al. (2004). The uncertainties in column 3 are those derived from the Price et al. (2004) global mean fits; column 4 is our calculation of the spectral irradiance at the effective wavelength of the MSX bandpass. DIRBE (Diffuse Infrared Background Experiment) instrument aboard COBE (Cosmic Background Explorer) obtained measurements of several bright stars, including Sirius (Burdick and Murdock 1997).

Like Vega, Sirius A's angular diameter is measured interferometrically. Early the first measurements were made by Hanbury Brown and Twiss (1958). The current value is 6.041 ± 0.017 milliarcsec (Davis et al. 2011). Armed with this datum and the observed flux, the emergent flux can be calculated from about 1390 smaller:

$$F_e = 4F_{obs}/\theta^2$$

Table 8-6
Calibration experiments for Sirius

Wavelength range (μm)	Method	References
4–22	SI-traceable – MSX	Price et al. (2004) Burdick and Murdock (1997)
1–5	Transfer standard – COBE/DIRBE	Burdick and Murdock (1997)
0.5–1.0	Vega HST/STIS	Bohlin and Gilliland (2004)

Table 8-7
Irradiance values and the measured uncertainties for Sirius

Wavelength (μm)	Irradiance (W cm^{-2})	Uncertainty (%)	Spectral Irradiance (W m^{-2}μm^{-1})	Reference
4.29	1.146×10^{-15}	0.8	1.102E−1	Band B1; Price et al. (2004)
4.35	1.874×10^{-15}	0.6	1.047E−1	Band B2; Price et al. (2004)
8.28	2.833×10^{-15}	0. 2	8.432E−1	Band A; Price et al. (2004)
12.13	3.195×10^{-16}	0.3	1.858E−1	Band C; Price et al. (2004)
14.65	1.959×10^{-16}	0.1	8.785E−1	Band D; Price et al. (2004)
21.34	1.22×10^{-16}	0.8	1.955E−1	Band E; Price et al. (2004)

where F_e is the emergent flux at the stellar surface, F_{obs} is the flux observed at Earth, and θ is the limb-darkened corrected angular diameter. A stellar atmosphere model for the star can then be constructed and subsequently utilized to extrapolate the calibration to other wavelengths. In the case of Sirius A, the atmosphere model appears to be that of a well-behaved AV star. Hence, it is desirable to undertake direct measurements of Sirius' irradiance in the near ultraviolet, visible, and near infrared to complement the MSX data.

3.2.3 The Sun

Not surprisingly, the Sun has been and continues to be the most well-observed star at all wavelengths. The solar spectral and total irradiance has been directly measured against SI-traceable standards in the ultraviolet, visible, and infrared by several space missions including UARS (Upper Atmosphere Research Satellite), ATLAS (Atmospheric Laboratory for Applications and Science) on the Space Shuttle (1992, 1993, 1994), EURECA (European Retrieval Carrier, 1992–1993), and SORCE (Solar Radiation and Climate Experiment, 2003-present). Ground-based experiments are those of Neckel and Labs (1984), Lockwood et al. (1992), and Burlov-Vasiljev et al. (1995). Arvesen et al. (1969) carried out observations from airplanes. Thuillier et al. (1997, 1998a, b, 2003) discuss the pre-2000 results of the experiments in the context of the observations and solar models. Formal uncertainties in the visible and infrared ATLAS data are between 1.2% and 2.2% (Thuillier et al. 2003) and in the UV are 2–4%. Formal uncertainties in the SORCE data are less than 0.1% (Rottman 2006). Colina et al. (1996) present an absolute flux calibrated spectrum of the Sun between 0.12 and 2.5 μm with a maximum uncertainty of 5% between 0.4 and 2.5 μm, and up to 20% below 0.4 μm. Solar variability is highest at short wavelengths. Long ward of ~800 nm the solar variability is less than 1%, irrespective of phase in the solar cycle (Rottman 2006). (see *Table 8-8*)

Because the Sun is a variable star, with significant solar "weather," detailed physical models are less accurate than the solar flux measurements. Solar models are less accurate precisely where the variability is greatest – at high energies. However, in the near infrared and beyond, the Sun can be used as an absolute calibrator. The solar analog method for calibration, where the flux of GV stars is averaged together and compared to the Sun, works quite well, with errors within 2% (cf. Rieke et al. 2008) because at longer wavelengths the variability is small.

Table 8-8
Some recent direct calibration experiments for the sun

Mission	Experiment	Wavelength	Method	Accuracy absolute/ relative	References
ATLAS	SUSIM	120–410 nm	Direct: NIST/SURF	6%/2%	Atlas Website
	SOLSPEC	120–300 nm	Direct: NIST/SURF and lamps	3%/0.25%	Thuillier et al. (2003)
	ACR and SOLCON	Total irradiance monitor	Direct	0.1%/<0.06%	Crommelynck et al. (1996)
SORCE	SIM	300–2,400 nm	Direct: NIST SIRCUS, and electric substitution radiometer	2%/0.03% per year	Rottman (2006) SORCE website
	TIM	Total irradiance monitor	Direct: NIST and Electrical Substitution Radiometer	0.035%/0.0001% per year	
	SOLSTICE	115–320 nm	Direct: NIST/SURF Indirect: stable stars	<5%/0.5% per year	
	XPS	0.1–34 nm	Direct: NIST/SURF and reference photometer	12–24%/1% per year	
EURECA	SOSP	180–3,200 nm	Direct: blackbody	3%	Thuillier et al. (2003)
UARS	ACRIM	Total irradiance monitor	Direct: radiometer	?/0.0005%	UARS website
	SUSIM	115–410 nm	Direct: NIST/SURF	1–10%/2%	Brueckner et al. (1993)
	SOLSTICE	120–300 nm	Direct: NIST/SURF Indirect: stable stars	5–10%/2%	Woods et al. (1996)

3.2.4 White Dwarfs

In the ultraviolet, white dwarfs (and their models) have been adopted as primary standards. Kruk et al. (1997) and Kruk (1997) outline the main difficulties of obtaining accurate absolute flux calibration of UV instruments, among which are the limitations of in situ laboratory measurements and the challenges of obtaining reliable measurements above the atmosphere. Observations of hot stars with sounding rockets (Strongylis and Bohlin 1979), satellites (e.g., Code and Meade 1979), Apollo 17, and the Voyager missions (Holberg et al. 1982) led to the establishment of a common absolute scale, and the development of the white dwarf models (cf. Koester 2010; Lanz et al. 1996) and subsequently the establishment of the hot DA white dwarfs, GD 153, GD 71, HZ 43, and G 191B2B (BD + 52 913), as primary standards (Bohlin et al. 1995). We note a caveat: the uncertainties shortward of ~1,200 Å are several times greater than the published uncertainties at longer wavelengths.

In ➋ *Fig. 8-1*, we show the spectra of the principal primary standards discussed in this section. The Sun is the brightest. There is a ten-order magnitude difference in brightness between it and the next set of standards, Vega and Sirius, represented by Alpha Lyrae in the figure. Approximately 10^5–10^6 times fainter are the white dwarfs. We also show a solar analog, P330E (GSC 02581–02323), and two late-type stars, LDS 749B and VB 8 that are used as transfer standards. Note that all spectra follow Raleigh-Jeans law long ward of ~2 μm, irrespective of temperature and spectral type.

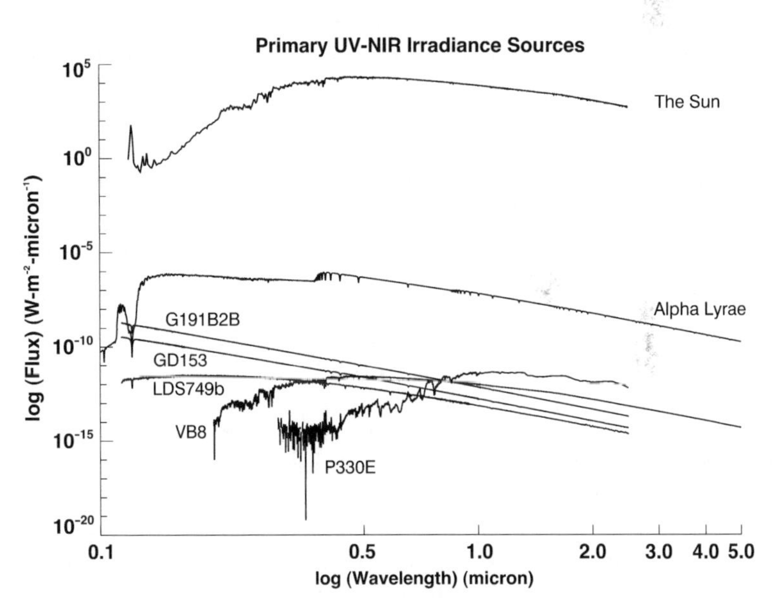

Fig. 8-1

Primary stellar irradiance standards in the ultraviolet through infrared. Note the difference in brightness between the brightest standards, Alpha Lyrae and the Sun, the more commonly used secondary standards. Also note that beyond 2 μm stellar spectral energy distribution follows the Rayleigh-Jean law, irrespective of spectral type (temperature). Spectral data are from the HST/CALSPEC database

3.3 Primary Flux Standards in the Gamma Ray, X-Ray, Submillimeter, and Radio

3.3.1 Gamma Ray and X-Ray Standards

Gamma ray and X-ray astronomy are primarily carried out in space, as the Earth's atmosphere is an effective shield against high-energy photons. As is usual, instruments are calibrated on the ground before launch to characterize the behavior of detectors, electronics, and the energy response of the total optical path. Once deployed, onboard calibration hardware is used in conjunction with celestial sources for spectral and flux calibration and to check for drift in the instrument response. Hardware, particularly detectors, "age" in the harsh, extraterrestrial environment, whereas the primary celestial sources are presumed to be stable. Space missions Swift, Integral, BeppoSAX, Exosat, MIR XEXE, and RXTE use or used the Crab[5] as a primary calibrator (Kirsch et al. 2005). Fermi relies on its onboard systems for calibration (Abdo et al. 2009). At lower X-ray energies, a white dwarf like HZ 44 is also used as a primary flux standard, as does the Chandra X-ray Observatory.

The Crab's flux output consists of pulsed emission from the pulsar and diffuse emission from the surrounding nebula. The integrated flux of the Crab from the very high energies to the radio region is shown in ➲ *Fig. 8-2*. The diffuse emission is, however, variable (Greiveldinger and Aschenbach 1999; Ling and Wheaton 2003) by as much as 20% near 1 keV (see Fig. 7 of Kirsch et al. 2005), though Kirsch et al. (2005) attribute the deviations to poor corrections. The Crab's pulsar's X-ray flux also exhibits (regular) variation in its X-ray emission (Weisskopf et al. 2004). Wilson-Hodge et al. (2011) find that the Crab's X-ray and gamma-ray flux has declined by as much as 7% in 4 years, as reproduced in ➲ *Fig. 8-2*. It is therefore not an ideal calibration source, though it is not obvious which celestial sources can serve as absolute flux standards at these high-energy ranges as most objects with sufficient flux at high energies are variable (➲ *Fig. 8-3*).

At the lower X-ray energies, HZ 44 remains a good flux calibrator. Its SI heritage is through stellar atmosphere models developed from suborbital observations of white dwarfs (Kruk 1997) and ultimately tied to the Vega photometric system through Landolt's (1992) measurements. HZ 43's emission is not known to have temporal variation.

3.3.2 Millimeter and Radio

Absolute flux calibration at radio frequencies faces similar challenges to that in the visible, as discussed in ➲ Sect. 2. Additional challenges are resolved sources, as occurs with the larger arrays; sources that are too bright for the instrument (e.g., Baars et al. 1977) a common problem for new telescopes that are both larger and more sensitive than their predecessors; and sources that are polarized. Obtaining 1% accuracy in flux calibration is as technically challenging at these long wavelengths (e.g, Yen et al. 1998), as it is in the uv-o-IR.

Cas A is the primary flux standard in the radio, despite its known frequency-dependent secular decrease in flux density. Additional primary flux standards are Cyg A, Tau A (Crab Nebula), 3C 58, and 3C 274 (Virgo A), which have uncertainties between 1% and 3% (ALMA calibration webpages). Spectral energy distributions for these sources are shown in ➲ *Fig. 8-4*.

[5] A common flux unit is the crab, equivalent to $2.4 \times 10^{-11}\ \mathrm{W\,m^{-2}}$ between 2 and 10 keV.

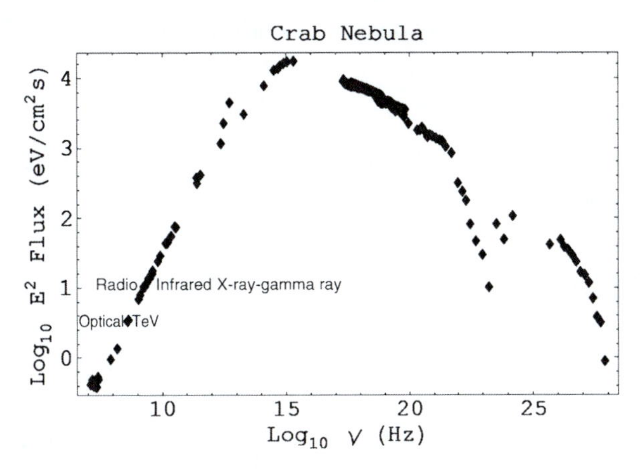

Fig. 8-2

The integrated flux of the Crab from the X-ray to the radio (Reproduced Fig. 16.4 from Kirk et al. (2009). Radio data from Baars et al. (1977), infrared from Green et al. (2004), optical from Veron-Cetty and Woltjer (1993), and X-ray through gamma-ray (EGRET, COMPTEL and BeppoSAX) from Kuiper et al. (2001) and TeV data points (>1,025 Hz) from H.E.S.S. observations Aharonian et al. (2006))

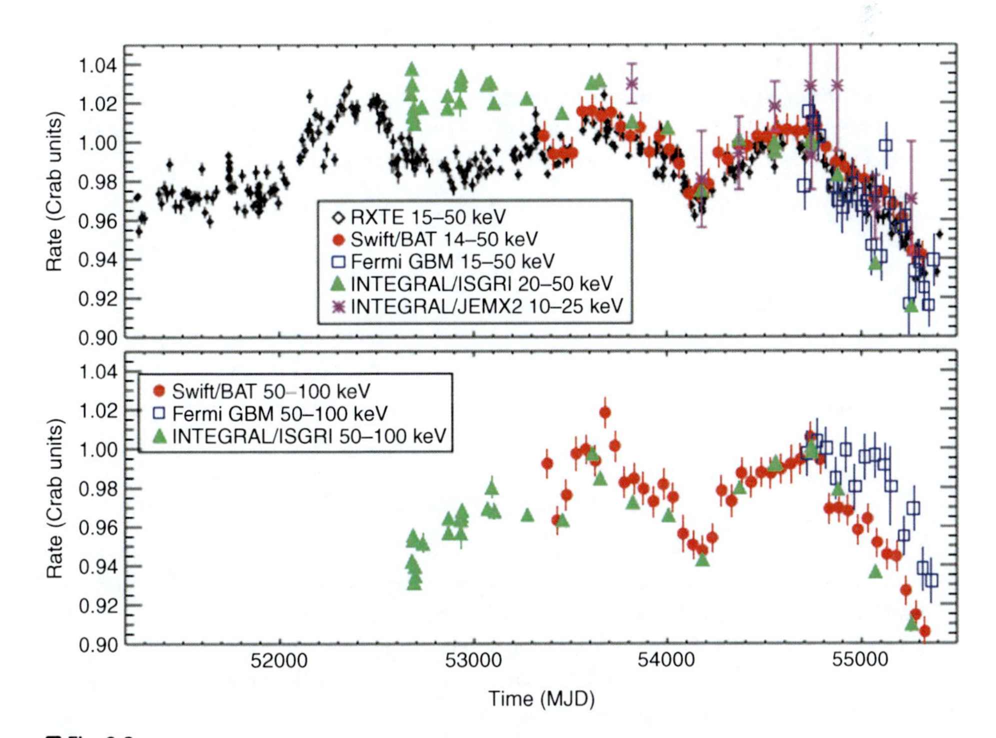

Fig. 8-3

Variation in the flux from the Crab Nebula over 24 months (2008–2010) measured by Swift, Fermi, RXTE, and Integral (Fig. 5 of Wilson-Hodge et al. 2011)

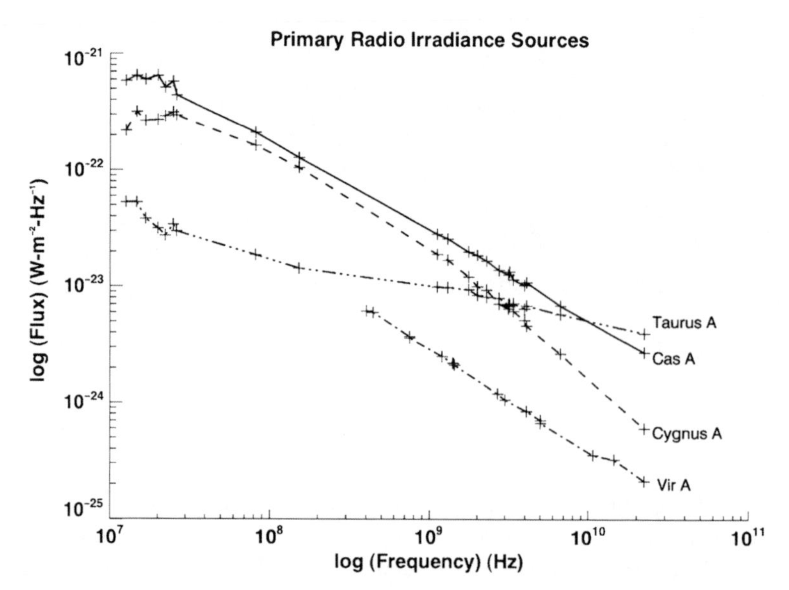

Fig. 8-4
The primary radio irradiances. Plotted is the flux for Taurus A, Cas A, Cygnus A, and Virgo A using data from Baars et al. (1977)

At millimeter and submillimeter wavelengths, these sources are resolved; hence, solar system objects are the primary flux standards. Mars and Uranus are primary calibrators, with uncertainties in the flux densities <3% (cf. Weiland et al. 2011, and ALMA calibration webpages).

3.4 Secondary Flux Standards

Secondary standards generally have higher uncertainties in their flux determinations due to systematic errors (e.g., extinction by astrophysical dust, different spectral types, linearity effects) as well as to specific calibration methods or incomplete knowledge (e.g., through synthetic photometry). By and large, most observational work relies on secondary standards for a variety of reasons:

1. Accessibility: primary standards are few in number so established networks of secondary standards are distributed around the sky, often near the celestial equator where they are accessible throughout the year and to most observatories.
2. Brightness: secondary standards are often much fainter than the primary standards and therefore accessible with large telescopes.
3. Variety: particularly in the ultraviolet and visible portion of the electromagnetic spectrum where having secondary standard stars of different spectral types, i.e., colors, is useful to match target colors and thereby minimize color effects.

Secondary standards are calibrated against the primary flux standards either directly or with models. In the former, observations are taken of the primary and the secondary with the same telescope/instrument. The latter case is used when the primary is too bright or when the wavelength regime is different. A model is fit to the observed spectral energy distribution of the primary (which can be used to predict the flux at additional, unobserved wavelength regions). To set the scale, the observed energy distribution of the secondary is also fit with a model, which is then compared at a common wavelength with the primary standard observation. Compilations of secondary standards across the electromagnetic spectrum are readily available. In the visible, the most commonly used are the extensive catalogs of Landolt (1983, 1992) and in the infrared those of Hawarden et al. (2001) and Elias et al. (1982). Catalogs of standard stars serve an additional purpose as well, in that they provide a means of transforming from one instrument's photometric system to another, e.g., Elias et al. (1983). Further discussions of photometric systems and color transformations are available in the literature; cf. Bessell (2005).

4 The Earth's Atmosphere

The atmosphere blocks over half of the electromagnetic spectrum between the radio and gamma-ray regions, and for those portions, observations can only be made from space. Of the remainder, essentially all are affected at some level by the atmosphere (➋ *Fig. 8-5*). Thus, most ground-based observations must be corrected in some way for atmospheric effects.

At frequencies below 30 MHz, the Earth's ionosphere absorbs essentially all radio waves. Although not strictly relevant for calibration, the ionosphere causes scintillation of sources up to frequencies of about 1.5 GHz. At higher frequencies, absorption by the atmosphere becomes increasingly important. Water vapor introduces a spike in opacity at about 20 GHz. A complex of oxygen lines around 60 GHz blocks out a window on either side of about 8 GHz. Above this frequency, the dominant opacity is due to absorption by oxygen, water vapor, and hydrosols (suspended water droplets) (Liebe 1985), and the dryness of a site is all-important. Additional blockages occur around 120 GHz, 180 GHz, 320 GHz, and so on. Between about 450 and 900 GHz (700–300 μm), only a few windows admit any signal at all. After that (300 to ~25–30 μm wavelength), the atmosphere is completely opaque (Traub and Stier 1976). Windows begin opening up again at shorter wavelengths. The main absorbers at wavelengths longer than about 14 μm are CO_2, CH_4, and H_2O. Between 8 and 14 μm, ozone is also a contributor. Between 1 and 8 μm, the main absorbers are H_2O, CO_2, and NH_3. At wavelengths shorter than 2 μm, the atmosphere becomes increasingly more transparent, albeit with a few more blockages from water vapor down to 0.9 μm (Lord 1992). At wavelengths shorter than 0.9 μm, the dominant sources of opacity become oxygen (A Band 0.7615 μm; B Band 0.6875 μm), aerosols, Rayleigh scattering, and ozone the atmosphere becomes opaque once again at wavelengths shorter than ~0.3 μm.

4.1 Calculating Opacity

The impact of the atmospheric opacity affects calibration of astronomical data in two ways. First, the overall flux recorded by a detector decreases. Second, if the opacity varies in strength across a bandpass, the effective wavelength of the bandpass changes. In optical photometry,

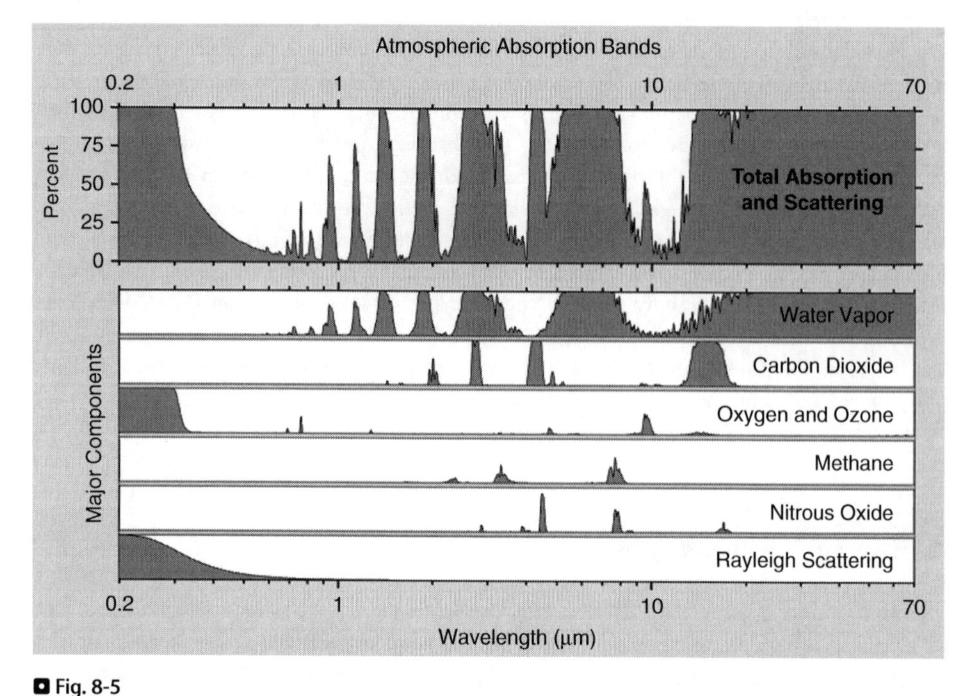

Fig. 8-5

The *upper panel* shows the transmission windows of the Earth's atmosphere (in *white*), and the major components of absorption and scattering in the *bottom panel*. This image is created by Robert A. Rohde, Global Warming Art (http://www.globalwarmingart.com/wiki/File:Atmospheric_Transmission_png), who use date from the Spectral Calculator of GATS, Inc. which implements the LINEPAK system of calculating absorption spectra (Gordley et al. 1994) from the HITRAN2004 (Rothman et al. 2005) spectroscopic database

the latter effect is sometimes called "second-order extinction," since the correction to a photometric measurement depends on both the amount of extinction and on the spectral energy distribution (i.e., color) of an object. In a broadband filter such as Johnson B, red objects, e.g., suffer less extinction than blue objects.

Opacity at a given wavelength can be expressed as an optical depth τ_ν such that the transmission through the atmosphere is given as $T_\nu = e^{\tau\nu}$. For an absorber whose opacity is constant with wavelength across a bandpass, the total opacity scales linearly with the integrated number density of absorbers along a particular path through the atmosphere: $\tau \propto \int \rho ds$, where ρ is the density of absorber and the integral is along a path through the atmosphere. One often expresses the integral in terms of "air mass" X, which is the integral of the total atmospheric density along some path relative to that at zenith. The proportionality constant needed to convert X to an extinction will thus be dependent on the location of an observatory, in particular its altitude. One often treats the atmosphere as a flat slab – the plane-parallel approximation. In this approximation, the distance of a star from the zenith (Z; the zenith angle) is determined, and the air mass of the observation is given as $X = sec(Z)$. At high air masses (Z > 60°), a more accurate expression that accounts for the curvature of the Earth (Hardie 1962) is preferred.

For portions of the spectrum where the opacity changes rapidly with wavelength across a bandpass, this simple treatment breaks down. Many of the strong features in the absorption spectrum of the atmosphere, e.g., the A- Band due to oxygen, are actually composed of a large number of narrow lines arising from electronic transitions in the oxygen molecule (including additional lines for molecules that incorporate different isotopes) (Wark and Mercer 1965). The overall opacity (integrated over the line profile as a function of wavelength) requires a standard curve-of-growth type analysis. For weak lines, the integrated opacity follows the opacity law given above, being linearly proportional to the integrated number of absorbers. For strong lines, the damping wings of the line profile dominate the opacity, and the transmission across the line profile is either *all* or *nothing*, with the width of the line (where *nothing* is transmitted) being proportional to the square root of the integrated number of absorbers. Thus, the mean opacity through a broadband will depend on X according to some law between X and $X^{1/2}$.

4.2 Measurement of Opacity

The most straightforward way to handle absorption is to observe a known calibration source that is near a science target such that the atmosphere affects both sources equally. If second-order corrections are required, they are usually small enough in magnitude that one needs only an approximate estimate of the extinction coefficient in order to compute the correction.

In the simplest case where the opacity varies linearly with X and the absorption constituents within the atmosphere are stable with time, one can observe a reference object (which need not have been calibrated a priori) multiple times and at different air masses during a given night, determine the slope (extinction coefficient) of the extinction-air mass relation, and use this calibration to correct all observations back to their zero-air mass values that would be measured above the atmosphere. This simple model is often sufficient for doing relative photometry, where one compares science targets with existing calibration sources, particularly if one accounts for an extinction coefficient that varies during a night.

For more accurate photometry, however, one needs to apply a more sophisticated model of the atmosphere and make measurements that are sensitive to the time-variable components of the atmosphere that are contributing to extinction at the wavelengths of interest.

4.3 Atmospheric Models

Several programs that model atmospheric radiation transfer are currently in use or under development. One of the big issues is few of them are reliable across the entire spectral range of interest in astronomy. The model with the widest astronomical application is the Moderate Resolution Transmittance (MODTRAN) code, which is useable for the region from UV to microwave (Berk et al. 1998). However, even this code has limitations in the UV region of the spectrum. Nevertheless, because of its versatility, MODTRAN is the default model of choice for high precision atmosphere fitting requirements. Another program that works (mostly) across the entire astronomical spectral range is RRTM (Mlawer et al. 1997), a Rapid Radiative Transfer Model. This code is based upon the Discrete Ordinates Radiative Transfer Program (DISORT; Stamnes et al. 1988) that uses a plane-parallel approximation of the atmosphere, and therefore it does not offer the high precision required for many astronomy applications.

The current version of the MODTRAN code in use for component analysis of the atmosphere is MODTRAN-4 (v3r1), developed by the Air Force Research Laboratories (AFRL).

This state-of-the-art atmospheric band radiation transport code calculates atmospheric transmittance and radiance for frequencies from to 50,000 cm^{-1} ($\lambda > 0.2\ \mu$m) at moderate spectral resolution and has been available (for a fee) to the public since 2000 (Berk et al. 1999; Anderson et al. 2001). The LSST project (Large Synoptic Survey Telescope) is using this code to develop the atmospheric correction techniques by fitting MODTRAN-4 components stellar spectra to examine portions of the Earth's atmosphere. The regions of fitting include absorption and scattering (Rayleigh), telluric components (O_2, O_3, water vapor, trace elements), Mie scattering by macroscopic aerosols, and intra-cloud shadowing due to ice crystals and water droplets. The initial efforts show the atmospheric effects on astronomical observations can be determined (and corrected) to a few millimagnitudes across the optical-NIR portion of the spectrum (Burke et al. 2010). AFRL are currently using MODTRAN-5 which supports increased spectral resolution and increased trace element components, and allows for treatment of spectral line wings as well as varying line strength distributions.

For some observational programs, other modeling programs may be more useful. For example, the Harvard-Smithsonian Center for Astrophysics has developed "am," a tool for radiative transfer computations at microwave to submillimeter wavelengths (Paine 2011). The primary application for this code is radio astronomy, and the spectra computed with it can include thermal emission, absorption, transmission, and excess delay in signal propagation.

4.4 Time-Dependent Components

In certain situations, it is advantageous to utilize dedicated instruments to monitor the atmospheric transmission or its constituent components in order to track the time-dependent components that contribute to the overall opacity.

Water vapor is the most pernicious single component of the atmosphere, affecting observations from millimeter wavelengths through the optical bands. Because it is highly time-variable, special efforts are made to monitor its column density. In general, one's goal is to measure the amount of *precipitable water vapor* along a line of sight. One can do so by either measuring thermal emission in some particularly strong water vapor line or (in the optical, at least) use narrow band photometry to measure some background source in bands that are centered on and offset from a particular water vapor absorption feature. The Atacama Large Millimeter Array uses the former method. A radiometer mounted in the focal plane monitors water vapor emission in the 183 GHz line to provide a measure of the total precipitable water vapor along a particular line of sight. (In addition to absorption, water vapor introduces delays in signal propagation that affect coherence critical for interferometer operation; Hills 2010.) SOFIA (Stratospheric Observatory for Infrared Astronomy) carries a similar type radiometer onboard (Cooper et al. 2003). A device was constructed to use the second method (Hansen and Caimanque 1975) at the Cerro Tololo Interamerican Observatory in Chile to monitor the water vapor absorption feature at 1.87 μm. More recently, GPS receivers are being used to monitor water vapor (Blake and Shaw 2011).

Some optical surveys have used a dedicated telescope to monitor transmission. For example, the SDSS used a separate telescope to monitor standard stars in a set of broadband filters that matched the main survey bandpasses (Tucker et al. 2006). The LSST survey (Ivezic et al. 2008) is planning to go further and monitor a set of stars spectroscopically and use the data as input to a MODTRAN model (Stubbs et al. 2007; Burke et al. 2010).

5 Validation of the Calibration Process

It is one thing to perform a calibration of an astronomical source to a stated level of precision, but quite another to verify that the accuracy has, indeed, been reached.

The uncertainty in a particular observational experiment is often due to the combination of a large number of possible errors, both statistical and systematic. An example of an "error tree" for an experiment might include uncertainties in the:

- SI measurement of a transfer standard
- Aperture size and shape of an integrating sphere
- Distance between, e.g., sphere and telescope
- Temperature of a blackbody
- Atmospheric transmission
- Temperature of detector
- Wavelength measurement or filter transmission curve

As well as errors due to:

- On or off – axis misalignments of optical elements
- Detector characterization: linearity, dark current, and gain
- Geometric distortion

The level of detail on the list or how many elements are required is highly dependent on the level of accuracy required; the more demanding the requirement, the more detailed an error tree is needed.

However, even the most careful error tree analysis is seldom complete or exact. No detector can ever be characterized perfectly – time, cost, and technology all impose limits. A filter bandpass as measured in the lab might be different from what one would measure at the telescope because of temperature and/or humidity changes. The horizontal and vertical distribution of water vapor in the atmosphere can never be measured continuously in time. An effect may not even be known to exist before a spacecraft is launched. A good example is the count-rate nonlinearity found in spectra obtained with the HST NICMOS (Near Infrared Camera and Multi-Object Spectrometer), showing a discrepancy such that bright stars (high count rate) were too bright, but fainter stars with lower count rates were too faint relative to spectra obtained with the HST Space Telescope Imaging Spectrograph (STIS) (Bohlin et al. 2006). As a consequence, measuring the "error in the error" is an important part of the calibration process.

Furthermore, errors inevitably increase when the calibration process involves multiple steps. For example, the South Pole Telescope (SPT) flux calibration comes from observations of the H II region RWC 38. This source was calibrated against the WMAP5 survey. That survey, in turn, was calibrated by observing the cosmic microwave background (CMB) dipole anisotropy. The dipole was measured by the DMR (Differential Microwave Radiometer) on the COBE (Cosmic Background Explorer) satellite. The DMR, finally, was calibrated before launch using blackbody references in a test chamber; the blackbody temperatures were measured by a thermocouple (Bennet et al. 1992). In traveling from the laboratory to the SPT, the calibration accuracy degraded from a fraction of a percent to about 10% (Staniszewski et al. 2009).

As a consequence, accurate calibration of astronomical sources benefits from having multiple, independent measurements that serve to validate the correctness of any single one. For multistage calibrations, independent measurements that serve to validate even a subset of the process are valuable and important. There is no hard and fast rule for determining how many independent measurements are needed to achieve any particular level of accuracy; an error-tree type analysis can help but may not be sufficient. Furthermore, sometimes new methods are developed "on-the-fly" as a particular experiment runs. We can find many instances, however, where two measurement methods were indispensible, and three were preferable. For example, the COBE mission eventually developed three independent calibration methods to measure the absolute temperature of the cosmic microwave background (Mather et al. 1999). The MSX (Midcourse Space Experiment) Mission used three independent methods to calibrate its data (Price et al. 2004). Sometimes even three is not enough – a comparison of measurements from multiple instruments onboard three X-ray satellites shows that the temperatures measured for the gas in a set of galaxy clusters observed in common agree to within 1% when using 2–7 keV data but disagree by an order of 10% when using 0.5–2 keV data; conversely, the fluxes agree in the low-energy band but disagree in the high-energy band (Nevalainen et al. 2010).

Glossary

ALMA Atacama Large Millimeter Array, international Radio array in Chile
BeppoSAX Italian-Dutch X-ray Satellite
Chandra Chandra Xray Observatory
CMB Cosmic Microwave Background
COBE Cosmic Background Explorer, a NASA satellite that carried DIRBE, DMR and FIRAS
CTIO Cerro Tololo Interamerican Observatory, in northern Chile and operated by NOAO.
DIRBE Diffuse Infrared Background Experiment, instrument on COBE
DMR Differential Microwave Radiometer, instrument on COBE
EXOSAT European X-ray Observatory
FIRAS Far Infrared Absolute Spectrophotometer, instrument on COBE
GPS Global Positioning System, a collection of satellites in Earth orbit used to provide accurate position and time information.
HST Hubble Space Telescope
Integral International Gamma-Ray Astrophysics Laboratory, gamma ray imager and spectrometer, and X-ray and visible imagers
LSST Large Synoptic Survey Telescope, a future 8 m telescope in Chile
MIR HEXE High Energy X-ray Experiment, a German payload on the Mir space station
MSX Mid Course Space Experiment
NICMOS Near Infrared Camera and Multiobject Spectrograph, an HST instrument
NIST National Institute of Standards and Technology, United States
NOAO National Optical Astronomy Observatory, United States
RXTE Rossi X-ray Timing Explorer
SDSS Sloan Digital Sky Survey
SOFIA Stratospheric Observatory for Infrared Astronomy, carried on a Boeing 747 airplane
SORCE Solar Radiation and Climate Experiment, carries SIM, SOLSTICE, TIM and XPS
Spitzer Spitzer Infrared Space Telescope
SPT South Pole Telescope
STIS Space Telescope Imaging Spectrograph, an HST instrument
SURF a calibration facility at NIST
Swift Swift Gamma-Ray Burst Mission, gamma ray, X-ray and ultraviolet instruments
WMAP Wilkinson Microwave Anisotropy Probe
SIM Spectral Irradiance Monitor, on SORCE

SOLSTICE Solar Stellar Irradiance Comparison Experiment on UARS
TIM Total Irradiance Monitor, on SORCE
XPS XUV Photometer System, on SORCE
EURECA European Retrievable Carrier, carried SOSP
SOSP Solar Spectrum spectrometer, on EURECA
UARS Upper Atmosphere Research Satellite, carried SOLCON, ACRIM, SUSIM, SOLSPEC
ATLAS-1 Atmospheric Laboratory for Applications and Science on space shuttle Atlantis
SOLCON Measurement of the Solar Constant
ACRIM Active Cavity Radiometer Irradiance Monitor on UARS, ATLAS-1
SOLSPEC Solar specrum Measurement
SUSIM Solar Ultraviolet Spectral Irradiance Monitor on UARS, ATLAS-1

References

Abdo, A. A., et al. 2009, Astroparticle Physics, Vol 32, Issue 3–4, pp 193–219, SLAC-PUB-14765, Corresponding Author: E. do Couto e Silva

Aharonian, F., Akhperjanian, A. G., Bazer-Bachi, A. R., et al. 2006, A&A, 457, 899

Albrecht, A., et al. 2006, arXiv: astro-ph/0609591

Anderson, G. P., et al. 2001, Proc SPIE, 4381, 455

Arp, U., et al. 2007, Appl Opt, 46, 25

Arvesen, J. C., et al. 1969, Appl Opt, 8, 2215

Baars, J. W. M., Genzel, R., Pauliny-Toth, I. I. K & Witzel, A. 1977, A&A, 61, 99

Baldini, L. et al. 2007, AIP Conf Proc, 921, 190

Bennet, C., et al. 1992, ApJ, 391, 466

Berk, A., Bernstein, L. S., Anderson, G. P., Acharya, P. K., Robertson, D. C., Chetwynd, J. H., & Adler-Golden, S. M., 1998, Remote Sens Environ, 65, 367–375

Berk, A., et al. 1999, Proc SPIE, 3756, 348

Bessell, M. S. 2005, ARAA 43, 293

Blackwell, D. E., & Shallis, M. J. 1977, MNRAS, 180, 177

Blackwell, D. E., Shallis, M. J., & Selby, M. J. 1979, MNRAS, 188, 847

Blackwell, D. E., Petford, A. D., & Shallis, M. J. 1980, A&A, 82, 24

Blackwell, D. E., Leggett, S. K., Petford, A. D., Mountain, C. M., & Selby, M. J., 1983, MNRAS, 205, 897

Blake, C. H., & Shaw, M. M. 2011, PASP, 123, 1302

Bohlin, R. 2010

Bohlin, R. 2007, ASP Conf Ser, 364, 315

Bohlin, R. C., & Gilliland, R. 2004, AJ, 127, 3508

Bohlin, R. C., Colina, L., & Finley, D. S. 1995, AJ, 110, 1316

Bohlin, R. C., Riess, A., & De Jong, R. 2006, ISR NICMOS 2006-002

Booth, A. J., Selby, M. J., Blackwell, D. E., Petford, A. D., & Arribas, S. 1989, A&A, 218, 167

Brueckner, G. E., Edlow, K. L., Floyd, L. E., Lean, J. L., & VanHoosier, M. E., 1993, JGR, 98, 695

Burdick, V., & Murdock, T. L, 1997, Technical report, General Research Corp. Danvers, MA United States. COBE Final Report: DIRBE Celestial Calibration

Burke, D. L., et al. 2010, ApJ, 720, 811

Burlov-Vasiljev, K. A., et al. 1995, Sol Phy, 157, 51

Campins, H., Rieke, G. H., & Lebofsky, M. J. 1985, AJ, 90, 896

Ciardi, D. R., van Belle, G. T., Thompson, R. R., Akeson, R. L., & Lada, E. A., 2000, BAAS, 32, 1476

Code, A. 1960, AJ, 65, 278

Code, A., & Meade, M. R. 1979, ApJS, 39, 195

Cohen, M., et al. 1992, AJ, 104, 1650

Colina, L., et al. 1996, AJ, 112, 307

Cooper, R., Roellig, T., Lunming, Y., Shiroyama, B., & Meyer, A. W. 2003, Proc SPIE, 4857, 11

Davis, J., et al. 2011, PASA, 28, 58

Defrère, D., et al. 2011, A&A, 534, 5

Elias, J. H., Frogel, J. A., Matthews, K., & Neugebauer, G., 1982, AJ, 87, 1029

Elias, J. H., Frogel, J. A., Hyland, A. R., & Jones, T. J. 1983, AJ, 88, 1027

Findlay, A. 1966, ARAA, 4, 77

Fixsen, D. J., et al. 1994, ApJ, 420, 457

Glushneva, I. N. 1964, Sov Astron, 8, 163

Green, D. A., Tuffs, R. J., & Popescu, C. C. 2004, MNRAS, 355, 1315

Greiveldinger, C., & Aschenbach, B. 1999, ApJ, 510, 305–311

Hanbury Brown, R., & Twiss, R. Q. 1958, Proc R Soc Lond, 248, 222

Hanbury Brown, R., Davis, J., & Allen, L. R. 1974, MNRAS, 167, 121
Hansen, O. L., & Caimanque, L. 1975, PASP, 87, 935
Hardie, R. H. 1962, in Astronomical Techniques, ed. W. A. Hiltner (Chicago: University of Chicago Press), 184
Hawarden, T., Leggett, S. K., Letawsky, M. B., Ballantyne, D. R., & Casali, M. M. 2001, MNRAS, 325, 563
Hayes, D. S. 1970, 159, 165
Hayes, D. S., & Latham, D. W. 1975, ApJ, 197, 593
Hayes, D. S., Latham, D. W., & Hayes, S. H. 1971, BAAS, 3, 501
Hayes, et al. 1975, APJ, 197, 587
Hill, G., Gulliver, A. F., & Adelman, S. J. 2010, ApJ, 712, 250
Hills, R., 2010, in ALMA Newsletter No. 6, 2
Holberg, J., & Bergeron, P. 2006, AJ, 132, 1221
Holberg, J. B., Forrester, W. T., Shemansky, D. E., & Barry, D. C. 1982, ApJ, 257, 656
Holberg, J. B., Bergeron, P., & Gianninas, A. 2008, AJ, 135, 1239
Ivezic, Z., et al. 2008, astro-ph/0805. 2366
Johnson, H., 1965, Commun Lunar Planet Lab, 3, 73
Kent, S. M., et al. 2009, astro-ph/0903. 2799
Kharitonov, A. V., et al. 1980, AZh, 57, 287
Kirk, J. G., Y. Lyubarsky, Y., & Petri, J. 2009, in Neutron Stars and Pulsars, (Berlin: Springer), 421
Kirsch, M. G., et al. 2005, Proc SPIE, 5898, 22
Knyazeva, L. N., & Kharitonov, A. V. 1990, Sov. Astron, 34, 626
Koester, D. 2010, MmSAI, 81, 921
Kravtsov, A. V., Vikhlinin, A., & Nagai, D. 2006, ApJ, 650, 128
Kruk, J. W., et al. 1997, ApJ, 482, 546
Kruk, J. W. 1997, IAUS, 189, 67
Kuiper, L., Hermsen, W., & Cusumano, G., et al. 2001, A&A, 378, 918
Labs, D., & Neckel, H. 1968, Z Astrophys, 69, 1
Landolt, A. 1983, AJ, 88, 439
Landolt, A. 1992, AJ, 104, 340
Lanz, T., Barstow, M. A., Hubeny, I., & Holberg J, B. 1996, ApJ, 473, 1089
Larason, T. C., & Houston, J. M. 2008, NIST Special Publication 250–41 http://www.nist.gov/calibrations/upload/sp250-41a.pdf
Leggett, S. K. 1985, A&A, 153, 273
Liebe, H. J. 1985, Radio Sci, 20, 1069
Ling, J. C., & Wheaton, Wm. A. 2003, ApJ, 598, 334
Livigini, D. J. 2003, NIST Special Publication 250–62 http://www.nist.gov/calibrations/upload/sp250-62.pdf
Lockwood, G. W., Tüg H., & White, N. M. 1992, ApJ, 390, 668
Lord, S. 1992, NASA Technical Memorandum 103957
Mather, J., et al. 1999, ApJ, 512, 511
Megessier, C., 1995, A&A, 296, 771
Mlawer, E. J., et al. 1997, J Geophys Res, 102, 663–682
Morrissey, P., et al. 2007, ApJS, 173, 682
Mountain, C. M., et al. 1985. A&A, 151, 399–402
Müller, S., Löhne, T., & Krirov, A. V. 2010, ApJ, 708, 1728
Neckel, H., & Labs, D. 1984, SoPh, 90, 205
Neugebauer, G., et al. 1971, AJ, 71, 719
Nevalainen, J., et al. 2010, A&A, 523, A33
Oke, J. B., & Schild, R. 1970, ApJ, 161, 1015
Paine, S., 2011, SMA Technical Memo No. 152, Rev 7
Petford, A. D., et al. 1985. A&A, 146, 195
Price, S., et al. 2004, AJ, 128, 889
Reese, E., et al. 2010, ApJ, 721, 653
Rieke, G. H., Lebofsky, M. J., & Low, F. J. 1985, AJ, 90, 900
Rieke, G. H., et al. 2008, AJ, 135, 2245
Rothman, L. S., et al. 2005, JQSRT, 96(2), 139
Rottman, G. 2006, Space Sci Rev, 125(1–4), 39
Selby, M. J., et al. 1983. MNRAS 203, 795
Selby, M. J., et al. 1980. MNRAS, 193, 111
Sinton, W. H., & Strong, J. 1960, ApJ, 131, 459
Sirk, M. M., et al. 1997, ApJ, S, 110, 347
Stamnes, K., Tsay, S. C., Wiscombe, W., & Jayaweera, K. 1988, Trans Antennas Propag, AP-28, 367
Staniszewski, Z., et al. 2009, ApJ, 701, 32
Strongylis, G. J., & Bohlin, R. C. 1979, PASP, 91, 205
Stubbs, C., et al. 2007, PASP, 119, 1163
Su, K. Y. L., et al. 2005, ApJ, 628, 487
Terez, E. I. 1985, Photometric and Polarimetric Investigations of Celestial Bodies (Kiev: Naukova Dumka), 55
Terez, G. A., & Terez, E. I. 1979, Sov Astron 23, 449
Thuillier, G., et al. 1997, Sol Phys, 171, 283
Thuillier, G., et al. 1998a, Metrologia, 35, 689
Thuillier, G., et al. 1998b, Sol Phys, 177, 41
Thuillier, G. et al., 2003. Sol Phys, 214, 1
Traub, W. A., & Stier, M. T. 1976, Appl Opt, 15, 364
Tucker, D., et al. 2006, Astr Nach, 327, 821
Tug, H., White, N. M., & Lockwood, G. W. 1977, A&A, 61, 679
Ulich, B., Davis, J. H., Rhodes, P. J., & Hollis, J. M. 1980, Trans Antennas Propag, AP-28, 367
Veron-Cetty, M. P., &Woltjer, L. 1993, A&A, 270, 370
Wark, D. Q., & Mercer, D. M. 1965, Appl Opt, 4, 839
Weiland, J. L., et al. 2011, ApJS, 192, 19
Weisskopf, M. C., et al. 2004, ApJ, 601, 1050
Willstrop, R. V. 1960, MNRAS, 121, 117
Wilson-Hodge, C. A., et al. 2011, ApJ, 727, L40
Woods, T. N., et al. 1996, JGR, 101, 9541
Wright, E. L. 1976, ApJ, 210, 250
Yun, M. S., Mangum, J., Bashan, T., Holdaway, M., & Welch, J. 1998, ALMA article on calibration: http://www.alma.nrao.edu/ memos/ html-memos/ alma211/ memo211.html

9 Virtual Observatories, Data Mining, and Astroinformatics

Kirk Borne
George Mason University, Fairfax, VA, USA

T.D. Oswalt, H.E. Bond (eds.), *Planets, Stars and Stellar Systems. Volume 2: Astronomical Techniques, Software, and Data*, DOI 10.1007/978-94-007-5618-2_9, © Springer Science+Business Media Dordrecht 2013

Abstract: The historical, current, and future trends in knowledge discovery from data in astronomy are presented here. The story begins with a brief history of data gathering and data organization. A description of the development of new information science technologies for astronomical discovery is then presented. Among these are e-Science and the virtual observatory, with its data discovery, access, display, and integration protocols; astroinformatics and data mining for exploratory data analysis, information extraction, and knowledge discovery from distributed data collections; new sky surveys' databases, including rich multivariate observational parameter sets for large numbers of objects; and the emerging discipline of data-oriented astronomical research, called astroinformatics. Astroinformatics is described as the fourth paradigm of astronomical research, following the three traditional research methodologies: observation, theory, and computation/modeling. Astroinformatics research areas include machine learning, data mining, visualization, statistics, semantic science, and scientific data management. Each of these areas is now an active research discipline, with significant science-enabling applications in astronomy. Research challenges and sample research scenarios are presented in these areas, in addition to sample algorithms for data-oriented research. These information science technologies enable scientific knowledge discovery from the increasingly large and complex data collections in astronomy. The education and training of the modern astronomy student must consequently include skill development in these areas, whose practitioners have traditionally been limited to applied mathematicians, computer scientists, and statisticians. Modern astronomical researchers must cross these traditional discipline boundaries, thereby borrowing the best of breed methodologies from multiple disciplines. In the era of large sky surveys and numerous large telescopes, the potential for astronomical discovery is equally large, and so the data-oriented research methods, algorithms, and techniques that are presented here will enable the greatest discovery potential from the ever-growing data and information resources in astronomy.

Keywords: Astroinformatics, bayesian classification, classification, clustering, data management, data mining, data preparation, data profiling, data science, data transformation, databases, decision tree, distance metrics, e-Science, exploratory data analysis, fourth paradigm, informatics, K-means, machine learning, neural network, outlier detection, semantic science, semisupervised learning, similarity metrics, sky surveys, supervised learning, survey science, unsupervised learning, virtual observatory, visualization, VOEvent

List of Abbreviations: *2MASS* 2-Micron All-Sky Survey; *AAO* Anglo-Australian Observatory; *ADAC* Astronomical Data Archives Center (Japan); *ADASS* Astronomical Data Analysis Software and Systems; *ADS* Astronomical Data Center; *ApJS* Astrophysical Journal Supplement; *ANN* Artificial neural network; *BD* Bonner Durchmusterung; *CADC* Canadian Astronomy Data Center; *CDS* Center de Donnees astronomique de Strasbourg (France); *GCVS* General Catalog of Variable Stars; *DDM* Distributed data mining; *DMD* Distributed mining of data; *DOE* Department of Energy; *DSS* Digital Sky Survey; *EDA* Exploratory data analysis; *HD* Henry Draper; *HEASARC* High Energy Astrophysics Science Archive Research Center; *IPAC* Infrared Processing and Analysis Center; *IRSA* Infrared Science Archive; *IVAO* International Virtual Observatory Alliance; *KDD* Knowledge Discovery in Databases; *KNN* K-nearest neighbors; *LEDAS* Leicester Database and Archive Service (UK); *LSST* Large Synoptic Survey Telescope; *MAST* Multimission Archive at Space Telescope; *MDD* Mining of distributed data; *ML* Machine learning; *NASA* National Aeronautics and Space Administration; *NED* NASA/IPAC Extragalactic Database; *NGC* New General Catalog; *NSF* National Science Foundation; *NVO* National

Virtual Observatory; *Pan-STARRS* Panoramic Survey Telescope and Rapid Response System; *PB* Petabyte; *PDMP* Project Data Management Plan; *PI* Principal investigator; *RA/Dec* Right ascension and declination; *RDF* Resource Description Framework; *SAO* Smithsonian Astrophysical Observatory *SIMBAD* Set of Identifications, Measurements, and Bibliography for Astronomical Data *SDSS* Sloan Digital Sky Survey; *SVM* Support vector machine; *TB* Terabyte; *VAO* Virtual Astronomy Observatory; *TMSS* Two-Micron Sky Survey; *VO* Virtual observatory; *WWW* World Wide Web; *XML* eXtensible Markup Language

URLs for Major Data Tools and Services

ADS http://www.adsabs.harvard.edu/; *CDS* http://cdsweb.u-strasbg.fr/; *HEASARC* http://heasarc.gsfc.nasa.gov/; *IRSA* http://irsa.ipac.caltech.edu/; *IVOA* http://ivoa.net/; *MAST* http://archive.stsci.edu/; *NED* http://ned.ipac.caltech.edu/; *SDSS* http://www.sdss.org/; *SIMBAD* http://simbad.u-strasbg.fr/simbad/; *VAO* http://www.usvao.org/

Somewhere, something incredible is waiting to be known.

Carl Sagan

1 Introduction

Data are the source of astronomical discovery and knowledge. Astronomers have always relied on observational data for their science, with relatively minor dependence on experimental data (from experiments in the lab). As astronomers have collected data from observatories of many sizes and wavelength domains, so too have astronomers published data collections for use by others. As far back as Ptolemy's Almagest (and undoubtably further back to the ancient Chinese), data on astronomical objects have been compiled and disseminated. In the modern era (beginning in the eighteenth century), formal journal publications have provided a means for such data publication. Specifically, it is not the data that are published (though this is sometimes true), but the derived results are published, in the form of catalogs, which are derived from the analysis of observational data obtained at telescopes. The derived results presented in published catalogs include object tables, observation logs, parameter lists, and more. Historically, in addition to published books (e.g., the BD catalog, the HD catalog, the NGC, the Yale Bright Star Catalog, the GCVS, or the SAO star catalog), the Astrophysical Journal Supplement (ApJS) series was the home for such catalogs and published data collections. It was not uncommon for beginning astronomy students to spend days in their university's library, compiling lists of astronomical catalogs from various publications – the ApJS was the primary source of such information. A young astronomer became familiar with these information resources since these were the building blocks of knowledge in the student's chosen scientific discipline.

Astronomy is therefore primarily a forensic science, in the sense that it involves the collection, systematic investigation, and subsequent analysis of physical evidence in order to extract clues regarding the nature of some phenomenon or event. Some aspects of astronomical research are experimental (such as laboratory astrophysics or in situ planetary exploration), in the classic sense of experimental science, but this is not typical – astronomers do not build a star or galaxy or planet in their labs, under varying conditions, and then probe the properties of that astronomical object and its responses to stimuli as a function of those changing conditions. Modern theoretical astrophysics research does have this experimental character through modeling, computation, and simulation of objects and their astrophysical processes. This is essential research, but it has a subjective character – it is only as accurate as the underlying hypotheses and algorithms that are invoked in the model or simulation. For example, a recipe for star formation might be an essential ingredient in modern numerical simulations of galaxy evolution, but it is still just a recipe, which at best reproduces some of the observed characteristics of star formation in a broad context or in very specific cases. In contrast to experimental and theoretical astrophysics, fundamental observational astronomy depends on the collection of scientific evidence from remote inaccessible radiation sources that are not affected by scientist's probing inquiries. But now, with the growth in increasingly large sky surveys, it becomes possible to perform astrophysical experiments within the collected data themselves – the testing of diverse hypotheses and performing what-if experiments are carried out through data exploration and analysis of complete samples of different object classes. These exploratory data analysis inquiries experimentally probe the behaviors of astrophysical objects and processes under varying conditions that are enabled by the sheer large size of the data collections, which ostensibly include representative examples of many different classes of astronomical objects and physical behaviors. This data-oriented approach to scientific research (generally called informatics and called astroinformatics for data-oriented astronomy research) is objective, evidence-based, and data-driven (not model-driven). The collection of evidence (observational data) has always been fundamental to the conduct of productive astronomical research, and this is becoming even more emphatically true as more and more large sky survey databases are being collected. In the modern era, not only is the collection of data a central activity, but also the processing, archiving, mining, analysis, curation, and long-term preservation of data are critical science-enabling activities.

Quickly fading are the days where the astronomer acquires photographic plates or personal data tapes from isolated trips to the observatory, followed by the one-time analysis of those data, ending with those data being stored in some drawer or file cabinet to collect dust. Most data are now acquired on portable reproducible media, including the production of archival copies of the data at the originating observatory. It is not uncommon for secondary analyses of the data to be conducted by astronomers who were not associated with the original observer or principal investigator, through the use of archival data and online data services (White et al. 2009).

When examining data-oriented astronomical research, it is helpful to recognize that scientific data can be viewed at increasing levels of abstraction. All of these levels represent research-ready data (and information) resources. At the most fundamental level, data are acquired from scientific instruments. These data can have a variety of modalities: images, spectra, time series, photon counts, particle counts (e.g., cosmic rays or neutrinos), interferograms, fringe visibility maps, and data cubes (e.g., integral field spectra). These data may be raw, processed, or calibrated. From these fundamental data sources, scientists extract information, usually in the form of catalogs. Information, in this sense, represents a higher level of abstraction from the original data. Information may include extracted parameters from data (e.g., magnitudes and positions for objects in images, redshifts or line ratios from spectra, or flux maps from photon counts or visibility maps). Information is usually represented in tables, catalogs, or standard graphics products (such as contour diagrams or spectral plots). Frequently, information is called "metadata," which can be defined as "data about data," though in the field of scientific data management, metadata usually refers to structural, syntactic, and administrative information about a data product.

From information, scientists infer knowledge. Knowledge, in this sense, represents a still higher level of abstraction from data and information. Knowledge may include classifications of objects or of objects' behavior, correlations between extracted information parameters (e.g., the Hubble law of cosmic expansion, scaling relations in galaxies, the H-R diagram, or stellar population heuristics in star-forming regions), or other patterns in the data (e.g., voids in the cosmic distribution of galaxies or characteristic supernova light curve shapes). Knowledge is usually represented in graphical form, in physics-based equations, or in text (i.e., journal articles). It is not common in the field of astronomy to represent knowledge in structured form (e.g., in knowledge-oriented databases, using factual assertions such as "blue stars are hot," or "red stars are cool," using RDF semantics or formal ontologies; Raskin and Pan 2005), though other science disciplines are starting to develop such structured knowledge bases that are nearly as easily queried as standard databases consisting of tables and catalogs (Hendler 2003). The latter are more accurately described as information bases.

Finally, at the highest level of data abstraction is the scientific understanding that scientists derive from data, information, and knowledge. Scientific understanding is captured in the scientific literature, in formal and informal communications, and in the collective beliefs of practicing scientists. Among the four levels of data abstraction, scientific understanding is the hardest to capture in a structured, searchable form. It is usually conveyed through graduate training, courses, textbooks, conferences, review articles, and those sections in journal articles that are reserved for interpretation of the results of data analyses. But it is scientific understanding that all scientists aim to achieve and to make fundamental contributions to.

As an illustrative example of this data-information-knowledge-understanding abstraction hierarchy, consider the case of galaxies. Data are obtained for galaxies: images and spectra. Information is extracted from data: positions, magnitudes (fluxes), and redshifts. Knowledge is discovered from information that there exists a correlation between magnitudes and redshifts (the Hubble Diagram). From this knowledge, Hubble inferred one of the greatest leaps of scientific understanding in the twentieth century that the universe is expanding. The "understanding" was in some sense "present" in the original data, though in fact the information and knowledge were encoded in the original data, and it was the human intellect that produced the scientific understanding through the synthesis of all of the other pieces.

Astronomical data and metadata (information) are now readily accessible in structured form through online services: search engines, queryable databases, and data archives. Astronomical knowledge and understanding are also readily accessible, though in unstructured form, through research publications, journal articles, preprint servers, conference proceedings, and personal web content. A great leap forward in astronomical knowledge search and discovery, including links and associations among scientific facts, is yet to be realized through the establishment of formal astronomical knowledgebases and fact-oriented repositories.

New modes of discovery are enabled by the growth of data and computational resources in the sciences. This cyberinfrastructure includes databases, virtual observatories (distributed data), high-performance computing (clusters and petascale machines), distributed computing (the grid, the cloud, and peer-to-peer networks), intelligent search and discovery tools, and innovative visualization environments. Data volumes from multiple sky surveys have grown from gigabytes into terabytes during the past decade and will grow from terabytes into tens (or hundreds) of petabytes in the next decade. This plethora of new data both enables and challenges effective astronomical research, requiring new approaches. Thus far, astronomy has tended to address these challenges in an informal and ad hoc manner, with the necessary special expertise being assigned to e-Science (Hey and Trefethen 2002) or survey science. However, there is an even wider scope and therefore a broader vision of this data-driven revolution in astronomical research. The solutions to many of the problems posed by massive astronomical databases exist within disciplines that are far removed from astronomy, whose practitioners do not normally interface with astronomy. For astronomy to effectively cope with and reap the maximum scientific return from existing and future large sky surveys, facilities, and data-producing projects, astronomical research now includes a major new subdiscipline, *astroinformatics*. Astroinformatics includes a set of naturally related specialties including data organization, data description, astronomical classification taxonomies, astronomical concept ontologies, data mining, machine learning, visualization, and statistics (Borne 2010).

The focus of this chapter is primarily on the machine learning algorithms that enable astronomical data mining – more specifically, the focus is on algorithms that enable scientific discovery from large astronomy data collections. Machine learning is a special research subdiscipline of artificial intelligence (involving computer science and applied mathematics) that focuses on the development of algorithms that learn from data – these algorithms may be implemented in machines (e.g., robotics) or applied as software tools (e.g., in data processing and analysis pipelines). The application of machine learning to large datasets is specifically referred to as data mining. Data mining has many definitions – those which are most relevant for astronomy are (a) knowledge discovery in databases (KDD); (b) an information extraction activity whose goal is to discover hidden facts (knowledge) contained in large databases; and (c) methods by which knowledge is tranformed from a "data format representation" into a "rule format representation." The first definition goes directly to the heart of any application of data mining: the discovery of new scientific knowledge. The second definition is the most common, and it describes the activity and its goals: data mining is a proactive data exploration research task. This also most clearly reflects the data-to-information-to-knowledge research process that was described earlier. The final definition may seem the most obscure, but it makes most sense in the context of rule mining algorithms, such as decision trees (➋ Sect. 3.2.3) and association rule mining, both of which produce a set of rules (e.g., classification decision points or "if-then" rules) that are derived from data. From any of these points of view, astronomers who apply data mining in their research are aiming to make new discoveries and to reveal previously unknown knowledge about astronomical objects and astrophysical processes through exploration of large data and information resources (Borne 2009a; Ball and Brunner 2010).

➋ Section 2 will review the world of astronomy data, including data resources, the virtual observatory, sky surveys, data science, and data mining. ➋ Section 3 will review data mining, including a discussion of methods, algorithms, and astronomy applications. ➋ Section 4 will introduce the new discipline of astroinformatics, including research topics, challenge areas, and a vision for the future of data-intensive astronomical research.

2 The Evolving Astronomical Data Environment

Data streams from experiments, sensors, and simulations are increasingly complex and growing in volume: growing from gigabytes into terabytes during the past decade and growing from terabytes into tens (or hundreds) of petabytes in the next decade (Bell et al. 2006). This is true in most sciences, including time-domain sky surveys, climate simulations, remote sensing sensor networks, fusion science, and particle physics experiments (Hey et al. 2009). Each discipline is producing its own "*tonnabytes*" data volume challenge (Borne 2009b, c). In response to these developments, a transformation in scientific research is occurring that is more revolutionary than evolutionary. The exponential growth in scientific data in all disciplines is changing the practice of science and the pathways to discoveries. In astronomy, this revolutionary transformation was inevitable.

In the one to two decades prior to the WWW revolution, there appeared Internet-accessible astronomical data sources, catalogs, and information resources. Astronomers began to learn new digital tools of data discovery and access, such as telnet and ftp (and e-mail). The emergence of digital versions of astronomical data and catalogs thus inspired the growth of astronomical data centers, such as the ADC (Astronomical Data Center) in the US and the CDS in France. The CDS' SIMBAD database aggregated the data, metadata, and information from numerous stellar catalogs into one master digital data repository. Electronically accessible digital catalogs soon sped up the process of discovery and access to published astronomical data (again, primarily catalogs of tabular data). Astronomers could then download their own copy of the HD or SAO catalogs, or just about any of the major catalogs that had appeared in ApJS or other journals (e.g., the Abell (1958) catalog of rich clusters of galaxies, the Sharpless (1959) catalog of H II regions, the 3CR radio source catalog (Bennett 1962), or the Lynds (1962) catalog of dark nebulae).

It was during this period (late 1960s through early 1990s) that three other "members" of the scientific and astronomical data environments appeared. The first two were generic concepts – data management (including data archives) and relational databases – and the third was NASA's ADS (Astrophysics Data System).

Data management (specifically, science data management) became an essential activity of large data-producing projects and enterprises. For example, NASA required (and still does require) PDMPs (Project Data Management Plans) from their funded missions. Data management includes many activities that are now taken for granted, but which were novel concepts for most scientists – description (metadata), indexing, access, curation, and preservation – functions that were most familiar to librarians. These processes became essential in the era of digital data and digital accessibility. During the same time period, the relational data model was developed to manage data in databases. The paper by Codd (1970) is a nice little introduction to the relational data model – this one paper alone has had over 6,000 citations (an enviable career for most researchers)! Both data management and databases have enabled sustained growth and use of large astronomical data repositories for use by astronomers. They became essential cyberinfrastructure for scientific research, long before scientists introduced the term "cyberinfrastructure."

One of the early examples of true cyberinfrastructure for astronomy was NASA's ADS, which first appeared in 1988 as a proof of concept with 40 astronomical papers. It now has links to nearly ten million records and is used on a frequent and regular basis by nearly all astronomers worldwide – it is the online repository of all astronomy publications, plus much more (including access to data and advanced search indices). In its earliest incarnation, ADS was envisioned as an Internet portal to access digital astronomy data (not only published articles). The ADS was then perceived by astronomers as "the interface" to online data. The ADS was ahead of its time – what the initial ADS tried to be is what the WWW became: the interface, the network, the indexer, the search engine, the source of data and information everywhere. What ADS subsequently became (i.e., the online database of all astronomy publications, with linked data, citations, and references) is perhaps the most used online research tool in all of astronomy.

Following the appearance of the WWW and web browsers, numerous astronomy data centers emerged on the web. These included major NASA facility data archives, such as the Hubble Data Archive, MAST, the "new" ADS (see above), HEASARC, IRSA, and NED at IPAC, plus major astronomical data centers worldwide, such as ADAC, CADC, and CDS, in addition to countless personal and institutional web pages with links to data. The larger data centers provided sophisticated search and analysis data services, while the smaller data providers offered little more than links to data files. One type of service became commonly used – name resolvers – both NED and SIMBAD provided name resolver services. Essentially, these services act in a very straightforward manner: the user enters the common name of an astronomical object, and the name resolver service returns the RA/Dec position of the object. The online data centers used name resolvers as a web service – i.e., the user simply typed in the object name in the data center's user interface, and a background process invoked the name resolver, captured the RA/Dec, and then inserted the RA/Dec into their search tool – the user simply had to submit the query. These name resolvers automated the search and retrieval functions for lists of objects as well. This was a great improvement in efficiency of the search experience, but the user still had to visit numerous sites in order to find all data for a given object or class of objects. Fortunately, this limitation would be short-lived.

The proliferation of astronomical information services online was similar to the rapid expansion of online business information, online stores, and online government resources. In response to user demand within those latter domains, order eventually emerged from the chaos in the forms of e-business, e-commerce, and e-government, respectively. The protocols for these e-experiences enabled rapid discovery of and access to distributed information sources, automatic aggregation and integration of such online information into a single portal, one-stop shopping for the end-user, and greater user search efficiency. Scientists saw the value in this and adopted these same web service protocols. Consequently, distributed science resources were stitched into one user experience – e-Science was born. It was only a matter of time before astronomers also adopted these web protocols – the adoption of e-Science led to yet another "game changer" in astronomical research – the VO (virtual observatory), thus making astronomy the "poster child" for e-Science around the world. Other science disciplines have even referred to their distributed federated data environments as "virtual observatories" also. For astronomy, the VO is often called an "efficiency amplifier" (A. Szalay, private communication), and that is still the best way to understand e-Science's utility and impact. The VO and its major features are described in ➲ Sect. 2.1, followed by a brief introduction to survey science (➲ Sect. 2.2), a review of data science (➲ Sect. 2.3) in astronomy within the context of mining large data collections, and a taxonomy for data mining (➲ Sect. 2.4) – all of which represent major steps toward data-driven scientific discovery in the era of data-intensive science (➲ Sect. 2.5).

2.1 The Virtual Observatory: Data Discovery and Access

As the previous comments have emphasized, the astronomical data environment has evolved rapidly in the past 20 years through the advent of web-based online services. Even very large data archives and information repositories are now readily accessible through the VO by anyone with a web browser. New research on previously analyzed data is easy, and this secondary use of data is producing many new scientific results (White et al. 2009).

The VO was the most significant technological development for astronomy in the first decade of the twenty-first century. The VO is science-enabling cyberinfrastructure for discovery and access to all geographically distributed astronomical data collections. The current VO operational framework within the US is called the Virtual Astronomy Observatory (VAO;

www.us-vo.org), which was developed initially as an NSF-funded ITR (Information Technology Research) project, but it actually spans multiple international projects that have invested significant resources (IVOA; www.ivoa.net). The IVOA partners have developed standards and protocols for data modeling, metadata, data description, data discovery, data access, data sharing, and integration of distributed data. While the VO has been remarkably successful and productive, its focus during the initial development years was primarily on infrastructure, upon which a variety of application frameworks are being built. Two of the most significant application domains that are now emerging for the VO (for which a VO-enabled framework is required) are time-domain astronomy and astronomical data mining. The focus is primarily on data mining in the remainder of this chapter, though time-domain astronomy is discussed in the next section.

One of the significant VO developments has been the establishment of registries for data discovery, access, and integration. Various data portals take advantage of these registries in order to provide "one-stop shopping" for all of the astronomer's data needs. One particularly useful data portal for data mining specialists is OpenSkyQuery (Greene et al. 2008). This site enables multi-database queries, even when (i.e., specifically when) the databases are geographically distributed around the world and when the databases have a variety of schema and access mechanisms. This facility enables distributed data mining of joined catalog data from numerous astrophysically interesting databases. This facility does not mine astronomical images (i.e., the pixel data, stored in image archives), nor does it mine all VO-accessible databases, but it does query and join on many of the most significant astronomical object catalogs. In this context, "distributed data mining" is taken to mean the "mining of distributed data" (➋ Sect. 3.3.1), as opposed to the algorithmically interesting and very challenging context of "distributed mining of data" (➋ Sect. 3.3.2). A set of VO-inspired astronomical contexts and use cases for data mining are presented that are applicable to either type of distributed data mining (➋ Sect. 3.4).

While the VO has been remarkably successful and productive, its initial focus historically was primarily on infrastructure, upon which a variety of application frameworks are being built (Euchner et al. 2004; Grosbl et al. 2005; Graham et al. 2005; Liu et al. 2006; Gardner et al. 2007; Graham 2009; Becciani et al. 2010; Oreiro et al. 2011). Time-domain astronomy is the next great application domain for which a VO-enabled framework is now required (Becker 2008; Djorgovski et al. 2008; Drake et al. 2009; Graham 2010). As already described, another important application domain for VO is scientific data mining, specifically distributed data mining across multiple geographically dispersed data collections (Borne 2001a, b, 2003; Djorgovski et al. 2001; Fabbiano et al. 2010).

The VO makes available to any researcher anywhere a vast selection of data for browsing, exploration, mining, and discovery (Graham et al. 2008). The VO protocols enable a variety of user interactions: query distributed databases through a single interface (e.g., VizieR (Ochsenbein et al. 2000) or the next generation OpenSkyQuery.net (A. Thakar, private communication)); search distributed image archives with a single request (e.g., DataScope (McGlynn 2008)); discover data collections and catalogs by searching their descriptions within a comprehensive registry (e.g., the IVOA searchable registries (Plante et al. 2004; Schaaf 2007)); and integrate and visualize data search results within a single GUI or across a set of linked windows (e.g., VIM (Visual Integration and Mining; Williams 2008), Aladin (Boch et al. 2008), or SAMP (Simple Application Messaging Protocol; Taylor et al. 2010)). The scientific knowledge discovery potential of the worldwide VO-accessible astronomical data and information resources is almost limitless – the sky is the limit! The VO has consequently been described as an "efficiency amplifier" for data-intensive astronomical research (A. Szalay, private communication) due to the fact that the efficiency of data discovery, access, and integration has been greatly amplified by the VO protocols, standards, tools, registries, and resources. This has been a great leap forward in the ability for astronomers to do multiarchive distributed data research. However, while the data discovery process has been made much more efficient by the VO, the knowledge discovery from data (KDD) process can also be made much more efficient (measured by "time-to-solution") and more effective (measured by depth and breadth of solution). Improved KDD from VO-accessible data can be achieved particularly through the application of sophisticated data mining and machine learning algorithms (Ball and McConnell 2011), whether operating on the distributed data in situ (distributed mining of data) or operating on data subsets retrieved to a central location from distributed sources (mining of distributed data). Numerous examples and detailed reviews of these methods, applications, and science results can be found in the review papers by Borne (2009a) and Ball and Brunner (2010).

2.2 Sky Surveys: Science Discovery at the Frontier of Information Science

Astronomers now systematically study the sky with large sky surveys. These surveys make use of uniform calibrations and well-engineered pipelines for the production of a comprehensive set of quality-controlled data products. Surveys are used to measure and collect data from all objects that are visible within large regions of the sky, in a systematic, controlled, and repeatable fashion. These statistically robust procedures thereby generate very large unbiased samples of many classes of astronomical objects. Thus, sky surveys provide some of the most sought-after quality-controlled astronomical data content that is accessible through the virtual observatory. It should not be surprising that sky surveys "done right," such as SDSS, are generating at least as many research publications (based on project-produced archived data) as even the most popular large telescopes generate through PI-led observations (Trimble and Ceja 2010).

A common feature of modern astronomical sky surveys is that they are producing massive databases. Surveys produce hundreds of terabytes (TB) up to 100 (or more) petabytes (PB) both in the image data archive and in the object catalogs (databases), which motivate and inspire countless data mining and exploratory data analysis archival research programs. These surveys include DSS (Digitized Sky Survey; generated from the first and second generation Palomar Observatory Sky Surveys and from the UK Schmidt AAO Southern Sky Survey), SDSS (Sloan Digital Sky Survey), 2MASS (2-Micron All-Sky Survey), and Pan-STARRS (Panoramic Survey Telescope and Rapid Response System), plus the future LSST (Large Synoptic Survey Telescope). Such large sky surveys have enormous potential to enable countless astronomical discoveries (see Djorgovski et al., this volume ➋ Chap. 3, Time-Domain Astronomy). Such discoveries will span the full spectrum of statistics: from rare one-in-a-billion (or one-in-a-trillion) type objects to the complete statistical and astrophysical specification of a class of objects (based upon millions of instances of the class).

2.2.1 Time-Domain Sky Surveys: The New Frontier

One of the key features of these surveys is that the main telescope facility will be dedicated to the primary survey program, with no specific plans for follow-up observations. This is emphatically true for the LSST project (Mould 2004). Paradoxically, the follow-up observations are scientifically essential – they contribute significantly to new scientific discovery, to the classification and characterization of new astronomical objects and sky events, and to rapid response to short-lived transient phenomena. Since it is anticipated that LSST will generate many thousands (probably millions) of new astronomical event alerts during each night of observations, there is a critical need for innovative machine learning, knowledge discovery, and intelligent follow-up observation decision support. Consequently, the rapid real-time detection, classification, characterization, assessment of time-criticality, and determination of astronomical relevance for a plethora of exotic astrophysical phenomena are scientifically imperative and extreme data challenges. A data mining classification broker for LSST and other time-domain surveys is one possible approach toward coping with this challenge, but research on the scalable machine learning algorithms that are needed to drive this engine still remains to be done (Borne 2008a, 2009; see also Bloom et al. 2008).

LSST is the most impressive sky survey being planned for the next decade (lsst.org; Tyson 2004; Strauss 2004). Compared to other sky surveys, the LSST survey is expected to deliver time-domain coverage for orders of magnitude more objects. Extensive science concepts and astronomical research problems that would benefit from the use of LSST survey data have been well documented in a major compendium – the LSST Science Book (LSST Science Collaborations and the LSST Project 2009). The project proposes to produce ~15–30 TB of data per night of observation for 10 years. The final image archive is anticipated to be ~100 PB, and the final LSST astronomical object catalog (object-attribute database) is expected to be ~20–40 PB, comprising over 200 attributes for 50 billion objects and ~50 trillion source observations. In addition to time series for all of these objects, LSST is planning to maintain an alerts database for all of the nightly events – perhaps one million per night.

2.2.2 Data Mining in Sky Survey Databases

Many astronomical use cases for data mining are anticipated with the LSST database and from similar sky survey databases (with a particular emphasis on time-domain surveys below; Borne et al. 2008), including the following:

- Provide rapid probabilistic classifications for approximately one million LSST events each night
- Find new correlations and associations of all kinds from the 200+ science attributes
- Discover voids in multidimensional parameter spaces (e.g., period gaps)
- Discover new and exotic classes of astronomical objects or new properties of known classes
- Discover new and improved rules for classifying known classes of objects
- Identify novel, unexpected behavior in the time domain from time series data
- Hypothesis testing – verify existing (or generate new) astronomical hypotheses with strong statistical confidence, using millions of training samples
- Serendipity – discover the rare one-in-a-billion type of objects through outlier detection
- Quality assurance – identify glitches and image processing errors through deviation detection

Consider this example: astronomers currently discover a few hundred new supernovae per year. Since the beginning of human history, significantly fewer than 10,000 supernovae have been recorded. The identification, classification, and analysis of supernovae are among the key requirements for understanding the ubiquitous cosmic dark energy. Since supernovae are temporal phenomena, it is imperative for astronomers to respond quickly (within minutes) to each new event with rapid follow-up observations in many measurement modes (e.g., light curves, spectroscopy, or imaging of the host galaxy). Historically, with <10 new supernovae being discovered each week, such follow-up observations have been feasible. However, LSST promises to produce a list of *1,000 new supernovae each night* for 10 years! How can such a staggering number of new objects be efficiently classified and properly prioritized for follow-up observation? Astronomers will be faced with the enormous challenge of efficiently mining, correctly classifying, and intelligently prioritizing this huge number of new events for rapid follow-up observation each night for a decade. Incoming event alert data will be subjected to a suite of machine learning (ML) algorithms for event classification, outlier detection, object characterization, and novelty discovery. These algorithms will contribute to probabilistic ML models that will produce rank-ordered lists of the most significant and/or most unusual events. This will enable rapid information extraction, knowledge discovery, and scientific decision support for real-time astronomical research facility operations – to prioritze the most time-critical follow-up observations.

2.2.3 Enabling Technologies for Time-Domain Sky Surveys

Time-domain astronomy has advanced significantly in recent years through one of the VO's major enabling technologies – the VOEvent protocol. As a consequence of this development, the future promises even greater advances as time-domain astronomy takes center stage in the coming years (Bloom et al. 2008; Williams et al. 2010). VOEvent is used as a messaging standard and information payload for notification of temporal astronomical events (Williams and Seaman 2008). For LSST, the VOEvent protocol is a de facto requirement for publishing and distributing the nightly event stream of about one million sky alerts (e.g., see skyalert.org).

The revolution in time-domain astronomy in the next 10 years (brought on by Palomar Quest, Palomar Transient Factory, Catalina Real-time Transient Survey, Pan-STARRS, LSST, the Dark Energy Survey, and others) will create an extreme-scale data mining challenge. From time-domain astronomy, a new vision of the night sky will emerge, as time series are collected for a plethora of objects and object classes (e.g., supernovae, novae, accreting black holes, microquasars, gamma-ray bursts, gravitational microlensing events, extrasolar planetary transits across galactic stars, new comets, incoming asteroids, trans-Neptunian objects, dwarf planets, optical transients, variable stars of all classes, and anything that changes in brightness or celestial position). Temporal variations that are novel, unexpected, previously unknown, or outside the bounds of existing classification schemes will generate an alert to the worldwide astronomical community. The event alert notification packet will include a characterization and a probabilistic classification of the event, with some measure of the confidence of the classification (Bloom et al. 2007, 2008; Borne 2008a). Without good event characterization and classification information in those alert packets, and hence without some means with which to prioritize the anticipated huge number of events (~1,000,000 each night) from the new time-based instruments, the astronomy community will consequently be buried in the data deluge and may miss some of the most important and exciting astronomical discoveries of the coming decade.

Even in the seemingly simple case of variable stars for which their variability is known, well studied, and well characterized already, the situation is not so simple. Suppose that one of these stars' eigenmodes of variability was to change, then that would be extremely interesting – this could perhaps be a signal that some potentially exotic astrophysical processes are involved (Sarro et al. 2009). Astronomers will definitely want to be notified promptly about these types of variations, which are essentially *changes in the stationarity of the source*. Detection and characterization of nonstationarity can be measured through changes in the eigenvectors and eigenvalues of the light curve (Debosscher et al. 2007; Rebbapragada et al. 2009). Because these eigenstates provide convenient and efficient short-hand representations of the features contained in the full data stream, the eigenstate information can be easily stored, monitored, searched, and flagged as interesting and/or changing.

2.3 Astronomical Data Science

Data science (or data-oriented science) represents the fourth paradigm of scientific research, following experiment (observational science), theory, and computational science (modeling and simulation) (Hey et al. 2009; Mahootian and Eastman 2009). Data science is similar to data management in the sense that it is a broad term that encompasses many aspects, activities, and processes that focus on data. Unlike data management, data science is first and foremost a field of scientific research. Data management includes some important and significant research elements, but it also incorporates a set of practical functions. Some of the more interesting (research-oriented) aspects of scientific data management include data quality assessment (which often involves advanced statistics), data provenance (which seeks to identify descriptors and methods for tracing the origin, creation, curation, and change histories of data products), data organization (including data models, data structures, compression algorithms, and access protocols), data indexing (including advanced sorting and hashing methods), and advanced data archiving concepts (e.g., Ramapriyan et al. 2002; Harberts et al. 2003; Rotem and Shoshani 2009; Plante et al. 2010). In contrast, data science (often referred to as "informatics") includes research in several disciplines: statistics, data mining, machine learning, visualization, and semantic science. Astronomical data mining is described (with some examples) in ➋ Sect. 3, and the new field of astronomical data science (i.e., astroinformatics) is described in ➋ Sect. 4. Astroinformatics includes data mining, machine learning, visualization, and semantic science – research focused on discovery from large datasets through applied mathematics and advanced computer science algorithms. Astrostatistics is a separate research discipline that focuses on discovery from complex data through advanced statistical methods and algorithms.

In the above discussion, there was no mention of traditional data analysis and data reduction methods that have been used by astronomers since the early days of formal research. These research methods are not the same thing as data science. Data reduction and analysis refer to methods of deriving scientific results from data – these methods can be automated as well as manually conducted. They may include statistical methods and knowledge discovery methods that are now associated with the field of data science. But most of those traditional analysis methods do not scale when the data volumes expand from a few dozen CCD images per night to a few thousand, and when the number of objects being analyzed grows from a few hundred (or a few thousand) to a few tens of billions or trillions. Data science techniques and algorithms (from the fields of statistics, data mining, machine learning, visualization, and semantic science) are absolutely essential in the new era of "big data." The summation of these data science methodologies for astronomical research is now called astroinformatics, which is described in ➋ Sect. 4 of this chapter.

2.4 A Taxonomy for Data Mining

Some of the very early efforts to mine distributed data through VO data discovery and access protocols were more proof-of-concept than novel astronomical research projects. Nevertheless, they pointed the way and demonstrated the discovery potential of distributed data mining in VO-accessible world of astronomy data. Traditionally, astronomers think of distributed data mining specifically as the mining of distributed data, though the computer science machine learning research community thinks of distributed data mining as the distributed mining of data. The latter is a major research challenge, as it involves approximating the result at different sites, then sharing the results between the sites, and then repeating the mining calculations until a solution is converged upon. In this research domain, it is assumed that the data are either too large to move around (because of tight bandwidth bottlenecks) or that the data may be proprietary (or secure) with restricted local access (no remote downloads permitted). Two examples of distributed data mining are examined in ➋ Sect. 3.3: one of the first type (mining of distributed data) and one of the second type (distributed mining of data).

In general, astronomers will want to apply data mining algorithms on distributed multiwavelength (multimission) VO-accessible data within one of these four distinct astronomical research contexts (Borne 2003), where "events" refer to a class of astronomical objects or phenomena (e.g., supernovae, planetary transits, or microlensing):

1. *Known events/known classification rules* use existing models (i.e., descriptive models or classification algorithms) to locate known phenomena of interest within collections of large databases (frequently from multiple sky surveys).
2. *Known events/unknown rules* apply machine learning algorithms to learn the classification rules and decision boundaries between different classes of events or find new astrophysical properties of known events (e.g., known classes) by examining previously unexplored parameter spaces (e.g., wavelength or temporal domains) in distributed databases.
3. *Unknown events/known rules* use expected relationships (from predictive algorithms or physics-based models) among observational parameters of astrophysical phenomena to predict the presence of previously unseen events within various databases or sky surveys.
4. *Unknown events/unknown rules* use pattern detection, correlation analysis, or clustering properties of data to discover new observational (in this case, astrophysical) relationships (i.e., unknown connections) among astrophysical phenomena within multiple databases or use thresholds or outlier detection techniques to identify new transient or otherwise unique ("one-of-a-kind") events, and thereby discover new astrophysical phenomena (the "unknown unknowns") – some researchers prefer to call this "surprise detection" (Borne and Vedachalam 2012).

2.5 Steps to Discovery with the Virtual Observatory

The above list represents four distinct contexts for astronomical knowledge discovery from large distributed databases. Another view of the same problem space is to examine the basis set of data mining functions (use cases) that astronomers apply to their data – these are the functional primitives, not the actual algorithms (some of which are described in ➋ Sect. 3.2), which might be quite numerous for any given data mining function (Borne 2001a, b). These basic VO-driven data mining use cases include the following:

(a) *Object cross-identification* (or *cross-matching*) – It is a well-known, notorious "problem" in astronomy that individual objects are often called by many names. This is an even more serious problem now with the advent of all-sky surveys in many wavelength bands – the same object will appear in a multitude of VO-accessible catalogs, observation logs, and databases, and the object will almost certainly be cataloged uniquely in each case according to a survey-specific naming convention. Identifying the same object across these distributed data collections is not only crucial to understanding its astrophysical properties and its physical nature, but this use case is also one of the most basic inquiries that an astronomer will make of these all-sky surveys. The problem of isolating these object cross-IDs across multiple surveys reduces in its simplest form to finding high-likelihood *spatial associations* of given objects among a collection of catalogs. Some databases, such as the Sloan Digital Sky Survey, provide cross-matching services, as does OpenSkyQuery.net (Budavari et al. 2009).
(b) *Object Cross-correlation* – After the cross-database identifications of the same object have been established, there are a wealth of astrophysical "What if?" queries that can be applied to the observational parameters of objects that are present in multiple databases. In the most general sense, these correspond to the application of *classification*, *clustering*, and *regression* (cross-correlation or anticorrelation) algorithms among the various database parameters. These correlations need not be confined to two parameters; they may be highly dimensional (e.g., the fundamental plane of elliptical galaxies; Djorgovski and Davis 1987; Dressler et al. 1987). Correlating the observational properties of a class of objects in one astronomical database with a different set of properties for the same class of objects in another database is the essence of VO-style distributed data mining. The results of these inquiries correspond to *pattern associations*.
(c) *Nearest-neighbor identification* – In addition to identifying spatial associations (in real space) across multiple distributed astronomical data archives, it is also scientifically desirable to find "spatial" associations (*nearest neighbors*) in the highly dimensional space of complex astronomical observational parameters – to search for dense *clusterings* and *associations* among observational parameters in order to find new classes of objects or new properties of known classes. The subsets of objects that have such matching sets of parameters correspond to *coincidence associations*. They offer a vast potential for new scientific exploration and will provide a rich harvest of astronomical knowledge discovery from VO-accessible distributed databases.
(d) *Systematic data exploration* – Within very large distributed databases, there are likely to be significant subsets of data (regions of observational parameter space) that have gone largely unexplored, even by the originating scientific research teams that produced the data (Djorgovski et al. 2001). Archival data researchers accessing the databases will want to apply a wide variety of *parameter-based* and *correlation-based* constraints to the data. The exploration process involves numerous iterations on "What if?" queries. Some of these queries will not be well constrained or targeted, and will thus produce very large output sets. For most situations, this will be too large to be manageable by a typical user's desktop data analysis system. So, systematic data exploration includes browse and preview functions (to reveal qualitative properties of the results prior to retrieval) and allows iterations on preceding queries (either to modify or to tighten the search constraints).

(e) *Surprise detection (= outlier/anomaly/novelty discovery)* – This corresponds to one of the more exciting and scientifically rewarding uses of large distributed databases: finding something totally new and unexpected. Outlier detection comes in many forms. One expression of outlyingness may be "interestingness," which is a concept that represents how "interesting" (nontypical) are the parameters that describe an object within a collection of different databases. These may be outliers in individual parameter values, or they may have unusual correlations or patterns among the parameter values, or they have a significantly different distribution of parameter values from its peers (Borne and Vedachalam 2012).

3 Astronomical Data Mining

It is not unreasonable to assert that astronomers have always been data miners (Djorgovski et al. 2001; Borne 2001a, b, 2009; Ball and Brunner 2010) since the three fundamental astronomy research activities map exactly onto the three major categories of ML (machine learning; data mining) research. These ML categories are as follows:

(a) Characterize the data (unsupervised learning, clustering, class discovery)
(b) Assign class labels (supervised learning, classification)
(c) Discover the unknown (semisupervised learning, outlier/novelty detection)

Such ML research methodologies are now becoming more essential than ever as the acquisition of data is rapidly accelerating and causing a nearly insurmountable data avalanche (Bell et al. 2006; Szalay 2008; Borne 2009a, 2010; Hey et al. 2009). Consequently, the landscape of astronomical research is changing in dramatic and transformative ways. Astronomy is now a data-intensive science, and it will become even more data-intensive in the coming decade (Brunner et al. 2002; Szalay et al. 2002; Becla et al. 2006). ML algorithms that are capable of handling extreme-scale data will enable data mining, information retrieval, and knowledge discovery from huge astronomy datasets, thus enabling new realms of knowledge discovery within massively growing data collections.

3.1 Mining Astronomical Data: Images and Catalogs

Astronomical data mining is exploratory data analysis (EDA): it invokes machine learning algorithms to explore data through data mining, surprise detection, and scientific discovery in extreme-scale data. While it is possible to "mine the pixels" within image archives, using machine vision, pattern detection, and feature recognition algorithms, that is not the focus of this chapter – the primary focus here is on mining of catalogs (structured tables and databases), against which many standard data mining and machine learning algorithms can be applied. To illustrate the potential for discovery, two simple examples of astronomical data mining are presented. In each case, structured parameter sets (i.e., multivariate databases or streams of multivariate arrays of parameters) are mined. Following these examples, a summary is provided of the major machine learning algorithms that are invoked in data mining applications (➋ Sect. 3.2), two orthogonal examples of distributed data mining in astronomy are then described (➋ Sect. 3.3), and finally a summary of astronomical data mining research challenges is presented (➋ Sect. 3.4).

3.1.1 Cosmological Data Mining

First, cosmology catalogs can be mined for novel features and surprising correlations, using parameters that correspond to the measured physical characteristics (e.g., size, shape, flux, color, orientation, location, group membership) of the myriads of galaxies that are detected within large sky images. Cosmology catalogs (e.g., SDSS and 2MASS) consist of aggregated (and organized) collections of all the structured information content (hundreds of attributes) representing hundreds of millions of galaxies detected within these sky surveys. Identifying new and unexpected correlations, patterns, and outliers among these parameter sets may reveal new physics and/or new classes of objects. One example of the type of correlation that could be discovered is the so-called fundamental plane of elliptical galaxies. The class of ellipticals has been known for more than 20 years to show dimension reduction among a subset of observed attributes, such that the three-dimensional distribution of three of those astrophysical parameters (velocity dispersion, surface brightness, and radius) reduces to a two-dimensional plane (Djorgovski and Davis 1987; Dressler et al. 1987) that is not parallel to any of the observed parameter planes. The causes of this are essentially gravitational physics (the virial theorem) and star formation physics (i.e., a nearly universal stellar initial mass function).

3.1.2 Time-Domain Data Mining

As a second example, event streams from time-domain astronomical sky surveys (e.g., those aggregated at SkyAlert.org) can be mined for new discoveries, change detection, new classes of objects, new behaviors of known objects, and surprises. Up until recently, these event streams have been small (<100 events per week), but the data rates will soon reach epidemic proportions when the new time-domain sky surveys begin operations (e.g., LSST (Large Synoptic Survey Telescope)) – predicted to generate up to one million events each and every night! For example, during *every 5 min* of its 10-year sky survey duration (~30,000 h total), the LSST will generate more new sky events than have been previously detected in all of recorded human history.

3.1.3 The Role of Machine Learning in Large Sky Surveys

The characterization (unsupervised ML) and classification (supervised ML) of both types of astronomical data sets in the above examples (massive multivariate data catalogs and high-rate parameter streams) are often identified as major research challenges for data-intensive astronomy (Tyson et al. 2008; Ivezic et al. 2008; Bloom et al. 2008; Kegelmeyer et al. 2008; Borne 2008a, b, 2010). Specifically for time-domain astronomy (and large sky surveys that do repeat observations of the sky), the rapid detection, characterization, and analysis of interesting phenomena and emergent behavior in high-rate data streams are critical aspects of the science, enabling data-driven decision support, instrument-steering, and prompt follow-up observation of the most interesting phenomena.

While there are many data mining algorithms for classification (supervised learning), the real discoveries occur through the application of semisupervised and unsupervised learning. One author has stated the opportunity this way: "Unsupervised exploratory analysis plays an important role in the study of large, high-dimensional datasets" (Shabalin et al. 2009). Consequently, this is where some of the most surprising and interesting discoveries will be made in data-intensive scientific disciplines. Specifically, unsupervised learning (data characterization) will play a significant role in time-domain astronomy, particularly for the LSST project, which will publish millions of astronomical event alert messages nightly. One of the major departures from prior thinking on this topic (e.g., Borne 2008a; Bloom et al. 2008) is the focus on *characterization*, which must be distinguished from the typical astronomer's goal: specific object *classification*. Since classification is a form of supervised learning, it relies on human-provided class labels – the result is a set of human-driven subjective assertions about the object or event. Conversely, characterization is a form of unsupervised learning, which relies only on the observed features of the object/event. Consequently, the result is a set of entirely data-driven objective *features*. This feature vector becomes the "knowledge vector" that is exactly what the time-domain astronomy community needs most urgently: it identifies what are the key features of the transient object, its behavior, and its characteristics. For example, a set of characterizations for a single event may indicate that an object increased in brightness, that the increase was a factor of >100, and that this increase occurred within less than 1 day. This is a critical piece of information since it immediately excludes certain classes of astronomical objects and certain types of astrophysical behaviors.

It is inevitable that some feature vectors (data characterizations) will be compromised in this way, either because of faulty input data or of flaws in the characterization engine. Nevertheless, characterizations need not be perfect in order to be useful. The main point here is that characterization (as a form of unsupervised learning) is significantly more robust and repeatable (by different astronomers) than detailed classification. It is also true that one faulty element of a feature vector for a given object will likely make it stand out as an outlier to a machine-learning algorithm – i.e., it will be flagged as inconsistent with other features – this provides a form of quality assurance and error flagging for end-users.

3.2 Knowledge Discovery from Data: Short Tutorial on Algorithms

Two major reviews of data mining in astronomy are Ball and Brunner (2010) and Borne (2009a). A summary of major machine learning algorithms and data mining methods is presented here, but the reader is referred to one of the above articles for more details on astronomy-specific applications and to any of the countless books on the subject of data mining and machine learning for more details on the algorithms (e.g., *Introduction to Data Mining* by Tan et al. 2006; *The Top Ten Algorithms in Data Mining* by Wu and Kumar 2009; *Handbook of Statistical Analysis and Data Mining Applications* by Nisbet et al. 2009; and *Data Mining: Practical Machine Learning Tools and Techniques* by Witten et al. 2011).

A broad division of data mining methods and techniques would include classification (supervised learning), clustering (unsupervised learning), and outlier detection (semisupervised learning). As stated earlier (➋ Sect. 3), these are already familiar concepts and activities for astronomers. Their meaning is essentially the same within the field of machine leaning, with the addendum that the machine learning community assigns a much broader definition to classification and clustering: (a) classification refers to any method that learns from existing labeled data and then assigns labels to new previously unseen data – classification includes such diverse techniques as linear regression, Bayesian classification, neural networks, or decision trees – and (b) clustering refers to partitioning the data in any parameter space, in any number of dimensions, and it also refers to class discovery (i.e., finding the groupings within the dataset without prior knowledge or bias to guide the results) – clustering may be partitional (in which one set of clusters is found) or hierarchical (in which multiple sets of clusters are found), which can be either divisive (top-down) or agglomerative (bottom-up).

A menu of data mining methods would reflect a wide range of applied mathematics techniques, including discrete math, linear algebra, combinatorics, geometry, graph methods, and more. Some common data mining methods are (a) clustering (group together similar items and separate dissimilar items); (b) classification (assign class labels to new data items using the known classes and groups; (c) association mining (find unusually frequent co-occurring associations of parameter values among database objects; (d) regression analysis (predict a numeric parameter value for a dependent variable using the value of an object's independent variable, based upon known values for existing data); (e) link (or affinity) analysis (identify and quantify linkages between data items based upon features that are shared in common); (f) feature detection and pattern recognition (identify known patterns in the data based upon objective features); and (g) outlier detection (find the surprising and unexpected, outside the bounds of prior data or model predictions).

Some of the most common data mining algorithms for classification include Bayesian classification, decision trees, neural networks, regression, support vector machines, Markov models, and K-nearest neighbors. Some of the most common data mining algorithms for clustering include K-means, nearest neighbor, squared error, and mixture models. Other techniques for unsupervised learning include principal component analysis, independent component analysis, association mining, and link analysis. Brief descriptions for a few of these algorithms are given in the following paragraphs. Note that even within one of these categories, there may exist many dozens or hundreds of diverse algorithms within that one category.

3.2.1 Unsupervised Learning: Mixture Models

Mixture models are used to find the optimal number and representative sizes of partitions that best describe the complete data set. The mixtures may be Gaussians or some other kernel function, which can be univariate or multivariate depending on the dimensionality of the parameter space in which one is clustering the data. The actual choice of kernel function is not important since the method aims to characterize (describe) the behavior of the data, not to explain it, in the same way that a Fourier or wavelet decomposition of a data stream is used to characterize the data's behavior. Mixture modeling is a statistical clustering technique in that it attempts to divide the full data sample into a mixture of smaller subsamples and then derives a set of distribution parameters for each component: width (dispersion or variance), mean, and amplitude. One particularly powerful application of mixture modeling is to measure the parameters for the varying mixtures of components as a function of some other independent variable and then examine how the parameters (mean, variance, amplitude) for the various mixture components vary as a function of that independent variable. This technique is useful for finding and quantifying dependencies in multiple dimensions. For an astronomy example, see Pimblett (2011).

3.2.2 Unsupervised Learning: K-Means Clustering

K-means clustering is by far the most popular clustering algorithm, primarily owing to its simplicity and robustness. It requires a nontrivial amount of computing for two reasons: (a) it requires multiple passes through the entire data set, and (b) in each pass, the distance between each data point and the centroids of the tentative clusters needs to be calculated. Nevertheless, it is very effective at finding a good set of K clusters, where K is the number of clusters and is arbitrarily assigned by the researcher. In addition to the selection of K, the algorithm begins with the researcher arbitrarily assigning a center for each of the K clusters in the chosen feature space (see ➋ Sect. 3.5.4). Then, on each successive pass through the entire data set, these steps are performed: (a) the distance between each data object and each cluster center is calculated; (b) the data object is assigned to be a member of the cluster whose center is nearest; (c) after all data objects are assigned to their nearest clusters, the cluster means are recalculated using its updated set of assigned data objects; (d) if the cluster means have not changed since the last iteration, then the process stops and the cluster means are now used as descriptors for the distribution of clusters, but if the cluster means have changed since the last iteration, then the process goes back to step (a) using the new cluster means and then continues. For an astronomy example, see Hojnacki et al. (2007).

3.2.3 Supervised Learning: Decision Tree Classification

Decision trees are applied in two steps: first, the algorithm learns the rules from known (labeled) data, and second, it applies the rules in order to classify new previously unseen data. Learning the rules consists of a progressive process of finding which attribute does the best job in separating the different classes of objects (e.g., star-galaxy separation). After the best attribute (parameter) is found, then the algorithm searches for the best attribute that classifies those objects that were initially misclassified by the first rule that was learned, and the process continues until either there are no more training data from which to learn new rules, or there are no more attributes to test, or the accuracy of the classification rules applied to the test data set indicates that overfitting is taking place (see ➋ Sect. 3.5.8). The method by which the algorithm selects the best parameter at each step is very similar to the game of 20 questions, in which a person tries to guess what another person is thinking by asking a series of questions. In this game, a wise person will start by asking a question that separates the universe of possibilities into nearly equal halves and continues to do this at each step (e.g., Is it an organic entity? Is it a human? Is it a woman? Is she dead? Was she a writer? Did she write poetry? Is it Emily Dickinson?). Conversely, a child may start by asking a question that does nothing to improve her/his chances of discovering the correct answer (e.g., Is it my mommy?) – in this case, the universe of possibilities was reduced from infinity to infinity minus one. Similarly, the decision tree rule learning algorithm uses an entropy measure (specifically, information gain) to find which attribute creates the most order in the data (separating class A from class B) when that attribute is applied as the distinguishing feature of the data set. Decision trees are powerful and very popular, primarily because they are easy to use and understand – a series of questions (yes/no, if/then, attribute greater than or less than X) that are straightforward to interpret and to apply. For an astronomy example, see Ball et al. (2006).

Two interesting modifications of the standard decision tree model-building and rule-learning algorithms are oblique decision trees (Murthy et al. 1994) and random forests

(Breiman 2001). In oblique decision trees, the decision point (at one of the internal nodes of the tree) is not based upon testing values for a single attribute, but it can be based upon a linear combination of attributes (e.g., instead of testing if $X > n$, the oblique tree could test if $aX - cY > n$). In other words, the decision plane is not parallel to any of the parameter axes, but it is oblique relative to any of those. This is very sensible for scientific datasets, particularly when linear combinations of the variables may contain some scientific significance. For an astronomy example, see White (2008).

For random forests, the tree-building phase tends to be more creative. If the database contains N different scientific parameters (attributes), then N decision trees are built. Each one of these N trees uses all of the attributes, but with a different one of the attributes randomized in each distinct tree. Specifically, all of the values for a particular attribute are mixed up, in totally random order, and then reassigned to the objects in the source list (e.g., the B-V color of objects A, B, and C might actually be 0.5, 1.2, and 2.3, but the randomized list may assign their B-V colors as 1.2, 2.3, and 0.5, respectively). This is done for each of the N attributes to be tested in the decision tree model. If the classification accuracy of the decision tree for a given attribute X is not noticeably worse when X is randomized versus the classification accuracy when the X values are assigned to the correct objects, then it is clear that the attribute X carries no significance or weight in the actual classification of the objects, and consequently the attribute X is dropped from the list of attributes that the final decision tree needs to contain. For an astronomy example, see Carliles et al. (2010).

3.2.4 Supervised Learning: K-Nearest Neighbor Classification (KNN)

KNN classification is one of the most simple classification algorithms. It relies on the user's ability to calculate distances between database objects (see ➋ Sect. 3.5.7). In this usage, distance does not mean spatial distance, but it refers to distance in any multivariate parameter space. It is a very expensive algorithm in spite of its simplicity. For each item in the database, its distance is measured to every other item in the training set. Then this set of distances is sorted – the object is then classified into the class that corresponds to the class of the majority of nearest K objects in the training set. K is usually chosen to be the square root of the total number of objects in the training set, with K deliberately chosen to be an odd number (as a tie-breaker). After the first item has been classified in the above manner, the user then reads the second item from the database and repeats the above steps: measure all distances to all objects in the training set, sort all of the distances, find the nearest K neighbors, count how many of the K neighbors are in each of the known classes, and finally assign the new item into the majority class. This continues until all new objects in the database are classified. Clearly, the procedure is expensive as all distances are recalculated with every new item, and the list of distances must be sorted every time. Nevertheless, the simplicity of the algorithm makes it popular. Some users simplify it even further by setting $K = 1$ – an object is assigned to the same class as its nearest neighbor in parameter space. For an astronomy example, see Wang et al. (2010).

3.2.5 Supervised Learning: Support Vector Machine (SVM)

SVM classification is one of the most powerful and most complex classification algorithms. The technique comes from optimization theory in mathematics. It searches for the data transformation that does the best job at maximizing the separation between class A objects and class B objects. The kernel function that performs this data transformation (X,Y) → F(X,Y) may be highly nonlinear and nonobvious (see ➋ Sect. 3.5.6). An objective function is used to find the optimal kernel function. The objective function is essentially the cumulative separation distance between those class A objects and those class B objects that are closest to their separating hyperplane – those specific objects are known as the support vectors for their own respective classes (see ➋ Sect. 3.5.4). SVM is not for the faint-hearted – entire courses on SVM are taught in data science and applied mathematics departments. But it is very accurate and objective. For an astronomy example, see Wadadekar (2005).

3.2.6 Supervised Learning: Artificial Neural Net Classification (ANN)

Similar to SVM, entire courses (even two-semester courses) are taught on ANN. ANNs attempt to emulate the biological neural network within the brains of living organisms. Biological neural networks learn from experience and establish neural pathways that trigger certain responses when subjected to particular inputs. Similarly, an ANN can learn to classify a set of training data based upon the various attribute values of each training object. Then, given the set of attribute values for a new data item, it can calculate the output class of the object from the previously learned nonlinear regression model. ANNs are very accurate and popular, but their biggest shortcoming is that they are a black box – it is almost totally unclear how a given model arrives at a particular classification result – consequently, ANNs are not very popular when one needs to present results to upper management since explaining how the ANN arrived at its result is not particularly straightforward. Nevertheless, if accuracy and efficiency are more important than understanding the algorithm's inner workings, then ANN is a great choice for classification. For an astronomy example, see Bazell and Peng (1998).

3.2.7 Supervised Learning: Bayesian Classification

Bayes' theorem (1763) is a straightforward equation derived from the mathematical tautology: $P(C \text{ and } X) = P(C|X)\,P(X) = P(X|C)\,P(C)$, where $P(C \text{ and } X)$ is the probability of both C and X occurring, $P(X)$ is the probability of X occurring, $P(X|C)$ is the probability of X given a particular value for C, and similarly for $P(C)$ and $P(C|X)$. A simple rearrangement of the terms yields $P(C|X) = P(X|C)\,P(C)/P(X)$, where $P(C|X)$ is interpreted in data mining classification applications to signify the probability that an object has class C given the set of attribute (parameter) values X. It is easy to estimate $P(C|X)$ from a database: (a) estimate $P(X|C)$ by counting how often the set of attribute values $\{X\}$ occurs in class C objects in the training data set; (b) estimate $P(C)$ by counting the frequency of occurrence of each class C in the training set; (c) estimate $P(X)$ by counting the frequency of occurrence of each attribute value combination $\{X\}$ in the training set; and (d) apply Bayes' theorem. Bayesian modeling is also very common for model estimation (e.g., in cosmology). Bayesian classification specifically uses complete samples of classified objects as the training model for classifying previously unseen data items. For an astronomy example, see Sebok (1979).

3.3 Distributed Data Mining in Astronomy

Two examples of distributed data mining (DDM) are presented here. These illustrate the two approaches to DDM: mining of distributed data (MDD), and distributed mining of data (DMD). In its most general form, DDM follows the process flow that is depicted in ➋ *Fig. 9-1* (which is patterned after the data-information-knowledge-understanding paradigm discussed in the Introduction).

3.3.1 Mining of Distributed Data: Finding VLIRGs in Multiwavelength Archives

An early example of mining of distributed data (MDD) was carried out by Borne (2003) – the basic elements of that investigation are presented here. In this case study, some of the key aspects

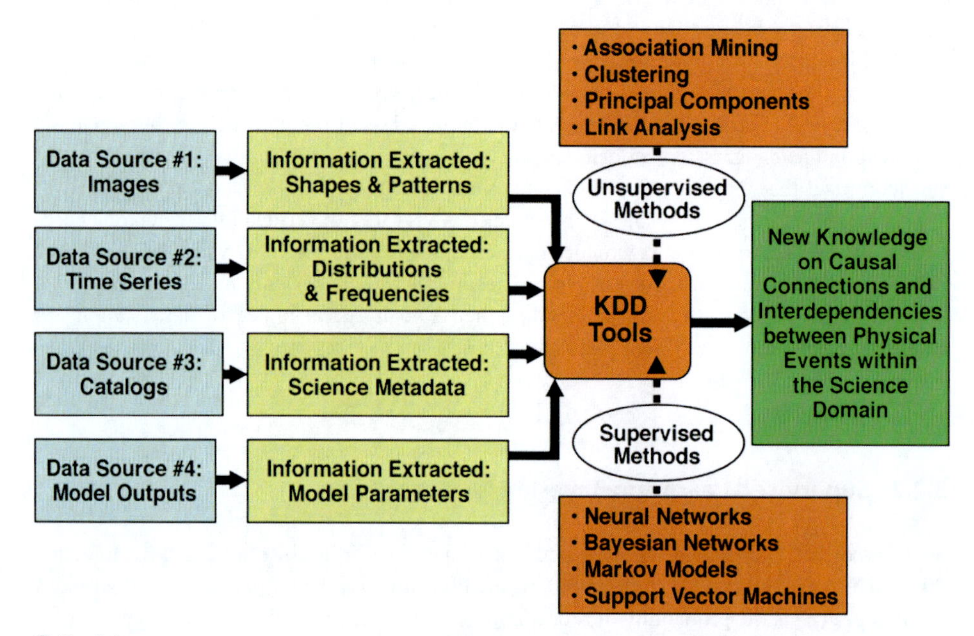

Fig. 9-1

Schematic illustration of the knowledge discovery in databases (*KDD*) process flow for scientific discovery. Numerous astronomical databases, generated from large sky surveys, provide a wealth of data and a variety of data types for mining. Data mining proceeds first with data discovery and access from multiple distributed data repositories, then continues with the information extraction step, and then is enriched through the application of machine learning algorithms and data mining methods for scientific knowledge discovery. Finally, the scientist evaluates, synthesizes, and interprets the newly discovered knowledge in order to arrive at a greater scientific understanding of the universe and its constituent parts, processes, and phenomena. *Supervised methods* are those based on training data – these are primarily classification methods. *Unsupervised methods* are those that do not rely on training data or preknown classes – these include traditional clustering methods

of MDD that relate to astrophysics research problems were presented. The case study focused on one particularly interesting rare class of galaxies that had not been well studied up to that time: the very luminous infrared galaxies (VLIRGs; with infrared luminosities between 10^{11} and 10^{12} solar luminosities). It was anticipated prior to the study that mining across a variety of multiproject multiwavelength multimodal (imaging, spectroscopic, catalog) databases would yield interesting new scientific properties and understanding of the VLIRG phenomenon.

VLIRGs represent an important phase in the formation and evolution of galaxies. On one hand, they offer the opportunity to study how the fundamental physical and structural properties of galaxies vary with IR (infrared) luminosity, providing a possible evolutionary link between the extremely chaotic and dynamic Ultraluminous IR galaxies (ULIRGs; with infrared luminosities between 10^{12} and 10^{13} solar luminosities) and the "boring" set of normal galaxies. On the other hand, VLIRGs are believed to be closely related to other significant cosmological populations in the sense that VLIRGs could be either the low-redshift analogs or the direct result of the evolution of those cosmologically interesting populations of galaxies that typically appear only at the earliest stages of cosmic time. Several such cosmologically interesting populations have been identified by astronomers. These include (1) the galaxies that comprise the cosmic infrared background (CIB); (2) the high-redshift submillimeter sources at very large distance; (3) infrared-selected quasars; and (4) the "extremely red objects" (EROs) found in deep sky surveys. The primary questions pertaining to these cosmological sources are What are they? Are they dusty quasars, or dusty starbursting galaxies, or massive old stellar systems (big, dead, and red galaxies)? Some of these classes of objects are undoubtedly members of the extremely rare class of ULIRGs, while the majority is probably related to the more numerous, though less IR-luminous, class of galaxies: the VLIRGs.

The Borne (2003) data mining research project was a proof-of-concept for a VO-enabled science search scenario, aimed at identifying potential candidate contributors to the CIB (cosmic infrared background). This was significant since it was known that fully one-half of all of the radiated energy in the entire universe comes to us through the CIB. The approach involved querying several distributed online databases and then examining the linkages between those databases, the image archives, and the published literature. As typical for most multimission multiwavelength astronomical research problems, this search scenario started by finding object cross-identifications across various distributed source lists and archival data logs. In a very limited sample of targets that were investigated in order to validate this VO-enabled distributed data mining (MDD) approach to the problem, one object was found in common among three distributed databases. This object was a known hyperluminous infrared galaxy (HyLIRG; with infrared luminosity greater than 10^{13} solar luminosities) at moderate redshift, harboring a Quasar, which was specifically imaged by the Hubble Space Telescope because of its known HyLIRG characteristics. In this extremely limited test scenario, the result was in fact what was searched for: a distant IR-luminous galaxy that is either a likely contributor to the CIB or else it is an object similar in characteristics to the more distant objects that likely comprise the CIB.

While the above MDD case study provides an example of distributed database querying, it is not strictly an example of data mining. It primarily illustrates the power of cross-identifications across multiple databases (see ➋ Sect. 2.5) – i.e., joins on object positions across various catalogs, subject to constraints on various observable properties (such as optical-to-IR flux ratios). After such query result lists are generated, any appropriate machine learning algorithm can then be applied to the data in order to find new patterns, new correlations, principal components, or outliers; or to learn classification rules or class boundaries from labeled (already classified) objects; or to classify new objects based upon previously learned rules or boundaries.

3.3.2 Distributed Mining of Data: The Fundamental Plane of Elliptical Galaxies

In contrast to the above example, the distributed mining of data (DMD) is a major research challenge area in the field of machine learning and data mining. There is a large and growing research literature on this subject (Bhaduri et al. 2008). Despite this fact, there has been very little application of DMD algorithms and techniques to astronomical data collections, and even those few cases have not begun to tap the richness and knowledge discovery potential from VO-accessible data collections. DMD algorithms have been applied to distributed PCA (principal components analysis; Giannella et al. 2006; Das et al. 2009), distributed outlier detection (Dutta et al. 2007), distributed classification of tagged documents (Dutta et al. 2009), and systematic data exploration of astronomical databases using the PADMINI (peer-to-peer astronomy data mining) system (Mahule et al. 2010). In order to provide a specific detailed illustration of one of the astronomical research problems solved by these methods, the distributed PCA results will be presented here (from Bhaduri et al. 2011).

The class of elliptical galaxies has been known for more than 20 years to show dimension reduction among a subset of physical attributes, such that the three-dimensional distribution of three of those astrophysical parameters reduces to a two-dimensional plane (Djorgovski and Davis 1987; Dressler et al. 1987). The normal to that plane represents the principal eigenvector of the distribution, and it is found that the first two principal components capture significantly more than 90% of the variance among those three parameters.

By analyzing existing large astronomy databases (such as the Sloan Digital Sky Survey [SDSS] and the 2-Micron All-Sky Survey [2MASS]), a very large data set of elliptical galaxies with SDSS-provided spectroscopic redshifts was generated. The data set consists of 102,600 cross-matched objects within an empirically determined volume-limited completeness cut-off: redshift $z < 0.23$. Each galaxy in this large data set was then assigned (labeled with) a new "local galaxy density" attribute, calculated through a volumetric Voronoi tessellation of the total galaxy distribution in space. The inverse of the Voronoi volume represents the local galaxy density (i.e., each galaxy occupies singly a well-defined volume that is calculated by measuring the distance to its nearest neighbors in all directions and then generating the three-dimensional convex polygon whose faces are the bisecting planes along the direction vectors pointing toward the nearest neighbors in each direction – the enclosed volume of the polygon is the Voronoi volume). Note that this is even possible only because every one of the galaxies in the chosen sample has a measured redshift from the SDSS spectroscopic survey. Other groups who have tried to do this have used photometric redshifts (photo-z's; Rossi and Sheth 2008), which have (at best!) an uncertainty of $\Delta z = 0.01$ (especially when using a large number of smaller bandpasses; Wolf et al. 2004), corresponding to a distance uncertainty of $\sim 18\,h^{-1}$ Mpc, which is far above the characteristic scale of the large-scale structure ($\sim 5\,h^{-1}$ Mpc comoving). The work described here does not suffer from this deficiency since spectroscopic redshifts are at least one to two orders of magnitude better than photo-z's, and so the distance estimates and Voronoi volume calculations (hence, local galaxy density estimates) are well determined. It is also interesting to note that since the dynamical timescale (age) of a gravitating system is inversely proportional to the square root of the local galaxy density (e.g., Fellhauer and Heggie 2005), consequently the dynamical timescale is directly proportional to the square root of the calculated Voronoi volume. Therefore, studying the variation of galaxy parameters and relationships as a function of each galaxy's local Voronoi volume is akin to studying the time evolution of the ensemble population of galaxies.

For the initial correlation mining work, the entire galaxy data set was divided into 30 equal-sized partitions as a function of the inverse-Voronoi volume independent variable: the local galaxy density. Consequently, each bin contains more than 3,000 galaxies, thereby generating statistically robust estimators of the fundamental parameters in each bin. From the SDSS and 2MASS catalogs, about four dozen measured parameters were extracted for each galaxy (out of a possible 800+ from the combined two catalogs): numerous colors (computed from different combinations of the five-band SDSS $u'g'r'i'z'$ photometry and the three-band 2MASS JHK_s photometry), various size and radius measures, concentration indices, velocity dispersions (from SDSS), and various surface brightness measures. For those parameters that depend on distance, the SDSS redshift was used to recalculate physical measures for those parameters (e.g., radii, surface brightness).

As a result of the data sampling criteria, one is able to study eigenvector changes of the fundamental plane of elliptical galaxies as a function of density. This distributed data mining problem has also uncovered some new astrophysical results: it was found that the variance captured in the first two principal components increases systematically from low-density regions to high-density regions (➋ *Fig. 9-2*), and it was found that the direction of the principal eigenvector also drifts systematically in the three-dimensional parameter space from low-density regions to the highest-density regions (➋ *Fig. 9-3*).

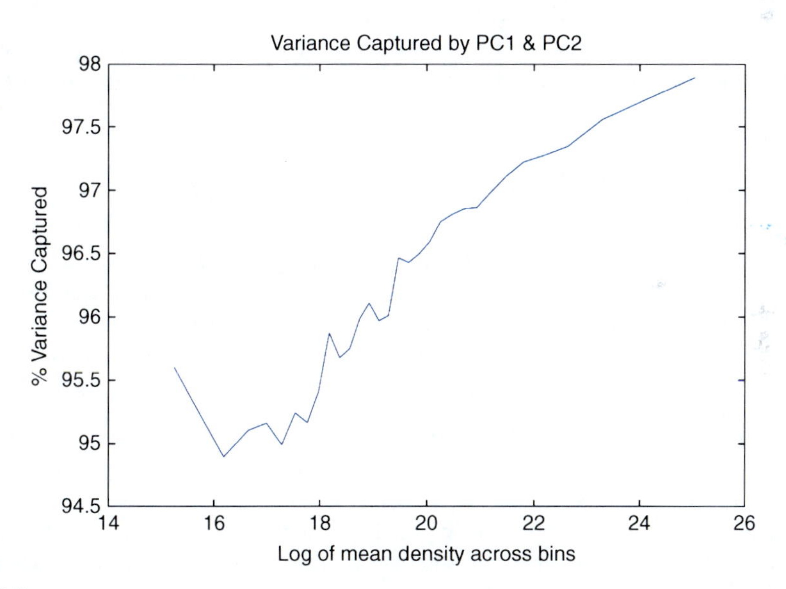

Fig. 9-2

Variance captured by the first two principal components with respect to the log of the mean galaxy local density (for 30 different bins containing ~3,000 galaxies each). The sample parameters used in this analysis are i-band Petrosian radius containing 50% of the galaxy flux (from SDSS), velocity dispersion (SDSS), and K-band mean surface brightness (2MASS). This plot clearly shows that the fundamental plane relation becomes tighter with increasing local galaxy density (inverse Voronoi volume) (Das et al. 2009)

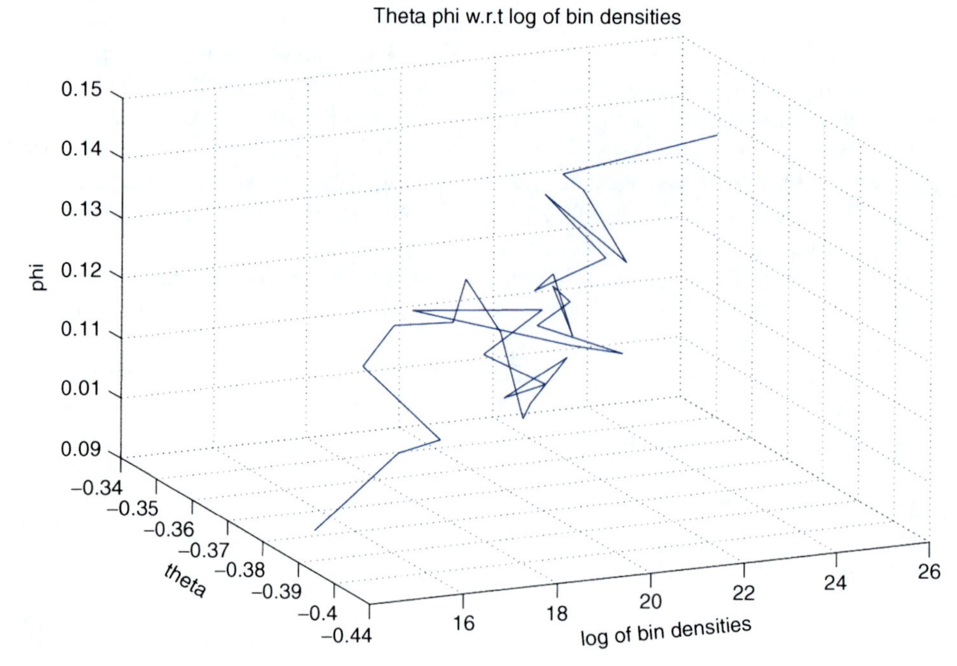

Fig. 9-3
Direction cosines for the normal vector to each fundamental plane calculated as a function of local galaxy density (inverse Voronoi volume). Though there are some chaotic effects, in general, there is a trend for the tilt of the fundamental plane to drift nearly systematically (Das et al. 2009)

3.4 Research Challenges in Astronomical Data Mining

For young researchers or for anyone starting a research program in this area, it is quite informative and instructive to examine (and to take direction from) the data mining research challenges that have already been documented in other reports. Research challenges have been distilled below from three reports in particular, all of which were sponsored by the Department of Energy Office of Science (DOE-1 2007; DOE-2 2008; DOE-3 2008):

- Develop analysis methodologies for discovering and characterizing emergent behavior in complex systems and systems of systems, to aid in the discovery of unexpected dynamics.
- Develop efficient methods for the statistical analysis of large, heterogeneous data sets.
- Develop rigorous but computationally feasible methods for dimensional reduction of data.
- Develop reliable and mathematically sound techniques for deducing information from data and knowledge sources that are high dimensional and heterogeneous in nature and quality.
- Re-engineer machine learning and statistics algorithms to scale with the size of datasets.
- Develop algorithms for detection of outliers, anomalies, and nonstationarity in data streams.
- Develop new mathematics to extract novel insights from complex data.
- Additional research areas include characterizing areas of interest in massive time-varying data, identification of areas in massive datasets which scientists should investigate in-depth, statistical analysis, feature detection and tracking, creation of condensed representations

of data (for visualization and exploration), summarization and abstraction of petascale datasets, parametric exploration of complex scientific datasets, interactive mining of petascale time-varying multivariate datasets, near real-time identification of anomalies in streaming data, knowledge capture from petascale data, and mechanisms for massive data management and analysis.

The above list represents a remarkably rich, inviting, and comprehensive research agenda.

3.5 Steps to Discovery with Data Mining

It is sometimes said that data mining is not auto-magic. In other words, there is a lot of manual intervention required, particularly at the earliest stages of a data mining project. This intervention is usually dominated by the data preparation (or data cleaning) phase, but it also includes several additional data-focused activities as identified below – some of these steps are inherent in most forms of scientific data analysis, not unique to data mining applications. Summarized below are some of the key steps and basic concepts for achieving success in a data mining project.

3.5.1 Concept: Key Steps in the Data Mining Process

First, it is essential to recognize that data mining is a process, not a haphazard application of subroutines, scripts, and free software. The data mining process follows a sequence of key steps. The key steps that are invoked in any data mining project are (a) data browse, preview, and selection; (b) data cleaning and preparation; (c) feature selection; (d) data normalization and transformation; (e) similarity and/or distance metric selection; (f) select the data mining method and appropriate algorithm; (g) apply the data mining algorithm to the data; (h) gather and analyze the data mining results; (i) estimate the accuracy of the results; (j) verify that over-fitting has been avoided; and (k) take action (since data mining should be used as a means to an end: decision support through actionable intelligence from data; e.g., write a paper on the results; write a proposal to get more data; retrieve more data from the archive; obtain more data at the telescope).

3.5.2 Concept: Data Previewing (Data Profiling)

Examining one's data prior to the application of any analysis tool is always a good idea. In data mining, this is especially true since often the data mining tool is a black box (i.e., the scientist does not actually see what is happening nor sees intermediate steps). Data previewing allows the researcher to assess the good, bad, and ugly parts of the database. Data previewing activities may include (a) histograms of parameter value distributions; (b) scatter plots of parameter combinations; (c) min-max value checks (especially check these against expected min-max values); (d) summarizations and aggregations (e.g., mean, median, mode, variance, skewness, kurtosis, sums ordered by groups of parameters); (e) find all of the unique parameter values (and check these against expected values); (f) check physical units; (g) check that the data are scaled properly; (h) external checks (ideally with other databases, to cross-validate some of the values); and (i) verify that the output from the data retrieval request is consistent with the input query.

3.5.3 Concept: Data Preparation (Cleaning the Data)

It is often said that the data preparation phase of a data mining project may take 40–60% of the total effort. Experience has shown that this is usually a significantly higher percentage. Data preparation includes many steps, some of which are typically called data reduction in astronomy. The steps include (a) dealing with NULL (missing) values (remember: absence of evidence is not the same as evidence of absence); (b) dealing with measurement error and other data errors; (c) dealing with (or simply characterizing) noise; (d) dealing with outliers (unless finding outliers is the goal of the science research); (e) transformations (e.g., units, scale factors, projections of the data, linear combinations or ratios of data values); (f) data normalization (to put parameters with wildly different numerical values onto equal weighting); (g) relevance analysis (i.e., feature selection – identifying the scientifically valid and interesting parameters to mine); (h) remove redundant attributes (e.g., recession velocity and redshift of galaxies – choose one); and (i) dimensionality reduction (especially if several large databases are used, such as SDSS and 2MASS, which could generate several hundred parameters to correlate, visualize, and mine).

3.5.4 Concept: Feature Selection (Building the Feature Vector)

A feature vector is the attribute (parameter) vector of an object, containing only those attributes that describe, characterize, (ideally) define, and uniquely identify the object. The set of feature vectors for all objects to be mined provides the fundamental data for the data mining algorithm. The feature vector is usually only a subset of all possible parameter values within the database (e.g., one may choose to find correlations only between color and luminosity of the sources, so the feature vector may consist of four attributes: RA, Dec, color, luminosity). Ideally, for classification applications, the feature set should be the smallest subset of scientific parameters that are sufficient to predict class labels well, without overfitting (see ➋ Sect. 3.5.8). There are objective means for selecting the optimal set of features: PCA (principal component analysis), correlation analysis, or information gain (used in decision tree calculations; ➋ Sect. 3.2.3). There are also subjective means for selecting the feature set: good old-fashioned domain knowledge (i.e., scientists often know which scientific parameters are most useful, most meaningful, and most effective in classifying the objects that they know best).

3.5.5 Concept: Data Types

Being aware of the different data types to be mined is critical, in order to feed the correct data to the algorithm. For example, a regression algorithm will require continuous numeric data; logistic regression works with discrete data; and Bayesian classification can work with categorical data. The various data types (not mutually exclusive) are (a) continuous (e.g., any scientifically measured numerical value); (b) discrete (i.e., binary (0/1, or star/galaxy, or yes/no); boolean (true/false), cardinal numbers (number of objects in a multiple system), or a specific list of allowed values (e.g., constellation names, instrument modes); and (c) categorical (i.e., non-numeric character or text data, such as the name of the object), which can be ordinal (ordered values, such as the stellar classification sequence OBAFGKM) or nominal (not ordered, such as a list of possible classifications for an object).

3.5.6 Concept: Data Normalization and Transformation

Data normalization transforms data values for different parameters into a uniform set of units or into a uniform scale (e.g., a common min-max range). Data normalization can also be used to assign the correct weighting to the different parameters in the feature vector (e.g., weight the parameters in inverse proportion to their measurement error). Examples include (a) transforming all numerical values from min to max to a 0–1 scale, or −1 to 1, or 0–100; (b) normalizing data to zero mean and unit variance; (c) converting (encode) discrete or character (categorical) data into numeric values (specifically for use in a data mining algorithm that requires numeric inputs); (d) transforming ordinal data to a ranked list (numeric values); and (e) discretizing continuous data into bins (specifically for use in a data mining algorithm that requires discrete values as input).

3.5.7 Concept: Similarity and Distance Metrics

Similarity between complex data objects is one of the central notions in data mining, especially in clustering analyses. The fundamental problem that the researcher faces is to determine whether any selected pair of data objects exhibit similar characteristics. The problem is both interesting and difficult because the similarity measures should be applicable to nonnumeric data and because the similarity meaures should allow for imprecise matches. Similarity and its inverse (distance) provide the basis for all of the fundamental data mining clustering techniques and for many data mining classification techniques (such as nearest-neighbor classification; Sect. 3.2.4). Clustering algorithms depend on a distance (or similarity) measure in order to determine (a) the closeness, similarity, or "alikeness" among members of the same cluster, and (b) the distance, dissimilarity, or "unlikeness" between members from different clusters. The biggest challenge in defining a similarity or distance metric arises with categorical data. For example, what is the distance from hot to warm? Is it larger than the distance from warm to cold? Or consider this: Is the distance from blue to red larger or smaller than the distance from round to square?

Any similarity metric sim(A,B) or distance metric dist(A,B) must satisfy several general mathematical requirements when comparing two database objects A and B: (a) identity: dist(A,B) = 0 and sim(A,B) = 1 if and only if A and B are identical; (b) positivity: dist(A,B) ≥ 0 and sim(A,B) ≥ 0; (c) symmetry: dist(A,B) = dist(B,A) and sim(A,B) = sim(B,A); and (d) the triangle inequality: dist(A,C) $\leq$ dist(A,B) + dist(B,C). The latter is not strictly required, but metrics that violate the triangle inequality (called semimetrics) can produce very unusual (perhaps quite interesting) results (e.g., Missaoui et al. 2005). Note that in order to calculate the "distance" between different parameter values, those values must typically be transformed or normalized, either to the same units or to a similar scale (or both). The normalization of both categorical (nonnumeric) and numerical data with different units generally requires domain expertise (i.e., it cannot be automated entirely). This is part of the preprocessing (data preparation) step in any data mining project.

Some of the most popular and widely used similarity and distance metrics are (a) the general Lp-norm distance = $(\mathrm{sum}(|x - y|)^p)^{1/p}$; (b) Euclidean distance (L2-norm); (c) Manhattan distance (L1-norm) = $|x1 - y1| + |x2 - y2| + |x3 - y3| + \ldots$, where {x1,x2,x3,...} is the feature vector for object x and {y1,y2,y3,...} is the feature vector for object y; (d) cosine similarity = cosine of the angle between two feature vectors = the dot product of the two vectors divided by the product of their norms; and (e) general similarity and distance functions derived from the other: sim = 1/(1 + dist), or dist = (1/sim) – 1. In the latter case, similarity varies from 1 to 0 as distance varies from 0 to infinity.

3.5.8 Concept: Overfitting

Overfitting is very common in regression analyses or any other curve-fitting application. Specifically it occurs when the curve to be fitted has more power than it should have for noisy or nonmonotonic data, such as fitting a 99th-order polynomial through a time series of 100 noisy measurements. Such a fit is unlikely to represent reality, but it is more likely to be a reflection of either overzealousness, or extreme optimism, or serious misunderstanding in the researcher. Similarly, when applying a classification algorithm in data mining to a set of data, the researcher will use a set of training data to train the algorithm and then validate the accuracy with a set of test data. If the training is allowed to reach perfect accuracy on the training data, then it is likely to have bad error properties on the test data and seriously bad error properties on new data yet to be classified. The remedy for avoiding overfitting is to monitor continuously the accuracy of the classification results against the test data, as the classification model is being learned. As long as the test data accuracy continues to improve, in concert with improving accuracy on the training data set, then all is well. But, as soon as the error rate on the test data begins to worsen, even if the accuracy on the training data is still improving, then it is time to stop the training – the model is as good as the data will allow it to be. Of course, if more data are available for additional training and testing, then further refinements in the model are warranted, insofar as the test data accuracy continues to improve.

Overfitting is one of the most common and easily overlooked errors in data mining experiments. To monitor this and thereby to avoid it, researchers have tried a number of data partitioning techniques on the initial set of labeled data: (a) 50–50 partitioning of the initial data into training data and test data or (b) tenfold validation, which includes ten separate training experiments with 90% of the data being used for training and with ten different permuted selections of the other 10% of the data being used for testing. In the latter case, if some of the experimental test results are good and some are bad, then it is possible that overfitting is the culprit, and consequently the model is probably overconstrained, thus requiring fewer parameters or lower-power terms in the fit.

3.5.9 Concept: Accuracy

Classification accuracy has been mentioned several times previously (e.g., ➋ Sects. 3.2.3, ➋ 3.2.6, and ➋ 3.5.8). Having a robust objective measure of accuracy is critical to the classification process, particularly during the training and testing phases of learning the classifier. The simplest measure of accuracy is to estimate the precision and recall of a classifier when applied to the test data set. (Note: one does not test the accuracy of the classifier on the training set since this was the set that was used to learn the classifier – this would be an example of circular logic.) When applied to the test data set, the precision metric measures the number of objects classified correctly relative to the total number of objects assigned to a particular classification, and the recall metric measures the number of objects classified as a particular class relative to the total number of objects that are truly in that class within the original test dataset. Consider this example: assume that the test dataset consists of 1,000 galaxies, of which 50 are known starbursts and 100 are known dry mergers. If the classifier identified 200 of the test set galaxies as starbursts, of which 48 are truly starbursts, then the precision of the starburst classifier is only 24% (48/200), while its recall is 96% (48/50). In this first example, the precision of the classifier is poor (low accuracy), but the recall is high. Suppose now that the classifer identified 40 of the test set galaxies as dry mergers, of which 36 are truly dry mergers, then the precision of the dry merger classifier is 90% (36/40), but the recall is only 36% (36/100). One of these examples appears to have done well from the perspective of precision, while the other appears to be better from the perspective of recall. It is desirable to have both high-precision and high-recall accuracy. If that is not possible, then some combination of the two is needed in order to find the optimal solution (i.e., optimal classifier). To accommodate this in the accuracy calculation, one can use different estimators for the accuracy: (a) overall accuracy, (b) producer's accuracy, and (c) user's accuracy. To illustrate these, consider again the starburst classification example. In this example, the number of true positives (TP) identified was 48, the number of false positives (FP) identified was 152 (200–48), the number of true negatives (TN) was 798 (800–2), and the number of false negatives (FN) was 2 (the two misclassified starbursts). For these numbers, we find the following measures of accuracy:

Overall accuracy = (TP+TN)/(TP + FP + TN + FN) = 48+798/1,000 = 84.6%
Producer's accuracy = (TP)/(TP + FN) = 48/(48+2) = 96% = RECALL
User's accuracy = 4 (TP)/(TP + FP) = 48/(48+152) = 24% = PRECISION

Therefore, besides their implicit meaning, the terms "precision" and "recall" also have these meanings: producer's accuracy (recall) measures the percentage of cases in which the classifier (the producer) correctly classifies a particular class of objects (i.e., recalling them from the database with the correct class label), and user's accuracy (precision) measures the confidence that the user has in the accuracy (i.e., precision) of the classification of any particular object assigned to that class. As a result of this analysis, we see that the overall accuracy can be used as an estimator of combined (precision + recall) accuracy during the training phase of data mining classification algorithms. This is particularly useful for those algorithms that are trained iteratively, for which an estimate of accuracy is needed following each iteration. Finally, we note that the overall accuracy of the dry merger classifier in the above example is as follows:

$$\text{Overall accuracy} = (36 + 896)/1,000 = 93.2\%,$$

where TP = 36, FP = 4, FN = 64, and TN = 896. Thus, in this simple example, the dry merger classifier can be considered as more accurate than the starburst classifier.

3.5.10 Concept: Bagging and Boosting

Overfitting is occasionally induced when the researcher is aware of outlier or rare classes in the dataset. In these cases, it is difficult to train the classifier very accuractly since there are insufficient examples of the rare classes in the training set to provide an accurate representation of the properties of these rare classes. Consequently, the researcher tries to force the classifier to include these rare instances in the training phase. As a consequence of this overzealousness, the researcher may continue to extend the training of the classifier beyond the point at which overfitting sets in. Two techniques are available in such cases, which take into account these rare classes without leading to overfitting. These two approaches are bagging and boosting (Quinlan 1996) – they are used to improve the predictive accuracy of the classifiers on rare classes. Boosting refers to the process of weighting rare classes and rare instances more heavily when calculating the accuracy of the classifier – i.e., poor classification performance on the rare classes is penalized more heavily than poor accuracy on the more common classes (which are probably more easily classified by any number of other methods anyway). Bagging refers to the process of adding statistically more training examples of the rare objects during the classifier training phase. Bagging is achieved through a procedure called "bagging with replacement" – in this case, when any object is selected for the training sample via random sampling of the original dataset, that object is placed back into the original sample. In this way, each object has an equal chance of being selected more than once, and the rare objects (representing rare classes) are similarly likely to have a higher incidence of occurrence in the final training set than in a nonbagging training set selection process.

4 Astroinformatics: The Science of Data-Oriented Astronomical Research

Astronomy and many other scientific disciplines are developing subdisciplines that are information-rich and data-intensive to such an extent that these are now becoming (or have already become) recognized stand-alone research disciplines and full-fledged academic programs on their own merits (e.g., bioinformatics, and geoinformatics). The late Jim Gray emphasized the development of this new science paradigm (data-intensive science) by naming it *X-Informatics*, where X refers to any science (Gray 2003). In this context, informatics specifically means data science (including information science), which is *the discipline of organizing, accessing, integrating, and mining data from multiple sources for discovery and decision support.* Thus, astroinformatics is the new data-oriented paradigm for twenty-first century astronomy research and education. A detailed summary and research roadmap for the emerging field of astroinformatics appears in the paper by Borne (2010), which was derived from an ASTRO2010 position paper on the same topic (Borne et al. 2009). Some of the major topics in astroinformatics are presented here, starting with a high-level summary of the data science (informatics) research discipline within the context of astronomy (➋ Sect. 4.1), followed by an outline of the major astroinformatics research themes and challenge areas (➋ Sect. 4.2) and ending with a vision for astronomical discovery using data-oriented methodologies (the fourth paradigm of science; ➋ Sect. 4.3).

4.1 Astroinformatics: Data Science Research for Astronomy

Astroinformatics includes a broad spectrum of informatics specialties including data modeling, data organization, data description, transformation, and normalization methods for data integration and information visualization and knowledge extraction, indexing techniques, information retrieval methods, data mining and knowledge discovery methods, content-based and context-based information representations, consensus semantic annotation tags, astronomical classification taxonomies, astronomical concept ontologies, data-intensive computing, and astrostatistics (Borne 2007). These methodologies enable data integration, data mining, information retrieval, knowledge discovery, and scientific decision support (e.g., robotic telescope operations and object selection) across heterogeneous massive data collections.

➲ *Figure 9-1* illustrates the connection between data mining and astroinformatics. This graphic schematically represents the traditional data-to-information-to-knowledge-to-understanding flow from data to meaning (from sensors to sense) specifically within a scientific context (Borne 2007). As described earlier (➲ Sect. 3.3), this also illustrates the KDD (knowledge discovery in databases; i.e., data mining) process flow for scientific knowledge discovery: (a) from distributed heterogeneous data sources, information of different types is extracted; (b) from the information, new knowledge nuggets are mined using a variety of KDD (machine learning) algorithms; and (c) with this knowledge, one applies human reasoning to attain greater understanding of the universe. What is most interesting in ➲ *Fig. 9-1* is the informatics layer. The *informatics layer* is indicated by the boxes labeled "information extracted." On the left is the "data layer" – standardization is not required nor is it feasible, in this layer. Astronomy has excellent discipline-wide agreement on data syntax (the standard data format FITS), but patchy agreement on context and semantics: how the data are stored, organized, indexed, queried, or presented to the user; and how to repurpose and independently interpret the data. The middle layer is the informatics layer – this is where standardized representations of the "information extracted" are needed, for use in the "KDD layer" (data mining layer). The informatics layer could be described as the "metadata layer" since it captures "information about the data" (as do traditional metadata), but it is a *rich metadata layer*. The informatics layer includes standardized scientific metadata, but it also includes taxonomies, ontologies, UCDs (Uniform Content Descriptors; Dolensky 2004), astronomical vocabularies, and XML-based (self-documenting) representations of the information content extracted from the data (Borne 2007).

Metadata-enabled science (e.g., astroinformatics) applications include data description, data discovery, data integration, intelligent data understanding, data retrieval recommendations, archival data self-description and self-discovery, data quality tagging, relevance analysis, specification of science catalog parameters, dimension reduction, science database feature selection, constraint-based mining, data provenance (history, tracking, and re-creation), semantic data discovery and information integration, and more. Consequently, a major emphasis in astroinformatics development will be in the science data middle layer (➲ *Fig. 9-1*), which (a) is more astronomy-specific than the general data management structures and policies at the physical layer and (b) has more semantic "understanding" and concept linkages (associations) than typical astronomy-specific query engines at the interface layer. The informatics layer is thus used to represent (through all of the various types of metadata) the information content that is extracted from data archives, which is then used in data integration, machine learning (data mining), and visualization algorithms to facilitate scientific knowledge discovery and experimental process decision support (such as robotic telescopes and automatic follow-up observations of real-time alerts).

4.2 Research Challenges in Astroinformatics

Borne (2010) outlined the general themes and challenges in astroinformatics research and education. These could represent a career of research topics for young astroinformatics researchers:

- Innovative uses of information technology in astronomical applications, including decision support (for robotic telescopes and responding to targets of opportunity) and dynamic data-driven applications (such as autonomous follow-up observations of real-time sky alerts)

- Tools for organizing, storing, and indexing very large datasets, including large sky surveys and numerical simulations
- Efficient management and utilization of information and data that are discoverable and accessible through VO portals, including knowledge acquisition and management, science workflow tools, data mining, acquisition and dissemination, novel visual presentations, and stewardship of large-scale data repositories and archives
- Information and knowledge processing, natural language processing, information extraction, integration of data from heterogeneous sources, event detection, and feature recognition
- Knowledge representation, including astronomical folksonomies, vocabularies, taxonomies, ontologies, and related semantic metadata
- Metadata for linkage between experimental and model results to benefit astronomical research (e.g., the matching of numerical simulation outputs with observed astrophysical phenomena)
- Human-machine interaction, including interface design, use and understanding of science discipline-specific information, intelligent agents, and information needs and uses, all of which are tailored to specific astronomical research problems and projects
- High-performance computing and communications relating to scientific applications, including efficient machine-machine interfaces, transmission and storage, real-time decision support
- Innovative uses of information technology and authentic scientific data in the classroom to enhance learning, retention, and understanding of science discipline-specific information

4.3 A Vision for Discovery with Astroinformatics

Machine learning and data mining algorithms, when applied to very large data streams, can generate the classification labels (tags) automatically and autonomously. Generally, scientists do not want to leave this decision making to machine intelligence alone – scientists prefer to have human intelligence in the loop also. When humans and machines work together to produce the best possible classification label(s), this is collaborative annotation (an example of this is citizen science, such as the Galaxy Zoo project; Fortson et al. 2011). Collaborative annotation is a form of human computation (von Ahn 2007). Human computation refers to the application of human intelligence to solve complex difficult problems that cannot be solved by computers alone. Humans can see patterns and semantics (context, content, and relationships) more quickly, accurately, and meaningfully than machines. Human computation therefore applies to the problem of annotating, labeling, and classifying voluminous data streams. Of course, the application of autonomous machine intelligence (data mining and machine learning) to the annotation, labeling, and classification of data granules is also valid and efficacious. The combination of both human and machine intelligence is critical to the success of event characterization and classification in enormous data-intensive astronomy sky survey projects, especially time-domain surveys like the LSST (see ➲ Sect. 3).

Astroinformatics achieves its most useful form when the various methodologies of the discipline (i.e., data organization, data description, data characterization, astronomical classification taxonomies, astronomical concept ontologies, data mining, machine learning, visualization, semantic science, and statistics) are all brought together to solve critical astronomical research problems. Consequently, it is important to issue a challenge to astronomy departments everywhere: astroinformatics is a new paradigm for graduate training in data-intensive astronomy that must be taught. In astronomy, the Data-Sensor-Computing-Model synergy is self-perpetuating (Eastman et al. 2005), as rapid advances in the two technology areas (sensors [telescopes and detectors] and computation) have continued unabated (Gray and Szalay 2004), leading to (a) still more data (Becla et al. 2006); (b) amazingly rich numerical simulation outputs (e.g., the Millennium Simulation, Springel et al. 2005); and (c) the need for increasingly scalable and more robust models (Kegelmeyer et al. 2008). With the anticipated accelerating advance in data generation capabilities over the coming years, the field of astronomy will demand an increasingly skilled workforce in the areas of computational and data sciences in order to confront these challenges. Such skills are more critical than ever. It is imperative that astronomy graduate programs advance their offerings into the twenty-first century: the next generation of astronomers' experiences with data are now increasingly more virtual (through online databases and the VO) than physical (through trips to mountaintop observatories).

Knowing how to mine, analyze, visualize, and derive scientific knowledge from large complex data collections are essential skills for future astronomers. A new model for interdisciplinary astronomy graduate education is envisioned, one that provides unique training in the fundamental astronomical and astrophysical topic areas required for research success, plus a rigorous suite of graduate courses in high-performance computing, data mining, statistics, time-series analysis, and information science. This cyber-oriented entree into astronomical research can provide a pathway and springboard for diverse career opportunities and (at the undergraduate level) for broadening participation in STEM disciplines.

Clearly, this is an ambitious astroinformatics agenda. It will not be fully accomplished in just a year or two. It will require several years of research and development. This is fortunate since the most dramatic need for knowledge discovery from large data sets in astronomy will come with the start-up of the very large time-domain surveys in the next decade, such as the LSST sky survey. The research community has a few years to get these astroinformatics methodologies right – all of those years are needed in order to complete the challenging research program described in this review. In so doing, astronomers can hope to mine effectively and efficiently the vast data repositories in astronomy for the wealth of scientific discoveries hidden therein.

Acknowledgments

This research has been supported in part by NASA AISR grant number NNX07AV70G. The author thanks numerous colleagues for their significant and invaluable contributions to the ideas expressed in this chapter: Jogesh Babu, Douglas Burke, Andrew Connolly, Timothy Eastman, Eric Feigelson, Matthew Graham, Alexander Gray, Norman Gray, Suzanne Jacoby, Thomas Loredo, Ashish Mahabal, Robert Mann, Bruce McCollum, Misha Pesenson, M. Jordan Raddick, Keivan Stassun, Alex Szalay, Tony Tyson, and John Wallin. The author is grateful to Dr. Hillol Kargupta and his research associates for many years of productive collaborations in the field of distributed data mining in virtual observatories.

References

Abell, G. O. 1958, ApJS, 3, 211

Ball, N. M., & Brunner, R. J. 2010, Data mining and machine learning in astronomy. Int. J. Mod. Phys. D, 19(7), 1049

Ball, N. M., & McConnell, S. 2011, IVOA KDD-IG: A User Guide for Data Mining in Astronomy, downloaded from http://www.ivoa.net/cgi-bin/twiki/bin/view/IVOA/IvoaKDDguide

Ball, N. M., et al. 2006, ApJ, 650, 497

Bayes, Rev. T. 1763, An essay toward solving a problem in the Doctrine of chances. Philos. Trans. R. Soc. Lond., 53, 370

Bazell, D., & Peng, Y. 1998, ApJS, 116, 47

Becciani, U., et al. 2010, Publ. ASP, 122, 119

Becker, A. C. 2008, AN, 329, 280

Becla, J., et al. 2006, Designing a multi-petabyte database for LSST, in Observatory Operations: Strategies, Processes, and Systems, Proc. SPIE, Vol. 6270, ed. D. R. Silva, & R. E. Doxsey. doi:10.1117/12.671721

Bell, G., Gray, J., & Szalay, A. 2006, Petascale computational systems. IEEE Comput, 39(1), 110

Bennett, A. S. 1962, Mem. R. Astron. Soc., 68, 163

Bhaduri, K., Das, K., Liu, K., Kargupta, H., & Ryan, J. 2008 (Release 1.8), downloaded from http://www.cs.umbc.edu/~hillol/DDMBIB/

Bhaduri, K., et al. 2011, J. Stat. Anal. Data Min., 4(3), 336

Bloom, J. S., Butler, N. R., & Perley, D. A. 2007, Gamma-ray bursts, classified physically, in AIP Conf. Proc., Vol. 1000 (Melville, NY: American Institute of Physics), Gamma-Ray Bursts, 11

Bloom, J. S., et al. 2008, Towards a real-time transient classification engine. Astron. Nach., 329, 284

Boch, T., Fernique, P., & Bonnarel, F. 2008, Astronomical Data Analysis Software and Systems (ADASS) XVI, ASP Conf. Ser. 394 (Chicago: Astronomical Society of the Pacific), 217

Borne, K. 2001a, Science user scenarios for a VO design reference mission: science requirements for data mining, in Virtual Observatories of the Future, ASP Conf. Ser. 225 (Chicago: Astronomical Society of the Pacific), 333

Borne, K. 2001b, Data mining in astronomical databases, in Mining the Sky (Berlin/Heidelberg: Springer-Verlag), 671

Borne, K. 2003, SPIE Data Mining and Knowledge Discovery, Vol. 5098 (Bellingham: SPIE), 211

Borne, K. D. 2007, Astroinformatics: the new eScience paradigm for astronomy research and education. Microsoft eScience Workshop at RENCI, downloaded from http://research.microsoft.com/en-us/um/redmond/events/escience2007/escienceagenda_posters.aspx

Borne, K. 2008a, A machine learning classification broker for the LSST transient database. Astron. Nach., 329, 255

Borne, K. 2008b, Data science challenges from distributed petascale astronomical sky surveys, in DOE Workshop on Mathematical Analysis of Petascale Data, downloaded from http://www.orau.gov/mathforpetascale/slides/Borne.pdf

Borne, K. 2009a, Scientific data mining in astronomy, in Next Generation Data Mining (Chapman and Hall/Boca Raton: CRC), 91

Borne, K. 2009b, The VO and Large Surveys: What More Do We Need? downloaded from http://www.astro.caltech.edu/~george/AIworkshop/Borne.pdf

Borne, K. 2009c, The Zooniverse: Advancing Science through User-Guided Learning in Massive Data Streams, downloaded from http://www.kd2u.org/NGDM09/schedule_NGDM/schedule.htm

Borne, K. 2010, Astroinformatics: data-oriented astronomy research and education. Earth Sci. Inform., 3, 5

Borne, K., & Vedachalam, A. 2012, Surprise detection in multivariate astronomical data, in Statistical Challenges in Modern Astronomy V, ed. E. D. Feigelson, & G. J. Babu (New York: Springer), 275–290

Borne, K., Becla, J., Davidson, I., Szalay, A., & Tyson, J. A. 2008, The LSST data mining research agenda, in AIP Conference Proceedings for Classification and Discovery in Large Astronomical Surveys, Vol. 1082 (Melville, NY: American Institute of Physics), 347

Borne, K., et al. 2009, Astroinformatics: a 21st century approach to astronomy, in ASTRO2010 Decadal Survey in Astronomy and Astrophysics position paper, arXiv:0909.3892v1

Breiman, L. 2001, Mach. Learn., 45(1), 5

Brunner, R., Djorgovski, S. G., Prince, T. A., & Szalay, A. S. 2002, Massive datasets in astronomy, in The Handbook of Massive Data Sets, ed. J. Abello, P. M. Pardalos, & M. Resende (Norwell: Kluwer), 931–979

Budavari, T., et al. 2009, ApJ, 694, 1281

Carliles, S., et al. 2010, ApJ, 712, 511

Codd, E. F. 1970, Commun. ACM, 13(6), 377

Das, K., et al. 2009, in SIAM Conference on Data Mining SDM09, 247–258, downloaded from http://www.siam.org/proceedings/datamining/2009/dm09.php

Debosscher, J., et al. 2007, Automated supervised classification of variable stars. I. Methodology. A&A, 475, 1159

Drake, A. J., et al. 2009, ApJ, 696, 870

Djorgovski, S. G., et al. 2008, AN, 329, 263

Djorgovski, S. G., & Davis, M. 1987, Fundamental properties of elliptical galaxies. ApJ, 313, 59

Djorgovski, S. G., et al. 2001, Exploration of parameter spaces in a virtual observatory, in Mining the Sky, Proc. SPIE, Vol. 4477, ed. J.-L. Starck, & F. Murtagh (Bellingham: SPIE), 43

DOE-1, 2007, Visualization and Knowledge Discovery: Report from the DOE/ASCR Workshop on Visual Analysis and Data Exploration at Extreme Scale, downloaded from http://www.sc.doe.gov/ascr/ProgramDocuments/Docs/DOE-Visualization-Report-2007.pdf

DOE-2, 2008, Mathematics for Analysis of Petascale Data Workshop Report, downloaded from http://www.sc.doe.gov/ascr/ProgramDocuments/Docs/PetascaleDataWorkshopReport.pdf

DOE-3, 2008, Applied Mathematics at the U.S. Department of Energy: Past, Present and a View to the Future, http://www.sc.doe.gov/ascr/ProgramDocuments/Docs/Brown_Report_May_08.pdf

Dolensky, M. 2004, Applicability of emerging resource discovery standards to the VO, in Toward an International Virtual Observatory, ed. P. J. Quinn, & K. M. Gorski (Berlin: Springer), 265

Dressler, A., et al. 1987, Spectroscopy and photometry of elliptical galaxies. I – a new distance estimator. ApJ, 313, 42

Dutta, H., et al. 2007, in SIAM Conference Data Mining SDM07, 473–476, downloaded from http://www.siam.org/proceedings/datamining/2007/dm07.php

Dutta, H., et al. 2009, in IEEE International Conference on Data Mining, Workshops, 495–500, downloaded from http://ieeexplore.ieee.org/xpl/freeabs_all.jsp?arnumber=5360457

Eastman, T., Borne, K., Green, J, Grayzeck, E., McGuire, R., Sawyer, D. 2005, eScience and archiving for space science. Data Sci. J., 4, 67–76

Euchner, F., et al. 2004, Astronomical Data Analysis Software and Systems (ADASS) XIII, ASP Conf. Ser. 314 (Chicago: Astronomical Society of the Pacific), 578

Fellhauer, M., & Heggie, D. 2005, A&A, 435, 875

Fabbiano, G., et al. 2010, Recommendations of the VAO Science Council, arXiv:1006.2168v1, http://www.aui.edu/vao.php?q=science.council

Fortson, L., et al. 2011, Galaxy zoo: morphological classification and citizen science, in Advances in Machine Learning and Data Mining for Astronomy, ed. M. J. Way, J. D. Scargle, K. M. Ali, & A. N. Srivastava (Chapman and Hall/Boca Raton: CRC)

Gardner, J. P., Connolly, A., & McBride, C. 2007, Astronomical Data Analysis Software and Systems (ADASS) XVI, ASP Conf. Ser. 376 (Chicago: Astronomical Society of the Pacific), 69

Giannella, C., Dutta, H., Borne, K., Wolff, R., & Kargupta, H. 2006, in SIAM Conference on Data Mining SDM06, Workshop on Scientific Data Mining, downloaded from http://www.siam.org/meetings/sdm06/workproceed/Scientific%20Datasets/

Graham, M. J. 2009, Astronomical Data Analysis Software and Systems (ADASS) XVIII, ASP Conf. Ser. 411, 165

Graham, M. J., et al. 2005, Astronomical Data Analysis Software and Systems (ADASS) XIV, ASP Conf. Ser. 347 (Chicago: Astronomical Society of the Pacific), 394

Graham, M. J., Fitzpatrick, M. J., & McGlynn, T. A. (eds) 2008, The National Virtual Observatory: Tools and Techniques for Astronomical Research, ASP Conf. Ser. 382 (Chicago: Astronomical Society of the Pacific)

Graham, M. J. 2010, Hot-Wiring the Transient Universe, 119, available from http://hotwireduniverse.org/

Gray, J. 2003, Online Science, downloaded from http://research.microsoft.com/en-us/um/people/gray/JimGrayTalks.htm

Gray, J., & Szalay, A. 2004, Where the Rubber Meets the Sky: Bridging the Gap Between Databases and Science, Microsoft Technical Report MSR-TR-2004-110, IEEE Data Engineering Bulletin, 27(4), 3–11

Greene, G., et al. 2008, The National Virtual Observatory: Tools and Techniques for Astronomical Research, ASP Conf. Ser. 382 (Chicago: Astronomical Society of the Pacific), 111

Grosbl, P., et al. 2005, Astronomical Data Analysis Software and Systems (ADASS) XIV, ASP Conf. Ser. 347, 124

Harberts, R., et al. 2003, Intelligent Archive Visionary Use Case: Virtual Observatories, downloaded from http://disc.sci.gsfc.nasa.gov/intelligent_archive/presentations/presentations.shtml

Hendler, J. 2003, Science, 299(5606), 520

Hey, T., & Trefethen, A. 2002, Future Gen. Comput. Syst., 18, 1017

Hey, T., Tansley, S., & Tolle, K. (eds) 2009, The Fourth Paradigm: Data-Intensive Scientific Discovery, downloaded from http://research.microsoft.com/en-us/collaboration/fourthparadigm/

Hojnacki, S. M., et al. 2007, ApJ, 659, 585

Ivezic, Z., et al. 2008, Parameterization and classification of 20 Billion LSST objects: lessons from SDSS, in Classification and Discovery in Large Astronomical Surveys, AIP Conf. Proc., Vol. 1082 (Melville, NY: American Institute of Physics), 359

Kegelmeyer, P., et al 2008, Mathematics for Analysis of Petascale Data: Report on a Department of Energy Workshop, downloaded from http://www.sc.doe.gov/ascr/ProgramDocuments/Docs/PetascaleDataWorkshopReport.pdf

Liu, C., et al. 2006, Advanced Software and Control for Astronomy, Proc. SPIE, Vol. 6274 (Bellingham: SPIE), 627415

LSST Science Collaborations and the LSST Project 2009, LSST Science Book, Version 2.0, arXiv:0912.0201, http://www.lsst.org/lsst/scibook

Lynds, B. T. 1962, ApJS, 7, 1

Mahootian, F., & Eastman, T. 2009, World Futures, 65, 61

Mahule, T., et al. 2010, in NASA Conference on Intelligent Data Understanding, downloaded from https://c3.ndc.nasa.gov/dashlink/resources/220/, pp. 243-257

McGlynn, T. 2008, in The National Virtual Observatory: Tools and Techniques for Astronomical Research, ASP Conf. Ser. 382 (Chicago: Astronomical Society of the Pacific), 51

Missaoui, R., et al. 2005, Similarity measures for efficient content-based image retrieval. IEEE Proc. Vision Image Signal Process., 152(6), 875

Mould, J. 2004, LSST Followup, downloaded from http://www.lsst.org/Meetings/CommAccess/abstracts.shtml

Murthy, S. K., Kasif, S., & Salzberg, S. 1994, J Artif. Intell. Res., 2, 1

Nisbet, R., Elder, J., IV, & Miner, G. 2009, Handbook of Statistical Analysis and Data Mining Applications (Amsterdam/Boston: Academic)

Ochsenbein, F., Bauer, P., & Marcout, J. 2000, The VizieR database of astronomical catalogues. A&ASS, 143, 23

Oreiro, R., et al. 2011, A&A, 530, A2

Pimblett, K. A. 2011, MNRAS, 411, 2637

Plante, R., et al. 2004, Astronomical Data Analysis Software and Systems (ADASS) XIII, ASP Conf. Ser. 314 (Chicago: Astronomical Society of the Pacific), 585

Plante, R., et al. 2010, Building Archives in the Virtual Observatory Era in Software and Cyberinfrastructure for Astronomy, Proc. SPIE, Vol. 7740 (Bellingham: SPIE), 77400K

Quinlan, J. R. 1996, Bagging, boosting, and c4.5, in the Proceedings of the 13th National Conference on Artificial Intelligence, AAAI Press (Portland, OR: Association for the Advancement of Artificial Intelligence), 725

Ramapriyan, H. K., et al. 2002. Conceptual Study of Intelligent Archives of the Future, downloaded from http://disc.sci.gsfc.nasa.gov/intelligent_archive/presentations/presentations.shtml

Raskin, R., G. & Pan, M. J. 2005, Knowledge representation in the semantic web for earth and environmental terminology (SWEET). Comput. Geosci., 31(9), 1119

Rebbapragada, U., et al. 2009, Finding anomalous periodic time series: an application to catalogs of periodic variable stars. Mach. Learn., 74(3), 281

Rossi, G., & Sheth, R. K. 2008, MNRAS, 387, 735

Rotem, D., & Shoshani, A. 2009, Scientific Data Management: Challenges, Technology, and Deployment (Chapman and Hall/Boca Raton: CRC)

Sarro, L., et al. 2009, Automated supervised classification of variable stars. II. Application to the OGLE database. A&A, 494, 739

Sebok, W. 1979, AJ, 84, 1526

Schaaf, A. 2007, Web Information Systems Engineering, WISE 2007 Workshop, Lecture Notes in Computer Science, Vol. 4832 (Heidelberg: Springer), 52

Shabalin, A. A., Weigman, V. J., Perou, C. M., & Nobel, A. B. 2009, Finding large average submatrices in high dimensional data. Ann. Appl. Stat., 3(3), 985

Sharpless, S. 1959, ApJS, 4, 257

Springel, V., et al. 2005, Simulations of the formation, evolution and clustering of galaxies and quasars. Nature, 435, 629

Strauss, M. 2004, Towards a Design Reference Mission for the LSST, downloaded from http://www.lsst.org/Meetings/CommAccess/abstracts.shtml

Szalay, A. 2008, Preserving digital data for the future of eScience. Science News (from the August 30, 2008 issue)

Szalay, A., Gray, J., & vandenBerg, J. 2002, Petabyte scale data mining: dream or reality? in Astronomy Telescopes and Instruments, Proc. SPIE, Vol. 4836 (Bellingham: SPIE), 333

Tan, P.-N., Steinbach, M., & Kumar, V. 2006, Introduction to Data Mining (Boston: Addison Wesley)

Taylor, M., et al. 2010, IVOA Recommendation: Simple Application Messaging Protocol Version 1.2, downloaded from http://www.ivoa.net/Documents/latest/SAMP.html

Trimble, V., & Ceja, J. A. 2010, Astron. Nach., 331, 338

Tyson, J. A. 2004, The Large Synoptic Survey Telescope: Science & Design, downloaded from http://www.lsst.org/Meetings/CommAccess/abstracts.shtml

Tyson, J. A., and LSST collaboration 2008, LSST Petascale Data R&D Challenges, downloaded from http://universe.ucdavis.edu/docs/LSST_petascale_challenge.pdf

von Ahn, L. 2007, Human computation, in The proceedings of the 4th International Conference on Knowledge Capture. doi:10.1145/1298406.1298408

Wadadekar, Y. 2005, Publ. ASP, 117, 79

Wang, D., Zhang, Y., & Zhao, Y. 2010, in Software and Cyberinfrastructure for Astronomy, Proc. SPIE, Vol. 7740 (Bellingham: SPIE), 701937.1

White, R. L. 2008, Astronomical applications of oblique decision trees, in AIP Conference Proceedings for Classification and Discovery in Large Astronomical Surveys, Vol. 1082 (Melville, NY: American Institute of Physics), 37

White, R. L. et al. 2009, The High Impact of Astronomical Data Archives, ASTRO2010 Decadal Survey in Astronomy and Astrophysics position paper, downloaded from http://adsabs.harvard.edu/abs/2009astro2010P..64W

Witten, I. H., Frank, E., & Hall, M. A. 2011, Data Mining: Practical Machine Learning Tools and Techniques (3rd ed.; Amsterdam/Boston: Morgan Kaufmann)

Williams, R., 2008, Astronomical Data Analysis Software and Systems (ADASS) XVI, ASP Conf. Ser. 394 (Chicago: Astronomical Society of the Pacific), 173

Williams, R., & Seaman, R. 2008, in The National Virtual Observatory: Tools and Techniques for Astronomical Research, ASP Conf. Ser. 382 (Chicago: Astronomical Society of the Pacific), 425

Williams, R., Bunn, S., & Seaman, R. 2010, Hot-Wiring the Transient Universe, available from http://hotwireduniverse.org/

Wolf, C., et al. 2004, A&A, 421, 913

Wu, X., & Kumar, V. 2009, The Top Ten Algorithms in Data Mining (Chapman and Hall/Boca Raton: CRC)

10 Statistical Methods for Astronomy

Eric D. Feigelson[1,3] · *G. Jogesh Babu*[2,3]
[1]Department of Astronomy & Astrophysics, The Pennsylvania State University, University Park, PA, USA
[2]Department of Statistics, The Pennsylvania State University, University Park, PA, USA
[3]Center for Astrostatistics, The Pennsylvania State University, University Park, PA, USA

T.D. Oswalt, H.E. Bond (eds.), *Planets, Stars and Stellar Systems. Volume 2: Astronomical Techniques, Software, and Data*, DOI 10.1007/978-94-007-5618-2_10, © Springer Science+Business Media Dordrecht 2013

Abstract: Statistical methodology, with deep roots in probability theory, provides quantitative procedures for extracting scientific knowledge from astronomical data and for testing astrophysical theory. In recent decades, statistics has enormously increased in scope and sophistication. After a historical perspective, this review outlines concepts of mathematical statistics, elements of probability theory, hypothesis tests, and point estimation. Least squares, maximum likelihood, and Bayesian approaches to statistical inference are outlined. Resampling methods, particularly the bootstrap, provide valuable procedures when distributions functions of statistics are not known. Several approaches to model selection and goodness of fit are considered.

Applied statistics relevant to astronomical research are briefly discussed. Nonparametric methods are valuable when little is known about the behavior of the astronomical populations or processes. Data smoothing can be achieved with kernel density estimation and nonparametric regression. Samples measured in many variables can be divided into distinct groups using unsupervised clustering or supervised classification procedures. Many classification and data mining techniques are available. Astronomical surveys subject to nondetections can be treated with survival analysis for censored data, with a few related procedures for truncated data. Astronomical light curves can be investigated using time-domain methods involving the autoregressive models, frequency-domain methods involving Fourier transforms, and state-space modeling. Methods for interpreting the spatial distributions of points in some space have been independently developed in astronomy and other fields.

Two types of resources for astronomers needing statistical information and tools are presented. First, about 40 recommended texts and monographs are listed covering various fields of statistics. Second, the public domain **R** statistical software system has recently emerged as a highly capable environment for statistical analysis. Together with its ~3,000 (and growing) add-on CRAN packages, **R** implements a vast range of statistical procedures in a coherent high-level language with advanced graphics. Two illustrations of **R**'s capabilities for astronomical data analysis are given: an adaptive kernel estimator with bootstrap errors applied to a quasar dataset, and the second-order *J* function (related to the two-point correlation function) with three edge corrections applied to a galaxy redshift survey.

Keywords: Astrostatistics; Bayesian inference; Bootstrap resampling; Censoring (upper limits); Data mining; Goodness of fit tests; Hypothesis tests; Kernel density estimation; Least squares estimation; Maximum likelihood estimation; Model selection; Multivariate classification; Nonparametric statistics; Probability theory; R (statistical software); Regression; Spatial point processes; Statistical inference; Statistical software; Time series analysis; Truncation

1 Role and History of Statistics in Astronomy

Through much of the twentieth century, astronomers generally viewed statistical methodology as an established collection of mechanical tools to assist in the analysis of quantitative data. A narrow suite of classical methods were commonly used, such as model fitting by minimizing a χ^2-like statistic, goodness of fit tests of a model to a dataset with the Kolmogorov–Smirnov statistic, and Fourier analysis of time series. These methods are often used beyond their range of applicability. Except for a vanguard of astronomers expert in specific advanced techniques (e.g., Bayesian or wavelet analysis), there was little awareness that statistical methodology had progressed considerably in recent decades to provide a wealth of techniques for wide range of problems. Interest in astrostatistics and associated fields like astroinformatics is rapidly growing today.

The role of statistical analysis as an element of scientific inference has been widely debated (Rao 1997). "Statistics" originally referred to the collection and rudimentary characterization of empirical data. In recent decades, it has accrued other meanings: inference beyond the dataset to the underlying populations under study, quantification of uncertainty in the data, induction of mechanisms causing patterns in the data, and assistance in making decisions based on the data.

Some statisticians feel that, while statistical characterization can be effective, statistical modeling is often unreliable. Statistician G. E. P. Box famously said "Essentially, all models are wrong, but some are useful," and Sir D. R. Cox (2006) wrote "The use, if any, in the process of simple *quantitative* notions of probability and their numerical assessment [to scientific inference] is unclear." Others are more optimistic. Astrostatistician P. C. Gregory (2005) writes:

> Our [scientific] understanding comes through the development of theoretical models which are capable of explaining the existing observations as well as making testable predictions. ... Fortunately, a variety of sophisticated mathematical and computational approaches have been developed to help us through this interface, these go under the general heading of statistical inference.

Astronomers might distinguish cases where the model has a strong astrophysical underpinning (such as fitting a Keplerian ellipse to a planetary orbit) and cases where the model does not have a clear astrophysical explanation (such as fitting a power law to the distribution of stellar masses). In all cases, astronomers should carefully formulate the question to be addressed, apply carefully chosen statistical approaches to the dataset to address this question with clearly stated assumptions, and recognize that the link between the statistical findings and reality may not be straightforward.

Prior to the twentieth century, many statistical developments were centered around astronomical problems (Stigler 1986). Ancient Greek, Arabic, and Renaissance astronomers debated how to estimate a quantity, such as the length of a solar year, based on repeated and inconsistent measurements. Most favored the middle of the extreme values, and some scholars feared that repeated measurement led to greater, not reduced, uncertainty. The mean was promoted by Tycho Brahe and Galileo Galilei, but did not become the standard procedure until the mid-eighteenth century. The eleventh century Persian astronomer al-Biruni and Galileo discussed propagation of measurement errors. Adrian Legendre and Carl Friedrich Gauss developed the "Gaussian" or normal error distribution in the early nineteenth century to address discrepant measurement in celestial mechanics. The normal distribution was intertwined with the least-squares estimation technique developed by Legendre and Pierre-Simon Laplace. Leading astronomers throughout Europe contributed to least-squares theory during the nineteenth century.

However, the close association of statistics with astronomy atrophied during the beginning of the twentieth century. Statisticians focused their attention on biological sciences and human affairs: biometrics, demography, economics, social sciences, industrial reliability, and process control. Astronomers made great advances by applying newly developed fields of physics to explain astronomical phenomena. The structure of a star, for example, was powerfully explained by combining gravity, thermodynamics, and atomic and nuclear physics. During the middle of the century, astronomers continued using least-squares techniques, although heuristic procedures were also common. Astronomers did not adopt the powerful methods of maximum likelihood estimation (MLE), formulated by Sir R. A. Fisher in the 1920s and widely promulgated in other fields. MLE came into use during the 1970s–1980s, along with the nonparametric Kolmogorov–Smirnov statistic. Practical volumes by Bevington (1969) and Press et al. (1986) with Fortran source codes promulgated a suite of classical methods.

The adoption of a wider range of statistical methodology began during the 1990s and is still growing. Sometimes developments were made independently of mathematical statistics; for example, development of the two-point correlation function to study galaxy clustering was largely independent of the closely related mathematical developments relating to Ripley's K function for non-random spatial point processes. Progress was made in developing procedures for analyzing unevenly spaced time series and populations from truncated (flux-limited) astronomical surveys, problems that rarely arose in other fields. As computational methods advanced, parametric modeling using Bayesian analysis became attractive for a variety of problems that are not satisfactorily addressed by traditional frequentists methods.

The 1990s also witnessed the emergence of cross-disciplinary interactions between astronomers and statisticians including collaborative research groups, conferences, didactic summer schools, and software resources. A small but growing research field of astrostatistics was established. The activity is propelled both by the sophistication of data analysis and modeling problems, and the exponentially growing quantity of publicly available astronomical data. Both primary and secondary data products are provided by organizations such as NASA's Science Archive Research Centers, ground-based observatory archives, and databases such as NASA's Extragalactic Database and SIMBAD. The NASA/Smithsonian Astrophysics Data System, a billion-hit Web site, is an effective portal to the astronomical research literature and databases. The International Virtual Observatory that facilitates access to distributed databases will only increase the need for statistical resources to analyze multiwavelength datasets selected by the astronomer for examination. These datasets are reaching petabyte scales with multivariate catalogs providing measured properties of billions of astronomical objects.

Thus, the past two centuries have witnessed the astronomical foundations of statistics, a growing distance between the fields, and a recent renaissance in astrostatistics. The need for, and use of, advanced statistical techniques in the service of interpreting astronomical data and testing astrophysical theory is now ascendent.

The purpose of this review is to introduce some of the concepts and results of a broad scope of modern statistical methodology that can be effective for astronomical data and science analysis. ➲ Section 2 outlines concepts and results of statistical inference that underlie statistical analysis of astronomical data. ➲ Section 3 reviews several fields of applied statistics relevant for astronomy. ➲ Section 4 discusses resources available for the astronomer to learn appropriate statistical methodology, including the powerful **R** software system. Key terminology is noted in quotation marks (e.g., "bootstrap resampling"). The coverage is abbreviated and not complete in any fashion. Topics are mostly restricted to areas of established statistical methodology relevant to astronomy; recent advances in the astronomical literature are mostly not considered. A more comprehensive treatment for astronomers can be found in Feigelson and Babu (2012), and the volumes listed in ➲ *Table 10-1* should be consulted prior to serious statistical investigations.

Table 10-1
Selected statistics books for astronomers

Broad scope	
Adler (2010)	R in a Nutshell
Dalgaard (2008)	Introductory Statistics with R
Feigelson and Babu (2012)	Modern Statistical Methods for Astronomy with R Applications
Rice (1994)	Mathematical Statistics and Data Analysis
Wasserman (2005)	All of Statistics: A Concise Course in Statistical Inference
Statistical inference	
Conover (1999)	Practical Nonparametric Statistics
Evans et al. (2000)	Statistical Distributions
Hogg and Tanis (2009)	Probability and Statistical Inference
James (2006)	Statistical Methods in Experimental Physics
Lupton (1993)	Statistics in Theory and Practice
Ross 2010	A First Course in Probability
Bayesian statistics	
Gelman et al. (2004)	Bayesian Data Analysis
Gregory (2005)	Bayesian Logical Data Analysis for the Physical Sciences
Kruschke (2011)	Doing Bayesian Data Analysis: A Tutorial with R and BUGS
Resampling methods	
Efron and Tibshirani (1993)	An Introduction to the Bootstrap
Zoubir and Iskander (2004)	Bootstrap Techniques for Signal Processing
Density estimation	
Bowman and Azzalini (1997)	Applied Smoothing Techniques for Data Analysis
Silverman (1998)	Density Estimation
Takezawa (2005)	Introduction to Nonparametric Regression
Regression and multivariate analysis	
Kutner et al. (2004)	Applied Linear Regression Models
Johnson and Wichern (2007)	Applied Multivariate Statistical Analysis
Nondetections	
Helsel (2004)	Nondetects and Data Analysis
Klein and Moeschberger (2010)	Survival Analysis
Lawless (2002)	Statistical Models and Methods for Lifetime Data
Spatial processes	
Bivand et al. (2008)	Applied Spatial Data Analysis with R
Fortin and Dale (2005)	Spatial Analysis: A Guide for Ecologists
Illian et al. (2008)	Statistical Analysis and Modelling of Spatial Point Patterns
Martínez and Saar (2002)	Statistics of the Galaxy Distribution
Starck and Murtagh (2006)	Astronomical Image and Data Analysis

Table 10-1
(Continued)

Data mining, clustering, and classification	
Everitt et al. (2001)	Cluster Analysis
Duda et al. (2001)	Pattern Classification
Hastie et al. (2009)	The Elements of Statistical Learning
Way et al. (2011)	Advances in Machine Learning and Data Mining for Astronomy
Time series analysis	
Chatfield (2004)	The Analysis of Time Series
Cowpertwait and Metcalfe (2009)	Introductory Time Series with R
Nason (2008)	Wavelet Methods in Statistics with R
Shumway and Stoffer (2006)	Time Series Analysis and Its Applications with R Examples
Graphics and data visualization	
Chen et al. (2008)	Handbook of Data Visualization
Maindonald and Braun (2010)	Data Analysis and Graphics using R
Sarkar (2008)	Lattice: Multivariate Data Visualization with R
Wickham (2009)	ggplot2: Elegant Graphics for Data Analysis

2 Statistical Inference

2.1 Concepts of Statistical Inference

A "random variable" is a function of potential outcomes of an experiment. A particular realization of a random variable is often called a "data point." The magnitude of stars or redshifts of galaxies are examples of random variables. An astronomical dataset might contain photometric measurements of a sample of stars, spectroscopic redshifts of a sample of galaxies, categorical measurements (such as radio-loud and radio-quiet active galactic nuclei), or brightness measurements as a function of sky location (an image), wavelength of light (a spectrum), or of time (a light curve). The dataset might be very small so that large-N approximations do not apply, or very large making computations difficult to perform. Observed values may be accompanied by secondary information, such as estimates of the errors arising from the measurement process. "Statistics" are functions of random variables, ranging from simple functions, such as the mean value, to complicated functions, such as an adaptive kernel smoother with cross-validation bandwidths and bootstrap errors (*Fig. 10-2* below).

Statisticians have established the distributional properties of a number of statistics through formal mathematical theorems. For example, under broad conditions, the "Central Limit Theorem" indicates that the mean value of a sufficiently large sample of independent random variables is normally distributed. Astronomers can invent statistics that reveal some scientifically interesting properties of a dataset, but can not assume they follow simple distributions unless this has been established by theorems. But in many cases, Monte Carlo methods such as the bootstrap can numerically recover the distribution of the statistic from a particular dataset under study.

Statistical inference, in principle, helps reach conclusions that extend beyond the immediate data to derive broadly applicable insights into the underlying population. The field of statistical inference is very large and can be classified in a number of ways. "Nonparametric inference" gives probabilistic statements about the data which do not assume any particular distribution (e.g., Gaussian, power law) or parametric model for the data, while "parametric inference" assumes some distributions or functional relationships. These relationships can be simple heuristic relations, as in linear regression, or can be complex functions derived from astrophysical theory. Inference can be viewed as the combination of two basic branches: "point estimation" (such as estimating the mean of a dataset) and the "testing of hypotheses" (such as a 2-sample test on the equality of two medians).

It is important that the scientist be aware of the range of applicability of a given inferential procedure. Several problems relevant to astronomical statistical practice can be mentioned. Various statistics that resemble Pearson's χ^2 do permit a weighted least squares regression, but the statistic often does not follow the χ^2 distribution. The Kolmogorov–Smirnov statistic can be a valuable measure of difference between a sample and a model, but tabulated probabilities are incorrect if the model is derived from that sample (Lilliefors 1969) and the statistic is ill-defined if the dataset is multivariate. The likelihood ratio test can compare the ability of two models to explain a dataset, but it cannot be used if an estimated parameter value is consistent with zero (Protassov et al. 2002).

2.2 Probability Theory and Probability Distributions

Statistics is rooted in probability theory, a branch of mathematics seeking to model uncertainty (Ross 2010). Nearly all astronomical studies encounter uncertainty: observed samples represent only small fractions of underlying populations, properties of astrophysical interest are measured indirectly or incompletely, errors are present due to the measurement process. The theory starts with the concept of an "experiment," an action with various possible results where the actually occurring result cannot be predicted with certainty prior to the action. Counting photons at a telescope from a luminous celestial object, or waiting for a gamma-ray burst in some distant galaxy are examples of experiments. An "event" is a subset of the (sometimes infinite) "sample space," the set of all outcomes of an experiment. For example, the number of supermassive black holes within 10 Mpc is a discrete and finite sample space, while the spatial distribution of galaxies within 10 Mpc can be considered as an infinite sample space.

Probability theory seeks to assign probabilities to elementary outcomes and manipulate the probabilities of elementary events to derive probabilities of complicated events. Three "axioms of probability" are: the probability $P(A)$ of an event A lies between 0 and 1, the sum of probabilities over the sample space is 1, and the joint probability of two or more events is equal to the sum of individual event probabilities if the events are mutually exclusive. Other properties of probabilities flow from these axioms: additivity and inclusion-exclusion properties, conditional and joint probabilities, and so forth.

"Conditional probabilities" where some prior information is available are particularly important. Consider an experiment with m equally likely outcomes and let A and B be two events. Let $\#A = k$, $\#B = n$, and $\#(A \cap B) = i$ where $\cap$ means the intersection ("and"). Given information that B has happened, the probability that A has also happened is written $P(A \mid B) = i/n$, this is the conditional probability and is stated "The probability that A has occurred given B is i/n." Noting that $P(A \cap B) = \frac{i}{m}$ and $P(B) = \frac{n}{m}$, then

$$P(A \mid B) = \frac{P(A \cap B)}{P(B)}. \tag{10.1}$$

This leads to the "multiplicative rule" of probabilities, which for n events can be written

$$\begin{aligned} P(A_1 \cap A_2 \cap \ldots A_n) = P(A_1)\, P(A_2 \mid A_1) \ldots P(A_{n-1} \mid A_1, \ldots A_{n-2}) \\ \times P(A_n \mid A_1, \ldots A_{n-1}). \end{aligned} \tag{10.2}$$

Let $B_1, B_2, \ldots, B_k$ be a partition of the sample space. The probability of outcome A in terms of events B_k,

$$P(A) = P(A \mid B_1)P(B_1) + \cdots + P(A \mid B_k)P(B_k), \tag{10.3}$$

is known as the "Law of Total Probability." The question can also be inverted to find the probability of an event B_i given A:

$$P(B_i \mid A) = \frac{P(A \mid B_i)P(B_i)}{P(A \mid B_1)P(B_1) + \cdots + P(A \mid B_k)P(B_k)}. \tag{10.4}$$

This is known as "Bayes' Theorem" and, with a particular interpretation, it serves as the basis for Bayesian inference.

Other important definitions and results of probability theory are important to statistical methodology. Two events A and B are defined to be "independent" if $P(A \cap B) = P(A)P(B)$. Random variables are functions of the sample or outcome space. The "cumulative distribution function" (c.d.f.) F of a random variable X is defined as

$$F(x) = P(X \leq x). \tag{10.5}$$

In the discrete case where X takes on values $a_1, a_2, \ldots, a_n$, then F is defined through the probability mass function (p.m.f.) $P(a) = P(x = a_i)$ and

$$F(x) = \sum_{a_x \leq x} P(x = a_i). \tag{10.6}$$

Continuous random variables are often described through a "probability density distribution" (p.d.f.) f satisfying $f(y) \geq 0$ for all y and

$$F(x) = P(X \leq x) = \int_{-\infty}^{x} f(y)dy. \tag{10.7}$$

Rather than using the fundamental c.d.f.s, astronomers have a tradition of using binned p.d.f.s, grouping discrete data to form discontinuous functions displayed as histograms. Although valuable for visualizing data, this is an ill-advised practice for statistical inference: arbitrary decisions must be made concerning binning method, and information is necessarily lost within bins. Many excellent statistics can be computed directly from the c.d.f. using, for example, maximum likelihood estimation (MLE).

"Moments" of a random variable are obtained from integrals of the p.d.f. or weighted sums of the p.m.f. The first moment, the "expectation" or mean, is defined by

$$E[X] = \mu = \int x f(x)dx \tag{10.8}$$

in the continuous case and $E[X] = \sum_a aP(X = a)$ in the discrete case. The variance, the second moment, is defined by

$$Var[X] = E[(X - \mu)^2]. \tag{10.9}$$

A sequence of random variables $X_1, X_2, \ldots, X_n$ is called "independent and identically distributed (i.i.d.)" if

$$P(X_1 \le a_1, X_2 \le a_2, \ldots, X_n \le a_n) = P(X_1 \le a_1)P(X_2 \le a_2)\ldots P(X_n \le a_n), \tag{10.10}$$

for all n. That is, $X_1, X_2, \ldots, X_n$ all have the same c.d.f., and the events $(X_i \le a_i)$ are independent for all a_i. The "Law of Large Numbers" is a theorem stating that

$$\frac{1}{n}\sum_{i=1}^{n} X_i \approx E[X] \tag{10.11}$$

for large n for a sequence of i.i.d. random variables.

The continuous "normal distribution," or Gaussian distribution, is described by its p.d.f.:

$$\phi(x) = \frac{1}{\sqrt{2\pi}\sigma} \exp\left\{-\frac{(x-\mu)^2}{2\sigma^2}\right\}. \tag{10.12}$$

When X has the Gaussian density in (➋ 10.12), then the first two moments are

$$E(X) = \mu \quad Var(X) = \sigma^2. \tag{10.13}$$

The normal distribution, often designated $N(\mu, \sigma^2)$, is particularly important as the Central Limit Theorem states that the distribution of the sample mean of any i.i.d. random variable about its true mean approximately follows a normal.

The Poisson random variable X has a discrete distribution with p.m.f.

$$P(X = i) = \lambda^i e^{-\lambda}/i! \tag{10.14}$$

for integer i. For the "Poisson distribution," the mean and variance are equal:

$$E(X) = Var(X) = \lambda. \tag{10.15}$$

If $X_1, X_2, \ldots, X_n$ are independent random variables with the Poisson distribution having rate λ, then $(1/n)\sum X_i$ is the best, unbiased estimator of λ. If $x_1, x_2, \ldots, x_n$ is a particular sample drawn from the Poisson distribution with rate $\lambda > 0$, then the sample mean $\overline{x} = \sum x_i/n$ is the best unbiased estimate for λ. Here the x_i are the realizations of the random variables X_i's. The difference of two Poisson variables, and the proportion of two Poisson variables, follows no simple known distribution and their estimation can be quite tricky. This is important because astronomers often need to subtract background from Poisson signals, or compute ratios of two Poisson signals. For faint or absent signals in a Poisson background, MLE and Bayesian approaches have been considered (Cowan 2006; Kashyap et al. 2010). For Poisson proportions, the MLE is biased and unstable for small n and/or p near 0 or 1, and other solutions are recommended (Brown et al. 2001). If background subtraction is also present in a Poisson proportion, then a Bayesian approach is appropriate (Park et al. 2006).

The power law distribution is particularly commonly used to model astronomical random variables. Known in statistics as the Pareto distribution, the correctly normalized p.d.f. is

$$f(x) = \frac{\alpha b^\alpha}{x^{\alpha+1}}, \tag{10.16}$$

for $x > b$. The commonly used least-squares estimation of the shape parameter α and scale parameter b from a binned dataset is known to be biased and inefficient even for large n. This procedure is not recommended and the "minimum variance unbiased estimator" based on the MLE is preferred (Johnson et al. 1994):

$$\alpha^* = \left(1 - \frac{2}{n}\right)\hat{\alpha}_{MLE} \quad \text{where} \quad \hat{\alpha}_{MLE} = \frac{n}{\sum_{i-1}^{n} \ln(x_i/\hat{b}_{MLE})}$$

$$b^* = \left(1 - \frac{1}{(n-1)\hat{\alpha}_{MLE}}\right)\hat{b}_{MLE} \quad \text{where} \quad \hat{b}_{MLE} = x_{min}. \tag{10.17}$$

MLE and other recommended estimators for standard statistics of some dozens of distributions are summarized by Evans et al. (2000, also in Wikipedia) and are discussed comprehensively in volumes by Johnson et al. (1994).

2.3 Point Estimation

Parameters of a distribution or a relationship between random variables are estimated using functions of the dataset. The mean and variance, for example, are parameters of the normal distribution. Astronomers fit astrophysical models to data, such as a Keplerian elliptical orbit to the radial velocity variations of a star with an orbiting planet or the Navarro–Frenk–White distribution of Dark Matter in galaxies. These laws also have parameters that determine the shape of the relationships.

In statistics, the term "estimator" has a specific meaning. The estimator $\hat{\theta}$ of θ is a function of the random sample that gives the value estimate when evaluated at the actual data, $\hat{\theta} = f_n(X_1, \ldots, X_n)$, pronounced "theta-hat." Such functions of random variables are also called "statistics." Note that an estimator is a random variable because it depends on the observables that are random variables. For example, in the Gaussian case (➋ 10.12), the sample mean $\hat{\mu} = \bar{X} = \frac{1}{n}\sum_{i=1}^{n} X_i$ and sample variance defined as $\hat{\sigma}^2 = S^2 = \frac{1}{n-1}\sum_{i=1}^{n}(X_i - \overline{X})^2$ are estimators of μ and σ^2, respectively. If the X_i's are replaced by actual data, $x_1, x_2, \ldots, x_n$, then $\hat{\mu} = \frac{1}{n}\sum_{i=1}^{n} x_i$ is a point estimate of μ.

Least squares (LS), method of moments, maximum likelihood estimation (MLE), and Bayesian methods are important and commonly used procedures in constructing estimates of the parameters. Astronomers often refer to point estimation by "minimizing χ^2"; this is an incorrect designation and cannot be found in any statistics text. The astronomers' procedure is a weighted LS procedure, often using measurement errors for the weighting, that, under certain conditions, will give a statistic that follows a χ^2 distribution.

The choice of estimation method is not obvious, but can be guided by the scientific goal. A procedure that gives the closest estimate to the true parameter value (smallest bias) will often differ from a procedure that minimizes the average distance between the data and the model (smallest variance), the most probable estimate (maximum likelihood), or estimator most consistent with prior knowledge (Bayesian). Astronomers are advised to refine their scientific questions to choose an estimation method, or compare results from several methods to see how the results differ.

Statisticians have a number of criteria for assessing the quality of an estimator including unbiasedness, consistency, and efficiency. An estimator $\hat{\theta}$ of a parameter θ is called "unbiased" if the expected value of $\hat{\theta}$, $E(\hat{\theta}) = \theta$. That is, $\hat{\theta}$ is unbiased if its overall average value for all potential datasets is equal to θ. For example, in the variance estimator S^2 of σ^2 for the Gaussian distribution, $n-1$ is placed in the denominator instead of n to obtain an unbiased estimator. If $\hat{\theta}$ is unbiased estimator of θ, then the variance of the estimator $\hat{\theta}$ is given by $E((\hat{\theta}-\theta)^2)$. Sometimes, the scientist will choose a minimum variance estimator to be the "best" estimator, but often a bias is accepted so that the sum of the variance and the square of the bias, or "mean square error" (MSE), is minimized:

$$MSE = E[(\hat{\theta}-\theta)^2] = Var(\hat{\theta}) + (\theta - E[\hat{\theta}])^2. \tag{10.18}$$

An unbiased estimator that minimizes the MSE is called the "minimum variance unbiased estimator" (MVUE). If there are two or more unbiased estimators, the one with smaller variance is usually preferred. Under some regularity conditions, the "Cramér-Rao inequality" gives a lower bound on the lowest possible variance for an unbiased estimator.

2.4 Least Squares

The LS method, developed for astronomical applications 200 years ago (➋ Sect. 1), is effective in the regression context for general linear models. Here "linear" means linear in the parameters, not in the variables. LS estimation is thus appropriate for a wide range of complicated astrophysical models. Suppose X_i are independent but not identically distributed, say, $E(X_i) = \sum_{j=1}^{k} a_{ij}\beta_j$, (the mean of X_i is a known linear combination of parameters $\beta_1, \ldots, \beta_k$), then the estimators of the parameters β_j can be obtained by minimizing the sum of squares of $(X_i - \sum_{j=1}^{k} a_{ij}\beta_j)$. In some cases, there is a closed form expression for the least squares estimators of $\beta_1, \ldots, \beta_k$. If the error variances σ_i^2 of X_i are also different (heteroscedastic), then one can minimize the weighted sum of squares

$$\sum_{i=1}^{n} \frac{1}{\sigma_i^2}\left(X_i - \sum_{j=1}^{k} a_{ij}\beta_j\right)^2 \tag{10.19}$$

over $\beta_1, \ldots, \beta_k$. If $Y_1, Y_2, \ldots, Y_n$ are independent random variables with a normal distribution $N(\mu, 1)$, then the sum of squared normals $\sum_{i-1}^{n}$ is distributed as the χ^2 distribution.

The method of moments is another classical approach to point estimation where the parameters of the model are expressed as simple functions of the first few moments and then replace the population moments in the functions with the corresponding sample moments.

2.5 Maximum Likelihood Method

Building on his criticism of both least-squares method and the method of moments in his first mathematical paper as an undergraduate, R. A. Fisher (1922) introduced the method of maximum likelihood. The method is based on the "likelihood" where the p.d.f. (or probability mass function for a discrete random variable) is viewed as a function of the data given the model and specified values of the parameters. In statistical parlance, if the data are an i.i.d. random sample $X_1, \ldots, X_n$, with a common p.d.f. or p.m.f. $f(., \theta)$, then the likelihood L and loglikelihood ℓ are given by

$$\ell(\theta) = \ln L(\theta) = \sum_{i=1}^{n} \ln f(X_i, \theta). \tag{10.20}$$

In Fisher's formulation, the model parameters θ are treated as fixed and the data are variable.

The "maximum likelihood estimator" (MLE) $\hat{\theta}$ of θ is the value of the parameter that maximizes $\ell(\theta)$. The results for some common probability distributions treated in astronomy are easily summarized:

1. If $X_1, X_2, \ldots, X_n$ are i.i.d. random variables distributed as Poisson with intensity λ, $f(t,\lambda) = e^{-\lambda}\lambda^t/t!$, then the mean $\hat{\lambda} = \bar{X} = \sum_{i=1}^n X_i/n$ is the LS and method of moments estimator of λ. In this case, the MLE does not exist if all the realizations of $X_i = 0$.
2. If X is distributed as normal (Gaussian) with mean μ and variance σ^2, then $(\hat{\mu}, \hat{\sigma}^2) = (\bar{X}, ((n-1)/n)S^2)$ is the MLE of (μ, σ^2). Here the MLE $\hat{\theta}$ is consistent but not unbiased; however, this can often be overcome by multiplying $\hat{\theta}$ by a constant.
3. If X is distributed as power law (Pareto) with slope (shape) parameter α and location parameter b, $f(t) = \alpha b^\alpha/t^{\alpha+1}$ with $t > b$, then the MLE for the slope is $\hat{\alpha}_{MLE} = n/\sum_i \ln(X_i/X_{min})$ as given in (➋ 10.17). For asymmetric distributions like the power law, the MLE, LS, and MVUE estimators often differ. The MVUE estimator for the power law is $\alpha^* = (1-2/n)\hat{\alpha}$. The LS estimator for the power law slope commonly used by astronomers does not have a closed expression, is biased, and converges very slowly. Astronomers are advised to use the MVUE estimator rather than LS fits to binned data to estimate power law slopes.

While these examples are for rather simple situations, MLE estimators can often be numerically calculated for more complex functions of the data. For many useful interesting functions g of the parameters θ, $g(\hat{\theta})$ is the MLE of $g(\theta)$ whenever $\hat{\theta}$ is the MLE of θ. Computing the maximum likelihood is usually straightforward. The "EM Algorithm" is an easily implemented and widely used procedure for maximizing likelihoods (Dempster et al. 1977; McLachlan and Krishnan 2008). This procedure, like many other optimization calculations, may converge to a local rather than the global maximum. Astronomers use the EM Algorithm for an MLE in image processing where it is called the Lucy-Richardson algorithm (Lucy 1974). In some cases, the MLE may not exist and, in other cases, more than one MLEs exist.

Knowledge of the limiting distribution of the estimator is often needed to obtain confidence intervals for parameters. In many commonly occurring situations with large n, the MLE $\hat{\theta}$ has an approximate normal distribution with mean θ and variance $1/I(\theta)$ where

$$I(\theta) = nE\left(\frac{\partial}{\partial\theta}\log f(X_1,\theta)\right)^2. \tag{10.21}$$

This is the "Fisher information matrix." Thus, 95% (or similar) confidence intervals can be derived for MLEs.

With broad applicability, efficient computation, clear confidence intervals, and strong mathematical foundation, maximum likelihood estimation rose to be the dominant method for parametric estimation in many fields. Least squares methodology still predominates in astronomy, but MLE has a growing role. An important example in astronomy of MLE with Fisher information confidence intervals was the evaluation of cosmological parameters of the concordance Λ Cold Dark Matter model based on fluctuations of the cosmic microwave background radiation measured with the Wilkinson Microwave Anisotropy Probe (Spergel et al. 2003).

2.6 Hypotheses Tests

Along with estimation, hypotheses tests are a major class of tools in statistical inference. Many astronomical problems such as source detection and sample comparison can be formulated as Yes/No questions to be addressed with statistical hypotheses testing.

Consider the case of source detection where the observed data Y consists of signal μ and noise ϵ, $Y = \mu + \epsilon$. The problem is to quantitatively test the "null hypothesis" $H_0 : \mu = 0$ representing no signal against the "alternative hypothesis" $H_a : \mu > 0$. Statistical hypothesis testing resembles a court room trial where a defendant is considered innocent until proven guilty. A suitable function of the data called the "test statistic" is chosen, and a set of test statistic values or "critical region" is devised. The decision rule is to reject the null hypothesis if the function of the data falls in the critical region. There are two possible errors: a false positive that rejects the null hypothesis, called a "Type I error"; and a false negative that fails to reject the null hypothesis when the alternative hypothesis is true, or "Type II error." Ideally, one likes to minimize both error types, but this is impossible to achieve. Critical regions constructed to keep Type I error under control, say at 5% level are called "levels of significance." One minus the probability of Type II error is called the "power of the test." High power tests are preferred.

A result of a hypothesis test is called "statistically significant" if it is unlikely to have occurred by chance, that is, the test rejects the null hypothesis at the prescribed significance level α where $\alpha = 0.05, 0.01$, or similar value. Along with the results of a statistical test, often the so-called p-value is reported. The "p-value" of the test is the smallest significance level at which the statistic is significant. It should be important to note that the null hypothesis and the alternative hypothesis are not treated symmetrically: the null hypothesis can be rejected at a given level of significance, but the null hypothesis can not formally be accepted.

2.7 Bayesian Estimation

Conceptually, Bayesian inference uses aspects of the scientific method that involves evaluating whether acquired evidence is consistent or inconsistent with a given hypothesis. As evidence accumulates, the degree of belief in a hypothesis ought to change. With enough evidence, it should become very high or very low. Thus, Bayesian inference can be used to discriminate between conflicting hypotheses: hypotheses with very high support should be accepted as true and those with very low support should be rejected as false. However, this inference method is influenced by the prior distribution, initial beliefs that one holds before any evidence is ever collected. In so far as the priors are not correct, the estimation process can lead to false conclusions.

Bayesian inference relies on the concept of conditional probability to revise one's knowledge. Prior to the collection of sample data one had some (perhaps vague) information on θ. Then combining the model density of the observed data with the prior density one gets the posterior density, the conditional density of θ given the data. Until further data are available, this posterior distribution of θ is the only relevant information as far as is concerned.

As outlined above, the main ingredients for Bayesian inference are the likelihood function, $L(\theta \mid X)$ with a vector θ of parameters and a prior probability density, $\pi(\theta)$. Combining the two via Bayes' theorem (➋ 10.4) yields the posterior probability density

$$\pi(\theta \mid X) = \frac{\pi(\theta)\, L(X \mid \theta)}{\int \pi(u)\, L(X \mid u) du} \tag{10.22}$$

when the densities exist. In the discrete case where $\pi(\theta)$ is the probability mass function, the formula becomes

$$\pi(\theta \mid X) = \frac{\pi(\theta)\, L(X \mid \theta)}{\sum_{j=2}^{k} \pi(\theta_j)\, L(X \mid \theta_j)}. \tag{10.23}$$

If there is no special information on the parameter θ except that it lies in an interval, then one often assumes θ is uniformly distributed on the interval. This is a choice of a "non-informative prior" or reference prior. Often, Bayesian inference from such a flat prior coincides with classical frequentist inference.

The estimator $\hat{\theta}$ of θ defined as the mode of $\pi(\theta|X)$, the value of θ that maximizes the posterior $\pi(\theta|X)$, is the most probable value of the unknown parameter θ conditional on the sample data. This is called the "maximum a posteriori" (MAP) estimate or the "highest posterior density" (HPD) estimate.

The mean of the posterior distribution gives another Bayes estimate by applying least squares on the posterior density. Here the $\hat{\theta}_B$ that minimizes the posterior dispersion

$$E[(\theta - \hat{\theta}_B)^2 \mid X] = \min E[(\theta - a)^2 \mid X] \tag{10.24}$$

is given by $\hat{\theta}_B = E[\theta \mid X]$. If $\hat{\theta}_B$ is chosen as the estimate of θ, then a measure of variability of this estimate is the posterior variance, $E[(\theta - E[\theta \mid x])^2|X|$. This gives the posterior standard deviation as a natural measure of estimation error; that is, the estimate is $\hat{\theta}_B \pm \sqrt{E[(\theta - E[\theta \mid x])^2|X|}$. In fact, for any interval around $\hat{\theta}_B$, the posterior probability containing the true parameter can be computed. In other words, a statement such as

$$P(\hat{\theta}_B - k_1 \leq \theta \leq \hat{\theta}_B + k_2|X) = 0.95 \tag{10.25}$$

gives a meaningful 95% "credible region." These inferences are all conditional on the given dataset x.

Bayesian inference can be technically challenging because it requires investigating the full parameter space. For models with many parameters and complex likelihood functions, this can involve millions of calculations of the likelihood. Sophisticated numerical methods to efficiently cover the parameter space are needed, most prominently using Markov chains with the Gibbs sampler and Metropolis-Hastings algorithm. These are collectively called "Markov Chain Monte Carlo" calculations. Once best-fit $\hat{\theta}$ values have been identified, further Monte Carlo calculations can be performed to examine the posterior distribution around the best model. These give credible regions in parameter space. Parameters of low scientific interest can be integrated to remove them from credible region calculations, this is called "marginalization" of nuisance parameters.

The Bayesian framework can also effectively choose between two models with different vectors of parameters. This is the "model selection" problem discussed in ➋ Sect. 2.9. Suppose the data X has probability density function $f(x \mid \theta)$ and the scientist wants to compare two models, $M_0 : \theta \in \Theta_0$ versus $M_1 : \theta \in \Theta_1$. A prior density that assigns positive prior probability to Θ_0 and Θ_1 is chosen, and the posterior "odds ratio" $P\{\Theta_0|x\}/P\{\Theta_1|X\}$ is calculated. A chosen threshold like 1/9 or 1/19 will decide what constitutes evidence against a null hypothesis. The "Bayes factor" of M_0 relative to M_1 can also be reported:

$$BF_{01} = \frac{P(\Theta_0|X)}{P(\Theta_1|X)} \bigg/ \frac{P(\Theta_0)}{P(\Theta_1)} = \frac{\int_{\Theta_0} f(x|\theta) g_0(\theta) d\theta}{\int_{\Theta_1} f(x|\theta) g_1(\theta) d\theta} \tag{10.26}$$

The smaller the value of BF_{01}, the stronger the evidence against M_0. Unlike classical hypothesis testing, the Bayesian analysis treats the hypotheses symmetrically. The method can be extended to compare more than two models.

2.8 Resampling Methods

Astronomers often devise a statistic that measures a property of interest in the data, but find it is difficult or impossible to determine the distribution of that statistic. The classical statistical methods concentrate on statistical properties of estimators that have a simple closed form, but these methods often involve unrealistically simplistic model assumptions. A class of computationally intensive procedures known as "resampling methods" address this limitation, providing inference on a wide range of statistics under very general conditions. Resampling methods involve constructing hypothetical datasets derived from the observations, each of which can be analyzed in the same fashion to see how the chosen statistic depends on plausible random variations in the observations. Resampling the original data preserves whatever distributions are truly present, including selection effects such as truncation and censoring.

The "half-sample method" is an old resampling method dating to the 1940s. Here one repeatedly chooses at random half of the data point, and estimates the statistic for each resample. The inference on the parameter can be based on the histogram of the resampled statistics. An important variant is the Quenouille–Tukey "jackknife method" where one constructs exactly n hypothetical datasets each with $n - 1$ points, each one omitting a different point. It is useful in reducing the bias of an estimator as well as estimating the variance of an estimator. The jackknife method is effective for many statistics, including LS estimators and MLEs, but is not consistent for discrete statistics such as the sample median.

The most important of resampling methods is the "bootstrap" introduced by Bradley Efron in 1979 (Efron and Tibshirani 1993). Here one generates a large number of datasets, each randomly drawn from the original data such that each drawing is made from the entire dataset, so a simulated dataset is likely to miss some points and have duplicates or triplicates of others. This "resampling with replacement" can be viewed as a Monte Carlo simulation from an existing data without any assumption on the underlying population.

The importance of the bootstrap emerged during the 1980s when mathematical study demonstrated that it gives nearly optimal estimate of the distribution of many statistics under a wide range of circumstances (Babu 1984; Babu and Singh 1983). For example, theorems using Edgeworth expansions establish that the bootstrap provides a good approximation for a Studentized smooth functional model (Babu and Singh 1984). A broad class of common statistics can be expressed as smooth functions of multivariate means including LS estimators (means and variances, t-statistics, correlation coefficients, regression coefficients) and some MLEs. The bootstrap is consequently widely used for a vast range of estimation problems.

While bootstrap estimators have very broad application, they can fail for statistics with heavy tails, some non-smooth and nonlinear situations, and some situations where the data are not independent. A lack of independence can occur, for example, in proximate pixels of an astronomical image due to the telescope point spread function, or in proximate observations of a time series of a variable celestial object. The bootstrap may also be inapplicable when the data have "heteroscedastic" measurement errors, that is, the variances that differ from point to point. Bootstrap confidence intervals also require that the statistic be "pivotal" such that the limiting distribution is free from the unknown parameters of the model. Fortunately, methods are available to construct approximately pivotal quantities in many cases, and in the dependent case such as autocorrelated images or time series, a modification called the "block bootstrap" can be applied. Loh (2008) describes an application to the galaxy two-point correlation function.

The most popular and simple bootstrap is the "nonparametric bootstrap" where the resampling with replacement is based on the "empirical distribution function" (e.d.f.) of the original data. The "parametric bootstrap" uses a functional approximation, often a LS or MLE fit, rather than the actual dataset to obtain random points. This is a well-known simulation procedure (Press et al. 1986). Bootstrapping a regression problem requires a choice: one can bootstrap the residuals from the best fit function (classical bootstrap), or one can bootstrap multivariate data points (paired bootstrap). The paired bootstrap is robust against heteroscadasticity in the errors.

2.9 Model Selection and Goodness of Fit

The aim of model fitting is to provide most parsimonious "best" fit of a parametric model to data. It might be a simple heuristic model to phenomenological relationships between observed properties in a sample of astronomical objects, or a more complex model based on astrophysical theory. A good statistical model should be parsimonious yet conforming to the data, following the principle of Occam's Razor. A satisfactory model avoids underfitting which induces bias, and avoids overfitting which induces high variability. A model selection criterion should balance the competing objectives of conformity to the data and parsimony.

The statistical procedures for parameter estimation outlined above, such as LS and MLE, can link data with astrophysical models, but they do not by themselves evaluate whether the chosen model is appropriate for the dataset. The relevant methods fall under the rubrics of statistical "model selection," and "goodness of fit." The common procedure in astronomy based on the reduced chi-squared $\chi^2_\nu \simeq 1$ is a primitive technique not used by statisticians or researchers in other fields.

Hypothesis testing discussed above can be used to compare two models which share a common structure and some parameters, these are "nested models." However, it does not treat models symmetrically. A more general framework for model selection will be based on likelihoods. Let D denote the observed data and $M_1, \ldots, M_k$ denote models for D under consideration. For each model M_j, let $f(D|\theta_j, M_j)$ denote the likelihood, the p.d.f. (or p.m.f. in the discrete case) evaluated at the data D, and let $\ell(\theta_j) = \ln f(D|\theta_j, M_j)$ denote the loglikelihood where θ_i is a p_j dimensional parameter vector.

Three classical hypothesis tests based on MLEs for comparing two models were developed during the 1940s. To test the null hypothesis $H_0 : \theta = \theta_0$, the Wald Test uses $W_n = (\hat{\theta}_n - \theta_0)^2 / Var(\hat{\theta}_n)$, the standardized distance between θ_0 and the maximum likelihood estimator $\hat{\theta}_n$ based on a dataset of size n. The distribution of W_n is approximately chi-square with one degree of freedom. In general, the variance of $\hat{\theta}_n$ is not known; however, a close approximation is $1/I(\hat{\theta}_n)$, where $I(\theta)$ is the Fisher's information. Thus $I(\hat{\theta}_n)(\hat{\theta}_n - \theta_0)^2$ has a chi-square distribution in the limit, and the Wald test rejects the null hypothesis H_0, when this quantity is large. The "likelihood ratio test" uses the logarithm of ratio of likelihoods,

$\ell(\hat{\theta}_n) - \ell(\theta_0)$, and Rao's score test uses the statistic $S(\theta_0) = (\ell'(\theta_0))^2/(nI(\theta_0))$, where ℓ' denotes the derivative of ℓ. The likelihood ratio is most commonly used in astronomy. Protassov et al. (2002) warn about its common misuse.

If the model M_1 happens to be nested in the model M_2, the largest likelihood achievable by M_2 will always be larger than that achievable by M_1. This suggests that the addition of a penalty on models with more parameters would achieve a balance between overfitting and underfitting. Several penalized likelhood approaches to model selection have been actively used since the 1980s. The "Akaike's Information Criterion" (AIC), based on the concept of entropy, for model M_j is defined to be

$$AIC = 2\ell(\hat{\theta}_j) - 2p_j. \tag{10.27}$$

Unlike hypothesis tests, the AIC does not require the assumption that one of the candidate models is correct, it treats models symmetrically, and can compare both nested and non-nested models. Disadvantages of the AIC include the requirement of large samples and the lack of consistency in giving the true number of model parameters even for very large n. The "Bayesian Information Criterion" (BIC) is a popular alternative model selection criterion defined to be

$$BIC = 2\ell(\hat{\theta}_j) - p_j \ln n. \tag{10.28}$$

Founded in Bayesian theory, it is consistent for large n. The AIC penalizes free parameters less strongly than does the BIC.

Goodness of fit can be estimated using nonparametric tests similar to the Kolmogorov–Smirnov statistic discussed in ➋ Sect. 2.10. However, the goodness of fit probabilities derived from these statistics are usually not correct when applied in model fitting situations when the parameters are estimated from the dataset under study. An appropriate approach is bootstrap resampling that gives valid estimates of goodness of fit probabilities under a very wide range of situations. Both the nonparametric and parametric bootstrap can be applied for goodness of fit tests. The method cannot be used for multivariate data due to identifiability problems.

A more difficult problem is comparing best-fit models derived for non-nested model families. One possibility is using the "Kullback–Leibler information," a measure of proximity between data and model arising from information theory.

2.10 Nonparametric Statistics

Nonparametric statistical inference gives insights into data which do not depend on assumptions regarding the distribution of the underlying population. Most standard statistics implicitly assume, through the Central Limit Theorem, that all distributions are normal, measurement uncertainties are constant and increase as $\sqrt{N}$ as the sample size increases, and chosen parametric models are true. But our knowledge of astronomical populations and processes – Kuiper Belt Objects, galactic halo stellar motions, starburst galaxies properties, accretion onto supermassive black holes, and so forth – is very limited. The astronomer really does not know that the observed properties using convenient units are in fact normally distributed or that relationships between properties are in fact (say) power law. Nonparametric approaches to statistical inference should thus be particularly attractive to astronomers, and can precede more restrictive parametric analysis.

Some nonparametric statistics are called "distribution-free" because they are valid for any underlying distribution. Some methods are particularly "robust" against highly skewed distributions or outliers due to erroneous measurements or extraneous objects. Some are based on the rankings of each object within the dataset. However, many nonparametric methods are restricted to univariate datasets; for example, there is no unique ranking for a bivariate dataset.

Nonparametric analysis often begins with "exploratory data analysis" as promoted by statistician John Tukey. The "boxplot" is a compact and informative visualization of a univariate dataset (➋ *Fig. 10-1*). It displays the five-number summary (minimum, 25% quartile, median, 75% quartile, maximum) with whiskers, notches, and outliers. There is a broad consensus that the "median," or central value, of a dataset is the most reliable measure of location. "Trimmed means" are also used. The spread around the median can be evaluated with the "median absolute deviation" (MAD) given by

$$MAD(X) = \text{Median}\,|X - \text{Median}(X)|. \tag{10.29}$$

The cumulative distribution of a univariate distribution is best estimated by the "empirical distribution function" (e.d.f.):

$$\hat{F}_n(x) = \frac{1}{n}\sum_{i=1}^{n} I[X_i \le x], \tag{10.30}$$

for i.i.d. random variables, where I is the indicator function. It ranges from 0.0 to 1.0 with step heights of $1/n$ at each observed value. For a significance level $\alpha = 0.05$ or similar value, the approximate confidence interval for $F(x)$.

The true distribution function at x, is given by

$$\widehat{F}_n(x) \pm z_{1-\alpha/2}\sqrt{\widehat{F}_n(x)[1-\widehat{F}_n(x)]/n}, \tag{10.31}$$

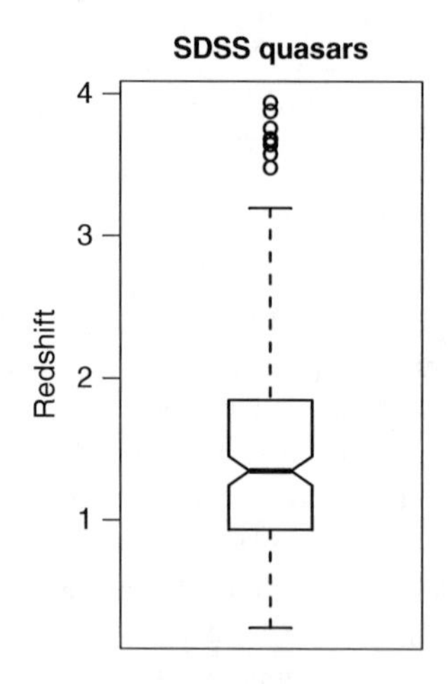

Fig. 10-1
Boxplot for the redshift distribution of 200 quasars from the Sloan Digital Sky Survey

where z_α are the quantiles of the Gaussian distribution.

Important nonparametric statistics are available to test the equality of two e.d.f.'s, or for the compatibility of an e.d.f. with a model. The three main statistics are:

$$\begin{aligned}
\text{Kolmogorov–Smirnov (KS)} \quad M_{KS} &= \max_x |\widehat{F}_n(x) - F_0(x)| \\
\text{Cramer–von Mises (CvM)} \quad W^2_{CvM} &= n \sum_{i-1}^{n} (\widehat{F}_n(X_i) - F_0(X_i))^2 \\
\text{Anderson–Darling (AD)} \quad A^2_{AD} &= n \sum_{i=1}^{n} \frac{(\widehat{F}_n(X_i) - F_0(X_i))^2}{F_0(X_i)(1 - F_0(X_i))}.
\end{aligned} \tag{10.32}$$

The KS statistic is most sensitive to large-scale differences in location (i.e., the median value) and shape between the two distributions. The CvM statistic is effective for both large-scale and small-scale differences in distribution shape. But both of these measures are relatively insensitive to differences near the ends of the distribution. This deficiency is addressed by the AD statistic, a weighted version of the C-vM statistic to emphasize differences near the ends. The AD test is demonstrably the most sensitive of the e.d.f. tests; this was confirmed in a recent astronomical study by Hou et al. (2009). The distributions of these statistics are known and are distribution-free for all continuous F.

But all these statistics are no longer distribution-free under two important and common situations: when the data are multivariate, or when the model parameters are estimated using the dataset under study. Although astronomers sometimes use two-dimensional KS-type tests, these procedures are not mathematically validated to be distribution-free. Similarly, when comparing a dataset to a model, the e.d.f. probabilities are distribution-free only if the model is fully specified independently of the dataset under study. Standard tables of e.d.f probabilities thus do not give a mathematically correct goodness of fit test. Fortunately, a simple solution is available: the distribution of the e.d.f. statistic can be established for each dataset using bootstrap resampling. Thus, a recommended nonparametric goodness of fit procedure combines the sensitive Anderson–Darling statistic with bootstrap resampling to establish its distribution and associated probabilities.

Nonparametric statistics includes several distribution-free rank-based hypothesis tests. The "Mann-Whitney-Wilcoxon statistic" tests the null hypothesis that two samples are drawn from the same population. This is an effective alternative to the "t test" for normal populations. Extensions include the "Hodges-Lehmann test for shift" and the "Kruskal-Wallis test" for $k > 2$ samples. "Contingency tables" are very useful when the categorical, rather than continuous, variables are present. The "χ^2 test" and "Mantel-Haenszel test" are used as 2- and k-sample tests, respectively. "Kendall's τ," "Spearman's ρ," and "Cox-Stuart test" are extremely useful rank tests for independence between paired variables, (X_i, Y_i). Kendall's and Spearman's statistics are nonparametric versions of Pearson's linear correlation coefficient. These tests rest on the assumption of i.i.d. random variables, they are thus not appropriate when one of the variables has a fixed order as in astronomical lightcurves, spectra, or images.

3 Applied Fields of Statistics

3.1 Data Smoothing

Density estimation procedures smooth sets of individual measurements into continuous curves or surfaces. "Nonparametric density estimation" makes no assumption regarding the underlying distribution. A common procedure in astronomy is to collect univariate data into histograms giving frequencies of occurrences grouped into bins. While useful for exploratory examination, statisticians rarely use histograms for statistical inference for several reasons. The choice of bin origin and bin width is arbitrary, information is unnecessarily lost within the bin, the choice of bin center is not obvious, multivariate histograms are difficult to interpret, and the discontinuities between bins does not reflect the continuous behaviors of most physical quantities.

"Kernel density estimation," a convolution with a simple unimodal kernel function, avoids most of these disadvantages and is a preferred method for data smoothing. For an i.i.d. dataset, either univariate or multivariate, the kernel estimator is

$$\hat{f}_{kern}(x,h) = \frac{1}{nh(x)} \sum_{i=1}^{n} K\left(\frac{X - X_i}{h(x)}\right), \tag{10.33}$$

where $h(x)$ is the "bandwidth" and the kernel function K is normalized to unity. The kernel shape is usually chosen to be a Gaussian or Epanechikov (inverted parabola) function. Confidence intervals for $f(x)$ for each x can be readily calculated around the smoothed distribution, either by assuming asymptotic normality if the sample is large or by bootstrap resampling.

The choice of bandwidth is the greatest challenge. Too large a bandwidth causes oversmoothing and increases bias, while too small a bandwidth causes undersmoothing and increases variance. The usual criterion is to choose the bandwidth to minimize the "mean integrated square error" (MISE):

$$MISE(\hat{f}_{kern}) = E\left[\left(\int \hat{f}_{kern}(x) - f(x)\right)^2 dx\right]. \tag{10.34}$$

A heuristic bandwidth for unimodal distributions, known as Silverman's rule-of-thumb, is $h = 0.9\sigma n^{-1/5}$ where σ is the standard deviation of the variable and n is the number of data points. A more formal approach is "cross-validation" which maximizes the log-likelihood of estimators obtained from jackknife simulations. A variety of adaptive smoothers are used where $h(x)$ depends on the local density of data points, although there is no consensus on a single optimal procedure. One simple option is to scale a global bandwidth by the local estimator value according to $h(x_i) = h/\sqrt{\hat{f}(X)}$. An important bivariate local smoother is the "Nadaraya-Watson estimator." Other procedures are based on the distance to the "k-th nearest neighbor" (k-nn) of each point; one of these is applied to an astronomical dataset in *Fig. 10-2*.

A powerful suite of smoothing methods have recently emerged known as "semi-parametric regression" or "nonparametric regression." The most well-known variant is William Cleveland's LOESS method that fits polynomial splines locally along the curve or surface. Extensions include local bandwidth estimation from cross-validation, projection pursuit, and kriging. Importantly, numerically intensive calculations in these methods give confidence bands around the estimators. These methods have been introduced to astronomy by Miller et al. (2002) and Wang et al. (2005).

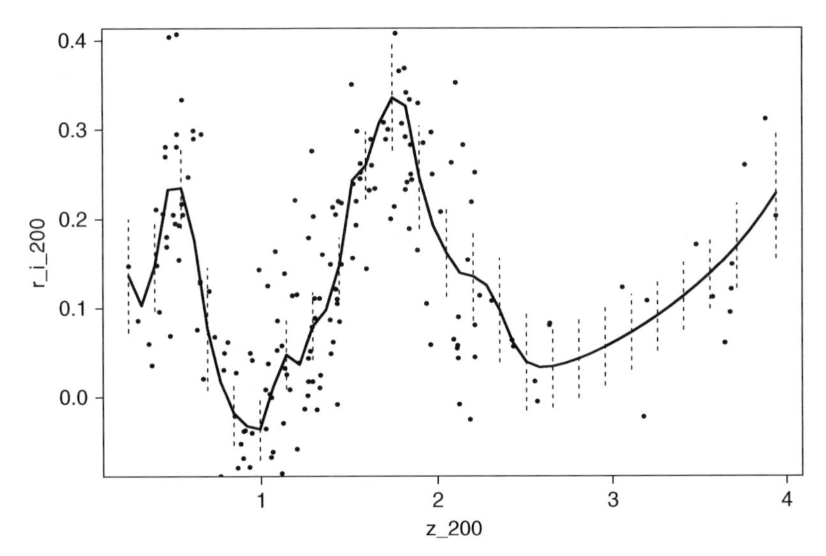

Fig. 10-2
Adaptive kernel density estimator of quasar $r - i$ colors as a function of redshift with bootstrap confidence intervals, derived using the np **package, one of ~3,000 add-on** CRAN **packages**

3.2 Multivariate Clustering and Classification

Many astronomical studies seek insights from a table consisting of measured or inferred properties (columns) for a sample of celestial objects (rows). These are multivariate datasets. If the population is homogeneous, their structure is investigated with methods from multivariate analysis such as principal components analysis and multiple regression. But often the discovery techniques capture a mixture of astronomical classes. Multi-epoch optical surveys such as the planned Large Synoptic Survey Telescope (LSST) will find pulsating stars, stellar eclipses from binary or planetary companions, moving asteroids, active galactic nuclei, and explosions such as novae, supernovae, and gamma-ray bursts. Subclassifications are common. Spiral galaxy morphologies were divided into Sa, Sb, and Sc categories by Hubble, and later were given designations like SBab(rs). Supernovae were divided into Types Ia, Ib, and II, and more subclasses are considered.

However, astronomers generally developed these classifications in a heuristic manner with informed but subjective decisions, often based on visual examinations of two-dimensional projections of the multivariate datasets. A popular method for unsupervised clustering in low-dimensions is the "friends-of-friends algorithm," known in statistics as single linkage hierarchical clustering. But many astronomers are not aware that this clustering procedure has serious deficiencies and many alternatives are available. As an astronomical field matures, classes are often defined from small samples of well-studied prototypes that can serve as "training sets" for supervised classification. The methodologies of unsupervised clustering and supervised classification are presented in detail by Everitt et al. (2001), Hastie et al. (2009), and Duda et al. (2001). Many of the classification procedures have been developed in the computer science, rather than statistics, community under the rubrics of "machine learning" and "data mining."

As with density estimation, astronomers often seek nonparametric clustering and classification as there is no reason to believe that stars, galaxies, and other classes have multivariate normal (MVN) distributions in the observed variables and units. However, most nonparametric methods are not rooted in probability theory, the resulting clusters and classes can be very sensitive to the mathematical procedure chosen for the calculation, and it is difficult to evaluate statistical significance of purported structures. Trials with different methods, bootstrap resampling for validation, and caution in interpretation are advised.

Most clustering and classification methods rely on a metric that defines distances in the p-space, where p is the number of variables or "dimensionality" of the dataset. A Euclidean distance (or its generalization, a Minkowski m-norm distance) is most often adopted, but the distances then depend on the chosen units that are often incompatible (e.g., units in a stellar astrometric catalog may be in degrees, parsecs, milliarcsecond per year, and kilometers per second). A common solution in statistics is to standardize the variables,

$$X_{std} = \frac{X - \bar{X}}{\sqrt{Var(X)}}, \tag{10.35}$$

where the denominator is the standard deviation of the dataset. Astronomers typically choose a logarithmic transformation to reduce range and remove units. A second choice needed for most clustering and classification methods is the definition of the center of a group. Centroids (multivariate means) are often chosen, although medoids (multivariate medians) are more robust to outliers and classification errors. A third aspect of a supervised classification procedure is to quantify classificatory success with some combination of Type 1 errors (correct class is rejected) and Type II errors (incorrect class is assigned).

Unsupervised agglomerative hierarchical clustering is an attractive technique for investigating the structure of a multivariate dataset. The procedure starts with n clusters each with one member. The clusters with the smallest value in the pairwise "distance matrix" are merged, their rows and columns are removed and replaced with a new row and column based on the center of the cluster. This merging procedure is repeated n times until the entire dataset of n points is contained in a single cluster. The result is plotted as a classification tree or dendrogram. The structure depends strongly on the definition of the distance between a cluster and an external data point. In single linkage clustering, commonly used by astronomers, the nearest point in a cluster is used. However, in noisy or sparse data, this leads to spurious "chaining" of groups into elongated structures. For this reason, single linkage is discouraged by statisticians, although it may be appropriate in the search for filamentary patterns. Average linkage and Ward's minimum variance method give a good compromise between elongated and hyperspherical clusters.

"k-means partitioning" is another widely used method that minimizes the sum of within-cluster squared distances. It is related both to Voronoi tesselations and to classical MANOVA methods that rely on the assumption of MVN clusters. k-means calculations are computationally efficient as the distance matrix is not calculated and cluster centroids are easily updated as objects enter or depart from a cluster. A limitation is that the scientist must choose in advance the number k of clusters present in the dataset. Variants of k-means, such as robust k-medoids and the Linde-Buzo-Gray algorithm, are widely used in computer science for pattern recognition in speech and for image processing or compression.

MLE clustering based on the assumption of MVN structures are also used with model selection (i.e., choice of number of clusters in the best model) using the Bayesian Information

Criterion. This is an implementation of "normal mixture models" and uses the "EM Algorithm" for maximizing the likelihood. Other methods, such as DBSCAN and BIRCH, have been recently developed by computer scientists to treat more difficult situations like clusters embedded in noise, adaptive clustering and fragmentation, and efficient clustering of megadatasets.

Techniques for supervised classification began in the 1930s with Fisher's "linear discriminant analysis" (LDA). For two classes in a training set, this can be viewed geometrically as the projection of the cloud of p-dimensional points onto a 1-dimensional line that maximally separates the classes. The resulting rule is applied to members of the unclassified test set. LDA is similar to principal components analysis but with a different purpose: principal components find linear combinations of the variables that sequentially explain variance for the sample treated as a whole, while LDA finds linear combinations that efficiently separate classes within the sample. LDA can be formulated with likelihoods for MLE and Bayesian analysis, and has many generalizations. Classes of machine learning techniques including "Support Vector Machines."

"Nearest neighbor classifiers" (k-nn) are a useful class of techniques whereby cluster membership of a new object is determined by a vote among the memberships of the k nearest neighboring points in the training set. As in kernel density estimation, the choice of k balances bias and variance, and can be made using cross-validation. Bootstrap resampling can assist in evaluating the stability of cluster number and memberships. In "discriminant adaptive nearest neighbor" classification, the metric is adjusted to the local density of points, allowing discovery of subclusters in dense regions without fragmenting low density regions.

Astronomers often define classes by rules involving single variables, such as "Class III pre-main sequence stars have mid-infrared spectral indices $[5.8] - [8] < 0.3$" or "Short gamma-ray bursts have durations < 2 seconds." These partition the datasets along hyperplanes parallel to an axis, although sometimes oblique hyperplanes are used. Such criteria are usually established heuristically by examination of bivariate scatterplots, and no guidance is provided to estimate the number of classes present in the dataset. In statistical parlance, these rule-based classifications are "classification trees." Mature methodologies called "classification and regression trees" (CART) have been developed by Leo Breiman and colleagues to grow, prune, and evaluate the tree. Sophisticated variants like bootstrap aggregation ("bagging") to quantify the importance and reliability of each split and "boosting" to combine weak classification criteria have proved very effective in improving CART and other classification procedures. Important methods implementing these ideas include AdaBoost and Random Forests. CART-like methods are widely used in other fields and could considerably help astronomers with rule-based classification.

Many other multivariate classifiers are available for both simple and complex problems: naive Bayes, neural networks, and so forth. Unfortunately, as most methods are not rooted in mathematical statistics, establishing probabilities for a given cluster or pattern in a training set, or probabilities for assigning new objects to a class, is difficult or impossible to establish. Indeed, a formal "No Free Lunch Theorem" has been proved showing that no single machine learning algorithm can be demonstrated to be better than another in the absence of prior knowledge about the problem. Thus, the astronomer's subjective judgments will always be present in multivariate classification; however, these can be informed by quantitative methodologies which are not yet in common use.

3.3 Nondetections and Truncation

Astronomical observations are often subject to selection biases due to limitations of the telescopes. A common example is the magnitude-limited or flux-limited survey where many fainter objects are not detected. In an unsupervised survey, this leads to "truncation" in the flux variable, such that nothing (not even the number) is known about the undetected population. In a supervised survey, the astronomer seeks to measure a new property of a previously defined sample of objects. Here nondetections produce "left-censored" data points: all of the objects are counted, but some have upper limits in the newly measured property. The statistical treatment of censoring is well established under the rubric of "survival analysis" as the problem arises (usually in the form of right-censoring) in fields such as biomedical research, actuarial science, and industrial reliability (Feigelson and Nelson 1985). The statistical treatment of truncation is more difficult as less is known about the full population, but some relevant methodology has been developed for astronomy.

The "survival function" $S(x)$ for a univariate dataset is defined to be the inverse of the e.d.f.:

$$S(x) = P(X > x) = \frac{\#\text{observations} \geq x}{n} = 1 - F(x). \tag{10.36}$$

A foundation of survival analysis was the derivation of the nonparametric maximum likelihood estimator for a randomly censored dataset by Kaplan and Meier in the 1950s:

$$\hat{S}_{KM}(x) = \prod_{x_i \geq x} \left(1 - \frac{d_i}{N_i}\right), \tag{10.37}$$

where N_i is the number of objects (detected or undetected) $\geq x_i$ and d_i are the number of objects at value x_i. If no ties are present, $d_i = 1$ for all i. The ratio d_i/N_i is the conditional probability that an object with value above x will occur at x. This "product-limit estimator" has discontinuous jumps at the detected values, but the size of the jumps increases at lower values of the variable because the weight of the nondetections are redistributed among the lower detections. For large samples, the "Kaplan–Meier (KM) estimator" is asymptotically normal with variance

$$\widehat{Var}(\hat{S}_{KM}) = \hat{S}_{KM}^2 \sum_{x_i \geq x} \frac{d_i}{N_i(N_i - d_i)}. \tag{10.38}$$

This nonparametric estimator is valid only when the censoring pattern is not correlated with respect to the variable x. Note that when the estimator is used to obtain a luminosity function of a censored dataset, $S(L)$ and the censoring occur due to a flux limit $f_0 = L/4\pi d^2$, the censoring pattern is only partially randomized depending on the distribution of distances in the sample under study. There is no general formulation of an optimal nonparametric luminosity function in the presence of non-random censoring patterns. However, if the parametric form of the luminosity function is known in advance, then estimation using maximum likelihood or Bayesian inference is feasible to obtain best-fit parameter values.

It is possible to compare two censored samples with arbitrary censoring patterns without estimating their underlying distributions. Several nonparametric hypothesis tests evaluating the null hypothesis $H_0 : S_1(x) = S_2(x)$ are available. These include the Gehan and Peto-Peto tests (generalizations of the Wilcoxon two-sample test for censored data), the logrank test, and weighted Fleming–Harrington tests. They all give mathematically correct probabilities that the two samples are drawn from the same distributions under different, reasonable treatments of the nondetections.

A truly multivariate survival analysis that permits censoring in all variables has not been developed, but a few limited methods are available including generalizations of Kendall's τ rank correlation coefficient and various bivariate linear regression models. Cox regression, which relates a single censored response variable to a vector of uncensored covariates, is very commonly used in biometrical studies.

Truncation is ubiquitous in astronomical surveys as, except for a very few complete volume-limited samples, only a small portion of huge populations are available. Unlike controlled studies in social sciences, where carefully randomized and stratified subsamples can be selected for measurement, the astronomer can identify only the closest and/or brightest members of a celestial population.

As in survival analysis treating censored data, if the parametric form of the underlying distribution of a truncated dataset is known, then likelihood analysis can proceed to estimate the parameters of the distribution. The nonparametric estimator equivalent to the Kaplan–Meier estimator was formulated by astrophysicist Donald Lynden-Bell in 1971, and studied mathematically by statistician Michael Woodroofe and others. The Lynden-Bell–Woodroofe (LBW) estimator for a truncated dataset is similar to the Kaplan–Meier estimator, again requiring that the truncation values be independent of the true values. A generalized Kendall's τ statistic is available to test this assumption. A few extensions to the LBW estimator have been developed including a nonparametric rank test for independence between a truncated variable and a covariate, and a two-step least-squares regression procedure.

The KM and LBW estimators, as nonparametric maximum likelihood estimators of censored and truncated distributions, are powerful tools. Their performance is superior to heuristic measures often used by astronomers, such as the detection fraction or Schmidt's $1/V_{max}$ statistic. However, care must be taken: the dataset can not have multiple populations, the upper limit pattern may bias the result, results are unreliable when the censoring or truncation fraction is very high, and bootstrap resampling is often needed for reliable error analysis. Furthermore, nonparametric survival methods cannot simultaneously treat nondetections and measurement errors in the detected points, which are linked by the instrumental process. An integrated approach to measurement errors and nondetections is possible only within a parametric framework, as developed by Kelly (2007).

3.4 Time Series Analysis

The sky is filled with variable objects: orbits and rotations of stars and planets, stochastic accretion processes, explosive flares, supernovae, and gamma-ray bursts. Some behaviors are strictly periodic, others are quasi-periodic, autocorrelated (including $1/f$-type red noise), or unique events. Gravitational wave observatories are predicted to reveal all of these types of variations from high-energy phenomena, but to date only noise terms have been found. Wide-field multi-epoch optical surveys of the sky, culminating with the planned LSST, make movies of the sky and will emerge with hundreds of millions of variable objects.

Time series analysis is a vast, well-established branch of applied mathematics developed mostly in the fields of statistics, econometrics, and engineering signal processing (Chatfield 2004; Shumway and Stoffer 2006). Astronomers have contributed to the methodology to address problems that rarely occur in these other fields: unevenly spaced observations, heteroscedastic measurement errors, and non-Gaussian noise. Note that the mathematics applies to any random variable x that is a function of a fixed time-like variable t, this includes astronomical spectra (intensities as a function of fixed wavelengths) and images (intensities as a function of fixed location).

Unless periodicities are expected, analysis usually begins by examining the time domain observations, $x(t_i)$ where $i = 1, 2, \ldots, n$. If the scientific goals seek large-scale changes in the values of x, then parametric regression models of the form $x(t) = f(t) + \epsilon(t)$ can characterize the trend, where the noise term is commonly modeled as a standard normal, $\epsilon = N(0, \sigma^2)$. If trends are present but are considered uninteresting, they can be removed by a differencing filter, such as $y(t_i) = x(t_i) - x(t_{i-1})$, or spline fits.

Very often, autocorrelation is present due either to astrophysical or instrumental causes. The "autocorrelation function" as a function of lag time t_k is

$$ACF(t_k) = \frac{\sum_{i=1}^{n-k} [x(t_i) - \bar{x}][x(t_{i+k}) - \bar{x}]}{\sum_{i=1}^{n} [x(t_i) - \bar{x}]^2}. \tag{10.39}$$

The "partial autocorrelation function" is very useful, as its amplitude gives the autocorrelation at lag k removing the correlations at shorter lags. Normal "white" noise produces ACF values around zero, while positive values indicate correlated intensities at the specified t_k lag times. When autocorrelation is present, the number of independent measurements is less than the number of observations n. One effect on the statistics is to increase the uncertainty of standard statistics, such as the variance of the mean which is now

$$\widehat{Var}[\bar{x}(t)] = \frac{\sigma^2}{n} \left[1 + 2 \sum_{k=1}^{n-1} \left(1 - \frac{k}{n} \right) ACF(k) \right]. \tag{10.40}$$

Autocorrelated time series are usually modeled with stochastic "autoregressive models." The simplest is the random walk,

$$x(t_i) = x(t_{i-1}) + \epsilon_i \tag{10.41}$$

which gives an autocorrelation function that slowly declines with k, $ACF = 1/\sqrt{1 + k/i}$. The random walk is readily generalized to the linear autoregressive (AR) model with dependencies on p past values. If the time series has stochastic trends, then a moving average (MA) term is introduced with dependencies on q past noise values. The combined autoregressive moving average, ARMA(p, q), model is

$$x(t_i) = \alpha_1 x(t_{i-1}) + \ldots + \alpha_p x(t_{i-p}) + \epsilon(t_i) + \beta_1 \epsilon(t_{i-1}) + \ldots + \beta_q \epsilon(t_{i-q}). \tag{10.42}$$

This has been generalized to permit stronger trends, long-term autocorrelations, multivariate time series with lags, non-linear interactions, heteroscedastic variances, and more (ARIMA, FARIMA, VARIMA, ARCH, GARCH, and related models). Model parameters are usually obtained by maximum likelihood estimation with the Akaike Information Criterion (➋ 10.27) for model selection. This broad class of statistical models have been found be effective in modeling a wide range of autocorrelated time series for both natural and human-generated phenomena.

A few methods are available for autocorrelated time series that are observed with uneven spacing. A correlation function can be defined that involves binning either in time or in lag: the "discrete correlation function" (DCF) in astronomy (Edelson and Krolik 1988) and the "slot autocorrelation function" in physics. Binning procedures and confidence tests are debated, one recommendation is to apply Fisher's z transform to the DCF to obtain an approximately normally-distributed statistic. The "structure function" is a generalization of the autocorrelation function originally developed to study stochastic processes that can also can be applied to unevenly spaced data (Simonetti et al. 1986). Structure functions, and the related singular measures, can give considerable insight into the nature of a wide variety of stochastic time series.

A major mode of time domain analysis that is rarely considered in astronomy is "state-space modeling." Here the time series is modeled as a two-stage parametric regression problem with an unobserved state vector defining the temporal behavior, a state matrix describing changes in the system, an observation matrix linking the data to the underlying state, and noise terms. The model can be as complex as needed to describe the data: deterministic trends, periodicities, stochastic autoregressive behaviors, heteroscedastic noise terms, break points, or other components. The parameters of the model are obtained by maximum likelihood estimation, and are updated by an efficient algorithm known as the "Kalman filter." Kitagawa and Gersch (1996) illustrate the potential of state space modeling for problems in astronomy and geology.

Spectral analysis (also called harmonic or Fourier analysis) examines the time series transformed into frequency space. Periodic signals that are distributed in time but concentrated in frequency is now clearly seen. The Fourier transform can be mapped to the ACF according to

$$f(\omega) = \frac{\sigma_x^2}{\pi}\left[1 + 2\sum_{k=1}^{\infty} ACF(k)\cos(\omega k)\right]. \tag{10.43}$$

This shows that, for white noise with $ACF = 0$, the power spectrum $f(\omega)$ is equal to the signal variance divided by π. If a phase-coherent sinusoidal signal is superposed, the $ACF(k) = cos(\omega_0 k)/(2\sigma_x^2)$ is periodic and the power spectrum is infinite at ω_0. An autoregressive $AR(1)$ process, the spectral density is large at low frequencies and decline at high frequencies, but may be more complicated for higher-order ARMA processes.

Fourier analysis is designed for a very limited problem: a stationary time series of evenly spaced observations, infinite duration, with sinusoidal periodic signals superposed on white noise. Any deviation from these assumptions causes aliasing, spectral leakage, spectral harmonics and splitting, red noise, and other problems. For realistic datasets, the classical "Schuster periodogram" is a biased estimator of the underlying power spectrum, and a number of techniques are used to improve its performance. Smoothing (either in the time or frequency domain) increases bias but reduces variance; common choices are the Daniell (boxcar), Tukey–Hanning or Parzen windows. Tapering, or reducing the signal at the beginning and end of the time series, decreases bias but increases variance. The cosine or Hanning taper is common, and multitaper analysis is often effective (Percival and Walden 1993).

For unevenly spaced data, the empirical power spectrum is a convolution of the underlying process and the sequence of observation times. The "Lomb–Scargle periodogram" (Scargle 1982) is a generalization of the Schuster periodogram for unevenly spaced data. It can be formulated as a modified Fourier analysis, a least-squares regression to sine waves, or a Bayesian solution assuming a Jefferys prior for the time series variance. An alternative to the Lomb–Scargle periodogram, which may have better performance at high frequencies, is the Direct Quadratic Spectrum Estimator (Marquardt and Acuff 1984). Astronomers have also developed a suite of non-Fourier periodograms for periodicity searches in unevenly spaced data. These are statistics of the time domain data folded modulo a range of frequencies. They include Dworetsky's "minimum string length," Lafler-Kinman's truncated autocorrelation function, Stellingwerf's "phase dispersion minimization," Schwarzenberg-Czerny's ANOVA statistic, and Gregory-Loredo's Bayesian periodogram. Specialized methods are now being developed for the specific problem of uncovering faint planetary transits in photometric time series of stars.

Evaluation of the significance of spectral peaks, both for traditional spectral analysis and for unevenly spaced data, is difficult. Analytic approximations are available based on gamma, chi-square or F distributions, but these are applicable only under ideal conditions when the noise is purely Gaussian and no autocorrelation is present. Monte Carlo permutations of the data, simulations with test signals, and other methods (e.g., Reegen 2007) are needed to estimate false alarm probabilities and other characteristics of real-data power spectra.

3.5 Spatial Point Processes

Spatial point datasets are a type of multivariate datasets where some of the variables can be interpreted as spatial dimensions. Examples include locations of Kuiper Belt Objects in the 2-dimensional sky, galaxies in a 3-dimensional redshift survey (where recessional velocity is mapped into radial distance), photons on a 4-dimensional X-ray image (two sky dimensions, energy, and arrival time), and galactic halo stars in 6-dimensional phase space.

Simple "spatial point processes" are stationary (properties invariant under translation) and isotropic (properties invariant under rotation). A stationary Poisson point process generates random locations; this pattern is called "complete spatial randomness" (CSR). More complex models can be developed using non-stationary processes. Data points may have associated non-spatial "mark variables": fluxes in some band, masses, velocities, classifications, and so forth. Extensive methodology for interpreting such spatial point processes has been developed for applications in geology, geography, ecology, and related sciences (Fortin and Dale 2005; Illian et al. 2008). Some of this work has been independently pursued in astronomy, particularly in the context of galaxy clustering (Martínez and Saar 2002).

One class of statistical methods is hypothesis tests to determine whether the observed pattern is consistent with CSR. These tests can be based on the distribution of inter-point distances, or nearest-neighbor distances from observed points or random locations. Other methods treat spatial autocorrelation, measures of variance as a function of scale. Moran's I, for example, developed in the 1950s is a second-order statistic for binned stationary spatial data that gives a spatial "correlogram" that is similar to the temporal autocorrelation function. The "semi-variogram" is a widely used unbinned counterpart of Moran's I, for example, it can be decomposed into a scale-free variance attributable to measurement error and a scaled variance attributable to spatial autocorrelation in the observed phenomena. "Markov random fields" provide another important avenue to characterizing and modeling complicated but stationary spatial patterns. In a Bayesian treatment, astronomers could associate the smoothness prior based on the telescope point spread function, and analyze the remaining structure for noise, texture, sources, and other spatial features.

Astronomers have long experience in analyzing the "two-point correlation function" for characterize global autocorrelation in a stationary and isotropic spatial point process. Let P_{12} be the joint probability that two objects lie in "infinitesimal spheres" of volume (area in two dimensions) $dV_1 dV_2$ around two points at locations $\mathbf{x}_1$ and $\mathbf{x}_2$. This probability can be modeled as the sum of a CSR process with spatial density $\bar{\rho}$ and a correlated process depending only on the distance d:

$$dP_{12} = \bar{\rho}^2[1 + \xi(d)]dV_1 dV_2. \tag{10.44}$$

Several estimators for ξ have been investigated, often involving the ratio of observed densities to those in CSR simulations that have the same survey selection effects as the observed dataset. Edge effects are particularly important for small surveys. The variance of the ξ estimator is often estimated using a normal approximation, but this is biased due to the spatial autocorrelation. An alternative approach based on the block bootstrap is proposed by Loh (2008).

Statisticians and researchers in other fields use integrals of the two-point correlation function to avoid arbitrary choices in binning. "Ripley's K function" measures the average number of objects within a circle around each point:

$$K(d) = \frac{1}{\hat{\lambda} n} \sum_{i=1}^{n} \#[S \text{ in } C(\mathbf{s}_i, d)], \tag{10.45}$$

where C denotes a circle of radius s centered on the data point locations $\mathbf{x}_i$ and λ measures the global density. For a CSR process, $\lambda = n/V$ where V is the total volume of the survey, $E[K(d)] = \pi d^2$ and $Var[K(d)] = 2\pi d^2/(\widehat{\lambda^2} V)$. To remove the rapid rise as circle radii increase, the stabilized function $L = \sqrt{K(d)/\pi} - d$ (for two dimensions) with uniform variance is often considered. These statistics can easily be extended to measure clustering interactions between two or more populations.

As with the two-point correlation function, the K function is biased due to edge effects as the circle radii become comparable to the survey extent. Various edge-corrections have been developed. Other related functions have different sensitivities to clustering and edge effects. The $G(d)$ function is the cumulative distribution of nearest neighbor distances computed from the data points, and $F(d)$ is the similar distribution with respect to random locations in the space. F is called the "empty space function" and serves the same role as the CSR simulations in astronomers' estimates of ξ. "Baddeley's J function," defined by $J(d) = [1 - G(d)]/[1 - F(r)]$, is highly resistant to edge effects while still sensitive to clustering structure: $J(d) = 1$ for CSR patterns, $J(r) < 1$ for clustered patterns out to the clustering scale, and $J(r) > 1$ for patterns with spatial repulsion (e.g., a lattice). ➋ *Figure 10-3* shows an application of Baddeley's J to a galaxy spatial distribution.

All of the above statistics measure global patterns over the entire space, and thus are applicable only to stationary spatial processes. In a nonstationary point process, the clustering pattern varies across the observed space. Moran's I can be applied locally to the k-nearest neighbors, giving an exploratory mapping tool for spatially variable autocorrelation. Probabilities for statistical testing are now difficult to obtain, but block bootstrap methods that maintain small-scale structure but erase large-scale structure might be useful.

In some situations, the spatial variables are themselves less interesting than the mark variables which represent an unseen continuous process that has been sampled at distinct spatial locations. For example, individual background galaxies distorted by gravitational lensing trace an unseen foreground Dark Matter distribution, and individual star radial velocities trace a global differential rotation in the galactic disk. An important method for interpolating mark variables was developed in geology under the rubric of "kriging." For a stationary Gaussian spatial point process, kriging gives the minimum square error predictor for the unobserved continuous distribution. Kriging estimates are based on semi-variogram autocorrelation measures, usually using maximum likelihood methods.

4 Resources

The level of education in methodology among astronomers (and most physical scientists) is far below their needs. Three classes of resources are needed to advance the applications of statistics

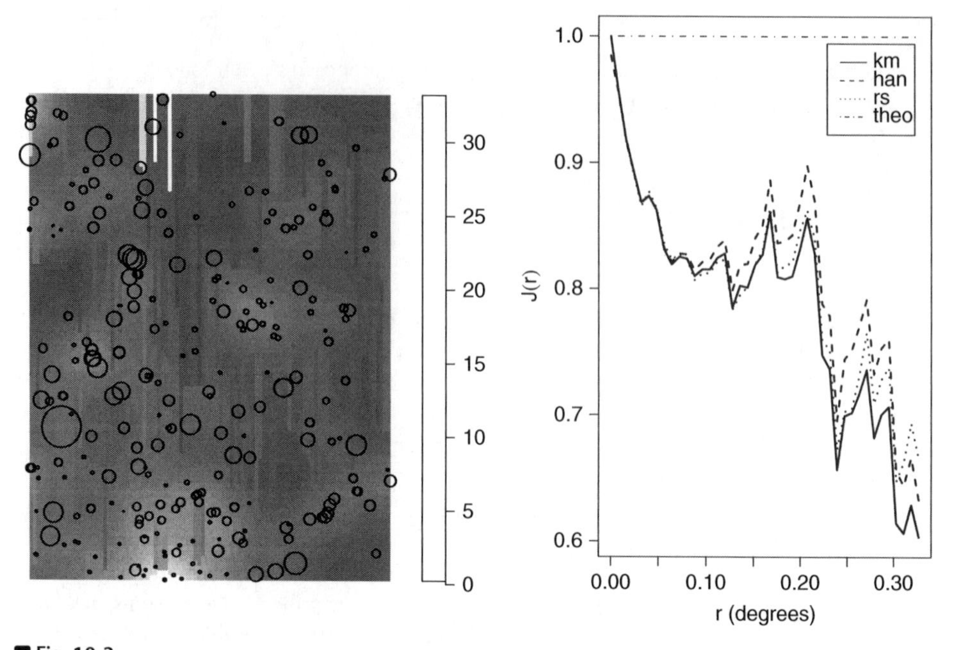

Fig. 10-3

Analysis of a small galaxy redshift survey as a spatial point process using CRAN's *spatstat* package. *Left*: Sky locations of the galaxies with circle sizes scaled to their redshifts, superposed on a kernel density estimator of the sky positions. *Right*: Baddeley's *J* function derived from this galaxy sample. *J* is related to the integral of the two-point correlation function, and is shown with three edge corrections

for astronomical research: education in existing methodology, research in forefront astrostatistics, and advanced statistical software. In the USA, most astronomy students take no classes in statistics while the situation is somewhat better in some other countries. Brief summer schools exposing students to methodology are becoming popular, and partially alleviate the education gap.

Self-education in statistics is quite feasible as many texts are available and astronomers have the requisite mathematical background for intermediate-level texts. The Wikipedia Web site is quite effective in presenting material on statistical topics in a compact format. But astronomers confronting a class of statistical problems would greatly benefit from reading appropriate textbooks and monographs. Recommended volumes are listed in *Table 10-1*.

Research in astrostatistics has rapidly increased since the mid-1990s. Dozens of studies to treat problems in unevenly spaced time series, survey truncation, faint source detection, fluctuations in the cosmic background radiation, and other issues have been published in the astronomical literature. Some of these efforts develop important new capabilities for astronomical data analysis. However, many papers are not conversant with the literature in statistics and other applied fields, even at the level of standard textbooks. Redundant or inadequate treatments are not uncommon.

The paucity of software within the astronomical community has, until recently, been a serious hurdle to the implementation of advanced statistical methodology. For decades, advanced methods were provided only by proprietary statistical software packages like *SAS* and *S-Plus* which the astronomical community did not purchase. The commonly purchased *IDL* package for data analysis and visualization incorporates some statistical methods, particularly those in *Numerical Recipes* (Press et al. 1986), but not a full scope of modern methods.

The software situation has enormously improved in the past few years with the growth of **R**, an implementation of the *S* language released under GNU Public License. **R** is similar to *IDL* in style and operation, with brief user commands to implement both simple and complex functions and graphics. Base **R** implements some dozens of standard statistical methods, while its user-supplied add-on packages in the Comprehensive **R** Archive Network (**CRAN**) provide thousands of additional functionalities. Together, **R/CRAN** is a superb new resource for promulgation of advanced statistical methods into the astronomical community. Its use is illustrated below. Dozens of books have recently emerged on use of **R** for various purposes.

4.1 Web Sites and Books

The field of statistics is too vast to be summarized in any single volume or Web site. *Wikipedia* covers many topics, often in an authoritative fashion (see http://en.wikipedia.org/wiki/List_of_statistics_articles). Additional on-line resources specifically oriented toward astrostatistics are provided by Penn State's Center for Astrostatistics (http://astrostatistics.psu.edu), the California-Harvard Astrostatistics Collaboration (http://www.ics.uci.edu/~dvd/astrostat.html), and the International Computational AstroStatistics Group (http://www.incagroup.org). The International Statistics Institute has formed a new Astrostatistics Network to foster cross-disciplinary interaction (http://isi-web.org/com/ast).

The greatest resource are the hundreds of volumes written by statisticians and application experts. ➋ *Table 10-1* gives a selection of these books recommended for astronomers. They include undergraduate to graduate level texts, authoritative monographs, and application guides for the **R** software system.

4.2 The R Statistical Software System

R (R Development Core Team 2010) is a high-level software language in the public domain with **C**-like syntax. It provides broad capabilities for general data manipulation with extensive graphics, but its strength is the dozens of built-in statistical functionalities. Compiled code for **R** can be downloaded for Windows, MacOS, and Unix operating systems from http://r-project.org. The user community of **R** is estimated to be one to two million individuals, and over 100 books have been published to guide researchers in its use.

The **Comprehensive R Archive Network** (**CRAN**) of user-supplied add-on packages has been growing nearly exponentially for a decade, currently with ~3,000 packages. Some **CRAN** packages have narrow scope, while others are themselves large statistical analysis systems for specialized analyses in biology, ecology, econometrics, geography, engineering, and other research communities. **CRAN** packages, as well as user-provided code in other languages (C, C++, Fortran, Python, Ruby, Perl, Bugs, and XLisp), can be dynamically brought into an **R** session. Some **CRAN** packages can be productively used in astronomical science analysis. At the time of writing, only one package has been specifically written for astronomy: **CRAN**'s *fitsio* package for input of FITS (Flexible Image Transport System, http://fits.gsfc.nasa.gov/) formatted data used throughout the astronomical community.

The **R** scripts below, and accompanying figures, illustrate the analysis and graphical output from **R** and **CRAN** programs based on an astronomical dataset. Similar **R** scripts for a wide variety of statistical procedures applied to astronomical datasets are given in Feigelson and Babu (2012).

The first **R** script inputs a multivariate dataset of photometric and other measurements in ASCII using **R**'s *read.table* function. The data are extracted from the Sloan Digital Sky Survey quasar survey (Schneider et al. 2010). The *dim* (dimension) function tells us there are 17 columns for 33,845 quasars, *names* give the column headings, and *summary* gives the minimum, quartiles, mean, and maximum value for each variable. The next lines select the first 200 quasars and define redshift and $r - i$ color index variables.

The second script makes a boxplot summarizing the redshift distribution (➋ *Fig. 10-1*). It gives the median with a horizontal line, 25% and 75% quartiles with "hinges," "whiskers," outliers, and notches representing a nonparametric version of the standard deviation. It shows an asymmetrical distribution with a heavy tail toward higher redshifts around $z \simeq 3 - 4$. This is one of the common plots made within **R**.

The third script calculates a sophisticated, computationally intensive, nonparametric kernel smoother to the two-dimensional distribution of the $r - i$ color index dependence on redshift z. This calculation is provided by the **CRAN** package *np*, *Nonparametric kernel methods for mixed datatypes* (Hayfield and Racine 2008). Its *npregbw* function computes an adaptive kernel density estimator for a p-dimensional matrix of continuous, discrete, and/or categorical variables. The options chosen here give a local-linear regression estimator with a k-nearest neighbor adaptive bandwidth obtained using least-squares cross-validation. ➋ *Figure 10-2* shows the quasar color-redshift data points superposed on the kernel smoother with error bars representing the 95% confidence band based on bootstrap resampling. The methods are outlined in the **CRAN** help file and are described in detail by Hall et al. (2007).

```
# 1. Construct a sample of 200 SDSS quasar redshifts and r-i
     colors

qso <- read.table('http://astrostatistics.psu.edu/datasets/
    SDSS_QSO.dat',  header=T)
dim(qso)
names(qso)
summary(qso)
z_200 <- qso[1:200,4]
r_i_200 <- qso[1:200,9] - qso[1:200,11]

# 2. Univariate distribution of redshifts: boxplot

boxplot(z_200, varwidth=T, notch=T, main='SDSS quasars',
   ylab='Redshift', pars=list (boxwex=0.3,boxlwd=1.5,
   whisklwd=1.5,staplelwd=1.5,outlwd=1.5,font=2))

# 3. Bivariate adaptive kernel estimator with bootstrap errors

install.packages('np')
library(np)
```

```
citation('np')
bw_adap <- npregbw(z_200, r_i_200, regtype='ll',
bwtype='adaptive_nn')
npplot(bw_adap, plot.errors.method="bootstrap")
points(z_200, r_i_200, pch=20)
```

To illustrate an analysis of galaxy clustering, the fourth **R** script inputs 4,215 galaxies from a redshift survey of the Shapley supercluster by Drinkwater et al. (2004). A subsample of 286 galaxies is selected from the original sample by specifying a small location in the sky.

The fifth **R** script starts by installing the **CRAN** *spatstat, Spatial Statistics,* package described by Baddeley (2010). The functions *owin* and *as.ppp* convert the data table into a special format used by *spatstat*. The galaxies are then plotted on a smoothed spatial distribution with symbol sizes scales to the recessional velocity as a "mark" variable. The second-order *J* function, calculated by the function *Jest*, is then plotted showing sensitivity at large angles to different edge correction algorithms (➋ *Fig. 10-3*). The horizontal line is the predictor for "complete spatial randomness." Baddeley's *J* function is related to Ripley's *K* function, the cumulative (unbinned) two-point correlation function, and is designed to reduce edge effects.

```
# 4. Input and examine Shapley galaxy dataset

shap <- read.table('http://astrostatistics.psu.edu/datasets/
   Shapley_galaxy.dat', header=T)
attach(shap)
dim(shap)
summary(shap)
shap_lo <- shap[(R.A.<214) & (R.A.>209) & (Dec.>-34) &
   (Dec.<-27),]
dim(shap_lo)

# 5. Display galaxy distribution and calculate Baddeley
   J function
# using spatstat package

install.packages('spatstat')
library(spatstat)
citation(spatstat)
shap_lo_win <- owin(range(shap_lo[,1]), range(shap_lo[,2]))
shap_lo_ppp <- as.ppp(shap_lo[,c(1,2,4)], shap_lo_win)
summary(shap_lo_ppp)

par(mfrow=c(1,2))
plot(density(shap_lo_ppp, 0.3), col=gray(5:20/20), main='')
plot(shap_lo_ppp, lwd=2, add=T)
plot(Jest(shap_lo_ppp), lwd=2, col='black', cex.lab=1.3,
  cex.axis=1.3, main='',xlab='r (degrees)',
  legendpos='topright')
par(mfrow=c(1,1))
```

Acknowledgments

The work of Penn State's Center for Astrostatistics is supported by NSF grant SSE AST-1047586.

References

Adler, J. 2010, R in a Nutshell: A Desktop Quick Reference (O'Reilly Media)

Babu, G. J. 1984, Bootstrapping statistics with linear combinations of chi-squares as weak limit, Sankhya, Ser A, 46, 85–93

Babu, G. J., & Singh, K. 1983, Inference on means using the bootstrap, Ann Stat, 11, 999–1003

Babu, G. J., & Singh, K. 1984, On one term Edgeworth correction by Efron's bootstrap, Sankhya, Ser A, 46, 219–232

Baddeley, A. 2010, Analysing Spatial Point Patterns in R, http://www.spatstat.org

Bevington, P. R. 1969, Data Reduction and Error Analysis for the Physical Sciences (McGraw-Hill)

Bivand, R. S., Pebesma, E. J., & Gómez-Rubio, V. 2008, Applied Spatial Data Analysis with R (New York: Springer)

Bowman, A. W., & Azzalini, A. 1997, Applied Smoothing Techniques for Data Analysis (Clarendon)

Brown, L. D., Cai, T. T., & DasGupta, A. 2001, Interval estimation for a binomial proportion (with discussion), Stat Sci, 16, 101–133

Chatfield, C. 2004, The Analysis of Time Series: An Introduction (6th ed.; London: Chapman and Hall)

Chen, C.-h., Härdle, W., & Unwin, A. (eds.), 2008, Handbook of Data Visualization (New York: Springer)

Conover, W. J. 1999, Practical Nonparametric Statistics (New York: Wiley)

Cowan, G. 2006, The small-N problem in high energy physics, in Statistical Challenges in Modern Astronomy IV, ed. G. J. Babu, & E. D. Feigelson (New York: Springer), 75–86

Cowpertwait, P. S. P., & Metcalfe, A. V. 2009, Introductory Time Series with R (New York: Springer)

Cox, D. R. 2006, Principles of Statistical Inference (Cambridge, UK: Cambridge University Press)

Dalgaard, P. 2008, Introductory Statistics with R (New York: Springer)

Dempster, A. P., Laird, N. M., & Rubin, D. B. 1977, Maximum likelihood from incomplete data via the EM algorithm, J R Stat Soc B, 39, 1–38

Drinkwater, M. J., Parker, Q. A., Proust, D. Slezak, E., & Quintana, H., 2004, Publ Astron Soc Aust, 21, 89

Duda, R. O., Hart, P. E., & Stork, D. G. 2001, Pattern Classification (2nd ed.; New York: Wiley)

Edelson, R. A., & Krolik, J. H. 1988, The discrete correlation function: A new method for analyzing unevenly sampled variability data, Astrophys J 333, 646–659

Efron, B., & Tibshirani, R. J. 1993, An Introduction to the Bootstrap (London: Chapman and Hall)

Evans, M., Hastings, N., & Peacock, B. 2000, Statistical Distributions (3rd ed.; New York: Wiley)

Everitt, B. S., Landau, S., & Leese, M. 2001, Cluster Analysis (4th ed.; Arnold)

Feigelson, E. D., & Babu, G. J. 2012, Modern Statistical Methods for Astronomy with R Applications (Cambridge, UK: Cambridge University Press)

Feigelson, E. D., & Nelson, P. I. 1985, Statistical methods for astronomical data with upper limits. I – Univariate distributions, Astrophys J 293, 192–206

Fortin, M.-J., & Dale, M. R. T. 2005, Spatial Analysis: A Guide for Ecologists (Cambridge, UK: Cambridge University Press)

Fisher, R. A. 1922, On the mathematical foundations of theoretical statistics, Philos Trans R Soc A, 222, 309–368

Gelman, A., Carlin, J. B., Stern, H. S., & Rubin, D. B. 2004, Bayesian Data Analysis (2nd ed.; London: Chapman and Hall)

Gregory, P. C. 2005, Bayesian Logical Data Analysis for the Physical Sciences: A Comparative Approach with Mathematica Support (Cambridge, UK: Cambridge University Press)

Hall, P., Li, Q., & Racine, J. S. 2007, Nonparametric estimation of regression functions in the presence of irrelevant regressors, Rev Econ Stat, 89, 784–789

Hastie, T., Tibshirani, R., & Friedman, J. 2009, The Elements of Statistical Learning: Data Mining, Inference, and Prediction (2nd ed.; New York: Springer)

Hayfield, T., & Racine, J. S. 2008, Nonparametric Econometrics: The np Package, J Stat Softw, 27(5), 1–32

Helsel, D. R. 2004, Nondetects and Data Analysis: Statistics for Censored Environmental Data (Wiley-Interscience)

Hogg, R. V., & Tanis, E. 2009, Probability and Statistical Inference (10th ed.; Prentice Hall)

Hou, A., Parker, L. C., Harris, W. E., & Wilman, D. J. 2009, Statistical tools for classifying galaxy group dynamics, Astrophys J, 702, 1199–1210

Illian, J., Penttinen, A., Stoyan, H., & Stoyan, D. 2008, Statistical Analysis and Modeling of Spatial Point Processes (Wiley-Interscience)

James, F. 2006, Statistical Methods in Experimental Physics (2nd ed.; World Scientific)

Johnson, N. L., Kotz, S., & Balakrishnan, N. 1994, Continuous Univariate Distributions, Vols. 1 and 2 (2nd ed.; Wiley-Interscience)

Johnson, R. A., & Wichern, D. W. 2007, Applied Multivariate Statistical Analysis (6th ed.; Prentice-Hall)

Kashyap, V. L., van Dyk, D. A., Connors, A., Freeman, P. E., Siemiginowska, A., Xu, J., & Zezas, A. 2010, On computing upper limits to source intensities, Astrophys J, 719, 900–914

Kelly, B. C. 2007, Some aspects of measurement error in linear regression of astronomical data, Astrophys J, 665, 1489–1506

Kitagawa, G., & Gersch, W. 1996, Smoothness Priors Analysis of Time Series (New York: Springer)

Klein, J. P., & Moeschberger, M. L. 2010, Survival Analysis: Techniques for Censored and Truncated Data (New York: Springer)

Kruschke, J. K. 2011, Doing Bayesian Data Analysis: A Tutorial with R and BUGS (Academic)

Kutner, M. H., Nachtsheim, C. J., & Neter, J. 2004, Applied Linear Regression Models (4th ed.; McGraw-Hill)

Lawless, J. F. 2002, Statistical Models and Methods for Lifetime Data (2nd ed.; New York: Wiley)

Lilliefors, H. W. 1969, On the Kolmogorov-Smirnov test for the exponential distribution with mean unknown, J Am Stat Assoc, 64, 387–389

Loh, J. M. 2008, A valid and fast spatial bootstrap for correlation functions, Astrophys J, 681, 726–734

Lucy, L. B. 1974, An iterative technique for the rectification of observed distributions, Astron J, 79, 745–754

Lupton, R. 1993, Statistics in Theory and Practice (Princeton University Press)

Maindonald, J., & Braun, W. J. 2010, Data Analysis and Graphics Using R: An Example-Based Approach (3rd ed.; Cambridge, UK: Cambridge University Press)

Marquardt, D. W., & Acuff, S. K. 1984, Direct quadratic spectrum estimation with irregularly spaced data, in Time Series Analysis of Irregularly Observed Data, ed. E. Parzen, Lecture Notes in Statistics, Vol. 25 (New York: Springer)

Martínez, V. J., & Saar, E. 2002, Statistics of the Galaxy Distribution (London: Chapman and Hall)

McLachlan, G. J., & Krishnan, T. 2008, The EM Algorithm and Extensions (2nd ed.; Wiley-Interscience)

Miller, C. J., Nichol, R. C., Genovese, C., & Wasserman, L. 2002, A nonparametric analysis of the cosmic microwave background power spectrum, Astrophys J, 565, L67–L70

Nason, G. 2008, Wavelet Methods in Statistics with R (New York: Springer)

Park, T., Kashyap, V. L., Siemiginowska, A., van Dyk, D. A., Zezas, A., Heinke, C., & Wargelin, B. J. 2006, Bayesian estimation of hardness ratios: Modeling and computations, Astrophys J, 652, 610–628

Percival, D.B., & Walden, A.T. 1993, Spectral Analysis for Physical Applications (Cambridge, UK: Cambridge University Press)

Press, W. H., Teukolsky, S. A., Vetterling, W. T., & Flannery, B. P. 1986, Numerical Recipes: The Art of Scientific Computing (Cambridge, UK: Cambridge University Press)

Protassov, R., van Dyk, D. A., Connors, A., Kashyap, V. L., & Siemiginowska, A. 2002, Statistics, handle with care: Detecting multiple model components with the likelihood ratio test, Astrophys J, 571, 545–559

R Development Core Team 2010, R: A Language and Environment for Statistical Computing (Vienna: R Foundation for Statistical Computing)

Rao, C. R. 1997, Statistics and Truth: Putting Chance to Work (2nd ed.; World Scientific)

Reegen, P. 2007, SigSpec. I. Frequency- and phase-resolved significance in Fourier space, Astron Astrophys, 467, 1353–1371

Rice, J. 1994, Mathematical Statistics and Data Analysis (2nd ed.; Duxbury Press)

Ross, S. M. 2010, A First Course in Probability (10th ed.; Prentice Hall)

Sarkar, D. 2008, Lattice: Multivariate Data Visualization with R (New York: Springer)

Scargle, J. D. 1982, Studies in astronomical time series analysis. II: Statistical aspects of spectral analysis of unevenly spaced data, Astrophys J 263, 835–853

Schneider, D. P. et al. 2010, The Sloan Digital Sky Survey Quasar Catalog. V. Seventh Data Release, Astron J, 139, 2360–2373

Shumway, R. H., & Stoffer, D. S., 2006, Time Series Analysis and Its Applications with R Examples (2nd ed.; New York: Springer)

Silverman, B. W. 1998, Density Estimation (London: Chapman and Hall)

Simonetti, J. H., Cordes, J. M., & Heeschen, D. S. 1986, Flicker of extragalactic radio sources at two frequencies, Astrophys J 296, 46–59

Spergel, D. N., Verde, L., Peiris, H. V., Komatsu, E., Nolta, M. R., Bennett, C. L., Halpern, M., Hinshaw, G., Jarosik, N., Kogut, A., Limon, M., Meyer, S. S., Page, L., Tucker, G. S., Weiland, J. L., Wollack, E., & Wright, E. L. 2003, First-year Wilkinson Microwave Anisotropy Probe (WMAP) observations: Determination of cosmological parameters, Astrophys J, 148, 175–194

Starck, J.-L., & Murtagh, F. 2006, Astronomical Image and Data Analysis (2nd ed.; New York: Springer)

Stigler, S. M. 1986, The History of Statistics: The Measurement of Uncertainty Before 1900 (Harvard University Press)

Takezawa, K. 2005, Introduction to Nonparametric Regression (New York: Wiley)

Wang, X., Woodroofe, M., Walker, M. G., Mateo, M., & Olszewski, E. 2005, Estimating Dark Matter distributions, Astrophys J, 626, 45–158

Way, M. J., Scargle, J. D., Ali, K., & Srivastava, A. N. (eds.), 2011, Advances in Machine Learning and Data Mining for Astronomy (London: Chapman and & Hall)

Wasserman, L. 2005, All of Statistics: A Concise Course in Statistical Inference (New York: Springer)

Wickham, H. 2009, ggplot2: Elegant Graphics for Data Analysis (2nd ed.; New York: Springer)

Zoubir, A. M., & Iskander, D. R. 2004, Bootstrap Techniques for Signal Processing, (Cambridge, UK: Cambridge University Press)

11 Numerical Techniques in Astrophysics

Matt Wood
Department of Physics and Space Sciences, Florida Institute of Technology, Melbourne, FL, USA

T.D. Oswalt, H.E. Bond (eds.), *Planets, Stars and Stellar Systems. Volume 2: Astronomical Techniques, Software, and Data*, DOI 10.1007/978-94-007-5618-2_11, © Springer Science+Business Media Dordrecht 2013

Abstract: We discuss selected numerical techniques in astrophysics, including classes of partial differential equations, fundamental representations of partial differential equations as finite difference equations, N-body techniques, and Lagrangian and Eulerian hydrodynamics techniques.

Keywords: Methods: numerical – hydrodynamics – N-body

1 Introduction

Numerical simulations have come to be an essential third approach in our search for a deeper understanding of astrophysical phenomena, alongside observation and analytical theory. We receive information from the universe largely in the form of electromagnetic radiation. Historically, visible-light images of planets, nebulae, and galaxies piqued interest and spurred developments in mathematics, physics, and celestial mechanics. Ever-larger telescopes probed fainter objects, and spectrographs opened a window onto the chemical compositions and radial velocities of objects observed. Radio and space-based telescopes opened windows to the full electromagnetic spectrum. We learned that the laws of physics deduced from careful observation and experimentation applied everywhere in the observable universe. We also learned that the universe contained more extreme states of matter than we could probe in terrestrial laboratories; for example, Sir Arthur Eddington (1926) noted in discussing Adams' early observations of Sirius B "... that matter 2000 times denser than platinum is not only possible, but is actually present in the universe."

Early models of stellar structure, including those of white dwarf stars (Chandrasekhar 1939, and see also Schwarzschild 1958), were polytropes computed by hand. The development of digital computers has meant that researchers could program far more detailed approximations to the objects under study, and in so doing study stellar structure, astrophysical hydrodynamics, and the dynamics of large-N collections of point-masses such as galaxies or star clusters. The exponential pace of increasing processing speed and available RAM has followed "Moore's law" of a doubling of processor capabilities every 2 years for over half a century. Currently and for the foreseeable future, numerical astrophysics is an indispensable means of studying the evolution universe and its contents.

Here will be presented a brief overview of the numerical techniques most commonly used to model astrophysical objects and phenomena. Most important are numerical implementations of the equations of orbital mechanics, hydrodynamics, and radiative transport; however, the constitutive physics calculations of opacities, equation of state, nuclear physics, and magnetic field effects on plasmas are essential components to our numerical models. Many excellent texts have been written on numerical techniques in astrophysics, including Bowers and Wilson (1991), Hockney and Eastwood (1988), and Bodenheimer et al. (2007). In addition, there is an even larger library of texts devoted to the more general subject of computational physics and computational techniques in related physical sciences, such as Hamming (1962), Gould et al. (1988), Garcia (2000), Allen and Tildesley (1989), Burnett (1987), Harrison (2001), Cramer (2004), Giordano and Nakanishi (2005), Pang (2006), Press et al. (2007), Landau et al. (2007, 2008), Thijssen (2007), and DeVries and Hasbun (2010), to name just a few.

In particular, the reader may find Bodenheimer et al. (2007) to be of particular utility as it is still current and far more comprehensive than this short review can be, and also contains a

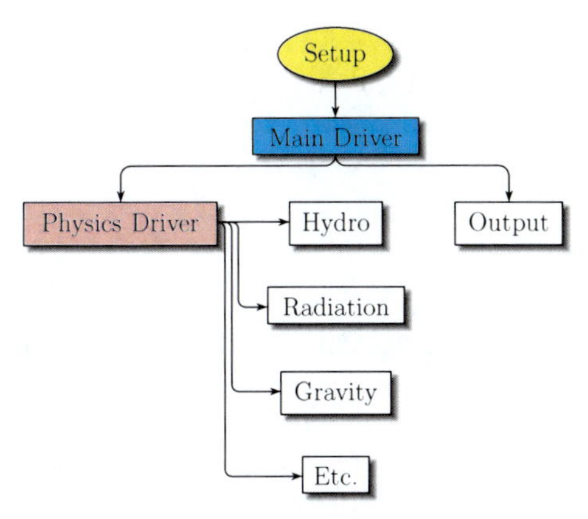

Fig. 11-1
Schematic flowchart of a typical astrophysical numerical simulation code

CDROM with source codes that allow the reader to set up and explore his or her own problems in orbital dynamics, Eulerian grid-based hydrodynamics, Lagrangian smoothed particle hydrodynamics, radiative transfer, and stellar evolution.

In many cases, the flowchart of a numerical simulation code for astrophysical problems will resemble *Fig. 11-1* (Bowers and Wilson 1991). As an example, hydrodynamics is a common physics module. The three conservation Euler equations for inviscid gas dynamics may be written in Lagrangian (co-moving) form and are given by the equations of mass, momentum, and energy conservation, respectively (e.g., Shu 1992):

$$\frac{d\rho}{dt} + \rho\nabla\cdot\vec{\mathbf{v}} = 0, \tag{11.1}$$

$$\frac{d\vec{\mathbf{v}}}{dt} + \frac{\nabla P}{\rho} = 0, \tag{11.2}$$

$$\frac{de}{dt} + \frac{P}{\rho}\nabla\cdot\vec{\mathbf{v}} = 0, \tag{11.3}$$

where $d/dt = \partial/\partial t + \vec{\mathbf{v}}\cdot\nabla$ is the convective derivative. These three coupled nonlinear partial differential equations represent five equations containing the six unknowns ρ, e, P, and $\vec{\mathbf{v}}$ which denote the mass density, specific internal energy, pressure, and fluid velocity, respectively. When these equations are supplemented with the applicable boundary and initial conditions and equation of state, the problem can in principle be solved. Note that the equation of state $P = P(\rho, e)$ may be assumed to be a simple ideal gas relation $P = (\gamma - 1)\rho e$ for some applications (where γ is the ratio of specific heats), but in general will be a complicated function (often pre-computed and tabulated) giving P in terms of ρ, e, and the composition. The latter might include the use of the Saha equation to determine the ionization fraction. Thus, the hydrodynamics physics module would be made up of one or more subroutines implementing a finite difference solution scheme to equations (11.2) and (11.3), and a subroutine which returns the needed quantities from the equation of state.

In ➋ Sect. 2, classes of partial differential equations and simple difference equations are discussed. N-Body techniques are discussed in ➋ Sect. 3. In ➋ Sect. 4, smoothed particle hydrodynamics techniques are discussed, and in ➋ Sect. 5, Eulerian hydrodynamics techniques including shock treatments and Godunov methods are presented. Conclusions are presented in ➋ Sect. 6.

2 Numerics: Approximating Differential Equations

The equations governing most physical phenomena are continuous partial differential equations. The problems of interest can usually be classified as initial value problems, boundary value problems, or a combination of the two. The astrophysical problems of interest are generally too complex to be amenable to analytic solutions. Computers in general cannot be programmed to solve multidimensional coupled systems of equations directly, but they can solve algebraic relations very, very fast. The researcher generally will take his or her partial differential equations and recast them as difference equations (see, e.g., Press et al. 2007). The continuous variables of space and time will often be computed on a grid. The simplest is to represent a derivative as a finite difference, and this is treated first. Care must always be taken when doing numerical simulations that the numerical approximations capture the physics that is to be studied, and do not introduce spurious results that are unphysical. The researcher must check that the numerical solution is not strongly dependent on the time and space resolution of the simulation, and that the physics of interest is either resolved by the differencing or treated in such a way that it need not be.

2.1 Classes of Partial Differential Equations

Numerical simulations in an astrophysical context usually model initial value and/or boundary value problems. If the former, we follow the state of a system evolved through time, and if the alter, the equations describing the system at a given point in time are solved (Bodenheimer et al. 2007). In general, partial differential equations (PDEs) are classified as elliptical, parabolic, or hyperbolic, where the classification of the PDE determines how information must propagate, and thus how it might or might not be parallelized. Elliptic equations are common in boundary value problems in which the solution at each point in the domain of interest depends on the values at the boundary enclosing the domain. Examples of elliptical equations are Poisson's equation

$$\nabla^2 u = 4\pi G\rho(x, y, z), \tag{11.4}$$

where in 3D Euclidean space we have:

$$\nabla^2 = \frac{\partial^2 u}{\partial x^2} + \frac{\partial^2 u}{\partial y^2} + \frac{\partial^2 u}{\partial z^2}, \tag{11.5}$$

and where u is the gravitational potential, G is the gravitational constant, and ρ is the mass density. If $\rho = 0$, then (➋ 11.4) reduces to the Laplace equation

$$\nabla^2 u = 0. \tag{11.6}$$

Examples of parabolic equations are initial value problems that also require one or more boundary conditions to be specified. The classic example of a parabolic equation is the diffusion equation:

$$\frac{\partial u}{\partial t} = \nabla \cdot (D\nabla u), \tag{11.7}$$

where the positive-valued D is the diffusion coefficient, and, for example, u might represent the temperature or density. If D is a function of $u = u(x, y, z, t)$ then the equation is nonlinear. In the 1D case where D is not a function of u, we have

$$\frac{\partial u}{\partial t} = D\frac{\partial^2 u}{\partial x^2}. \tag{11.8}$$

As an example, if treating heat conduction in a medium, then the initial temperature values in the domain must be specified, as well as the time-independent values at the boundaries of the domain.

Hyperbolic equations are typically initial value problems, where the initial state of a system is evolved forward in time via the governing equations. Boundary conditions may be applied and external forces specified, but they are not in general required. The standard example for the class of hyperbolic equations is the wave equation

$$\frac{\partial^2 u}{\partial t^2} = c^2\nabla^2 u, \tag{11.9}$$

where c is the speed of propagation of the wave.

In astrophysical simulations of complex phenomena, equations of two or more of these types may be present, and the numerical experimenter must be careful that the experiment (i.e., the code) is carefully written so that the solutions represent reality.

2.2 Simple Difference Equations

The first derivative of a function $f(x)$ is defined as

$$f' = \frac{df}{dx} \equiv \lim_{h\to 0}\frac{f(x+h) - f(x)}{h}, \tag{11.10}$$

so numerically it is natural to approximate

$$\frac{df}{dx} \approx \frac{f(x+h) - f(x)}{h}, \tag{11.11}$$

where h is small enough. Indeed (➋ 11.11) expresses the forward finite difference equation. If we generalize to a multidimensional $f(x, y, z)$ given on a regular equally spaced Cartesian grid with grid points (x_i, y_j, z_k) with $x_i \in (x_1, \ldots, X_{N_x})$, $y_j \in (y_1, \ldots, X_{N_y})$, and $z_k \in (z_1, \ldots, X_{N_z})$, then we write the forward difference approximation of the partial derivative as

$$\left.\frac{\partial f}{\partial x}\right|_{x_i,y_j,z_k} \approx \frac{f(x_{i+1}, y_j, z_k) - f(x_i, y_j, z_k)}{h}. \tag{11.12}$$

Similarly, the backward difference approximation to the partial derivative is

$$\left.\frac{\partial f}{\partial x}\right|_{x_i,y_j,z_k} \approx \frac{f(x_i, y_j, z_k) - f(x_{i-1}, y_j, z_k)}{h}, \tag{11.13}$$

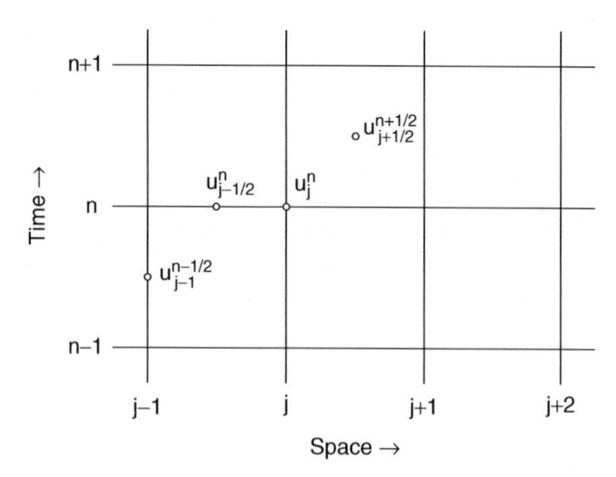

Fig. 11-2
Finite differences on a simple space-time grid. ***Points*** **show differing choices for variable centering**

and the centered difference approximation is

$$\left.\frac{\partial f}{\partial x}\right|_{x_i,y_j,z_k} \approx \frac{f(x_{i+1},y_j,z_k) - f(x_{i-1},y_j,z_k)}{2h}. \tag{11.14}$$

A finite difference approximation to the second-order partial derivative is

$$\left.\frac{\partial^2 f}{\partial x^2}\right|_{x_i,y_j,z_k} \approx \frac{f(x_{i+1},y_j,z_k) - 2f(x_i,y_j,z_k) + f(x_{i-1},y_j,z_k)}{h^2}. \tag{11.15}$$

Fig. 11-2 shows a space-time grid, and possible locations for the values of the discretized function $u(x,t)$. The spatial variable x will span some defined range $x_{\min}$ to $x_{\max}$, and time will run from $t = 0$ to $t = t_{\max}$. In the simplest case, the spatial variable x is discretized in M equal points, where $x_j = x_{\min} + j\Delta x$. In like fashion, time is discretized into N points with values $t^n = n\Delta t$. The region between the defined grid points is typically called the grid cell or zone. As shown schematically in *Fig. 11-2*, the values of the discretized function $u(x,t)$ can be defined at cell corners (e.g., u^n_j), cell edges (e.g., $u^{n-1/2}_{j-1}$ or $u^n_{j-1/2}$), or cell centers (e.g., $u^{n+1/2}_{j+1/2}$).

The *time step* Δt can be constant in time or may vary globally or even individually per object being simulated. The spatial grid can similarly be held fixed as initialized at the beginning of the simulation, or the code may implement the ability to expand or contract in some optimized way (if, e.g., simulating a supernova explosion or protostellar collapse, respectively), or the code may implement the ability to add or delete grid points to the spatial mesh as needed (e.g., to resolve shocks). *Fig. 11-3* shows a space-time grid that is both contracting in the spatial dimension, and using progressively shorter time steps.

Derivatives with respect to time are handled similarly, and again there are a few choices that can be made for how to step a numerical solution through time. Consider the partial differential equation

$$\frac{\partial u(x,t)}{\partial t} = g(u,x,t). \tag{11.16}$$

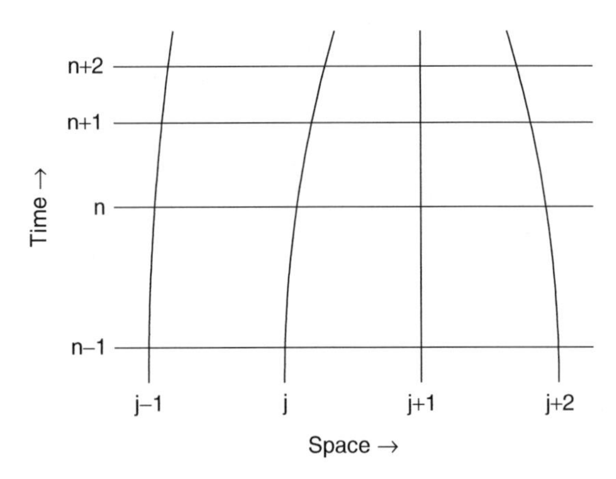

Fig. 11-3
Figure shows a grid with changing dimensions

To advance the solution from time t^n to time $t^{n+1} = t^n + \Delta t$, the simplest choice is clearly the explicit scheme:

$$u_i^{n+1} = u_i^n + g^n \Delta t, \tag{11.17}$$

which is an approximation to the differential equation that is first-order accurate. For certain classes of problems (specifically, functions that behave monotonically) and a sufficiently small choice of time step, this simple scheme is perfectly adequate.

Equation 11.17 uses forward time differencing. It is also possible to cast the equation using backward time differencing as

$$u_i^{n+1} = u_i^n + g^{n+1} \Delta t, \tag{11.18}$$

or centered time differencing as

$$u_i^{n+1} = u_i^n + \frac{g^n + g^{n+1}}{2} \Delta t. \tag{11.19}$$

Equations 11.18 and 11.19 are *implicit* schemes, since they include g^{n+1}, which is not known at time t^n, but must be determined along with u_j^{n+1}. The choice of forward or backward time differencing can have significant ramifications for the numerical stability of a solution. Equation 11.18 is first-order accurate, and (11.19) is second-order accurate. In general, errors in numerical solutions are smaller when higher-order accuracy is used, and as a result it is possible to use larger time steps that can be taken for a fixed error goal.

A given numerical code may contain a mixture of forward, backward, and centered difference equations, as given by needs for expediency and stability. Volumes have been written on the subject, and we direct the interested reader to explore further as interested (Bodenheimer et al. 2007; Strikwerda 2007; Thomas 2010).

2.3 The Critical Role of Stability

With the finite difference approximations above, we could directly recast our PDEs into FDEs for any simulation. Doing so would likely lead to problems, however, if no thought is given to how exactly variables and the derivatives of those variables are represented on the grid. As a simple example (see Giordano and Nakanishi 2005, Chapter 3.), consider a simple pendulum of length L with small angular amplitude θ, governed by the equation

$$\frac{d^2\theta}{dt^2} \cong -\left(\frac{g}{L}\right)\theta, \tag{11.20}$$

where g is the gravitational acceleration. The analytical solution is $\theta = \theta_0 \cos(\omega t + \phi)$, where θ_0 is the amplitude of the oscillation, $\omega = \sqrt{g/L}$, and ϕ is the phase at $t = 0$. To solve this numerically, we first write (➋ 11.20) as two first-order equations

$$\frac{d\omega}{dt} = -\left(\frac{g}{L}\right)\theta, \tag{11.21}$$

$$\frac{d\theta}{dt} = \omega. \tag{11.22}$$

When cast directly as forward difference equations, we have Euler's method:

$$\omega_{i+1} = \omega_i - \left(\frac{g}{L}\right)\theta_i \Delta t, \tag{11.23}$$

$$\theta_{i+1} = \theta_i + \omega_i \Delta t. \tag{11.24}$$

It is recommended that the reader who is new to numerical astrophysics actually code this up, to find that no matter how small Δt is chosen, the method is not acceptable because it does not conserve energy. Consider time step n where $\theta = 0$. Over the previous time step $n - 1$ the angular acceleration was constant and equal to that at t^{n-1}, even though at t^n the angular acceleration should be zero. Over the next time step, the angular acceleration is zero, even though it is in reality negative over the rest of the interval. Thus, the amplitude of the oscillations in this example will grow without bound for all choices of Δt, and so the scheme is *inherently unstable* when applied to this problem. The scheme can be modified slightly by replacing ω_i with ω_{i+1} in (➋ 11.24), which is the Euler-Cromer method. Cromer (1981) noted that this simple modification conserves energy for oscillatory problems.

Stability of the numerical method is even more important than the accuracy of the method. For example, as above a method might produce results that are very good initially, but that diverge from the analytical solution the longer the simulation runs. Such a method is said to be *unstable* for the particular problem. In brief, a method is (1) unconditionally stable when applied to a particular problem if the errors decrease with time, (2) conditionally stable when the errors decrease with time as long as the time step Δt is less than some critical value, and (3) unconditionally unstable if the errors grow with time regardless of the time step used.

3 *N*-Body Simulations

N-body codes are concerned with predicting the motions of a small to large number N ($i = 1, \ldots, N$) of objects that interact gravitationally as point masses. The acceleration of mass i is simply the net force on i from the other $N - 1$ particles in the system:

$$\frac{d^2\vec{\mathbf{r}}_i}{dt^2} = -\sum_{i=1}^{N}\sum_{j\neq i}^{N} \frac{Gm_j(\vec{\mathbf{r}}_i - \vec{\mathbf{r}}_j)}{|\vec{\mathbf{r}}_i - \vec{\mathbf{r}}_j|^3}. \tag{11.25}$$

The problem is initialized by specifying the positions $\vec{\mathbf{r}}_i$ and velocities $\vec{\mathbf{v}}_i$ at time $t = 0$. The N-body problem consists of two distinct parts. First, the net force on each particle must be calculated, and second, the system must be advanced in time. A brief overview of the subject is given below, and for a detailed treatment, we recommend Chapter 3 of Bodenheimer et al. (2007).

The first N-body simulation in the astrophysical literature was the analog simulation of Holmberg (1941) who used 37 light bulbs representing two interacting model galaxies. The first N-body experiments that were simulated on computers were those of von Hoerner (1960, 1963), who used first 16 and later 25 particles.

3.1 Particle-Particle Methods

The simplest method for coding N-body dynamics is the particle-particle (PP) scheme. It is simple enough that it can be coded with little time and effort (and is a useful exercise for the reader). The pseudocode is listed in ➋ Algorithm 1, adapted from Hockney and Eastwood (1988). For each mass m_i the net gravitational force from the other $N-1$ bodies is calculated directly. Once all net forces $\vec{\mathbf{F}}_\mathbf{i}$ are calculated, the velocities and positions of all particles are updated in time using, for example, a second-order leap-frog integrator or higher-order Runge-Kutta integrator (e.g., Press et al. 2007). In practice, the time steps δt must be allowed to vary, as close particle-particle interactions will yield large accelerations, and these accelerations in turn would yield unacceptably large velocity changes unless the time step is sufficiently small. On the other hand, if no particles are strongly interacting, it would be inefficient to use a time step that was so small that none of the velocities changed. For research codes where N might be $\sim 10^8$, computational efficiency further requires that particles have their own individual times steps. With this implemented, a small time step for one pair of closely interacting particles does not slow down the entire simulation.

Algorithm 1: TIME STEP LOOP OF THE PP METHOD($\vec{\mathbf{x}}, \vec{\mathbf{v}}$)

comment: Clear force accumulators

for $i \leftarrow 1$ **to** N
  do $\vec{\mathbf{F}}_i \leftarrow 0$

comment: Accumulate forces

for $i \leftarrow 1$ **to** $N-1$

$$\textbf{do} \begin{cases} \textbf{for } j \leftarrow i+1 \textbf{ to } N \\ \quad \textbf{do} \begin{cases} \vec{\mathbf{F}}_i \leftarrow \vec{\mathbf{F}}_i + \vec{\mathbf{F}}_{ij} \\ \vec{\mathbf{F}}_j \leftarrow \vec{\mathbf{F}}_j - \vec{\mathbf{F}}_{ij} \end{cases} \end{cases}$$

comment: Integrate equations of motion

for $i \leftarrow 1$ **to** N

$$\textbf{do} \begin{cases} \vec{\mathbf{v}}_i^{\text{new}} \leftarrow \vec{\mathbf{v}}_i^{\text{old}} + \dfrac{\vec{\mathbf{F}}_i}{m_i}\delta t \\ \vec{\mathbf{x}}_i^{\text{new}} \leftarrow \vec{\mathbf{x}}_i^{\text{old}} + \vec{\mathbf{v}}_i \delta t \end{cases}$$

comment: Update time counter

$t \leftarrow t + \delta t$

comment: Return updated positions and velocities

return $(\vec{\mathbf{x}}, \vec{\mathbf{v}})$

The PP method scales as $\mathcal{O}(N^2)$, and so is really only a reasonable choice for problems with either $N \lesssim 1,000$ particles, or if special purpose hardware is available. The first widely used such hardware was the GRAPE[1] ("GRAvity PipE") computer that made use of specialized pipeline processors to compute the gravitational forces between particles (Makino et al. 1997, 2003; Sugimoto et al. 1990). The next step by Makino (2008) was GRAPE-DR which is hardware that is faster than a nominal CPU and designed to be more general purpose but still highly optimized for computing N-body problems. Currently, general-purpose graphical processing units (GPGPUs) are replacing CPUs and GRAPE hardware because they are both faster and far less expensive than comparable CPU or GRAPE hardware[2] (Belleman et al. 2008; Konstantinidis and Kokkotas 2010; Portegies Zwart et al. 2007). Currently, the PP method is used primarily in production runs of globular cluster simulations, as these require that each particle be followed as accurately as possible.

3.2 Particle-Mesh Methods

An early technique to reduce the computational requirements of N-body simulations of collisionless systems is the particle-mesh (PM) technique (Hockney and Eastwood 1988; Klypin and Shandarin 1983; White et al. 1983). In this technique, mesh points are assigned masses from the point mass positions, Poisson's equation

$$\nabla^2 \phi = 4\pi G \rho \tag{11.26}$$

is solved at the M mesh points, and forces are computed from the mesh-defined potentials and interpolated to the particle positions. Poisson's equation can be solved very efficiently on a regular grid using Fourier transform techniques (e.g., Couchman 1991). The code scales roughly as $\mathcal{O}(N \times M)$. Although the mesh can be adaptively refined (Couchman 1991; Knebe et al. 2001; Kravtsov et al. 1997; Teyssier 2002; Truelove et al. 1998), the PM method alone will not resolve close interactions since particles only move according to the average potential and as a result is rarely used in modern codes.

3.3 Particle-Particle Particle-Mesh (P^3M) Methods

A hybrid scheme that takes the best of both of the previous two methods is the Particle-Particle Particle-Mesh (P^3M) technique. This technique begins with the PM method, but particles that are physically close have their interactions added as well using a particle-particle computation over nearest neighbors (Efstathiou et al. 1985; Hockney and Eastwood 1988). The key to the P^3M method is to split the force summation into two parts. The first part is that from the smoothly varying potential calculated on the mesh, and the second, short-range part is from the rapidly varying potential arising from particles that are physically close:

$$\vec{\mathbf{F}}_{ij} = \vec{\mathbf{F}}_{ij}^{m} + \vec{\mathbf{F}}_{ij}^{sr}, \tag{11.27}$$

where $\vec{\mathbf{F}}_{ij}^{sr} \sim 0$ for all but a few nearest neighbors. The details of the implementation of the P^3M method are beyond the scope of the present work, but the interested reader should refer to Chapter 8 of Hockney and Eastwood (1988) for an in-depth presentation. The method is not currently in wide use.

[1] The GRAPE project website is www.astrogrape.org.
[2] Visit gpgpu.org.

3.4 Tree Methods

The tree algorithm has a similar physical justification as the P^3M algorithm, namely, that the $1/r^2$ force law yields a relatively large effect on the force summation for nearby particles and little effect on an individual basis for distant particles. So the forces of nearby particles are computed directly, but those of distant particles are computed as summations over volumes. Tree algorithms were pioneered by Appel (1985). The hierarchical $\mathcal{O}(N \log N)$ force calculation introduced by Barnes and Hut (1986, 1989) has been used extensively in the intervening years, as it is relatively simple to implement in numerical codes.

The Barnes-Hut algorithm works by grouping particles in a hierarchy of cubical spaces called an oct-tree, as follows. The entire simulation is placed within a single cube (or node or cell), called the root of the tree. This cell is then subdivided into eight subcells, which will each contain a subset of the original particles. If a subcell contains a single particle, it is not further divided, but if there are two or more particles, it is again divided into eight subcells. Thus the process continues until each subcell has at most a single particle within it. Once this process is complete, then for each cell in the simulation that is occupied by one or more particles, the total mass, center of mass, and quadrupole moment is calculated.

To calculate the net force on a particular body, start at the root of the tree and traverse the nodes of the tree. If the center of mass of a particular node is sufficiently distant from the body, then treat the collection of particles in that node as a single body at the center of mass and with the total mass of that node. Usually what is done to determine a value of "sufficiently far" is to calculate the ratio of the linear size s of the volume given by the node to the distance d to the node's center of mass. For values of s/d less than some typical value (typically ~0.5), the node is considered to be sufficiently far and the masses within can be treated as a single mass at the center of mass distance.

As a nearly trivial example, consider a quadruple star system consisting of two close binaries ($a \sim 10$ AU) both of which orbit the common center of mass for the four stars with a significantly larger separation for the system. In this case, when calculating the net force on a given star, the close binary companion's contribution would be calculated directly, but the more distant binary pair's contribution would be computed using the mass and center of mass of the binary, not the individual stars.

Other variations on the tree algorithm are of course possible. The tree may be built bottom-up by grouping nearest-neighbor pairings (Jernigan and Porter 1989), or the algorithm may implement multipole expansions for the gravitational field in a region of space (Greengard and Rokhlin 1987). Dehnen (2000) implemented mutual cell-cell interactions thus further reducing the required simulation times by a factor of ~4 over standard tree codes.

There are freely available N-body codes that use the tree algorithm. The VINE code (Nelson et al. 2008; Wetzstein et al. 2009) is freely available[3] and designed to simulate galaxy interactions, galactic dynamics, and star and planet formations, and is being adapted by the author to simulate cataclysmic variable accretion disks. GADGET is another freely available code (Springel 2005; Springel et al. 2001), designed to compute cosmological N-body/SPH simulations on massively parallel computers with distributed memory. A limitation of tree algorithms is that they can be considerably slower than PM methods that make use of Fourier techniques when dealing with mass distributions that have a low density contrast. Xu (1995) implemented a hybrid TreePM method which tried to combine the best features of both methods by restricting the

[3]www.usm.uni-muenchen.de/people/mwetz/Download.html.

tree algorithm to computing short-range interactions, and the PM algorithm to computing the long-range gravitational forces. GADGET offers the option of using the TreePM approach.

Fortin et al. (2011) recently published a comparison of different codes for galactic N-body simulations with the goal of helping potential users (particularly less experienced ones) determine the appropriate code for their specific problem. The authors compare different implementations of three different algorithms, the Barnes-Hut algorithm (i.e., GADGET), the fast multipole method (e.g., Cheng et al. 1999), and the Dehnen (2000) algorithm implemented in falcON. They find that falcON is the fastest code, often by a wide margin, and is well behaved in terms of conserving linear momentum and in terms of being relatively insensitive to concentration. GADGET-2 is the second fastest, and can outperform falcON since it is parallel. The fast multipole method offered the highest parallel efficiency and scalability, but otherwise was found to be generally slower and less suited for galactic work.

4 Lagrangian Hydrodynamics: Smoothed Particle Hydrodynamics

The method of smoothed particle hydrodynamics (SPH) approximates continuum hydrodynamics using a staggered grid, where the grid points are effectively Lagrangian point markers in the fluid (Lucy 1977; Monaghan 1992, 2005). It is a powerful technique for simulating physical systems that are highly dynamic and bounded by vacuum; thus, for many astrophysical systems, SPH can be a more computationally efficient approach than more conventional Eulerian techniques. The SPH method has been reviewed in depth recently by Abel (2011), Springel (2010a), Price (2010), and Rosswog (2009, 2010).

SPH replaces the fluid continuum with a finite number of particles which interact pairwise with each other, where the strength of that interaction is a function of the interparticle distance. The function is the *kernel* W, and many choices for the kernel function are possible. The simplest choice would be a Gaussian as in fact used by Gingold and Monaghan (1977), but the consequence of this choice is the interaction force is nonzero for all other particles in the system. It is common to use a twice-differentiable piecewise polynomial function which approximates a Gaussian, but which has compact support, being identically zero beyond twice the *smoothing length* h (Monaghan and Lattanzio 1985). Specifically, normalized for 3D applications:

$$W(q,h) = \frac{1}{\pi h^3} \times \begin{cases} 1 - \frac{3}{2}q^2 + \frac{3}{4}q^3 & \text{if } 0 \le q < 1; \\ \frac{1}{4}(2-q)^3 & \text{if } 1 \le q \le 2; \\ 0 & \text{otherwise;} \end{cases} \tag{11.28}$$

where $q = |\vec{\mathbf{r}}|/h$. With this choice, interparticle forces need only be calculated for neighbors within $2h$ of a given particle. In the continuum limit, a field quantity can be estimated by the integral

$$\langle A(\vec{\mathbf{r}})\rangle = \int W(\vec{\mathbf{r}} - \vec{\mathbf{r}}', h) A(\vec{\mathbf{r}}') d\vec{\mathbf{r}}', \tag{11.29}$$

where the integration is over all space and where W is defined such that $\int W\, dV \equiv 1$. To compute a field quantity using the finite number of SPH particles within $2h$ of $\vec{\mathbf{r}}$, we have

$$A(\vec{\mathbf{r}}) = \sum_j m_j \frac{A_j}{\rho_j} W(\vec{\mathbf{r}} - \vec{\mathbf{r}}_j, h), \tag{11.30}$$

where m_j is the mass of the j-th particle, ρ_j is its density, and $\vec{\mathbf{r}}_j$ is its position. A general form for the estimate of the derivative of a field quantity is given by

$$\nabla A(\vec{\mathbf{r}}) = \sum_j m_j \frac{A_j}{\rho_j} \nabla W(\vec{\mathbf{r}} - \vec{\mathbf{r}}_j, h). \tag{11.31}$$

Note that these equations allow the field quantity and its derivative to be estimated everywhere, and not just at the positions of the particles.

The general form of the Lagrangian momentum and energy equations relevant for accretion disk studies in stellar binaries are (cf. (➋ 11.2) and (➋ 11.3) above):

$$\frac{d^2\vec{\mathbf{r}}}{dt^2} = -\frac{\nabla P}{\rho} + \vec{\mathbf{f}}_{visc} - \frac{GM_1}{r_1{}^3}\vec{\mathbf{r}}_1 - \frac{GM_2}{r_2{}^3}\vec{\mathbf{r}}_2, \tag{11.32}$$

$$\frac{du}{dt} = -\frac{P}{\rho}\nabla \cdot \vec{\mathbf{v}} + \epsilon_{visc}, \tag{11.33}$$

where u is the specific internal energy, $\vec{\mathbf{f}}_{visc}$ is the viscous force, ϵ_{visc} is the energy generation from viscous dissipation, and $\vec{\mathbf{r}}_{1,2} \equiv \vec{\mathbf{r}} - \vec{\mathbf{r}}_{M1,M2}$ are the displacements from stellar masses M_1 and M_2, respectively.

The usual SPH form of the momentum equation for a given particle i is (Monaghan 1992):

$$\frac{d\vec{\mathbf{v}}_i}{dt} = -\sum_j m_j \left(\frac{P_i}{\rho_i{}^2} + \frac{P_j}{\rho_j{}^2} + \Pi_{ij} \right) \nabla_i W_{ij} - \frac{GM_1}{r_{i1}{}^3}\vec{\mathbf{r}}_{i1} - \frac{GM_2}{r_{i2}{}^3}\vec{\mathbf{r}}_{i2}, \tag{11.34}$$

where the notation $\nabla_i W_{ij}$ denotes the gradient of $W(\vec{\mathbf{r}}_i - \vec{\mathbf{r}}_j, h)$, taken with respect to the coordinates of particle i. The most common prescription for the artificial viscosity is (Gingold and Monaghan 1983)

$$\Pi_{ij} = \begin{cases} (-\alpha c_{ij}\mu_{ij} + \beta\mu_{ij}^2)/\rho_{ij} & \vec{\mathbf{v}}_{ij} \cdot \vec{\mathbf{r}}_{ij} \leq 0; \\ 0 & \text{otherwise;} \end{cases} \tag{11.35}$$

where

$$\mu_{ij} = \frac{h\vec{\mathbf{v}}_{ij} \cdot \vec{\mathbf{r}}_{ij}}{r_{ij}^2 + \eta^2 h_{ij}^2}. \tag{11.36}$$

Here $c_{s,ij}$ is the average of the sound speeds for particles i and j, $\eta \approx 0.01$ is a parameter introduced to protect against singularities for nearly coincident particles, $\rho_{ij} = (\rho_i + \rho_j)/2$, and we use the notation $\vec{\mathbf{v}}_{ij} = \vec{\mathbf{v}}_i - \vec{\mathbf{v}}_j$, $\vec{\mathbf{r}}_{ij} = \vec{\mathbf{r}}_i - \vec{\mathbf{r}}_j$. The μ^2 term was introduced to help prevent particle penetration in high Mach number shocks. The magnitude of the viscous dissipation is controlled by parameters α and β, with typically $\alpha \approx 0.5$ to 1.0 and $\beta = 2\alpha$. Note that as is typical in artificial viscosity formalisms, only approaching particles are subject to a viscous force.

The viscosity approximation as given in (➋ 11.35) is essentially a combination of a bulk and von Neumann-Richtmyer viscosity (von Neumann and Richtmyer 1950). The viscosity vanishes for solid-body rotation, but not for shear flows as might be found in accretion disks. To address this, Balsara (1995) introduced a correction factor that is multiplied by Π_{ij}, which reduces the magnitude of the computed viscosity when there is a strong shearing flow. The correction factor for particle i is given by

$$f_i^{\mathrm{AV}} = \frac{|\nabla \cdot \vec{\mathbf{v}}|_j}{|\nabla \cdot \vec{\mathbf{v}}|_i + |\nabla \times \vec{\mathbf{v}}|_i}, \tag{11.37}$$

and the correction factor is taken as the average of the ij pair $(f_i^{\mathrm{AV}} + f_j^{\mathrm{AV}})/2$.

The internal energy of each particle is typically integrated using

$$\frac{du_i}{dt} = \frac{1}{2}\sum_j m_j \left(\frac{P_i}{\rho_i{}^2} + \frac{P_j}{\rho_j{}^2} + \Pi_{ij} \right) \vec{\mathbf{v}}_{ij} \cdot \nabla_i W_{ij}. \tag{11.38}$$

SPH as detailed above – including constant h particles and using the same time step for all particles – is relatively straightforward to code, but may run inefficiently and/or may under-resolve some regions. Because the method provides an interpolated estimate of the field quantities, there must be a sufficient number of particles in the sum in order for the estimate to be useful. Typically for 3D problems, some 30–70 neighbors are desired. Much less than 30 and the field is under sampled, and more than $\sim 10^2$ neighbors and the incremental change in the estimate per additional particle is negligible.

Because SPH is often used to model systems which may have strong shocks as well as tenuous regions, it was quickly realized that the smoothing length should be variable. Several authors began using smoothing lengths that varied both in space and time (Benz et al. 1990; Evrard 1988; Hernquist and Katz 1989, and see the recent reviews noted above). So what is needed is a method to calculate h on a per-particle basis, where h will be small in regions where many particles are tightly clustered, and h will be large in regions that have few particles per unit volume (Price 2010). An obvious choice for constant-mass particles is thus to let h scale with the inverse of the local number density of particles $h(\vec{\mathbf{r}}) \propto n(\vec{\mathbf{r}})^{1/3}$, which then feeds back into the equation for the density summation. We then have at the location of particle i two simultaneous equations for the local density and smoothing length

$$\rho(\vec{\mathbf{r}}_i) = \sum_j m_j W(\vec{\mathbf{r}}_i - \vec{\mathbf{r}}_j, h_i); \qquad h(\vec{\mathbf{r}}_i) = \zeta \left(\frac{m_i}{\rho_i} \right)^{1/3}, \tag{11.39}$$

where ζ is a parameter specifying the scaling between the smoothing length and the mean particle spacing $(m/\rho)^{1/3}$. In most recent codes, these equations are solved simultaneously using, for example, a Newton-Raphson root-finding method (Price 2010). As noted in, for example, Springel (2010a), this is equivalent to keeping the mass in the kernel volume constant for all particles.

As discussed more fully in Springel (2010a), SPH has been modified in recent years to include self-gravity (as discussed earlier), magnetic fields (Dolag and Stasyszyn 2009; Dolag et al. 1999; Price 2010; Price and Monaghan 2004a, b, 2005; Rosswog 2007), and radiation in the flux-limited-diffusion approximation (Forgan et al. 2009; Whitehouse and Bate 2006; Whitehouse et al. 2005). Alternative approaches to implementing radiative transfer into the SPH scheme include (1) that of Petkova and Springel (2009) who simplify the transfer equation such that it can be computed in terms of the local energy density of the radiation, (2) Pawlik and Schaye (2008) who compute radiative transfer by using emission and absorption cones between particles, and (3) Nayakshin et al. (2009) who use a Monte-Carlo technique wherein the photons are treated as virtual particles.

Although not SPH, it deserves mention that recently Springel (2010b) has introduced a novel Lagrangian scheme which goes a long way toward eliminating the weaknesses of both the Eulerian grid and Lagrangian SPH techniques. The new method is based on Voronoi tessellation of a set of discrete points. The mesh is used to solve via a finite volume approach and exact Riemann solver the hyperbolic conservation laws of ideal hydrodynamics. The mesh-generating points can be moved arbitrarily to follow the local motion of the fluid, but if held stationary the method is equivalent to an ordinary Eulerian method. The new method can also adjust the spatial resolution as required by the local dynamics, which is one of the principle advantages of the SPH method. A similar, first-order method was originally introduced by Whitehurst (1995) in the form of his FLAME code based on Delaunay and Voronoi tessellations, but no application publications followed.

5 Eulerian Hydrodynamics: Hydrodynamics on a Grid

5.1 Introduction

In standard grid-based approaches to solving hydrodynamical problems, a volume of space is subdivided into finite-sized cells, and the physical quantities such as temperature, density, fluid velocity, etc., are computed inside those cells or at cell faces using finite difference or finite volume techniques as discussed above (see, e.g., Bodenheimer et al. 2007; Mignone et al. 2007; O'Shea et al. 2004; Ryu et al. 1993; Stone and Norman 1992a). As discussed in Bodenheimer et al. (2007), grid methods as a rule are preferable when shocks in the fluid are an important component of the problem being simulated. However, for many kinds of astrophysical problems SPH codes that implement adaptive particle sizes may be optimal, as SPH naturally increases spatial resolution where needed.

5.2 Shocks and the Rankine-Hugoniot Conditions

Shocks are discontinuities in the flow of the fluid. A shock front is a surface that is characterized by a nearly discontinuous change in P, ρ, $v_\perp$, and T, where $v_\perp$ is the component of the velocity perpendicular to the shock front. Related to shocks but generally less problematic to simulate are contact discontinuities. A contact discontinuity is characterized by a nearly discontinuous change in ρ, T, or $v_\parallel$, but a continuous P and $v_\perp$.

Shocks often form naturally in physical systems when sound waves propagate – because the speed of sound increases with density, a compression wave which is initially symmetric about the maximum density will have its leading edge steepen to a shock front as it propagates through the fluid. The shock front narrows to a scale of just a few mean free paths, but we can use conservation laws to relate the physical properties on the two sides of the shock. The resulting relations are called the Rankine-Hugoniot ("jump") conditions (Landau and Lifshitz 1959).

Let us consider a shock front in a 1D inviscid polytropic flow in the reference frame where the front is stationary and there are no external forces. In this case, the 1D Euler equations can be written as

$$\frac{\partial \rho}{\partial t} = -\frac{\partial}{\partial x}(\rho u), \tag{11.40}$$

$$\frac{\partial(\rho u)}{\partial t} = -\frac{\partial}{\partial x}(\rho u^2 + P), \tag{11.41}$$

$$\frac{\partial \mathcal{E}}{\partial t} = -\frac{\partial}{\partial x}\left[u(\mathcal{E} + P)\right], \tag{11.42}$$

where

$$\mathcal{E} \equiv \frac{1}{2}\rho u^2 + e \tag{11.43}$$

is the total energy density, e is the internal energy density, and u is the fluid velocity relative to the stationary shock front, with $u_1 > c_{s,1}$ supersonic in the pre-shock region. Conservation of fluxes of mass ρu, momentum $\rho u^2 + P$, and total energy $u(\frac{1}{2}\rho u^2 + e + P)$ across the shock front results in

$$\rho_1 u_1 = \rho_2 u_2, \tag{11.44}$$

$$\rho_1 u_1^2 + P_1 = \rho_2 u_2^2 + P_2, \tag{11.45}$$

$$u_1\left(\frac{1}{2}\rho_1 u_1^2 + e_1 + P_1\right) = u_2\left(\frac{1}{2}\rho_2 u_2^2 + e_2 + P_2\right). \tag{11.46}$$

Using these equations and assuming a polytropic equation of state $P = (\gamma - 1)e$, it is easy to show that the ratio of densities in the post- and pre-shock regions is given by (e.g., Bodenheimer et al. 2007)

$$\frac{\rho_2}{\rho_1} = \frac{P_2 + m^2 P_1}{P_1 + m^2 P_2}, \tag{11.47}$$

where

$$m^2 = \frac{\gamma - 1}{\gamma + 1}. \tag{11.48}$$

Even if the shock is very strong ($P_2 \gg P_1$), the density contrast is only $\rho_2/\rho_1 = 4$ for the case of $\gamma = 5/3$, but for an isothermal shock able to radiate instantaneously the energy dissipated in the shock front, $m^2 \rightarrow 0$ and the density contrast may be arbitrarily high.

Alternatively, the shock may be characterized by the value of the Mach number $\mathcal{M} = |u_1|/c_{s,1}$ of the pre-shock flow, in which case it can be shown (Balachandran 2007) that

$$\frac{\rho_2}{\rho_1} = \frac{(\gamma + 1)\mathcal{M}^2}{(\gamma - 1)\mathcal{M}^2 + 2}, \tag{11.49}$$

$$\frac{P_2}{P_1} = \frac{2\gamma\mathcal{M}^2 - (\gamma - 1)}{\gamma + 1}, \tag{11.50}$$

and that in the frame of reference of the shock, the post-shock medium is subsonic:

$$\mathcal{M}_2^2 = \frac{\frac{2}{\gamma-1} + \mathcal{M}_1^2}{\frac{2\gamma}{\gamma-1}\mathcal{M}_1^2 - 1}. \tag{11.51}$$

In the limit as $\mathcal{M}_1 \rightarrow \infty$, we find $\mathcal{M}_2 \rightarrow \sqrt{(\gamma - 1)/2}$.

There are two ways of treating shocks in Eulerian codes. The first is to implement a Riemann solver which makes use of the results obtained in this section, and the second is to implement an artificial viscosity. We discuss each in turn.

5.3 The Riemann Problem

A Riemann problem consists of a conservation law applied to the case of piecewise constant data that includes a single discontinuity. Methods to solve the Riemann problem are called Riemann solvers. For a detailed treatment of this subject, see Toro (1999) or Knight (2006).

The simplest application is the shock tube. Consider a 1D system of two volumes of fluid initially separated by a diaphragm located at x_0. Let the initial pressures and densities of the

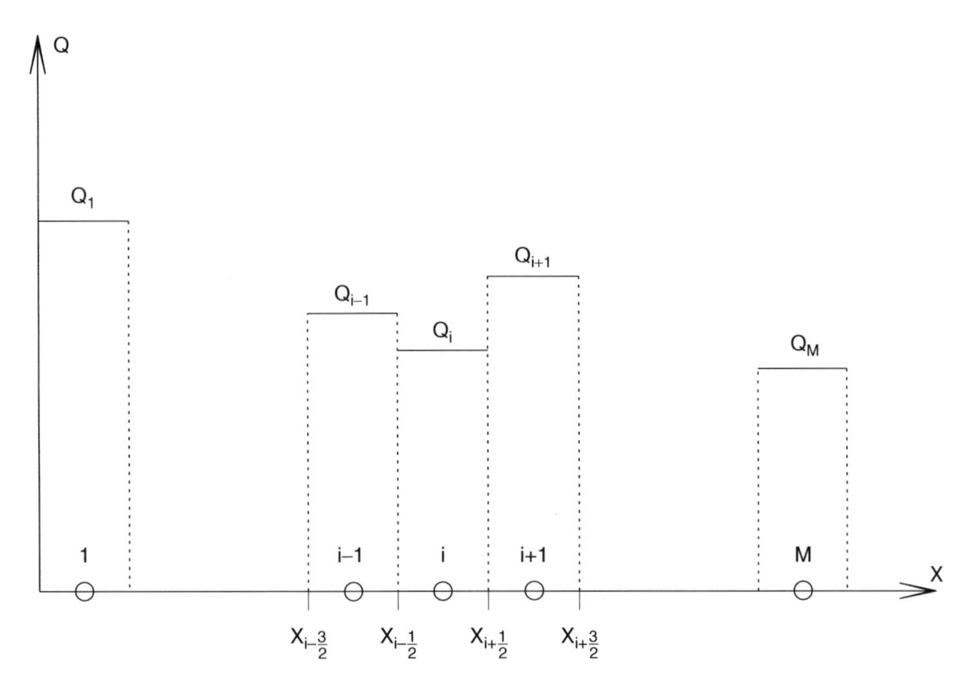

Fig. 11-4
A schematic representation of component $\mathcal{Q}_i$ at time t^n

fluids to the left and right of the diaphragm be initially (P_1, ρ_1) and (P_5, ρ_5), respectively, with $P_1 > P_5$, and the fluids originally at rest. At $t = 0$ the diaphragm is ruptured, resulting in the left fluid expanding into the right. Five regions subsequently develop: (1) the original undisturbed left fluid, (2) rarefaction wave, (3) uncompressed left fluid with constant velocity and pressure and bounded at the right by a contact discontinuity, (4) compressed right fluid bounded at the right by a shock, and (5) undisturbed right fluid (Bodenheimer et al. 2007; Hawley et al. 1984). Given the initial conditions for pressure and densities in the two initial regions, it is possible to solve for the subsequent pressures, densities, and velocities of the fluid in the middle three regions. This result can then be applied to a typical piecewise constant array representing some physical variable in a simulation.

Godunov (1959) was the first to implement the conservative upwind method to nonlinear systems of conservation laws. Godunov's first-order upwind method computes interface numerical fluxes $\mathcal{F}_{i+\frac{1}{2}}$ using solutions of local Riemann problems, under the assumption that at time step n in the simulation the values $\mathcal{Q}_i$ on the grid are piecewise constant (see Fig. 11-4), where Q is the vector of dependent variables $\mathcal{Q} = (\rho, \rho u, \mathcal{E})^T$, and $\mathcal{F}$ is the vector of fluxes $\mathcal{F} = (\rho u, \rho u^2 + P, (\mathcal{E} + P)u)^T$. This allows the values of $\mathcal{Q}_i$ to be updated from one time step to the next using the integral form of the Euler equations

$$\mathcal{Q}_i^{n+1} = \mathcal{Q}_i^n - \frac{1}{\Delta x} \int_{t^n}^{t^{n+1}} \left(\mathcal{F}_{i+\frac{1}{2}} - \mathcal{F}_{i-\frac{1}{2}} \right) dt. \tag{11.52}$$

For example, following Knight (2006), assume the piecewise constant function for some variable $\mathcal{Q}_i$ is known for $i = 1, \ldots, M$ at time t^n. In general, there will be a discontinuity in Q_i at

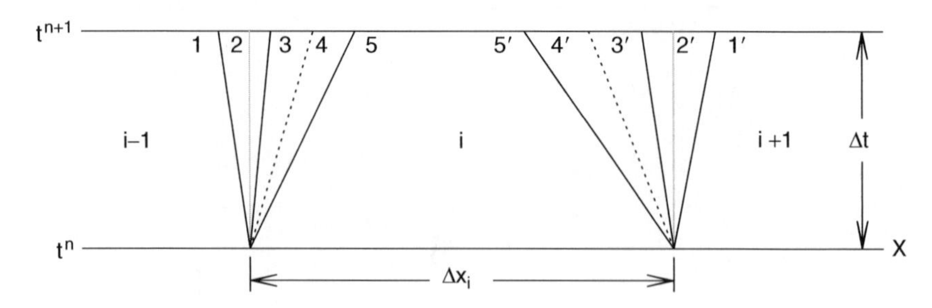

Fig. 11-5

A schematic flow structure for Godunov's Second Method. The contact discontinuities are represented as *dotted lines*

each interface. The flux $\mathcal{F}_{i+\frac{1}{2}}$ can be determined from the solution of the local general Riemann problem at interface $x_{i+\frac{1}{2}}$, and similarly for flux $\mathcal{F}_{i-\frac{1}{2}}$ at the left bound of cell i (see *Fig. 11-5*). In the solution the states for regions 1 and 5 can be taken as

$$Q^5_{i+\frac{1}{2}} = Q_i, \tag{11.53}$$

$$Q^1_{i+\frac{1}{2}} = Q_{i+1}, \tag{11.54}$$

and

$$Q^1_{i-\frac{1}{2}} = Q_{i-1}, \tag{11.55}$$

$$Q^5_{i-\frac{1}{2}} = Q_i, \tag{11.56}$$

where we use superscripts to denote the region, and where we assume that as shown in *Fig. 11-4* $Q_{i-1} > Q_i < Q_{i+1}$. The solution to the Riemann problem for Q at the interface is dependent upon $(x - x_{i+\frac{1}{2}})/(t - t^n)$ and not on x and t separately, and therefore the solution for Q at the interface is independent of time. We then have

$$\int_{t^n}^{t^{n+1}} \mathcal{F}_{i+\frac{1}{2}}\, dt = \mathcal{F}_{i+\frac{1}{2}} \Delta t = \mathcal{F}(Q^{RP}_{i+\frac{1}{2}})\Delta t, \tag{11.57}$$

where $Q^{RP}_{i+\frac{1}{2}}$ is the solution of the Riemann problem at interface $x_{i+\frac{1}{2}}$. Using a similar approach for $\mathcal{F}_{i-\frac{1}{2}}$, we have

$$Q^{n+1}_i = Q^n_i - \frac{\Delta t}{\delta x}\left(\mathcal{F}(Q^{RP}_{i+\frac{1}{2}}) - \mathcal{F}(Q^{RP}_{i-\frac{1}{2}})\right). \tag{11.58}$$

This is the essence of Godunov's Second Method, which is most often used in practice.

5.4 Artificial Viscosity

The second approach to handling shocks is to implement an artificial viscosity. Shocks are nearly discontinuous physically, with a characteristic width of a few particle mean free paths. This in general is many times smaller than the typical grid spacing in simulations. By introducing an artificial viscosity, the shock front may be artificially spread over a few grid cells, as required to maintain a stable solution. First introduced by von Neumann and Richtmyer (1950), artificial viscosities are now widely used to help make flows more diffusive near shock fronts so the simulation maintains stability.

The von Neumann-Richtmyer artificial viscosity is a bulk viscosity which acts as an additional pressure:

$$\Pi = \begin{cases} q\rho(\Delta x)^2 \left(\frac{\partial v}{\partial x}\right)^2 & \text{if } \frac{\partial v}{\partial x} < 0; \\ 0 & \text{if } \frac{\partial v}{\partial x} > 0; \end{cases} \quad (11.59)$$

where q is a parameter varied to produce the desired shock width. Note that the artificial viscosity only acts during compression. Throughout the equations describing the system, one replaces $P \rightarrow P + \Pi$. The artificial viscosity so defined has the desirable properties that it only operates at a significant level in the vicinity of shock fronts and can be tuned so as to be only as diffusive as needed. Additional details can be found in Bodenheimer et al. (2007).

5.5 Available Codes

Stone and Norman (1992a) introduced the code ZEUS-2D to the community. ZEUS-2D is a finite difference hydrodynamics code implemented in a covariant formalism. In addition, ZEUS-2D incorporated magnetohydrodynamics (Stone and Norman 1992b), and radiation hydrodynamics (Stone et al. 1992). ZEUS is in fact several different numerical codes, and each version is freely available. ZEUS-2D and the MPI-enabled version ZEUS-MP can be obtained from the Laboratory for Computational Astrophysics website,[4] and ZEUS-3D is make available by David Clarke[5] (Clarke 1996, 2010). With over 700 referred citations to Stone and Norman (1992a), the ZEUS codes are arguably the best-tested astrophysical HD/MHD codes in existence.

Recently, Stone and collaborators have been developing a new code for astrophysical magnetohydrodynamics (MHD) called Athena (Stone et al. 2008). The authors designed Athena primarily for solving problems in the areas of the interstellar medium, star formation, and accretion flows, and it is freely available to the community.[6] The version at the time of this writing includes, for example, compressible hydrodynamics and MHD in 1D, 2D, and 3D, special relativistic dynamics, self-gravity or a defined static gravitational field, both Navier-Stokes and anisotropic viscosity, and optically thin radiative cooling. In addition, several choices are available for the Riemann solvers and spatial reconstruction methods.

Also freely available to the community is the PLUTO[7] code (Mignone et al. 2007). PLUTO is a modular Godunov-type code designed to be used in astrophysical problems characterized by high Mach numbers. PLUTO offers several choices of hydrodynamic solvers, and can be used to simulate Newtonian, relativistic, MHD, or relativistic MHD flows in Cartesian, cylindrical, or spherical coordinates. The code can run on single-processor machines, but can also run on large parallel machines using MPI, if available. Computational grids may be static or adaptive.

The FLASH code is available from the Flash Center at the University of Chicago.[8] It is a multiphysics multiscale simulation code with a large base of users. The FLASH code is intended primarily for applications involving high-energy density physics (e.g., supernovae). A large selection of physics solvers is available, including various hydrodynamics, MHD, equation

[4] lca.ucsd.edu/portal/software/.
[5] www.ap.smu.ca/~dclarke/zeus3d/.
[6] trac.princeton.edu/Athena.
[7] plutocode.ph.unito.it/.
[8] See flash.uchicago.edu/site/.

of state, radiative transfer, diffusion and conduction, nuclear burning, gravity, and magnetic resistivity and conductivity solvers. The code is in its fourth version.

6 Conclusion

Numerical Astrophysics is one of the youngest scientific fields, with a history spanning only a few decades, but one which is currently vibrant and growing. Numerical experimentation allows us to study the physics of astronomical processes at a level of detail undreamed of a century ago. In this short manuscript, only a few of the many areas of numerical astrophysics have been discussed. The interested reader has available a vast literature, including above-cited texts and conference proceedings, as well as many public domain codes that are suitable for projects ranging from self-education to cutting-edge research. Many if not most code authors are generous with their codes and time, and approachable with suggestions of collaborative efforts. Roughly half a century ago, researchers were computing stellar structure and evolution by hand with a slide rule (Schwarzschild 1958), computing one time step per several hours of clock time. Today we can use hundreds of CPU *centuries* to complete a multimillion particle simulation in a matter of calendar weeks, and by the end of the decade will likely be able to simulate galactic dynamics with a particle representation of each of the $\sim 10^{11}$ stars present in a large spiral galaxy like our own.

Acknowledgments

Thanks to Terry Oswalt and Howard Bond for suggesting that I write this manuscript, to Dean Townsley for commenting on a draft of the manuscript, and to my wife Jane E. Wood for her encouragement and patience over the many hours spent in the office.

References

Abel, T. 2011, MNRAS, 413, 271
Allen, M. P., & Tildesley, D. J. 1989, Computer Simulat ion of Liquids (Oxford: Oxford University Press)
Appel, A. W. 1985, SIAM J Sci Stat Comp, 6, 85
Balachandran, P. 2007, Fundamentals of Compressible Fluid Dynamics (New Delhi: Prentice-Hall)
Balsara, D. S. 1995, J Comput Phys, 131, 357
Barnes, J., & Hut, P. 1986, Nature, 324, 44
Barnes, J. E., & Hut, P. 1989, ApJS, 70, 389
Belleman, R. G., Bédorf, J., & Portegies Zwart, S. F. 2008, New Astron, 13, 103
Benz, W., Bowers, R. L., Cameron, A. G. W., & Press W. H. 1990, ApJ, 348, 647
Bodenheimer, P., Laughlin, G. P., Różyczka, M., & Yorke, H. W. 2007, Numerical Methods in Astrophysics: An Introduction (Boca Raton, FL: Taylor & Francis)
Bowers, R. L., & Wilson, J. R. 1991, Numerical Modeling in Applied Physics and Astrophysics (Boston: Jones and Bartlett)
Burnett, D. S. 1987, Finite Element Analysis From Concepts to Applications (Reading, MA: Addison Wesley)
Chandrasekhar, S. 1939, An Introduction to the Study of Stellar Structure (Chicago: University of Chicago Press)

Cheng, H., Greengard, L., & Rokhlin, V. 1999, J Comput Phys, 155, 468
Clarke, D. A. 1996, ApJ, 457, 291
Clarke, D. A. 2010, ApJS, 187, 119
Couchman, H. M. P. 1991, ApJ, 368, L23
Cramer, C. J. 2004, Essentials of Computational Chemistry: Theories and Models (2nd ed.; Chichester, UK: Wiley)
Cromer, A. 1981, AJP, 49, 455
Dehnen, W. 2000, ApJ, 536, L39
DeVries, P. L., & Hasbun, J. E. 2010, A First Course in Computational Physics (2nd ed.; Sudbury, MA: Jones & Bartlett)
Dolag, K., & Stasyszyn, F. 2009, MNRAS, 398, 1678
Dolag, K., Bartelmann, M., & Lesch, H. 1999, A&A, 348, 351
Eddington, A. 1926, The Internal Constitution of the Stars (Cambridge: Cambridge University Press)
Efstathiou, G., Davis, M., White, S. D. M., & Frenk, C. S. 1985, ApJS, 57, 241
Evrard, A. E. 1988, MNRAS, 235, 911
Forgan, D., Rice, K., Stamatellos, D., & Whitworth, A. 2009, MNRAS, 394, 882
Fortin, P., Athanassoula, E., & Lambert, J.-C. 2011, A&A, 531, A120
Garcia, A. 2000, Numerical Methods for Physics (2nd ed.; Englewood Cliffs, NJ: Prentice-Hall)
Gingold, R. A., & Monaghan, J. J. 1977, MNRAS, 181, 375
Gingold, R. A., & Monaghan, J. J. 1983, MNRAS, 204, 715
Giordano, N. J., & Nakanishi, H. 2005, Computational Physics (2nd ed.; New Jersey: Prentice Hall)
Godunov, S. K. 1959, Math Sbornik, 47, 357
Gould, H., Tobochnik, J., & Christian, W. 1988, An Introduction to Computer Simulation Methods (3rd ed.; Reading, MA: Addison Wesley)
Greengard, L., & Rokhlin, V. 1987, J Comp Phys, 73, 325
Hamming, R. W. 1962, Numerical Methods for Scientists and Engineers (New York, NY: McGraw-Hill)
Harrison, P. 2001, Computational Methods in Physics, Chemsitry and Biology: An Introduction (West Susses, UK: Wiley)
Hawley, J. F., Smarr, L. L., & Wilson, J. R. 1984, ApJ, 277, 296
Hernquist, L., & Katz, N. 1989, ApJS, 70, 419
Hockney, R. W., & Eastwood, J. W. 1988, Computer Simulation Using Particles (Bristol, UK: Adam Hilger)
Holmberg, E. 1941, ApJ, 94, 385
Jernigan, J. G., & Porter, D. H. 1989, ApJS, 71, 871
Klypin, A. A., & Shandarin, S. F. 1983, MNRAS, 204, 891
Knebe, A., Green, A., & Binney, J. 2001, MNRAS, 325, 845
Knight, D. D. 2006, Elements of Numerical Methods for Compressible Flows (New York: Cambridge University Press)
Konstantinidis, S., & Kokkotas, K. D. 2010, A&A, 522, A70
Kravtsov, A. V., Klypin, A. A., & Khokhlov, A. M. 1997, ApJS, 111, 73
Landau, L. D., & Lifshitz, E. M. 1959, Fluid Mechanics (London: Pergamon Press)
Landau, R. H., Páez, M. J., & Bordeianu, C. C. 2007, Computational Physics: Problem Solving with Computers (2nd ed; Weinheim, Germany: Wiley-VCH)
Landau, R. H., Páez, M. J., & Bordeianu, C. C. 2008, A Survey of Computational Physics: Introductory Computational Science (Princeton, NJ: Princeton University Press)
Lucy, L. B. 1977, AJ, 82, 1013
Makino, J. 2008, in Dynamical Evolution of Dense Stellar Systems, Proceedings of the International Astronomical Union, IAU Symp., 246, 457, doi:10.1017/S1743921308016165
Makino, J., Fukushige, T., Koga, M., & Namura, K. 2003, PASJ, 55, 1163
Makino, J., Taiji, M., Ebisuzaki, T., & Sugimoto, D. 1997, ApJ, 480, 432
Mignone, A., Bodo, G., Massaglia, S., Matsakos, T., Tesileanu, O., Zanni, C., & Ferrari, A. 2007, ApJS, 170, 228
Monaghan, J. J. 1992, ARA&A, 30, 543
Monaghan, J. J. 2005, Rep Prog Phys, 68, 1703
Monaghan, J. J., & Lattanzio, J. C. 1985, A&A, 149, 135
Nayakshin, S., Cha, S.-H., & Hobbs, A. 2009, MNRAS, 397, 1314
Nelson, A. F., Wetzstein, M., & Naab, T. 2009, ApJS, 184, 326
O'Shea, B. W., Bryan, G., Bordner, J., et al. 2004, arXiv:astro-ph/0403044
Pang, T. 2006, An Introduction to Computational Physics (2nd ed.; Cambridge, UK: Cambridge University Press)
Pawlik, A. H., & Schaye, J. 2008, MNRAS, 389, 651
Petkova, M., & Springel, V. 2009, MNRAS, 396, 1383
Portegies Zwart, S. F., Belleman, R. G., & Geldof, P. M. 2007, New Astron, 12, 641
Press, W. H., Teukolsky, S. A., Vetterling, W. T., & Flannery, B. P. 1992, Numerical Recipes in Fortran: The Art of Scientific Computing (2nd ed.; Cambridge: Cambridge University Press)
Price, D. J. 2010, J Comp Phys (in press) (arXiv:1012.1885)
Price, D. J., & Monaghan, J. J. 2004a, MNRAS, 348, 123

Price, D. J., & Monaghan, J. J. 2004b, MNRAS, 348, 139
Price, D. J., & Monaghan, J. J. 2005, MNRAS, 364, 384
Rosswog, S. 2007, Astron Nachr, 328, 663
Rosswog, S. 2009, New A Rev, 53, 78
Rosswog, S. 2010, J Comput Phys, 229, 8591
Ryu, D., Ostriker, J. P., Kang, H., & Cen, R. 1993, ApJ, 414, 1
Schwarzschild, M. 1958, Structure and Evolution of the Stars (Princeton, NJ: Princeton University Press)
Shu, F. H. 1992, The Physics of Astrophysics, Vol. II (Mill Valley, CA: University Science Books)
Springel, V. 2005, MNRAS, 364, 1105
Springel, V. 2010a, ARA&A, 48, 391
Springel, V. 2010b, MNRAS, 401, 791
Springel, V., Yoshida, N., & White, S. D. M. 2001, New A, 6, 79
Stone, J. M., & Norman, M. L. 1992a, ApJS, 80, 753
Stone, J. M., & Norman, M. L. 1992b, ApJS, 80, 791
Stone, J. M., Mihalas, D., & Norman, M. L. 1992, ApJS, 80, 819
Stone, J. M., Gardiner, T. A., Teuben, P., Hawley, J. F., & Simon, J. B. 2008, ApJS, 178, 137
Strikwerda, J. 2007, Finite Difference Schemes and Partial Differential Equations (2nd ed.; Pacific Grove, CA: Wadsworth & Brooks/Cole)
Sugimoto, D., Chikada, Y., Makino, J., Ito, T., Ebisuzaki, T., & Umemura, M. 1990, Nature, 345, 33
Teyssier, R. 2002, A&A, 385, 337
Thijssen, J. M. 2007, Computational Physics (2nd ed.; New York: Cambridge University Press)
Thomas, J. W. 2010, Numerical Partial Differential Equations: Finite Difference Methods (New York, NY: Springer)
Toro, E. F. 1999, Riemann Solvers and Numerical Methods for Fluid Dynamics: A Practial Introduction (3rd ed.; New York: Springer-Verlag)
Truelove, J. K., Klein, R. I., McKee, C. F., Holliman, J. H., II, Howell, L. H., Greenough, J. A., & Woods, D. T. 1998, ApJ, 495, 821
von Hoerner, S. 1960, Z Astrophys, 50, 184
von Hoerner, S. 1963, Z Astrophys, 57, 47
von Neumann, J., & Richtmyer, R. D. 1950, J Appl Phys, 21, 232
Wetzstein, M., Nelson, A. F., Naab, T., & Burkert, A. 2009, ApJS, 184, 298
White, S. D. M., Frenk, C. S., & Davis, M. 1983, ApJ, 274, L1
Whitehouse, S. C., & Bate, M. R. 2006, MNRAS, 367, 32
Whitehouse, S. C., Bate, M. R., & Monaghan, J. J. 2005, MNRAS, 364, 1367
Whitehurst, R. 1995, MNRAS, 277, 655
Xu, G. 1995, ApJS, 98, 355

Index

D

E

F

Printed by Publishers' Graphics LLC